DISCARD

Grzimek's
Animal Life Encyclopedia

Second Edition

••••

Grzimek's
Animal Life Encyclopedia

Second Edition

●●●●

Volume 2
Protostomes

Sean F. Craig, Advisory Editor
Dennis A. Thoney, Advisory Editor
Neil Schlager, Editor

Joseph E. Trumpey, Chief Scientific Illustrator

Michael Hutchins, Series Editor
In association with the American Zoo and Aquarium Association

Detroit • New York • San Diego • San Francisco • Cleveland • New Haven, Conn. • Waterville, Maine • London • Munich

THOMSON

GALE

Grzimek's Animal Life Encyclopedia, Second Edition

Volume 2: Protostomes
Produced by Schlager Group Inc.
Neil Schlager, Editor
Vanessa Torrado-Caputo, Associate Editor

Project Editor
Melissa C. McDade

Editorial
Stacey Blachford, Deirdre S. Blanchfield, Madeline Harris, Christine Jeryan, Kate Kretschmann, Mark Springer

Permissions
Margaret Chamberlain

Indexing Services
Synapse, the Knowledge Link Corporation

Imaging and Multimedia
Mary K. Grimes, Lezlie Light, Christine O'Bryan, Barbara Yarrow, Robyn V. Young

Product Design
Tracey Rowens, Jennifer Wahi

Manufacturing
Wendy Blurton, Dorothy Maki, Evi Seoud, Mary Beth Trimper

For permission to use material from this product, submit your request via Web at http://www.gale-edit.com/permissions, or you may download our Permissions Request form and submit your request by fax or mail to: The Gale Group, Inc., Permissions Department, 27500 Drake Road, Farmington Hills, MI, 48331-3535, Permissions hotline: 248-699-8074 or 800-877-4253, ext. 8006, Fax: 248-699-8074 or 800-762-4058.

Cover photo of land snail by JLM Visuals. Back cover photos of sea anemone by AP/Wide World Photos/University of Wisconsin-Superior; land snail, lionfish, golden frog, and green python by JLM Visuals; red-legged locust © 2001 Susan Sam; hornbill by Margaret F. Kinnaird; and tiger by Jeff Lepore/Photo Researchers. All reproduced by permission.

While every effort has been made to ensure the reliability of the information presented in this publication, The Gale Group, Inc. does not guarantee the accuracy of the data contained herein. The Gale Group, Inc. accepts no payment for listing; and inclusion in the publication of any organization, agency, institution, publication, service, or individual does not imply endorsement of the editors and publisher. Errors brought to the attention of the publisher and verified to the satisfaction of the publisher will be corrected in future editions.

ISBN   0-7876-5362-4 (vols. 1–17 set)
          0-7876-5778-6 (vol. 2)

This title is also available as an e-book.
ISBN   0-7876-7750-7 (17-vol set)

Contact your Gale sales representative for ordering information.

LIBRARY OF CONGRESS CATALOGING-IN-PUBLICATION DATA

Grzimek, Bernhard.
    [Tierleben. English]
    Grzimek's animal life encyclopedia.— 2nd ed.
      v. cm.
Includes bibliographical references.
Contents: v. 1. Lower metazoans and lesser deuterosomes / Neil Schlager, editor — v. 2. Protostomes / Neil Schlager, editor — v. 3. Insects / Neil Schlager, editor — v. 4-5. Fishes I-II / Neil Schlager, editor —v. 6. Amphibians / Neil Schlager, editor — v. 7. Reptiles / Neil Schlager, editor — v. 8-11. Birds I-IV / Donna Olendorf, editor — v. 12-16. Mammals I-V / Melissa C. McDade, editor — v. 17. Cumulative index / Melissa C. McDade, editor.
ISBN 0-7876-5362-4 (set hardcover : alk. paper)
    1. Zoology—Encyclopedias. I. Title: Animal life encyclopedia. II. Schlager, Neil, 1966- III. Olendorf, Donna IV. McDade, Melissa C. V. American Zoo and Aquarium Association. VI. Title.
QL7 .G7813 2004

                 590'.3—dc21
                 2002003351

Printed in Canada
10 9 8 7 6 5 4 3 2 1

Recommended citation: *Grzimek's Animal Life Encyclopedia,* 2nd edition. Volume 2, *Protostomes,* edited by Michael Hutchins, Sean F. Craig, Dennis A. Thoney, and Neil Schlager. Farmington Hills, MI: Gale Group, 2003.

Contents

Contents

Foreword

Earth is teeming with life. No one knows exactly how many distinct organisms inhabit our planet, but more than 5 million different species of animals and plants could exist, ranging from microscopic algae and bacteria to gigantic elephants, redwood trees and blue whales. Yet, throughout this wonderful tapestry of living creatures, there runs a single thread: Deoxyribonucleic acid or DNA. The existence of DNA, an elegant, twisted organic molecule that is the building block of all life, is perhaps the best evidence that all living organisms on this planet share a common ancestry. Our ancient connection to the living world may drive our curiosity, and perhaps also explain our seemingly insatiable desire for information about animals and nature. Noted zoologist, E. O. Wilson, recently coined the term "biophilia" to describe this phenomenon. The term is derived from the Greek *bios* meaning "life" and *philos* meaning "love." Wilson argues that we are human because of our innate affinity to and interest in the other organisms with which we share our planet. They are, as he says, "the matrix in which the human mind originated and is permanently rooted." To put it simply and metaphorically, our love for nature flows in our blood and is deeply engrained in both our psyche and cultural traditions.

Our own personal awakenings to the natural world are as diverse as humanity itself. I spent my early childhood in rural Iowa where nature was an integral part of my life. My father and I spent many hours collecting, identifying and studying local insects, amphibians and reptiles. These experiences had a significant impact on my early intellectual and even spiritual development. One event I can recall most vividly. I had collected a cocoon in a field near my home in early spring. The large, silky capsule was attached to a stick. I brought the cocoon back to my room and placed it in a jar on top of my dresser. I remember waking one morning and, there, perched on the tip of the stick was a large moth, slowly moving its delicate, light green wings in the early morning sunlight. It took my breath away. To my inexperienced eyes, it was one of the most beautiful things I had ever seen. I knew it was a moth, but did not know which species. Upon closer examination, I noticed two moon-like markings on the wings and also noted that the wings had long "tails", much like the ubiquitous tiger swallow-tail butterflies that visited the lilac bush in our backyard. Not wanting to suffer my ignorance any longer, I reached immediately for my *Golden Guide to North American Insects* and searched through the section on moths and butterflies. It was a luna moth! My heart was pounding with the excitement of new knowledge as I ran to share the discovery with my parents.

I consider myself very fortunate to have made a living as a professional biologist and conservationist for the past 20 years. I've traveled to over 30 countries and six continents to study and photograph wildlife or to attend related conferences and meetings. Yet, each time I encounter a new and unusual animal or habitat my heart still races with the same excitement of my youth. If this is biophilia, then I certainly possess it, and it is my hope that others will experience it too. I am therefore extremely proud to have served as the series editor for the Gale Group's rewrite of *Grzimek's Animal Life Encyclopedia*, one of the best known and widely used reference works on the animal world. *Grzimek's* is a celebration of animals, a snapshot of our current knowledge of the Earth's incredible range of biological diversity. Although many other animal encyclopedias exist, *Grzimek's Animal Life Encyclopedia* remains unparalleled in its size and in the breadth of topics and organisms it covers.

The revision of these volumes could not come at a more opportune time. In fact, there is a desperate need for a deeper understanding and appreciation of our natural world. Many species are classified as threatened or endangered, and the situation is expected to get much worse before it gets better. Species extinction has always been part of the evolutionary history of life; some organisms adapt to changing circumstances and some do not. However, the current rate of species loss is now estimated to be 1,000–10,000 times the normal "background" rate of extinction since life began on Earth some 4 billion years ago. The primary factor responsible for this decline in biological diversity is the exponential growth of human populations, combined with peoples' unsustainable appetite for natural resources, such as land, water, minerals, oil, and timber. The world's human population now exceeds 6 billion, and even though the average birth rate has begun to decline, most demographers believe that the global human population will reach 8–10 billion in the next 50 years. Much of this projected growth will occur in developing countries in Central and South America, Asia and Africa—regions that are rich in unique biological diversity.

Finding solutions to conservation challenges will not be easy in today's human-dominated world. A growing number of people live in urban settings and are becoming increasingly isolated from nature. They "hunt" in supermarkets and malls, live in apartments and houses, spend their time watching television and searching the World Wide Web. Children and adults must be taught to value biological diversity and the habitats that support it. Education is of prime importance now while we still have time to respond to the impending crisis. There still exist in many parts of the world large numbers of biological "hotspots"—places that are relatively unaffected by humans and which still contain a rich store of their original animal and plant life. These living repositories, along with selected populations of animals and plants held in professionally managed zoos, aquariums and botanical gardens, could provide the basis for restoring the planet's biological wealth and ecological health. This encyclopedia and the collective knowledge it represents can assist in educating people about animals and their ecological and cultural significance. Perhaps it will also assist others in making deeper connections to nature and spreading biophilia. Information on the conservation status, threats and efforts to preserve various species have been integrated into this revision. We have also included information on the cultural significance of animals, including their roles in art and religion.

It was over 30 years ago that Dr. Bernhard Grzimek, then director of the Frankfurt Zoo in Frankfurt, Germany, edited the first edition of *Grzimek's Animal Life Encyclopedia*. Dr. Grzimek was among the world's best known zoo directors and conservationists. He was a prolific author, publishing nine books. Among his contributions were: *Serengeti Shall Not Die*, *Rhinos Belong to Everybody* and *He and I and the Elephants*. Dr. Grzimek's career was remarkable. He was one of the first modern zoo or aquarium directors to understand the importance of zoo involvement in *in situ* conservation, that is, of their role in preserving wildlife in nature. During his tenure, Frankfurt Zoo became one of the leading western advocates and supporters of wildlife conservation in East Africa. Dr. Grzimek served as a Trustee of the National Parks Board of Uganda and Tanzania and assisted in the development of several protected areas. The film he made with his son Michael, *Serengeti Shall Not Die*, won the 1959 Oscar for best documentary.

Professor Grzimek has recently been criticized by some for his failure to consider the human element in wildlife conservation. He once wrote: "A national park must remain a primordial wilderness to be effective. No men, not even native ones, should live inside its borders." Such ideas, although considered politically incorrect by many, may in retrospect actually prove to be true. Human populations throughout Africa continue to grow exponentially, forcing wildlife into small islands of natural habitat surrounded by a sea of humanity. The illegal commercial bushmeat trade—the hunting of endangered wild animals for large scale human consumption—is pushing many species, including our closest relatives, the gorillas, bonobos and chimpanzees, to the brink of extinction. The trade is driven by widespread poverty and lack of economic alternatives. In order for some species to survive it will be necessary, as Grzimek suggested, to establish and enforce a system of protected areas where wildlife can roam free from exploitation of any kind.

While it is clear that modern conservation must take the needs of both wildlife and people into consideration, what will the quality of human life be if the collective impact of short-term economic decisions is allowed to drive wildlife populations into irreversible extinction? Many rural populations living in areas of high biodiversity are dependent on wild animals as their major source of protein. In addition, wildlife tourism is the primary source of foreign currency in many developing countries and is critical to their financial and social stability. When this source of protein and income is gone, what will become of the local people? The loss of species is not only a conservation disaster; it also has the potential to be a human tragedy of immense proportions. Protected areas, such as national parks, and regulated hunting in areas outside of parks are the only solutions. What critics do not realize is that the fate of wildlife and people in developing countries is closely intertwined. Forests and savannas emptied of wildlife will result in hungry, desperate people, and will, in the long-term lead to extreme poverty and social instability. Dr. Grzimek's early contributions to conservation should be recognized, not only as benefiting wildlife, but as benefiting local people as well.

Dr. Grzimek's hope in publishing his *Animal Life Encyclopedia* was that it would "...disseminate knowledge of the animals and love for them", so that future generations would "...have an opportunity to live together with the great diversity of these magnificent creatures." As stated above, our goals in producing this updated and revised edition are similar. However, our challenges in producing this encyclopedia were more formidable. The volume of knowledge to be summarized is certainly much greater in the twenty-first century than it was in the 1970's and 80's. Scientists, both professional and amateur, have learned and published a great deal about the animal kingdom in the past three decades, and our understanding of biological and ecological theory has also progressed. Perhaps our greatest hurdle in producing this revision was to include the new information, while at the same time retaining some of the characteristics that have made *Grzimek's Animal Life Encyclopedia* so popular. We have therefore strived to retain the series' narrative style, while giving the information more organizational structure. Unlike the original *Grzimek's*, this updated version organizes information under specific topic areas, such as reproduction, behavior, ecology and so forth. In addition, the basic organizational structure is generally consistent from one volume to the next, regardless of the animal groups covered. This should make it easier for users to locate information more quickly and efficiently. Like the original Grzimek's, we have done our best to avoid any overly technical language that would make the work difficult to understand by non-biologists. When certain technical expressions were necessary, we have included explanations or clarifications.

Considering the vast array of knowledge that such a work represents, it would be impossible for any one zoologist to have completed these volumes. We have therefore sought specialists from various disciplines to write the sections with which they are most familiar. As with the original *Grzimek's*, we have engaged the best scholars available to serve as topic editors, writers, and consultants. There were some complaints about inaccuracies in the original English version that may have been due to mistakes or misinterpretation during the complicated translation process. However, unlike the original *Grzimek's*, which was translated from German, this revision has been completely re-written by English-speaking scientists. This work was truly a cooperative endeavor, and I thank all of those dedicated individuals who have written, edited, consulted, drawn, photographed, or contributed to its production in any way. The names of the topic editors, authors, and illustrators are presented in the list of contributors in each individual volume.

The overall structure of this reference work is based on the classification of animals into naturally related groups, a discipline known as taxonomy or biosystematics. Taxonomy is the science through which various organisms are discovered, identified, described, named, classified and catalogued. It should be noted that in preparing this volume we adopted what might be termed a conservative approach, relying primarily on traditional animal classification schemes. Taxonomy has always been a volatile field, with frequent arguments over the naming of or evolutionary relationships between various organisms. The advent of DNA fingerprinting and other advanced biochemical techniques has revolutionized the field and, not unexpectedly, has produced both advances and confusion. In producing these volumes, we have consulted with specialists to obtain the most up-to-date information possible, but knowing that new findings may result in changes at any time. When scientific controversy over the classification of a particular animal or group of animals existed, we did our best to point this out in the text.

Readers should note that it was impossible to include as much detail on some animal groups as was provided on others. For example, the marine and freshwater fish, with vast numbers of orders, families, and species, did not receive as detailed a treatment as did the birds and mammals. Due to practical and financial considerations, the publishers could provide only so much space for each animal group. In such cases, it was impossible to provide more than a broad overview and to feature a few selected examples for the purposes of illustration. To help compensate, we have provided a few key bibliographic references in each section to aid those interested in learning more. This is a common limitation in all reference works, but *Grzimek's Encyclopedia of Animal Life* is still the most comprehensive work of its kind.

I am indebted to the Gale Group, Inc. and Senior Editor Donna Olendorf for selecting me as Series Editor for this project. It was an honor to follow in the footsteps of Dr. Grzimek and to play a key role in the revision that still bears his name. *Grzimek's Animal Life Encyclopedia* is being published by the Gale Group, Inc. in affiliation with my employer, the American Zoo and Aquarium Association (AZA), and I would like to thank AZA Executive Director, Sydney J. Butler; AZA Past-President Ted Beattie (John G. Shedd Aquarium, Chicago, IL); and current AZA President, John Lewis (John Ball Zoological Garden, Grand Rapids, MI), for approving my participation. I would also like to thank AZA Conservation and Science Department Program Assistant, Michael Souza, for his assistance during the project. The AZA is a professional membership association, representing 215 accredited zoological parks and aquariums in North America. As Director/William Conway Chair, AZA Department of Conservation and Science, I feel that I am a philosophical descendant of Dr. Grzimek, whose many works I have collected and read. The zoo and aquarium profession has come a long way since the 1970s, due, in part, to innovative thinkers such as Dr. Grzimek. I hope this latest revision of his work will continue his extraordinary legacy.

Silver Spring, Maryland, 2001
Michael Hutchins
Series Editor

How to use this book

Grzimek's Animal Life Encyclopedia is an internationally prominent scientific reference compilation, first published in German in the late 1960s, under the editorship of zoologist Bernhard Grzimek (1909–1987). In a cooperative effort between Gale and the American Zoo and Aquarium Association, the series has been completely revised and updated for the first time in over 30 years. Gale expanded the series from 13 to 17 volumes, commissioned new color paintings, and updated the information so as to make the set easier to use. The order of revisions is:

Volumes 8–11: Birds I–IV
Volume 6: Amphibians
Volume 7: Reptiles
Volumes 4–5: Fishes I–II
Volumes 12–16: Mammals I–V
Volume 3: Insects
Volume 2: Protostomes
Volume 1: Lower Metazoans and Lesser Deuterostomes
Volume 17: Cumulative Index

Organized by taxonomy

The overall structure of this reference work is based on the classification of animals into naturally related groups, a discipline known as taxonomy—the science in which various organisms are discovered, identified, described, named, classified, and catalogued. Starting with the simplest life forms, the lower metazoans and lesser deuterostomes, in volume 1, the series progresses through the more complex classes of animals, culminating with the mammals in volumes 12–16. Volume 17 is a stand-alone cumulative index.

Organization of chapters within each volume reinforces the taxonomic hierarchy. In the case of the volume on Protostomes, introductory chapters describe general characteristics of all organisms in these groups, followed by taxonomic chapters dedicated to Phylum, Class, Subclass, or Order. Species accounts appear at the end of the taxonomic chapters. To help the reader grasp the scientific arrangement, each type of chapter has a distinctive color and symbol:

■ = Phylum Chapter (lavender background)

◆ = Class Chapter (peach background)

⬠ = Subclass Chapter (peach background)

● = Order Chapter (blue background)

Introductory chapters have a loose structure, reminiscent of the first edition. Chapters on taxonomic groups, by contrast, are highly structured, following a prescribed format of standard rubrics that make information easy to find. These chapters typically include:

Opening section
Scientific name
Common name
Phylum
Class (if applicable)
Subclass (if applicable)
Order (if applicable)
Number of families
Thumbnail description
Main chapter
Evolution and systematics
Physical characteristics
Distribution
Habitat
Behavior
Feeding ecology and diet
Reproductive biology
Conservation status
Significance to humans

Species accounts
Common name
Scientific name
Order (if applicable)
Family
Taxonomy
Other common names
Physical characteristics
Distribution
Habitat
Behavior
Feeding ecology and diet
Reproductive biology
Conservation status
Significance to humans

Resources
Books
Periodicals
Organizations
Other

Color graphics enhance understanding

Grzimek's features approximately 3,000 color photos, including nearly 110 in the Protostomes volume; 3,500 total color maps, including approximately 115 in the Protostomes volume; and approximately 5,500 total color illustrations, including approximately 280 in the Protostomes volume. Each featured species of animal is accompanied by both a distribution map and an illustration.

All maps in *Grzimek's* were created specifically for the project by XNR Productions. Distribution information was provided by expert contributors and, if necessary, further researched at the University of Michigan Zoological Museum library. Maps are intended to show broad distribution, not definitive ranges.

All the color illustrations in *Grzimek's* were created specifically for the project by Michigan Science Art. Expert contributors recommended the species to be illustrated and provided feedback to the artists, who supplemented this information with authoritative references and animal specimens from the University of Michigan Zoological Museum library. In addition to illustrations of species, *Grzimek's* features drawings that illustrate characteristic traits and behaviors.

About the contributors

Virtually all of the chapters were written by scientists who are specialists on specific subjects and/or taxonomic groups. Sean F. Craig reviewed the completed chapters to insure consistency and accuracy.

Standards employed

In preparing the volume on Protostomes, the editors relied primarily on the taxonomic structure outlined in *Invertebrates*, edited by R. C. Brusca, and G. J. Brusca (1990). Systematics is a dynamic discipline in that new species are being discovered continuously, and new techniques (e.g., DNA sequencing) frequently result in changes in the hypothesized evolutionary relationships among various organisms. Consequently, controversy often exists regarding classification of a particular animal or group of animals; such differences are mentioned in the text. Readers should note that even though insects are protostomes, they are treated in a separate volume (Volume 3).

Grzimek's has been designed with ready reference in mind, and the editors have standardized information wherever feasible. For **Conservation Status**, *Grzimek's* follows the IUCN Red List system, developed by its Species Survival Commission. The Red List provides the world's most comprehensive inventory of the global conservation status of plants and animals. Using a set of criteria to evaluate extinction risk, the IUCN recognizes the following categories: Extinct, Extinct in the Wild, Critically Endangered, Endangered, Vulnerable, Conservation Dependent, Near Threatened, Least Concern, and Data Deficient. For a complete explanation of each category, visit the IUCN web page at <http://www.iucn.org/themes/ssc/redlists/categor.htm>.

In addition to IUCN ratings, chapters may contain other conservation information, such as a species' inclusion on one of three Convention on International Trade in Endangered Species (CITES) appendices. Adopted in 1975, CITES is a global treaty whose focus is the protection of plant and animal species from unregulated international trade.

In the Species accounts throughout the volume, the editors have attempted to provide common names not only in English but also in French, German, Spanish, and local dialects.

Grzimek's provides the following standard information on lineage in the **Taxonomy** rubric of each Species account: [First described as] *Epimenia australis* [by] Thiele, [in] 1897, [based on a specimen from] Timor Sea, at a depth of 590 ft (180 m). The person's name and date refer to earliest identification of a species. If the species was originally described with a different scientific name, the researcher's name and the date are in parentheses.

Readers should note that within chapters, species accounts are organized alphabetically by order name, then by family, and then by genus and species.

Anatomical illustrations

While the encyclopedia attempts to minimize scientific jargon, readers will encounter numerous technical terms related to anatomy and physiology throughout the volume. To assist readers in placing physiological terms in their proper context, we have created a number of detailed anatomical drawings that are found within the particular taxonomic chapters to which they relate. Readers are urged to make heavy use of these drawings. In addition, many anatomical terms are defined in the **Glossary** at the back of the book.

Appendices and index

In addition to the main text and the aforementioned **Glossary**, the volume contains numerous other elements. **For further reading** directs readers to additional sources of information about protostomes. Valuable contact information for **Organizations** is also included in an appendix. An exhaustive **Protostomes family list** records all orders of protostomes as recognized by the editors and contributors of the volume. And a full-color **Geologic time scale** helps readers understand prehistoric time periods. Additionally, the volume contains a **Subject index.**

Acknowledgements

Gale would like to thank several individuals for their important contributions to the volume. Dr. Sean F. Craig and Dr. Dennis A. Thoney, topic editors for the Protostomes volume, oversaw all phases of the volume, including creation of the topic list, chapter review, and compilation of the appendices. Neil Schlager, project manager for the Protostomes volume, and Vanessa Torrado-Caputo, associate editor at Schlager Group, coordinated the writing and editing of the text. Dr. Michael Hutchins, chief consulting editor for the series, and Michael Souza, program assistant, Department of Conservation and Science at the American Zoo and Aquarium Association, provided valuable input and research support.

Advisory boards

Series advisor

Michael Hutchins, PhD
Director of Conservation and Science/William Conway
Chair
American Zoo and Aquarium Association
Silver Spring, Maryland

Subject advisors

Volume 1: Lower Metazoans and Lesser Deuterostomes

Dennis A. Thoney, PhD
Director, Marine Laboratory & Facilities
Humboldt State University
Arcata, California

Volume 2: Protostomes

Sean F. Craig, PhD
Assistant Professor, Department of Biological Sciences
Humboldt State University
Arcata, California

Dennis A. Thoney, PhD
Director, Marine Laboratory & Facilities
Humboldt State University
Arcata, California

Volume 3: Insects

Arthur V. Evans, DSc
Research Associate, Department of Entomology
Smithsonian Institution
Washington, DC

Rosser W. Garrison, PhD
Research Associate, Department of Entomology
Natural History Museum
Los Angeles, California

Volumes 4–5: Fishes I– II

Paul V. Loiselle, PhD
Curator, Freshwater Fishes
New York Aquarium

Brooklyn, New York

Dennis A. Thoney, PhD
Director, Marine Laboratory & Facilities
Humboldt State University
Arcata, California

Volume 6: Amphibians

William E. Duellman, PhD
Curator of Herpetology Emeritus
Natural History Museum and Biodiversity Research Center
University of Kansas
Lawrence, Kansas

Volume 7: Reptiles

James B. Murphy, DSc
Smithsonian Research Associate
Department of Herpetology
National Zoological Park
Washington, DC

Volumes 8–11: Birds I–IV

Walter J. Bock, PhD
Permanent secretary, International Ornithological Congress
Professor of Evolutionary Biology
Department of Biological Sciences,
Columbia University
New York, New York

Jerome A. Jackson, PhD
Program Director, Whitaker Center for Science,
Mathematics, and Technology Education
Florida Gulf Coast University
Ft. Myers, Florida

Volumes 12–16: Mammals I–V

Valerius Geist, PhD
Professor Emeritus of Environmental Science
University of Calgary
Calgary, Alberta
Canada

Contributing writers

Charles I. Abramson, PhD
Oklahoma State University
Stillwater, Oklahoma

Tatiana Amabile de Campos, MSc
Universidade Estadual de Campinas
Campinas, Brazil

Alberto Arab, PhD

William Arthur Atkins
Atkins Research and Consulting
Normal, Illinois

Michela Borges, MSc
Universidade Estadual de Campinas
Campinas, Brazil

Geoffrey Boxshall, PhD
The Natural History Museum
London, United Kingdom

Sherri Chasin Calvo
Independent Science Writer
Clarksville, Maryland

David Bruce Conn, PhD
Berry College
Mount Berry, Georgia

Sean F. Craig, PhD
Humboldt State University
Arcata, California

Henri Jean Dumont, PhD, ScD
Ghent University
Ghent, Belgium

Gregory D. Edgecombe, PhD
Australian Museum
Sydney, Australia

Igor Eeckhaut, PhD
Mons, Belgium

Christian C. Emig, Dr. es-Sciences
Centre d'Oceanologie
Marseille, France

Kevin F. Fitzgerald, BS
Independent Science Writer

Steven Mark Freeman, PhD
ABP Marine Environmental Research
Ltd.
Southampton, United Kingdom

Rick Hochberg, PhD
Smithsonian Marine Station at Fort
Pierce
Fort Pierce, Florida

Samuel Wooster James, PhD
University of Kansas
Lawrence, Kansas

Gregory C. Jensen, PhD
University of Washington
Seattle, Washington

Reinhardt Møbjerg Kristensen, PhD
Zoological Museum
University of Copenhagen
Copenhagen, Denmark

David Lindberg, PhD
Museum of Paleontology
University of California, Berkeley
Berkeley, California

Estela C. Lopretto, PhD
Museo de La Plata
Buenos Aires, Argentina

Tatiana Menchini Steiner, PhD
Universidade Estadual de Campinas
Campinas, Brazil

Leslie Ann Mertz, PhD
Wayne State University
Detroit, Michigan

Paula M. Mikkelsen, PhD
American Museum of Natural History
New York, New York

Elizabeth Mills, MS
Portland, Maine

Katherine E. Mills, MS
Cornell University
Department of Natural Resources
Ithaca, New York

Peter B. Mordan, PhD
The Natural History Museum
London, United Kingdom

Paulo Ricardo Nucci, PhD
Universidade Estadual de Campinas
Campinas, Brazil

Erica Veronica Pardo, PhD

Amanda Louise Reid, PhD
Bulli, Australia

Patrick D. Reynolds, PhD
Hamilton College
Clinton, New York

John Riley, PhD
University of Dundee
Dundee, Scotland

Johana Rincones, PhD
Universidade Estadual de Campinas
Campinas, Brazil

Greg W. Rouse, PhD
South Australian Museum
Adelaide, Australia

Michael S. Schaadt, MS
Cabrillo Marine Aquarium
San Pedro, California

Ulf Scheller, PhD
Jarpas, Sweden

Horst Kurt Schminke, PhD
Carl von Ossietzky Universität Olden-
burg
Oldenburg, Germany

Contributing writers

Anja Schulze, PhD
Harvard University
Cambridge, Massachusetts

Mark Edward Siddall, PhD
American Museum of Natural History
New York, New York

Martin Vinther Sørensen, PhD
Zoological Museum
University of Copenhagen
Copenhagen, Denmark

Eve C. Southward, PhD, DSc
Marine Biological Association
Plymouth, United Kingdom

Tatiana Menchini Steiner, PhD
Universidade Estadual de Campinas
São Paulo, Brazil

Per A. Sundberg, PhD
Göteborg University
Göteborg, Sweden

Michael Vecchione, PhD
National Museum of Natural History
Washington, DC

Les Watling, PhD
University of Maine
Darling Marine Center

Walpole, Maine

Tim Wood, PhD
Wright State University
Dayton, Ohio

Jill Yager, PhD
Antioch College
Yellow Springs, Ohio

Contributing illustrators

Drawings by Michigan Science Art

Joseph E. Trumpey, Director, AB, MFA
Science Illustration, School of Art and Design, University of Michigan

Wendy Baker, ADN, BFA

Ryan Burkhalter, BFA, MFA

Brian Cressman, BFA, MFA

Emily S. Damstra, BFA, MFA

Maggie Dongvillo, BFA

Barbara Duperron, BFA, MFA

Jarrod Erdody, BA, MFA

Dan Erickson, BA, MS

Patricia Ferrer, AB, BFA, MFA

George Starr Hammond, BA, MS, PhD

Gillian Harris, BA

Jonathan Higgins, BFA, MFA

Amanda Humphrey, BFA

Emilia Kwiatkowski, BS, BFA

Jacqueline Mahannah, BFA, MFA

John Megahan, BA, BS, MS

Michelle L. Meneghini, BFA, MFA

Katie Nealis, BFA

Laura E. Pabst, BFA

Amanda Smith, BFA, MFA

Christina St.Clair, BFA

Bruce D. Worden, BFA

Kristen Workman, BFA, MFA

Thanks are due to the University of Michigan, Museum of Zoology, which provided specimens that served as models for the images.

Maps by XNR Productions

Paul Exner, Chief cartographer
XNR Productions, Madison, WI

Tanya Buckingham

Jon Daugherity

Laura Exner

Andy Grosvold

Cory Johnson

Paula Robbins

• • • • •

Topic overviews

What is a protostome?

Evolution and systematics

Reproduction, development, and life history

Ecology

Symbiosis

Behavior

Protostomes and humans

• • • • •

What is a protostome?

Origin of Protostomia

The term Protostomia (from the Greek "proto," meaning first, and "stoma," meaning mouth) was coined by the biologist Karl Grobben in 1908. It distinguishes a group of invertebrate animals based upon the fate of the blastopore (the first opening of the early digestive tract) during embryonic development. Animals in which the blastopore becomes the mouth are called protostomes; those in which the mouth develops after the anus are called deuterostomes (from the Greek "deutero," meaning second, and "stoma," meaning mouth).

Protostomia and Deuterostomia are considered superphyletic taxa, each containing a variety of animal phyla. Traditionally, the protostomes include the Annelida, Arthropoda, and Mollusca, and the deuterostomes comprise the Echinodermata and Chordata. Grobben was not the first biologist to recognize the distinction between these two groups, but he was the first to place importance on the fate of the blastopore as a major distinguishing criterion. Historically, the two groups are distinguished by the following criteria:

1. Embryonic cleavage pattern (that is, how the zygote divides to become a multicellular animal)

2. Fate of the blastopore

3. Origin of mesoderm (the "middle" embryonic tissue layer between ectoderm and endoderm that forms various structures such as muscles and skeleton)

4. Method of coelom formation

5. Type of larva

These developmental features are different in the two groups and can be summarized as follows:

Developmental features of protostomes

1. Cleavage pattern: spiral cleavage

2. Fate of blastopore: becomes the mouth

3. Origin of mesoderm arises from mesentoblast (4d cells)

4. Coelom formation: schizocoely

5. Larval type: trochophore larva

Developmental features of deuterostomes

1. Cleavage pattern: radial cleavage

2. Fate of blastopore: becomes the anus

3. Origin of mesoderm: pouches off gut (endoderm)

4. Coelom formation: enterocoely

5. Larval type: dipleurula

Cleavage pattern refers to the process of cell division from one fertilized cell, the zygote, into hundreds of cells, the embryo. In protostomes, the developing zygote undergoes spiral cleavage, a process in which the cells divide at a 45° angle to one another due to a realignment of the mitotic spindle. The realignment of the mitotic spindle causes each cell to divide unequally, resulting in a spiral displacement of small cells, the micromeres, that come to sit atop the border between larger cells, the macromeres. Another superphyletic term used to describe animals with spiral cleavage is Spiralia. Spiral cleavage is also called determinate cleavage, because the function of the cells is determined early in the cleavage process. The removal of any cell from the developing embryo will result in abnormal development, and individually removed cells will not develop into complete larvae.

In deuterostomes, the zygote undergoes radial cleavage, a process in which the cells divide at right angles to one another. Radial cleavage is also known as indeterminate cleavage, because the fate of the cells is not fixed early in development. The removal of a single cell from a developing embryo will not cause abnormal development, and individually removed cells can develop into complete larvae, producing identical twins, triplets, and so forth.

The fate of the blastopore has classically been used as the defining characteristic of protostomes and deuterostomes. In protostomes, the blastopore develops into the mouth, and the anus develops from an opening later in development. In deuterostomes, the blastopore develops into the anus, and the mouth develops secondarily.

Mesoderm and coelom formation are intimately tied together during development. In protostomes, the mesoderm originates from a pair of cells called mesentoblasts (also called

Protostomes have exoskeletons. When they grow, they shed their outer layer. (Photo by A. Captain/R. Kulkarni/S. Thakur. Reproduced by permission.)

4d cells) next to the blastopore, which then migrate into the blastocoel, the internal cavity of the embryo, to become various internal structures. In coelomates, the mesentoblasts hollow out to become coeloms, cavities lined by a contractile peritoneum, the myoepithelium. In protostomes, the process of coelom formation is called schizocoely. In deuterostomes, the mesoderm originates from the wall of the archenteron, an early digestive tract formed from endoderm. The archenteron pouches out to form coelomic cavities, in a process called enterocoely.

Protostomia and Deuterostomia are also characterized by different larvae. In most protostomes, the larval type is a trochophore, basically defined by the presence of two rings of multiciliated cells (prototroch and metatroch) surrounding a ciliated zone around the mouth. Most deuterostomes have a dipleurula-type larva, defined by the presence of a field of cilia (monociliated cells) surrounding the mouth.

Contemporary reexamination of Protostomia

For more than a century, biologists have divided the bilateral animals into two main lineages (the diphyletic origin of Bilateria), the most well known of which is the Protostomia/Deuterostomia split. Similar divisions include the Zygoneura/Ambulacralia-Chordonia split proposed by the German invertebrate embryologist Hatschek in 1888, the Hyponeuralia/Epineuralia split proposed by the French zoologist Cuenot in 1940, and a Gastroneuralia/Notoneuralia split proposed by the German zoologist Ulrich in 1951, among others. These divisions often emphasized different developmental and adult features, thereby leading to different names and hypotheses about animal relationships. Although none of these groups have been granted formal taxonomic rank (for example, as a subkingdom or superphylum) by the International Code of Zoological Nomenclature, the names nevertheless remain active in the literature.

Contemporary research on protostome relationships utilizes a host of methods and technologies that were unavailable to biologists in the early twentieth century, such as Grobben. Modern biologists employ electron microscopy, fluorescent microscopy, biochemistry, and a collection of molecular techniques to sequence the genome, trace embryonic development, and gain insight into the origin of various genes and gene clusters, such as *Hox* genes. Other fields of research, including cladistic analysis and bioinformatics, continue to make important contributions. The latter fields are computer-based technologies that employ algorithms and sta-

A fire worm (*Eurythoe complanata*) with venomous bristles. (Photo by A. Flowers & L. Newman. Reproduced by permission.)

than a handful of species from any phylum that meet all the traditional protostome criteria. Questions have been raised about their relationships, even among "typical" protostomes such as arthropods. For example, the only known arthropods with typical spiral cleavage are the cirripede crustaceans (barnacles such as *Balanus balanoides*) and some primitive chelicerates (such as horseshoe crabs and spiders), and these are but a few compared to the millions of species in the phylum. Moreover, some phyla, such as Ectoprocta, Nematomorpha, and Priapulida, share no developmental characters with the typical protostome, and for others, such as Gnathostomulida and Loricifera, very little developmental information exists. This has called into question the validity of the Protostomia as a natural, monophyletic group. In fact, whether or not Protostomia is accepted by biologists as monophyletic often depends upon the type of data collected, such as molecular sequences, embryology, and morphology, and how the data are analyzed.

The evolutionary origin of the Protostomia, and of the groups it includes, remains a major challenge to modern biologists. Although proof of the monophyly of the Protostomia is elusive, many of the phyla are clearly related, and make up clades that some biologists consider monophyletic. For example, in 1997 Aguinaldo et al. proposed the establishment of two clades within the Protostomia based on molecular sequence data: Ecdysozoa (the molting animals, including the Arthropoda, Nematoda, Priapulida, and Tardigrada), and Lophotrochozoa (the ciliated animals, including the Annelida, Echiura, and Sipuncula). Biologists continue to debate these hypotheses and test them with independent biochemical, developmental, molecular, and morphological data.

tistics to handle and analyze large data sets, such as lists of morphological characters and nucleotide sequences. Together with new paleontological discoveries in the fossil realm, these novel techniques and technologies provide modern biologists with a useful way to reexamine the traditional protostome relationships and to develop new hypotheses on animal relationships and evolution.

With the arrival of new information and a more encompassing examination of all the animal phyla, the modern view of Protostomia has broadened from that originally proposed by Grobben, which, at one time or another, included the following phyla: Brachiopoda, Chaetognatha, Cycliophora, Ectoprocta, Entoprocta, Echiura, Gastrotricha, Gnathostomulida, Kinorhyncha, Loricifera, Nemertinea, Nematoda, Nematomorpha, Onychophora, Phoronida, Platyhelminthes, Priapulida, Rotifera, Sipuncula, and Tardigrada. Many of these phyla contain species that display one or more developmental characters outlined by Grobben; however, it is rare to find more

A land snail crawling on grass. (Photo by JLM Visuals. Reproduced by permission.)

Resources

Books

Brusca, R. C., and G. J. Brusca. *Invertebrates,* 2nd ed. New York: Sinauer Associates, 2003.

Nielsen, C. *Animal Evolution: Interrelationships of the Living Phyla,* 2nd ed. New York: Oxford University Press, 2001.

Periodicals

Aguinaldo, A. M. A., et al. "Evidence for a Clade of Nematodes, Arthropods and Other Moulting Animals." *Nature* 387 (1997): 483–491.

Løvtrup, S. "Validity of the Protostomia-Deuterostomia Theory." *Systematic Zoology* 24 (1975): 96–108.

Winnepenninckx, B., T. Backeljau, L. Y. Mackey, J. M. Brooks, R. de Wachter, S. Kumar, and J. R. Garey. "Phylogeny of Protostome Worms Derived from 18S rRNA Sequences." *Molecular Biology and Evolution* 12 (1995): 641–649.

Rick Hochberg, PhD

Evolution and systematics

Roots and methods of systematics and classification

There is only one figure in the 1859 first edition of Charles Darwin's *On the Origin of Species*; it is what biologists now call a phylogenetic tree. A phylogenetic tree is a diagram that shows how animals have evolved from a common ancestor by branching out from it. Darwin himself did not use the term "phylogeny," but he referred to his tree as a "diagram of divergence of taxa." Darwin wrote primarily about evolution in *Origin*, but he devoted parts of Chapter XIII to classification, in which he gave a clear account of what he considered a natural system for classifying organisms: "I hold the natural system is genealogical in its arrangement, like a pedigree; but the degrees of modification which the different groups have undergone, have to be expressed by ranking them under different so-called genera, sub-families, families, sections, orders, and classes." The science of systematics, which includes taxonomy, is the oldest and most encompassing of all fields of biology, and in 1859 natural history was largely a matter of classifying. Darwin's statement may not sound particularly revolutionary to modern adherents of the theory of evolution; however, biological classification before the mid-nineteenth century had essentially been a matter of imposing some kind of order on a complex nature created by God. (Obviously, classifications at that time did not reflect any underlying process simply because the process of evolution was unknown then.) Darwin's concept quickly won acceptance among biologists, and phylogenetic trees became the standard way to depict the evolution of recent taxa and how taxa have originated from a common ancestor.

If scientists wish to classify living animals to reflect their evolutionary relationships, they must first investigate the phylogeny of the organisms in question. Darwin did not devise a method for determining phylogenetic relationships other than in very general terms, however. Although phylogenies began to appear in the late nineteenth century, they were based on subjective assessments of the morphological similarities and differences that were then regarded as indications of kinship. Even though many authors have used phylogenetic terms in discussing their systems of classification, one must bear in mind that many of the classifications found in textbooks are not based on any explicit phylogenetic analysis. In fact, it was not until the mid-twentieth century that theoretical as well as

methodological advances in the field of systematics led biologists to better supported phylogenetic hypotheses. As of 2003, most systematists and evolutionary biologists use these methods, which are known as phylogenetic systematics or cladistics, to infer relationships among various animals and present the results in the form of a cladogram or phylogenetic tree.

The basic concept in phylogenetic systematics is monophyly. A monophyletic group of species is one that includes the ancestral species and all of its descendants. Thus, a monophyletic taxon is a group of species whose members are related to one another through a shared history of descent; that is, a single evolutionary lineage. There are several taxa and names still in use that are not monophyletic; some have survived because they are still in common use—for example, "invertebrates"—and others because we know little about their evolutionary history. The basis for determining evolutionary relationships is homology, a term that refers to similarities resulting from shared ancestry. The cladistic term for this similarity is synapomorphy. It is these homologous characters that point to a common ancestry. For example, the presence of a backbone in birds, lizards and humans indicates that these three groups share a common ancestor and are thus related. Similarity, however, does not always reflect common ancestry; sometimes it points to convergent evolution. The advanced octopus eye, which in many ways resembles the human eye, is not an indication of a relationship between octopods and humans. Cladistic methods are used to distinguish between similarities resulting from a common ancestry and similarities due to other causes.

In the past, biologists used morphological characters as the primary source for investigating relationships. Most of the current taxonomic classification is based on assessment of morphological similarities and differences. Morphological characters alone, however, have obvious limitations in determining phylogeny within the animal kingdom. It is difficult to find similarities between, say, a flatworm and a sponge if the researcher must rely on gross morphology and anatomy. Some taxonomists therefore turned to embryological characteristics; many relationships among animals have been established on the basis of sperm morphology or larval biology. The advent of polymerase chain reaction (PCR) technology and direct nucleotide sequencing has brought about immense changes in the amount of information available for phylogeny evaluation. The finding that animals at all levels share large portions of the genome makes it possible to compare taxa as far apart as vertebrates and nematodes at homologous gene loci. Biologists can feel confident that they are comparing the "same" thing when they look at the base composition of a gene such as 18s rDNA, because there are so many overall genetic similarities in this gene between taxa. A staggering amount of new information has been collected from DNA; at present, all new gene sequences are deposited in banks like GenBank, which makes them available to the worldwide scientific community. Researchers are thus getting closer to finding the actual Tree of Life; electronic databases and information sharing have led them much closer to realizing Darwin's vision of a classification based on genealogy. On the other hand, these recent advances mean that many traditional views of relationships among various groups of animals are open to question. It is not easy to write about metazoan phylogeny today, knowing that so many established "truths" have already been overturned, and more will certainly be challenged in the near future.

As of 2003, the classification of animals is defined by the International Code of Zoological Nomenclature, established on January 1, 1758—the year of publication of the tenth edition of Linnaeus's *Systema Naturae*. This code regulates the naming of species, genera, and families. It states that a species name should refer to a holotype, which is a designated specimen deposited in a museum or similar institution. In theory, the holotype should be available to anyone who wishes to study it. The genus is defined by the type (typical) species, and the family name is defined by the type genus. Although the convention has developed of using a series of hierarchical categories, these other ranks are not covered by the code and are not defined in the same explicit way. The most inclusive category is kingdom, followed by phylum. Although phyla are not formally defined in the Code, and authors disagree about their definition in some instances, "phylum" is probably the category most easily recognized by nonspecialists. Members of a phylum have a similar baüplan (the German word for an architect's ground plan for a building) or body organization, and share some obvious synapomorphies, or specialized characters that originated in the last common ancestor. These similarities and shared characters are not always obvious, however; the rank of phylum for some taxa is open to debate. Molecular data have also challenged the monophyletic status of some phyla that were previously unquestioned. It is evident that the cladistic approach to systematics, combined with an ever increasing amount of data from molecular genetics, has ushered in a period of taxonomic turmoil.

Kingdoms of life

The world of living organisms can be divided into two major groups, the prokaryotes and the eukaryotes. The prokaryotes lack membrane-enclosed organelles and a nucleus, while the eukaryotes do possess organelles and a nucleus inside their membranes, and have linear chromosomes. (By 2003, however, an organelle was found in a bacterium—which overturns the assumption that these specialized compartments are unique to the eukaryotes.) The prokaryotes have been subdi-

vided into two kingdoms: Eubacteria (bacteria) and Archea (archaebacteria). The eukaryotic kingdom has been subdivided into the Animalia or Metazoa and the "animal-like organisms" or Protozoa. As of 2003, however, the protozoans are most often referred to as the Kingdom Protista; that is, eukaryotic single-celled microorganisms together with certain algae. This kingdom contains around 18 phyla that include amoebas, dinoflagellates, foraminiferans and ciliates. Kingdom Animalia contains about 34 phyla of heterotrophic multicellular organisms. The number is approximate because there is currently no consensus regarding the detailed classification of taxa into phyla. About 1.3 million living species have been described, but this number is undoubtedly an underestimate. Estimates of undescribed species range from lows of 10–30 million to highs of 100–200 million.

The beginnings of life

Clearly the prokaryotes are the most ancient living organisms, but when did they first appear? There is indirect evidence of prokaryotic organisms in some of the oldest sediments on earth, suggesting that life first appeared in the seas as soon as the planet cooled enough for life as we know it today to exist. There are three popular theories regarding the origin of life on earth. The classic theory, which dates from the 1950s, suggests that self-replicating organic molecules first appeared in the atmosphere and were deposited in the seas by rain. In the seas, these molecules underwent further reactions in the presence of energy from lightning strikes to make nucleic acids, proteins, and the other building blocks of life. More recently, the second theory has proposed that the first synthesis of organic molecules took place near deep-sea hydrothermal vents that had the necessary heat energy and chemical activity to form these molecules. The third theory maintains that organic molecules came to Earth from another planet.

The data suggest that the first eukaryotic cells appeared several billion years ago, but we know very few details about the early evolution of these eukaryotes. Although they appeared early, they probably took a few hundred million more years to develop into multicellular organisms. Eukaryotic cells are appreciably larger than prokaryotic cells and have a much higher degree of organization. Each cell has a membrane-bound nucleus with chromosomes, and a cytoplasm containing various specialized organelles that carry out different functions, including reproduction. An example of an organelle is the mitochondrion. Mitochondria serve as the sites of cell respiration and energy generation. The presence of these organelles, and their similarity to the structures and functions of free-living bacteria, suggest that bacteria were incorporated into the precursors of eukaryotic cells and lost their autonomy in the process. This scenario is referred to as the theory of endosymbiosis; in essence, it defines the eukaryotic cell as a community of microorganisms. The first endosymbionts are believed to have been ancestral bacteria incorporating other bacteria that could respire aerobically. These bacteria subsequently became the mitochondria. The accumulation of free oxygen in the oceans from photosynthesis may have triggered the evolution of eukaryotes; this hypothesis is supported by the coincidental timing of the first eukaryotic cells and a rise

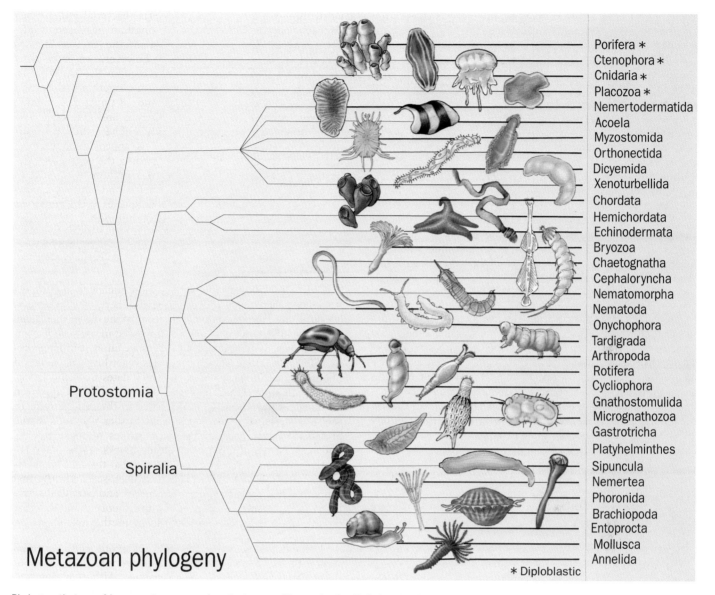

Porifera *
Ctenophora *
Cnidaria *
Placozoa *
Nemertodermatida
Acoela
Myzostomida
Orthonectida
Dicyemida
Xenoturbellida
Chordata
Hemichordata
Echinodermata
Bryozoa
Chaetognatha
Cephaloryncha
Nematomorpha
Nematoda
Onychophora
Tardigrada
Arthropoda
Rotifera
Cycliophora
Gnathostomulida
Micrognathozoa
Gastrotricha
Platyhelminthes
Sipuncula
Nemertea
Phoronida
Brachiopoda
Entoprocta
Mollusca
Annelida

Protostomia

Spiralia

Metazoan phylogeny

* Diploblastic

Phylogenetic tree of lower metazoans and protostomes. (Illustration by Christina St. Clair)

in the levels of free oxygen in the oceans. On the basis of fossil findings, scientists think that the forming of these early complex cells took place rather quickly, probably between 2800 and 2100 million years ago (mya), even though the oldest known eukaryote (*Grypania*, a coiled unbranched filament up to 1.18 in [30 mm] long) comes from rocks that are more than 2100 million years old. The earliest eukaryotic cells known to belong to any modern taxon are red algae, thought to be about 1000 million years old.

Data from the molecular clock suggest that the last common ancestor of plants and animals existed about 1.6 billion years ago, which is long after the first appearance of eukaryotes and long before any definite fossil records of metazoans. The Ediacaran fauna (600–570 mya) contains the first evidence of the existence of many modern phyla. This evidence,

however, is largely a matter of trace fossils, which result from animals moving through sediment. The relation of these trace fossils to modern phyla is therefore a matter of debate. As of 2003, biologists tend to regard the entire fauna as including many species now viewed as primitive members of extant phyla. The modern phyla thought to be represented among the Ediacaran fauna include annelid-like forms, Porifera, Cnidaria, Echiura, Onychophora, and Mollusca. There are many fossils from this period that cannot be assigned with certainty to any recent phyla; these forms probably represent high-level taxa that later became extinct. Although most of the Ediacaran organisms were preserved as shallow-water impressions in sandstone, there are around 30 sites worldwide representing deepwater and continental slope communities.

An illustration depicting what the ocean may have looked like during the Jurassic Period; present are an ammonite (*Titanites anguiformis*), based on fossils from Portland, Dorset, England, and ichthyosaurs (*Stenopterygius* sp.), based on fossils from Holtzmanaden, Germany. (Photo by Chase Studios, Inc./Photo Researchers, Inc. Reproduced by permission.)

The Ediacaran fauna was almost entirely soft-bodied, although it also included some animals that palaeontologists place among the mollusks and early arthropod-like organisms. They draw this conclusion from the fossilized remains of chitinous structures thought to be the jaws of annelid-like animals and the radulae (rows of teeth functioning as scrapers) of mollusks. Many of the animals from this period appear to have lacked complex internal structures, but by the late Ediacara period larger animals appeared that probably had internal organs, considering their size. For example, the segmented sheet-like *Dickinsonia*, which was probably a polychaete, grew as long as 39.3 in (1 m). It is unlikely that an animal of that size could survive without internal structures that digested and metabolized food. It is clear from these fossils that large and complicated metazoan animals already existed 540 mya.

The amount of fossil evidence for bilaterally symmetrical metazoans increased exponentially during the transition be-

tween the Precambrian and Cambrian Periods, about 544 mya. This transition is called the Cambrian explosion. The question is whether this sudden appearance of a number of phyla is evidence for a rapid radiation (diversification) of animal forms. Some researchers have suggested that the absence or lack of metazoan life in the early fossils is due to the simple fact that the first animals were small organisms lacking structures (like shells) that fossilized well. Some findings support the view that the first animals were microscopic; however, as has already been mentioned, there were also large animals in this period. Although there are problems with using the molecular clock method of measurement (calibrating the nodes in a phylogenetic tree based on assumptions about the rate of mutations in a molecule), metazoan phylogenies based on molecular data indicate that many recent phyla existed before the Cambrian explosion but did not leave fossil evidence until later. Some researchers have proposed that a more complicated life style, more complex interactions among animals, and especially the advent of predation were a strong selective force for developing such features as shells as antipredation devices. While the Ediacaran fauna seems to have consisted of suspension and detritus feeders who were largely passive as well as a very few active predators, animal communities during the Early Cambrian Period included most of the trophic levels found in modern marine communities. According to some authors, it is this second set of interactions that led to structures that could be fossilized. There have also been explanations based on such abiotic factors as atmospheric or geochemical changes. In either case, it is possible that although abundant fossils from the major animal phyla are found in Cambrian strata, the organisms originated in an earlier period. In other words, the so-called "Cambrian explosion" may simply reflect the difficulty of preserving soft-bodied or microscopic animals. Many paleontologists hold that these phyla originated instead during the Neoproterozoic Period during the 160 million years preceding the Cambrian explosion. Their opinion is based on findings from Ediacaran fauna.

Dating the origins of the metazoan phyla is thus controversial. The resolution of this controversy has been sought in DNA sequence data and the concept of a molecular clock. The molecular clock hypothesis assumes that the evolutionary rates for a particular gene are constant through time and across taxa, or that we can compensate for disparate rates at different times. The results from such studies differ; however, one relatively recent result indicates that protostomes and deuterostomes diverged around 544–700 mya, and that the divergence between echinoderms and chordates took place just before the Cambrian period. This study shows that mollusks, annelids, and arthropods had existed for over 100 million years before the Cambrian explosion, and echinoderms and early chordates may have arisen 50 million years prior to this explosion.

Protists

Although the term "protozoa" has been used for a long time, and ranked as a phylum for a hundred years, it is now clear that the name does not define a monophyletic group. "Protozoa" is really a name attached to a loose assemblage of primarily single-celled heterotrophic eukaryotic organisms. The Kingdom Protista contains both organisms traditionally called protozoa as well as some autotrophic groups. (The distinction between heterotrophy and autotrophy is, however, blurred in these organisms.) There are no unique features, or synapomorphies, that distinguish this kingdom from others; protists can be defined only as a grouping of eukaryotes that lack the organization of cells into tissues and organs that is seen in animals (or in fungi and plants for that matter). Current understandings of protist phylogeny and classification are in a state of constant flux. Recent molecular studies have overturned so many established classification schemes that any attempt to describe taxa within this kingdom as of 2003 risks becoming obsolete in a matter of months. In any event, however, the protists are the first eukaryotic organisms, and the forerunners of the multicellular animals known as metazoans.

One example of protists is phylum Ciliophora, the ciliates, which are very common in benthic (sea bottom) and plank-

Dorsal reconstruction of a crustacean (*Waptia fieldensis*) from the Middle Cambrian Burgess Shale of British Columbia. This small, fairly common, shrimp-like arthropod had gill branches for swimming, tail flaps for steering, and legs for walking on the seafloor. It averaged about 3 in. (7.5 cm) in length. (Photo by Chase Studios, Inc./Photo Researchers, Inc. Reproduced by permission.)

tonic communities in marine, brackish, and freshwater habitats as well as in damp soils. Several ciliates are important mutualistic endosymbionts of such ruminants as goats and sheep, in whose digestive tracts they convert plant material into a form that can be absorbed by the animal. Other examples include the euglenids and their kin. They are now placed in the phylum Kinetoplastida, but used to be part of what was called the phylum Sarcomastigophora, which, at that time, also contained the dinoflagellates (now placed in their own phylum Dinoflagellata). The phylum Kinetoplastida includes two major subgroups. The trypanosomes are the better known of the two, since several species in this group cause debilitating and often fatal diseases in humans. Species of *Leishmania* cause a variety of ailments collectively known as leishmaniasis, transmitted by the bite of sand flies. Leishmaniasis kills about a thousand people each year and infects over a million worldwide. More serious diseases are caused by members of the genus *Trypanosoma* which live as parasites in all classes of vertebrates. Chagas' disease, for example, is caused by a *Trypanosoma* species transmitted to humans by a group of hemipterans (insects with sucking mouth parts) known as assassin or kissing bugs. These insects feed on blood and often bite sleeping humans (commonly around the mouth, whence the nickname). They leave behind fecal matter that contains the infective stages of the trypanosome, which invades the body through mucous membranes or the insect's bite wound.

Other protists that cause serious diseases in humans belong to the phylum Apicomplexa. Members of the genus *Plasmodium* cause malaria, which affects millions of people in over a hundred countries. Malaria has been known since antiquity; the relationship between the disease and swampy land led to the belief that it was contracted by breathing "bad air" (*mal aria* in Italian). Nearly 500 million people around the world

A trilobite fossil (*Ceraurus pleurexanthemus*) from the Ordovician Period, found in Quebec, Canada. (Photo by Mark A. Schneider/Photo Researchers, Inc. Reproduced by permission.)

are stricken annually with malaria; of these, 1–3 million die from the disease, half of them children. The most deadly species of this genus, *Plasmodium falciparum*, causes massive destruction of red blood cells, which results in high levels of free hemoglobin and various breakdown products circulating in the patient's blood and urine. These broken-down cellular fragments lead to a darkening of the urine and a condition known as blackwater fever.

There are around 4000 described species of dinoflagellates, now assigned to their own phylum, Dinoflagellata. Many of these species are known only as fossils. There are fossils that unquestionably belong to this group dating back 240 million years; in addition, evidence from Early Cambrian rocks indicates that they were abundant as early as 540 mya. Some planktonic dinoflagellates occasionally undergo periodic bursts of population growth responsible for a phenomenon known as red tide. During a red tide, the density of these dinoflagellates may be as high as 10–100 million cells per 1.05 quarts (1 liter) of seawater. Many of the organisms that cause red tides produce toxic substances that can be transmitted to humans through shellfish. Another well-known group of protists are the amoebas (phylum Rhizopoda), a small phylum of around 200 described species. The most obvious feature of rhizopodans is that they form temporary extensions of their cytoplasm known as pseudopodia, which are used in feeding and locomotion.

Earliest metazoans

The origin of the metazoan phyla is a matter of debate; several theories are presently proposed. The syncytial theory suggests that the metazoan ancestor was a multinucleate, bi-

laterally symmetrical, ciliated protist that began to live on sea bottoms. A syncytium is a mass of cytoplasm that contains several or many nuclei but is not divided into separate cells. The principal argument in support of the syncytial theory is the presence of certain similarities between modern ciliates and acoel flatworms. Most of the objections to this hypothesis concern developmental matters and differences in general levels of complexity among the adult animals. Another proposal known as the colonial theory suggests that a colonial flagellated protist gave rise to a planuloid (free-swimming larva) metazoan ancestor. The ancestral protist, according to this theory, was a hollow sphere of flagellated cells that developed some degree of anterior-posterior orientation related to its patterns of motion, and also had cells that were specialized for separate somatic and reproductive functions. This theory has been modified over the years by various authors; most evidence as of 2003 points to the protist phylum Choanoflagellata as the most likely ancestor of the Metazoa. Choanoflagellates possess collar cells that are basically identical to those found in sponges. There are a number of choanoflagellate genera commonly cited as typifying a potential metazoan precursor, for example *Proterospongia* and *Sphaeroeca*.

The differences between the two theories may be summarized as follows. There is a ciliate ancestor in the syncytial theory; this ancestor gave rise to one lineage leading to "other protists" and the Porifera (sponges), and another lineage leading to flatworms, cnidarians, ctenophorans, flatworms, and "higher metazoans." The colonial theory posits three separate lineages: one leading to other protists, a second to Porifera, and a third to the rest of the metazoans. Both theories place Porifera at the base of the phylogenetic tree, probably because sponges are among the simplest of living multicellular organisms. They are sedentary filter feeders with flagellated cells that pump water through their canal system. Sponges are aggregates, or collections, of partially differentiated cells that show some rudimentary interdependence and are loosely arranged in layers. These organisms essentially remain at a cellular grade of organization. Porifera is the only phylum representing the parazoan type of body structure, which means that the sponges are metazoans without true embryological germ layering. Not only are true tissues absent in sponges, most of their body cells are capable of changing form and function.

Diploblastic metazoans

The next step in the direction of more complex metazoans was the evolution of the diploblastic phyla, the cnidarians and the comb jellies. "Diploblastic" refers to the presence of two germ layers in the embryonic forms of these animals. Both Cnidaria and Ctenophora are characterized by primary radial symmetry and two body layers, the ectoderm and the endoderm. One should note that some authors argue for the presence of a third germ layer in the ctenophores that is embryologically equivalent with the other two. Phylum Cnidaria includes jellyfish, sea anemones and corals, together with other less known groups. Cnidarians lack cephalization, which means that they do not reflect the evolutionary tendency to locate important body organs in or near the head.

In addition, cnidarians do not have a centralized nervous system or discrete (separate) respiratory, circulatory, and excretory organs. The primitive nature of the cnidarian bauplan is shown by the fact that they have very few different types of cells, in fact fewer than a single organ in most other metazoans. The essence of the cnidarian bauplan is radial symmetry, a pattern that resembles the spokes of a wheel, and places limits on the possible modes of life for a cnidarian. Cnidarians may be sessile, sedentary, or pelagic, but they do not move in a clear direction in the manner of bilateral cephalized creatures. The cnidarians, however, have one of the longest metazoan fossil histories. The first documented cnidarian fossil is from the Ediacaran fauna, which contains several kinds of medusae and sea pens that lived nearly 600 mya. There are two major competing theories about the ancestral cnidaria, focusing on whether the first cnidarian was polyploid or medusoid in form. According to one theory, modern Hydrozoa lie at the base of the cnidarian phylogeny (planuloid theory); other theories are inconclusive as to whether the modern Anthozoa or Hydrozoa were the first cnidarians.

Ctenophores, commonly called comb jellies, sea gooseberries, or sea walnuts, are transparent gelatinous animals. Like the cnidarians, the ctenophores are radially symmetrical diploblastic animals that resemble cnidarians in many respects. They differ significantly from cnidarians, however, in having a more organized digestive system, mesenchymal musculature, and eight rows of ciliary plates at some stage in their life history, as well as in some other features. Although the ctenophores and cnidarians are similar in their general construction, it is difficult to derive ctenophores from any existing cnidarian group; consequently, the phylogenetic position of ctenophores is an open question. Traditional accounts of ctenophores describe them as close to cnidarians but separating from them at a later point in evolutionary history. Ctenophores, however, are really quite different from cnidarians in many fundamental ways; many of the apparent similarities between the two groups may well reflect convergent adaptations to similar lifestyles rather than a phylogenetic relationship. Ctenophores have a pair of anal pores that have sometimes been interpreted as homologous with the anus of bilaterian animals (worms, humans, snails, fish, etc.). Furthermore, a third tissue layer between the endoderm and ectoderm may be a characteristic reminiscent of the Bilateria. These findings would support the phylogenetic position of ctenophores in comparison to that of the cnidarians, but recent molecular studies in fact point to a plesiomorphic position. "Plesiomorphic" refers to primitive or generalized characteristics that arose at an early stage in the evolution of a taxon. As a result of these studies, the relationship between cnidarians and ctenophores is still unsettled and is an active area of research.

Bilateral symmetry, triploblastic metazoans and the protostomes

Although there are currently several different views regarding the details of metazoan phylogeny, all analyses make a clear differentiation between the lower and diploblastic animals on the one hand and the triploblastic animals on the other. Some time after the radiate phyla evolved, animals with bilateral symmetry (a body axis with a clear front end and back end) and a third germ layer (the mesoderm) appeared. The appearance of bilateral symmetry was associated with the beginning of cephalization as the nervous system was concentrated in the head, and was accompanied by the development of longitudinal nerve cords. There are two fundamentally different patterns of mesoderm development, which are mirrored in the two major lineages of the Bilateria, the Deuterostomia and the Protostomia. This section will discuss the evolution of the protostomes. It must be emphasized, however, that metazoan phylogeny is undergoing continual revision. The molecular data are inconclusive not only regarding the relationships of some metazoan taxa to one another, but also which phyla belong to the two major clades. The following discussion of protostome evolution is derived from the most recent analyses based on combinations of DNA and morphological data.

Most authors assign about 20 phyla to the Protostomia; however, recent molecular data and cladistic analyses based on extended sets of morphological data do not agree as to whether the brachiopods and Phoronida should be included among the deuterostomes. These analyses are also incongruent when it comes to the position of several phyla; in addition, they call into question the monophyly of traditionally well-recognized taxa. Still, most authors have so far identified the flatworms (Platyhelminthes) as the first phylum to emerge among the protostomes. Platyhelminths are simple, wormlike animals lacking any apomorphies that distinguish them from the hypothetical protostome ancestor. An apomorphy is a new evolutionary trait that is unique to a species and all its descendants. The absence of apomorphies among the platyhelminths, then, means that there is no unique feature, such as the rhyncocoel found in ribbon worms, that can be used to identify the group as monophyletic. Various hypotheses regarding the origin of flatworms, their relationship to other taxa, and evolutionary patterns within the group have been hotly debated over the years. Recent DNA data even suggests that the phylum Platyhelminthes is not monophyletic and that one of the orders (Acoela) should be placed in a separate phylum. These analyses furthermore place the flatworms in a more apomorphic position on the tree, indicating that what were regarded as primitive and ancestral conditions are really either secondary losses, or that the ancestor was quite different from present notions of it.

Morphological characters still suggest that the first protostomes were vermiform (worm-shaped) animals like the ribbon worms in phylum Nemertea. Other taxa with wormlike ancestral features include the sipunculans (phylum Sipuncula) and the echiurans (phylum Echiura), which are animals that burrow into sediments on the ocean floor by using the large trunk coelom for peristalsis. The other major protostome taxa escaped from infaunal life, perhaps in part through the evolution of exoskeletons or the ability to build a tube. The emergence of the mollusks may be an instance of this transition. The most primitive mollusks are probably the vermiform aplacophorans, or worm mollusks. It seems likely that these animals arose from an early wormlike protostome, indicating that the sipunculans are related to the mollusks. This hypothesis is also supported by morphological data and 18S rDNA sequences. Except for the aplacophorans, all other mollusks have solid calcareous shells produced by glands in their mantles. These shells, which provide structural support and serve as defense mechanisms, vary greatly in size and shape.

The most diverse group of animals with exoskeletons is phylum Arthropoda, which includes over one million described species, most of them in the two classes Crustacea and Insecta. The first arthropods probably arose in Precambrian seas over 600 mya, and the true crustaceans were already well established by the early Cambrian period. The arthropods have undergone a tremendous evolutionary radiation and are now found in virtually all environments on the planet. The arthropods constitute 85% of all described animal species; this figure, however, is a gross underestimate of the actual number of arthropod species. Estimates of the true number range from 3–100 million species. The arthropods resemble the annelids in being segmented animals; in contrast to the annelids, however, they have hard exoskeletons. This feature provides several evolutionary advantages but clearly poses some problems as well. Being encased in an exoskeleton of this kind puts obvious limits on an organism's growth and locomotion. The fundamental problem of movement was solved by the evolution of joints in the body and appendages, and sets of highly regionalized muscles. The intricate problem of growth within a constraining exoskeleton was solved through the complex process of ecdysis, a specific hormone-mediated form of molting. In this process, the exoskeleton is periodically shed to allow for increases in body size. It may be that ecdysis is unique to Arthropoda, Tardigrada and Onychophora and not homologous to the cuticular shedding that occurs in several metazoan phyla. This question is unresolved as of 2003. In the phylogeny included here, the arthropods are assigned to the same clade (a group of organisms sharing a specific common ancestor) as other cuticular-shedding taxa. Some authors refer to this clade as Ecdysozoa, a classification that is receiving increasing support from various sources of information.

The development of an exoskeleton clearly conferred a great selective advantage, as evidenced by the spectacular success of arthropods with regard to both diversity and abundance. This success took place despite the need for several coincidental changes to overcome the limitations of the exoskeleton. One of the key advantages of an exoskeleton is protection—against predation and physical injury, to be sure, but also against physiological stress. The morphological diversity among arthropods has resulted largely from the differential specialization of various segments, regions, and appendages in their bodies. It is clear that segmentation, in which body structures with the same genetic and developmental origins arise repeatedly during the ontogeny of an organism, is advantageous in general and leads to evolutionary plasticity. The segmented worms of phylum Annelida exemplify this evolutionary plasticity. This taxon comprises around 16,500 species; annelids have successfully invaded virtually all habitats that have sufficient water. Annelids are found most commonly in the sea, but are also abundant in fresh water. In addition, many annelid species live comfortably in damp terrestrial environments.

Segmentation as expressed in the arthropods and annelids has traditionally been considered a character that indicates a close relationships between these two taxa, and they are often placed in the same clade. Several recent analyses dispute this picture, however; the phylogeny included here assigns Annelida and Arthropoda to two different clades. It also places the flatworms in a much more apomorphic position and not in their customary location at the base of the phylogenetic tree. Systematics and classification are unsettled as of 2003, and several "truths" about evolutionary relationships are likely to be overturned in the near future. These new phylogenies will also lead to the revision of theories regarding the evolution of behavior and characters.

Resources

Books

Brusca, Richard C., and Gary J. Brusca. *Invertebrates*, 2nd ed. Sunderland, MA: Sinauer Associates, 2003.

Clarkson, Euan N. K. *Invertebrate Palaeontology and Evolution*, 4th ed. Malden, MA: Blackwell Science, Ltd., 1999.

Felsenstein, Joseph. *Inferring Phylogenies.* Sunderland, MA: Sinauer Associates, 2003.

Futuyama, Douglas J. *Evolutionary Biology*, 2nd ed. Sunderland, MA: Sinauer Associates, 1998.

Nielsen, Claus. *Animal Evolution, Interrelationships of the Living Phyla.* Oxford, U.K.: Oxford University Press, 1995.

Periodicals

Giribet, Gonzalo, et al. "Triplobastic Relationships with Emphasis on the Acoelomates and the Position of Gnathostomulida, Cycliophora, Plathelminthes, and Chaetognatha: A Combined Approach of 18S rDNA Sequences and Morphology." *Systematic Biology* 49 (2000): 539–562.

Nielsen, Claus, et al. "Cladistic Analyses of the Animal Kingdom." *Biological Journal of the Linnaean Society* 57 (1996): 385–410.

Zrzavý, Jan S., et al. "Phylogeny of the Metazoa Based on Morphological and 18S Ribosomal DNA Evidence." *Cladistics* 14 (1998): 249–285.

Per Sundberg, PhD

Reproduction, development, and life history

Ontogeny and phylogeny

All animals must reproduce, passing copies of their genes into separate new bodies in future generations. These genetic copies may be genetically identical, produced by asexual processes, or genetically distinct, produced by sexual processes. In sexual processes, which are by far the more common among animals, the initial result of reproduction is a single cell, known as a zygote, containing the new and unique set of genes. Yet, by definition animals are multicellular, and generally consist of hundreds, thousands, or millions of cells. Even more important is the fact that the assemblage of cells that we recognize as a given animal species must be organized into a specific pattern. This pattern, when viewed as a whole, defines the morphology of the animal. The morphology, in turn, underlies a complex functional organization of the animal in which the cells are grouped into tissues (such as epidermis), organs (such as kidneys), and organ systems (such as digestive systems). Each separate group of cells within an animal's body performs a specific function in what is called the division of labor among body cells. The processes through which the single zygote becomes the complex multicellular adult animal, with many tissues, organs, and systems in their proper places, all functioning in a coordinated manner, are referred to as the animal's ontogeny.

Clearly, the ontogeny of an animal is critical to determining what kind of creature the animal is and what it can become. Thus, it should be no surprise that specific ontogenetic patterns tend to be characteristic of particular groups of animals. For example, the leg of a crab and the leg of a mouse are very different structures; the crab has an epidermal exoskeleton around muscles, whereas the mouse has muscles and epidermis around an internal bony skeleton. The crab's structural characteristics define it as a member of the phylum Arthropoda; the mouse's structural characteristics define it as a member of the phylum Chordata. Since structure must come about through ontogeny, it stands to reason that each morphologically distinct phylum of animals must also have a distinctive ontogenetic pattern. So, taking it one step further, we can reason that an animal's ontogeny (the developmental history of the individual), correlates with its phylogeny (the evolutionary history of the phylum).

Some nineteenth-century naturalists and biologists were so struck by this relationship that they argued that an ani-

mal's entire evolutionary history was repeated during the course of its embryonic development as an individual. Later studies have shown that this exact repetition does not occur. However, the pattern of ontogeny is so important to an animal's formation that a basic correlation between ontogeny and phylogeny does exist. For this reason, comparative zoologists have long regarded patterns of embryonic development as being crucial to the understanding of where each group of animals fits into the larger phylogenetic scheme.

Protostomes vs. deuterostomes

Given the long-standing recognition that embryonic developmental patterns reflect evolutionary relationships, it comes as no surprise that the two major branches of the animal kingdom are defined by differences in specific embryonic attributes. Early metazoans, such as sponges (phylum Porifera) and jellyfishes (phylum Cnidaria), have rather simple bodies that exhibit a rather high degree of developmental plasticity. However, with the advent of flatworms (phylum Platyhelminthes), we see greater complexity, and generally less plasticity. In flatworms and all higher animals, the body forms early into a three-layered embryo, and thus is said to be triploblastic. Each of the three layers is known as a germ layer, because it will form into all the organs of that body layer. The outer ectoderm layer will develop into external structures such as the epidermis, skin, exoskeleton, nervous system, and sensory structures. The middle mesoderm layer becomes internal organs such as kidneys, reproductive systems, circulatory systems, and muscles. The inner endoderm layer will develop into the gut cavity and its derivatives, such as the stomach, intestine, and liver.

Above the level of flatworms, all higher animals possess an additional mesodermal feature—a membrane-lined body cavity, or coelom. This feature is so important that these higher animals, which constitute more than 85% of animal species, are known collectively as coelomates. But the coelom forms in two very different ways, each of which corresponds generally with two very different sequences of basic embryonic events. Thus, the higher animals fall into two great branches, the Deuterostomia and the Protostomia, each defined by a unique set of embryonic characteristics. The variations relate to (1) the pattern of cleavage; (2) the fate of the embryonic

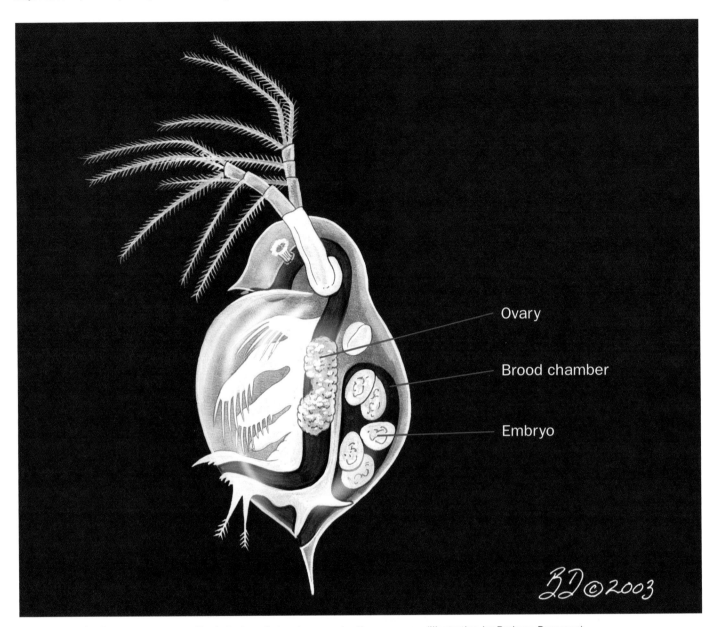

The common freshwater crustacean, *Daphnia*, brooding embryos under the carapace. (Illustration by Barbara Duperron)

blastopore formed during gastrulation; and (3) the method of coelom formation. The relevant characteristics among the protostomes are described in the following sections.

Gametogenesis

As higher animals, all sexually reproducing protostomes form gametes, generally in the form of sperm (in male systems) and oocytes (in female systems). (Oocytes are sometimes ambiguously called "eggs.") Most protostomes are gonochoristic, meaning that they have separate male and female individuals. However, hermaphroditism, or the formation of male and female gametes in the same individual, is very common in many protostome phyla, reaching high levels in groups such as the leeches and earthworms (phylum Annelida), and the pulmonate snails (phylum Mollusca). Among hermaphrodites, sperm and oocytes can be produced by the same gonad, or by separate male and female gonads, depending on the species. In any case, the basic processes of gamete formation, or gametogenesis, are fundamentally similar in all phyla. In all cases, it begins with reduction of the chromosome number from paired sets to single sets, so that subsequent joining of pairs from the male and female results in restoration of paired sets. After the reductional division, each gamete must take on structural and functional characteristics that enable it to engage in pairing with the gamete of the opposite sex.

Spermatogenesis, the formation of sperm, thus begins with reductional division, then proceeds to development of a generally motile and diminutive cell, capable of positioning itself in physical contact with the oocyte. Although it is technically true that most sperm can swim, in some species sperm can crawl, slither, or glide. In each of these cases, however, it is important to note that sperm cannot travel great distances; the various propulsion devices, therefore, are more important for small-scale positioning than for actually seeking out the oocyte. This is especially true of the many marine protostomes that spawn their naked gametes directly into the seawater. The formation of an individual sperm generally involves extreme condensation of the chromosomes, and elaboration of motility devices such as flagella, oocyte-encounter and oocyte-manipulation devices, and energy stores.

An important aspect of spermatogenesis in most species is the close synchronization of sperm development and release. The primary basis for synchronization is the maintenance of close physical contact between the spermatogenic cells throughout their development. In virtually all cases, this contact involves the actual sharing of a common cell membrane and cytoplasm among large clusters of cells. These clusters are known as spermatogenic morulae. In annelid worms, velvet worms (phylum Onychophora), and some other protostome groups, these morulae have the appearance of balls of sperm, all within a large saclike gonad or the body cavity, with the heads pointed inward and the tails pointing outward. In shrimp, lobsters, and other crustaceans, as well as insects (phylum Arthropoda), the morulae occupy individual chambers in the gonad. In snails, clams, and their relatives (phylum Mollusca), the morulae often occur in concentric rings, with the less-developed cells in the outer rings, near the gonad wall, and the more-developed cells in the central rings, near to the ducts leading to the outside.

Oogenesis, the formation of oocytes, also begins with a reductional division of the chromosomes, but then proceeds to the formation of a generally large, spherical, nonmotile cell. The oocyte does not generally contribute to preliminary positioning with the sperm, but it does play a vital role in bringing the two gametes to a point of fusing to form a single composite cell, the fertilized zygote. In fact, contrary to popular belief, it is more correct to say that the oocyte fertilizes the sperm, rather than that the sperm fertilizes the oocyte. In reality, both gametes make vital contributions to this union, but it is clearly the oocyte that is responsible for most of what happens after that. During oogenesis, the oocyte is equipped with special structures and regulatory enzymes for internalizing the sperm nucleus, directing the fusion of the two nuclei, setting up the rapid sequence of cell divisions that follow, and even establishing the patterns of division and subsequent embryonic events. Following fertilization, most protostomes develop rapidly into a fully functional feeding larva or juvenile, and the oocyte must take care of all the needs of the developing embryo until it is capable of feeding on its own. Thus, in addition to the mechanical and regulatory apparatus, the oocyte generally must contain large nutrient stores in the form of lipid- and protein-rich yolk.

Some protostomes regularly engage in sexual reproduction, yet do not require the development of both sperm and oocytes. In many of these cases, the species are technically gonochoristic, but males are rarely or never produced. However, if the offspring develop from true oocytes, with the re-

duction of chromosome number, even without subsequent fertilization, this is a form of sexual reproduction. If no true oocytes are formed by the reductional division of the chromosomes, the reproduction is asexual, even though the progeny cells look like oocytes. Whether sexual or asexual, this type of reproduction by female-only species is known as parthenogenesis. Among protostomes, some insects (phylum Arthropoda) are well known for their parthenogenetic capabilities.

Copulation, spawning, and fertilization

Because gametes are capable of limited or no motility relative to the vast habitat in which the animals live, each species must have a way of bringing the sperm and oocytes close to each other so that fertilization can occur. The mechanisms for doing this are numerous, and involve a dazzling diversity of behavioral and anatomical modifications across the spectrum of protostome life. Despite the diversity, all can be grouped generally into two broad categories, copulation and spawning.

Copulation involves various mechanisms by which one member of a mating pair physically introduces sperm into the body of its partner. In hermaphroditic species, this insemination is usually reciprocal. The precise mechanism of insemination varies among protostome groups, as does the site of insemination. Many snails (phylum Mollusca), especially marine prosobranchs, possess a large penis that can extend all the way out of the shell of the male and into the mantle cavity of the female, depositing sperm directly in the genital opening. Many male crustaceans and insects (phylum Arthropoda) have complex exoskeletal structures, derived from specific appendages or body plates, which lock mechanically with complementary plates surrounding the genital opening of the female. Some protostomes transfer special packets of sperm, known as spermatophores, to their mating partner, so that the individual sperm can be released into the female's system some time after copulation has ended. For example, male squids (phylum Mollusca) use a modified arm to place a loaded spermatophore inside the mantle cavity of a female. Some hermaphroditic leeches (phylum Annelida) actually spear their mating partner through the skin with a dartlike spermatophore, which slowly injects the sperm through the body wall following copulation. In almost all cases, whether by sperm or spermatophore transfer, copulation is followed by internal fertilization, and at least some degree of internal development. The benefits of internal fertilization and development are especially great in terrestrial environments, so virtually all terrestrial protostomes copulate. Likewise, the freshwater environments are not generally hospitable for gametes and embryos, so most freshwater protostomes are copulators, although there are some exceptions.

The vast majority of marine invertebrates are broadcast spawners, meaning that they broadcast their gametes freely into the open seawater. A few freshwater species, such as the well-known invasive zebra mussel (phylum Mollusca) also engage in broadcast spawning. In most cases of broadcast spawning, both the sperm and oocytes are spawned so that fertilization is external. But in a few groups, such as some

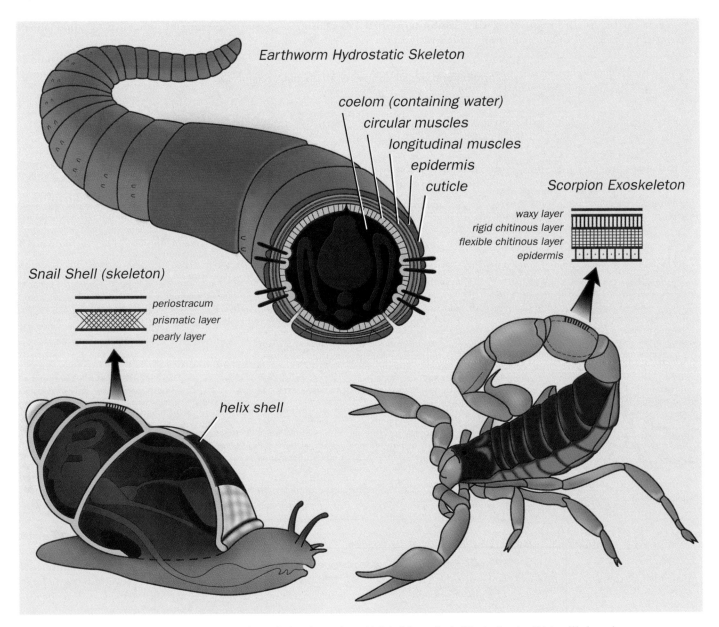

Different types of skeletons: external (snail), hydrostatic (earthworm), and jointed (scorpion). (Illustration by Kristen Workman)

clams and other bivalve mollusks, only the males spawn, leaving the adult females to draw sperm into their bodies for internal fertilization. Following internal fertilization, many species brood their young for some period of time, either internally, as in some snails, or externally in egg masses, as in some decapod crustaceans. Even among broadcast spawners with external fertilization, some species take up embryos or larvae from the open water and brood them internally, or brood them externally on the body surface.

For most protostomes, sexual reproduction is highly periodic, so copulatory and spawning behavior are also periodic. Focusing all gamete-releasing into defined periods of time is yet another way that the fully formed gametes can achieve higher rates of success in encountering one another. Among terrestrial and freshwater species, the periodicity is generally annual, occurring only at certain seasons of the year. The same may be true for marine species, particularly in near-shore environments, where seasonal runoff of rainwater from rivers provides seasonal cues for sexual activity, as well as seasonal surges in nutrients to feed the resulting larvae. In other marine environments, reproductive periodicity is often influenced more by lunar or tidal rhythms, and so may occur in monthly rather than in annual cycles.

Cleavage

Following successful fertilization, the zygote must commence formation of a multicellular embryo, a process known

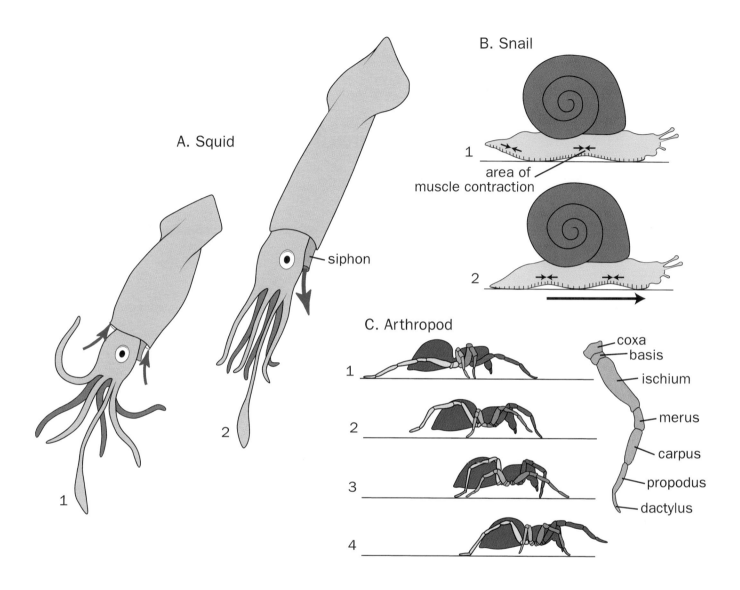

Locomotion in different animals: A. Squid propulsion; B. A snail's muscular foot; C. Leg extension in arthropods. (Illustration by Patricia Ferrer)

as embryogenesis. The actual establishment of multicellularity from the unicellular zygote involves a process known as cleavage. Cleavage involves more than simple cell division, for example, mitosis. True multicellularity involves the division of labor among cells, so each cell has to take on a special identity and developmental fate shortly after becoming independent of its progenitor cell. The process of acquiring a distinctive function is known as differentiation, and acquiring a specific developmental fate is known as determination. Protostomes generally undergo differentiation and determination very early in development, in many cases at the very first cell division of cleavage. This is easily visible under a standard microscope for some phyla, but is hidden from view by the highly modified cleavage patterns of insects, spiders, and some other arthropods.

The first thing that distinguishes protostomes from deuterostomes is this early determination. Thus, protostomes are often said to undergo determinate cleavage, or mosaic development, in contrast to the indeterminate cleavage, or regulative development, of deuterostomes. These two cleavage patterns are so different that they can be distinguished easily with a microscope. The determinate cleavage of protostomes results from a plane of cell division, usually visible after the second division, that cuts diagonally across the original zygote axis, thus compartmentalizing different regulative and nutritive chemicals in each of the resulting cells. This is referred to as spiral cleavage, since the cells dividing diagonally appear under the microscope to spiral around the original axis. In contrast, the indeterminate cleavage of deuterostomes results from planes of cell division that cut alternatively lon-

When scorpion offspring are born, the mother assists them in climbing onto her back, where they stay until their first molt. They then climb down, and live independently. (Photo by A. Captain/R. Kulkarni/S. Thakur. Reproduced by permission.)

gitudinally along the zygote axis, then transversely across the axis, thus leaving each resulting tier of cells with similar regulative and nutritive chemicals. This is referred to as radial cleavage, since the cells dividing at alternating parallel and right angles to the original axis appear under the microscope to radiate in parallel planes from that axis. The most important thing is not whether the resulting cell masses appear to spiral or to radiate, but that the spiraling cells of the protostomes show determination of specific germ layers as early as the first cell division, and almost universally by the third. Thus, at the very earliest stages of cleavages, specific cells of protostomes have already been determined to a fate of forming one of the three germ layers.

Within these basic functional forms of cleavage, there are many variations in the specific spatial configurations and the extent of cell division. Most protostomes undergo some type of holoblastic cleavage, in which the two daughter cells become completely separated, each with its own complete cell membrane. This type of cleavage may be described as either equal cleavage or unequal cleavage, depending on whether the daughter cells are equal in size. Most protostomes exhibit unequal holoblastic cleavage. In all these, the large cells are called macromeres, and they usually form the endoderm and the mesoderm. Small micromeres at the other end of the em-

bryo generally form the ectoderm. Some animals have mesomeres of an intermediate size, which may contribute to either the ectoderm or the mesoderm, depending on the species. In contrast, many arthropods with very large, heavily yolked oocytes undergo a form of incomplete cleavage known as superficial cleavage, in which the incompletely divided daughter cells ultimately reside as a layer surrounding a shared yolk mass. This appears similar to the meroblastic cleavage seen in large yolky eggs of birds and reptiles, but true superficial cleavage in arthropods begins with multiple divisions of the nuclei prior to the division of the cytoplasm.

Blastulation, gastrulation, and coelom formation

During and after cleavage, embryonic development continues with a series of rearrangements among the cells and cell layers. In the first of these, known as blastulation, the cells in the solid mass resulting from cleavage simply arrange themselves in preparation for the establishment of the spatially segregated germ layers. Blastulation begins during the middle-to-late stages of cleavage, and varies in the degree of layer organization. The final blastula stage of most protostomes is a solid mass of cells, known as a stereoblastula. Typical examples of this can be seen among many marine mollusks and annelids. In some protostomes, the blastula stage, known as a coeloblastula, is a hollow ball of cells that are arranged in a single layer around the central cavity, known as a blastocoel. The ribbon worms (phylum Nemertea) are not considered coelomates by most biologists, and therefore are not technically protostomes. However, they undergo typical spiral cleavage and develop through a coeloblastula stage and exhibit other protostome characteristics, so most biologists consider them to be closely related to the protostomes.

After blastulation, the blastula is now ready to undergo a critical process in which the three embryonic germ layers are established. This process is known as gastrulation, since it is characterized by the internalization of the endodermal cells to form the archenteron, which is the ancestral gastrointestinal tract. Gastrulation involves a specific set of cell movements that vary widely, depending on the animal group. These mechanisms range from invagination, to inward migration, to inward growth and proliferation. The end result, however, is the same. The endodermal cells are now internal, forming the archenteron gut tube, while the mesodermal cells take up residence between the endoderm and the ectoderm, which comprises cells that remained on the outside of the embryo. Regardless of how the gastrulation process takes place, the embryo is left with an opening to the outside; this opening, the blastopore, is encircled by a rim that forms the boundary between endoderm and ectoderm, and will develop into an opening into the gut in the adult animal. The precise nature of the opening is the second major defining attribute of the protostomes, in which the fate of the blastopore is to form the adult mouth. Conversely, the fate of the blastopore in deuterostomes is to form the anus.

Shortly or immediately after gastrulation is complete, protostomes form their body cavity, the coelom. By definition, a true coelom is always a body cavity within mesodermal tissue. The mechanism by which the coelom is formed is the third primary distinction between protostomes and deuterostomes. In most deuterostomes, the coelom forms by outpocketing from the original archenteron, a process known as enterocoely, since the coelomic cavities are thus derived directly from embryonic enteric cavities. In protostomes, the coelom forms from a split in the previously solid mass of mesodermal cells, a process thus known as schizocoely. There are some exceptions to this rule, but it holds true in most cases. Some protostomes lack a coelom as adults, but even these typically go through a coelomate embryonic and/or larval stage.

Larval and postlarval development

The gastrula stage is technically the last stage of embryonic development, so every stage following, up to the adult, is postembryonic. Many protostomes undergo postembryonic development that is direct. In these cases, the gastrula develops directly into a juvenile, which is typically a miniature, but sexually immature, version of the adult. The juvenile then has simply to grow and mature to become an adult. The vast majority of protostomes take a very different approach, engaging in a more complex pattern known as indirect development.

This involves the development of the gastrula into some sort of distinctive larva, which is both immature and quite different from the adult. Typically, larvae have functions in the life history that are critical to the species, yet differ from that of the adult. In most marine protostomes, the primary function of the larval form is to provide for the dispersal of the species to colonize new habitats. Larvae are generally well suited for this since they are very small, and thus easily carried freely floating in the water as plankton. This planktonic dispersal of larvae is especially well developed among the marine annelids, mollusks, and crustaceans, but also occurs in the minor protostome phyla, such as Echiura and Sipuncula.

Larvae occur in many types, depending on the phylum and species, and each of these types has been given a specific name. In the simplest forms, such as with marine mollusks and annelids, the trochophore is little more than a gastrula with bands of cilia for swimming. At the other end of the spectrum of complexity, marine crustaceans may go through a succession of anatomically distinct larval stages, such as the nauplius, zoea, or megalopa. Larvae of all groups rely on considerable nutrients as they disperse and develop, but they acquire them in different ways. Depending on the species, they are either planktotrophic, feeding on plankton as they drift, or lecithotrophic, relying on stored yolk material obtained from the mother. Regardless of the number or type of larval stages, each species will eventually undergo metamorphosis, a dramatic change of morphology into the adult form. In some species, there is an intermediate juvenile stage, so that postembryonic development is mixed, having indirect and direct components. Insects are especially variable in this regard. In the case of freshwater insects, the larval and juvenile stages are often the dominant stage in the life cycle. In some of these, such as caddisflies and mayflies, the larvae may live for one to several years, whereas the adult lives for only days. In some terrestrial insects, such as cicadas, the larvae may live up to 17 years, with the adults living only a few weeks. In cases such as these, the larva actually defines the species ecologically, and the adult is simply a short-lived stage necessary for sexual reproduction.

Sexual maturation

The final stage of postembryonic development is sexual maturation. This is preceded by the final development of critical body parts, and even of the fundamental body framework, as in the segmentation, or metamerism, of annelids and arthropods. Sexual maturation may occur immediately following embryonic development, or may be arrested for many years. Many protostomes undergo sequential cycles of sexual maturation, growing gonads and/or gametes during certain seasons, and completely losing them in others. During nonreproductive periods, such an animal may appear to be a large juvenile. Notable among these are the many marine polychaete worms (phylum Annelida) that lack distinct gonads, but whose gametes form from mesodermal peritoneal cells lining the coelom only when the proper environmental cues induce them to transform.

The most important variation among postembryonic ontogenetic strategies involves the degree to which animals can

Opalescent squid (*Loligo opalescens*) mating off of southern California, USA. (Photo by Gregory Ochocki/Photo Researchers, Inc. Reproduced by permission.)

modify their cells to perform various functions, that is, to change the developmental fate of their cells. Some species, especially among the lower metazoans, have cells that are developmentally plastic, in that the cells retain the ability to become, or to form through cell division, many different cell types, depending on the needs of the animal. Other animals tend to have cells with very limited developmental plasticity. For example, a skin cell in adult humans can only produce other skin cells; any departures from this would result in a malformation, such as skin cancer. No coelomates have as much developmental plasticity as do lower metazoans such as sponges and cnidarians. However, as a general rule among the coelomates, protostomes tend to have a lower degree of developmental plasticity than do deuterostomes.

Asexual reproduction

Asexual reproduction is defined as any reproduction that does not involve meiosis, which involves genetic recombination and reduction of chromosomal number during cell division. Asexual reproduction is not as common among protostomes as it is among the lower metazoans such as sponges, hydroids, and flatworms. However, it is quite common and well developed among some oligochaete and poly-

chaete annelids. In these, it usually involves some type of fission, in which the adult body splits into two or more pieces, each of which reconstitutes the missing parts. Although rare, asexual reproduction does occur among some groups, such as the asexual parthenogenesis of some aphids and some freshwater snails. Another type of asexual reproduction, the polyembryony of some wasps, involves a rare asexual proliferation of embryonic masses to form several individuals.

Phylum summaries

Brief summaries of the primary reproductive and developmental strategies of each generally recognized protostome phylum follow. However, the variations are great, and the short summaries below are intended only to place each phylum within the overall context of strategies and processes discussed above.

Phylum Annelida

The segmented worms are quite variable in their reproductive and developmental strategies. Most of this variation occurs along class-specific lines. For example, the vast majority of polychaetes (including clamworms, scaleworms, and fanworms), most of which are marine, are gonochoristic. In marked contrast, clitellates (earthworms, leeches, and their relatives), many of which are terrestrial and freshwater, are hermaphroditic. Most polychaetes are broadcast spawners; clitellates generally exchange sperm by mutual insemination, through copulation or spermatophore insertion. Even among the copulators, most annelids leave their young at an early embryonic stage. Some polychaetes and leeches engage in external brooding, with the young developing either directly on the outside of their body or in their burrows or tubes. Most polychaetes possess rather simple gonads, and some have dramatic synchronized spawning events that culminate in rupture of the body wall to release the gametes into the surrounding seawater. Clitellates generally have complex gonads, commensurate with their copulatory behavior and deposition of offspring within cocoons in freshwater and terrestrial environments. Cleavage is generally holoblastic, and the typical spiral pattern may be obscured by modifications in some clitellates. Most annelids of all classes produce a coeloblastula that undergoes gastrulation by an invagination process. Marine polychaetes develop into trochophore larvae that are similar to those of mollusks, possibly indicating a phylogenetic relationship. Clitellates generally undergo direct development. Asexual reproduction does not occur in most annelids, but is utilized by a few polychaetes and some freshwater oligochaete clitellates.

A tiger flatworm (*Maritigrella fuscopunctata*) using hypodermic insemination. (Photo by A. Flowers & L. Newman. Reproduced by permission.)

Phylum Onychophora

The exclusively tropical and terrestrial velvet worms are gonochoristic, with well-developed gonads and a copulatory behavior that involves the deposition of spermatophores by the male externally, onto the body of the female. Fertilization is internal, following movement of the sperm directly through the body wall into the female. Cleavage is holoblastic in most species, and superficial in others, but always resembles that of various arthropods. Some species brood their young internally. Postembryonic development is direct, and there is no record of asexual reproduction.

Phylum Tardigrada

Water bears are mostly gonochoristic, but a few hermaphroditic species are known. They have rather simple gonads, and their copulation results in internal or external fertilization, with zygotes retained within the shed cuticle of the mother. They go through holoblastic cleavage leading to a coeloblastula, and followed by direct development into a juvenile. The juvenile is smaller than the final adult, but consists of the same number of cells. Thus, maturation involves growth of cells without new cellular reproduction. Sexual parthenogenesis occurs in some, but asexual reproduction is unknown.

Phylum Arthropoda

As the overwhelmingly largest phylum in the animal kingdom, arthropods exhibit a diversity of reproductive and developmental strategies that cannot be easily summarized. Most are gonochoristic, but there are exceptions. Most arthropods are insects, and the primarily terrestrial nature of this group has resulted in reproductive systems and development patterns that have adapted to the desiccating effects of this environment. In all habitats, including the marine environment where the majority of crustaceans live, copulation is

the general rule. This occurs by numerous and varied means. Brooding and maternal care is also well developed in the phylum, with the complex societies of wasps and ants being at the very pinnacle of its development in the animal kingdom. The gonads are generally tubular throughout the phylum, but the tubes are generally regionally modified to perform a variety of sophisticated functions. Sperm and oocytes typically develop inside complex follicles from which they derive materials prior to fertilization, which is typically internal in all environments. Cleavage may be holoblastic, as in most spiders, millipedes, and crustaceans. Insects and some other groups, however, have such large yolk reserves that holoblastic cleavage is impossible, and these exhibit superficial cleavage around the central yolk mass. Marine crustaceans usually have planktonic larval forms, and larvae or nymph juveniles of many insects are important feeding stages in the life cycle. Asexual reproduction occurs in very few species, and is limited to asexual parthenogenesis and some reported cases of asexual division of embryos.

Phylum Mollusca

Reproduction is extremely varied in this second largest of all animal phyla. Among the large classes, examples of gonochorism and hermaphroditism abound, with striking sexual dimorphism in some of the former. Broadcast spawning or copulation occurs among the marine species, but copulation is more common among those of terrestrial and freshwater environments. Several species engage in sometimes-sophisticated brooding behavior. The reproductive systems are generally quite sophisticated, especially among the snails that lay complex egg masses. The vast majority of mollusks undergo holoblastic cleavage, but squids and octopuses have large yolky eggs that cleave incompletely, similar to the eggs of birds and reptiles. Blastulae may be either hollow or solid, depending on the species. Most species produce a trochophore larva, and in many, especially marine snails and bivalves, this is followed by a distinctive veliger that begins to acquire juvenile characteristics leading up to settlement. Some species brood their trochophores and veligers. The most extreme example of this involves freshwater unionid clams, which brood their modified glochidia veligers prior to releasing them to become parasites on fish. Asexual reproduction is rare, and involves only a few known cases of asexual parthenogenesis. Some populations in these cases have no males at all.

Phylum Echiura

The spoon worms are gonochoristic, and have simple gonads similar to those of annelids, to which they are closely related. With few exceptions, they are broadcast spawners, whose unequal holoblastic cleavage leads to an annelidlike tro-

chophore larva. One oddity among certain spoon worms involves a tiny dwarf male living inside the genital sac of the dramatically larger female, resulting in internal fertilization. Asexual reproduction is not known to occur.

Phylum Sipuncula

The peanut worms are gonochoristic, with rather simple gonads from which they broadcast spawn their gametes into the open seawater. Their reproductive systems are similar to those of many polychaetes, but their development is more like that of mollusks; they are thought to be related to both, but more closely to the mollusks. Holoblastic cleavage leads through gastrulation to a trochophore larva, similar to that of mollusks. Some species have more than one larval stage. Asexual reproduction has been reported to occur, but has been poorly studied.

Resources

Books

Arthur, Wallace. *The Origin of Animal Body Plans: A Study in Evolutionary Developmental Biology.* Cambridge, U.K.: Cambridge University Press, 1997.

Conn, David Bruce. *Atlas of Invertebrate Reproduction and Development,* 2nd ed. New York: Wiley-Liss, 2000.

Conn, David Bruce, Richard A. Lutz, Ya-Ping Hu, and Victor S. Kennedy. *Guide to the Identification of Larval and Postlarval Stages of Zebra Mussels, Dreissena spp. and the Dark False Mussel, Mytilopsis leucophaeata.* Stony Brook: New York Sea Grant Institute, 1993.

Gilbert, Scott F., and Anne M. Raunio, eds. *Embryology: Constructing the Organism.* Sunderland, MA: Sinauer Associates, 1997.

Strathmann, Megumi F., ed. *Reproduction and Development of Marine Invertebrates of the Northern Pacific Coast.* Seattle: University of Washington Press, 1987.

Wilson, W. H., Stephen A. Sticker, and George L. Shinn, eds. *Reproduction and Development of Marine Invertebrates.* Baltimore, MD: Johns Hopkins University Press, 1994.

Young, Craig M., M. A. Sewell, and Mary E. Rice, eds. *Atlas of Marine Invertebrate Larvae.* San Diego, CA: Academic Press, 2002.

Periodicals

Conn, David Bruce, and Colleen M. Quinn. "Ultrastructure of the Vitellogenic Egg Chambers of the Caddisfly, *Brachycentrus incanus* (Insecta: Trichoptera)." *Invertebrate Biology* 114 (1995): 334–343.

Hodgson, Alan N., and Kevin J. Eckelbarger. "Ultrastructure of the Ovary and Oogenesis in Six Species of Patellid Limpets (Gastropoda: Patellogastropoda) from South Africa." *Invertebrate Biology* 119 (2000): 265–277.

McHugh, Damhnait, and Peter P. Fong. "Do Life History Traits Account for Diversity of Polychaete Annelids?" *Invertebrate Biology* 121 (2002): 325–349.

Pernet, Bruno. "Reproduction and Development of Three Symbiotic Scaleworms (Polychaeta: Polynoidae)." *Invertebrate Biology* 119 (2000): 45–57.

Ruppert, Edward E. "The Sipuncula: Their Systematics, Biology and Evolution." *Bulletin of Marine Science* 61 (1997): 1–110.

David Bruce Conn, PhD

· · · · ·

Ecology

Introduction

Protostomes are one of the most diverse and abundant groups in the animal kingdom. Their distribution, variety, and abundance are largely the result of evolutionary adaptations to climatic changes in their respective environments. At present, they inhabit a wide range of terrestrial and marine environments. Many protostomes are familiar to most people, including spiders, earthworms, snails, mussels, and squid, to mention just a few. Their lifestyles, origins and diversity all have underlying structures and functions that depend on the interactions between abiotic (nonliving) and biotic (living) components of the environment both past and present. From the fossil record, protostomes first appeared about 600 million years ago, although researchers believe that many of the early members of this group became extinct. Those few that did survive, however, evolved and radiated, or diversified, into the variety of protostomes that biologists recognize today.

Diversity of protostomes

Protostomes are presently classified into annelids, arthropods, mollusks, brachiopods and bryozoans. The annelids are thought to include about 9,000 species known to be living in marine, freshwater, or moist soil environments. Perhaps the best-known annelid genera are *Hirudo* (leeches), *Nereis* (clamworms) and *Lumbricus* (earthworms).

By contrast, the largest group in the animal kingdom is the arthropods, which account for almost three-quarters of all living animal species, and have adapted successfully to most terrestrial and aquatic habitats around the world. Early arthropods known as trilobites are an extinct group that have been extensively described from the fossil record. With regard to present-day species, some arthropods are free-living while others are parasitic. This extremely large group of protostomes includes the crustaceans (e.g., *Astacus*, crayfish; *Carcinus*, shore crab); the myriapods (e.g., *Lulus*, millipedes); the insects (e.g., *Periplaneta*, cockroaches; *Apis*, bees); and the arachnids (e.g., *Scorpio*, scorpions; *Epeira*, web-spinning spiders).

Mollusks are the second largest group in the animal kingdom, comprising around 100,000 known living species. Most are marine (e.g., *Mytilus*, mussels; *Loligo*, squid; and *Octopus*),

although some are such well-known terrestrial animals as *Helix*, the land snail, and *Limax*, the garden slug.

Brachiopods (lampshells) and bryozoans are marine organisms that are distinguished by a feeding structure called a lophophore. Bryozoans are colony-forming animals attached at the base to the substrates on which they live. The most common bryozoans include such encrusting species as *Bowerbankia* and gelatinous colonies like *Alcyonidium*, some of which provide food and shelter to many small benthic organisms. Understanding how protostomes interact with one another and their physical environment, yet continue to survive from generation to generation is a central theme in their ecology.

Major themes in ecology

Ecology is a broad topic, yet it has a number of themes that apply to all living organisms. At its most basic level, ecology is the study of interactions between animals and the abiotic and biotic factors in their environment through the acquisition and reallocation of energy and nutrients. Ecologists also examine the cyclic transfer of these elements to sustain life processes. The major themes in ecology include limiting factors; ecosystems; population issues; ecological niches; species interactions; competition; predator and prey dynamics; feeding strategies; reproductive strategies; and biodiversity. Research related to these themes has produced some of the most complex and diverse findings in the animal kingdom when it is focused on protostomes.

Limiting factors

The concept of limiting factors in ecology is related to the control or regulation of population growth. These factors include abiotic as well as biotic aspects of the environment. For protostomes, the availability and consumption of food is an important biotic limiting factor. During the planktonic stages of many benthic (ocean bottom) protostomes, for example, the seasonal abundance of phytoplankton in the water column will directly affect the mortality of protostomes, and hence the number or success of individuals during recruitment.

Other limiting factors are abiotic, particularly temperature, salinity and light. These factors affect the type and number

A crab feeds on a dead fish, aiding in decomposition. (Illustration by Barbara Duperron)

of protostomes that can exist within a given environment. For many species, the prevailing abiotic conditions are strongly associated with the characteristics of their habitat. These features are usually well defined. The spatial distribution and abundance of protostomes in either a freshwater pond or rocky shore, for example, will tend to show clear patterns defined by both physical and biological factors.

Protostomes are poikilothermic, or cold-blooded, which means that they do not regulate their body temperature; consequently, they are at the mercy of ambient conditions. Poikilothermy does have, however, a few advantages. Poikilotherms can conserve energy needed for warmth and reallocate it to such other important body functions as flight in insects; maintaining water-jet propulsion in squid and cuttlefish; and molting of the exoskeleton in crustaceans. The chief disadvantage of poikilothermy is that at low temperatures, many protostomes either reduce or restrict their activity. Temperature gradients in aquatic environments tend to determine the distribution of certain species; for example, high temperatures are better tolerated by the shore crab *Necora puber* when exposed to desiccation (drying out) by a retreating tide, than by such subtidal species as the masted crab *Corystes cassivelaunus*.

The salt concentration of sea water is critical for most marine protostomes. Most marine species tolerate only a relatively narrow salinity range (33–35 psu); under normal conditions, however, seasonal fluctuations in salinity are usually gradual. These fluctuations may be associated with changes in seawa-

ter temperature, freshwater input from estuaries, or evaporation in enclosed water bodies. High or low salinity outside the normal range of a species will affect its ability to regulate the body's osmotic pressure relative to its surroundings. Failure to regulate this pressure can result in death. In estuaries, salinity gradients are quite pronounced and can directly influence the distribution of protostomes year-round.

The length of daylight can have a profound effect on protostome behavior by modulating internally controlled rhythms, such as physiological responses to feeding or daily locomotory activity. Timely emergence of prey coincides with dawn and dusk activities, or strictly nocturnal existence. Generally, these activities are to avoid predators that use visual cues to detect prey.

Ecosystems

In the broadest sense, an ecosystem is a functional unit made up of all the organisms or species in a particular place that interact with one another and with their environment to provide a continuous flow of energy and nutrients. The success of a species in interacting with its environment will affect the long-term success of its population as well as the functioning of the ecosystem in which it lives.

The biotic component of an ecosystem consists of autotrophic and heterotrophic organisms. All living organisms fall into one of these two categories. Autotrophic organisms (e.g., plants, many protists, and some bacteria) are essentially self-sufficient, synthesizing their own food from simple inorganic material; whereas heterotrophic organisms, including the protostomes, require the organic material produced by the autotrophs. The organisms in any ecosystem are linked by their potential to pass on energy and nutrients to others, usually in the form of waste products and either living or dead organisms. Protostomes that obtain their energy from living organisms are called consumers, of which there are two basic types, herbivores and carnivores. Herbivores consume plant material, whereas carnivores eat both herbivores and other carnivores. Detritivores obtain energy from either dead organisms or from organic compounds dispersed in the environment.

The transfer of energy from one organism to another and their feeding relationships (e.g., producers and consumers) is called a food web. The various stages of the web are called trophic levels; for example, the first level is occupied by autotrophic organisms or primary producers, and the subsequent levels are occupied in turn by heterotrophic organisms or consumers. An ecosystem constantly recycles energy and nutrients as organisms are consumed, die, and decompose, only to be assimilated by detritivores and utilized as nutrients by plants ready to produce food for consumers.

Populations

A population is defined as a group of individuals of the same species inhabiting the same area, which can be defined as a local, regional, island, continental, or marine area. In ecology, the size and nature of a population reflect the dynamic relationships among reproductive rates, survivorship, migration, and immigration. One measure of reproduction is fecundity, which is defined as the number of offspring produced by a single sexually mature female. Fecundity in protostomes is usually expressed as the average number of fertilized eggs produced in a breeding cycle or over a lifetime. Most female protostomes produce high numbers of fertilized eggs over the course of their lives. An oyster, for example, is estimated to produce on average 100 million fertilized eggs during its maturity.

Survivorship refers to the percentage of individuals that survive to reach sexual maturity; it may well have a greater impact on the size of a particular population than fecundity. In the aquatic environment, planktonic larvae are often highly vulnerable to predators; although fecundity is high in females of these species, it is offset by relatively low survivorship resulting from heavy predation. Other factors that increase mortality include starvation, diseases of various types, and cannibalism. In addition, the size of a population may increase or decrease because of migration and immigration. The migratory monarch butterfly *Danaus* is a well-known example of a species with a well-defined migratory pattern. A slightly less familiar example is the greenfly *Aphis*, which produces several generations of wingless parthenogenetic females through the summer months until autumn, when the wingless aphids produce winged females. These winged females are produced specifically for migration and dispersal. Immigration usually involves the recruitment of individuals into an existing population. In some circumstances, individuals that are displaced following the loss of a habitat may move into the habitat of a neighboring population.

Population stability is often determined by the success of reproduction and recruitment. This success will largely depend on a given species' reproductive strategy. For example, broadcast spawners rely on a high number of fertilized eggs being widely dispersed, whereas brooders produce a smaller number of eggs with limited dispersal potential. The objective of both strategies is long-term survival. Ecologists use numerical models to understand and explain the interactional dynamics of a population. At an elementary level, these models describe the growth pattern of a population in terms of the number of individuals surviving over time. When the number of individuals in a population or its growth rate neither increases nor decreases, the population is said to be stable. Several factors may be responsible for maintaining this stability, as individuals will continue to grow and reproduce. Depleted food supplies or increased predation are common examples of factors limiting population growth. The rate at which a species reproduces in order to maintain its population size is related to its population strategy.

There are two basic strategies for achieving population maintenance: *r*-strategy and *K*-strategy. *R*-strategy refers to a rapid exponential growth in population followed by overuse of resources and an equally rapid population decline. Species that rely on *r*-strategy tend to be short-lived organisms that mature quickly and often die shortly after they reproduce. They have many young with little or no investment in rearing them; few defensive strategies; and an opportunistic tendency to invade new habitats. Most pest species exhibit *r*-strategy reproduction. Species that rely on *K*-strategy, by

Zebra mussels (*Dreissena polymorpha*), seen here encrusted on a dock, have become pests to humans in some areas. (Photo by A. Flowers & L. Newman. Reproduced by permission.)

contrast, tend to grow slowly, to have relatively long life spans, and to produce very few young. They usually invest care in the rearing of their young. Most endangered species fall into this second category. The lesser octopus *Eledone*, for example, attaches its egg masses to rocks, where the adults give the eggs a certain amount of parental care until they hatch; this species is an example of an animal that exhibits the *K*-strategy. Protostomes have many reproductive strategies that are intermediate between the extremes of the *r*- and *K*-strategy.

Individuals of every species must survive long enough to reproduce successfully. Reproduction takes additional energy, and so species that can efficiently capture and ingest their food resource are most likely to reproduce at their optimal level and leave more descendants. This is a fundamental characteristic of the concept of the ecological niche.

Ecological niches

The concept of an ecological niche is fundamental to understanding the ways in which a species interacts with the abiotic and biotic factors in its environment. From a broad

Golden apple snails (*Pomacea canaliculata*) were originally introduced into Taiwan to start an escargot industry. When the snails did not do well commercially, many were released in the wild, where they have caused devastation to the local rice paddies. (Photo by Holt Studios/Nigel Cattlin/Photo Researchers, Inc. Reproduced by permission.)

perspective, the ecological niche of a species is its relative position within the community in which it lives, often referred to as its habitat. More specifically, the ecological niche also includes the ability of a species to successfully employ life history strategies that allow it to produce the maximum number of offspring. Life history strategies are behavioral responses that enable members of a species to adapt to and make use of their habitat; forage successfully for food; avoid and defend themselves against predators; and find mates and reproduce. Successful breeding will produce descendants or offspring that will carry the genetic makeup of the individual into the next generation. The second generation will continue to interact with the environment frequently repeating the life cycle of its parents.

Species interactions

A species represents the lowest taxonomic group that can be clearly defined by characteristics that separate it from another at the same taxonomic level. For example, members of a species must be capable of breeding among themselves to produce offspring; possess a genetically similar makeup; and have unique structural and functional characteristics that equip the species to survive. Some species exhibit distinctive variations in structure and function among their members. Honeybees are perhaps the best known example of morphological and functional differentiation, with individuals classi-

fied as drones, workers and queens. While all of these individuals are both members of a single colony and members of the same species, they display slight variations in their individual genetic composition, which creates major differences in their functional roles. All individual bees are adapted to performing specific tasks that ensure the survival of the colony as a whole.

There are two basic types of interactions among individual animals, namely intraspecific and interspecific. "Intraspecific" refers to interactions among animals of the same species. It incorporates the concepts of completion (for food, shelter, territory, and breeding partners) and social organization (e.g., the interactions among individuals within a colony of insects). "Interspecific" refers to interactions among members of different species, including predator and prey relationships; competition for food, shelter, and space; and such different associations as parasitism. Parasites have negative effects on their hosts, from which they obtain food, protection, and optimal conditions for survival. They often cause disease and deprive their hosts of nutrients. For example, tapeworms live inside the gut of their host, which provides them with nutrients that allow the tapeworms to grow by adding segments to their body; each segment is essentially a factory for the production of more offspring.

Competition

Competition develops when two or more species seek to acquire the same resources within a given habitat. For example, field experiments have shown that two species of barnacle, *Chthamalus stellatus* and *Balanus balanoides*, compete directly for space, substrates (surfaces to live on), and elevation within the intertidal zone of a rocky shoreline. The competition between these two species is so effective that *C. stellatus* is usually found only on the upper shore, where it is better adapted to surviving such conditions, while *B. balanoides* is generally confined to the lower shore, where it outcompetes *C. stellatus* for space. Ecologists refer to this type of interaction as competitive exclusion, or Gause's principle, named for the Russian biologist C. F. Gause, who first discovered the occurence in 1934. Gause used two species of paramecia in a series of laboratory experiments. Since then the principle of competitive exclusion, which states that only one species in a given community can occupy a given ecological niche at any one time, has been extensively documented in both laboratory and field investigations.

The concept of competition is closely linked to the concept of the ecological niche. When a given ecological niche is filled, there may be competition among species for that niche. The chances of a species becoming established depends on many factors, including migration; availability of food; the animal's ability to find suitable food; and its ability to defend itself or compete for the ecological niche. The more specialized a species, the lower its chances of finding itself in direct competition with another species. This general rule is related to the concept of resource partitioning, which refers to the sharing of available resources among different species to reduce opportunities of competition and ensure a more stable community.

The stages in the life cycles of some species appear to have evolved in order to minimize competition and thus maximize survival rates by enabling members of a species to occupy completely different habitats at different points in their life cycle. Many benthic protostomes, for example, begin their relatively brief early life as planktonic larvae feeding on other small planktonic organisms in the water column before settling to the seabed, where they undergo metamorphosis into a juvenile form. A number of insects have a short adult life span of only a few hours, following a much longer period of 2–3 years as nymphs living on weeds or decaying matter. These strategies may serve to reduce competition for resources between adults and juveniles (or adults and larvae).

Predator and prey dynamics

Predation is the flow of energy through a food web; it is an important factor in the ecology of populations, the mortality of prey, and the birth of new predators. Predation is an important evolutionary force in that the process of natural selection tends to favor predators that are more effective and prey that is more evasive. Some researchers use the term "arms race" to describe the evolutionary adaptations found in some predator-prey relationships. Certain snails, for example, have developed heavily armored shells, while their predator crabs have evolved powerful crushing claws. Prey defenses can have a stabilizing effect in predator-prey interactions when the predator serves as a strong selective agent to favor better defensive adaptations in the prey. Easily captured animals are eliminated, while others with more effective defenses may rapidly dominate a local population. Other examples of prey defenses include the camouflage of the peppered moth, and behavior such as the nocturnal activity of prey to avoid being seen by predators.

Feeding strategies

The feeding strategies of protostomes have evolved to locate, capture, and handle different types of food. These strategies can be broadly classified into four groups: 1) Suspension or filter feeding. Organisms in this group obtain their food by filtering organic material from the water column directly above the sea floor, river bed, or lake bottom. 2) Deposit feeding. These protostomes consume particles of food found on the surface of the sediment. 3) Scavenging. Scavengers are organisms that eat carrion, or dead and decaying animal matter. 4) Predation. Predators capture and ingest prey species.

Suspension feeding in mollusks, for example, involves drawing organic particles into the mantle cavity in respiratory currents and trapping them on their ciliated gills. This type of feeding is usually associated with sessile protostomes, many of which have feather-like structures adapted for collecting material from the passing water currents. Brachiopods and bry-

ozoans are suspension feeders with a specialized lophophore, which is a ring of ciliated tentacles that is used to gather food. In lowland river beds, the larval forms of the insect *Diptera* (blackflies) feed by attaching themselves to plants with hooks anchored in secreted pads of silk. The larvae trap organic particles using paired head fans.

Deposit feeders ingest detritus and other microscopic organic particles. These animals are commonly associated with muddy sediments and selectively pluck organic particles from the sediment surface, although some are less selective. The semaphore crab *Heloecius cordiformis* has a specialized feeding claw shaped like a spoon for digging through the surface layers of muddy sediments. The crab's mouthparts then sift through the sediment and extract the organic matter.

Scavengers are unselective feeders that eat when the opportunity arises. Many amphipods scavenge for animal and plant debris on the sea floor. When leeches are not sucking the blood of a host, they often feed on detritus or decaying plant and animal material. Such scavengers as octopus, shrimp, and isopods often swarm in large numbers to a piece of flesh that has fallen to the sea bed.

Predators often have highly sophisticated structures to help them locate, capture, and restrain prey whether mobile or sedentary. Often these structures include mandibles (modified jaws) that are able to crush and hold prey items. The larvae of Dobson flies are predatory; they feed on aquatic macroinvertebrates in rivers and streams. In reef habitats, such cone shells as *Conus geographus* are predators that capture their prey using a harpoon-shaped radula, which is a flexible tongue or ribbon lined with rows of teeth present in most mollusks. When the prey comes within striking range, the radula shoots out and penetrates any exposed tissue. The cone shell then releases a deadly venom known as conotoxin that paralyzes its prey (such as fish or marine polychaete worms).

Biodiversity

Biodiversity is a measurement of the total variety of life forms and their interactions within a designated geographical area. Ecologists use the concept to evaluate genetic diversity, species diversity and ecosystem diversity. Areas high in biological diversity are strongly associated with habitat complexity, for the simple reason that complex habitats provide numerous hiding places for prey, opportunities for predators to ambush their prey, and a range of substrates suitable for sessile organisms. Moreover, habitats high in biodiversity are thought to be more stable and less vulnerable to environmental change. For these reasons, conservation of biologically diverse areas is considered important to maintaining the longevity and health of an ecosystem. An instructive example of a complex habitat that provides shelter and feeds a broad range of species (including protostomes), and yet is threatened at the same time, is the coral reef.

Resources

Books

Allan, J. D. *Stream Ecology: Structure and Function of Running Waters.* New York: Chapman and Hall, 1995.

Emberlin, J. C. *Introduction to Ecology.* Plymouth, MA: Macdonald and Evans Handbooks, 1983.

Giller, P. S., and B. Malmqvist. *The Biology of Streams and Rivers.* Oxford, U.K.: Oxford University Press, 1998.

Green, N. P. O., G. W. Stout, D. J. Taylor, and R. Soper. *Biological Science, Organisms, Energy and Environment.* Cambridge, U.K.: Cambridge University Press, 1984.

Hayward, P. J., and J. S. Ryland. *Handbook of the Marine Fauna of North-West Europe.* Oxford, U.K.: Oxford University Press, 1995.

Lalli, C. M., and T. R. Parsons. *Biological Oceanography, An Introduction.* Oxford, U.K.: Elsevier Science, 1994.

Lerman, M. *Marine Biology: Environment, Diversity, and Ecology.* Redwood City, CA: Benjamin Cummings Publishing Company, 1986.

Mann, K. H. *Ecology of Coastal Waters with Implications for Management.* Malden, MA: Blackwell Science, 2000.

McLusky, D. S. *The Estuarine Ecosystem.* New York: Halsted Press, 1981.

Price, P. W. *Insect Ecology.* New York: John Wiley and Sons, 1997.

Robinson, M. A., and J. F. Wiggins. *Animal Types 1, Invertebrates.* London: Stanley Thornes, 1988.

Steven Mark Freeman, PhD

·····

Symbiosis

Introduction

Without symbiosis, living organisms would be quite different from what they are today. This is true not only because symbiotic relationships were fundamental to the separation of eukaryotes (organisms whose cells have true nuclei) from prokaryotes (cellular organisms lacking a true nucleus), but also because they represent a unique biological process without which many organisms could not exist. All herbivorous mammals and insects, for example, would starve without their cellulose-digesting mutualists; coral reefs could not form if corals were not associated with algae; and the human immune system would not be as complex as it is today if humans had not been infested so often by parasites over the course of their evolution.

The English word "symbiosis" is derived from two Greek words, *sym*, meaning "with," and *bios*, meaning "life." The term was introduced into scientific usage in 1879 by Heinrich Anton de Bary, professor of botany at the University of Strasbourg. De Bary used symbiosis in a global sense to refer to any close association between two heterospecific organisms. He explicitly referred to parasitism as a type of symbiosis, but excluded associations of short duration. According to de Bary's definition of symbiosis, the infection of humans by *Plasmodium falciparum*, the agent that causes malaria, is an instance of symbiosis whereas the pollenization of flowering plants by insects is not. As of 2003, "symbiosis" is used in two basic ways. First, the term can be used to refer to a close association between two organisms of different species that falls into one of three categories—mutualism, commensalism, or parasitism. Second, symbiosis may be used in a more restricted sense as a synonym of mutualism, to identify a relationship in which the two associated species derive benefits from each other.

Terminology

Symbiosis always implies a biological interaction between two organisms. The first organism is called the host, and is generally larger than the other or at least supports the other. The second organism is called the symbiont. It is usually the smaller of the two organisms and always derives benefits from its host. The symbiont is called an ectosymbiont when it lives on the surface of the host; it is called an endosymbiont when it is internal— that is, when it lives inside the host's digestive system, coelom, gonads, tissues or cells. The symbiosis is beneficial to the host in mutualism, neutral in commensalism, and harmful in parasitism.

These three categories are often abbreviated by the signs "+/+" for mutualism; "0/+" for commensalism; and "-/+" for parasitism, where the symbols to the left and right of the slash represent the primary effect of the association on the host and symbiont respectively. The "+" indicates that the relation is beneficial, "0" that it is neutral, and "-" that it is harmful. A symbiosis is defined as facultative if the host is not necessary over the full course of the symbiont's life cycle; it is defined as obligatory if the symbiont is dependent on the host throughout its life cycle. In addition, symbionts are described as either specific or opportunistic. A specific symbiont is associated with a few host species, the highest specificity being assigned to symbionts that infest only one species. Opportunistic symbionts, on the other hand, are associated with a wide range of hosts belonging to a wide variety of different taxonomic groups.

Some symbiotic associations are difficult to place within one of these three categories. Biologists often prefer to speak of the existence of a symbiotic continuum along which mutualism, commensalism, and parasitism shade into one another without strict dividing lines. Suckerfishes are an illustrative example of the problem of precise categorization. The suckerfish is an organism that attaches itself to large marine vertebrates (a host) by means of an anterior sucker. Some authors consider this symbiosis an example of mutualism because the suckerfishes eat ectoparasites located on the skin of the vertebrates to which they are attached; the suckerfishes are also able to conserve energy because while they are attached to a host, they allow their hosts to swim for them. But other researchers regard suckerfishes as ectocommensals because they eat the remains of their hosts' prey. They are even considered inquilines (symbionts that live as "tenants" in a host's nest, burrow, fur, etc. without deriving their nourishment from the host) on occasion because some of them live inside the buccal (cheek) cavities of certain fishes.

Life cycles are another factor that complicates the categorization of symbiotic relationships, in that some organisms move from one symbiotic state to another over the course of their life cycle. Myzostomids, for example, are tiny marine

An example of parasitism: a gastroid of the genus *Stilifer* living on the arm of a sea star. (Photo by Igor Eeckhaut. Reproduced by permission.)

worms associated with the comatulid crinoids, organisms related to sea stars. Most myzostomids are parasitic when they are young and cause deformities on the skin of their hosts. They develop, however, into ectocommensals that do not harm the crinoids except for stealing their food. Myzostomids are the oldest extant animal parasites currently known; deformities attributed to these strange worms have been identified on fossil crinoids from the Carboniferous Period, 360–286 million years ago.

The problem of categorization becomes even more complicated when organisms change their symbiotic relationships according to environmental conditions. This is the case in the mutualism between the freshwater cnidarian *Hydra*, which lives in ponds and slowly moving rivers, and the alga *Chlorella*, which lives in the cnidarian's cells. Under normal environmental conditions, the algae perform photosynthesis and release substantial amounts of carbon to the animal's cells in the form of a sugar known as maltose. In darkness, however, the flow of carbon-based compounds is reversed, with the nutrients coming from the feeding of *Hydra* being diverted by the algae. As a result, the growth of the cnidarians is reduced and the mutualist algae have become parasites.

Commensalism

Commensalism (from the Latin *com*, or "with," and *mensa*, or "table") literally refers to "eating together" but encompasses a wide range of symbiotic interactions. A commensal symbiont feeds at the same place as its host or steals the food of its host. This narrower definition is restricted to a very few organisms; most of the time, commensalism covers all associations that are neutral for the hosts, in which the commensal organisms benefit from the acquisition of a support, a means of transport, a shelter, or a food source. There are three major types of commensal relationships: phoresy (from the Greek *phoros*, "to carry"), in which the host carries or trans-

ports the phoront; aegism (from the Greek *aegidos*, or aegis, the shield of Athena), in which the host protects the aegist; and inquilism (from the Latin *incolinus*, "living inside"), in which the host shelters the inquiline in its body or living space without negative effects. Inquilism has been described by some researchers as a form of "benign squatting."

The loosest symbiotic associations are certainly the facultative phoresies. The modified crustacean *Lepas anatifera*, which is often attached to the skins of cetaceans (whales, dolphins, and porpoises) or the shields of turtles, is an instructive example of a phoresy. These crustaceans can be found hanging from floating pieces of wood as well as from members of other species. If other organisms often serve *L. anatifera* as a substrate, the association is not at all obligatory over the course of the crustacean's life cycle. The polychaete worm *Spirorbis* is a similar instance of a phoresy; its tube can be found sticking either to various types of organisms or to rocks in intertidal zones.

A stronger association exists between aegist symbionts and their hosts. Aegist coeloplanids are tiny flat marine invertebrates found on the spines of sea urchins or the skin of sea stars. These organisms are related to comb jellies or sea gooseberries, which are planktonic organisms. The coeloplanids are protected from potential predators by their host's defensive organs or structures. They eat plankton and organic materials from the water column trapped by their sticky fishing threads. The coeloplanids do not harm their echinoderm hosts even when hundreds of individuals are living on a single host. Many aegist relationships involve marine invertebrates, especially poriferans (sponges), cnidarians, and echinoderms. The associated organisms include polychaete worms; such crustaceans as crabs and shrimps; brittle stars; and even fishes.

The relationships between aegists and their hosts are often quite close. For example, the snapping shrimp *Synalpheus* lives and mates on comatulid crinoids. The crinoids have feathery rays or arms that hide the shrimps from predators. The shrimps often leave their hosts in order to feed, but are able to relocate them by smell as well as sight; they recognize the odor of a substance secreted by their hosts. This behavior is also found in some crabs, like *Harrovia longipes*, which also lives on comatulid crinoids.

Inquiline symbionts are particularly interesting to researchers. The most extraordinary inquilines, however, are the symbiotic pearlfishes. Pearlfishes belong to the family Carapidae, which includes both free-living and symbiotic fishes. The latter are associated with bivalves and ascidians. They can also be found in the digestive tubes of sea stars and the respiratory trees of sea cucumbers. Pearlfishes that have been extracted from their hosts cannot live more than a few days. They are totally adapted to their symbiotic way of life: their bodies are spindle-shaped; their fins are reduced in size; and their pigmentation is so poorly developed in some species that their internal organs are visible to the naked eye. Pearlfishes are often specific, and make use of olfaction (sense of smell) as well as vision to recognize their hosts. They must have some type of physiological adaptation that protects them against their hosts' internal fluids, such as the digestive juices of sea stars; however, these adaptations are not understood as of 2003.

Parasitism

In parasitic relationships (from the Greek *para*, "beside," and *sitos*, "food"), the parasitic organism first acquires a biotic substrate where it lives for part of its life cycle. This biotic substrate is often a food source for the parasite and sometimes a source of physiological factors essential to its life and growth. The parasite then seeks out a host.

Most parasites do not kill the hosts they infest even if they are pathogenic and cause disease. Diseases are alterations of the healthy state of an organism. Parasitic diseases may be divided into two types: structural diseases, in which the parasite damages the structural integrity of the host's tissues or organs; and functional diseases, in which the parasite affects the host's normal growth, metamorphosis, or reproduction. Parasitism is by far the most well-known symbiotic category, as many parasites have a direct or indirect impact on human health and economic trends. The causes of disease have always fascinated people since ancient times. Early humans thought that diseases were sent by supernatural forces or evil spirits as punishment for wrongdoing. It was not until the nineteenth century that scientific observations and studies led to the germ theory of disease. In 1807, Bénédicte Prévost demonstrated that the bunt disease of wheat was produced by a fungal pathogen. Prévost was the first to demonstrate the cause of a disease by experimentation, but his ideas were not accepted at that time as most people clung to the notion of spontaneous generation of life. By the end of the nineteenth century, however, Anton de Bary's work with fungi, Louis Pasteur's with yeast, and Robert Koch's with anthrax and cholera closed the debate on spontaneous generation and established the germ theory of disease.

In parasitic associations, animals are either parasites or hosts. There are three animal phyla, Mesozoa, Acanthocephala, and Pentastomida, that are exclusively parasitic. In addition, parasites are commonly found in almost all large phyla, including Platyhelminthes, Arthropoda and Mollusca. Many gastropod mollusks, for example, are parasites of such echinoderms as brittle stars and sea stars. *Stilifer linckiae* buries itself so deeply in the body wall of some sea stars that only the apex of the shell remains visible at the center of a small round hole. The gastropod's proboscis pierces the sea star's tissues and extends into the body cavity of its host, where it sucks the host's internal fluids and circulating cells. The morphology of *Stilifer linckiae* closely resembles that of free-living mollusks. In most cases, however, the body plans of parasites have been modified from those of their free-living relatives. The parasites tend to lose their external appendages and their organs of locomotion; in addition, their sense organs are commonly reduced or absent.

The rhizocephalan sacculines are a remarkable example of the evolutionary modification of a crustacean body plan. They are so profoundly adapted to parasitism that only an understanding of their early larval form allows them to be recognized as crustaceans. *Sacculina carcini*, for example, can often be observed as an orange sac on the ventral side of a crab. The early larval stages of this organism are free-living nauplii that move into the water column until they find a crab. Only female larvae seek out and attach themselves to the base

An example of commensalism: the crab *Harrovia longipes* associated with a crinoid. (Photo by Igor Eeckhaut. Reproduced by permission.)

of one of the crab's bristles. Once attached, the female larvae molt and lose their locomotory apparatus, giving rise to new forms known as kentrogon larvae. Kentrogon larvae are masses of cells, each armed with a hollow stylet or thin probe. The stylet pierces the body wall of the crab as far as the body cavity; the cell mass then passes through the stylet into the host's body. In this way the kentrogon injects itself into the crab. The internal mass proceeds to grow and differentiate into two main parts: an internal sacculina that absorbs nutrients through a complex root system gradually extending throughout the crab's body; and an external sacculina that forms after the root system, emerges from the ventral side of the crab, and develops into the true body of the female parasite. The female reproductive system opens to the outside through a pore that allows the entry of a male larva. The male larva injects its germinal cells, which eventually become spermatozoa capable of fertilizing the ova. *S. carcini* reproduces throughout the year on the crab *Carcinus maenas*, but more frequently between August and December.

Mutualism

In mutualism (from the Latin *mutuus*, "reciprocal"), the interactions between the symbiotic organisms can be as simple as a service exchange or as complex as metabolic exchanges. Mutualistic animals are associated with a range of different organisms, including bacteria, algae, or other animals. Many marine fishes, for example, are cleaned regularly of ectoparasites and damaged tissues by specialized fishes or shrimps called cleaners. The cleaners provide a valuable service by keeping the fishes free of parasites and disease; in turn, they acquire food and protection from predators. Cleaning mutualisms occur throughout the world, but are most commonly found in tropical waters. The cleaning fishes or shrimps involved in this type of mutualism establish cleaning stations on such exposed parts of the ocean floor as pieces of coral. The cleaner organisms are generally brightly colored and stand out against the background pattern of the coral. The bright colors, along with the cleaners' behavioral

Sea anemones of the genus *Heteractis* use stinging cells to capture prey such as small fish. The commensal anemonefish (*Amphiprion* sp.) have developed a way to mimic the anemone's own membrane, so that the anemone does not know that the anemonefish is a foreign animal. The fish gets protection from other fish, and since the anemonefish is territorial and tidy, it appears to keep the anemone safe and clean as well. (Photo ©Tony Wu/www.silent-symphony.com. Reproduced by permission.)

displays, attract fishes to the cleaning stations. The cleaners are then allowed to enter the mouth and gill chambers of such species as sharks, parrotfishes, grunts, angelfishes, and moray eels. Most cleaning fishes belong to the genus *Labroides*. Parasites that are removed from the cleaned fishes include copepods, isopods, bacteria, and fungi. Beside fishes, cleaning shrimps are also common in the tropics. The best-known species are the Pederson cleaner shrimp, *Periclimenes pedersoni*, and the banded coral shrimp, *Stenopus hispidus*. When the fishes approach their cleaning stations, these shrimps wave their antennae back and forth until the fishes get close enough for the shrimps to climb on them. Experiments have shown that the cleaners control the spread of parasites and infections among members of their host species. Cleaning symbioses also occur between land organisms: for example, various bird species remove parasites from crocodiles, buffalo and cattle; and the red rock crab *Grapsus grapsus* cleans the iguana *Amblyrhynchus subcristatus*.

The most evident mutualism between oceanic species is the one that exists between sea anemones and clown fishes. Fishes of the genera *Amphiprion*, *Dascyllus*, and *Premnas* are commonly called clown fishes due to their striking color patterns. The symbiosis is obligatory for the fish but facultative for the anemones. The brightly colored clown fishes attract larger predator fishes that sometimes venture too close to the anemones; they can be stung by the anemone tentacles, killed, and eaten. Clown fishes share in the meal and afterward remove wastes and fragments of the prey from the anemone. A number of experiments have been conducted in order to understand why the clown fishes are immune to the stinging tentacles of the sea anemones when other fishes are not. It is known that the clown fishes must undergo a period of acclimation before they are protected from the anemones. Further studies showed that the mucous coating of the clown fishes changes during this period of acclimation, after which the anemones no longer regard them as prey. The change in the mucous coating was first thought to result from fish secretions, but researchers were able to demonstrate that it results from the addition of mucus from the anemones themselves.

Resources

Books

Ahmadjian, Vernon, and Surindar Paracer. *Symbiosis: An Introduction to Biological Associations.* Hanover and London: University Press of New England, 1986.

Baer, Jean G. *Animal Parasites.* London: World University Library, 1971.

Douglas, Angela E. *Symbiotic Interactions.* Oxford, U.K.: Oxford University Press, 1994.

Morton, Bryan. *Partnerships in the Sea: Hong Kong's Marine Symbioses.* Leiden, The Netherlands: E. J. Brill, 1989.

Noble, Elmer R., and Glenn A. Noble. *Parasitology: The Biology of Animal Parasites.* Philadelphia: Lea and Febiger, 1982.

Periodicals

Eeckhaut, I., D. van den Spiegel, A. Michel, and M. Jangoux. "Host Chemodetection by the Crinoid Associate *Harrovia longipes* (Crustacea: Brachyura: Eumedonidae) and a Physical Characterization of a Crinoid-Released Attractant." *Asian Marine Biology* 17 (2000): 111–123.

Van den Spiegel, D., I. Eeckhaut, and M. Jangoux. "Host Selection by the Shrimp *Synalpheus stimpsoni* (De Man 1888), an Ectosymbiont of Comatulid Crinoids, Inferred by a Field Survey and Laboratory Experiments." *Journal of Experimental Biology and Ecology* 225 (1998): 185–196.

Igor Eeckhaut, PhD

Behavior

The large number and sheer diversity of protostomes necessitates a restriction on the kinds of behaviors (and species) that can be discussed in this chapter. The behaviors highlighted here are based in part on their importance to the survival of an individual organism.

Protostomes are some of the most morphologically complex, ecologically diverse, and behaviorally versatile organisms in the animal kingdom. They consist of more than one million species divided into approximately 20 phyla. Major representative phyla include the Platyhelminthes (flukes, planarians, and tapeworms), Nematoda (roundworms), Mollusca (chitons, clams, mussels, nautiluses, octopods, oysters, snails, slugs, squids, and tusk shells), Annelida (bristleworms, earthworms, leeches, sandworms, and tubeworms), and Arthropoda (ants, centipedes, cockroaches, crabs, crayfish, lobsters, millipedes, scorpions, spiders, and ticks).

When considering what a protostome is, it is important to note that the answer appears to be changing with the accumulation of new information. Evidence from studies of morphology and the fossil record generally support the view that animals in the phyla Annelida, Mollusca, and Arthropoda are indeed protostomes. However, new data based on rRNA analysis suggests that some animals in the Pseudocoelomate phyla (gastrotrichs, rotifers, and roundworms) and in the Acoelomate phyla (flukes, planarians, tapeworms, and ciliated worms) are also protostomes.

Despite the impressive diversity of organisms in this group, the vast majority of protostomes share certain basic characteristics of embryonic development. Indeed, the very name protostomes means "first mouth" and nicely illustrates the common characteristic that the initial opening to the digestive tract in the embryo develops into a mouth. Additional protostome characteristics include an embryonic stage known in the literature as mosaic development. Mosaic development produces a series of cell divisions (cleavage patterns) in which the fate of individual cells following the first cell division is fixed (determinate cleavage), while subsequent cell divisions are arranged spirally (spiral cleavage). Moreover, in protostomes the origin of the mesoderm (the germ layer producing such structures as the heart, muscles, and circulatory organs) is created from both the ectoderm (the germ layer producing the skin or integument, nervous system, mouth and anal canals) and endoderm (the germ layer producing the linings of the digestive tract and related glands) in a region known as the 4d cell. Protostomes also have an internal body cavity situated between the digestive tract and the body wall known as the coelom. Two cylindrical masses constructed from mesodermal cells split and the resulting cavities enlarge and combine to form an internal body cavity (coelom) that is surrounded on all sides by mesoderm cells (schizocely).

All protostomes must engage in activities that lead to survival and reproduction. The honey bee and ant, for example, must find and digest food and protect the colony. The planarian and crab must also meet nutritional requirements, reproduce, and defend themselves, but do so often in an aquatic environment. An earthworm is faced with similar problems of survival, but usually solves them underground. Protostomes that fly, swim, or burrow are all faced with the same set of problems. The solutions to these problems represent an interaction of environment and morphology, and here lies the differences in what is called behavior.

The word "behavior" is ambiguous. A physiologist, for example, may be comfortable describing the "behavior of a neuron," while a behavioral scientist might find this objectionable. Moreover, among behavioral scientists there are often discrepancies in the definition of behavior. John B. Watson, who popularized an early form of "behaviorism" in the early twentieth century, once defined behavior as muscle contractions and glandular secretions. Other behavioral scientists such as B. F. Skinner have used several definitions of behavior, including "the movement of an organism in space in relation either to its point of origin or to some other object." Many of these definitions give a novice the impression that, for behavioral scientists, the subject matter consists of bodily movements and mechanical responses. Many define behavior not as movement of an organism (which is the proper study of kinesiology), but as an act. Defining behavior in terms of actions captures the notion that behavior has consequences, in other words, scientists are primarily interested in what an organism "does." By defining behavior in terms of actions and consequences, the focus of a behavioral analysis is not on the individual movements that constitute a behavior (as important as this is), but what the behavior "accomplishes." For instance, how an organism acts in a social situation, responds to threats, or captures food is

A clam siphon at work. (Photo by Nancy Sefton/Photo Researchers, Inc. Reproduced by permission.)

intimately related to its body plan. Protostomes have a symmetrical body plan (e.g., planarians, earthworms, lobsters, or ants). One of the more interesting body plans is radial symmetry. Animals with radial symmetry have no front or back and take the general form of a cylinder (e.g., sea stars, and sea anemones) with various body parts connected to a main axis. Such animals have feeding structures and sensory systems that interact with their environment in all directions. Such a body plan is most common among animals that are permanently attached to a substrate (e.g., sea anemones) or drifting in the open seas (e.g., jellyfishes).

Another type of symmetrical body plan found in protostomes is bilateral symmetry. Invertebrates with bilateral symmetry (e.g., planarians, earthworms, crustaceans, insects, and spiders) have a definite front and back, left and right, and backside and underside orientation. Animals with such a body plan generally can control their locomotion, unlike sessile or drifting species (radial symmetry). The front end (anterior) contains an assortment of feeding and sensory structures, often encapsulated in a head (cephalization) that confronts the environment first. Moreover, the underside (ventral surface) typically contains structures necessary for locomotion, and the backside (dorsal) becomes specialized for protection.

Behavior

Feeding behavior

Feeding behavior consists of several different types of acts associated with discovery, palatability, and ingestion. The expression of feeding behavior is a combination of evolutionary and environmental pressures. Depending on the species, protostomes consume an infinite variety of food ranging from microscopic organisms, vegetable matter, and other protostomes; some even grow their own food. Despite the large and varied number of protostomes, some generalizations can be found. First, the strategies for finding food can be reduced to those organisms that find food by living on it, foraging for it, waiting for it to pass by, growing it, and having other organisms

provide it. Second, the mechanisms associated with feeding can be reduced to those that singly and/or in combination feed by suspension, deposition, macroherbivory and predation.

SUSPENSION FEEDERS

Crustaceans such as daphnids, brine shrimp, copepods, and ostracods (i.e., those not considered in the class Malacostraca) are excellent examples of filter- or suspension-feeding protostomates. Suspension feeders obtain food by either moving through the water or by remaining stationary. In both cases, bacteria, plankton, and detritus flow through specially designed feeding structures.

Interestingly, because of the large amount of energy required to continuously filter water, there are relatively few protostomes that actually use continuous filtration. A less expensive and also the most commonly employed strategy are to develop specialized filter mechanisms that contain a "sticky" substance such as mucus. An example of this is found in some species of tube-dwelling polychetes that direct water through their burrows and trap food particles in mucus. The mucus is then rolled up into a "food pellet" and manipulated by ciliary action to the mouth where it is consumed.

DEPOSIT FEEDERS

Well-known examples of protostomes that obtain food from mud and terrestrial soils include most earthworms and some snails. Direct deposit feeders extract microscopic plant matter and other nourishment by swallowing sediment. Such feeders can be either burrowing or selective. In contrast to burrowing feeders such as earthworms, selective feeders obtain food from the upper layers of sediment.

MACROHERBIVORY FEEDERS

Macroherbivory feeders obtain food by consuming macroscopic plants. One of the best protostome examples of plant feeders is the order Orthoptera (crickets, locust, and grasshoppers). Members of this order have developed specialized mouthparts and muscle structures to bite and chew. The African Copiphorinae, for example, uses its large jaws to open seeds. Biting and chewing mouthparts are also seen in beetles and many orders of insects. Two other types of mouthparts common to macroherbivory feeders are sucking and piercing. Sucking mouthparts enable insects such as butterflies and honey bees to gather nectar, pollen, and other liquids. Protostomes such as cicadas feed by drawing blood or plant juices. The leaf cutter ants (*Atta cephalotes*) are interesting example of macroherbivory feeders. These ants cut leaves and flowers and transport them to their nests where they are used to grow a fungus that is their main food source. A related feeding behavior is also found in termites (Isoptera). Termites of the species *Longipeditermes longipes* forage for detritus such as rotting leaves that become a culture medium for the fungi on which they feed.

PREDATORY FEEDERS

Arguably the most sophisticated protostome feeders are those that obtain food by hunting, which requires the animal to locate, pursue, and handle prey. Most invertebrates locate prey by chemoreception; others use vision, tactile, or vibration, or some combination thereof. Predators can be classified as stalkers, lurkers, sessile opportunists, or grazers.

Planarians (Platyhelminthes) are an excellent example of animals that obtain food through hunting. The vast majority of planarians are carnivorous. They are active and efficient hunters because of their mobility and sensory systems. They feed on many different invertebrates, including rotifers, nematodes, and other planarians, and have several different methods of capture. One of the most common methods is to wrap their body around a prey item and secure it with mucus. An interesting example of this behavior can be found in terrestrial planarians. The terrestrial planarian *Microplana termitophaga* feeds on termites by living near termite mound ventilation shafts. The planarian stretches itself into the shaft and waves its head until a termite comes in contact, at which time the termite becomes stuck on the mucus produced by the worm. An interesting note is that it is not generally agreed upon that Platyhelminthes are protostomes.

Another interesting method that protostomes use to stalk prey can be found in members of the phylum Onychophora. These are wormlike animals that some scholars believe bridge the gap between annelids and arthropods. The velvet worm *Macroperipatus torquatus* forages nocturnally on crickets and other selected invertebrates and approaches its prey undetected by utilizing slow movements. When the potential prey is recognized as an item to be consumed, the worm attacks it by enmeshing the organism in a glue-like substance squirted from the oral cavity.

Perhaps the most well-known examples of hunting protostomes are the spiders in the phylum Arachnida. Members of the family Lycosidae, colloquially known as wolf spiders, can hunt by day, although some species hunt at night. Some wolf spiders pounce on prey from their burrow, while others actively leave the burrow on hunting trips. The jumping spiders of the family Salticidae and some lynx spiders of the family Oxyopidae also hunt for prey. Once found, the spiders can leap upon it from distances as much as 40 times their body length. Other examples of hunting behavior can be found in the metallic hunting wasp *Chlorion lobatum*, which specializes in capturing crickets, and in the army ant *Eciton burchelli*, which forms large colonies and searches for prey on the forest floor.

Defensive behavior

Protostomes must defend themselves against an impressive array of predators. To survive against an attack, various strategies have evolved. These strategies include active mimicry, flash and startle displays, and chemical/physical defense.

PHYSICAL AND CHEMICAL DEFENSE

A common behavior exhibited by protostomes in response to danger is adopting a threatening posture. When, for example, a specimen of *Brachypelma smithi* (mygalomorph spider) is threatened outside its burrow, it reacts by making itself appear larger by shifting weight onto the rear legs while simultaneously raising the front legs and exposing the fangs. Another physical defense mechanism of many species of my-

galomorphs is that they use the fine sharp hairs that cover them to pierce their predators. This is not only painful, but may be toxic (these hairs can pierce human skin to a depth of 0.078 in [2 mm]). A spider can release these hairs by rubbing the hind legs against the abdomen. In addition to making themselves appear larger and covered with sharp and, in some cases, toxic hairs, they can also squirt a liquid from their anus.

A novel form of defensive behavior in spiders is found in females and immature males of the black widow (*Latrodectus hesperus*). When threatened, the black widow emits strands of silk and manipulates the silk to cover its vulnerable abdomen and, sometimes, the aggressor. Especially interesting is the defensive behavior of the cerambycid beetles (genus *Hammaticherus*) that use spine-like appendages on their antennae to whip their aggressor. Equally fascinating is the behavior of arctiid moths that produce a series of clicks when detecting the sound made by hunting bats.

A well-known active defense system is found in social insects such as honey bees, termites, and ants. The latter two organisms actually maintain a caste of "soldiers" for colony defense, as do several species of aphids (*Colophina clematis, C. monstrifica, C. arma*). When threatened, these organisms attack by injecting venom into the aggressor and can use their powerful mandibles to incapacitate. In aquatic organisms such as those found in the order Decapoda, cuttlefish and squid defend themselves not only by an ability to escape, but also by discharging ink that temporarily disorientates the aggressor. Some decapods in the order Octopoda, which includes the octopus, have a similar ink defense system. At least one case has been observed in which *Octopus vulgaris* was recorded actually holding stones in its tentacles as a defensive shield against a moray eel.

In general, organisms during early ontogenetic development approach low-intensity stimulation and withdraw from high-intensity stimulation. Protostomes can always escape high-intensity stimulation offered by an aggressor by crawling, swimming, flying, or jumping. Such behavior is easily observed in grasshoppers and the decapod *Onychoteuthis*, popularly known as the "flying squid." The flying squid can escape aggressors by emitting strong water bursts from its mantle to propel the animal into the air where finlike structures allow it to glide for a brief period of time. Fleeing is not always effective. The katydid, *Ancistrocerus inflictus*, does not confront aggressors by an active defense system such as that found in spiders, ants, honey bees, and termites. Rather, *Ancistrocerus* may be found living near the nests of several wasp species. It is these wasps that provide protection for the katydid.

MIMICRY

There are various forms of mimicry. Some of the best-known protostomes that engage in mimicry are butterflies. The species *Zeltus amasa maximianus* (Lycaendae) increases its chances of surviving an attack by giving its enemy a choice of two heads—one of which is a decoy. By presenting a predator with a convincing false target, the probability of surviving an attack is increased. The decoy, or "false head," is created by morphological adaptations present on the wing

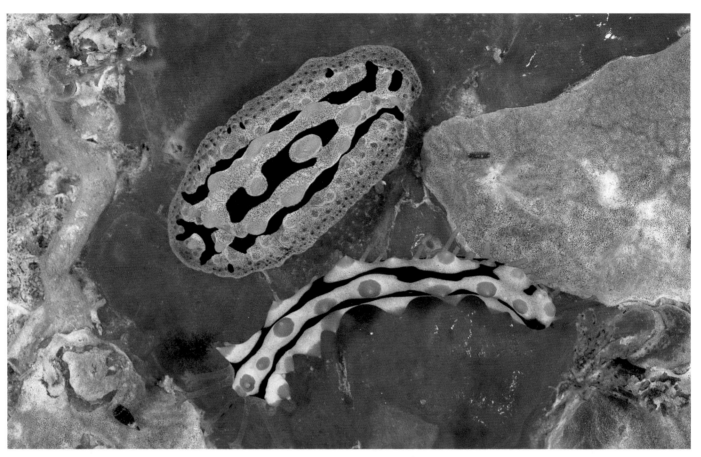

A toxic nudibranch (*Phyllidia coelestis*) and juvenile sea cucumber (*Bohadschia graeffei*), both displaying aposematic or warning coloration. (Photo by A. Flowers & L. Newman. Reproduced by permission.)

tips. False-head mimicry requires not only morphological adaptations, but also that the animal be able to engage in behavioral patterns that will focus a predator's attention on the decoy. One of the methods a butterfly might use to focus a predator's attention on their false head is to make their morphological adaptations seems more "attractive"; this process is accomplished by certain butterflies using the ribbon-like structures located near their wing tips. When the butterfly moves its wings the ribbons begin to resemble antennae, diverting attention away from the true head located on the opposite side of the butterfly.

A similar strategy is also common in caterpillars. In species of *Lirimiris* (Notodontidae), the animal actually inflates a head-like sac at its rear. The resulting fictitious appendage draws the attention of the predator away from the actual head to the comparatively tough rear end. Another version of the false head is found in crab spiders (*Phrynarachne* sp.), and longhorn beetles (*Aethomerus* sp.), whose mimicry resembles bird feces, and the *Anaea* butterfly caterpillar (Nymphalidae) that resembles dried leaf tips.

In contrast to false mimicry, some protostomes actually mimic the behavior of other species. Active mimicry is common in a wide range of invertebrates. An interesting example is found in *Acyphoderes sexualis* (Cerambyiid). This beetle mimics the behavior of two different animals, depending on the

threat. When the beetle is touched, it resembles some species of ponerine ants (Formicidae), and when threatened in flight, the behavior changes to resemble polybiine wasps. Another example is tephritid flies (*Rhagoletis zephyria*). At least two genera emit behaviors that resemble salticid spiders—their main predator. Many other fascinating examples can be found, including wasps (*Ropalidia* sp.) that create nests resembling fruit, and assassin bugs (*Hiranetis braconiformis*) that reduce the probability of serving as a parasitic host by imitating the walking pattern of its impregnator, complete with fake ovipositor.

STARTLE DISPLAYS AND FLASH COLORATION

When stimulated by an aggressor, some protostomes quickly modify their posture to make them appear larger and at the same time to quickly present a "flash" of color. The various postures and displays are characterized by their position, such as frontal displays and lateral displays. The colors associated with these displays are often effective forms of defense because aggressors learn to associate certain colors with results that may have occurred through prior interaction with the intended prey. For example, if the prey had exhibited a certain color to its attacker, and then the predator became sick after ingesting the prey, or the intended prey sprayed the aggressor with a disagreeable fluid its body produces, the predator learns to associate that outcome with the flash of color it had seen and will attempt to avoid repeating the situation. Rapid display of color is also effective because the display itself will often frighten aggressors away. An example of a flash display is found in the katydids (*Neobarrettia vannifera*). When disturbed, this animal quickly opens its wings to reveal a polka-dot pattern. A display resembling a large face awaits any aggressor who disturbs the peanut bug, *Laternaria laternaria* (Fulgoridae), and flag-legged insects (Coreidae) quickly wave a brightly colored leg that it can afford to lose.

Learned behavior

The reasons for studying learning in protostomes are varied. Some scientists hope to exploit the nervous system of invertebrates in an effort to reveal the biochemistry and physiology of learning. Other scientists are interested in comparing invertebrates with vertebrates in a hunt for the similarities and differences in behavior. Still other scientists use learning paradigms to explore applied and basic research questions such as how pesticides influence honey bee foraging behavior and if learning is used in defensive and social behaviors.

A prerequisite for the study of learning is that be clearly defined and that the phenomena investigated as examples of learning be clearly defined. When reviewing studies of learning, the scientist should be aware that definitions vary from researcher to researcher. For example, a researcher may consider behavior controlled by its consequences (i.e., behavior that is rewarded or punished) as an example of operant behavior, while others believe that it depends upon the type of behavior being modified (either operant or instrumental learning). Moreover, some believe that any association between stimuli represents examples of Pavlovian conditioning, while others believe that the "conditioned stimulus" must never elicit the response to be trained prior to any subsequent association.

Here, learning is defined as a relatively permanent change in behavior potential as a result of experience. Several important principles of this definition include the following:

- Learning is inferred from behavior.

- Learning is the result of experience; this excludes changes in behavior produced as the result of physical development, aging, fatigue, adaptation, or circadian rhythms.

- Temporary fluctuations are not considered learning; rather, the change in behavior identified as learned must persist as such behavior is appropriate.

- More often than not, some experience with a situation is required for learning to occur.

To better understand the process of learning in protostomes, many behavioral scientists have divided the categories of learning into non-associative and associative.

NON-ASSOCIATIVE LEARNING

This form of behavior modification involves the association of one event, as when the repeated presentation of a stimulus leads to an alteration of the frequency or speed of a response. Non-associative learning is considered to be the most basic of the learning processes and forms the building blocks of higher order types of learning in protostomes. The

organism does not learn to do anything new or better; rather, the innate response to a situation or to a particular stimulus is modified. Many basic demonstrations of non-associative learning are available in the scientific literature, but there is little sustained work on the many parameters that influence such learning (i.e., time between stimulus presentations, intensity of stimulation, number of training trials).

There are basically two types of non-associative learning: habituation and sensitization. Habituation refers to the reduction in responding to a stimulus as it is repeated. For a decline in responsiveness to be considered a case of non-associative learning, it must be determined that any decline related to sensory and motor fatigue does not exert an influence. Studies of habituation show that it has several characteristics, including the following:

- The more rapid the rate of stimulation is, the faster the habituation is.

- The weaker the stimulus is, the faster the habituation is.

- Habituation to one stimulus will produce habituation to similar stimuli.

- Withholding the stimulus for a long period of time will lead to the recovery of the response.

Sensitization refers to the augmentation of a response to a stimulus. In essence, it is the opposite of habituation, and refers to an increase in the frequency or probability of a response. Studies of sensitization show that it has several characteristics, including the following:

- The stronger the stimulus is, the greater the probability that sensitization will be produced.

- Sensitization to one stimulus will produce sensitization to similar stimuli.

- Repeated presentations of the sensitizing stimulus tend to diminish its effect.

ASSOCIATIVE LEARNING

This is a form of behavior modification involving the association of two or more events such as between two stimuli or between a stimulus and a response. In associative learning, the participant does learn to do something new or better. Associative learning differs from non-associative learning by the number and kind of events that are learned and how the events are learned. Another difference between the two forms of learning is that non-associative learning is considered to be a more fundamental mechanism for behavior modification than those mechanisms in associative learning. This is easily seen in the animal kingdom. Habituation and sensitization are present in all animal groups, but classical and operant conditioning is not. In addition, the available evidence suggests that the behavioral and cellular mechanisms uncovered for non-associative learning may serve as the building blocks for the type of complex behavior characteristic of associative learning. The term associative learning is reserved for a wide variety of classical, instrumental, and operant procedures in which responses are associated with stimuli, consequences, and other responses.

Classical conditioning refers to the modification of behavior in which an originally neutral stimulus—known as a conditioned stimulus (CS)—is paired with a second stimulus that elicits a particular response—known as the unconditioned stimulus (US). The response that the US elicits is known as the unconditioned response (UR). A participant exposed to repeated pairings of the CS and the US will often respond to the originally neutral stimulus as it did to the US. Studies of classical conditioning show that it has several characteristics, including the following:

- In general, the more intense the CS is, the greater the effectiveness of the training.

- In general, the more intense the US is, the greater the effectiveness of the training.

- In general, the shorter the interval is between the CS and the US, the greater the effectiveness of the training.

- In general, the more pairings there are of the CS and the US, the greater the effectiveness of the training.

- When the US no longer follows the CS, the conditioned response gradually becomes weaker over time and eventually stops occurring.

- When a conditioned response has been established to a particular CS, stimuli similar to the CS may elicit the response.

Instrumental and operant conditioning refer to the modification of behavior involving an organism's responses and the consequences of those responses. It may be helpful to conceptualize an operant and instrumental conditioning experiment as a classical conditioning experiment in which the sequence of stimuli and reward is controlled by the behavior of the participant.

Studies of instrumental and operant conditioning show that they have several characteristics, including the following:

- In general, the greater the amount and quality of the reward are, the faster the acquisition is.

- In general, the greater the interval of time is between response and reward, the slower the acquisition is.

- In general, the greater the motivation is, the more vigorous the response is.

- In general, when reward no longer follows the response, the response gradually becomes weaker over time and eventually stops occurring.

Non-associative and/or associative learning has been demonstrated in all the protostomes in which it has been investigated, including planarians (for some scientists, turbellarians are not considered protostomes), polychaetes, earthworms, leeches, water fleas, acorn barnacles, crabs, crayfish, lobsters, cockroaches, fruit flies, ants, honey bees, pond snails, freshwater snails, land snails, slugs, sea hares, and octopuses. While there is no general agreement, most behavioral scientists familiar with the literature would suggest that the most sophisticated examples of learning occur in Crustacea, social insects, gastropod mollusks, and cephalopods. Many of the organisms in these groups can solve complex and simple discrimination tasks, learn to use an existing reflex in a new context, and learn to control their behavior by the consequences of their actions.

Resources

Books

Abramson, C. I. *Invertebrate Learning: A Laboratory Manual and Source Book.* Washington, DC: American Psychological Association, 1990.

———. *A Primer of Invertebrate Learning: The Behavioral Perspective.* Washington, DC: American Psychological Association, 1994.

Abramson, C. I., and I. S. Aqunio. *A Scanning Electron Microscopy Atlas of the Africanized honey bee* (Apis mellifera*): Photographs for the General Public.* Campina Grande, Brazil: Arte Express, 2002.

Abramson, C. I., Z. P. Shuranova, and Y. M. Burmistrov, eds. *Contributions to Invertebrate Behavior.* (In Russian.) Westport, CT: Praeger, 1996.

Brusca, R. C., and G. J. Brusca. *Invertebrates.* Sunderland, MA: Sinauer Associates, Inc., 1990.

Lutz, P. E. *Invertebrate Zoology.* Menlo Park, CA: Benjamin/Cummings Publishing Company, Inc., 1986.

Preston-Mafham, R., and K. Preston-Mafham. *The Encyclopedia of Land Invertebrate Behavior.* Cambridge, MA: The MIT Press, 1993.

Other

"The Infography." *Fields of Knowledge.* [August 2, 2003.] <http://www.fieldsofknowledge.com/>.

Charles I. Abramson, PhD

• • • • •

Protostomes and humans

The influence of protostomes on humans

Many protostomes are small in size and often overlooked among larger natural features and more charismatic creatures of the surrounding world. However, protostomes have played a variety of important roles in human cultures, economies, and health since ancient times. Some protostome species have instilled a sense of fear in the minds of humans, while others have represented fortune. Some have provided nourishing sources of food, while others have advanced medical treatments. Still other protostomes influence ecosystem functions, both for the benefit and to the detriment of humans. Although protostomes have figured prominently in societies throughout the course of history, many are threatened by current human activities, and conservation efforts are important for ensuring the future survival of some species.

Cultural importance of protostomes

Both ancient and modern societies have used protostomes for a variety of purposes. Cowry shells, for example, became the first medium of exchange, or money. These shells were originally used for barter in China as early as 1200 B.C., and their function as currency continued in parts of Africa through the nineteenth century. Wealthy Chinese individuals were even buried with cowries in their mouths to ensure they would have the ability to make purchases in their afterlife. Cowries also have been worn as ornaments by members of many cultures, and women of the ancient Roman city of Pompeii believed that the shells served a functional role by preventing sterility.

Marine gastropods, particularly those of the genus *Murex*, were used to obtain a royal purple dye, known as Tyrian purple. References to this dye date back to 1600 B.C., and its production probably represents one of the earliest cultural and commercial uses of marine mollusks. In the battle of Actium of 31 B.C., Antony and Cleopatra displayed sails of Tyrian purple on their sailing vessels. In some ancient societies, only royalty and idols were allowed to wear cloth of this color. The industry to produce Tyrian purple dye flourished throughout the world until the decline of the Roman empire in the mid-fifteenth century, at which time it was replaced by cheaper dyes, including some obtained from other protostomes.

In addition, some species of protostomes have provided important sources of food in many societies. Shellfish such as clams, mussels, and oysters are consumed throughout much of the world, as are crustaceans such as shrimp, lobsters, and crabs. Snails are considered a delicacy in some countries, where they appear on menus as "escargot." Although they may resemble creatures from another era, horseshoe crabs are highly valued in many Asian markets.

Many protostomes are consumed by humans, and as such they support major fisheries and traditional livelihoods. In 2001 protostomes such as shrimp, clams, squids, and lobsters represented over 26% of all marine fishery harvests. In addition, more than 13 million tons of these species were produced in aquaculture operations. Both wild harvests and captive production provide livelihoods for many humans, and these fisheries are often deeply rooted in the culture of coastal areas. For example, Cajun communities in the bayous of the Mississippi River may include generations of shrimpers, while the coast of Maine is dotted with communities that are culturally centered around the lobster fishery.

Protostomes in mythology, religion, literature, and art

Spiders have frequently appeared in myths and literature throughout history. Biologically, spiders are classified as arachnids, a word that is derived from Greek (*arachne*). In Greek mythology, the mother of spiders was Arachne, a woman who was highly skilled in spinning and weaving. When Arachne boasted of her skills, the goddess Athena challenged her to a contest, after which she destroyed Arachne by transforming her into a spider. Arachne lived out the remainder of her life spinning thread from her belly and using it to weave a web—behaviors that would be carried on by all descending spiders. Spiders have also played prominent roles, both as benevolent and malevolent figures, in Native American myths and beliefs. Among Native American tribes, many viewed spiders as symbols of motherhood and rebirth, and the web was believed to afford protection from evil spirits; other tribes viewed spiders as demons and the web as a trap. In modern times, many children have been introduced to spiders through the "Little Miss Muffet" nursery rhyme, E. B. White's *Charlotte's Web*, or the Spider-Man comics.

Dungeness crab in Kodiak, Alaska, USA, being processed. (Photo by Vanessa Vick/Photo Researchers, Inc. Reproduced by permission.)

Other protostomes also appear in literature, religion, and art. In *Twenty Thousand Leagues under the Sea*, Jules Verne described a harrowing incident in which a giant squid attacked a submarine. Verne's literary description instilled a fear of sea creatures in the minds of many readers, but in reality, fewer than 250 sightings of giant squid have been confirmed. Aesop's fable *The Two Crabs* provided readers with a lesson on the importance of teaching by example. In Hindu culture, the conch symbolizes the sounds of creation. Lord Vinshu always carried a conch shell, which he used as a weapon against evil forces. Finally, protostomes occasionally appear in art, and one of the most widely recognized of these appearances occurs in Sandro Boticelli's fifteenth-century painting *The Birth of Venus*, which depicts the goddess Venus emerging from the sea on a scallop shell.

Medicinal uses of protostomes

Several species of protostomes have served important roles in the medical treatment of humans in historical and modern times. Leeches have been used for a wide variety of medicinal purposes as far back as 2,500 years ago. During Medieval times, leeches offered a popular approach to bloodletting, and doctors attempted to treat many illnesses by removing "im-

pure" blood or other substances from the body. Leeches secrete a substance that prevents blood from clotting, and they can consume five times their own body weight in blood. While leeches largely disappeared from medical facilities with the advent of modern techniques, they are reappearing as part of a limited number of medical treatments, including plastic and reconstructive surgery. In addition, leeches produce other beneficial compounds such as an anesthetic, antibiotics, and enzymes.

Horseshoe crabs have contributed to a number of medical advances. Studies of their eyes have led to treatments for human eye disorders. The blood of horseshoe crabs forms a substance, Limulus Amebocyte Lysate (LAL), that can identify gram-negative bacterial contamination in medical fluids, drugs, and on surgical devices; screening medical items with LAL has reduced complications and deaths related to septic shock and infections acquired in hospitals. Chitin from the shell of horseshoe crabs is extremely pure, nontoxic, and biodegradable. It is used in products such as contact lenses, surgical sutures, and skin lotions. In addition, chitin forms the chemical chitosan that removes metals and toxins from water, and its fat-absorbing properties help remove fat and cholesterol from human bodies.

Black-lipped pearl oysters (*Pinctada margaritifera*) are cultured for pearls. (Photo by Fred McConnaughey/Photo Researchers, Inc. Reproduced by permission.)

Compounds produced by other protostomes have potential medical uses as well. Several opisthobranch mollusks extract compounds from prey organisms; these compounds have shown promise as anticancer drugs. Additionally, compounds in the venom of cone shells are being considered as potential drugs for treating neurological disorders or acting as painkillers.

Ecological value of protostomes

Some species of protostomes, particularly those that burrow and feed in soft sediments, exert a major influence on the structure and functioning of their habitats. By disturbing sediment through a process called bioturbation, these organisms enhance decomposition, nutrient cycling, and oxygenation of the soil. On land, earthworms are major bioturbators, and their activities are essential for maintaining the health of soils used for agriculture and forestry. In marine and aquatic systems, shrimps, crabs, and polychaete worms contribute to bioturbation; the changes induced by these species greatly affect habitats such as mangroves, wetlands, and the seafloor.

In marine and aquatic systems, mussels, oysters, and barnacles heavily influence water quality. In efforts to obtain food, these organisms filter tremendous amounts of water through their bodies; in the process, they remove algae and waterborne nutrients from the water. In the past, oyster populations in the Chesapeake Bay filtered the estuary's entire water volume every three to four days. When humans began to harvest oysters in the bay, populations declined dramatically and water quality became very poor; by 2002, oysters filtered the bay's water only once a year.

Many protostomes, including bryozoans, earthworms, polychaete worms, crustaceans, and mollusks, have been introduced to areas outside of their native ranges as a result of human activities. When introduced to areas in which they have no natural predators, these species can cause major changes to ecosystems. For example, the zebra mussel, native to Eurasia, was introduced in the United States in 1988; it quickly spread throughout the Great Lakes and to other freshwater bodies. Because zebra mussels are extremely efficient at filtering water, they consume large portions of the algae and zooplankton that form the basis of the food chain in freshwater systems. Over time, the disrupted food chain can have a negative effect on other species, including native mussels and fishes.

More than 1,600 species of protostomes are recognized as Extinct, Endangered, or Vulnerable, according to the IUCN Red List. This number is based on an assessment of the status of less than 5% of all protostome species; in reality, the number of species at risk may be much larger. Some species such as conchs and oysters declined dramatically due to overharvesting. However, other species are threatened by habitat alterations or indirect ecosystem changes. Examples of human actions that indirectly affect protostomes include coastal development that impedes horseshoe crab nesting; hydrologic alterations that reduce shrimp in shallow marshes; and nutrient inputs that create anoxic waters in which shellfish and crabs cannot survive. Conservation efforts should minimize both the direct and indirect consequences of human actions that may adversely affect protostomes.

Shipworms (*Bankia setacea*) in fallen piling. Without being detected, shipworms can drill through lumber until it resembles a honeycomb and is rendered unfit for use. (Photo by Gary Gibson/Photo Researchers, Inc. Reproduced by permission.)

Resources

Books

Fusetani, Nobuhiro, ed. *Drugs from the Sea.* Basel, Switzerland: Karger, 2000.

Tanacredi, John T., ed. *"Limulus" in the Limelight: A Species 350 Million Years in the Making and in Peril?* New York: Kluwer Academic/Plenum Publishers, 2001.

Weigle, Marta. *Spiders and Spinsters: Women and Mythology.* Albuquerque: University of New Mexico Press, 2001.

Periodicals

Brussaard, Lijbert. "Soil Fauna, Guilds, Functional Groups and Ecosystem Processes." *Applied Soil Ecology* 9 (1998): 123–135.

Other

Gonzaga, Shireen, and Marc Airhart. "Seashells." August 2, 2003 [August 13, 2003]. <http://www.earthsky.com/Features/Articles/0800seashells.html>.

Katherine E. Mills, MS

Polychaeta
(Clam, sand, and tubeworms)

Phylum Annelida
Class Polychaeta
Number of families 86

Thumbnail description
Segmented worms with numerous bristles and
one pair of parapodia per segment

Photo: Bristle or fire worms (*Chloeia* sp.) primarily come out at night and are scavangers. (Photo ©Tony Wu/www.silent-symphony.com. Reproduced by permission.)

Evolution and systematics

Polychaetes do not fossilize very well as they are soft-bodied animals. There are few fossil records from an entire worm; these have been found from the Pennsylvanian fauna. The oldest fossil records, dating to the middle Cambrian of the Burgess Shale, include the aciculates *Wiwaxia* and *Canadia*, which have a prostomium with appendages, well-developed parapodia, and different kinds of chaetae.

A considerable diversification among polychaetes occurred in the Middle Cambrian, with six genera represented in Burgess Shale (Canada). The genera *Wiwaxia* and *Canadia* do not have jaws, but some later forms possess hard jaws that could mineralize with iron oxide. Such polychaete fossils are known as scolecodonts and have been described from the Ordovician, Silurian, and Devonian periods. Other polychaete fossils include tubes and burrow structures produced by some sedentary forms that secreted a mucous lining for their burrow.

Polychaeta and Clitellata, which includes the classes Oligochaeta and Hirudinoidea, are the two great lineages within the phylum Annelida. As the marine polychaetes fossils that appeared in Middle Cambrian are the earliest record of Annelida, they are considered to be the earliest derived group within the phylum. The exact nature of the ancestral group from which Polychaeta and Clitellata arose is still obscure, but it was likely a homonomous, metameric burrower that possessed a compartmentalized coelom, paired epidermal chaetae, and a head composed of a presegmental prostomium and a peristomium much like those found in modern polychaetes. Due to some recent molecular data and anatomical and developmental evidence, many scientists believe the pogonophorans and vestimentiferans should be placed within the Annelida as a specialized polychaete family, either Pogonophoridae or

Siboglionidae. The class Polychaeta is divided into 24 orders and 86 families, with more than 10,000 described species.

Physical characteristics

Polychaetes range in length from <0.078 in (<2 mm) to >9.8 ft (>3 m). The majority of them are <3.9 in (<10 cm) long and between 0.078–0.39 in (2–10 mm) wide. The morphology is greatly variable. The majority of polychaetes have a cylindrical and elongated body. The parapodia can be uniramous or biramous, with a dorsal lobe (notopodium) and ventral lobe (neuropodium).

The body morphology of polychaetes usually reflects their habits and habitats. Often, active forms have a more homonomous body construction than sedentary ones, which possess some degree of heteronomy, with differentiation in body regions that are utilized for particular functions.

Mobile forms usually have well-developed parapodia, eyes, and sensory organs. Additionally, in some species, the mouth has chitinous jaws and an eversible pharynx.

Sessile forms live in permanent burrows or tubes and have the parapodia reduced or absent; some of them possess special tentaclelike appendages projecting from the tube to collect food from the surface and also for aeration.

Polychaetes are extremely variable in color, varying from light tan to opaque, but most are colored red, pink, green, yellow, or a combination of colors. Some species are iridescent.

Polychaetes of the family Aphroditidae have the dorsum covered with scales (elytra) that can be overlaid by a hard hairlike layer. Because of these characteristics, one species is commonly called the "sea-mouse." Some planktonic species are

Eunice aphroditois is a predatory worm found in the Indo-Pacific. Not much research has been done on this particular species, though it was first described in the late 1700s. This and related worms are found all around the world in tropical and semitropical waters. It is a "sit and wait" predator, waiting in ambush for victims. (Photo ©Tony Wu/www.silent-symphony.com. Reproduced by permission.)

adapted to live in the water column by usually being transparent and flattened with fin parapodia.

It is very common to observe the gas exchange structures of some polychaetes, especially the colorful ones. The branchial morphology is greatly variable: filaments in cirratulids, anterior gills in terebellids, and a tentacular or branchial crown on the heads of sabellids, serpulids, and spirorbids. The more active polychaetes possess a highly vascularized portion of the parapodia, utilizing it for gas exchange.

Distribution

Polychaetes are found worldwide, living in every marine habitat from tropical to polar regions. Some species occur in brackish or freshwater environments. A few live on land, but in habitats completely inundated with water.

Habitat

Due to morphological variations, polychaetes occupy different kinds of habitats, both planktonic and benthic. They are found from the intertidal zone to the deepest depths of the ocean in sandy and muddy sediments, digging or in temporary or permanent mucous tubes that are part of the infaunal community, or crawling on the surface of the substrate. Some species live above the sediment surface as part of the epifaunal community; for example, some nereid species live between mussel beds attached to piers.

Polychaetes are also found in coral and rocky reefs, occupying crevices or beneath stones, and some construct sandy or calcareous tubes that are often attached to coral.

Planktonic forms have adapted structures to swim, and spend all their lives in the water column. Some species tolerate the low salinity of estuaries and a few live in freshwater environments. Some spionids and a group of nereids inhabit semi-terrestrial habitats where they can be covered by freshwater during the wet season and during tidal inundations.

There are some polychaetes that have been found in deep-sea thermal vents. Some species can be commensal or even parasitic. *Histriobdella homari* eats incrusted bacteria and blue-green algae found on the gills and branchial chamber of *Homarus americanus* and *Homarus gammarus*. One *Stratiodrilus* species feeds on microorganisms in the gill chamber of freshwater crayfish or anomuran crustaceans. In the family Oenonidae, there are parasitic forms that live in the coelomic cavity of other polychaetes, echiurans, or bivalves.

Behavior

Polychaetes exhibit few behavior patterns. Certain species are gregarious, forming dense aggregations, while others are solitary. Horizontal and vertical partitioning of space in the sediment occurs between conspecifics and closely-related species or species from the same trophic group (species that feed on the same food resources).

Some polychaetes respond to shadows passing overhead and retreat rapidly back into their tubes. Some species "smell" their prey in the sediment through chemical sensory organs. Species that are prey of birds and fishes can discard their posterior end to avoid predation and can regenerate lost parts.

Some species pair during the breeding season; however, following this phase, they are so aggressive that they often eat each other. During spawning, several species are luminescent in response to light as a reproductive strategy. Others react to changes in temperature, day length, and Moon phases. All these responses are coordinated to ensure successful spawning and reproduction.

Feeding ecology and diet

The morphological and functional diversity of polychaetes enables them to exploit food resources in almost all marine environments in different ways. Polychaetes are usually categorized into raptorial (including carnivores, herbivores, and scavengers), omnivores, surface and subsurface deposit feeders, suspension feeders, and filter feeders.

Polychaetes can be non-selective or selective deposit feeders. The non-selective deposit feeders ingest sand or mud grains, showing little or no discrimination for the size and nutritional value of the particles, assimilating any organic material in the ingested sediment. Selective deposit feeders, however, utilize structures such as palps, tentacles, or buccal organs to select particles with high nutritional value.

The raptorial species usually have a homonomous construction, well-developed parapodia, sensory organs on the head, and a pharynx armed with hard jaws (nereids, glycerids, phyllodocids, syllids). They make rapid movements across the substratum. They often prey on small invertebrates. The prey can be located through sensorial or chemical means. Certain forms (glycerids) have poison glands associated with the jaws. Some raptorial species are not active hunters; they lie in wait for passing prey.

Herbivorous polychaetes feed by scraping or tearing plant material with their pharyngeal structures. Some scavengers feed on any dead or organic material they encounter. Omnivorous species feed on any material they find, and there are some carnivores that feed on deposits when prey are scarce.

Surface and subsurface deposit feeders ingest sandy or muddy particles, feeding on organic material attached to them. They usually have a saclike pharynx that sucks the grains from the sediment when burrowing (capitellids). Some surface deposit feeders have grooved mucous tentacles that collect particles from the surface and carry them directly into the mouth (terebellids, cirratulids).

Suspension feeders possess specialized structures such as tentacular sulcated crowns, or palps on the head, that enable them to collect suspended material in the water column. Some forms are very active, frequently moving their palps (spionids), and others simply expose their crown, waiting for particles to fall onto their surface (Oweniidae).

Filter feeders have specialized crowns with pinnated radioles that create a water current through the pinnules, collecting the particles in suspension (sabellids, sabelariids, spirorbids, serpulids).

Reproductive biology

Polychaetes have different degrees of regeneration. They regenerate lost appendages such as palps, tentacles, cirri, and parapodia. The regeneration of posterior ends is common, but regeneration of a lost head end is uncommon. Many polychaetes utilize regeneration during asexual reproduction, producing a series of individuals, a bud that grows from an individual, or new individuals that develop from an isolated fragment.

The majority of polychaetes are dioecious (gonochoristic); hermaphroditism occurs in relatively few species. Gametes usually mature inside the coelom and are released by gonoducts, coelomoducts, nephridia, or through the rupture of the parental body wall. The majority of species release their gametes into the water, where fertilization takes place. The larvae are planktotrophic, but some species have lecitotrophic larvae, and a few have both. Species with internal fertilization brood their eggs or produce encapsulated eggs that float or are attached to the substratum.

As a rule, segments are generated from a posterior growth zone, arising and developing sequentially from the anterior to the posterior.

Some polychaetes have evolved methods to increase the chances of fertilization. Some sexually reproductive nereids, syllids, and eunicids form an epitokous individual, wherein various body parts or the whole body become a gamete-carrying bag capable of swimming from the bottom upward into the water column, where the gametes are spread. In many polychaetes, the larva is a trochophore that possesses a locomotory ciliary band near the mouth region. The lifespan can range from a few weeks to several years, depending on the reproductive strategy of the particular species.

Conservation status

Polychaetes are part of many investigations that contribute to the conservation and knowledge of the biodiversity of the marine environment. No species of polychaete are listed by the IUCN. However, the Palolo worm, *Eunice viridis*, could potentially be in need of conservation efforts.

Significance to humans

Polychaetes play an important role in the marine benthic food chain, not only serving as food for other organisms, but also recycling organic matter within the sediment and breaking down plant material. Some polychaetes, such as nereids, are known to be important food sources for birds and for economically important fishes; polychaetes are also used as bait for recreational fishing.

Polychaetes play an important role in monitoring marine environmental quality. They respond quickly to changes in the environment, promoted by anthropogenic compounds or chemical contaminants because of their direct contact with the sediments and water column.

Some species can provide an indication of the condition and health of the sediment they live in and often occur at high densities in polluted habitats.

Some polydorids (Spionidae) bore into oyster shells, affecting their appearance and hence their market value, and cause a decline in oyster cultures or become pests. There are numerous significant fouling species that can settle and grow on the hulls of ships.

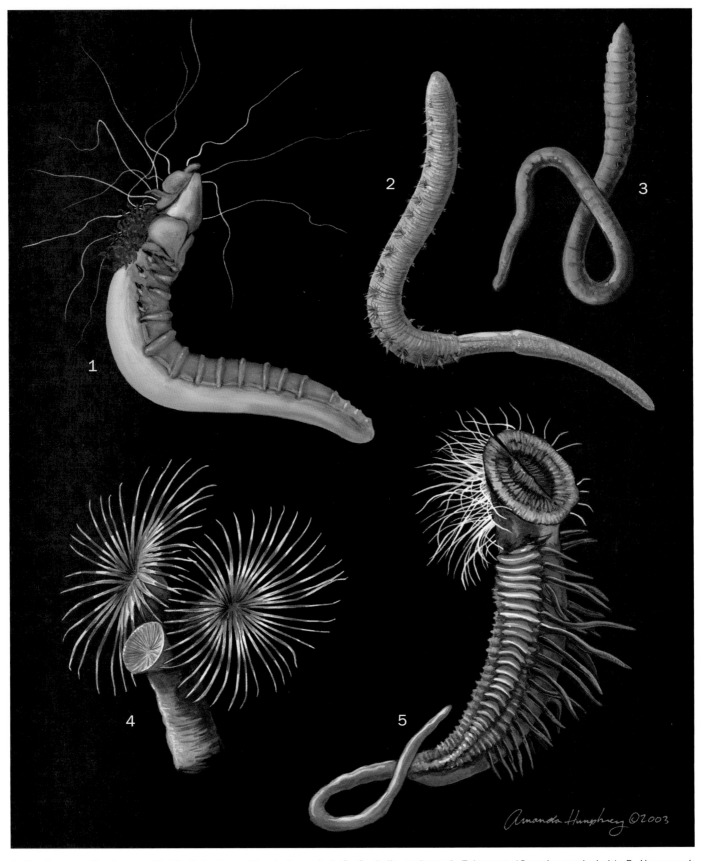

1. Sand mason (*Lanice conchilega*); 2. Lugworm (*Arenicola marina*); 3. *Capitella capitata*; 4. Tubeworm (*Serpula vermicularis*); 5. Honeycomb worm (*Sabellaria alveolata*). (Illustration by Amanda Humphrey)

1. Parchment worm (*Chaetopterus variopedatus*); 2. Ragworm (*Hediste diversicolor*); 3. Pile-worm (*Neanthes succinea*); 4. Fire worm (*Eurythoe complanata*); 5. Catworm (*Nephtys hombergii*). (Illustration by Amanda Humphrey)

Species accounts

Fire worm
Eurythoe complanata

ORDER
Amphinomida

FAMILY
Amphinomidae

TAXONOMY
Eurythoe complanata (Pallas, 1766), Caribbean Sea.

OTHER COMMON NAMES
English: Bristleworm.

PHYSICAL CHARACTERISTICS
Body elongated and flattened dorso-ventraly, wide prostomium, one pair of eyes. They have three antennae, one pair of palps on the head, and dorsal branchial filament tufts that provide them with a blood-red color; the head bears a flattened keel caruncule (structure projecting from the posterior end of the prostomium that carries chemosensory organs called nuchal organs). The parapodia are well developed with different kinds of chaetae and possess calcareous, glassy, hollow harpoon chaetae with neurotoxins that cause discomfort when they contact human skin, thus the reason for common name of "fire worm."

DISTRIBUTION
All tropical seas.

HABITAT
Inhabit cryptic intertidal and shallow subtidal areas, living in crevices, under and between rocks, or in dead coral substrata. Also found in sand and mud.

BEHAVIOR
Found intertidal areas under rocks, forming nests. Assumes a defensive posture, arching its body dorsally to display expansive fascicle of harpoon chaetae when disturbed. Active during the night and usually hidden during daytime.

FEEDING ECOLOGY AND DIET
Omnivorous and a scavenger. Ventral pharynx is eversible, unarmed, strongly muscular, and bears tranverse ridges. When feeding, it positions itself above the prey or food and uses mouth apparatus to rasp and squeeze food material into the mouth. After swallowing the food, the ridges carry it to the digestive tract. It can find prey by contact and also by chemosensory mechanisms.

REPRODUCTIVE BIOLOGY
Exhibits both asexual and sexual reproduction. Asexual reproduction occurs when individuals undergo fragmentation, dividing the body into one or more parts that regenerate to form heads, tails, or both, and grow into new individuals.

CONSERVATION STATUS
Not listed by the IUCN.

SIGNIFICANCE TO HUMANS
If individuals are handled, the bristly, calcareous harpoon chaetae breaks off and remains in contact with the skin. These

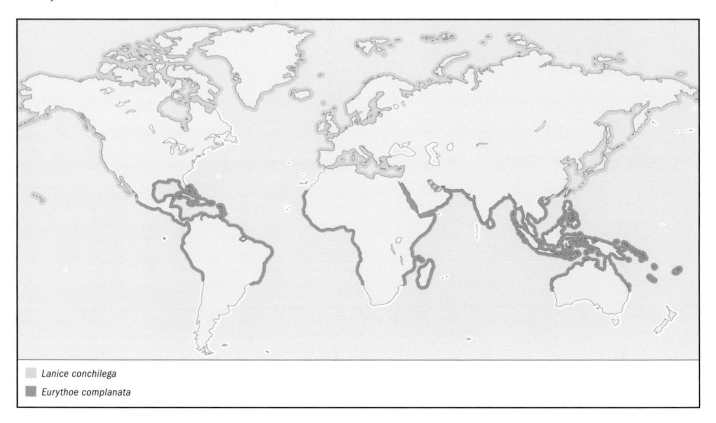

Lanice conchilega
Eurythoe complanata

chaetae release an acidic neurotoxin, causing reactions such as inflammation and swelling. ◆

Lugworm
Arenicola marina

ORDER
Capitellida

FAMILY
Arenicolidae

TAXONOMY
Arenicola marina Linnaeus, 1758, Sweden.

OTHER COMMON NAMES
English: Blow lug, bristleworm.

PHYSICAL CHARACTERISTICS
Has cylindrical and firm body divided into two regions: thorax and abdomen. The head is small and has no appendages or eyes. Has a rough proboscis that is usually visible. The thoracic region possesses 19 parapodia with chaetae, some of them bearing arborescent gills. Abdominal region is narrow and lacks chaetae and gills. Body segments are divided into five rings (annuli). Reaches between 4.7–7.8 in (120–200 mm) in length and varies in color from pink to dark pink, red, green, dark brown, or black.

DISTRIBUTION
Recorded from shores of western Europe, Norway, Spitzbergen, north Siberia, and Iceland. In the western Atlantic, it has been recorded from Greenland, along the northern coast from the Bay of Fundy to Long Island. Its southern limit is about 40°N.

HABITAT
Occurs all over intertidal region in sand and muddy sand, inhabiting a characteristic U- or J-shaped tube it forms by burrowing through the sediment.

BEHAVIOR
Digs U- or J-shaped burrow ranging from 7.8–15.7 in (20–40 cm) deep that has a depression at the head end and small amount of defecated material at the tail end; this defecated material has the color of clean sand. Feeding, defecation, and burrow irrigation behavior is cyclic and can last from 15–42 minutes, depending on the size of the animal. The burrow is irrigated and consequently aerated by intermittent cycles of peristaltic contractions of the body from the tail to the head end. Therefore, concentrated oxygenated water is taken in at the tail end and leaves by percolation through the feeding column.

FEEDING ECOLOGY AND DIET
Feeds on surface sedimentary material with the head end of its burrow, forming a feeding column and characteristic funnel or "blow hole" to the surface sediment. Sucks particles with the aid of the saclike proboscis. Small, ingested particles are attached to the proboscis papillae and the larger are rejected and accumulate around the burrow opening. Microorganisms such as bacteria, benthic diatoms, meiofauna, and detritus attached to particles are digested in the gut.

REPRODUCTIVE BIOLOGY
Gonochoristic, oviparous, with a lifespan from 5–10 years. The eggs and early larvae develop in the female burrow, but post larvae are capable of migration. Spawning takes place within the burrow and the release of gametes occurs by rhythmic contractions of the body wall. Gametes are released through the nephridia. Sperm is flushed out of the burrow by pumping activity of the male, while oocytes are retained in the horizontal shaft of the female's burrow.

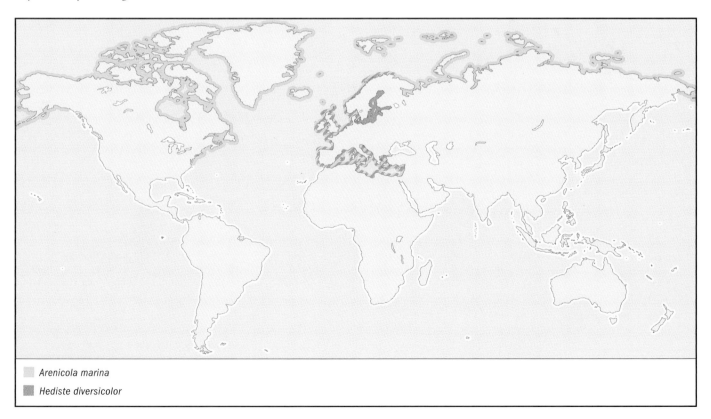

Arenicola marina

Hediste diversicolor

CONSERVATION STATUS
Not listed by the IUCN.

SIGNIFICANCE TO HUMANS
Utilized commercially as bait. Reworking of the sediment (bioturbation) by this and other species affects the cycling and retention of contaminants such as hydrocarbons and heavy metals within the sediment. It is presently used routinely as a standard bioassay organism for assessing the toxicity of marine sediments. ◆

No common name
Capitella capitata

ORDER
Capitellida

FAMILY
Capitellidae

TAXONOMY
Capitella capitata (Fabricius, 1870), Greenland.

OTHER COMMON NAMES
English: Bristleworm.

PHYSICAL CHARACTERISTICS
Body flexible, slender, elongated and blood red in color, earthwormlike appearance, 0.78–3.9 in (2–10 cm) long. The head is conical and shovel shaped. The parapodia are reduced with chaetae in both rami. Single genital pore between chaetigers eight and nine; in males this pore is surrounded by cross spines.

DISTRIBUTION
Found worldwide.

HABITAT
Inhabits muddy sand, gritty sand, fine sand, or mud on lower intertidal to sub-littoral habitats. Also found under pebbles or small stones, with burrows at or near the surface of the sediment. Occurs in high abundance, associated with organically enriched substrates.

BEHAVIOR
Burrower; often a solitary non-migratory species, living in high numbers in areas of organic enrichment where sewage inputs and fish farms are present. It is also found in sediments with high concentrations of metals and hydrocarbons.

FEEDING ECOLOGY AND DIET
Non-selective subsurface deposit feeder, feeding on microorganisms, phytoplankton, and detritus. Everts a saclike pharynx and aglutinates sand grains in order to feed.

REPRODUCTIVE BIOLOGY
Spawns once or twice a year, and in natural populations, sexual maturity is reached at about four months. Female produces from 100–1,000 eggs. Considered to be iteroparous, and the lecithotrophic larvae are brooded during part of their development within the adult tube. Potential longevity ranges from 45 days to two years.

CONSERVATION STATUS
Not listed by the IUCN.

SIGNIFICANCE TO HUMANS
Provides an important food source for the shrimp *Crangon crangon*. Has been accepted as pollution indicator. ◆

Capitella capitata
Sabellaria alveolata

Catworm
Nephtys hombergii

ORDER
Phyllodocida

FAMILY
Nephtyidae

TAXONOMY
Nephtys hombergii Savigny, 1818, coast of France.

OTHER COMMON NAMES
English: Bristleworm.

PHYSICAL CHARACTERISTICS
Thin worm reaching 3.9–7.8 in (10–20 cm) in length. Body rectangular in cross section and pearly iridescent white in color; head bears four small eyes and a papillated proboscis. Parapodia are well developed with golden bristles.

DISTRIBUTION
Found from the northern Atlantic, from such areas as the Barents Sea, the Baltic, and the North Sea, to the Mediterranean. Has been reported as far south as South Africa.

HABITAT
Common in lower intertidal zone, frequently found in high density in muddy substrata.

BEHAVIOR
During feeding activity, this solitary species burrows 1.9–.9 in (5–15 cm) beneath the sediment. Also swims with undulating movements of the body, but only short distances.

FEEDING ECOLOGY AND DIET
Scavenger or active carnivore, feeding on small invertebrates. Papillated proboscis is employed when capturing prey.

REPRODUCTIVE BIOLOGY
Gonochoristic. Mature individuals stimulated to spawn and release sperm and eggs on the sediment surface during low tide. Larvae are lecithotrophic at first stage and, after that, planktotrophic. Life cycle lasts 2–5 years, with individuals reaching maturity in two years.

CONSERVATION STATUS
Not listed by the IUCN.

SIGNIFICANCE TO HUMANS
Used as bait by fishermen and considered an important food source for birds. Its predation activity can decrease the biomass of prey in its habitat. ◆

Ragworm
Hediste diversicolor

ORDER
Phyllodocida

FAMILY
Nereididae

TAXONOMY
Nereis diversicolor (Müller, 1775), Baltic Sea.

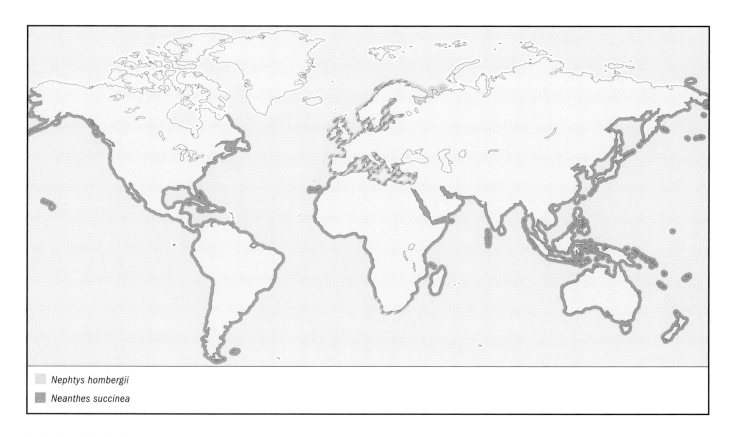

Nephtys hombergii
Neanthes succinea

OTHER COMMON NAMES
English: Bristleworm, clamworm.

PHYSICAL CHARACTERISTICS
Body elongated and flattened with conspicuous parapodia with bristles. The head has four eyes, two antennae, and two palps. Has eversible pharynx; armed with chitinous teeth. Color is greenish during spawning season and varies from red to light brown at other times.

DISTRIBUTION
Widely distributed throughout northwest Europe in the Baltic Sea, North Sea, and along Atlantic coast to the Mediterranean.

HABITAT
Euryhaline. Found in sandy and muddy sheltered environments in intertidal region.

BEHAVIOR
Solitary organism that builds conspicuous burrow that serves as a site for foraging and as refuge against predation by birds.

FEEDING ECOLOGY AND DIET
Has great variety of feeding modes: passive suspension feeder; surface and subsurface deposit feeder; active and passive omnivore; and scavenger preying on small invertebrates, plant material, or ingesting sand and mud particles to utilize the attached detritus. Utilizes eversible pharynx to capture prey.

REPRODUCTIVE BIOLOGY
Gonochoristic, oviparous, reproducing once in entire life. Larva is lecithotrophic. Maturation and spawning influenced by temperature in spring, and during this phase, males are brighter green and females are darker green. Reproductive pheromones coordinate processes such as mate location and synchronism for release of gametes when males discharge sperm around burrow of females. Adults reach maturity in one to three years.

CONSERVATION STATUS
Not listed by the IUCN.

SIGNIFICANCE TO HUMANS
Important food source for different species of wading birds and serves as bait for fishermen. ◆

Pile-worm
Neanthes succinea

ORDER
Phyllodocida

FAMILY
Nereidae

TAXONOMY
Nereis succinea (*Neanthes*) Frey and Leuckart, 1847, North Sea.

OTHER COMMON NAMES
English: Bristleworm.

PHYSICAL CHARACTERISTICS
Body is usually red in color, reaching up to 7.4 in (19 cm). Head with four large eyes, small antennae, and long palps. Pharynx bears chitinous jaws and small teeth (paragnaths). Parapodia are well developed with bristles.

DISTRIBUTION
Cosmopolitan in temperate and tropical waters.

HABITAT
Found from intertidal zone to subtidal regions in mud and sand and under rocks as infaunal organism, as well as in dock-fouling communities of marinas, hiding between mussel and oyster beds as part of the epifauna. Being an euryhaline species, it is also found in estuarine regions.

BEHAVIOR
Active crawler at night; hidden during daytime when it constructs a mucous-lined tube.

FEEDING ECOLOGY AND DIET
The jawed eversible proboscis is used to capture small invertebrates or graze on detritus and plant material. Usually feeds at night.

REPRODUCTIVE BIOLOGY
Sexes are separated. During breeding season, it becomes epitokous (modified forms filled with gametes called heteronereids) and individuals swim together stimulated by pheromones, releasing gametes in water where fertilization takes place.

CONSERVATION STATUS
Not listed by the IUCN.

SIGNIFICANCE TO HUMANS
Larva is preyed upon by fishes and birds that feed in water column. ◆

Tubeworm
Serpula vermicularis

ORDER
Sabellida

FAMILY
Serpulidae

TAXONOMY
Serpula vermicularis Linnaeus, 1767, western Europe.

OTHER COMMON NAMES
English: Bristleworm.

PHYSICAL CHARACTERISTICS
Permanent tube-dwelling polychaete; body length from 1.9–2.7 in (5–7 cm) with up to 200 segments. Head possesses two opposing circles of bipinnate radioles, forming a spiral crown when extended outside of its tube, and a calcareous operculum that is funnel-shaped and pinkish white in color. Tube is cylindrical with some irregular ridges and is made of calcium carbonate. Body can be pale yellow to red in color.

DISTRIBUTION
Northeast Atlantic and Mediterranean Sea.

HABITAT
Constructs calcareous tubes on hard substrata such as rocks, stones, and bivalve shells. Found from sublittoral fringe to circalittoral up to 820 ft (250 m) depth. In some very sheltered areas, the tubes aggregate together to form small reefs.

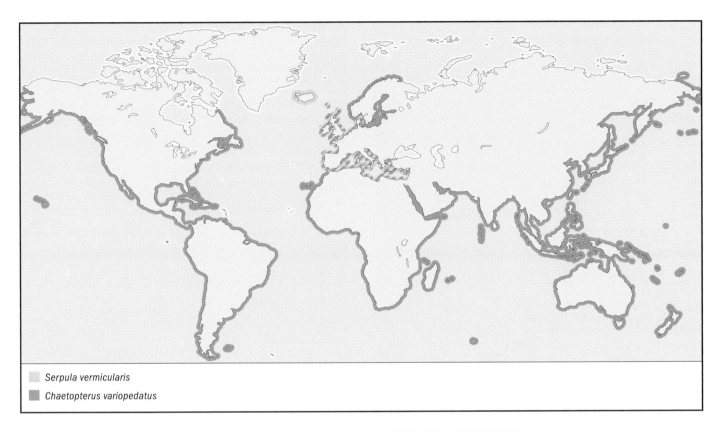

Serpula vermicularis

Chaetopterus variopedatus

BEHAVIOR
Tubes are permanently attached to the substrate. Can be gregarious if there are few substrates for larvae to settle on, but in open marine habitat, it is usually solitary.

FEEDING ECOLOGY AND DIET
Active suspension feeder, exposing its bipinnated crown out of its tube to feed. The pinnules of the radioles create water current into crown, which collects suspended material from the water column.

REPRODUCTIVE BIOLOGY
Gonochoristic; reproduces annually. Larva is planktotrophic and reaches maturity within one year. Lifecycle 2–5 years; spawning occurs in summer.

CONSERVATION STATUS
Not listed by the IUCN, but included in U.K. Biodiversity Action Plan.

SIGNIFICANCE TO HUMANS
Permanent tubes constructed by this species create microhabitats for other species, increasing local biodiversity. ◆

Parchment worm
Chaetopterus variopedatus

ORDER
Spionida

FAMILY
Chaetopteridae

TAXONOMY
Chaetopterus variopedatus (Renier, 1804), Mediterranean Sea.

OTHER COMMON NAMES
English: Bristleworm.

PHYSICAL CHARACTERISTICS
Body is irregularly segmented and divided into three regions with a dorsal ciliated groove running along its length and can be up to 9.8 in (25 cm) long; yellowish or greenish white in color. Anterior and first region is flattened on the ventral side and bears glands that secrete the tube. There are modified chaetae at the fourth chaetiger that are used to cut the tube wall. The second region of the body has winglike, modified notopodial lobes that secrete the mucus where the suspended particles are trapped. The third region is longer and distinctly segmented. Some parts of the body are phosphorescent.

DISTRIBUTION
Cosmopolitan in both tropical and temperate seas.

HABITAT
Found from intertidal to shelf depths in soft sediments such as sand and mud; can be found with their tubes attached to pier or any hard substratum. Their vacant tubes can be occupied by other species of polychaetes, and they share the tube with two crabs species from the genera *Pinnixia* and *Porcellana*; however, they do not occur in the same tube.

BEHAVIOR
Constructs a tough and flexible U-shaped tube in soft substrata with only the ends of the tube protruding from the surface. The tube has the consistency of paper and can be covered with sand, mud, or shell fragments.

FEEDING ECOLOGY AND DIET
It is a filter-feeding worm, dining on suspended organic material. Sea water is pumped through the tube by winglike notopodia, which secrete mucus that is drawn back, forming a bag.

The mucous bag filters the water to retain finer particles. Periodically, it is transported back to the mouth by the dorsal ciliated groove and ingested.

REPRODUCTIVE BIOLOGY
Gonochoristic; breeding all year round. Larva is long-lived and planktotrophic. The segments destined for the mid-body region develop precociously and more rapidly than anterior segments. This accelerated development (also called heterochrony) is not seen in other polychaetes.

CONSERVATION STATUS
Not listed by the IUCN.

SIGNIFICANCE TO HUMANS
Has long been used in developmental biology experiments because it is easy to maintain in laboratory settings and provides larvae and embryos *in vitro*. ◆

Sand mason
Lanice conchilega

ORDER
Terebellida

FAMILY
Amphitritinae

TAXONOMY
Lanice conchilega Pallas, 1766, Holland.

OTHER COMMON NAMES
English: Bristleworm.

PHYSICAL CHARACTERISTICS
Body up to 11.8 in (30 cm) long; yellow, pink, and brownish in color. Possesses 17 segments in front region with three pairs of blood-red colored arborescent gills and a white tentacular crown.

DISTRIBUTION
From the Arctic to the Mediterranean Sea, in the Arabian Gulf and the Pacific Ocean.

HABITAT
Found from intertidal zone down to 5,577 ft (1,700 m) in coarse, fine, clean sand, and mud.

BEHAVIOR
Can be found as solitary individual or in dense populations. Constructs tube from sand grains and shell fragments that have a frayed end protruding above the sand.

FEEDING ECOLOGY AND DIET
Active suspension feeder or surface deposit feeder, depending on local density; at low density, it is a deposit feeder, and at high density, it is suspension feeder to avoid competition with neighboring worms.

REPRODUCTIVE BIOLOGY
Gonochoristic; gametes occur from April to October in Northern Hemisphere. Larva is planktrotrophic; survives up to 60 days in the plankton, so dispersal distance may be great.

CONSERVATION STATUS
Not listed by the IUCN.

SIGNIFICANCE TO HUMANS
The tube constructed by this species increases the stability of sediment and consequently the stability of other species. Also, a mollusk species (*Acteon tornatilis*) and some birds eat this worm. ◆

Honeycomb worm
Sabellaria alveolata

ORDER
Terebellida

FAMILY
Sabelariidae

TAXONOMY
Sabellaria alveolata Linnaeus, 1767, England.

OTHER COMMON NAMES
English: Bristleworm.

PHYSICAL CHARACTERISTICS
Colonial segmented worm that builds tubes from cemented coarse sand and/or shell material, forming reefs. Adult size ranges from 1.1–1.5 in (30–40 mm). Thorax has three pairs of flattened chaetal sheaths; has chaetes that form an operculum utilized to close the tube opening. The color of the tube varies, depending on the color of the sand locally available for tube construction.

DISTRIBUTION
Mediterranean Sea, north Atlantic to south Morocco. Also found in the British Isles to its northern limit in the northeast Atlantic.

HABITAT
Open coasts, usually intertidal and occasionally subtidal. Settlement occurs on hard substrata, but species needs adequate sand and shell fragments to build up its tube.

BEHAVIOR
Permanently attached to the substratum and is gregarious. Adapted to close its operculum during low tide to avoid desiccation and predation. The tube can reach 7.8 in (20 cm) in length and around 0.19 in (5 mm) in diameter at the external opening.

FEEDING ECOLOGY AND DIET
Passive suspension feeder, ingesting seston during high tide.

REPRODUCTIVE BIOLOGY
Gonochoristic; reproductive frequency is annual. Female produces from 100,000–1,000,000 eggs. The larva is planktotrophic and has dispersal potential of more than 6.2 mi (10 km). The lifecycle ranges from 2–5 years; adults reach maturity within the first year.

CONSERVATION STATUS
Not listed by the IUCN, but included in U.K. Biodiversity Action Plan.

SIGNIFICANCE TO HUMANS
Provides substratum to other marine invertebrates to settle on, increasing the spatial heterogeneity and consequently local diversity. Fishermen collect them from reefs for use as bait. ◆

Resources

Books

Blake, James A., B. Hilbig, and P. H. Scott. *The Annelida Part 2—Polychaeta: Phyllodocida (Syllidae and Scale-bearing families), Amphinomida and Eunicida.* Taxonomic Atlas of the Benthic Fauna of the Santa Maria Basin and the Western Santa Barbara Channel series, Vol. 5. Santa Barbara, CA: Santa Barbara Museum of Natural History, 1995.

———. *The Annelida Part 4—Polychaeta: Flabelligerida to Sternaspidae.* Taxonomic Atlas of the Benthic Fauna of the Santa Maria Basin and the Western Santa Barbara Channel series, Vol. 7. Santa Barbara, CA: Santa Barbara Museum of Natural History, 2000.

Brusca, N. C., and G. J. Brusca. *Invertebrates.* 2nd edition. Sunderland: MA: Sinauer Associates Inc. Publishers, 2003.

Giese, Arthur. C., and John S. Pearse. *Reproduction of Marine Invertebrates.* London: Academic Press, Inc. LTD, 1975.

Glasby, Christopher J., et al. "Class Polychaeta." In *Polychaetes and Allies. The Southern Synthesis. Fauna of Australia. Polychaeta, Myzoztomida, Pogonophora, Echiura, Sipuncula,* Vol. 4. Melbourne, Australia: CSIRO Publishing, 2000.

Rouse, Greg, and Fredrik W. Pleijel. *Polychaetes.* New York: Oxford University Press Inc., 2001.

Periodicals

Arndt, C., and D. Schiedek. "*Nephtus hombergii*, A Free-living Predator in Marine Sediments: Energy Production under Environmental Stress." *Marine Biology* 129 (1998): 643–540.

Bat, L., and D. Raffaelli. "Sediment Toxicity Testing: A Bioassay Approach Using the Amphipod *Corophium volutator* and the Polychaete *Arenicola marina.*" *Journal of Experimental Marine Biology and Ecology* 226 (1998): 217–239.

Bridges, T. S. "Effects of Organic Additions to Sediment, and Maternal Age and Size, on Patterns of Offspring Investment and Performance in Two Opportunistic Deposit-feeding Polychaetes." *Marine Biology* 125 (1996): 345–357.

Dill, L. M., and A. H. G. Fraser. "The Worm Returns: Hiding Behavior of a Tube-dwelling Marine Polychaete, *Serpula vermicularis.*" *Behavioral Ecology* 8, no. 2 (1997): 186–193.

Fauchald, J., and P. A. Jumars. "The Diet of Worms: A Study of Polychaete Feeding Guilds." *Oceanography and Marine Biology: An Annual Review* 17 (1979): 193–284.

Giangrande, Adriana "Polychaete Reproductive Patterns, Life Cycles and Life Histories: An Overview." *Oceanography and Marine Biology: An Annual Review* 35 (1997): 323–389.

Gamenick, I., and O. Giere. "Ecophysiological Studies on the *Capitella capitata* Complex: Respiration and Sulfide Exposure." *Bulletin of Marine Science* 60 (1997): 613.

Grassle, J. F., and J. P. Grassle. "Sibling Species in the Marine Pollution Indicator (*Capitella capitata*) (Polychaete)." *Science* 192 (1976): 567–569.

Hardege, J. D., M. G. Bentley, and L. Snape. "Sediment Selection by Juvenile *Arenicola marina* (Polychaete)." *Marine Ecology Progress Series* 166 (1998): 187–195.

Mendez, N., J. Romero, and J. Flos. "Population Dynamics and Production of the Polychaete *Capitella capitata* in the Littoral Zone of Barcelona (Spain, NW Mediterranean)." *Journal of Experimental Marine Biology and Ecology* 218 (1997): 263–284.

Pearson, T. H., and R. Rosenberg. "Macrobenthic Succession in Relation to Organic Enrichment and Pollution of the Marine Environment." *Oceanography and Marine Biology: An Annual Review* 16 (1978): 229–311.

Riisgard, H. U., and G. T. Banta. "Irrigation and Deposit Feeding by the Lugworm *Arenicola marina*, Characteristics and Secondary Effects on the Environment. A Review of Current Knowledge." *Vie Milieu* 48 (1998): 243–257.

Suadicani, S. O., J. C. Freitas, and M. I. Sawaya. "Pharmacological Evidence for the Presence of a Beta-adrenoceptor-like Agonist in the Amphinomid Polychaete *Eurythoe complanata.*" *Comparative Bichemistry and Physiology* 104C, no. 2 (1993): 327–332.

Tsutsumi, H., and T. Kikuchi. "Study of the Life History of *Capitella capitata* Polychaeta: (Capitellidae) in Amakusa, South Japan, Including a Comparison with Other Geographical Regions." *Marine Biology* 80 (1984): 315–321.

Wilson, D. P. "Additional Observations on Larval Growth and Settlement of *Sabellaria alveolata.*" *Journal of the Marine Biological Association of the United Kingdom* 50, no. 1 (1970): 1–32.

Zebe, E., and D. Schiedek. "The Lugworm *Arenicola marina*: A Model of Physiological Adaptation to Life in Intertidal Sediments." *Helgoländer Meeresuntersuchungen* 50 (1996): 37–68.

Erica Veronica Pardo, PhD

Myzostomida
(Myzostomids)

Phylum Annelida

Class Myzostomida

Number of families 8

Thumbnail description
Minute, soft-bodied marine worms associated with echinoderms

Photo: A dorsal view of *Myzostoma polycyclus*, collected on a crinoid from Madagascar. (Photo by Igor Eeckhaut. Reproduced by permission.)

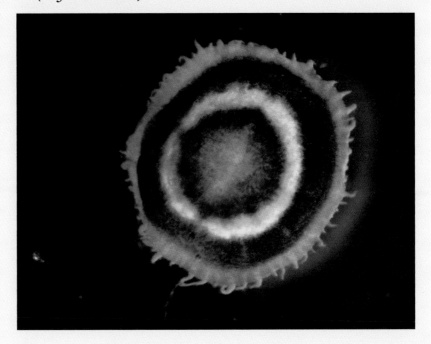

Evolution and systematics

The name "myzostomid" comes from the Greek *myzo*, meaning "to suck," and *stoma*, meaning "mouth." Leuckart described the first myzostomid, *Myzostoma parasiticum*, in 1827 and 1836. Since that discovery, the phylogenetic position of myzostomids within the metazoans has been a subject of controversy. The early assignments of myzostomids to Trematoda, then to Crustacea or Stelechopoda (a taxon grouping myzostomids with Tardigrada and Pentastomida), are no longer considered valid. Because myzostomids exhibit characters such as parapodia with chaetae, trochophore-type larvae, and segmentation (although incomplete), they are now classified in textbooks and encyclopedias as a family of Polychaeta or as a class of Annelida (the latter classification is followed here). However, phylogenetic analyses, including that of myzostomid DNA sequences, strongly support the view that they are not annelids but a phylum closely related to the flatworms.

The class Myzostomida is divided in two orders and eight families: the order Proboscidea (Myzostomatidae); and the order Pharyngidea (Protomyzostomatidae, Asteromyzostomatidae, Asteriomyzostomatidae, Endomyzostomatidae, Stelechopidae, Pulvinomyzostomatidae, and Mesomyzostomatidae). There are about 170 species.

Physical characteristics

The body of most myzostomids consists of an anterior cylindrical introvert (also called a proboscis) and a flat, disk-like trunk. The introvert is extended when the individual feeds, but it is retracted into an anteroventral pouch within the trunk most of the time. The trunk ranges in length from 0.118 in (3 mm) to 1.181 in (3 cm) for the largest species. There are usually five pairs of parapodia, located lateroventrally in two rows; each parapodium contains a protrusible hook, some replacement hooks, and a support rod, or aciculum. Most species have four pairs of slitlike or disklike lateroventral sense organs, commonly named lateral organs, and the trunk margin often bears needlelike cirri (more than 100 in some species). In a few species, humplike cirri occur at the base of each parapodium. Two male gonopores are located at the level of the third pair of parapodia; the female gonopore opens close to the anus, posteroventrally.

When myzostomids are parasitic, their body is often highly modified. The introvert, external appendages, and sensory organs are usually reduced or have disappeared. According to the location of myzostomids in the host, their trunk may be pleated dorsally, mushroom shaped, or irregular in shape.

Distribution

Myzostomids are found in all oceans from shallow waters to depths of over 9,840 ft (3,000 m). Most occur in tropical waters, but a few are found in the Arctic and Antarctic Oceans.

Habitat

These marine animals live in association with echinoderms.

Behavior

All myzostomids live in association with echinoderms. Most species (90% of myzostomids) are ectocommensals of crinoids, but a few are parasites of crinoids, asteroids, or ophiuroids, infesting their gonads, coelom, integumental, or digestive systems. The association between myzostomids and echinoderms is ancient, and signs of parasitic activities, similar to those induced by extant parasitic myzostomids, are found in fossilized crinoid skeletons dating back to the Carboniferous period.

Feeding ecology and diet

Very few myzostomids feed on host tissues or coelomic fluid, most feed on particles that they divert from the host using their introvert. Crinoids, the main hosts, are suspension feeders, catching food particles in the water column with podia on both sides of the ciliated grooves that run along their arms. When myzostomids want to feed, they insert their introvert into these ciliated grooves and suck up water and food particles into their mouth.

Reproductive biology

Most myzostomids are hermaphroditic. They are often functional simultaneous hermaphrodites, even though the male genital system develops a bit earlier than the female genital system during organogenesis. In some species, both males and females occur, and are often interpreted as the two stages of protandrous hermaphroditic species, the dwarf male apparently transforming into a female once it lives alone.

Reproduction in myzostomids takes place by the emission of spermatophores followed by the intradermic penetration of sperm cells. In ectocommensals, mating involves two mature individuals that contact each other, one individual ejecting one spermatophore that attaches to the integument of the other. Contact between the two individuals is very brief and they separate quickly after mating. Spermatophores generally attach to the back of the receiver, but they can be emitted successfully onto any body part.

Depending on the species, the emitted spermatophores may be white V-shaped, club-shaped, or ball-shaped baskets. After attachment, they pierce the integument and release all the sperm cells into it. Penetration can be observed, thanks to the presence of the white trails produced by the spermatophore contents extending into the translucent body of the receiver. These trails appear from 10 to 30 minutes after attachment of a spermatophore, and after from one to five hours, the spermatophores are reduced to an empty matrix. Fertilization of mature oocytes is internal, and eggs are accumulated into a uterus before being laid. Division is spiral, occurring in the water column, and gives rise to free-swimming trochophore larvae.

Conservation status

No species of myzostomids are listed by the IUCN.

Significance to humans

No parasitic myzostomids infest humans or animals reared by humans. They have no impact on human health.

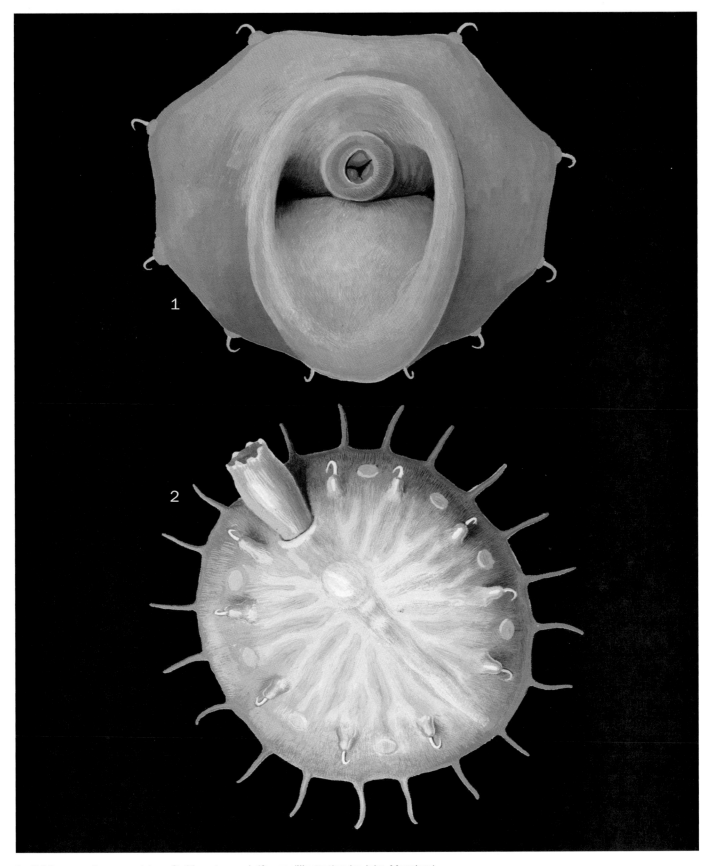

1. *Pulvinomyzostomum pulvinar*; 2. *Myzostoma cirriferum*. (Illustration by John Megahan)

Species accounts

No common name
Pulvinomyzostomum pulvinar

ORDER
Pharyngidea

FAMILY
Pulvinomyzostomatidae

TAXONOMY
Pulvinomyzostomum pulvinar von Graff, 1884, "in the Minch, from 60 to 80 fathoms."

OTHER COMMON NAMES
None known.

PHYSICAL CHARACTERISTICS
Females with stout trunk up to 0.16 in (4 mm) long with lateral margins folded upward and inward so dorsal surface is partly depressed; no introvert but a protrusive pharynx; both mouth and anus dorsal. No cirri, but 10 pairs of slitlike lateral organs with 5 pairs of reduced parapodia. Males with flattened, ovoid trunk up to 0.06 in (1.5 mm) long. Six pairs of lateral organs; 5 pairs of papilliform parapodia. No cirri.

DISTRIBUTION
Mediterranean Sea; northeast Atlantic Ocean along coasts of Europe.

HABITAT
Live in shallow and deep sea marine waters.

BEHAVIOR
Endosymbiont of shallow-water crinoids *Antedon bifida*, *Leptometra phalangium*, and *L. celtica*. Parasitic or commensal. Female generally infests anterior part of the crinoid digestive system with a male on its flank.

Myzostoma cirriferum

Pulvinomyzostomum pulvinar

FEEDING ECOLOGY AND DIET
Feeds on particles in crinoid digestive system.

REPRODUCTIVE BIOLOGY
Protandrous hermaphrodite. Males transform into females when they live alone. Reproductive process unknown.

CONSERVATION STATUS
Not listed by the IUCN.

SIGNIFICANCE TO HUMANS
None known. ◆

No common name
Myzostoma cirriferum

ORDER
Proboscidea

FAMILY
Myzostomatidae

TAXONOMY
Myzostoma cirriferum Leuckart, 1836, Mediterranean Sea.

OTHER COMMON NAMES
None known.

PHYSICAL CHARACTERISTICS
Flattened, ovoid trunk, up to 0.09 in (2.4 mm) long; cylindrical introvert of 0.04 in (1 mm) long (fully extended). Parapodia, lateral organs, and penises well developed, all ventrally located. Ten pairs of marginal cirri.

DISTRIBUTION
Mediterranean Sea; northeast Atlantic Ocean along coasts of Europe.

HABITAT
Live in shallow and deep sea marine waters.

BEHAVIOR
Ectocommensals of the shallow-water crinoids *Antedon bifida*, *A. petasus*, *A. adriatica*, *A. mediterranea*, and *Athrometra prolixa*. Several hundred can infest a single crinoid.

FEEDING ECOLOGY AND DIET
Feeds on particles diverted from the ambulacral grooves of crinoids.

REPRODUCTIVE BIOLOGY
Simultaneous hermaphrodite. Reproduces year round by transfer of spermatophores and intradermic penetration of sperm cells.

CONSERVATION STATUS
Not listed by the IUCN.

SIGNIFICANCE TO HUMANS
None known. ◆

Resources

Books

Grygier, M. J. "Class Myzostomida." In *Polychaetes and Allies: The Southern Synthesis. Fauna of Australia*, Vol. 4A, *Polychaeta, Myzostomida, Pogonophora, Echiura, Sipuncula*, edited by Pamela L. Beesley, Graham J.B. Ross, and Christopher J. Glasby. Melbourne, Australia: CSIRO Publishing, 2000.

Periodicals

Eeckhaut, I., and M. Jangoux. "Fine Structure of the Spermatophore and Intradermic Penetration of Sperm Cells in *Myzostoma cirriferum* (Annelida, Myzostomida)." *Zoomorphology* 111 (1991): 49–58.

———. "Life Cycle and Mode of Infestation of *Myzostoma cirriferum* (Annelida)" *Diseases of Aquatic Organisms* 15 (1993): 207–217.

Eeckhaut, I., D. McHugh, P. Mardulyn, R. Tiedemann, D. Monteyne, M. Jangoux, and M. Milinkovich. "Myzostomida: A Link Between Trochozoans and Flatworms?" *Proceedings of the Royal Society, London*, Series B, 267 (2000): 1383–1392.

Igor Eeckhaut, PhD

Oligochaeta
(Earthworms)

Phylum Annelida
Subphylum Clitellata
Class Oligochaeta
Number of families 17

Thumbnail description
Terrestrial worms that typically dwell in soil and that are characterized by a "tube within a tube" construction, with an outer muscular body wall surrounding a digestive tract that begins with the mouth in the first segment and ends with the anus in the last segment

Photo: A giant earthworm (*Haplotaxida*) burrows back into the moist leaf litter in the subtropical rainforest floor. (Photo by Wayne Lawler/Photo Researchers, Inc. Reproduced by permission.)

Evolution and systematics

Earthworms belong to a well-defined clade, the Clitellata, which includes leeches, branchiobdellids, many aquatic and small terrestrial worms with a single cell-layered clitellum, and the earthworms, most of which have a multi-layered clitellum. However, earthworms as a group lack a defining characteristic unique to earthworms. This is because they include the Moniligastridae, a south and east Asian earthworm family, which have a single-layered clitellum and prosoporous (male genital openings in front of the female genital openings). All other earthworms have a multi-layered clitellum and male genital openings behind the female pores (opisthoporous) and are called the Crassiclitellata. As soft-bodied invertebrates, earthworms lack a fossil record, other than burrow traces that may or may not have been created by earthworms. Their phylogenetic relationships have been a matter of controversy since the early twentieth century. Based on analysis of DNA sequence data, Jamieson et al. (2002) concluded that the large family Megascolecidae (in the broad sense, including the Acanthodrilidae and Octochaetidae, of some authors) is the sister-group of the Ocnerodrilidae, and that these in turn are together the sister-group of a clade composed of several families: Sparganophilidae, Komarekionidae, Almidae, Lutodrilidae, Hormogastridae, Lumbricidae, and Microchaetidae. The remaining two numerically important families, Glossoscolecidae and Eudrilidae, form a third major clade of Crassiclitellata, but relationships to the other two were not clear. Several small families, plus the Moniligastridae, were not included in the analysis. These families complete the family list of the Crassiclitellata: Ailoscolecidae, Alluroididae, Biwadrilidae, Diporochaetidae, and Kynotidae. Overall, there are 17 families, one order, and more than 4,000 species.

Physical characteristics

Earthworms have a "tube within a tube" construction, an outer muscular body wall surrounding a digestive tract that begins with the mouth in the first segment and ends with the anus in the last segment. Body wall musculature consists of an outer circular layer and an inner longitudinal layer, which respectively extend and shorten the body. Between the body wall and the gut is the body cavity, within which various other organs are arranged, generally segmentally. Segments are repeated units of the body, externally manifested as rings and internally separated by septa. In earthworms, each segment except the first bears setae, small chitinous bristles used for traction in the burrow.

A typical earthworm gut consists of the mouth, a muscular pharynx for taking in food, a gizzard for reducing food particles to smaller sizes, an esophagus, and an intestine. In the family Lumbricidae, the gizzard is located after the esophagus, just prior to the expansion of the intestine. Intestinal gizzards have evolved independently in other families and genera. The esophageal wall may secrete digestive enzymes, and in some earthworms, parts of the esophagus are modified as glands for the secretion of calcium carbonate into the gut contents. The intestine may be differentiated into digestive and absorptive regions, and often has a dorsal in-folding of the intestinal wall, called the typhlosole.

Small excretory organs, the nephridia, are arranged segmentally, from two per segment to many small nephridia per segment. Urine is excreted through nephropores to the outside, or is collected via systems of tubules and excreted into the intestine. In some families, nephridia of the anterior segments have been modified as glands for digestive secretions.

Close-up of worm casts in a lawn. (Photo by Holot Studios/Bob Gibbons/Photo Researchers, Inc. Rerproduced by permission.)

Earthworms are hermaphrodites. Reproductive organs are located in the anterior segments. The female reproductive system consists of paired ovaries in the 13th segment, ovarian funnels leading from the ovaries to an external female genital pore on the 14th segment, and depending on the family, there may be sperm receptacles called spermathecae. If present, these will generally be in some of segments 5–10. Spermathecae receive sperm from the mate during copulation. Alternatively, sperm may be deposited in packets called spermatophores, which will be found clinging to the exterior of the worm. The clitellum provides an outer casing for the ova and also secretes food used by the developing embryo.

Male organs consist of testes in one or both of segments 10 and 11, testicular funnels leading to sperm ducts through which sperm passes to the male genital openings, seminal vesicles in segments adjacent to the testicular segments (one or more of segments nine, 11, 12), and in some families, prostate glands that secrete fluids associated with the male genital pores. In other families, there are often glands associated with setae modified for use in copulation.

Distribution

Earthworms are globally distributed, but do not occur in deserts or regions where there is permafrost or permanent snow and ice. They may also be absent from the taiga biome and other cold climate vegetation types where soils are strongly acid (pH below 4). It has been shown that during the last 20,000 years, many glaciated areas have lacked the presence of earthworms, but in these and other places where they do not occur naturally, some species have been introduced by human activity. The Megascolecidae have the widest natural distribution, being present on all continents, except Europe. The Glossoscolecidae are confined to tropical South America, Central America, and a few Caribbean islands, while the Eudrilidae are found only in sub-Saharan Africa. The Lumbricidae are mainly in Europe, with a few species native to North America. Australian indigenous species are exclusively megascolecids. A few species have attained global temperate or tropical distributions with human assistance.

Habitat

The typical earthworm habitat is soil, but there are species living in freshwater mud, saltwater shorelines, and in suspended soils of tropical forests. The soil habitat can be divided into litter layer, topsoil, and deeper soil horizons, with different earthworms utilizing each.

Behavior

The three ecological categories of earthworms have very different behavior patterns. The anecic feeding behavior has been described. Their primary escape tactic is to rapidly withdraw into the burrow. Epigeic species crawl or burrow through organic matter deposits and feed on it. They have well-developed escape behavior that includes rapid motions, even the ability to jump and thrash about randomly, and to drop tail segments for the predator. Endogeics have little escape behavior, may just writhe or coil in the hand, and may exude some body cavity fluids. In some instances, these fluids may be noxious. Further details of earthworm behavior are poorly known, because they inhabit an opaque medium and are shy of light.

Feeding ecology and diet

Earthworms feed on dead and decomposing organic material such as fallen leaves, decaying roots, and soil organic matter. Epigeic worms are those feeding at or near the surface, or within accumulations of organic matter on or above the soil surface (e.g., logs, epiphyte root mats in trees, etc.). These will consume relatively freshly dead plant matter, as do anecic worms. Anecic earthworms maintain a deep burrow from which they emerge to ingest plant matter from the soil surface; the best known is the European night crawler, *Lumbricus terrestris*. Endogeic worms operate deeper in the soil and utilize organic matter that has already been somewhat or extensively modified from its original condition. Body size, coloration, and gut morphology are consistently different among these three categories. Epigeics are typically small, darkly colored, and have little secondary development of gut surface area. Anecics are large, colored only in the head, and have gut morphology similar to epigeics. Endogeic worms may be small or very large, but are usually un-pigmented, and show the greatest degree of gut surface area development.

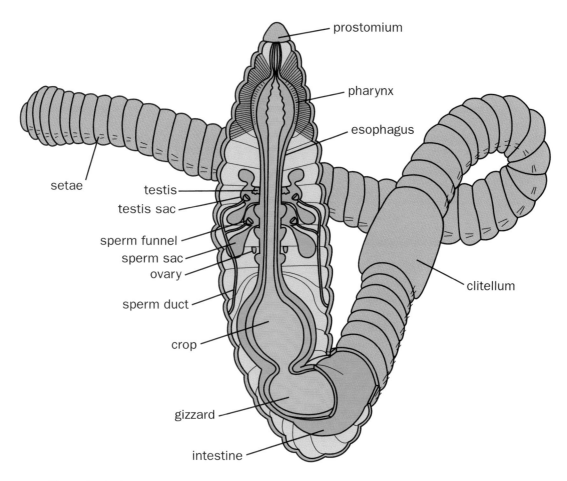

prostomium

pharynx

esophagus

setae

testis

testis sac

sperm funnel

sperm sac

ovary

sperm duct

crop

gizzard

intestine

clitellum

Earthworm anatomy. (Illustration by Laura Pabst)

Reproductive biology

Most earthworms are simultaneous hermaphrodites and exchange sperm during copulation. Sperm transfer may be external, in which the seminal fluid flows from male genital openings to the spermathecae, or there may be penis-like organs to insert the seminal fluid directly into the spermathecal openings. Sperm transfer by spermatophores is also known to occur. After copulation, fertilization takes place in the egg case. The case, or cocoon, is formed by the clitelum and passes over the female pores to receive one or more ova. It is then worked forward over the spermathecal pores, from which sperm are expelled into the case, and fertilization results. The cocoon is deposited in the soil or other substrate. The developing embryo feeds on clitellar and/or prostatic secretions, passes through larval stages, and emerges as a miniature earthworm. Growth and maturation may take months or years, depending on the species. In temperate zones, mating and cocoon deposition generally take place in the spring, with a secondary period possible in the autumn. In tropical areas, the peak of activity occurs during rainy seasons. However, the details of mating seasons in tropical earthworms are poorly known.

Some species of earthworms are clonal and reproduce by parthenogenesis. In this case, a diploid ovum is produced that is a genetic copy of the parent. No fertilization is necessary, so a single individual can reproduce unaided. This is important among the many species that have attained wide artificial distributions. In other instances, hermaphroditic species have been observed self-fertilizing. It is not known how common this is, or under what circumstances an individual may choose this course.

Conservation status

The 2002 IUCN Red List includes six species of earthworms; four are categorized as Vulnerable, one as Lower Risk/Near Threatened, and one (*Phallodrilus macmasterae*) as Critically Endangered. Only one is clearly protected, the Gippsland giant worm of Australia (*Megascolides australis*), which the IUCN classifies as Vulnerable. It has a very narrow range. *Driloleirus macelfreshi*, a giant worm from western Oregon in the United States, is suspected to be extinct, although the IUCN classifies it as Vulnerable. It is quite probable that many species are extinct because of habitat destruction, particularly in mountainous regions where the topography and earthworms' low dispersal rates contribute to high species diversity and small species ranges.

Significance to humans

Several species of earthworms (*Eisenia fetida, E. andrei, Eudrilus eugeniae,* and *Perionyx excavatus*) are used for production of vermicompost; some of these are used for fish bait as well. The use of earthworms as fish bait seems to be almost universal, and people use whatever worms they can find for this purpose. There are a few species commercially harvested and sold for bait: *Lumbricus terrestris* (Canada, northern United States), *Diplocardia riparia* (south-central United States), and *D. missippiensis* (Florida). As transformers of soil structure and organic matter, earthworms are significant to the maintenance and improvement of soils and plants growing in them, and thereby to humans who benefit from those plants.

1. River worm (*Diplocardia riparia*); 2. African worm (*Eudrilus eugeniae*); 3. Gippsland giant worm (*Megascolides australis*); 4. *Pontoscolex corethrurus*; 5. Common field worm (*Aporrectodea caliginosa*); 6. *Amynthas corticis*. (Illustration by Bruce Worden)

Species accounts

African worm
Eudrilus eugeniae

ORDER
Haplotaxida

FAMILY
Eudrilidae

TAXONOMY
Eudrilus eugeniae Kinberg, 1867, St. Helena Island.

OTHER COMMON NAMES
None known.

PHYSICAL CHARACTERISTICS
Dark mauve or pink throughout; 3.1–4.7 in (8–12 cm) length;
clitellum covering segments 14–18, prominent widely placed
female pores on segment 14, male pores on segments 17–18;
eight setae per segment in four pairs.

DISTRIBUTION
Sub-Saharan West Africa origin, now pantropical in rich organic microhabitats; cultured in earthworm farms globally.

HABITAT
Naturally occurring in savanna and forests; does well in vermi-
composting situations.

BEHAVIOR
Little is known apart from observations of seasonality of activity.

FEEDING ECOLOGY AND DIET
Cultured in pure organic media; in nature, produces piles of
fecal pellets (castings) during rainy season.

REPRODUCTIVE BIOLOGY
An outcrossing hermaphrodite; high enough reproductive rate
to make it economically useful.

CONSERVATION STATUS
Not listed by the IUCN. Status in native range not known;
probably stable. Now invasive in many tropical areas.

SIGNIFICANCE TO HUMANS
Cultured for fish bait, pet food, and the compost it produces.
Large numbers are grown annually in the United States,
mainly in southern states. ◆

Common field worm
Aporrectodea caliginosa

ORDER
Haplotaxida

FAMILY
Lumbricidae

TAXONOMY
Aporrectodea caliginosa Savigny, 1826, France.

Eudrilus eugeniae
Diplocardia riparia
Megascolides australis

OTHER COMMON NAMES
None known.

PHYSICAL CHARACTERISTICS
Slightly dusky pigmentation in the head segments, approximately 1.9–3.5 in (5–9 cm) length, with a clitellum (thickened band) covering segments 27–34, with small paired markings ventral side of segments 30, 32–34.

DISTRIBUTION
Western Europe, with peregrine distribution in New Zealand, northeast Asia, United States, and Canada and temperate South America, South Africa, and Australia. (Specific distribution map not available.)

HABITAT
Frequently found in human-influenced habitats, including arable land, pastures, and forests.

BEHAVIOR
In warmer temperate areas, it undergoes a temperature-induced summer dormancy, so that its peak activity periods are in the spring and autumn. Activity also requires moist soil conditions.

FEEDING ECOLOGY AND DIET
Flexible; may be found active to depths of 7.8 in (20 cm) or near the surface under litter or other decomposing organic material; seldom in purely organic substrates.

REPRODUCTIVE BIOLOGY
It is known to mate; most likely an outcrossing hermaphrodite. Reproduction is in the spring, with a secondary peak in the autumn.

CONSERVATION STATUS
Not listed by the IUCN. Abundant where it occurs, and may be considered an invasive exotic throughout much of its artificially acquired range.

SIGNIFICANCE TO HUMANS
Capable of persisting under agricultural tillage conditions, it may be of some importance in creating soil properties favorable to plant growth. ◆

No common name
Amynthas corticis

ORDER
Haplotaxida

FAMILY
Megascolecidae

TAXONOMY
Amynthas corticis Kinberg, 1867, Oahu, Hawaii, United States.

OTHER COMMON NAMES
None known.

PHYSICAL CHARACTERISTICS
Red to reddish brown dorsal pigmentation, lighter at segmental equators; 1.9–5.9 in (5–15 cm) length; clitellum on segments 14–16, single female pore ventral on segment 14, small paired male pores ventral side of segment 18, spermathecal pores at furrows 5/6/7/8/9, small oval discs ventrally in segments 6–9; setae in continuous rings of 10–50 per segment.

DISTRIBUTION
Subtropical China, with global warm temperate to cooler tropical peregrine distribution. (Specific distribution map not available.)

HABITAT
Frequently found in human-influenced habitats, including arable land, pastures, and forests.

BEHAVIOR
Capable of violent thrashing movements and jumping, may autotomize tail segments.

FEEDING ECOLOGY AND DIET
Flexible; may be found active to depths of 7.8 in (20 cm) or near the surface under litter or other decomposing organic material.

REPRODUCTIVE BIOLOGY
All known populations are probably asexual parthenogenetic morphs, a factor in its very wide artificial distribution.

CONSERVATION STATUS
Not listed by the IUCN. An invasive exotic throughout its artificially acquired range. The original homeland is not known, but may be in China.

SIGNIFICANCE TO HUMANS
Some attempts have been made to culture it for bait, and it has been blamed for damage in greenhouses. Like other peregrine species, it is able to tolerate disturbed soil conditions, so may be of some value if populations do not become too large. ◆

River worm
Diplocardia riparia

ORDER
Haplotaxida

FAMILY
Megascolecidae (Acanthodrilinae)

TAXONOMY
Diplocardia riparia Smith, 1895, Illinois, United States.

OTHER COMMON NAMES
None known.

PHYSICAL CHARACTERISTICS
Dark brown anterior dorsal pigmentation; 4.7–7.8 in (12–20 cm) length; clitellum covering segments 13–18, with small paired longitudinal grooves ventrally on segments 18–20.

DISTRIBUTION
Central United States, in Iowa, Illinois, Missouri, Kansas, and Nebraska.

HABITAT
Riparian forests and fine-textured alluvial soils of river banks; said to be most common under silver maple stands.

BEHAVIOR
Nothing is known.

FEEDING ECOLOGY AND DIET
Probably anecic or endogeic, feeding on organic debris buried in riverbank sediments.

REPRODUCTIVE BIOLOGY
Because specimens show evidence of sperm production, it is an outcrossing hermaphrodite.

CONSERVATION STATUS
Not listed by the IUCN. Locally abundant in banks of low-gradient streams in the central United States, but invasive species brought in as fish bait may threaten its populations in some areas.

SIGNIFICANCE TO HUMANS
Collected from natural populations and sold for fish bait. It is better able to withstand summer heat than other common bait species. ◆

Gippsland giant worm
Megascolides australis

ORDER
Haplotaxida

FAMILY
Megascolecidae

TAXONOMY
Megascolides australis McCoy, 1878, Australia.

OTHER COMMON NAMES
Aboriginal: Karmai.

PHYSICAL CHARACTERISTICS
Dark purple anterior pigmentation; approximately 31.5–39.3 in (80–100 cm) length; clitellum covering segments 27–34, with small paired markings ventral side of segments 30, 32–34.

DISTRIBUTION
Limited to the Bass River Valley, Victoria, Australia.

HABITAT
In clay soils along watercourses and other moderately wet places.

BEHAVIOR
The entire life is spent underground, including feeding, mating, and waste deposition.

FEEDING ECOLOGY AND DIET
Forms deep burrows reaching to the water table; feeds on organic matter in soil.

REPRODUCTIVE BIOLOGY
Egg cases are deposited underground, and embryos develop for 12–14 months, emerging at 7.8 in (20 cm) length.

CONSERVATION STATUS
Listed as Vulnerable by the IUCN. Has a very narrow range and is threatened by land and water use practices in the region.

SIGNIFICANCE TO HUMANS
It is of no economic importance, other than as a wonder of nature with some tourism drawing power. There is a Giant Worm Museum in Bass and a worm festival in the town of Korumburra. ◆

No common name
Pontoscolex corethrurus

ORDER
Haplotaxida

FAMILY
Glossoscolecidae

TAXONOMY
Pontoscolex corethrurus Müller, 1856, Brazil.

OTHER COMMON NAMES
English: Brushy-tail.

PHYSICAL CHARACTERISTICS
Un-pigmented, light pink in head, posterior section color depends on ingested soil; approximately 1.9–3.9 in (5–10 cm) length; clitellum covering segments 13–22; setae of tail enlarged, located in alternate positions between segments, either irregular or in 16 rows, but only eight per segment.

DISTRIBUTION
Originally from northeast South America, with pantropical peregrine distribution in disturbed habitats. (Specific distribution may not available.)

HABITAT
Frequently found in human-influenced habitats, including arable land, pastures, and second-growth forests; from low elevations to tropical montane cloud forests.

BEHAVIOR
When exploring, it extends a tentacle-like proboscis. Goes dormant when soils dry out.

FEEDING ECOLOGY AND DIET
Endogeic, flexible; may be found active to depths of 7.8 in (20 cm) or near the surface under litter or other decomposing organic material; able to utilize low-quality organic matter.

REPRODUCTIVE BIOLOGY
Possibly parthenogenetic, can reproduce as long as soil conditions permit.

CONSERVATION STATUS
Not listed by the IUCN. Abundant where it occurs, and may be considered an invasive exotic throughout much of its artificially acquired range.

SIGNIFICANCE TO HUMANS
As one of the few tropical earthworms capable of persisting under agricultural tillage conditions, it may be of some importance in creating soil properties favorable to plant growth, but it has been implicated in soil structure breakdown when population density becomes high. ◆

Resources

Books

Edwards, C. A., and P. J. Bohlen. *Biology and Ecology of Earthworms*. 3rd edition. New York: Chapman and Hall, 1996.

Lee, K. E. Earthworms, *Their Ecology and Relationships with Soils and Land Use*. Sydney, Australia: Academic Press, 1985.

Stephenson, J. *The Oligochaeta*. Oxford, U.K.: Clarendon Press, 1930.

Periodicals

Gates, G. E. "Burmese Earthworms. An Introduction to the Systematics and Biology of Megadrile Oligochaetes with Special Reference to Southeast Asia." *Transactions of the American Philosophical Society new series* 62, no. 7 (1972): 1–326.

Michaelsen, W. "Oligochaeta." *Tierreich* 10 (1900): 1–575.

Other

Jamieson, B. G. M. *Native earthworms of Australia (Megascolecidae, Megascolecinae)*. PDF Document on CD-ROM. Enfield, NH: Science Publishers, Inc., 2000.

Samuel Wooster James, PhD

Hirudinea
(Leeches)

Phylum Annelida
Class Hirudinea
Number of families 14

Thumbnail description
Annelids possessing a caudal sucker used for attachment to surfaces, and a fixed number of body segments with subdivided annuli (ring-like structures); best known for their blood-sucking members and their use in medicine

Photo: Newborn brood of aquatic leaf leeches. (Photo by Animals Animals ©K. Atkinson, OSF. Reproduced by permission.)

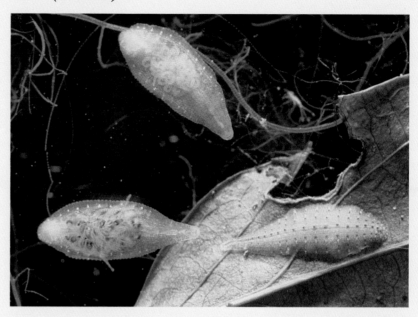

Evolution and systematics

Leeches are completely soft-bodied animals and could not be expected to leave a marked fossil record. Although there are two putative fossils from Bavarian deposits dating from the Upper Jurassic period (about 145 million years ago), *Epitrachys rugosus* (Ehlers, 1869) and *Palaeohirudo eichstaettensis* (Kozur, 1970), neither has both the definitive caudal sucker and the ring-shaped subdivisions of the body that would define them as leeches. The evolutionary relationships of leeches demonstrate that the ancestral hirudinid was a blood feeder in a freshwater environment. This finding suggests that leeches are no older than vertebrates and probably no older than amphibians.

The phylogenetic relationships of leeches have been the subject of several analyses based on morphological (structural) characters and DNA sequence data. Taken together, these analyses demonstrate beyond question that leeches, branchiobdellidans (crayfish worms) and acanthobdellidans form a monophyletic (descended from a common taxon) group of oligochaetes that is closely related to the lumbriculid oligochaetes, or earthworms.

Leeches are subdivided first into suborders based on anatomical adaptations for feeding. The Rhynchobdellida, as their name implies, have a muscular proboscis that allows them to feed on blood drawn from tissues beneath an organism's skin that are well supplied by blood vessels. The large worm-like Arhynchobdellida, of which *Hirudo medicinalis* is typical, have three muscular jaws, each of which may be armed with a row of teeth creating a serrated cutting edge.

The two principal families of proboscis-bearing Rhynchobdellida have pairs of medial cephalic eyespots that can detect two-dimensional movement. The Piscicolidae family comprises small elongate, mostly marine species that feed seasonally on fishes. The Glossiphoniidae tend to be strongly dorsoventrally flattened freshwater species typically preferring anuran (frogs and toads) or chelonian (turtles and tortoises) hosts, though a few are specific to fishes and others will feed on mammals.

The blood-feeding arhynchobdellids include the large aquatic Hirudinidae (medicinal leeches) and the smaller terrestrial Haemadipsidae (jungle leeches). The jungle leeches are more common in the humid forests of Southern Asia, India, Madagascar, Australia, and Indonesia than their aquatic cousins. Both of these groups are equipped with a parabolic arc of 10 eyespots that detect movement in three dimensions. Terrestrial leeches have the additional adaptation of respiratory auricles near their caudal sucker that permit gas exchange without excessive loss of fluid. These leeches also have well-developed sensory systems for detecting vibrations, carbon dioxide, and heat. In addition, there are several predatory arhynchobdellids like the slender Erpobdelliformes (families Erpobdellidae, Salifidae, and Americobdellidae); the larger amphibious Haemopidae; and several other poorly studied families.

Physical characteristics

Leeches have a clitellum, or specialized saddle-shaped glandular segment, that secretes cocoons or egg cases. These animals are simultaneous hermaphrodites. They have a body consisting of 34 body segments (somites) but lack the chaetae (stiff hairs or bristles) of other annelids. Though not all are sanguivorous, many leeches have special adaptations for blood feeding. Principal among these is the muscular caudal sucker made up of the last seven somites of the segmented body. The sucker is critical for attaching to and remaining on a host. It may double as a powerful swimming fluke (anchor) in larger and more active species. The six somites at the front of the leech also are modified into a region with a ventral sucker surrounding a muscular pharynx. Negative pressure applied by this anterior sucker aids in attachment and in encouraging blood to flow from a bite wound. Blood-feeding leeches usually are equipped with large branched gastric caeca, or pouches, that allow them to expand considerably during feeding; some leeches consume up to six times their unfed body weight. The fact that all present-day blood-feeding leeches have these features suggests that the ancestral leech also had them.

Distribution

Leeches can be found in one form or another in freshwater and terrestrial locations on all continents except Antarctica. Marine leeches are found in all oceans. The major families of leeches have global distributions with the exception of some species in the Haemadipsidae family. These species appear to have originated at the time of the late Gondwanan continental separation, after Africa and South America had parted from the remainder of the original supercontinent about 180–140 million years ago. Some minor families, including the Americobdellidae and Cylicobdellidae, are found only in South America. Most oceanic leeches seem to prefer temperate or arctic waters, with only a few species specific to elasmobranchs being found in tropical marine systems.

Habitat

The frequency with which leeches are encountered depends on geography. Most species that occur in European, African, or North and South American freshwater ecosystems can be found under submerged rocks and debris or along shorelines when they are not feeding on their respective hosts. Terrestrial leeches are found only in such perpetually moist environments as tropical or temperate coastal rainforests, where they remain in leaf litter unless they are seeking a meal. The habits of marine leeches are not well understood, as they are most commonly encountered while they are feeding on fishes or turtles. Few leeches appear to select specific sites on the hosts they feed from.

Behavior

Many leeches can swim by coordinating the depolarization of nerve cells along their ventral cord, which causes the lon-

gitudinal muscles to move the leech's body in a curving or wavelike pattern. The posterior sucker serves as an anchor to provide thrust. Leeches move along solid substrates by alternating attachment of their anterior and posterior suckers between periods of body stretching— much like the movement of an inchworm. When at rest during long periods of digestion or while brooding young, leeches lie under objects along the shorelines, often partially out of water. They frequently are observed performing their wavelike swimming movement in place as though to assist in ventilation. Leeches have several anterior eyespots and are able to detect movement from contrasting patterns of light and shadow.

Feeding ecology and diet

Sanguivorous (blood-feeding) leeches are capable of living on blood because of a number of bioactive chemicals in their saliva. Vertebrate blood, including human blood, is equipped with coagulation (clotting) factors. Most leeches require 30 minutes or more for feeding; however, vertebrate blood can clot in much less time. If the blood that the leech ingests were to clot inside its digestive tract, it could not mate, avoid its predators, or seek another meal. Leeches have circumvented the endpoints of the mammalian coagulation cascade, which is the medical term for the process of blood clotting. The coagulation cascade includes the clumping of platelets, the production of a fibrin matrix (network), and the cross-linking of that matrix into a firm clot. Leeches can interfere with this clotting system at a minimum of seven different points. Hirudin, a potent thrombin inhibitor, was the first anticoagulant (blood-thinning) compound to be isolated from leech saliva. Most other leech-derived anticoagulants are also protease inhibitors.

Some sanguivorous species are very host-specific; *Placobdelloides jaegerskioeldi* feeds exclusively on *Hippopotamus amphibius*, for example, while other leeches are less discriminating. Of the marine species, those that feed on cartilaginous fishes show no interest in teleosts (bony fishes), and *Ozobranchus* species feed only on turtles. In freshwater environments, the large and notorious hirudinid medicinal leeches more often acquire their nourishment from amphibians or fishes than from swimming humans.

Many leeches do not feed on blood at all. Glossiphoniids, like species of *Helobdella* and *Glossiphonia*, feed on aquatic oligochaetes and snails. The jawless Erpobdellidae feed on chironomid (midge) larvae, and the jawed haemopids consume whole earthworms, shredding them with jaws bearing two rows of large teeth. In addition, there are rarely encountered families like the South American Americobdellidae and Cylicobdellidae. The species in these families are terrestrial earthworm hunters and of uncertain phylogenetic affinities. Researchers have typically assumed that nonbloodfeeding varieties of leeches are more primitive than those with the "advanced" behavior of blood feeding. Recent phylogenetic work points to the existence of a blood feeding ancestor, however, and at least six transformations to predation in the history of leeches.

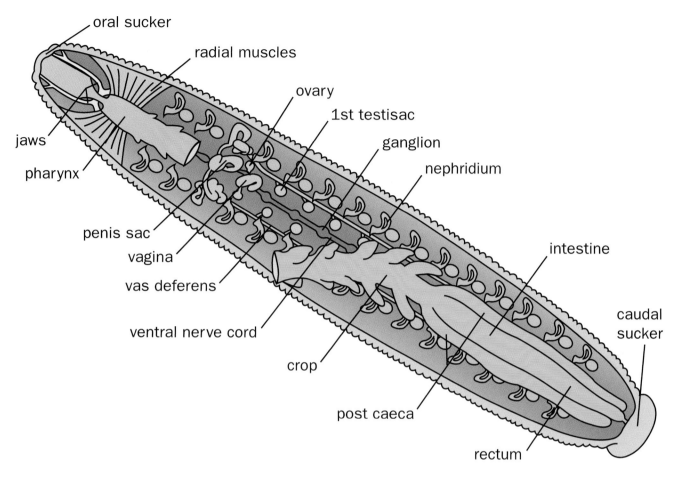

oral sucker

radial muscles

ovary

1st testisac

ganglion

nephridium

jaws

pharynx

penis sac

vagina

vas deferens

ventral nerve cord

crop

post caeca

rectum

intestine

caudal sucker

Leech anatomy. (Illustration by Laura Pabst)

Reproductive biology

Leeches are hermaphroditic animals with separate male and female reproductive systems. Male and female systems open independently to the exterior, each by means of a small ventral gonopore situated on the midventral line in the clitellar somites (normally somites XI through XIII). The most common form of sexual intercourse in leeches is traumatic insemination (direct injection into the female's body cavity). Male leeches randomly implant their spermatophores (small packets containing sperm) in the cuticle or outer covering of a recipient mate. The larger hirudinids and terrestrial haemadipsids have anatomical adaptations for internal fertilization, including a protrusible penis and a receptive vaginal sac permitting direct contact between male and female gonopores.

Reproducing individuals are usually distinguishable by the swelling of a certain number of annuli in the region that includes their gonopores. This swollen area constitutes the clitellum, a glandular segment that secretes the cocoons or egg cases. Unlike the homologous structure in terrestrial earthworms, the degree of prominence of the clitellum varies considerably in leeches.

The fish leeches, or Piscicolidae, exhibit an adaptation that helps their offspring to find an early blood meal. Rather than abandoning a secreted cocoon as the oligochaetes and many leeches do, the piscicolids cement their egg cases to the surface of crustaceans. When a fish eats the crustacean, the young leeches readily attach to the fish host's buccal (cheek) surfaces and migrate to its gills. Such Glossiphoniidae as *Haementeria ghilianii* are broad and flattened, and normally found feeding on turtles or amphibians. Species in this family secrete a membranous bag that holds their eggs on their underside in a brooding position underneath rocks and other debris. When the brood hatches, the young will turn and attach themselves to their parent's venter, or belly. When the parent finds its next blood meal, the young are carried to their first blood meal.

Conservation status

The European medicinal leech, *Hirudo medicinalis*, has been overexploited. In addition, its natural habitat has become fragmented and highly restricted. This species is listed as Lower Risk/Near Threatened by the IUCN, and is also listed in

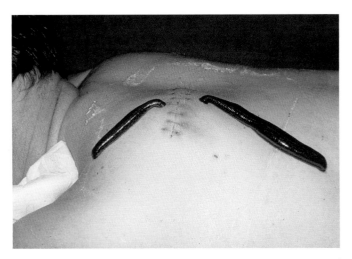

Use of the medicinal leech (*Hirudo medicinalis*) in the treatment of haemotoma, an accumulation of blood within the tissues that clots to form a solid swelling. The leech attaches its sucker near the injury, makes an incision and deposits an anticoagulant called hirudin, mixed with saliva, into the wound, breaking apart clots that are there, and preventing further clots from forming. (Photo by St. Bartholomew's Hospital/Science Photo Library/Photo Researchers, Inc. Reproduced by permission.)

CITES Appendix II. Many terrestrial rainforest species are as threatened as their specific habitats; examples include *Mesobdella gemmata* in the Valdivian coastal forests of Chile and *Haemadipsa sumatrana*.

Significance to humans

Leeches were well known for their medicinal applications in the eighteenth and nineteenth centuries, when they were used for such purposes as alleviating headaches and treating obesity. It is unlikely that any of these therapies were successful. Leeches are now, however, experiencing a renaissance of interest in medicine and pharmacology: they are the tools of choice for treating postoperative hematomas (localized collections of clotted blood) in microsurgery. In addition, their powerful salivary anticoagulants are being studied as possible treatments for heart disease and even cancer. Leeches are also potential indicator species for certain measurements of water quality, including heavy metal contamination and dissolved oxygen content.

1. Tiger leech (*Haemadipsa picta*); 2. North American medicinal leech (*Macrobdella decora*); 3. Giant Amazonian leech (*Haementeria ghilianii*); 4. Hippo leech (*Placobdelloides jaegerskioeldi*); 5. Euopean medicinal leech (*Hirudo medicinalis*). (Illustration by Bruce Worden)

Species accounts

Tiger leech
Haemadipsa picta

ORDER
Arhynchobdellae

FAMILY
Haemadipsidae

TAXONOMY
Haemadipsa picta Moore, 1929.

OTHER COMMON NAMES
English: Jungle leech; terrestrial leech.

PHYSICAL CHARACTERISTICS
Pale brownish or orange body with a prominent paramedial pair of dark stripes of pigmentation on dorsum and irregular chain-link patterns medially. In its unfed state, it is normally 0.39–0.78 in (1–2 cm) in length and 0.16–0.19 in (4–5 mm) in width. When the leech is foraging and extended, cephalic somites form an obvious piriform (pear-shaped) shape like the head of a snake. Parabolic arc of 10 cephalic eyes arranged in five pairs. Three jaws in pharynx with fine denticles. Respiratory slits (auricles) are located along the sides near the caudal sucker.

DISTRIBUTION
Malesian subregion from southern Burma and southern China as far as Sumatra and Borneo.

HABITAT
Terrestrial in rainforests, especially in leaf litter.

BEHAVIOR
Though poorly understood, the organism's foraging is thought to involve an excellent ability to respond to movement as detected by its 10 eyespots, as well as the detection of carbon dioxide given off by nearby hosts. Mechanoreceptors on the skin of these leeches allow them to pursue movements on land similar to footsteps. They move quite rapidly over surfaces in an inchworm fashion.

FEEDING ECOLOGY AND DIET
Haemadipsids like the tiger leech forage on the ground, along the branches and leaves of underbrush and other exposed surfaces. They attach themselves to these surfaces by their larger caudal sucker and extend their bodies up or out, with the oral sucker ready to attach to a passing host. Though this species probably subsists on frogs, any passing mammal will serve for a suitable blood meal. Following a rainstorm, these leeches may number in the hundreds along trails and on brush; they can be quite oppressive to humans in the region.

REPRODUCTIVE BIOLOGY
Mating takes place through gonopore-to-gonopore copulation. The eggs are laid in clitellar secretions surrounding the body, which are then slipped over the head to form a protective egg case or cocoon left under leaf litter. Development is direct.

CONSERVATION STATUS
Not listed by the IUCN. Not enough is known about the population size of this leech. The tiger leech exists only in endemic rainforests, however, all of which are under grave threat in the regions where this species is found.

SIGNIFICANCE TO HUMANS
Where they occur, jungle leeches are among the most self-assertive and omnipresent elements of a rainforest. Although it is unlikely that anyone would bleed to death from the dozens of leeches that find their way into socks or under shirts and trousers, many have been driven to distraction by their tenacity, the inability to escape their attentions, or the itchy welts that may arise from their bites. Occasionally featured in motion pictures (e.g., *Apocalypse Now*). ◆

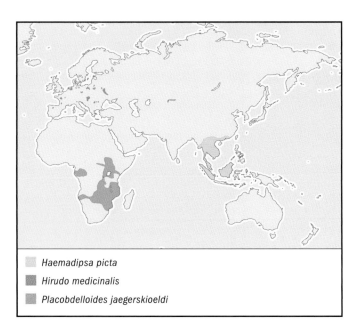

- Haemadipsa picta
- Hirudo medicinalis
- Placobdelloides jaegerskioeldi

European medicinal leech
Hirudo medicinalis

ORDER
Arhynchobdellae

FAMILY
Hirudinidae

TAXONOMY
Hirudo medicinalis Linnaeus, 1758.

OTHER COMMON NAMES
French: Sangsue médicinale; German: Medizinische Blutegel; Italian: Sanguisuga; Spanish: Sanguijuela medicinal; Swedish: Medicinsk blodigel.

PHYSICAL CHARACTERISTICS
Intricate red and green patterning on dorsum (upper surface), yellow medial lines, and tan venter (belly). Commonly grows up to 3.9 in (10 cm) or more in length and 0.39 in (1 cm) wide. Parabolic arc of 10 cephalic eyes arranged in five pairs.

Three jaws in pharynx with fine denticles (small toothlike structures).

DISTRIBUTION
British Isles, southern Scandinavia, continental Europe eastward to the Urals and to western Turkey. Distribution is very irregular and patchy as of 2003.

HABITAT
Found in naturally occurring freshwater lakes, ponds, streams, and marshes. Usually rests at air/water interface near shore.

BEHAVIOR
An excellent and agile swimmer, this leech can detect and move toward disturbances in the water from a distance of some yards. Vertical wavelike motions and use of the caudal sucker provide forward movement and thrust. Applies negative pressure to surfaces with its oral and caudal suckers for suction and attachment purposes.

FEEDING ECOLOGY AND DIET
Feeds principally on the blood of amphibians and fishes; occasionally feeds on mammals. Creates a three-part incision with its jaws and uses negative pressure of pharynx to draw out the upwelling blood. Can ingest quantities of blood several times greater than its unfed body weight. Can store blood in crop (enlarged area at the base of the esophagus) for months.

REPRODUCTIVE BIOLOGY
Mating takes place through gonopore-to-gonopore copulation involving an eversible penis and vaginal pouch. The eggs are laid in clitellar secretions surrounding the body, which are then slipped over the head to form a protective egg case or cocoon. Cocoons with approximately 10 eggs each are deposited on land near the edge of a body of water. Development is direct.

CONSERVATION STATUS
Overexploited; has seen its natural habitat become fragmented and highly restricted. Listed as Lower Risk/Near Threatened by the IUCN; also listed in CITES Appendix II.

SIGNIFICANCE TO HUMANS
Used medicinally for the purpose of phlebotomy (drawing blood) for millennia, and popularized for this use in the nineteenth century. Most historical uses are of dubious utility. Current use in microsurgery to reduce postoperative hematomas is quite legitimate and effective. Commercially available for medical purposes. Several anticoagulants, such as the antithrombin compound hirudin, have been extracted from salivary tissues and have biomedical/pharmacological use. Occasionally featured in motion pictures (e.g., *Speed 2*). ◆

North American medicinal leech
Macrobdella decora

ORDER
Arhynchobdellae

FAMILY
Macrobdellidae

TAXONOMY
Macrobdella decora Say, 1824.

OTHER COMMON NAMES
English: American leech.

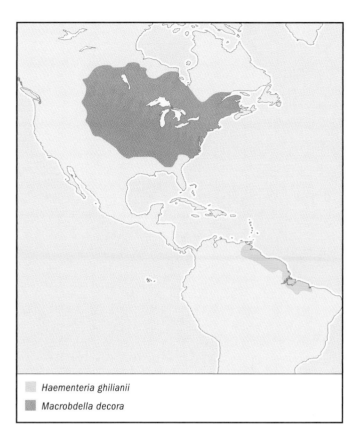

☐ *Haementeria ghilianii*
■ *Macrobdella decora*

PHYSICAL CHARACTERISTICS
Olive dorsum with a medial row of orange spots on every fifth annulus and paramedial rows of black dots. Orange venter with two rows of prominent sexual gland pores behind the gonopores. Commonly grows up to 3.9 in (10 cm) or more in length and 0.39 in (1 cm) wide. Parabolic arc of 10 cephalic eyes arranged in five pairs. Three jaws in pharynx with fine denticles.

DISTRIBUTION
North America; east of the Rocky Mountains from southern Canada to the Carolinas and along the Mississippi River drainages.

HABITAT
Found in naturally occurring freshwater lakes, ponds streams and marshes. When resting, usually found at the air/water interface near shore.

BEHAVIOR
An excellent and agile swimmer, this species can detect and move toward disturbances in water from a distance of some yards. Vertical wavelike motions and use of the caudal sucker provide forward movement and thrust. Oral and caudal suckers apply negative pressure to surfaces for suction and attachment purposes.

FEEDING ECOLOGY AND DIET
Feeds principally on the blood of amphibians and fishes; occasionally feeds on mammals. Creates a three-part incision with its jaws and uses negative pressure of the pharynx to draw out the upwelling blood. Can ingest quantities of blood several times greater than its unfed body weight. Can store blood in crop (enlarged area at the base of the esophagus) for months.

REPRODUCTIVE BIOLOGY

Mating is through gonopore-to-gonopore copulation but penis and vagina are quite small relative to those of its European counterpart. The eggs are laid in clitellar secretions surrounding the body, which are then slipped over head to form a protective egg case or cocoon. Cocoons with approximately 10 eggs each are deposited on land near the edge of a body of water. Development is direct.

CONSERVATION STATUS

Not listed by the IUCN. Populations are usually abundant when encountered.

SIGNIFICANCE TO HUMANS

Features prominently in creation mythology of the Osage tribe of Native Americans. Antiplatelet aggregation factor, decorsin, isolated from salivary secretions. Not widely used for phlebotomy. Occasionally featured in motion pictures (e.g., *Stand by Me*). ◆

Giant Amazonian leech
Haementeria ghilianii

ORDER
Rhynchobdellida

FAMILY
Glossiphoniidae

TAXONOMY
Haementeria ghilianii de Phillipi, 1849.

OTHER COMMON NAMES
French: Sangsue amazonienne.

PHYSICAL CHARACTERISTICS

Perhaps the largest freshwater leech, this species can exceed 11.8 in (30 cm) long and 3.93 in (10 cm) wide. The species is dorsoventrally flattened and shaped like a lance with a large caudal sucker. Adults often appear simply dark gray-brown in color, though younger specimens have a characteristic dorsal median broken stripe and other regular pigment patches on every third annulus. The somites in the head region carry one pair of eyes.

DISTRIBUTION

Found in locations near the mouth of the Amazon River, as far north as Venezuela and throughout the Guianas.

HABITAT
Coastal wetland marshes.

BEHAVIOR

A capable swimmer, this species is more often found on the underside of submerged rocks or debris where it hides while digesting a blood meal or brooding young.

FEEDING ECOLOGY AND DIET

Often found feeding on introduced cattle, this species otherwise feeds on endemic amphibians when young and on such local aquatic vertebrates as caimans, anaconda, and capybara. This leech feeds by inserting a muscular proboscis into the tissues of the host from which vascular blood is then pumped into its gastric pouches.

REPRODUCTIVE BIOLOGY

Mating is carried out by traumatic insemination of spermatophores into the integument (skin-like covering) of a recipient leech. Fertilized eggs are deposited on the underside of the parent and brooded until they hatch. The young are then carried to their first blood meal by the parent leech.

CONSERVATION STATUS

Not listed by the IUCN. Although this leech is difficult to find, it is not protected in any way.

SIGNIFICANCE TO HUMANS

The anticoagulant compound hementin, isolated from the salivary secretions of the giant Amazonian leech, is capable of breaking down a blood clot after it has formed. Hementin may be useful in medical treatment if it does not trigger allergic reactions from the patient's immune system during chronic use. Another anticoagulant compound known as tridegin has also been isolated from the secretions of *H. ghilianii*. ◆

Hippopotamus leech
Placobdelloides jaegerskioeldi

ORDER
Rhynchobdellida

FAMILY
Glossiphoniidae

TAXONOMY
Placobdelloides jaegerskioeldi Johansson, 1909.

OTHER COMMON NAMES
None known.

PHYSICAL CHARACTERISTICS

Dorsoventrally flattened and shaped like a lance with a large caudal sucker. Adults have a rough body surface due to the presence of large numbers of tubercles (small nodules). Color is dull brown to light gray with orange ringlets arranged in transverse rows. The somites in the head region carry one pair of eyes.

DISTRIBUTION

Found throughout sub-Saharan Africa wherever there are hippopotami.

HABITAT

In pools and pans frequented by hippos; in particular, inside the rectum of the animals themselves.

BEHAVIOR

Very sluggish, not a strong swimmer. Movement is mostly by alternate use of suckers.

FEEDING ECOLOGY AND DIET

This species is the only leech that is specific to a mammalian host. It feeds by inserting its proboscis into the rectal tissues of a hippopotamus, where it feeds to the point of fullness and mates with other leeches.

REPRODUCTIVE BIOLOGY

Mating is carried out by traumatic insemination of spermatophores into the integument of a recipient leech. Fertilized eggs are deposited on the underside of the parent and brooded until they hatch. Brooding is believed to occur outside the hippopotamus host. The young are then carried to their first meal in the rectum of the next hippopotamus by the parent.

CONSERVATION STATUS
Not listed by the IUCN.

SIGNIFICANCE TO HUMANS
So far this species is poorly studied but may prove interesting to medical researchers in light of its being specifically adapted to counteract the mammalian blood clotting system. ◆

Resources

Books

Harding, M. A., and J. P. Moore. *The Fauna of British India: Hirudinea.* London: Taylor and Francis, 1927.

Keegan, H. L., S. Tashioka, and H. Suzuki. *Bloodsucking Asian Leeches of Families Hirudinidae and Haemadipsidae.* Tokyo: United States Army Medical Command, Japan, 1968.

Mann, K. H. *Leeches (Hirudinea): Their Structure, Physiology, Ecology and Embryology.* London: Pergamon Press, 1962.

Oosthuizen, J. H., and M. E. Siddall. "The Freshwater Leeches (Hirudinea) of Southern Africa with a Key to All Species." In *Guide to Freshwater Invertebrates in Southern Africa.* Pretoria: South African Water Research Commission, 2001.

Sawyer, R. T. *Leech Biology and Behaviour.* Oxford, U.K.: The Clarendon Press, 1986.

Periodicals

Apakupakul, K., M. E. Siddall, and E. M. Burreson. "Higher-Level Relationships of Leeches (Annelida: Clitellata: Euhirudinea) Based on Morphology and Gene Sequences." *Molecular Phylogenetics and Evolution* 12 (1999): 350–359.

Elliott, J. M., and P. A. Tullett. "The Status of the Medicinal Leech *Hirudo medicinalis* in Europe and Especially in the British Isles." *Biological Conservation* 29 (1984): 15–26.

Light, J. E., and M. E. Siddall. "Phylogeny of the Leech Family Glossiphoniidae Based on Mitochondrial Gene Sequences and Morphological Data." *Journal of Parasitology* 85 (1999): 815–823.

Siddall, M. E., K. Apakupakul, E. M. Burreson, K. A. Coates, C. Erséus, S. R. Gelder, M. Källersjö, and H. Trapido-Rosenthal. "Validating Livanow: Molecular Data Agree That Leeches, Branchiobdellidans and *Acanthobdella peledina* Form a Monophyletic Group of Oligochaetes." *Molecular Phylogenetics and Evolution* 21 (2001): 346–351.

Wells, S., and W. Coombes. "The Status of and Trade in the Medicinal Leech." *Traffic Bulletin* 8 (1987): 64–69.

Mark Edward Siddall, PhD

Pogonophora
(Beard worms)

Phylum Annelida
Class Pogonophora
Number of families 13

Thumbnail description
Tube-living marine worms nourished by internal
chemoautotrophic bacteria.

Photo: *Ridgeia piscesae* at Middle Valley hydrothermal site in the northeast Pacific Ocean,
photographed through the porthole of *Alvin*, the
Navy-owned deep submergence vehicle. The tubes
are 0.20–0.39 in (5–10 mm) in diameter. (Photo
by E. C. Southward. Reproduced by permission.)

Evolution and systematics

The name Pogonophora comes from the Greek *pogon*,
meaning "beard," and *phora*, meaning "bearing," and refers to
the fact that many species have from one to many tentacles at
the anterior end. The earliest known fossil pogonophorans are
from the late Cambrian period. About 130 living species of
Pogonophora have been described (three subclasses, 13 families, and 29 genera). Their zoological position has been debatable for many years. Their possession of chaetae and
intersegmental septa is now considered good evidence of annelid affinity, and DNA analysis supports this, but more work
is needed to clarify the position of the Pogonophora within
the phylum Annelida. It has been suggested that pogonophores
represent a single family of polychaetes (Siboglinidae), but a
more conservative classification is used in this chapter.

Pogonophora is divided into the subclasses Frenulata (families Oligobrachiidae, Siboglinidae, Polybrachiidae, Lamellisabel-
lidae, and Spirobrachiidae) and Monilifera (family Sclerolinidae,
genus *Sclerolinum*). Some scientists classify Vestimentifera as a
subclass of Pogonophora as well, but it is treated separately in
this volume. Molecular studies indicate that Frenulata is a sister clade to Vestimentifera plus Monilifera.

Physical characteristics

The tubes the worms produce and live in are unbranched,
yellowish, brown, or black, sometimes ring patterned, sometimes segmented. They are made of chitin and protein.

Adult pogonophores have no digestive system. Their body
consists of a short anterior region bearing one or more tentaclelike appendages (or branchial filaments), a very long

trunk, and a small, segmented, opisthosoma. The body wall
consists of cuticle, epidermis, and circular and longitudinal
muscles. The coelom is divided into a small cephalic cavity
connected to the tentacle coeloms, paired cavities running
through the anterior region and trunk, and a series of cavities in the opisthosomal segments, separated by muscular
septa. The pogonophoran nervous system consists of an anterior mass of nerve cells, a ventral tract of fibers, and a general network of small fibers, all within the epidermis. The
trophosome, a central organ that contains specialized bacte-
ria-containing cells, occupies much of the trunk, sharing the
space with the reproductive organs. The vascular system consists of parallel dorsal and ventral vessels and various sinuses
connecting them. On the body surface there are cuticular
plaques and ridges. The chaetae are internally like those of
other annelids, but have unusual dentate heads. They are
arranged in low-lying bands rather than in projecting podia.

Frenulates have from one to 200 tentacles; a cuticular
ridge, or frenulum, encircling the short, cylindrical, forepart;
plaques only on the trunk; chaetae in two well-marked rings
about one-third of the way down the trunk; and only four
chaetae per segment on the small opisthosoma. Trophosome
tissue is absent from the anterior third of the trunk, which is
occupied by the reproductive organs.

Moniliferans are extremely thin. They have only two tentacles, a ring of plaques on the anterior body region, and scattered plaques on the long trunk. They have no obturaculum
or frenulum. Chaetae may occur in two sparse bands at the
hind of the trunk, but are more numerous on the opisthosomal segments. Trophosome tissue is restricted to the posterior half of the trunk.

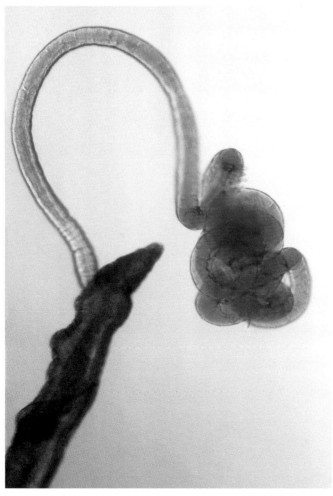

The anterior end and tentacles of *Siboglinum fiordicum*. (Photo by E. C. & A. J. Southward. Reproduced by permission.)

Distribution

Frenulates occur on continental slopes and deep trenches in the Atlantic (including Norwegian fjords), Pacific, Indian, Arctic, and Antarctic Oceans. Moniliferans occur in Norway, the Caribbean, the Southeast Pacific, and the Antarctic.

Habitat

Frenulates live in reducing sediments, with the top of the tube in the overlying oxygenated water. Moniliferans occur in decaying wood on volcanic seeps on the sea floor.

Behavior

The tentacles or branchial plume may be extended from the tube, particularly by vestimentiferans; the animals can retract quickly when disturbed by predators such as crabs. Main activities are gas exchange and tube building.

Feeding ecology and diet

An unusual type of hemoglobin in the blood binds both oxygen and sulfide and transports them to the trophosome. There, the symbiotic bacteria oxidize sulfide to produce chemical energy, which enables them to fix carbon dioxide and manufacture organic compounds. This chemosynthetic process makes pogonophorans virtually independent of light-based photosynthesis as a source of food. The bacteria produce glucose and amino acids, which can be transferred to the host cells directly, or derived from intracellular digestion of some of the bacteria. One frenulate species, *Siboglinum poseidoni*, lives in methane-rich conditions and utilizes methane with the aid of methanotrophic bacteria.

Reproductive biology

Frenulate males expel spermatophores that drift, become entangled on other tubes, then disintegrate to release long-headed sperm that fertilize the oocytes inside the female oviduct. Females produce either large eggs (around 400 microns) that are incubated in their tubes, or small eggs (around 150 microns) that may develop into pelagic larvae. Large eggs develop into ciliated larvae, which emerge from the maternal tube, then rapidly burrow deep into the sediment and form their own tubes. The symbiotic bacteria are acquired during the settlement stage, and the trophosome then starts to develop.

In moniliferans, *Sclerolinum brattstromi* males produce free sperm, and eggs are incubated by the female for a few days; later development is unknown. Asexual reproduction by fragmentation occurs.

Conservation status

No species of Pogonophora are listed by the IUCN.

Significance to humans

Pogonophorans are of great interest for scientific research.

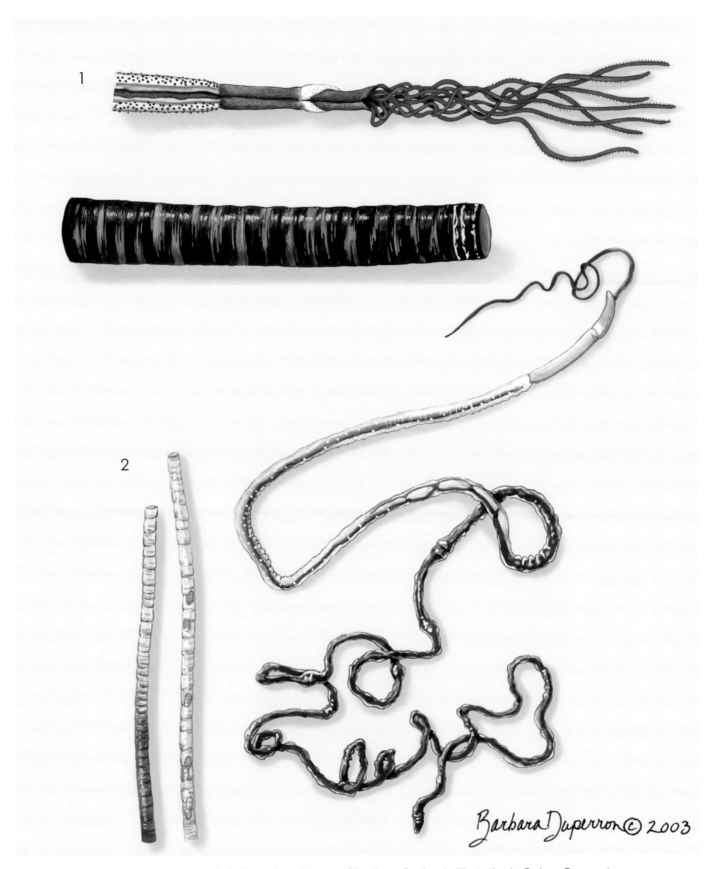

1. Black oligobrachia (*Oligobrachia ivanovi*); 2. Norwegian tubeworm (*Siboglinum fiordicum*). (Illustration by Barbara Duperron)

Species accounts

Norwegian tubeworm
Siboglinum fiordicum

ORDER
Athecanephria

FAMILY
Siboglinidae

TAXONOMY
Siboglinum fiordicum Webb, 1963.

OTHER COMMON NAMES
None known.

PHYSICAL CHARACTERISTICS
Body diameter 0.009 in (0.25 mm); maximum length 11.8 in (300 mm). One pinnulate tentacle. Tube diameter 0.01 in (0.27 mm); maximum length 11.8 in (300 mm); gray or brown ring pattern, no segmentation.

DISTRIBUTION
Norway.

HABITAT
Muddy sand or mud in fjords or continental shelf; depth range 82–656.7 ft (25 to 200 m), in muddy sand or mud.

BEHAVIOR
Uses opisthosoma to burrow into sediment.

FEEDING ECOLOGY AND DIET
Symbiosis with chemosynthetic bacteria.

REPRODUCTIVE BIOLOGY
Incubates embryos.

CONSERVATION STATUS
Not listed by the IUCN.

SIGNIFICANCE TO HUMANS
The most accessible frenulate species for scientific study. ◆

Black oligobrachia
Oligobrachia ivanovi

ORDER
Athecanephria

FAMILY
Oligobrachiidae

TAXONOMY
Oligobrachia ivanovi Southward, 1959.

OTHER COMMON NAMES
None known.

PHYSICAL CHARACTERISTICS
Body 0.015–0.02 in (0.4–0.6 mm) diameter, at least 7.87 in (200 mm) long; brown with white patches and black stripes. Five to 7 separate pinnulate tentacles; red and orange. Tube 0.015–0.04 in (0.4–1.1 mm) diameter; black with transparent anterior end, wrinkled surface, no segmentation.

DISTRIBUTION
Bay of Biscay (northeast Atlantic).

HABITAT
Mud on steep canyon slopes, at depths to 4,265–8,202 ft (1,300–2,500 m).

BEHAVIOR
Burrowing.

FEEDING ECOLOGY AND DIET
Symbiosis with chemosynthetic bacteria.

REPRODUCTIVE BIOLOGY
Incubates embryos.

CONSERVATION STATUS
Not listed by the IUCN.

SIGNIFICANCE TO HUMANS
None known. ◆

■ *Siboglinum fiordicum*
■ *Oligobrachia ivanovi*

Resources

Books

Desbruyères, D., and M. Segonzac, eds. *Handbook of Deep-Sea Hydrothermal Vent Fauna.* Brest, France: IFREMER, 1997.

Ivanov, A. V. *Pogonophora.* London: Academic Press, 1963.

Southward, E. C. "Class Pogonophora." In *Polychaetes and Allies: The Southern Synthesis. Fauna of Australia.* Vol. 4A, edited by P. L. Beesley, G. J. B. Ross, and C. J. Glasby. Melbourne, Australia: CSIRO Publishing, 2000.

———. "Pogonophora." In *Microscopic Anatomy of Invertebrates.* Vol. 12, edited by F. W. Harrison and M. E. Rice. New York: Wiley-Liss, 1993.

Van Dover, C. L. *The Ecology of Deep-Sea Hydrothermal Vents.* Princeton, NJ: Princeton University Press, 2000.

Periodicals

Halanych, K. M., R. A. Feldman, and R. C. Vrijenhoek. "Molecular Evidence that *Sclerolinum brattstromi* is Closely Related to Vestimentiferans, Not Frenulate Pogonophorans (Siboglinidae, Annelida)." *Biological Bulletin* 201 (2001): 65–75.

MacDonald, I. R., V. Tunnicliffe, and E. C. Southward. "Detection of Sperm Transfer and Synchronous Fertilization in *Ridgeia piscesae* at Endeavour Segment, Juan de Fuca Ridge." *Cahiers de Biologie Marine* 43 (2002): 395–398.

McHugh, D. "Molecular Evidence that Echiurans and Pogonophorans are Derived Annelids." *Proceedings of the National Academy of Science* 94 (1997): 8006–8009.

McMullin, E. R., S. Hourdez, S. W. Schaeffer, and C. R. Fisher. "Phylogeny and Biogeography of Deep Sea Vestimentiferan Tubeworms and Their Bacterial Symbionts." *Symbiosis* 34 (2003): 1–41.

Rouse, G. W. "A Cladistic Analysis of Siboglinidae Caullery, 1914 (Polychaeta, Annelida): Formerly the Phyla Pogonophora and Vestimentifera." *Zoological Journal of the Linnean Society* 132 (2001): 55–80.

Shillito, B., J. P. Lechaire, and F. Gaill. "Microvilli-like Structures Secreting Chitin Crystallites." *Journal of Structural Biology* 111 (1993): 59–67.

Southward, E. C. "Description of a New Species of *Oligobrachia,* (Pogonophora) from the North Atlantic, with a Survey of the Oligobrachiidae." *Journal of the Marine Biological Association of the U.K.* 58 (1978): 357–365.

———. "Development of Perviata and Vestimentifera (Pogonophora)." *Hydrobiologia* 402 (1999): 185–202.

Webb, M. "*Siboglinum fiordicum* sp. nov. (Pogonophora) from the Raunefjord, Western Norway." *Sarsia* 13 (1963): 33–44.

Eve C. Southward, PhD, DSc

Vestimentifera

(Hydrothermal vent and cold seep worms)

Phylum Vestimentifera

Number of families 8

Thumbnail description
Segmented worms that have an unusual anatomy and rely on symbiotic bacteria for nutrition. They are nearly always found in deep waters, some as members of hydrothermal vent communities and others found in association with reducing sediments such as "cold seeps"

Photo: This photo of *Riftia pachyptila* comes from a depth of 8,200 ft (2,500 m) at 9° 50 N. on the East Pacific Rise, and was shot on a dive in the submersible Alvin. (Photo by Craig M. Young, Oregon Institute of Marine Biology. Reproduced by permission.)

Evolution and systematics

A number of fossil hydrothermal vent and seep systems have been discovered containing tubular fossils that appear to be those of vestimentiferan tubes. These tubes have been found in sulfide ores dating to the Carboniferous and Cretaceous, and possibly to the Silurian period.

The varied and complex taxonomic history of Vestimentifera, containing 15 described species, represents one of the more fascinating tales in animal systematics. It has traditionally been regarded as a distinct phylum, probably due to its complex anatomy and bizarre lifestyle. Most scientists now consider it to be part of the polychaete family Siboglinidae (formerly phylum Pogonophora), within the phylum Annelida. This recent classification is based on its clear annelid features and on molecular sequence data. The more conservative classification is followed here.

Vestimentifera encompasses 10 genera, namely *Alaysia*, *Arcovestia*, *Escarpia*, *Lamellibrachia*, *Oasisia*, *Paraescarpia*, *Ridgeia*, *Riftia*, *Seepiophila*, and *Tevnia*. With the exception of *Escarpia* (two species) and *Lamellibrachia* (four species), each of these genera contains only one species.

Physical characteristics

Vestimentiferans have elongated cylindrical worm-like bodies and are always found living in tubes. In nearly all cases, the tube must be attached to a hard surface rather than lying free in the sediment. The animal secretes the tube and occupies most of it, extending the anterior portion of its body, the plume, out into the water. If disturbed, this vulnerable plume can be quickly retracted back into the tube. The plume is comprised of hundreds of branchial filaments that are clustered into lamellae. These filaments are filled with blood vessels, and the red hemoglobin in the blood gives the plumes their bright color. The plume and obturaculum are equivalent to the head of other annelid worms and there are no known sensory organs such as eyes.

A short region immediately behind the head is called the vestimentum and this has lateral flaps that fold over the top of the worm. The front part of the vestimentum forms a collar. It is this region that may be used for secretion of the tube. The next body region is generally referred to as the trunk and comprises most of the body. The trunk contains the reproductive organs and is also largely filled with an expanded gut tissue called the trophosome. There is no gut lumen as such and, in adults, there is no mouth or anus. The trophosome lies between the ventral and dorsal blood vessels and is filled with special cells called bacteriocytes that contain symbiotic chemoautotrophic bacteria. The remainder of the body is a short multi-segmented region called the opisthosoma. The anterior segments of the opisthosoma have rows of hook-like chaetae that act as anchors for the worm to retract into the tube.

Vestimentifera contains some of the largest of the annelids. *Riftia pachyptila* and *Ridgeia piscesae* grow to more than 4.9 ft (1.5 m) in length and live in tubes more 8.2 ft (2.5 m) long. Other vestimentifera such as *Lamellibrachia satsuma* reach 16 in (40 cm) in length as adults.

The tubes of Vestimentifera are whitish to gray-brown. The plume is usually bright red and surrounds a central white obturaculum. The body within the tube is generally green to brown and there are often large red blood vessels visible through the body surface.

Distribution

The particular requirements of vestimentiferans mean that they are restricted to deep-sea environments, with most found at depths >0.6 mi (>1 km), though one species was observed at <328 ft (<100 m) depth. Hydrothermal vents are found on active spreading ridges between continental plates. The major ones around the world where vestimentiferans have been found are the east Pacific Rise, the mid-Atlantic Ridge, and the Galapagos Rift. Hydrothermal vents and vestimentiferans are also found at sea-floor spreading centers in what are known as "back-arc basins" of the western Pacific, including the Okinawa Trough, Mariana Trough, and the Lau, Manus, and North Fiji Basins. Cold seeps are mainly located along subduction zones or continental margins, and vestimentiferans have been found at seeps in the Gulf of Mexico, off the coasts of North and South America, Spain, and in the Mediterranean Sea. Undoubtedly, many more Vestimentifera will be discovered as the deep sea is further explored.

Habitat

Most of the deep seafloor is soft sediment; vestimentiferans need a hard surface to attach their tubes to. Some vestimentiferans settle and grow on the chimneys of hydrothermal vents where the water temperature is around 68°F (20°C). They are also found on lava flows associated with vents. Often, they form large clusters, with younger worms making their tubes on those of larger worms. At cold seeps, the worms also tend to form clusters, with tubes growing on tubes.

Two dramatically different life histories are apparent in this group. It has been shown that *Riftia pachyptila* could colonize a new hydrothermal vent site, grow to sexual maturity, and have tubes of 4.9 ft (1.5 m) in length, in less than two years. This may represent the fastest growth rate of a marine invertebrate; this rapid growth appears to be essential because their habitat is ephemeral and lasts for only a few years or decades. In contrast, cold seeps such as the Louisiana slope (Gulf of Mexico) provide a stable supply of sulfide over centuries. *Lamellibrachia luymesi* that live in this environment grow very slowly and, while reaching more than 6.5 ft (2 m) in tube length, may take more than 100 years to do so.

Apart from hydrothermal vents, cold seeps, and whale carcasses, vestimentiferans have not been associated with any other habitat.

Behavior

Vestimentiferans form dense aggregations of both sexes at both hydrothermal and cold seep sites with worms at all stages of life. No other social organization is apparent.

Little is known of the mating system in vestimentiferans. It seems there is no contact between sexes and that sperm spawned by males makes its way to the tubes of females where they fertilize the eggs. There does not appear to be mate selection or parental behavior.

Feeding ecology and diet

Nutritional requirements for vestimentiferans are met through their symbiotic relationship with chemoautotrophic bacteria in the trophosome. A transitory mouth appears to be the pathway for bacteria to occupy the trophosome. The bacteria require carbon dioxide and either sulfide or thiosulphate,

all of which are supplied by the host. In return, the host obtains nutrition from the bacteria, or digests them (Southward 1993). All the chemoautotrophic bacteria found in vestimentiferans belong in the Proteobacteria and, where it has been studied, only one kind of bacterium is found in a given vestimentiferan species.

Little is know about predators of Vestimentifera. Possible predators include buccinid snails.

Reproductive biology

All vestimentiferans studied to date appear to have separate sexes with gametes that are produced in the trunk. No courtship behaviors have been documented to date. Males produce masses of sperm or sperm bundles that are spawned into the water and end up in the tubes of females. Fertilization appears to occur in or just outside the oviducts.

Females produce large numbers of eggs that are around 0.0039 in (0.1 mm) in diameter when mature.

Vestimentiferans face the problem of their favored habitats often being ephemeral. Thus, they need to have mechanisms that allow them to have some larvae continue to colonize the site where their parents are located, but also have others that are capable of dispersing long distances to other vents or seeps. How this is achieved is the subject of intense study.

The larvae of vestimentiferans are similar to that of many other polychaetes in that they have enough yolk to develop into small juveniles. The length of time this process takes is not presently known. Early larvae swim with the aid of a band of cilia at the front end. At a certain stage, to gain the nutrition the larvae need to continue development, symbiotic bacteria must colonize them. This seems to happen via the transitory mouth that appears in late-stage larva, allowing bacteria to enter the body.

Vestimentiferans probably release early embryos into the plankton. No larvae have ever been found in the tubes of females, nor is there any parental care. This is in contrast to other pogonophores in which brooding of larvae in the female tube is common.

Conservation status

No species of Vestimentifera are listed by the IUCN. Since their first discovery in 1969, vestimentiferans have been found on a regular basis as exploration of the deep sea continues. Presumably, there are many more species and genera to be discovered and named. At present, there is no known obvious threat to any vestimentiferans. When they are found, they tend to occur in large numbers, but the nature of their habitats means that they have restricted distributions and high levels of endemism.

Significance to humans

Vestimentifera attracted significant attention in 1977 when they were first discovered at hydrothermal vents and were shown to be a large and significant part of the animal communities found there. While they have no commercial value to humans, they have iconic status as deep-sea animals.

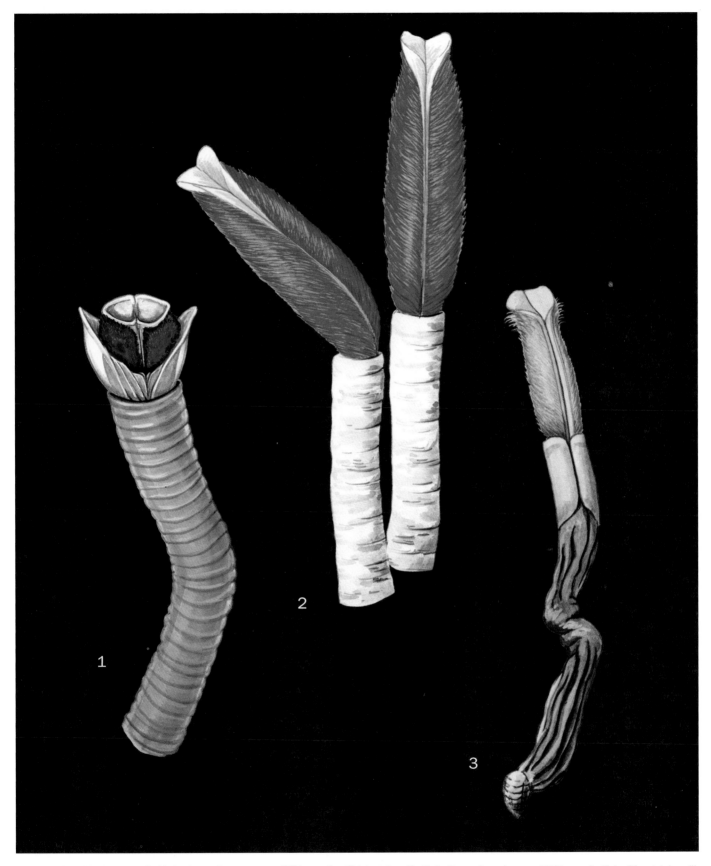

1. *Lamellibrachia luymesi*; 2. Hydrothermal vent worm (*Riftia pachyptila*) in tube; 3. Hydrothermal vent worm (*Riftia pachptila*) without tube. (Illustration by Joseph E. Trumpey and Barbara Duperron)

Species accounts

No common name
Lamellibrachia luymesi

ORDER
Basibranchia

FAMILY
Lamellibrachiidae

TAXONOMY
Lamellibrachia luymesi van der Land and Norrevang, 1977, Guyana Shelf.

OTHER COMMON NAMES
None known.

PHYSICAL CHARACTERISTICS
Body reaches 1.6 ft (0.5 m) in length, with tubes to 6.5 ft (2 m) in length. Plume dark red with ivory obturacula.

DISTRIBUTION
Off Guyana, Gulf of Mexico.

HABITAT
Found at depths of 0.3–0.6 mi (0.5–1 km); associated with cold seeps.

BEHAVIOR
Forms dense thickets on silty sediments where cold seeps occur. Presumably, one or more animals attach to a stone or piece of rock and gradually the tubes of the worms provide settling places for other worms. This allows for the large discrete thickets to form. Otherwise, little is known.

FEEDING ECOLOGY AND DIET
Symbiotic bacteria in trophosome provide nutrition for worm. Host provides bacteria with sulfide via its plume and carbon dioxide as a byproduct of its own respiration. Cold seeps provide a stable supply of sulfide over centuries. Worms found in this environment grow very slowly and adults may be 100 or more years old.

REPRODUCTIVE BIOLOGY
Separate sexes, with males spawning into the water and females releasing eggs after fertilizing them. Eggs 0.0039 in (0.1 mm) in diameter and larvae can disperse for several weeks without the need to feed.

CONSERVATION STATUS
Not listed by the IUCN.

SIGNIFICANCE TO HUMANS
None known. ◆

Hydrothermal vent worm
Riftia pachyptila

ORDER
Riftiida

FAMILY
Riftiidae

TAXONOMY
Riftia pachyptila Jones, 1981, East Pacific Rise.

OTHER COMMON NAMES
English: Deep-sea tubeworm, giant tubeworm, vestimentiferan tubeworm; French: Vers géant.

PHYSICAL CHARACTERISTICS
Largest of Vestimentifera; body reaches up to 4.9 ft (1.5 m) in length and white tubes to 8.2 ft (2.5 m). Bright red plume and white obturaculum.

DISTRIBUTION
East Pacific Rise, Galápagos Rift, and Guaymas Basin.

HABITAT
Found at depths of around 1 mi (1.5 km) associated with hydrothermal vents.

BEHAVIOR
Forms dense thickets on hydrothermal vent "chimneys." Otherwise, little is known.

FEEDING ECOLOGY AND DIET
Symbiotic bacteria in trophosome provide nutrition for the worm. The host provides bacteria with sulfide via its plume

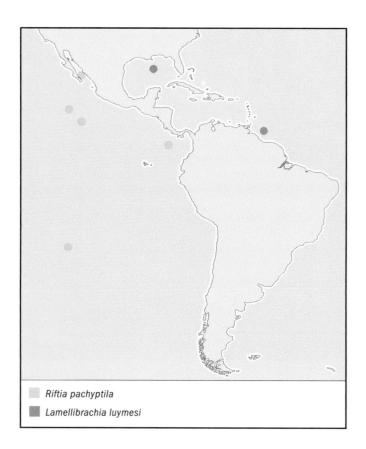

■ *Riftia pachyptila*
■ *Lamellibrachia luymesi*

and carbon dioxide as a byproduct of its own respiration. One of the fastest growing marine invertebrates, with tubes reaching 4.9 ft (1.5 m) in length in only 18 months.

REPRODUCTIVE BIOLOGY
Separate sexes, with males spawning into the water and females releasing eggs after fertilizing them. Larvae must be capable of dispersing to new vents since vents are ephemeral habitats.

CONSERVATION STATUS
Not listed by the IUCN.

SIGNIFICANCE TO HUMANS
"Posterchild" of hydrothermal vents. ◆

Resources

Books

Rouse, Greg W., and Fredrik Pleijel. *Polychaetes.* London: Oxford University Press, 2001.

Southward, Eve C. "Pogonophora." In *Microscopic Anatomy of Invertebrates, Volume 12, Onychophora, Chilopoda and Lesser Protostomata,* edited by Fredrik W. Harrison and Mary E. Rice. New York: Wiley-Liss, 1993.

———. "Pogonophora." In *Polychaetes and Allies: The Southern Synthesis. Fauna of Australia.* Volume 4A, *Polychaeta, Myzostomida, Pogonophora, Echiura, Sipuncula,* edited by Pam Beesely, Graham J. B. Ross, and Christopher J. Glasby. Melbourne, Australia: CSIRO Publishing, 2000.

Periodicals

Cavanaugh, C. M., S. L. Gardiner, M. L. Jones, H. W. Jannasch, and J. B. Waterbury. "Procaryotic Cells in the Hydrothermal Vent Tube Worm *Riftia pachyptila* Jones: Possible Chemoautotrophic Symbionts." *Science* 213 (1981): 340–342.

Fisher, C. R., I. A. Urcuyo, M. A. Simpkins, and E. Nix. "Life in the Slow Lane: Growth and Longevity of Cold-seep Vestimentiferans." *Marine Ecology-Pubblicazioni della Stazione Zoologica di Napoli* 18 (1997): 83–94.

Jones, M. L. "*Riftia pachyptila* Jones: Observations on the Vestimentiferan Worm from the Galápagos Rift." *Science* 213 (1981): 333–336.

Lutz, R. A., et al. "Rapid Growth at Deep-sea Vents." *Nature* 371 (1994): 663–664.

Rouse, G. W. "A Cladistic Analysis of Siboglinidae Caullery, 1914 (Polychaeta, Annelida): Formerly the Phyla Pogonophora and Vestimentifera." *Zoological Journal of the Linnean Society* 132 (2001): 55–80.

Tunnicliffe, V., A. G. McArthur, and D. McHugh. "A Biogeographical Perspective of the Deep-sea Hydrothermal Vent Fauna." *Advances in Marine Biology* 34 (1998): 354–442.

Webb, M. "*Lamellibrachia barhami,* gen. nov. sp. nov. (Pogonophora), from the Northeast Pacific." *Bulletin of Marine Science* 19 (1969): 18–47.

Greg W. Rouse, PhD

Sipuncula
(Peanut worms)

Phylum Sipuncula

Number of families 6

Thumbnail description
Unsegmented marine worm-like animals with a body divided into a trunk and retractable introvert

Photo: Peanut worm (*Phascolosoma* sp.) with partially extended introvert. (Photo by L. Newman & A. Flowers. Reproduced by permission.)

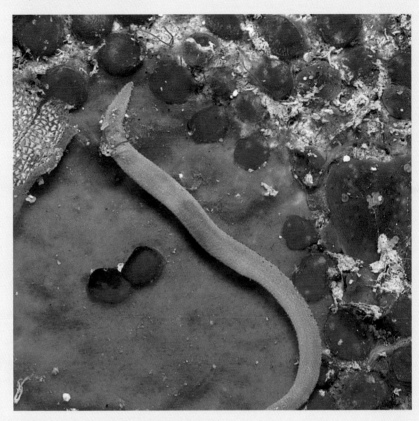

Evolution and systematics

No unambiguous fossil Sipuncula are currently known. *Ottoia prolifica* from the Burgess Shale has been proposed as a fossil sipunculan, but might also be an aschelminth or Priapulida. The paleozoic Hyolitha has a mix of attributes of sipunculans and mollusks, suggesting a close phylogenetic relationship with both. Fossilized burrows possibly created by sipunculans in soft sediments are known from early and mid-Paleozoic times. More recent Mesozoic and Cenozoic fossil burrows have also been attributed to sipunculan worms. Other sipunculans appear to have lived in association with corals and in vacated mollusk shells since the mid-Paleozoic, throughout the Mesozoic and Cenozoic.

In the early seventeenth century, Sipuncula were considered close relatives of holothurians. In 1847 Quatrefages erected the group Gephyrea, which he considered an intermediate between worms and holothurians and which also contained echiurans, sternaspids, and priapulids. Since the 1990s, there is general agreement that sipunculans are protostomes and closely related to annelids and mollusks, but their exact position still remains unresolved.

The phylum Sipuncula contains two classes, four orders, six families, 17 genera, and 147 species.

Physical characteristics

The sipunculan body is divided into trunk and retractable introvert. The ratio between introvert and trunk length varies among species. The mouth, at the anterior end of the introvert, is surrounded by an array of tentacles in the Sipunculidea. In the Phascolosomatidea, the tentacles are arranged in an arc around the nuchal organ, also located at the tip of the introvert. The anus lies dorsally, usually at the anterior end of the trunk, except in some species where it is shifted anteriorly onto the introvert. The nephridiopores lie ventrolaterally, typically at the level of the anus. Proteinaceous, nonchitinous hooks are often present on the distal part of the introvert and are either arranged in rings or scattered. Numerous papillae may be present on the trunk and introvert.

Trunk length varies from a few millimeters to about 11.8 in (30 cm) in exceptionally large specimens. Colors are usually shades of gray or brown, with occasional reddish, purple, or green pigment in the papillae and/or tentacles.

The body wall musculature is composed of an outer layer of longitudinal and an inner layer of circular muscles. One or two pairs of prominent introvert retractor muscles are present. A large coelom represents the main body cavity. The tentacles and the contractile vessel (fluid reservoir for tentacle extension) contain a second coelomic compartment. The intestine is characteristically U-shaped, with the ascending and descending branches coiled around each other in a double helix. A spindle muscle runs through the gut coil. It is attached anteriorly to the body wall near the anus and posteriorly to either the body wall or inside the gut coil. The sipunculan nervous system consists of a cerebral ganglion and a ventral nerve cord. Two nephridia are present, except in the genera *Phascolion* and *Onchnesoma*, which have only a single nephridium.

Distribution

Sipunculans occur in cold, temperate and tropical marine benthic habitats. They have been found in all depths from the intertidal zone to 22,510 ft (6,860 m).

Habitat

Some sipunculan species inhabit semi-permanent burrows in coarse or silty sand, and some live in crevices under rocks. A number of species bore into dead or, more rarely, live coral or other soft rocks, while one species even bores into a whale skull. Others inhabit empty mollusk shells, polychaete tubes, foraminiferan tests, or barnacles. Algal mats, large sponges, root mats of mangroves or sea grass, and byssal threads of bivalves also serve as habitats for some species.

Behavior

Relatively little is known about the behavior of sipunculans. Most species retract their tentacles and introvert quickly following a tactile stimulus. Many species are negatively phototactic and retreat into sediment or rock when given the opportunity. Burrowing and crawling are accomplished by utilizing the introvert hooks as anchors and the introvert musculature to pull the body forward. *Phascolion strombus*, an inhabitant of gastropod shells, is able to irrigate its shell to increase oxygen content by contractions of the body wall musculature. Swimming has only been reported in *Sipunculus* and consists of non-directional thrashing of the trunk.

Feeding ecology and diet

Most sipunculans are deposit feeders, except representatives of the genus *Themiste*, which have elaborately branched tentacles used for filter feeding. Sand-dwelling species ingest sediment and associated biomass that they collect with their tentacles. The tentacles are rarely visible above the seafloor

during the day, but may be extended at night to probe the surrounding sediment for food particles. Rock-dwelling species use their introvert hooks, mostly at nighttime, to scrape sediment and epifaunal organisms from the surrounding rock surface.

Reproductive biology

Most sipunculan species are dioecious. Only one species, *Nephasoma minutum*, is known to be hermaphroditic. *Themiste lageniformes* is facultatively parthenogenetic. Asexual reproduction by budding has been reported in *Aspidosiphon elegans*. No sexual dimorphism is known in Sipuncula. Gonads are only prevalent during the reproductive period. Gametes are released into the coelom where maturation proceeds. Mature gametes are taken up by the nephridia and released into the water through the paired nephridiopores.

The four developmental modes include:

- Direct lecithotrophic development without a pelagic stage.
- Indirect development with a lecithotrophic trochophore larva.
- Indirect development with two lecithotrophic larval stages: trochophore and pelagosphera.
- Indirect development with a lecithotrophic trochophore and a planktotrophic pelagosphera. The planktotrophic pelagosphera lasts up to six months in the water column before settling.

Conservation status

No sipunculan species are currently on the IUCN Red List. Because of their long-lived larval stages, many sipunculan species seem to be very widespread. Abundance ranges from rare to extremely common (e.g., the density of *Themiste lageniformes* can reach more than 2,000 individuals/10 ft^2 (m^2). Habitat destruction (e.g., mangroves, sea grass beds) can endanger regional populations.

Significance to humans

Fishermen in various parts of the world use sipunculan worms, mostly the larger sand-dwelling species, as bait. In Java, in the western Carolines, and in some parts of China, sipunculans are eaten by the locals.

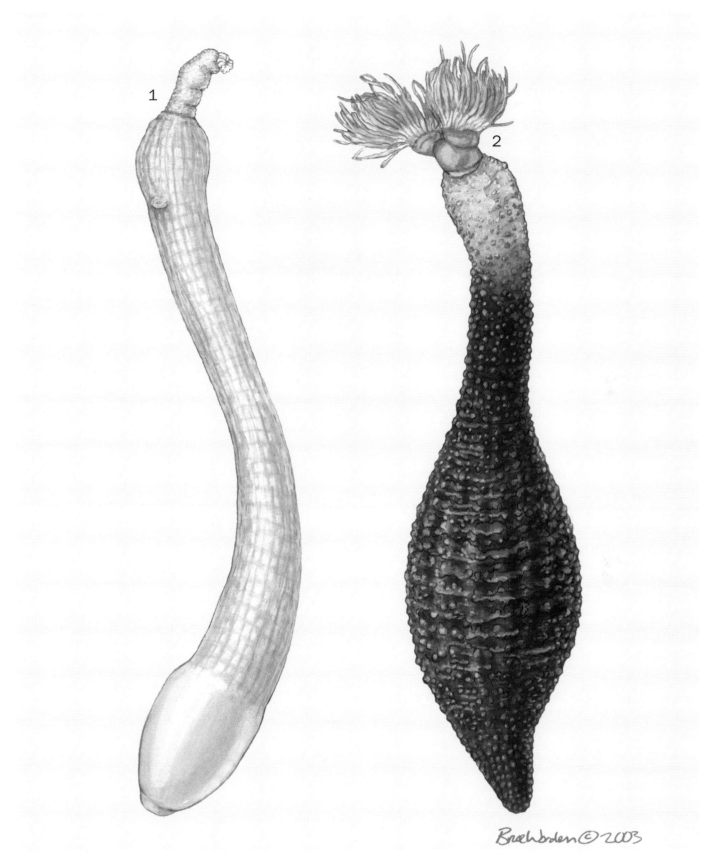

1. *Sipunculus nudus*; 2. *Antillesoma antillarum*. (Illustration by Bruce Worden)

Species accounts

No common name
Antillesoma antillarum

ORDER
Phascolosomatiformes

FAMILY
Phascolosomatidae

TAXONOMY
Antillesoma antillarum (Grübe and Oersted, 1858), West Indies.

OTHER COMMON NAMES
None known.

PHYSICAL CHARACTERISTICS
Up to 7.25 in (80 mm) long. Introvert 65–75% of trunk length. Body covered with large, dark papillae; numerous tentacles with purple or green pigment patches or stripes. Introvert hooks absent in adults.

DISTRIBUTION
Cosmopolitan in intertidal and shallow tropical and subtropical waters. (Specific distribution map not available.)

HABITAT
Lives in crevices or burrows into dead coral or soft rock.

BEHAVIOR
Nothing is known.

FEEDING ECOLOGY AND DIET
Nothing is known.

REPRODUCTIVE BIOLOGY
Dioecious; indirect developer with lecithotrophic trochophora and long-lived planktotrophic pelagosphera.

CONSERVATION STATUS
Not listed by the IUCN.

SIGNIFICANCE TO HUMANS
None known. ◆

No common name
Sipunculus nudus

ORDER
Sipunculiformes

FAMILY
Sipunculidae

TAXONOMY
Sipunculus nudus Linnaeus, 1766, type locality unknown, but perhaps Mediterranean Sea.

OTHER COMMON NAMES
None known.

PHYSICAL CHARACTERISTICS
Commonly up to 6 in (15 cm) long, sometimes reaching 10 in (25 cm); introvert up to one-third of trunk length. Longitudinal and circular body wall musculature in bands. Distinguished by number of longitudinal muscle bands: 24–34. Introvert hooks absent.

DISTRIBUTION
Cosmopolitan in temperate, subtropical, and tropical waters in subtidal zone to 2,953 ft (900 m) depth. (Specific distribution map not available.)

HABITAT
Semi-permanent sand borrows.

BEHAVIOR
During daytime, usually hides in burrow but might extend tentacles at night.

FEEDING ECOLOGY AND DIET
Ingests sediment and utilizes associated organic material.

REPRODUCTIVE BIOLOGY
Dioecious; indirect developer with lecithotrophic trochophore and long-lived planktotrophic pelagosphera.

CONSERVATION STATUS
Not listed by the IUCN.

SIGNIFICANCE TO HUMANS
Best examined sipunculan species; model organism for anatomy, physiology, biochemistry, and ecology. Used as bait in some parts of the world. ◆

Resources

Books

Cutler, Edward, B. *The Sipuncula. Their Systematics, Biology, and Evolution*. Ithaca, NY: Cornell University Press, 1994.

Edmonds, Stanley J. "Phylum Sipuncula." In *Fauna of Australia: Polychaetes and Allies, The Southern Synthesis*, Vol. 4A, edited by Pamela L. Beesley, Graham J. B. Ross, and Christopher J. Glasby. Melbourne, Australia: CSIRO Publishing, 2000.

Rice, Mary E. "Sipuncula." In *Microscopic Anatomy of Invertebrates: Onychophora, Chilopoda, and Lesser Protostomata*, Vol. 12, edited by Frederick W. Harrison and Mary E. Rice. New York: Wiley-Liss, 1993.

Stephen, A. C., and S. J. Edmonds. *The Phyla Sipuncula and Echiura*. London: Trustees of the British Museum (Natural History), 1972.

Periodicals

Maxmen, Amy B., Burnett F. King, Edward B. Cutler and Gonzalo Giribet. "Evolutionary Relationships Within the Protostome Phylum Sipuncula: A Molecular Analysis of Ribosomal Genes and Histone H3 Sequence Data." *Molecular Phylogenetics and Evolution* 27 (2003): 489–503.

Rice, Mary E. "Larval Development and Metamorphosis in Sipuncula." *American Zoologist* 16 (1976): 563–571.

Anja Schulze, PhD

Echiura
(Echiurans)

Phylum Echiura

Number of families 5

Thumbnail description
Worm-like invertebrates with non-segmented, bilaterally symmetrical bodies

Photo: A spoon worm (*Bonellia*) in Western Australia, with its forked tongue fully extended. (Photo by L. Newman & A. Flowers. Reproduced by permission.)

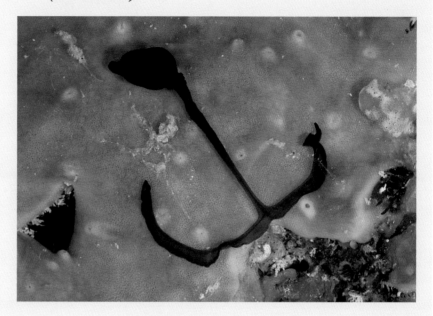

Evolution and systematics

As the echiurans' body has no large hard parts, fossils of these animals are rare. There are two fossils for this group: a fossil from Illinois, United States, dated from the Late Carboniferous and another fossil from Namibia, dated from the Late Cambrian.

A free-swimming trochophore larva is present in echiurans, sipunculans, mollusks, and annelidans, which suggests a phyllogenetic relationship. Echiurans and annelids have many features in common. The most important difference between them is the absence of segmentation in Echiura. Some authors consider echiurans to be properly placed within the phylum Annelida, though other studies have shown that echiurans and pogonophorans have a close affinity and both may be closer to mollusks than to annelidans or sipunculans.

The phylum Echiura encompasses about 160 species and 40 genera, divided into two classes: Echiuridea, with three orders and four families (Echiuridae, Bonellidae, Ikedidae, and Urechidae), and Sactosomatidea, with a single family, Sactosomatidae, with one species, *S. vitreum*.

Physical characteristics

Echiurans, also known as spoon worms, have a body divided in two distinct regions: a sausage-shaped saccular, non-segmented trunk and a ribbon-like proboscis at the anterior end. The length of the trunk may range from 0.39 in (1 cm) up to >19.6 in (>50 cm) and may be gray, dark green, reddish brown, rose, or red. It may be thick or thin, smoothed or roughened by glandular or sensory papillae. Internally, layers of muscles are responsible for peristaltic movements of the trunk. A pair of chitinous golden-brown chaetae usually occurs ventrally on the anterior part of the trunk. Some echiurans have one or two rings of chaetae around the anus.

The proboscis may be short or long, scoop- or ribbon-like, and flattened or fleshy and spatulate. It is generally white, rose, green, or brown. The distal end may be truncate or bifid. It is muscular, mobile, and highly extensible and contractile. It is able to extend 10 times its body length and can reach 3.2–6.5 ft (1–2 m). The ventral surface of the proboscis is ciliated, which helps in the feeding process. The mouth is located ventrally at the base of the proboscis and the anus is at the posterior extremity of the trunk.

Distribution

Echiurans are mainly marine, but some species live in brackish waters. The majority of spoon-worms are found in intertidal and shallow waters, but there are also species living at depths of 32,800 ft (10,000 m).

Habitat

Echiurans usually live in a U-shaped burrow with both ends of the burrow open. They are found mainly in soft benthic substrata such as sand, mud, or rubble, occupying burrows excavated by themselves or by other animals. Some species live in rock galleries excavated by boring invertebrates, whereas others live in empty shells, sand-dollar tests, coral or rock crevices, inside dead corals, or under stones. In general, some commensals are present inside the burrow, including polychaetes, crabs, mollusks, and fishes. The burrow provides

A spoon worm (*Bonellia*) with bilobed proboscis. (Photo by L. Newman & A. Flowers. Reproduced by permission.)

sand or mud. The movements by the peristalsis forces water through the tube, permitting the animal to obtain a supply of oxygen. In general, the burrow is kept clean and free from debris and fecal matter.

Feeding ecology and diet

The food of echiurans consists of dead organic matter and microorganisms that live on the substratum. Echiurans may be detritus feeders. They extend the proboscis out of the burrow onto the surface of the sediment. The tip of the ventrally ciliated surface collects particles, and glands produce mucus to adhere to these particles. Movements of the cilia conduct particles and mucus to the mouth. The proboscis is then extended in a new direction, and the procedure is repeated. A few species are filter feeders. The innkeeper worm, *Urechis caupo*, constructs a mucous net placed near the opening of the burrow. Peristaltic movements of the trunk draw water through the burrow, and particles and small organisms are trapped in the net. Ultimately, the worm eats both the net and the food.

Reproductive biology

Echiurans reproduce strictly by sexual means. Sexual dimorphism is pronounced only in the family Bonellidae, in which the male is much smaller than the female. Sexes are separate, and sperm and eggs are usually liberated at the same time in seawater where fertilization occurs, except in Bonellidae, in which individuals undergo internal fertilization.

Free-swimming and feeding trochophore larvae develop 22 hours to four days following fertilization. The larvae may drift in the plankton for up to three months, and during metamorphosis, it increases in length. The larva settles on the substratum and begins life as an adult.

a protected, ventilated home, and remains of food discarded by the spoon-worm may be eaten by the commensals.

Behavior

Echiurans are slow but not sedentary, and animals without a proboscis can swim. One of the most important movements is the peristalsis of the trunk, which allows the animal to move slowly over the surface and construct burrows in the

Conservation status

No species of Echiura are listed by the IUCN.

Significance to humans

Some species of echiurans are commonly used as laboratory animals for physiological, embryological, and biochemical studies. The substance bonellin has been studied because of its antibiotic properties.

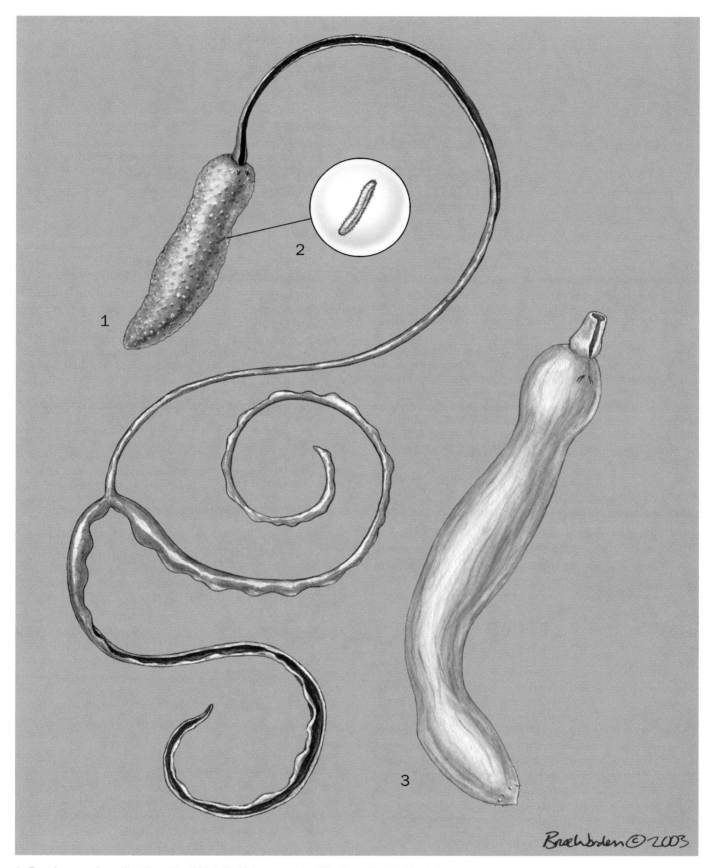

1. Female green bonellia (*Bonellia viridis*); 2. Male green bonellia (*Bonellia viridis*); 3. Innkeeper worm (*Urechis caupo*). (Illustration by Bruce Worden)

Species accounts

Green bonellia
Bonellia viridis

ORDER
Bonellioinea

FAMILY
Bonellidae

TAXONOMY
Bonellia viridis Rolando, 1821, Naples, Italy.

OTHER COMMON NAMES
English: Weenie worm; French: Bonnellie, bonnelie verte; Spanish: Gusano marino verde; Danish: Igelwurm.

PHYSICAL CHARACTERISTICS
Female's trunk ovoid to sausage shaped, and about 5.9 in (15 cm) long. Proboscis long and bifurcate at the end. When completely extended it can reach 4.9 ft (1.5 m). Trunk and proboscis pale to dark green, caused by presence of a dermal pigment, bonellin. One pair of ventral chaetae. Males 0.039–0.11 in (1–3 mm) long with a ciliated and planariform-like body, without pigment, proboscis, month, anus, or blood vascular system. Male's body occupied mainly by reproductive structures, and the male is often found living within the nephridia (genital sac) of the female.

DISTRIBUTION
Northeastern Atlantic Ocean, Mediterranean, Red Sea, and Indopacific.

HABITAT
Female does not make its own burrow, but inhabits burrows excavated in gravelly bottoms or burrows with multiple exits in rocky substrata or in clefts in rocks. It can be found in depths from 33 to 328 ft (10–100 m). A number of commensals inhabit the burrow. Male lives in the genital sac or on the body of the large female.

BEHAVIOR
Contraction of proboscis stem permits animal to move, and its bifurcate end has powerful cilia that help in locomotion and feeding. Moves back and forth inside burrow, as well as out of it. Contractions of the female's body wall renew the oxygen supply in the burrow.

FEEDING ECOLOGY AND DIET
Detritus feeder; food consists of detritus, small animals, organic material located at the root of vegetation and patches of sand between small rocks. Extends the proboscis from the burrow and grazes on the surrounding substratum with its bifid terminal lobes. Cilia on the ventral side of the proboscis move small particles and muscles pick up the larger ones. Cilia and muscles transfer the bolus mixed with mucus to the mouth. All metabolic needs of male are supplied by exchange with the female's body, indicating male has a parasitic mode of life.

REPRODUCTIVE BIOLOGY
Sexes are separate and fertilization occurs in the genital sac, where male often lives. Larva is free swimming. Bonellin has an important role in the determination of sex; it is considered

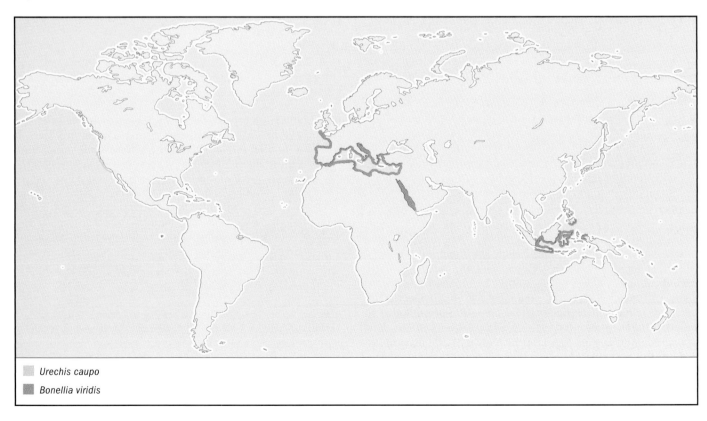

Urechis caupo
Bonellia viridis

to be a masculinizing factor. If the larva settles on ocean floor, it develops into a 3.9-in (10-cm) long female. If the larva settles on a female's body (particularly its proboscis), it develops into a 0.039–0.078-in (1–2-mm) long adult male in 1–2 weeks. Male lives as a parasite and produces a ready supply of sperm.

CONSERVATION STATUS
Not listed by the IUCN.

SIGNIFICANCE TO HUMANS
Specimens are commonly used as laboratory animals. The pigment bonellin has been studied as an important antibiotic. The substance has powerful and lethal effects on a number of organisms. ◆

Innkeeper worm

Urechis caupo

ORDER
Xenopneusta

FAMILY
Urechidae

TAXONOMY
Urechis caupo Fisher and MacGinitie, 1928, Elkhorn Slough, Monterey Bay, California, United States.

OTHER COMMON NAMES
English: Fat innkeeper worm.

PHYSICAL CHARACTERISTICS
Sausage-shaped trunk, and proboscis short and scoop-like. Pink body reaches length of 19.6 in (50 cm) and the cuticle of trunk is rugose. One pair of ventral anterior chaetae and one circle of stronger chaetae around the anus.

DISTRIBUTION
Pacific Ocean, west coast of United States (California).

HABITAT
Lives in U-shaped burrow in sand or sandy mud in intertidal or subtidal regions. Can be found in estuarine areas. Known as innkeeper worm because its burrow is occupied by several commensals.

BEHAVIOR
Able to construct its own tube. The proboscis starts a hole in the substratum and excavation process is continued until the burrow becomes a U-shaped tunnel with two exits. Both anterior ventral and posterior anal chetae are used during construction and maintenance of the tube. Can move over a smooth surface and spend the day obtaining food, cleaning the burrow, resting, and making respiratory movements with their tube. Can swim.

FEEDING ECOLOGY AND DIET
Produces a funnel-shaped mucous net, with its distal end attached to the internal wall of the burrow. By peristaltic movements, the trunk pumps seawater through the burrow and particles, and organisms larger than 0.00004 (1 µm) are trapped in net, which is detached from the body when loaded with food; animal eats net with food.

REPRODUCTIVE BIOLOGY
Spawning occurs in early summer. All individuals spawn at the same time, ensuring fertilization, which occurs externally; sperm and eggs are liberated in the seawater where the fertilization occurs. A free-swimming and feeding trochophore larva develops after fertilization.

CONSERVATION STATUS
Not listed by the IUCN.

SIGNIFICANCE TO HUMANS
Specimens are commonly used as laboratory animals and may also be used as bait. ◆

Resources

Books
Beesley, P. L., G. J. B. Ross, and C. J. Glasby, eds. *Polychaetes and Allies: The Southern Synthesis. Fauna of Australia.* Vol. 4A, *Polychaeta, Myzostomida, Pogonophora, Echiura, Sipuncula.* Melbourne, Australia: CSIRO Publishing, 2000.

Grassé, P. P. *Traitè de Zoologie.* Vol. 5. Paris: Masson et Cie, 1959.

Stephen, A. C., and S. J. Edmonds. *The Phyla Sipuncula and Echiura.* London: Trustees of the British Museum (Natural History), 1972.

Periodicals
Agius, L. "Larval Settlement in the Echiuran Worm *Bonellia viridis*: Settlement on Both the Adult Proboscis and Body Trunk." *Marine Biology* 53 (2002): 125–129.

Fisher, W. K., and G. E. MacGinitie. "The Natural History of an Echiuroid Worm." *Annals and Magazine of Natural History* 10 (1928): 204–213.

Jaccarini, V., L. Agius, P. J. Schembri, and M. Rizzo. "Sex Determination and Larval Sexual Interaction in *Bonellia viridis* Rolando (Echiura)." *Journal of Experimental Marine Biology and Ecology* 66 (1983): 25–40.

Nishikawa, T. "Comments on the Taxonomic Status of *Ikeda taenioides* (Ikeda, 1904) with Some Amendments in the Classification of the Phylum Echiura." *Zoological Science* 19 (2002): 1175–1180.

Other
"Introduction to the Echiura." [July 24, 2003]. <http://www.ucmp.berkeley.edu/annelida/echiura.html>.

Murina, V. V. "Phylum Echiura Stephen, 1965." 1998 [July 24, 2003]. <http://www.ibss.iuf.net/people/murina/echiura.html>.

Tatiana Menchini Steiner, PhD

Onychophora

(Onychophorans, velvet worms, and peripatus)

Phylum Onychophora

Number of families 2

Thumbnail description
Terrestrial carnivorous worms with legs along the whole length of the body, paired eyes, a conspicuous pair of antennae, and a velvety appearance

Photo: An onychophoran seen in Peru. (Photo by Danté Fenolio/Photo Researchers, Inc. Reproduced by permission.)

Evolution and systematics

A number of fossils have been linked to the Onychophora, but whether they really fit into this group is not known with certainty. *Aysheaia pedunculata* from the Middle Cambrian Burgess shale has many morphological similarities with extant onychophorans, but it differs fundamentally in having a terminal mouth. Another, possibly related, fossil is the Middle Pennsylvanian Mazon Creek *Helenodora inopinata*, which could have been marine or terrestrial. The connection between Priapulida (based on the fossil *Xenusion* from the Cambrian sandstone in Germany) and Onychophora is highly speculative. However, the marine Cambrian forms (e.g., *Microdictyon*, *Hallucigenia*, and *Luolishania*) have been definitely assigned to the Onychophora by some authors.

Although first mistakenly described in 1826 as a type of slug, the evolutionary history of onychophorans has long fascinated scientists. Onychophorans were thought to be a "missing link" between the Annelids (segmented worms) and the Arthropods (a group that includes the insects and spiders) because they share morphological characters with both these large phyla. However, molecular studies show that they are, in fact, more closely related to the arthropods. Because of their many unique characteristics, they are considered a separate phylum.

Two families, the Peripatopsidae and the Peripatidae, are recognized. It is assumed that the two families are sister taxa, but whether each family is a monophyletic group (evolved from a common ancestor) has yet to be determined. Compared to many other invertebrates, the gross morphology of

the Onychophora is remarkably similar over its wide and disjunct geographical range. The lack of distinguishing features has precluded a satisfactory higher-level classification and also causes difficulties at the species level, with cryptic species being common. No subfamilies are recognized.

Physical characteristics

The basic body structure of onychophorans is a simple one, but it is a design that works well. They have changed little in appearance over the last 500 million years. Onychophorans have a soft and flexible cylindrical body that is slightly flattened on the ventral side. The head bears a pair of annulated antennae with beady eyes located at their bases and a ventral circular mouth, equipped with fleshy lips and paired jaws. On either side of the mouth is an oral tube, sometimes called the "oral papillae." Paired legs, the oncopods, extend along the full length of the body and vary in number (depending on the species) from 13 to 43 pairs. The genital opening is on the ventral side at the posterior end of the body, and they have a terminal anus. Each oncopod is a conical, unjointed appendage with a foot that bears a pair of terminal claws. There are three to six spinous pads on the underside of each oncopod at the base of the feet, and often in males a crural, or pheromone-secreting, gland on the underside of some or all of the oncopods, close to their junction with the body. The body is covered by a thin cuticle that is molted at regular intervals. The cuticle is adorned with papillae, each composed of overlapping rows of scales. The largest papillae bear sensory

A velvet worm (phylum Onychophora) on a forest floor in Ecuador. These creatures have been described as the missing link between annelids and anthropods. They live only in humid, dimly lit places, and are primarily nocturnal. (Photo by Dr. Morley Read/Science Photo Library/Photo Researchers, Inc. Reproduced by permission.)

bristles. It is these sensory papillae that give onychophorans their velvety appearance and common name, "velvet worms."

Apart from a few white, cave-dwelling species, they are generally blue-gray or brownish in color, often intricately and beautifully patterned, with stripes, diamonds, spots, or chevrons.

Distribution

With the exception of Antarctica, Europe, and North America, extant onychophorans are found on all continents. The peripatids inhabit tropical America, and have also been found in tropical western Africa and southeastern Asia. Peripatopsids are found in countries associated with the previously connected southern super-continent, Gondwana, such as Chile, South Africa, Australia, New Guinea, and New Zealand. Their present distribution is patchy. In the northern hemisphere, the Onychophora are now restricted to tropical and subtropical regions, although fossil evidence suggests that they were much more widespread throughout Pangaea. The ancestors of peripatus moved from the sea to the land more than 400 million years ago.

Habitat

All onychophorans are confined to moist, humid microhabitats, such as tropical and subtropical forests, where they live in leaf litter or under stones, logs, or in the soil. During dry periods, or at low temperatures, some species migrate vertically within soil crevices and remain inactive until conditions improve.

Behavior

Velvet worms exhibit photonegative behavior, which means they hide from light and respond to air movements. These reactions make it difficult to observe their behavior, so very little is known about them. Covered with sensory bristles considered to be mechanoreceptors, onychophorans are highly sensitive to touch. This is undoubtedly the primary way in which they experience their environment, and touch probably also plays a significant role in sexual activity.

Feeding ecology and diet

Onychophora are carnivorous, and feed primarily on arthropods. They capture their prey by entangling it in threads of clear, sticky slime that is ejected from the oral tubes on either side of the head.

Very little is known about the feeding ecology of onychophorans in their natural environment. The slime is also used to deter predators and may be ejected up to 1.6 ft (0.5 m) in distance. In addition to other invertebrates, predators of onychophorans include birds and reptiles.

Reproductive biology

Female onychophorans are attracted to males by pheromonal secretions extruded from the males' crural glands. Reproduction takes place in an extremely curious manner. In some species, the males deposit packets of sperm (spermatophores) directly into the genital opening of the female, but in other species the spermatophores are placed somewhere on the body of the female. The skin tissue then collapses where the spermatophores are deposited, and the sperm migrate into the female's body, where they penetrate the ovaries to fertilize the eggs, or are stored for future use in paired sperm receptacles. Stored sperm remains viable for many months.

In some Australian species, the males place their spermatophores on their heads like tiny trophies in readiness to present them to a female. Some species have developed elaborate structures on their heads, including spikes, spines, hollow stylets, pits, and depressions, to either hold the sperm or assist in its transfer to the female. Mating has been observed in only a few species.

Embryonic development is extremely diverse. Some species lay large, yolk-filled eggs, while others retain yolky eggs within the female until they are ready to hatch. Some other species have small eggs without a yolky food source, and the young are retained in the body and obtain nourishment from the mother's body in a manner similar to that of placental mammals. In all species, the young are fully developed when born and, apart from lacking complete pigmentation, look like miniature adults. There is no parental care and the young forage independently soon after birth. Some species appear to give birth throughout the year, but seasonality has been recorded for other species.

Conservation status

The 2002 IUCN Red Lists includes 11 onychophoran species: three as Extinct; one as Critically Endangered; three as Endangered; two as Vulnerable; one as Lower Risk/Near Threatened; and one as Data Deficient. Because onychophorans are restricted to humid, cryptic habitats, they are highly susceptible to habitat disturbance, particularly deforestation. In some parts of the world, their relatively small populations could be endangered by excessive collecting. The very restricted distributions of many species also make them vulnerable to habitat alteration, and the destruction of indigenous forests will undoubtedly result in the extinction of many species. Most threats are a direct result of habitat disturbance by humans, including land clearing, pollution, and over-collecting.

Significance to humans

The Onychophora are excellent tools for biological research, being especially valuable for the study of evolution, phylogenetics, reproductive biology, physiology, discontinuous distribution, and continental drift.

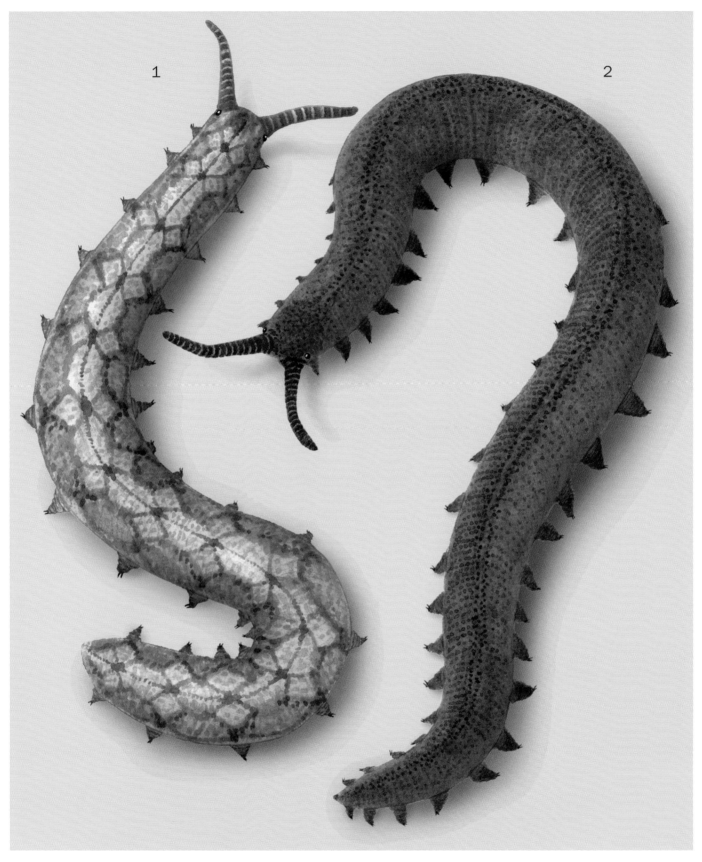

1. *Cephalofovea tomahmontis*; 2. *Epiperipatus biolleyi*. (Illustration by Dan Erickson)

Species accounts

No common name
Epiperipatus biolleyi

ORDER
No order designation

FAMILY
Peripatidae

TAXONOMY
Epiperipatus biolleyi Bouvier, 1902, Costa Rica.

OTHER COMMON NAMES
None known.

PHYSICAL CHARACTERISTICS
No modification of head papillae for reproduction. Number of oncopods variable, 30 pairs in females, 26–28 pairs in males; two anterior and one posterior distal foot papillae; nephridio-pore (dome-shaped urinary opening) between third and fourth spinous foot pad. Gonopore (genital opening) between penultimate pair of oncopods. Crural papillae present in males, opening at base of fourth last pair of oncopods. Anterior accessory gland papillae (function unknown) in males open at base of third last oncopod pair, posterior accessory glands open on anal segment. Females without spermathecae. Viviparous, young are fully developed when born. Body sienna brown or dusky pink with dark gray papillae and mid-dorsal line. Antennae and legs gray. Up to 2 in (52 mm) in length (males); 1.5 in (38 mm) in females (measured when walking).

DISTRIBUTION
The species occurs in Costa Rica.

HABITAT
Found at densities of 0.25 individuals per m² in sandy soil with an average humidity of 35% and pH 5.2–6.2, and under and inside rotting logs and microcaverns in the soil. They occur both in low montane moist forest and in areas that have been converted to pasture.

BEHAVIOR
They avoid light at around 470–600 nanometers. Walking speed is around 0.4 in (1 cm) per second. In nature, these animals often carry scars and mutilated oncopods.

FEEDING ECOLOGY AND DIET
Little is known about diet and feeding ecology in the wild.

REPRODUCTIVE BIOLOGY
Ultrastructural examination of the female genital system has shown that insemination occurs directly via the genital opening. The species is viviparous; the young are fully developed when born.

CONSERVATION STATUS
Not listed by the IUCN.

SIGNIFICANCE TO HUMANS
None known. ◆

☐ *Cephalofovea tomahmontis*
▨ *Epiperipatus biolleyi*

No common name
Cephalofovea tomahmontis

ORDER
No order designation

FAMILY
Peripatopsidae

TAXONOMY
Cephalofovea tomahmontis Ruhberg et al., 1988, Mt. Tomah, New South Wales, Australia.

OTHER COMMON NAMES
None known.

PHYSICAL CHARACTERISTICS
Thirty antennal rings, each with single row of bristles. Male with an eversible head structure consisting of dome-shaped, fleshy crown. Distal half of the crown has rounded, scaled papillae mediodorsally. Proximal part of crown unpigmented, with scattered, pigmented papillae. When inverted, the structure forms a depression, or pit. Female head papillae modified, reduced in size and crowded together in shallow pit. Fifteen pairs of oncopods in both sexes. Crural papillae present only on first pair, or first two pairs, of oncopods. Repeated Y-shaped segmental body pattern. Up to about 2 in (50 mm) in length, with females larger than males.

DISTRIBUTION
The species is found in the Blue Mountains, west of Sydney, NSW, Australia, from Mt. Tomah to Mt. Wilson.

HABITAT
Found in and under rotting logs and among leaf litter in eucalypt forest.

BEHAVIOR
They are highly secretive, so little is known about their behavior.

FEEDING ECOLOGY AND DIET
Little is known about their feeding ecology and diet. In captivity, they feed readily on slaters and *Drosophilia*.

REPRODUCTIVE BIOLOGY
Males have been found with spermatophores cupped within their partially everted head structure. Mating has not been observed. This species is ovoviviparous. Females give birth to fully developed young.

CONSERVATION STATUS
Not listed by the IUCN. This species can only be distinguished from its congeners using molecular methods, and the extent of its distribution range is not known. However, its status would seem to be secure at present.

SIGNIFICANCE TO HUMANS
None known. ◆

Resources

Books
Storch, Volker, and Hilke Ruhberg. "Onychophora." In *Microscopic Anatomy of Invertebrates*. Vol. 12, *Onychophora, Chilopoda and Lesser Protostomata*, edited by Frederick W. Harrison and Mary E. Rice. New York: Wiley-Liss, 1993.

Periodicals
Brockman, C., R. Mummert, H. Ruhberg, and V. Storch. "Ultrastructural Investigations of the Female Genital System of *Epiperipatus biolleyi* (Bouvier, 1902) (Onychophora, Peripatidae)." *Acta Zoologica* 80 (1999): 339–349.

Elliot, S., N. N. Tait, and D. A. Briscoe. "A Pheromonal Function for the Crural Glands of the Onychophoran *Cephalofovea tomahmontis* (Onychophora: Peripatopsidae)." *Journal of Zoology (London)* 231 (1993): 1–9.

Monge-Nágera, J., and J. P. Alfaro. "Geographic Variation of Habitats in Costa Rican Velvet Worms (Onychophora: Peripatidae)." *Biogeographica* 71, no. 3 (1995): 97–108.

Monge-Nágera, J., Z. Barrientos, and F. Aguilar. "Behavior of *Epiperipatus biolleyi* (Onychophora: Peripatidae) Under

Laboratory Conditions." *Revista de Biologia Tropical* 41, no. 3 (1993): 689–696.

Reid, A. L. "A Review of the Peripatopsidae (Onychophora) in Australia, with Descriptions of New Genera and Species, and Comments on Peripatopsid Relationships." *Invertebrate Taxonomy* 10, no. 4 (1996): 663–936.

Reid, A. L., N. N. Tait, and D. A. Briscoe. "Morphological, Cytogenetic and Allozymic Variation Within *Cephalofovea* (Onychophora: Peripatopsidae) with Descriptions of Three New Species." *Zoological Journal of the Linnean Society (London)* 114 (1995): 115–138.

Storch, V., R. Mummert, and H. Ruhberg. "Electron Microscopic Observations on the Male Genital Tract, Sperm Development, Spermatophore Formation, and Capacitation in *Epiperipatus biolleyi* (Bouvier) (Peripatidae, Onychophora)." *Mitteilungen aus dem Hamburgischen Zoologischen Museum und Institut* 92 (1995): 365–379.

Sunnucks, P., and N. Tait. "Tales of the Unexpected." *Nature Australia* 27, no. 1 (2001): 60–69.

Amanda Louise Reid, PhD

Tardigrada
(Water bears)

Phylum Tardigrada
Number of families 20

Thumbnail description
Microscopic, aquatic, multicellular, segmental animals with four pairs of legs

Photo: Colored scanning electron micrograph of a water bear (*Echiniscus testudo*) in its cryptobiotic tun, or barrel, state. (Photo by Andrew Syred/Science Photo Library/Photo Researchers, Inc. Reproduced by permission.)

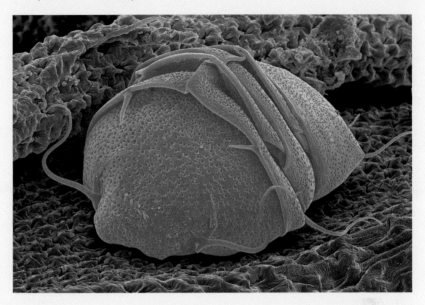

Evolution and systematics

The phylum Tardigrada belongs to the Panarthropoda group, together with the onychophorans (velvet worms) and arthropods, and comprises almost 1,000 described species. However, taxonomists expect that at least 10,000 species exist. The phylum is divided into three classes: Heterotardigrada, Eutardigrada, and Mesotardigrada. The latter was established on the basis of a single species, *Thermozodium esakii*, found in a hot sulfur spring in Nagasaki, Japan. However, the species has not been recorded since the end of World War II. The Heterotardigrada consists of two orders, Arthrotardigrada and Echiniscoidea, and is characterized by the presence of cephalic appendages, so-called cirri and clavae, that function as mechano- and chemoreceptors, respectively. The arthrotardigrades are marine forms that usually have median cirrus and telescopic legs, with or without toes, while the echiniscids are terrestrial armored or marine unarmored forms. The echiniscids have no median cirrus and the legs lack toes. All heterotardigrades have a separate gonopore and anus.

Eutardigrada consists of two orders, Parachela and Apochela, and is characterized by the absence or reduction of external sensory structures. The cuticle is unarmored and the legs have no toes. The so-called double claws of eutardigrades are differentiated into a primary and a secondary branch. Gametes, excretory products, and feces are released through a cloaca. True hetero- and eutardigrades are found in Cretaceous amber from Canada and the United States, and an aberrant tardigrade has recently been recorded in Siberian limestone from the Middle Cambrian.

Physical characteristics

Tardigrades are bilaterally symmetrical and vary in shape from cylindrical to extremely dorso-ventrally flattened. The majority of tardigrade species are white to translucent, but some terrestrial forms may exhibit strong colors such as yellow, orange, green, or red to olive-black. There are five distinct body segments, including a cephalic segment and four trunk segments, each bearing a pair of segmented legs with oblique- or cross-striated muscles. The terrestrial and limnic forms have reduced the segmentation in their stumpy legs that bear two to four claws, while the marine forms may have telescopic retractable legs, with up to 13 claws or four toes with complex claws. Other marine tardigrades have four to six toes with rod-shaped adhesive discs or round suction discs also inserted on the foot via toes.

The cuticle of tardigrades is very complex. Both the dorsal and ventral body cuticle may have segmental plates with different spines and appendages. The cuticle is frequently molted in juveniles and adults, much like in arthropods. In the beginning of the molting cycle, the tardigrade enters the simplex stage, which includes sheeting of stylets, stylet supports, buccal tubes, and pharyngeal cuticular rods, the so-called placoids. When the new cuticle is formed, the cuticle of the digestive system and toes/claws is also re-synthesized. The stylet apparatus is re-synthesized by two stylet glands ("salivary glands"), and cuticular claws and toes are formed in special claw glands in the legs.

The digestive system consists of three principal parts: the foregut (ectodermal origin), the midgut (mesodermal origin),

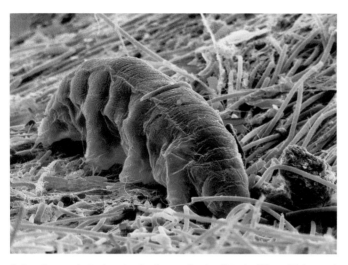

Colored scanning micrograph of a marine tardigrade (*Macrobiotus* sp.) showing four pairs of stumpy legs terminating in claws for clinging to sand or soil. (Photo by Andrew Syred/Science Photo Library/Photo Researchers, Inc. Reproduced by permission.)

and the hindgut (ectodermal origin). The foregut is a very complex feeding structure that consists of a mouth cavity, a stylet apparatus, a buccal tube, and a tri-radiate pharynx with placoids. The two stylets and the stylet supporters are probably homologous with mouth limbs in arthropods.

The nervous system consists of a three-lobed brain, a subpharyngeal ganglion, and four ventral trunk ganglia. Paired eyespots may be present inside the forebrain. Tardigrades lack respiratory organs, and gas exchange takes place through the epidermis. All tardigrades lack excretory proto- or metanephridia, which are common in many other invertebrates. Instead, eutardigrades have three Malpighian tubules at the junction between the mid- and hindgut. The Malpighian tubules may have both an excretory and osmoregulatory function. Heterotardigrades lacks these tubules, but some have segmental organs (coxal glands) that may have an excretory function.

The embryology of the tardigrades is still highly debated. Two theories exist: one theory postulates that the tardigrades have radial cleavage and an enterocoelic mode of coelom formation, whereas another theory suggests that the tardigrades have a modified spiral cleavage and schizocoelic mode of coelom formation. Schizocoely is also found among arthropods, which supports the close relationship between arthropods and tardigrades. The dispute about the tardigrade cleavage type has arisen because of disagreements among scientists about the observed cleavage pattern and cell fates. However, new and improved cell lineage studies appear to support the presence of radial cleavage.

Tardigrades are known to survive long periods of drying or freezing by cryptobiosis – a stage of latent life (ametabolic stage). In fact, it is only tidal tardigrades and the tardigrades that inhabit the interstitial water of mosses and lichens that are capable of cryptobiosis. There are four types of cryptobiosis: anhydrobiosis (dehydration), cryobiosis (very low temperature), osmobiosis (water potential and strong variations in salinity), and anoxybiosis (lack of oxygen). Anhydrobiosis and cryobiosis are well investigated. In these forms of cryptobiosis, the tardigrades can survive from a few months to several years. Neither Osmobiosis nor anoxybiosis are accepted by all scientists as true forms of cryptobiosis, but this is perhaps only a question of insufficient investigations. Usually, the terrestrial species will die after only short time (a day) in water without oxygen, but the tidal tardigrade *Echiniscoides sigismundi* has survived up to six months in seawater without oxygen. This species is currently the only known species that has the capacities to enter all four types of cryptobiosis.

Distribution

Some tardigrades have great migratory capacities, and eutardigrade species such as *Macrobiotus hufelandi* and *Milnesium tardigradum* may be true cosmopolitan species. The eggs of *Macrobiotus* species have been found in air-plankton collected by airplanes at several thousand feet. In Greenland, heterotardigrades have been found in rainwater samples after the powerful foehn storms. It is known that *Echiniscus testudo* is capable of migrating with the winds from one continent to another in its anhydrobiotic stage. However, the species has never been found in Australia, so it is not a true cosmopolitan. Dispersal in marine tardigrades is lesser known, but the tidal tardigrade, *Echiniscoides sigismundi*, may spread with the empty exuvia of the barnacles inside of which it lives. More than a hundred eggs of this tardigrade were found on one exuvium from the barnacle, *Semibalanus balanoides*. Ships may also help to disperse marine tardigrades. One example is the subspecies of *Echiniscoides sigismundi* in Australia. Along the coast of eastern Australia and in the Coral Sea, the subspecies *Echiniscoides sigismundi polynesiensis* is found, but in Nielsen Park in Sidney Harbor, the nominate form (*Echiniscoides sigismundi sigismundi*) was found on barnacles. The nominate form was described from Northern Europe and the subspecies, *E. sigismundi polynesiensis*, from the island of Tiahura in the Pacific Ocean. The only explanation for the presence of the nominate form in Sidney Harbor must be that ships from Europe have carried it from Europe to Australia.

Many species of tardigrades are not cosmopolitan. Species found in hot or warm springs may be endemic. The mesotardigrade, *Thermozodium esakii*, discovered in a hot sulfur spring in Nagasaki, was only found on one occasion, and may be extinct now. The very aberrant eutardigrade, *Eohypsibius nadjae*, was described from a cold mud volcano from west Greenland. Later, it was found in cold springs in the Faroe Islands and in the northern part of Italy. This unique species may be an Arctic relict that survived in cold springs. Several genera of Heterotardigrada show the old Gondwana (South America, Southern Africa, India, Antarctica, Australia, and New Zealand) distribution.

Habitat

Tardigrades are found in all different kinds of habitats, from the highest elevations in the Himalayas to the deepest trenches in the deep sea, and from hot, radioactive springs to the ice cathedrals inside the Greenland ice cap. Many of the so-called terrestrial species are semi-aquatic, because all tardigrades need a water film to be active. The arthrotardigrades are found in true marine habitats from the tidal beaches to the deep-sea mud. The terrestrial species live in mosses and lichens and tolerate desiccation for up to nine years.

Both the heterotardigrades and eutardigrades have independently invaded the terrestrial environment. One genus of eutardigrades, *Halobiotus*, has secondarily invaded the marine environment again. The species, *H. crispae*, is a strange tardigrade that cyclically changes form through the year—a transformation that usually is referred to as cyclomorphosis. The dark winter form is cyst-like and has a double cuticle, but it can still move around if it is not completely frozen. It may survive freezing for up to six months per year. The early spring form tolerates freshwater, has thin stylets, and lacks true placoids in the pharynx. The summer form is only active when the salinity is more than 30 parts per thousand, and it has a normal single layer cuticle and robust stylets with macroplacoids in the pharynx. In this stage, the gonads mature for reproduction.

Behavior

The name Tardigrada means "slow walker" and was given to the first described species of eutardigrades by the Italian scientist Spallanzani in 1776. He described their lumbering gait and in the capacity to form tuns in the cryptobiotic state, but already in 1773, the German priest Goeze had called a tardigrade *Kleiner Wasser Bär* (little water bear). The fascinating appearance of a moss-living eutardigrade, with its slow and bear-like gait, gives the observer associations to a miniature teddy bear. However, many tardigrades are not slow walkers at all. The carnivorous eutardigrade, *Milnesium tardigradum*, is the tiger among the water bears. When it attacks nematodes or rotifers, it moves very quickly and several of the eight legs do not touch the substrate during the jump. The fastest mowing tardigrades are found among the very specialized arthrotardigrades. In the genus *Batillipes*, the claws have been modified to form suction discs. The moving behavior of the species, *Batillipes noerrevangi* from Denmark, has been video-recorded. This species uses its toe discs for suction, as in the suction discs of geckoes. When it moves from a sand grain under the microscope to the glass slide, it moves faster than the human eye can follow.

Feeding ecology and diet

Tardigrades may be carnivorous, herbivorous, or bacteriovorous. Furthermore, a few marine tardigrade species are parasites on other marine invertebrates. *Tetrakentron synaptae* is found on the holothurian, *Leptosynapta galliennei*, where it punctures the epidermal cells of the holothurian and sucks out the cell contents. This species is the only tardigrade that has true adaptations for parasitism. It is dorso-ventrally flattened and all the sensory structures are reduced. As well, the claws are armed with three large hooks that are used to penetrate the epidermis of the holothurian. Females in particular are less mobile and are located in small depressions in the tegument of the holothurian. Another parasitic species is *Echiniscoides hoepneri* that lives on the embryos of the barnacle, *Semibalanus balanoides*.

The feeding ecology of the many marine species is not fully understood, but it is known that several species do not eat at all for long periods. Some species have symbiotic bacteria in special head vesicles. The genus *Wingstrandarctus* that lives in coral sand has three head vesicles containing thiobacteria (sulfur bacteria). The bacteria may give the tardigrade dissolved organic matter (DOM) products such as amino acids and glucose. In the deep-sea family Coronarctidae, the mid-gut may be filled with a white amorphous content that is very similar to gut contents found in bacteriovorous tardigrades.

The feeding ecology of terrestrial and freshwater tardigrades is much better known. The heterotardigrades of the family Echiniscidae seems to be adapted to suck out the cell contents of mosses. Many species have very long stylets to penetrate the thick cellulose walls of mosses. Large eutardigrade species such as *Milnesium tardigradum*, *Macrobiotus richtersi*, and *Amphibolus nebulosus* are carnivores and eat nematodes, rotifers, and other tardigrades. Smaller bryophilous species of eutardigrades may not always suck out the moss cells, but instead may eat the epiphytic diatoms and bacteria that live on the moss. Small eutardigrades living in soil or in the rhizoids of mosses have very thin and narrow buccal tubes. In the genus *Diphascon*, the buccal tube is flexible with spiral rings like a vacuum cleaner tube. This genus is also found in cryoconite ("star dust") on the Greenland ice cap. The species, *Diphascon recamieri*, has a rusty colored mid-gut and probably feeds on iron bacteria in the cryoconite.

Reproductive biology

The male reproductive system in all tardigrades seems to be relatively simple. The single testis is located dorsally and the two seminal vesicles open latero-ventrally via two seminal ducts in an oval gonopore papilla (in heterotardigrades) or in the cloaca (in eutardigrades). Penile structures have never been found in any male tardigrade, and it is still uncertain how the sperm transfer occurs. However, it seems that some females have structures that can be inserted into the male so that the female can actively grab the sperm. This method of sperm transfer is very unusual in the animal kingdom.

The female reproductive system in heterotardigrades consists of a single ovary; there is a single oviduct opening in a six-lobed rosette gonopore system anterior to the anus. In eutardigrades, the oviduct opens into the cloaca. Two cuticular seminal receptacles are present in many arthrotardigrades, and a single internal receptacle is present in several eutardigrades. The internal receptacle opens into the hindgut and it lacks a cuticle covering. The seminal receptacles are not homologous in heterotardigrades and eutardigrades. In *Milnesium tardigradum*, a ventral fourth Malpighian tubule has been described. However, this structure is actually a single seminal receptacle.

All tardigrades have been considered egg-laying, but there exists a single unpublished record from an arthrotardigrade (*Styraconyx* sp.) collected in the deep sea that shows a larva coming out of the gonopore. If this is true, other deep-sea tardigrades may be viviparous as well.

Eggshell morphology has great taxonomic importance in Eutardigrada. The egg of *Macrobiotus hufelandi* was the first to be observed in a scanning electron microscope, and the details of the egg sculpture have fascinated scientists ever since. All echiniscids form cysts before they lay eggs. When a female hatches from the cyst, she lays the unsculptured eggs in the old exuvium.

Although information on the mating behavior in Arthrotardigrada is extremely scarce, it has been observed that males and females in the species *Parastygarctus sterreri* mate venter to venter when the male ejects the sperms into the external seminal receptacles of the female.

In eutardigrades, the male clings with its first leg pair to the anterior part of the female. The claws on the first leg pair of the males may be strongly modified, as in *Milnesium tardigradum*. Internal fertilization is common in several eutardigrade species, and these species always lay free eggs. However, some species lay their eggs in the old exuvium right after molting, so that the male can afterward spread the sperm into the exuvium.

Most tardigrades are dioecious, but hermaphroditism also occurs. Hermaphroditism is especially common in many genera of limno-terrestrial eutardigrades, whereas it has only been recorded in a single arthrotardigrade species, *Orzeliscus* sp. Besides usual sexual reproduction, many species are capable of reproducing by parthenogenesis (reproduction without male fertilization). In some species, males have never been observed and all reproduction is solely by parthenogenesis. However, it has been shown that some apparently parthenogenetic species sometimes have populations with males. It is not known how the production of males is triggered in these populations.

It seems likely that the evolution of parthenogenesis in tardigrades is linked to the evolution of cryptobiosis. This postulate is supported by the strong correlation between the presence of parthenogenetic and cryptobiotic capacities. The parthenogenetic capacities in tardigrades have evolved independently in the echiniscid heterotardigrades and the parachelate eutardigrades.

Conservation status

No species are listed by the IUCN.

Significance to humans

The phenomenon of cryptobiosis has fascinated humans since it was discovered in tardigrades by Spallanzani in 1776. Today, researchers think that tardigrades could be used as test animals for traveling into outer space. Experiments in which tardigrades in anhydrobiosis (tun stage) are exposed to cosmic radiation, vacuum, and temperatures close to absolute zero have been very successful, and ongoing experiments with ionosphere balloons have shown that both the species *Echiniscus testudo* and *Richtersius coronifer* may be the right test animals for true outer space experiments in space shuttles. The pharmaceutical industry has been very interested in role of the sugar trehalose that tardigrades produce prior to anhydrobiosis and cryobiosis stages. Trehalose appears to protect the cellular membranes of tardigrades against damage from freezing and dehydration. Trehalose may be used in organ transplantation to avoid freeze damage. Recently, it has become clear that the phenomenon of cryptobiosis is much more complex than first thought. New results show that it is not only trehalose that is responsible for survival during cryobiosis and anhydrobiosis. In the species *Richtersius coronifer*, a very large protein (ice-nucleating agent) seems to protect the cellular structures from fast freezing in active animals. The phenomenon of cryptobiosis is a fascinating biological puzzle. By solving the puzzle of how a tardigrade can go into a reversible death (ametabolic stage) for many years, and after a few minutes of rehydration can climb around again, it might explain how life developed on earth.

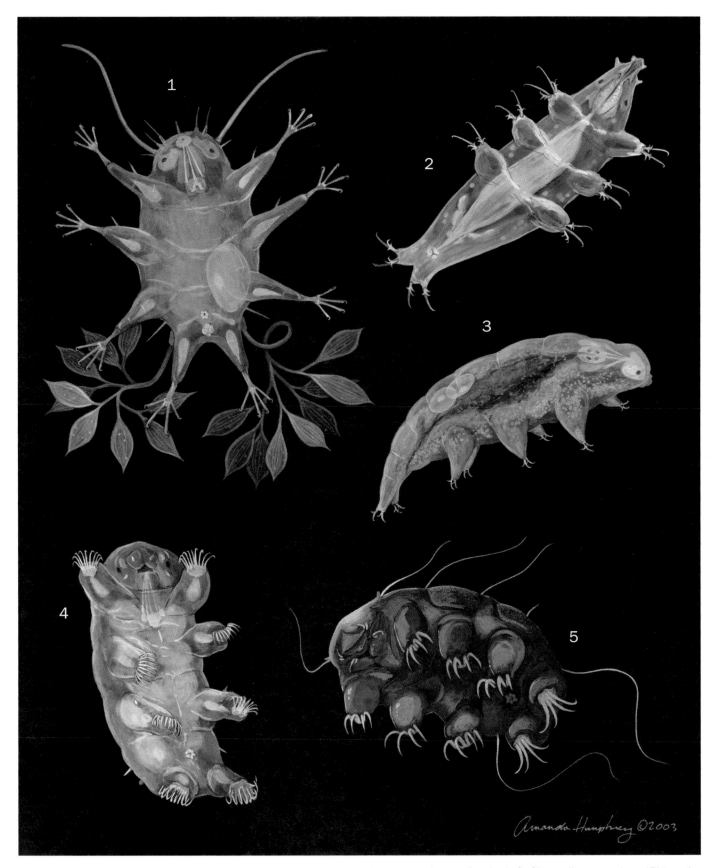

1. Balloon water bear (*Tanarctus bubulubus*); 2. Large carnivorous water bear (*Milnesium tardigradum*); 3. Giant yellow water bear (*Richtersius coronifer*); 4. Tidal water bear (*Echiniscoides sigismundi sigismundi*); 5. Turtle water bear (*Echiniscus testudo*). (Illustration by Amanda Humphrey)

Species accounts

Large carnivorous water bear
Milnesium tardigradum

ORDER
Apochela

FAMILY
Milnesiidae

TAXONOMY
Milnesium tardigradum Doyére, 1840, close to Paris, France.

OTHER COMMON NAMES
None known.

PHYSICAL CHARACTERISTICS
Measure 0.0197–0.0236 in (500–600 μm), females sometimes up to 0.0394 in (1,000 μm), males much smaller. Body is elongated to torpedo shaped; color varies from colorless to reddish or brownish and black eyes are posterior on each side of the pear-shaped pharynx; head region displays a number of unique characteristics. Terminal mouth opening is surrounded by six robust, triangular lamellae that serve as closing apparatus for mouth opening when it is not feeding; has six oral and two lateral papillae, which may have chemoreceptory function. Double claws unusual because primary and the secondary branches are completely separate; primary branch is very long and flexible, while secondary branch is a short and robust claw, usually with three spurs. Buccal tube is wide and short, and armed with short convex stylets and stylet supports.

DISTRIBUTION
Commonly cosmopolitan; from dry tropical deserts to polar freshwaters; *M. tardigradum* is actually a complex of several species; is also one of the tardigrade species that are present in the fossil records. A eutardigrade found in Cretaceous amber from United States, described as *M. swolenskyi*, is very similar to *M. tardigradum*, meaning they probably lived together with the dinosaurs. (Specific distribution map not available.)

HABITAT
Particularly common in drier temperate terrestrial habitats such as mosses and lichens; one of the first described species from mosses on roofs and in gutters of houses; species recorded from supralitoral lichens of the genus *Ramalina* and from ornitho-coprophilous lichens (grow in bird feces) such as the yellow *Xanthoria elegans*. In these lichens, it is colored reddish to brownish.

BEHAVIOR
A fast "runner," moving in a very characteristic way, and does not use its fourth pair of legs. Almost stands up, like a mini carnivore dinosaur, when it attacks prey.

FEEDING ECOLOGY AND DIET
Exclusively carnivorous, feeding on nematodes, rotifers, and other smaller eutardigrades. Large nematodes are attacked at the middle of their trunks and pierced by the two stylets; cell contents of nematode are sucked out by strongly muscular pharynx. Smaller nematodes swallowed like spaghetti. The mid-gut can be filled with jaws from rotifers; genus *Philodina* may be a favorite diet. Bucco-pharyngeal apparatus of smaller tardigrades also found in the mid-gut of the species.

REPRODUCTIVE BIOLOGY
Fertilization is internal; female has a single seminal vesicle, which has been mistaken for a fourth malpighian tubule. Males are much smaller than females; male claws on the first pair of legs are strongly modified; secondary branch of the double claw is a rough and robust hook; assumed that males use claws to attach to female during mating; up to 18 smooth-shelled eggs deposited in the cast exuvium. The size of the eggs ranges from 0.00276 to 0.00453 in (70–115 μm). Newly hatched juveniles resemble miniature adults.

CONSERVATION STATUS
Not listed by the IUCN.

SIGNIFICANCE TO HUMANS
Has been used for pest control of nematodes in soil, but without very good results. ◆

Balloon water bear
Tanarctus bubulubus

ORDER
Arthrotardigrada

FAMILY
Halechiniscidae

TAXONOMY
Tanarctus bubulubus Jørgensen and Kristensen, 2001, Faroe Bank, North Atlantic.

OTHER COMMON NAMES
English: Balloon animal.

PHYSICAL CHARACTERISTICS
Small species measuring 0.00346–0.00425 in (88–108 μm). Cephalic sensory structures consist of very long primary clavae (longer than the body), two lens-shaped buccal clavae, two long internal and external cirri, two short lateral cirri, and a single, median cirrus. Adults have telescopic legs with lance-like tibia, conic tarsus, and four toes with internal claws with small dorsal spurs; 18–20 balloon-like appendages are attached on the fourth leg pair. Balloons vary greatly in shape, and regulate the buoyancy of animal when it adheres to substrate. Females have two cuticular seminal receptacles that open lateral to the gonopore.

DISTRIBUTION
Faroe Bank and Bill Bailey Bank, North Atlantic. (Specific distribution map not available.)

HABITAT
Subtidal and lives interstitially in clean shell gravel. Is very rare and has only been found in water depths of 308–656 ft (94–200 m).

BEHAVIOR
Often attached to empty shells of tintinids (ciliates), but one specimen has been seen floating in the water column upside-down with the 18 balloons extended. Dorsal side of tardigrade is totally covered with empty pieces of coccoliths (calcareous algae); coccolith layer probably protects them from predators.

FEEDING ECOLOGY AND DIET
Four-lobed mid-gut is usually filled with a white amorphous content, similar to content in bacterivorous tardigrades. Stylets and buccal tube are very thin and narrow, indicating that it pierces bacteria and sucks out cell contents with its tri-lobed pharynx, which is armed with fused calcium carbonate-encrusted placoids. In one locality, the sediment was filled with methano-bacteria, and it has been suggested that species feeds on this type of bacteria.

REPRODUCTIVE BIOLOGY
Has internal fertilization; two seminal receptacles are often filled with thread-like spermatozoa. One egg can fill one-third of entire female. Two-clawed newly hatched larva lacks balloons, but already is length of 0.00303 in (77µm) when 18 balloons are present.

CONSERVATION STATUS
Not listed by the IUCN.

SIGNIFICANCE TO HUMANS
None known. ◆

Turtle water bear
Echiniscus testudo

ORDER
Echiniscoidea

FAMILY
Echiniscidae

TAXONOMY
Emydium testudo Doyére, 1840, close to Paris, France.

OTHER COMMON NAMES
None known.

PHYSICAL CHARACTERISTICS
Measures up to 0.0142 in (360 µm); brownish red, or reddish to yellow, with red eyespots. dorsal side covered with segmental sclerotized dorsal plates ornamented with rounded irregular pores. The cephalic sensory structures consist of two pairs of clavae, and three pairs of cirri; median cirrus is lacking. Lateral appendages on the trunk, also named cirri, are present. These filaments, called cirri a to e, probably have an adhesive function rather than a sensorial one; cirrus d is always lacking, and has instead two dorsal spikes. Four claws are robust and the internal claw has a miniscule basal spur.

DISTRIBUTION
Common; recorded from most of Europe, Canada, Greenland, South America, India, Turkey, and Afghanistan. (Specific distribution map not available.)

HABITAT
Lives in mosses and lichens, and appears to prefer insolated localities that often dry out. Common in urban areas on tile roofs with old moss populations.

BEHAVIOR
Has a slow bear-like gait.

FEEDING ECOLOGY AND DIET
Lives in green part of mosses. Has been postulated that it sucks out the moss cells, but in fact there are no observations of feeding behavior. The yellow color in coelomocytes and mid-gut cells is formed by carotene (same dye as in carrots); may come from the diet of mosses or lichens, but unfortunately the association between the diet and the color of cells has never been proven with labeled isotopes.

REPRODUCTIVE BIOLOGY
Males have never been observed; always suggested that species reproduces solely by parthenogenesis. However, small males have been found in other species of genus *Echiniscus*. If males present, they must be very rare. The two to five reddish eggs are laid in the old exuvium. Newly hatched juveniles have only two claws on each leg, and they have usually fewer trunk cirri than the adult.

CONSERVATION STATUS
Not listed by the IUCN.

SIGNIFICANCE TO HUMANS
None known, but species is sometimes found in house dust. ◆

Tidal water bear
Echiniscoides sigismundi sigismundi

ORDER
Echiniscoidea

FAMILY
Echiniscoididae

TAXONOMY
Echiniscus sigismundi Schultze, 1865, Helgoland, Germany.

OTHER COMMON NAMES
None known.

PHYSICAL CHARACTERISTICS
Adults measure 0.00618–0.0134 in (157–340 µm); lack dorsal plates, and the dorsal cuticle varies from smooth to a delicate mammilate sculpture. Large black eyes are always present; mouth opening is subterminal. Stylets are very long with large furcae; stylet supports lacking. Three straight placoids are encrusted with calcium carbonate. All sensory structures are reduced in length. Median cirrus is lacking, minuscule internal and external cirri are located around mouth cone. Key characteristic of genus *echiniscoides* is presence of numerous claws on the legs. In adults, the number of claws varies from 7–13; fourth pairs of legs have usually one claw less than other legs. All claws smooth and lack spurs.

DISTRIBUTION
Commonly cosmopolitan in the tidal zone; one record from soil samples at altitude of 3,280 ft (1,000 m) in the former Belgian Congo. Many subspecies described throughout the world.

HABITAT
Marine, intertidal species. Lives on *Enteromorpha* algae or as symbiont on barnacles. Always restricted to upper tidal zone and is capable of tolerating desiccation (anhydrobiosis).

BEHAVIOR

During low tide, animals are gregarious. More than 100 individuals have been observed in the sutures of barnacles; animal capable of tolerating all kinds of physiological stress; can survive in distilled as well as saturated seawater (osmobiosis). In polar regions, it may be frozen twice a day (low tide), or stay frozen for up to six months in areas without tides.

FEEDING ECOLOGY AND DIET

Herbivorous and pierces unicellular algae and cyanobacteria. Six-lobed mid-gut is usually green right after molting and turns black just before new molt; animal can only defecate during the molt, and it takes place inside the old exuvium.

REPRODUCTIVE BIOLOGY

Female lacks seminal vesicles; fertilization is external. Up to 12 mature eggs ready for ovoposition have been observed in single ovary. Several females lay free eggs in big clusters; afterward, males fertilize them. Newly hatched juveniles always have fewer claws than the adults.

CONSERVATION STATUS

Not listed by the IUCN.

SIGNIFICANCE TO HUMANS

None known. ◆

Giant yellow water bear
Richtersius coronifer

ORDER

Parachela

FAMILY

Macrobiotidae

TAXONOMY

Macrobiotus coronifer Richters, 1903, or *Adorybiotus coronifer* Richters, 1903, Spitsbergen Island.

OTHER COMMON NAMES

None known.

PHYSICAL CHARACTERISTICS

Large animals (up to 0.039 in [1 mm]), usually yellow to orange with large black eyes; buccal tube is rather narrow with hook-shaped appendices for the stylet muscle insertions; phar-

ynx is ovoid with two short and square macroplacoids; body lacks sensory appendages. Legs with two equally sized claws; each double claw has an enormous crescentic marking, named a lunula, with 10–18 denticles. The large, yellow eggs (larger than 0.00787 in [200 µm]) are more or less pliable and ornamented with rough thorns.

DISTRIBUTION

Common species in dry mosses on carbonate bedrock. Has been recorded from various lowland localities in Europe (Sweden), Turkey, Colombia, and the Arctic. Also found in the dry and high parts of the Himalayas at 18,300 ft (5,600 m) elevation (Nepal, Kala Pathar, Khumbu Himal). (Specific distribution map not available.)

HABITAT

Bryophilic (moss-living) tardigrade and lives mainly in alpine or arctic environments. Avoids acid bedrock, but is common in dry mosses growing on limestone or basalt. Prefers mosses of the genera *Grimia*, *Orthotrichum*, and *Tortula*. On Öland (Sweden), there may be more than 1,000 individuals in one moss-cushion.

BEHAVIOR

During dry spells, found in the anhydrobiotic tun stage. Capable of surviving severe desiccation and low temperatures down to -320°F (-196°C), both in anhydrobiotic and the hydrated states. During desiccation, accumulates trehalose. Molecules protect cell membrane against damage from dissociation. Can survive approximately nine years in dried conditions.

FEEDING ECOLOGY AND DIET

Herbivorous; it has been suggested that the two large stylets can penetrate the cell walls of the mosses and so animal can suck out cell contents with its strong pharyngeal apparatus. No direct observations made of feeding behavior, but the mid-gut can be filled with green to dark green chlorophyll-containing material.

REPRODUCTIVE BIOLOGY

Indications that it is capable of switching between sexual and parthenogenetic reproduction.

CONSERVATION STATUS

Not listed by the IUCN.

SIGNIFICANCE TO HUMANS

Used as a laboratory animal. Danish-Italian experiments with ionosphere balloons have shown that it can survive high temperatures, vacuum, and cosmic radiation. Will be used in further experiments in outer space. ◆

Resources

Books

Bertolani, Roberto. *Tardigradi. Guide per il Riconnoscimento delle Specie Animali delle Acque Interne Italiane.* Verona, Italy: Consiglio Nazionale Delle Ricerche, 1982.

Dewel, Ruth A., Diane R. Nelson, and William C. Dewel. "Tardigrada. Vol. 12: Onychophora, Chilopoda and Lesser Protostomata." In *Microscopic Anatomy of Invertebrates*, edited by F. W. Harrison and M. E. Rice. New York: Wiley-Liss, 1993.

Greven, Hartmut. *Die Bärtierschen. Die Neue Brehm-Bucherei*, Volume 537. Wittenberg, Germany: A. Ziemsen Verlag, 1980.

Kinchin, Ian M. *The Biology of Tardigrades.* London: Portland, 1994.

Nelson, Diane R., and Sandra J. McInnes. "Tardigrada." In *Freshwater Meiofauna: Biology and Ecology*, edited by S. D. Rundle, A. L. Robertson and J. M. Schmid-Araya. Leiden, The Netherlands: Backhuys Publishers, 2002.

Periodicals

Kristensen, R. M. "The First Record of Cyclomorphosis in Tardigrada Based on a New Genus and Species from Arctic Meiobenthos." *Zeitschrift für Zoologische Systematik und Evolutions-forschung* 20 (1982): 249–270.

Resources

————. "An Introduction to Loricifera, Cycliophora, and Micrognathozoa." *Integrative and Comparative Biology* 42 (2002): 641–651.

Rebecchi, L., V. Rossi, T. Altiero, R. Bertolani, and P. Menozzi. "Reproductive Modes and Genetic Polymorphism in the Tardigrade *Richtersius coronifer* (Eutardigrada, Macrobiotidae)." *Invertebrate Biology* 22 (2003): 19–27.

Wright, J. C., P. Westh, and H. Ramløv. "Cryptobiosis in Tardigrada" *Biological Reviews of the Cambridge Philosophical Society* 67 (1992): 1–29.

Reinhardt Møbjerg Kristensen, PhD
Martin Vinther Sørensen, PhD

Remipedia

(Remipedes)

Phylum Arthropoda

Subphylum Crustacea

Class Remipedia

Number of families 2

Thumbnail description
Small, marine, cave-dwelling crustaceans characterized by a short head and a long trunk bearing setose swimming appendages.

Photo: An individual of the species *Lasionectes entrichoma* from the Turks and Caicos Islands. (Photo by Jill Yager and Dennis Williams. Reproduced by permission.)

Evolution and systematics

The class Remipedia has two families, six genera, and 12 described species. At least three more species are undescribed. A single extinct species from a Carboniferous fossil has been placed in the class Remipedia.

Physical characteristics

Remipedes are troglobitic (cave-dwelling) crustaceans, lacking pigmentation and eyes. They are free-swimming and characterized by a short head and an elongate, segmented trunk. The head is covered with a cephalic shield. With the exception of the posterior-most segment(s), each trunk segment bears biramous, paddle-like, setose swimming appendages. Trunk segment number differs in each species. For example, the smallest species has no more than 16 segments as an adult; the largest has 29. Head appendages consist of a pair of small frontal filaments, long first antennae, and small paddle-like second antennae. The base of antenna 1 bears clusters of long aesthetascs for chemosensation. The setae of antenna 2 are long and plumose. The mandibles have a broad molar surface and cusped incisor processes. The first maxillae bear a terminal fang. The second maxillae and maxillipeds are nearly identical, and bear a terminal claw complex. These three feeding appendages are robust and prehensile. The largest species, *Godzillius robustus*, is about 1.8 in (45 mm) and the smallest, *Godzilliognomus frondosus*, is about 0.35 in (9 mm) in length.

Distribution

Remipedes are currently known from anchialine caves in most of the major islands of the Bahamas, several islands of the Turks and Caicos, the island of Cozumel, Mexico, Holguin and Matanzas Provinces, Cuba, and the island of Lanzarote in the Canary Islands. They are also known from continental coastal caves in the state of Quintana Roo, Mexico, and in Western Australia.

Habitat

Remipedes live exclusively in submerged caves (anchialine caves) found near the sea on islands or along the coast. Anchialine caves are caves that have surface openings on land and subsurface connections to the sea. This habitat is accessible only by trained cave divers using special scuba techniques and equipment. Anchialine caves typically have a fresh or brackish layer of water overlying deeper marine water. The density interface or halocline also delineates differences in temperature, pH, and dissolved oxygen. Remipedes live in the dark, beneath the density interface in waters that are low in dissolved oxygen. They live in an ecosystem that includes other troglobitic crustaceans, most commonly thermosbaenaceans, cirolanid isopods, caridean shrimp, and amphipods. Blind cave fish are occasionally found in the cave systems where remipedes live. With the exception of one Mexican cave system, all known locations of remipedes contain very few individuals. Most species are quite rare, and fewer than 10 individuals can be observed during a typical cave dive.

Behavior

Remipedes are typically observed swimming in the water column in most caves. However, in one cave in Mexico the remipedes are more frequently swimming along the surface of the sediment. Some of these remipedes have been observed

An individual of the species *Speleonectes gironensis* from Cuba. (Photo by Jill Yager and Dennis Williams. Reproduced by permission.)

resting motionless on the sediment and sometimes grooming. Several types of grooming behavior have been observed. The first antenna is flicked posteriorly and run along the setae of the swimming appendages. Circle grooming involves a remipede curling into a circle and cleaning trunk appendages with the feeding appendages.

Feeding ecology and diet

In their natural habitat remipedes have been observed feeding on cave shrimp and on blind cave fish. In laboratory experiments they will capture, subdue, and eat cave shrimp. Additionally they will eat non-native food items such as brine shrimp and bloodworms. They use the fang of the first maxilla to inject a secretory product into their prey that immobilizes and kills it. The fang is connected by a duct to a large secretory gland. The secretory product remains unanalyzed. Remipedes have a close association with the sediment in some caves. In captivity they have been observed to gather sedi-

ment into a small bolus, hold it over their mouth, and ingest material. Some scientists believe that they are using bacteria found in the sediment as either a food source or for some physiological purpose.

Reproductive biology

Remipedes are simultaneous hermaphrodites. The ovary originates in the head, and the oviducts extend to the base of the seventh swimming appendage where the female gonopore is found. The testes originate in the segment of the seventh swimming appendage, and the vas deferens extends to the fourteenth swimming appendage where the male gonopore is located. Sperm is flagellated and packaged into spermatophores. To date, nothing is known about remipede development. Tiny juveniles have been collected. Resembling adults, they are about one-third the length of adults and have at least 14 swimming appendages.

Conservation status

No remipedes are listed by the IUCN. One species, *Speleonectes lucayensis*, is protected by Lucayan National Park on Grand Bahama Island. Worldwide, anchialine caves are threatened ecosystems. Dangers include deforestation and development above or near the cave systems, use of biocides, improper sewage disposal, and destruction of nearby mangroves. The many caves along the coast of Quintana Roo, Mexico, are in imminent danger due to rapid development of tourist hotels, expansion of cities, and lack of enforcement of environmental laws. Most caves are formed in limestone, which is porous. Any kind of pollutant on the surface can eventually end up in the aquifer. One cave in Quintana Roo is home to thousands of remipedes, a number never recorded anywhere else in the world. This cave is in the path of development.

Significance to humans

Remipedes have no known commercial significance.

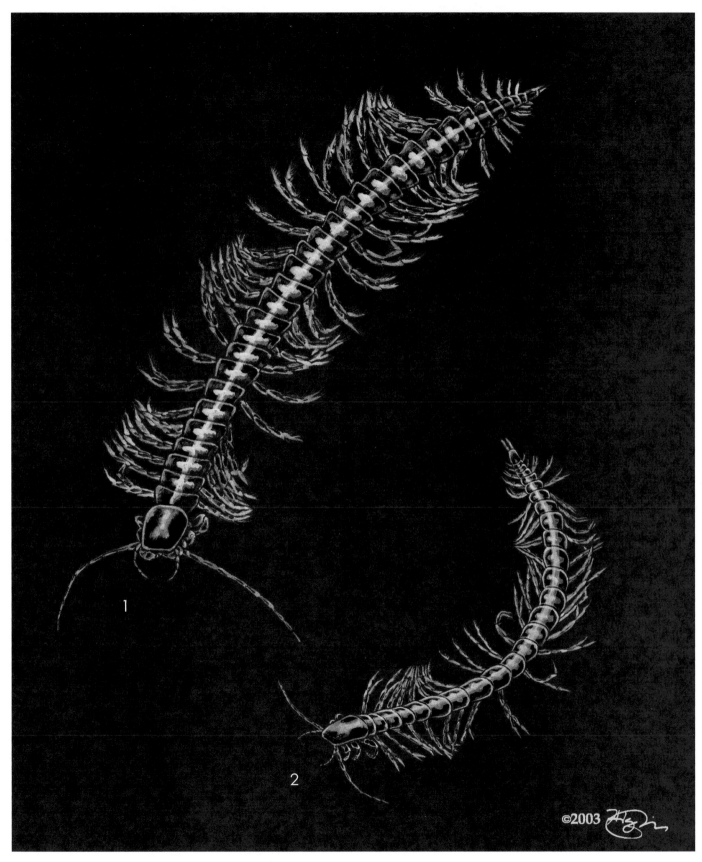

1. *Lasionectes entrichoma*; 2. *Speleonectes gironensis*. (Illustration by Jonathan Higgins)

Species accounts

No common name
Lasionectes entrichoma

ORDER
Nectiopoda

FAMILY
Speleonectidae

TAXONOMY
Lasionectes entrichoma

OTHER COMMON NAMES
None known.

PHYSICAL CHARACTERISTICS
Maximum length is 1.26 in (32 mm); maximum number of trunk segments is 32. The second maxillae and maxillipeds are robust and have thick rows of uniform short setae along the medial margins.

DISTRIBUTION
Anchialine caves of the Turks and Caicos Islands.

HABITAT
Found beneath the halocline from 16 ft (5 m) to deeper depths.

BEHAVIOR
Lasionectids swim in the water column.

FEEDING ECOLOGY AND DIET
Main diet is the caridean shrimp *Typhlatya garciai* Chace.

REPRODUCTIVE BIOLOGY
Hermaphroditic. Nothing is known about reproductive behavior.

CONSERVATION STATUS
Not listed by the IUCN.

SIGNIFICANCE TO HUMANS
None known. ◆

No common name
Speleonectes gironensis

ORDER
Nectiopoda

FAMILY
Speleonectidae

TAXONOMY
Speleonectes gironensis

OTHER COMMON NAMES
None known.

PHYSICAL CHARACTERISTICS
Maximum length is 0.55 in (14 mm). Maximum number of trunk segments is 25. First maxillae are robust. The caudal rami are at least two times the length of the anal segment.

DISTRIBUTION
Anchialine caves of Matanzas and Holguin Provinces, Cuba.

HABITAT
Found beneath the halocline from 39 ft (12 m) to deeper depths.

BEHAVIOR
Speleonectids swim in the water column.

FEEDING ECOLOGY AND DIET
Their main diet consists of small cave crustaceans.

REPRODUCTIVE BIOLOGY
Hermaphroditic. Nothing is known about reproductive behavior.

CONSERVATION STATUS
Not listed by the IUCN.

SIGNIFICANCE TO HUMANS
None known. ◆

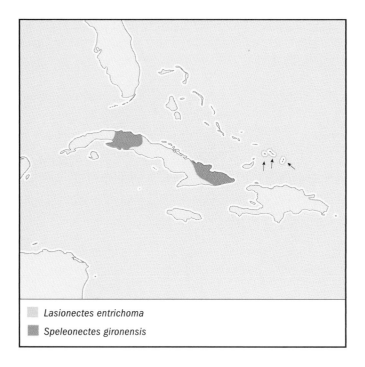

Lasionectes entrichoma
Speleonectes gironensis

Resources

Book

Yager, Jill. "The Reproductive Biology of Two Species of Remipedes." In *Crustacean Sexual Biology*, edited by R. T. Bauer and J. W. Martin. New York: Columbia University Press, 1991.

Periodicals

Yager, Jill. "*Cryptocorynetes haptodiscus*, New Genus, New Species, and *Speleonectes benjamini*, New Species, of Remipede Crustaceans from Anchialine Caves in the Bahamas, with Remarks on Distribution and Ecology." *Proceedings of the Biological Society of Washington* 100, no. 2 (1987): 302–320.

———. "*Pleomothra apletocheles* and *Godzillignomus frondosus*, Two New Genera of Remipede Crustaceans (Godzilliidae) from Anchialine Caves in the Bahamas." *Bulletin of Marine Science* 44 (1989): 1195–1206.

———. "Remipedia, a New Class of Crustacea from a Marine Cave in the Bahamas." *Journal of Crustacean Biology* 1 (1981): 328–333.

———. "*Speleonectes gironensis*, New Species (Remipedia: Speleonectidae), from Anchialine Caves in Cuba, with Remarks on Biogeography and Ecology." *Journal of Crustacean Biology* 14, no. 4 (1994): 752–762.

Yager, Jill, and Frederick R. Schram. "*Lasionectes entrichoma*, New Genus, New Species, (Crustacea: Remipedia) from Anchialine Caves in the Turks and Caicos, British West Indies." *Proceedings of the Biological Society of Washington* 99, no. 1 (1986): 65–70.

Jill Yager, PhD

● Cephalocarida
(Cephalocarids)

Phylum Arthropoda

Subphylum Crustacea

Class Cephalocarida

Number of families 2

Thumbnail description
Cephalocarids are small crustaceans with elongate bodies, short heads without carapace, and flattened, paddle-like appendages on the thorax

Illustration: *Hutchinsoniella macracantha.* (Illustration by John Megahan)

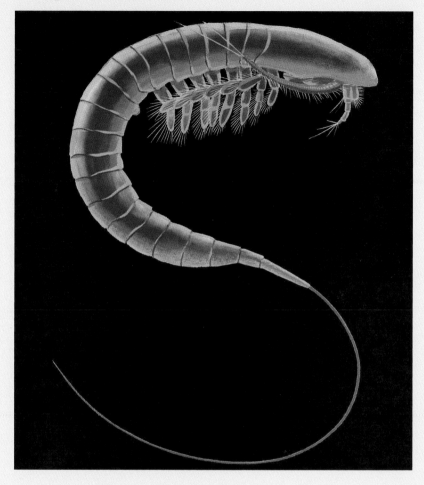

Evolution and systematics

Cephalocarids, when first discovered in Long Island Sound, were thought to be the most primitive of living crustaceans. This view, based on much detailed work on *Hutchinsoniella macracantha*, stemmed from the fact that cephalocarids possessed limbs that were phyllopodous (that is, flattened and lobe like, with the shape maintained by fluid pressure) and were similar in form from the back of the head to the end of the thorax, and were added sequentially during development. More recently cephalocarids have been recognized as an early group within the Crustacea, but most likely arose after the early line leading to remipedes. The subclass Cephalocarida comprises one extant order, Brachypoda; and two families, Hutchinsoniellidae and Lightiellidae. A total of nine species in four genera are currently known.

Physical characteristics

The cephalocarid head is short and wide, and is covered with a strong dorsal head shield. Ventrally, in front of the mouth, is a large, posteriorly directed labrum, which func-

tions to keep food particles—moving anteriorly as a result of a feeding current set up by the thoracic appendages—from going past the mouth opening. The mouth appendages posterior to the mouth, maxillules, and maxillae are built much like the following limbs of the thorax. That is, they have a basal protopod with endites on the inner margin and epipods and exopods on the outer margin. The endopod has a more or less ambulatory function and consists of 5–6 segments. Metachronal movements of these limbs cause an anteriorly directed feeding current in the mid-ventral groove between the paired appendages. There are 20 post-cephalic somites of which the first eight are considered to belong to the thorax. The abdominal somites do not bear appendages, except for the last somite (telson or anal somite) that has a pair of posteriorly directed uniramous appendages generally referred to as caudal rami.

Distribution

Cephalocarids are found in the upper few millimeters of very fine and often flocculent marine sediments in shallow to deep (5,250 ft [1,600 m]) waters. The nine species are known from east and west United States, Caribbean Islands, Brazil, Peru, southwest Africa, New Caledonia, New Zealand, and Japan.

Habitat

Most cephalocarids have commonly been found in flocculent surface muds with high organic content. A few, however, are known from coral rubble and substrates with fine sediment particles.

Behavior

Cephalocarids move through the upper few millimeters of the sediment by moving the thoracic limbs. They do appear to be able to swim or burrow. Occasionally, they will double up the body and use their appendages to groom the abdominal somites.

Feeding ecology and diet

Very small organic particles appear to be the primary food source of cephalocarids. Using the metachronal beat of the thoracic limbs, they pass these small organic particles to the mouth. As the limbs separate, particle-laden fluids are pulled through the interlimb space toward the mid-ventral food groove. As adjacent limbs come together, the fluid is pushed away from the body and particles are retained on the setae of the endites. Turbulence removes particles from the setae and into the food groove. It may be that mucus secreted from glands in the endites helps to bind the particles for ingestion.

Reproductive biology

Mating has not yet been observed, but cross-fertilization is likely since cephalocarids are functional hermaphrodites. Eggs are carried on the reduced appendages of the ninth post-cephalic somite. Young hatch as a metanauplius with three fully developed head appendages (antennules, antennae, and mandibles), two rudimentary appendages (maxillules and maxillae), and three post-cephalic somites without limbs. With each successive molt, the posterior-most appendage changes form from rudimentary to fully developed and the following limb appears in rudimentary form. In addition, at each molt one or two body somites are added until all twenty post-cephalic somites are present.

Conservation status

Very few cephalocarids are found in any abundance; most species so far being known only from one to four specimens. Conservation status is unknown, and no species are listed by the IUCN.

Significance to humans

Cephalocarids are of intellectual interest only, and have no other known significance.

Species accounts

No common name
Hutchinsoniella macracantha

ORDER
Brachypoda

FAMILY
Hutchinsoniellidae

TAXONOMY
Hutchinsoniella macracantha Sanders, 1955, Long Island Sound, United States.

OTHER COMMON NAMES
None known.

PHYSICAL CHARACTERISTICS
Head and thoracic somites wider than those of the abdomen. Eighth thoracic appendage is reduced to a simple lobe and an even smaller appendage is present on the ninth thoracic somite. (Illustration shown in chapter introduction.)

DISTRIBUTION
Virginia, Long Island Sound, and Buzzards Bay, in the United States; northwestern Atlantic Ocean continental slope.

HABITAT
Fine mud with flocculent organic matter.

BEHAVIOR
Lives in the upper few millimeters of the sediment.

FEEDING ECOLOGY AND DIET
Feeds on very fine organic detrital particles.

REPRODUCTIVE BIOLOGY
Developmental sequence consists of 19 molt stages.

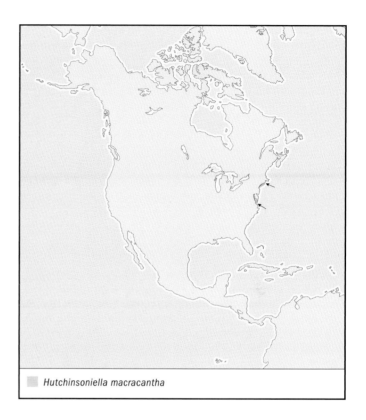

Hutchinsoniella macracantha

CONSERVATION STATUS
Not listed by the IUCN.

SIGNIFICANCE TO HUMANS
None known. ◆

Resources

Books
Schram, F. *Crustacea.* Oxford, U.K.: Oxford University Press, 1986.

Periodicals
Hessler, R. R., and H. L. Sanders. "Two New Species of *Sandersiella*, Including One from the Deep Sea." *Crustaceana* 13 (1973): 181–196.

Sanders, H. L. "The Cephalocarida, a New Subclass of Crustacea from Long Island Sound." *Proceedings of the National Academy of Sciences* 41 (1955): 61–66.

Les Watling, PhD

Anostraca
(Fairy shrimps)

Phylum Arthropoda

Subphylum Crustacea

Class Branchiopoda

Order Anostraca

Number of families 8

Thumbnail description
Lower crustaceans with elongated bodies and paired eyes on stalks; the body lacks a carapace (hard or bony shell)

Photo: Fairy shrimps are fresh water residents and have 11 pairs of legs. (Photo by Dr. William J. Johoda/Photo Researchers, Inc. Reproduced by permission.)

Evolution and systematics

On the basis of evidence from the fossil order Lipostraca, and the Upper Cambrian species *Rehbachiella kinnekullensis*, the anostracan line apparently split off at a very early stage from the rest of the Branchiopoda, about 500 million years ago. Fairy shrimps are widely considered the most primitive living crustaceans. Currently, scientists count eight families in two suborders within the Anostraca. This number is lower than the 11 families listed in older reference works. The change in taxonomy is the result of molecular investigations, which led researchers to group the families Artemiidae and Parartemiidae together to form the suborder Artemiina; and combine the families Chirocephalidae, Branchinectidae, Thamnocephalidae, Branchipodidae, Tanymastigidae, and Streptocephalidae to form the suborder Anostracina. DNA analysis also showed that the (sub)arctic genera *Polyartemia* and *Polyartemiella*, which have 17 and 19 pairs of limbs respectively as opposed to the usual 11, do not necessarily represent the most primitive anostracans. Rather, these genera may have resulted from mutations in homeobox genes, which are DNA sequences whose function is to divide an embryo into bands of tissue that will develop into specific organs. In females of the genus *Parartemia*, the eleventh pair of limbs may be reduced in size or completely missing.

The ancestral form *Rehbachiella* is known from horizontally banded limestone nodules found in the Orsten formation on the Baltic shore of Sweden. It has a rather large (more than 11) number of limbs, but no or few free abdominal segments. Unlike modern Anostraca, *Rehbachiella* had paired but sessile (located within the head) eyes. Even in extant fairy shrimps, however, rare cyclopic (one-eyed) mutants occasionally occur. The primitive ancestral anostracan may thus have had two sessile eyes that merged into a single median eye, which separated again at a later stage of evolution into two eyes on stalks.

Physical characteristics

Fairy shrimps are medium-sized branchiopods, usually 0.39–1.18 in (1–3 cm) long; but a few raptorial species, such as *Branchinecta gigas* may grow as long as 3.9 in (10 cm). The name of the order comes from two Greek words meaning "without" and "piece of hard tile." The fairy shrimp's thoracic limbs are flattened and leaflike, without true joints; the body lacks a carapace (hard or bony shell). Typical anostracans have 11 pairs of limbs, but some atypical species may have as many as 10, 17, or 19 pairs. One peculiar feature of all anostracan species is that they swim upside down. Some

Brine shrimps (*Artemia franciscana*) are a very important part of the ecosystem in saltwater and brackish habitats. (Photo by E. R.Degginger/Photo Researchers, Inc. Reproduced by permission.)

are largely translucent and hard to spot in the water; others, however, may develop bands or zones of bright color. The ovisac of females is often deep orange, red, or blue, and the rays in the branches of the tail may also have a distinctive color. The entire animal may develop a bright red or orange color.

The sexes are separate, except in some strains of *Artemia*, which may be parthenogenetic. Males have modified second antennae that serve to grasp females; the antennae may also demonstrate structural complexity. Females carry a ventral egg sac that may be either short and broad or long and thin, depending on the genus and species.

The eggs or cysts of anostracans are noteworthy because they are surrounded by a thick wall that allows them to resist drought and high temperatures. They develop into a gastrula containing about 4,000 cells, and then stop developing in order to survive adverse conditions. This stage of latency may continue for long periods of time, possibly more than a century in some strains of *Artemia*. On the other hand, *Artemia* is the only known genus in which viviparity may occur. In

some freshwater streptocephalids, as many as a third of the cysts produced may hatch shortly after being shed and bypass the resting stage.

Distribution

There are no extant marine fairy shrimps, but some species may occur in mountain lakes with almost pure water, while others—mainly *Artemia*—occur in saturated brine. In the Artemiina, the distribution of *Artemia* and *Parartemia* species used to be complementary. *Artemia* occurred in bodies of salt water on all continents except Australia, and *Parartemia* only in Australia. In the twentieth century, however, several species of *Artemia* were successfully introduced in various parts of Australia.

Most families of anostracans are found on three or four continents, but their ranges are often restricted to parts of a continent at the subfamily or genus level. For example, in the Thamnocephalidae, the genus *Thamnocephalus* is restricted to North America and northern South America, while *Dendrocephalus* is found exclusively in South America. A peculiar type of disjunct (widely separated) distribution is seen in the genus *Branchipodopsis*, which is represented in southern Africa by more than 15 species. Elsewhere in Africa, however, it has been reliably recorded only on the horn of Africa and the island of Socotra. The genus also occurs in such widely separated locations as Oman, the Caspian basin, and the Mongolian plateau. At the species (and sometimes genus) level, ranges may be extremely small, often restricted to the type locality. Such is the case with several species of Californian *Branchinecta*. Other species with extremely small ranges are found further south in Baja California. Another example is the genus *Dexteria*, limited to the area around Gainesville, Florida, and probably extinct by now as a result of the city's development.

Habitat

The main habitat of fairy shrimps is rain pools. The pools may be filled by periodic and predictable rains, or only erratically at long intervals. A number of anostracans, however, have adapted to two different types of water bodies: high mountain lakes and arctic ponds on the one hand, and saline lakes on the other. The life cycle of fairy shrimps in Arctic or Antarctic ponds is not regulated by alternation between wetting and drying, but by alternation between freezing and thawing. In saline lakes and ponds, such species as *Artemia* spp., *Parartemia* spp., and selected *Branchinecta* may occur from the first inundation, when the salt concentration is low, up to the point of saturation. All these habitats have one common feature: they are free of fishes and other vertebrate predators.

Behavior

Fairy shrimps form swarms that can be quite conspicuous if the animals are brightly colored. On the other hand, many species prefer to live in argillotrophic lakes or pools. The term "argillotrophic" means that the body of water in question produces low levels of phytoplankton because the water is clouded by high levels of suspended clay particles. The animals themselves are responsible for the turbidity, by stirring up sediment from the lake bottom. The turbidity serves a double purpose: it resuspends particles of sediment that are potentially nutritive, and it creates an environment that protects the shrimp from insect and bird predators that hunt by sight.

Beside swimming upside down, fairy shrimp, like zooplankton, migrate vertically over a 24-hour period if they live in sufficiently deep pools; they usually come to the surface at night. Females tend to live below the males in the water column.

Feeding ecology and diet

The vast majority of fairy shrimps are filter feeders. They use specialized endites on their legs to collect food in the ventral food groove. The food is then pushed towards the mouth. Filter feeding allows them to collect particles as small as bacteria and as large as algal cells. Fairy shrimps can consume even rotifers, nauplii (crustacean larvae), and nauplii of their own species.

A few species, like *Branchinecta ferox* and *B. gigas* are true raptorial predators, which means that they are adapted to seize prey. They pursue, catch and eat prey the size of large cladocerans and copepods, and other smaller-sized fairy shrimps that share their habitat.

Reproductive biology

Except for some members of the genus *Artemia*, fairy shrimp are bisexual and oviparous, usually with marked structural differences between the sexes. In males, the second antenna is modified into a remarkably complex clasping organ that is used to hold the female during copulation. In addition to the clasping organ, male anostracans have two penes. Copulation may take place so rapidly as to be hardly visible to the unaided eye, as in most streptocephalids; or last for many hours as the couple swims around in tandem formation, as in *Artemia*.

Following copulation and internal fertilization, the eggs are deposited in an external brood pouch of variable shape.

Each batch, or clutch, may contain several hundred eggs, and a female may produce up to forty clutches in her lifetime; thus total fertility may reach 4,000 eggs per female. The eggs are usually shed freely into the water. They may either sink to the bottom, as is usually the case in freshwater species, or float on the surface, to be deposited eventually along the lake shore.

Conservation status

Anostracan species with wide geographic ranges are usually under little or no threat. In such densely inhabited areas as California's central valley, however, where there is intense competition between urban and agricultural development on the one hand and conservation efforts on the other, many vernal pools have either been drained already or are threatened by obliteration. To a lesser extent, the same is true of endemic species in Baja California and other arid regions of Mexico. Such Florida endemics as *Dexteria floridana* may already be extinct. The 2002 IUCN Red List includes 28 anostracan species: six are categorized as Critically Endangered; nine as Endangered; 10 as Vulnerable; one as Lower Risk/Conservation Dependent; one as Lower Risk/Near Threatened; and one as Data Deficient. In the United States, five fairy shrimp species are listed as endangered; in California three of these species have been provided with habitat in an attempt to protect them.

Significance to humans

The genus *Artemia* is of considerable economic importance. The cysts of this species are harvested, cleaned, dried, packed and sold as fish food in the aquarium business. The cysts are also used in industrial aquaculture to feed fish larvae. The Libyan Fezzan desert contains several spring-fed dune lakes that have turned saline with time. Small communities living around these lakes use *Artemia* as their main source of animal protein. The women collect and dry the shrimp. These communities are called *dawada* (worm eaters) by the surrounding Arab tribes.

A species of *Streptocephalus* and a species of *Branchinella* are found in the hills of northeastern Thailand. These anostracans are fished by local tribespeople and used in a variety of local dishes.

1. Sudanese fairy shrimp (*Streptocephalus proboscideus*); 2. Brine shrimp (*Artemia salina*). (Illustration by Patricia Ferrer)

Species accounts

Brine shrimp
Artemia salina

FAMILY
Artemiidae

TAXONOMY
Artemia salina Linnaeus, 1758, England; and a cluster of about 10 related species.

OTHER COMMON NAMES
French: Crevette primitive, singe de mer; German: Salzwasser Feenkrebs, Salzkrebs, Urzeitkrebs.

PHYSICAL CHARACTERISTICS
Brine shrimp refers to *Artemia salina* and about 10 other related species. They are rather small anostracans, reaching only 0.6 in (15 mm) in length. The color varies from almost hyaline and transparent to bright red. The male antenna is strongly modified, but is not sufficient to identify the species. Microcharacters are required for identification, as well as biochemical and molecular methods.

DISTRIBUTION
The original specimens of *A. salina* were sampled from salt works at Lymington, England, but that population has long been extinct. The genus is widespread in bodies of salt water on all continents, and was introduced to Australia in the twentieth century. (Specific distribution map not available.)

HABITAT
Natural or artificial salt lakes and salinas (saltwater marshes) worldwide.

BEHAVIOR
The behavior of this species is similar to that of Sudanese fairy shrimp.

FEEDING ECOLOGY AND DIET
Artemia is a small-particle filter feeder. It can be grown on algae, yeasts, and a wide variety of micronized inert particles.

REPRODUCTIVE BIOLOGY
Like all fairy shrimps, artemia develops rapidly. The time between the hatching of the nauplius larva and maturation is slightly longer than a week. *Artemia* is also noted for its reproductive flexibility: under favorable conditions, it produces clutches of eggs at close intervals. Some species and/or strains may reproduce parthenogenetically, while others are viviparous. *Artemia* is the only known genus of fairy shrimp that shows this degree of versatility in reproductive tactics.

CONSERVATION STATUS
Not listed by the IUCN. The typical locality of the true brine shrimp has long disappeared; consequently, there is some uncertainty as to what constitutes the true habitat of *Artemia salina*, although it is likely geographically widespread. *Artemia monica* (Verrill, 1869) is one member of restricted occurrence. This member of the genus is limited to Mono Lake in California, where measures have been taken to prevent wide fluctuations in the salinity of the lake.

SIGNIFICANCE TO HUMANS
It has long been known that the presence of *Artemia* spp. (and of *Parartemia* as well) improves salt production in brine pools. That principle is still widely applied in salt works. In addition, an industry of *Artemia* cyst harvesting has developed around large salt lakes (Great Salt Lake in Utah, Kara Bogaz Gol in the Caspian basin, and others), where these cysts float in large masses on the surface of the water. The cysts can be collected in nets or scooped up from the lake shores where they accumulate. These cysts are later hatched to feed fish larvae, either in industrial aquaculture or by aquarium hobbyists. There are also a few instances of direct human consumption of brine shrimp, the best-known example being that of the Dawada (worm eaters) tribes in the Fezzan desert of Libya. ◆

Sudanese fairy shrimp
Streptocephalus proboscideus

FAMILY
Streptocephalidae

TAXONOMY
Streptocephalus proboscideus (Frauenfeld, 1873), Khartoum, Sudan.

OTHER COMMON NAMES
English: Freshwater fairy shrimp.

PHYSICAL CHARACTERISTICS
S. proboscideus is a medium-sized species that may reach 1.2 in (3 cm) in length. The color varies from almost translucent to almost black. The Latin name of the species, *proboscideus*, means "with a proboscis," and refers to a median appendage on the front of the head between the antennae. The male antennae are the main feature used to identify this species.

DISTRIBUTION
Arid and semiarid regions of eastern Africa, from northern Sudan to southern Africa and Namibia. (Specific distribution map not available.)

HABITAT
The Sudanese fairy shrimp is known primarily from shallow, turbid rain pools in which it may form huge swarms. Its cysts may lie dormant in dry mud for several years.

BEHAVIOR
The Sudanese fairy shrimp is an active swimmer requiring a water temperature of 77°F (25°C) or higher. It may filter as much as 2.11 quarts (2 liters) of water every 24 hours. *S. proboscideus* may live as long as 9 months under laboratory conditions.

FEEDING ECOLOGY AND DIET
S. proboscideus is an omnivore that can filter out particles as small as yeast cells and unicellular algae at the lower end of the spectrum, and as large as 0.008 in (0.2 mm) at the upper end. This fairy shrimp will eat its own nauplii (larvae) and those of related species indiscriminately.

REPRODUCTIVE BIOLOGY

Under optimum feeding conditions, an *S. proboscideus* nauplius turns into a mature individual in less than two weeks. The females then produce clutches of 100–300 eggs every 23 days. An average female may produce 35–40 such clutches in her lifetime. She must copulate and be fertilized after each clutch.

CONSERVATION STATUS

Not listed by the IUCN.

SIGNIFICANCE TO HUMANS

S. proboscideus is widely used in toxicity assays of water samples. It is cultivated as of 2003 for test kits used to measure the presence of heavy metal compounds, pesticides, ethylene glycol, PCP, and similar contaminants in bodies of water. Attempts at cultivating *S. proboscideus* as well as the American species *Thamnocephalus platyurus* to produce dry cysts for the home aquarium market, however, have failed to be commercially viable. ◆

Resources

Books

Dumont, H. J., and S. V. Negrea. *Introduction to the Class Branchiopoda*. Leiden, The Netherlands: Backhuys Publishers, 2002.

Erikson, C., and D. Belk. *Fairy Shrimp of California's Puddles, Pools and Playas*. Eureka, CA: Mad River Press, 1999.

Hamer, M. "Anostraca." In *Guides to the Freshwater Invertebrates of Southern Africa*, Vol. 1, *Crustacea*, edited by J. A. Day, B. A. Stewart, I. J. De Moor, and A. E. Louw. Pretoria, South Africa: Water Research Commission, 1999.

Persoone, G., et al. *The Brine Shrimp* Artemia, 3 vols. Wetteren, Belgium: Universa Press, 1980.

Periodicals

Belk, D., and J. Brtek. "Checklist of the Anostraca." *Hydrobiologia* 298 (1995): 315–353.

Brendonck, L. "Redescription of the Fairy Shrimp *Streptocephalus proboscideus* (Frauenfeld, 1873)." *Bulletin de l'Institut Royal des Sciences Naturelles de Belgique* 59 (1990): 4957.

Henri Jean Dumont, PhD, ScD

Notostraca

(Tadpole shrimps)

Phylum Arthropoda

Subphylum Crustacea

Class Branchiopoda

Order Nostostraca

Number of families 1

Thumbnail description
Pale-colored or translucent crustacean with a flattened shield-like carapace that covers two-thirds of the body and a thin, elongated abdomen ending in paired tail filaments

Photo: The tadpole shrimp species *Triops cancriformis* is one of the oldest known animal species to still exist. (Photo by Andreas Hart/OKAPIA/Photo Researchers, Inc. Reproduced by permission.)

Evolution and systematics

Tadpole shrimps are considered living fossils because their basic body plan has not changed in the last 300 million years. In fact, some fossil species known from the Paleozoic are basically indistinguishable from modern types. The oldest fossil of a tadpole shrimp dates from the carboniferous Paleozoic, and *Triops cancriformis* is the oldest living animal species on the planet, with fossils of this same species dating from the Triassic. Given the innumerous changes that Earth has undergone during the last 300 million years, tadpole shrimps are considered biological marvels of survival and adaptation.

Notostracans belong to the subclass Calamanostraca, which, together with the subclasses Diplostraca and Sarsostraca, form the three living groups of the class Branchiopoda. The fossil order Kazacharthra dates from the Jurassic and also belongs to the subclass Calamanostraca; they are distinguished from the Notostraca mainly by the shape of their carapace, which has spikes on the margin, rendering them a fierce appearance. Kazacharthran fossils were found exclusively in the now Republic of Kazakhstan, of the former USSR, and some authors consider them a very specialized group of Notostracans. Tadpole shrimps share with the other groups of the Branchiopoda the characteristic "gilled feet," which are leaf-like and divided into lobes, each containing a gill plate.

Modern Notostracans are grouped into a single family: Triopsidae. Only two genera are recognized (*Triops* and *Lepidurus*), comprising 15 living species with 11 subspecies.

Physical characteristics

Tadpole shrimps are among the largest Branchiopoda, with most species ranging in size from 0.4 to 1.6 in (10–40 mm). Nevertheless, some species may be larger; *Triops cancriformis* has been reported to reach a length of 4.0 in (11 cm). The head and all the limb-bearing segments of the trunk of tadpole shrimps are covered by a large and shield-like carapace, which is dorsoventrally flattened. They possess a pair of sessile compound eyes on the anterior portion of this carapace just behind a group of two to four ocelli; just behind the eyes there is a nuchal or dorsal organ that may act in chemoreception. The antennae are reduced or absent. The abdomen is thin, elongated, and flexible with very reduced appendages. The telson bears two long protuberances known as caudal rami. The genus *Triops* may be easily distinguished from *Lepidurus* because the latter possesses an extended supraanal plate on the telson between the caudal rami, while in *Triops* this plate is reduced or absent. The body of tadpole shrimps is generally pale or translucent, although it may present a pink or reddish coloration due to the presence of hemoglobin in the haemolymph. The carapace varies in color depending on the species, ranging from silvery gray, yellowish, olive, dark brown, and sometimes mottled, rendering them well camouflaged.

Distribution

Tadpole shrimps can be found almost exclusively in freshwater ephemeral pools worldwide, except for Antarctica. One species, *Lepidurus arcticus*, inhabits permanent lakes in Norway and Greenland.

Habitat

With the single exception of *Lepidurus arcticus*, which inhabits a few lakes in Norway and Greenland and may coexist with one fish species, all other notostracans can be found exclusively in ephemeral pools where fish predation can be avoided. These pools may last only a few weeks during the rainy season, and tadpole shrimps are specially adapted to this kind of niche, developing extremely quickly and producing drought-resistant eggs that may remain dormant for decades, and which may require a period of drought before they are capable of hatching. In general, *Triops* species are mostly found in warmer and short-lasting pools, while *Lepidurus* are more common in cooler and longer-lasting pools. Tadpole shrimps are usually found swimming in the benthos of these pools.

Behavior

The most interesting aspect of tadpole shrimp behavior is their numerous reproductive strategies. Notostracans may reproduce sexually, with species either having two separate sexes or being hermaphrodites. They may also reproduce parthenogenetically, and most species are both parthenogenetic and sexual. There are also cases in which hermaphrodites must cross-fertilize with other hermaphrodites, and they may reproduce by parthenogenesis as well. In fact, most species of tadpole shrimps use more than one of these strategies, and different populations of the same species may use different reproductive strategies in a population-dependent manner. This plasticity of reproduction in tadpole shrimps is one of the reasons they have survived through the millenia.

Feeding ecology and diet

One of the key adaptations of tadpole shrimps is their capability to eat almost anything available in their restricted habitat. This adaptation is important in order to maintain the rapid development needed to colonize temporary ponds, requiring about 40% of their body mass in food per day. Notostracans are facultative detritus feeders, scavengers, or even predators; they eat anything from bacteria, algae, protozoa, lower metazoans, insect larvae, tender plant roots and shoots, and even prey on smaller tadpole shrimps, fairy shrimps, and amphibian tadpoles. Tadpole shrimps use their numerous appendages to channel food down a ventral groove between the mandibles.

Reproductive biology

Tadpole shrimps may reproduce sexually with populations composed of separate sexes (males and females), males and hermaphrodites, or solely hermaphrodites. The latter either self-fertilize or cross-fertilize one another. In most species,

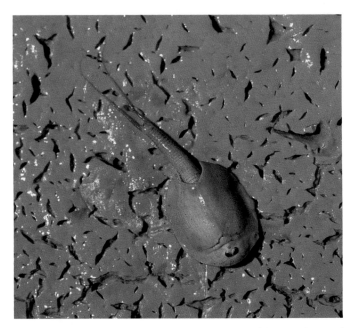

A shield shrimp (*Triops australiensis*) in drying mud in an ephemeral lake, near Charleville, western Queensland, Australia. (Photo by B. G. Thomson/Photo Researchers, Inc. Reproduced by permission.)

females or hermaphrodites also reproduce parthenogenetically, which is the most common means of reproduction in the Notostraca. During mating, the male holds the female above it while swimming in an upside-down position. Tadpole shrimps breed throughout their adult life.

The female or hermaphrodite carries the fertilized eggs in a brood pouch for several hours before dropping them in the water. These eggs are drought- and freeze-resistant and may remain dormant for up to several decades. Furthermore, through a mechanism that is not yet understood, a small percentage of these eggs may hatch shortly after being laid, while another portion may need only one period of drought before hatching, and some others may need two or more drought periods to hatch. In this way, tadpole shrimps increase the chances of their offspring surviving if any given pool should not last long enough to complete their development. In addition, the eggs are sensitive to light, osmotic pressure, and temperature, which allows them to hatch only in new (temporary) freshwater pools where predators can be avoided.

Tadpole shrimps hatch as nauplius larvae, and their development is extremely rapid, going through larval molts in as little as 24 hours if conditions are optimal. After each molt, the individual gains a pair of appendages. In most species, sexually mature tadpole shrimps may be found a couple of weeks after hatching.

Conservation status

The IUCN lists *Lepidurus packardi* as Endangered due to habitat destruction in California, where this species is endemic.

Significance to humans

Triops longicaudatus and *T. cancriformis* are considered pests of rice fields in countries where rice is directly planted. These tadpole shrimps may occur in enormous numbers, and they expose and eat the roots of rice seedlings while paddling through the mud in search of food. In Japan, however, the rice is planted as plantlets and *T. longicaudatus* is considered a biological control agent of weeds. Dormant cysts of some species of *Triops* are sold throughout the world in kits for rearing as aquatic pets.

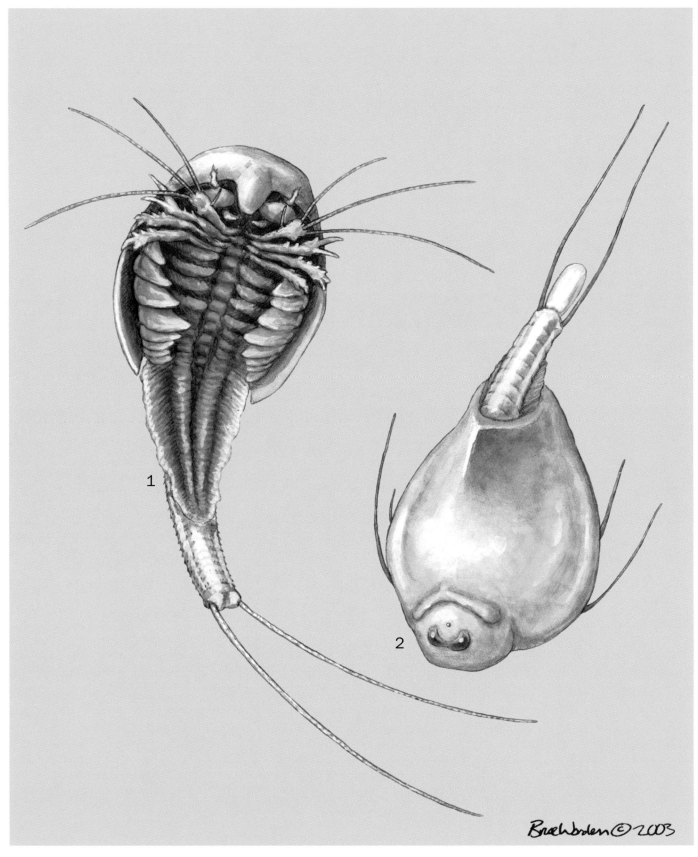

1. Vernal pool tadpole shrimp (*Lepidurus packardi*); 2. Longtail tadpole shrimp (*Triops longicaudatus*). (Illustration by Bruce Worden)

Species accounts

Vernal pool tadpole shrimp
Lepidurus packardi

FAMILY
Triopsidae

TAXONOMY
Lepidurus packardi Simon, 1886, California, United States. No subspecies recognized.

OTHER COMMON NAMES
None known.

PHYSICAL CHARACTERISTICS
Large tadpole shrimp, reaching up to 1 in (2.5 cm) in length, with a long pad-like supraanal plate on the telson between the caudal rami. Its carapace is olive or gray, sometimes mottled, allowing it to blend in with aquatic plants. Females can be distinguished from the males by the presence of ovisacs, also known as foot capsules, attached to the eleventh pair of gilled feet.

DISTRIBUTION
Endemic to the northern Central Valley of California (United States), where it is locally abundant and widespread in spite of the losses sustained to its vernal pool habitat.

HABITAT
Found in a variety of natural and artificial pond habitats, such as vernal pools, swales, ephemeral drainages, stock ponds, reservoirs, ditches, backhoe pits, and ruts caused by vehicular activity.

BEHAVIOR
During the evening hours, when the oxygen concentration of the water is low, *L. packardi* is usually found foraging at the water's surface on blades of grasses. During the day, it spends most of its time stirring the muddy pool bottom searching for prey.

FEEDING ECOLOGY AND DIET
Larval stages of the vernal pool tadpole shrimp are probably obligate filter feeders. However, adult *L. packardi* also prey actively on insect larvae, lower metazoans, other small crustaceans, and even smaller vernal pool tadpole shrimps. The food is collected by their gilled appendages while scavenging over vegetation or paddling through the mud.

REPRODUCTIVE BIOLOGY
Adult *L. packardi* are dioecius (males and females). The eggs carried by the female are orange in color. Cysts usually hatch after at least one drought and two to four days after rehydration. The nauplius larva develops into an adult after six or seven weeks, depending on food availability and temperature. This species reproduces only during the rainy season; the shrimps die out during the dry season or periods of drought.

CONSERVATION STATUS
Lepidurus packardi is listed as Endangered by the IUCN. The main reason for decline seems to be the elimination and degradation of its vernal pool habitat due to agricultural and urban development; other reasons include grazing, off-road vehicle

Lepidurus packardi
Triops longicaudatus

use, and hydrologic modification. On September 19, 1994, the U.S. Fish and Wildlife Service listed *L. packardi* as endangered because of its limited distribution, the small number of remaining populations, and the amount and nature of threats to this species. The U.S. Federal Government recognized this species as threatened once scientific research demonstrated that the vernal pool tadpole shrimp could go extinct without the protection afforded by the Endangered Species Act.

SIGNIFICANCE TO HUMANS
None known. ◆

Longtail tadpole shrimp
Triops longicaudatus

FAMILY
Triopsidae

TAXONOMY
Triops longicaudatus LeConte, 1846, United States. Some authors recognize two subspecies: *Triops longicaudatus longicaudatus* and *Triops longicaudatus intermedius*, but these have not been officially accepted.

OTHER COMMON NAMES
English: Rice tadpole shrimp, American tadpole shrimp; Spanish: Tortugueta ("little turtle").

PHYSICAL CHARACTERISTICS
Large tadpole shrimp reaching a length of up to 1.5 in (4.0 cm). In this species, the second maxilla is absent and there is no anal plate.

DISTRIBUTION
North America (including Hawaii but not Alaska), Central America, South America, Japan, West Indies, Galápagos Islands, and New Caledonia.

HABITAT
This is the most widespread species of notostracan, being found in a variety of temporary waters, including rice fields.

BEHAVIOR
Usually found scratching the soil surface in search of benthic food. When oxygen levels in the water are low, they will swim upside-down close to the surface of the water.

FEEDING ECOLOGY AND DIET
Omnivorous; it may eat detritus, scavenge dead organisms in its environment, or actively prey upon other animals, such as protozoa, insect larvae, other small crustaceans, and even cannibalize its siblings. In order to obtain food, adults actively paddle through the soil surface.

REPRODUCTIVE BIOLOGY
T. longicaudatus is known to exhibit several major reproductive strategies: individuals can be sexual (either male or female with populations that are either half male and half female or are female-biased), parthenogenetic, or hermaphroditic. What is interesting about *T. longicaudatus* is that different populations exhibit a different reproductive strategy or combination of strategies. This fact suggests that these different reproducing populations might be considered different subspecies or species in the future.

CONSERVATION STATUS
Not listed by the IUCN.

SIGNIFICANCE TO HUMANS
Triops longicaudatus is also known as the rice tadpole shrimp and is considered a pest species of rice fields where the rice is germinated in the field (mainly in the United States and Spain). *T. longicaudatus* damages the roots and leaves of seedling rice plants and also muddies the water, impeding the access of sunlight to the developing plants. In Japan, where the rice is transplanted, the long tail tadpole shrimp cannot harm the rice because it is too big and resistant to its attack. Instead, Japanese rice farmers use *T. longicaudatus* as a biological control agent against weeds. Dried cysts of *T. longicaudatus* are sold in kits to be bred as aquatic pets. ◆

Resources

Books
Bliss, Dorothy. E. *Biology of the Crustacea.* New York: Academic Press, 1982–1985.

Pennak, Robert W. *Fresh-Water Invertebrates of the United States.* New York: John Wiley and Sons, 1978.

Schram, Frederick R. *Crustacea.* New York: Oxford University Press, 1986.

Periodicals
Ahl, J. S. B. "Factors Affecting Contributions of the Tadpole Shrimp, *Lepiduris packardi,* to Its Oversummering Egg Reserves." *Hydrobiologia* 212 (1991): 137–143.

Goettle, Bradley. "Living Fossil in the San Francisco Bay Area?" *Tideline* 16, no. 4 (1996): 1–3.

Linder, F. "Contributions to the Morphology and Taxonomy of the Branchiopoda Notostraca, with Special Reference to

the North American Species." *Proceedings of the U.S. National Museum* 102 (1952): 1–69.

Longhurst, A. R. "A Review of the Notostraca." *Bulletin of the British Museum of Zoology* 3 (1955): 1–57.

Other
Eder, Erich. "Large Branchiopods—Living Fossils!" 17 April 2003 [27 July 2003]. <http//mailbox.univie.ac.at/erich.eder/UZK/index2.html>.

NatureServe. *NatureServe Explorer: An Online Encyclopedia of Life.* 1 July 2003 [27 July 2003]. <http//www.natureserve.org/explorer>.

University of California Museum of Paleontology. "Introduction to the Branchiopoda." 1 July 2003 [27 July 2003]. <http//www.ucmp.berkeley.edu>.

Johana Rincones, PhD
Alberto Arab, PhD

Conchostraca
(Clam shrimps)

Phylum Arthropoda

Subphylum Crustacea

Class Branchiopoda

Order Conchostraca

Number of families 5

Thumbnail description

Shrimps that possess a bivalved shell with a hinge that envelops the entire body and limbs; they are found exclusively in freshwater

Photo: Clam shrimps (*Limnadia lenticularis*) are found in temporary puddles, sometimes dug into the mud at the bottom of the puddle. (Photo by Andreas Hartl/OKAPIA/Photo Researchers, Inc. Reproduced by permission.)

Evolution and systematics

The fossil record of the Conchostraca is one of the oldest among the Branchiopoda, extending from the Lower Devonian. This record consists mostly of fossils of carapaces, and since the major diagnostic features of the clam shrimps relate to this carapace, these fossils are easily classified as conchostracans. Nevertheless, these characteristics do not provide details of the evolution within the group.

The Conchostraca and Cladocera belong to the subclass Diplostraca, which, together with the subclasses Calamanostraca and Sarsostraca, form the three living groups of the class Branchiopoda; two fossil groups, Kazacharthra and Lipostraca, also belong to this class. The term Branchiopoda literally means "gilled feet," and branchiopods are consequently said to breathe through their feet, which are leaflike and divided into lobes, each containing a gill plate. The presence of gills on the feet of the Branchiopoda is almost the only common characteristic among the diverse members of this group, although minor similarities can be found in the organization of the trunk segments and trunk limbs.

In 1986 the conchostracans were divided into five families of bivalve crustaceans: Lynceidae, Limnadiidae, Cyclestheriidae, Cyzicidae, and Leptestheriidae. However, the relationships among these families are not well understood and many authors believe that these relationships may not even exist. In fact, the validity of the term Conchostraca as a taxonomic name is debatable, since no evidence has been found that supports a monophyletic origin for clam shrimps. Nevertheless, up to the present, no author has presented a phylogenetically supported or generally accepted reclassification for this group.

Physical characteristics

Conchostracans are commonly known as clam shrimp due to their outward resemblance to bivalve mollusks. All conchostracans are laterally flattened and possess a bivalve carapace, joined by a dorsal hinge or fold with the two halves connected and controlled by a strong adductor muscle. This carapace covers the entire body and limbs (or almost), and, in most cases, is marked by concentric lines of growth. The trunk is divided into 10–32 segments, each with a pair of appendages. The second pair of antennae is well developed and biramous, and the compound eyes are sessile. Conchostracans can be distinguished from other groups of the Branchiopoda because their bodies are completely enclosed within the carapace and because they possess a comparatively reduced abdomen. They range in size from a few millimeters to up to 0.7 in (1.7 cm). Their color is usually translucent to pale, although some species might present a pink, reddish, or orange coloration because of the presence of hemoglobin in the haemolymph; the shell color varies from translucent to dark brown.

Distribution

Modern conchostracans occur exclusively in freshwater bodies on all continents, except Antarctica and northern polar regions. Some extinct species apparently inhabited marine habitats during the Devonian and Carboniferous.

Habitat

Clam shrimps are commonly found in temporary water bodies such as ephemeral ponds and troughs. Nevertheless, there have been reports of members of the family Lynceidae in prairie streams, while the family Cyclestheriidae mainly inhabits permanent bodies of water that are always associated with a thick algal mat. Clam shrimps spend most of their time on the bottom of these water bodies filtering nutritious particles, and they can frequently be found burrowed in the mud.

Behavior

Clam shrimps use their second antennae in addition to their legs for swimming, sometimes in an upside-down position, performing spiral or staggered movements. They can reproduce sexually, asexually (via parthenogenesis), or by both means. The females carry several hundred eggs attached to a specialized appendix; these eggs are usually shed when the female molts, although in some species, the eggs hatch in a brood pouch attached to the carapace. Some species produce drought-resistant eggs that can be dispersed by water or wind and are widely distributed.

Feeding ecology and diet

Conchostracans are generally acknowledged as filter feeders, but they can also scrape and tear at their food, and will scavenge almost any organism in their environment. Some species also scrape materials from rocks or other substrates. Conchostracans use their forefeet to collect food, while the hind appendages are modified as mandibles for grinding large food particles.

Reproductive biology

During copulation, the male clam shrimp grasps the ventral edge of the female shell and deposits a spermatophore in-side the female's carapace by extending his abdomen. The eggs are brooded inside the female's shell and are shed at the time of the female molt. Breeding is constant throughout the adult life of clam shrimps, and the female sheds eggs with every molt. Conchostracans usually produce two kinds of eggs: vegetative and asexual eggs, known as summer eggs, and more resistant, dry-season or winter eggs that are usually sexually produced. The summer eggs possess a thin shell and they might develop parthenogenetically or be fertilized; their development is rapid and they may hatch while still attached to the female. The winter eggs possess a thick shell and, in some species, they can remain dormant for long periods of time. Winter eggs are produced in lesser numbers and their production might be stimulated by external factors such as population density, temperature, and photoperiod. With the exception of the family Cyclestheriidae, the eggs of clam shrimps hatch as a nauplius larva that develops quickly into an adult without undergoing metamorphosis, although the acquisition of the carapace gives a false impression of metamorphosis. The family Cyclestheriidae reproduces mostly by parthenogenesis, with the eggs hatching in a brood pouch attached to the female and developing into replicas of the adult before being released without a naupliar stage.

Conservation status

No species of clam shrimp are listed by the IUCN. Some endemic species have limited distributions, but none have been considered endangered so far, probably because this group has been poorly studied.

Significance to humans

There is no known significance to humans.

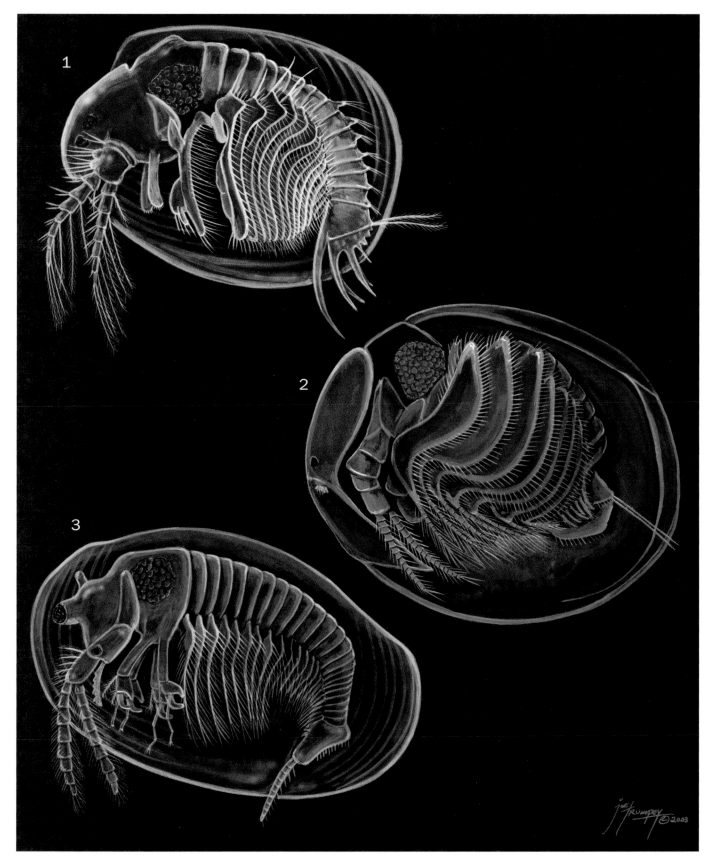

1. *Cyclestheria hislopi*; 2. Graceful clam shrimp (*Lynceus gracilicornis*); 3. Texan clam shrimp (*Eulimnadia texana*). (Illustration by Joseph E. Trumpey)

Species accounts

No common name
Cyclestheria hislopi

FAMILY
Ciclestheriidae

TAXONOMY
Estheria hislopi Baird, 1859, India.

OTHER COMMON NAMES
None known.

PHYSICAL CHARACTERISTICS
Shares some morphological features with the cladocerans, such as fused compound eyes, modified antennules, and a specialized brood pouch, leading some authors to believe that this species might be a missing link between conchostracans and cladocerans.

DISTRIBUTION
Presents a pantropical distribution, being restricted to a latitude between 30°N and 35°S in Asia, Africa, Australia, and Central and South America.

HABITAT
The natural habitat differs from that of other conchostracans because it mostly inhabit permanent bodies of water, where it is always associated with certain species of algae and other aquatic vegetation.

BEHAVIOR
Characterized by the rarity of males; reproduces primarily by parthenogenesis, and there have only been reports of males from four sites worldwide.

FEEDING ECOLOGY AND DIET
Feeds mainly by filtering detritus and plankton from its environment.

REPRODUCTIVE BIOLOGY
Populations are mostly parthenogenetic and composed almost entirely of hermophrodites. Males are extremely rare, arising only when physical conditions become unfavorable. The eggs undergo direct development in the brood chamber of the female and hatch as miniature replicas of the adults instead of the nauplius larva that is typical of other conchostracans.

CONSERVATION STATUS
Not listed by the IUCN.

SIGNIFICANCE TO HUMANS
None known. ◆

Cyclestheria hislopi
Lynceus gracilicornis
Eulimnadia texana

Texan clam shrimp
Eulimnadia texana

FAMILY
Limnadiidae

TAXONOMY
Eulimnadia texana Packard, 1871, Texas, United States.

OTHER COMMON NAMES
None known.

PHYSICAL CHARACTERISTICS
Present a pronounced sexual dimorphism; most notable characteristic is the modification of male's first pair of appendages into claw-like claspers that are used to hold onto the margins of a hermaphrodite's carapace during mating.

DISTRIBUTION
Restricted to southern United States, west of the Mississippi River and north of Mexico.

HABITAT
All types of ephemeral freshwater bodies.

BEHAVIOR
Common to all clam shrimps.

FEEDING ECOLOGY AND DIET
Omnivorous; able to filter feed as well as forage along pond bottoms.

REPRODUCTIVE BIOLOGY
Populations are usually composed mainly of hermaphrodites with the percentage of males ranging from 0–40% in natural populations. Hermaphrodites are able to produce drought-resistant cysts that are carried in a brood chamber inside the carapace. These eggs are usually released into a burrow dug by the hermaphrodites. Develops extremely fast, with an individual reaching reproductive size in 4–6 days under natural conditions.

CONSERVATION STATUS
Not listed by the IUCN.

SIGNIFICANCE TO HUMANS
None known. ◆

Graceful clam shrimp
Lynceus gracilicornis

FAMILY
Lynceidae

TAXONOMY
Lynceus gracilicornis Packard, 1871, Texas, United States.

OTHER COMMON NAMES
None known.

PHYSICAL CHARACTERISTICS
Large clam shrimp, presenting a body coloration ranging from orange to rose and a dark maroon shell; eggs carried by the female are yellow to orange. This species distinguished from other members of the Lynceidae because males bear a pair of dimorphic claspers, with the right clasper being much larger than the left; also characterized by the absence of growth marks on the carapace.

DISTRIBUTION
Texas and northern Florida; however, it is suspected to inhabit other regions in between.

HABITAT
Usually found in the shallow grassy parts of natural or artificial temporary ponds. Some individuals might be found in deep water when oxygen levels are high.

BEHAVIOR
Swims upside-down or on its side using its feet and antennae for backward propulsion.

FEEDING ECOLOGY AND DIET
Feeds on plankton that it collects while swimming.

REPRODUCTIVE BIOLOGY
Male clasps the lower border of the female's shell and swims while holding the female above him. Females may carry up to 200 eggs in a cohesive mass that is readily visible through the carapace.

CONSERVATION STATUS
Not listed by the IUCN.

SIGNIFICANCE TO HUMANS
None known. ◆

Resources

Books
Bliss, Dorothy E. *Biology of the Crustacea*. New York: Academic Press, 1982–1985.

Schram, Frederick R. *Crustacea*. New York: Oxford University Press, 1986.

Periodicals
Martin, J. W., B. E. Felgenhauer, and L. G. Abele. "Redescription of the Clam Shrimp *Lynceus gracilicornis* (Packard) (Branchiopoda, Conchostraca, Lynceidae) from Florida, with Notes on Its Biology." *Zoologica Scripta* 15, no. 3 (1986): 221–232.

Olesen, J., J. W. Martin, and E. W. Roessler. "External Morphology of the Male of *Cyclestheria hislopi* (Baird, 1859) (Crustacea, Branchiopoda, Spinicaudata) with a Comparison of Male Claspers Among the Conchostraca and Cladocera and Its Bearing on Phylogeny of the Bivalved Branchiopoda." *Zoologica Scripta* 25, no. 4 (1996): 291–316.

Other
Boyce, Sarah Lyn. "The Conchostraca." February 12, 2003 [July 24, 2003]. <http://crustacea.nhm.org/peet/conchostraca/index.html>

Eder, Erich. "Large Branchiopods—Living Fossils!" June 24, 2003 [July 24, 2003]. <http://mailbox.univie.ac.at/erich.eder/UZK/index2.html>

"Introduction to the Branchiopoda." University of California Museum of Paleontology. July 1, 2003 [July 24, 2003]. <http://www.ucmp.berkeley.edu/arthropoda/crustacea/branchiopoda.html>

Johana Rincones, PhD
Alberto Arab, PhD

•
Cladocera
(Water fleas)

Phylum Arthropoda
Subphylum Crustacea
Class Branchiopoda
Order Cladocera
Number of families 15

Thumbnail description
Small transparent zooplankton found in bodies
of freshwater throughout the world; a few
species inhabit marine waters

Photo: Photomicrograph of a gravid female water
flea (*Daphnia magna*) carrying eggs. (Photo by Holt
Studios/Nigel Cattlin/Photo Researchers, Inc. Reproduced by permission.)

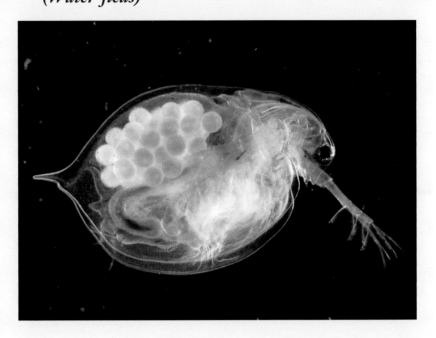

Evolution and systematics

Cladocerans remain well-preserved in aquatic sediments because their shells are composed of chitin, a white semi-transparent horny substance. Fossil evidence suggests that the major evolutionary development of the cladocerans occurred by the late Paleozoic or early Mesozoic era, approximately 250 million years ago. The taxonomy of the cladocerans, however, is still controversial. Most scientists recognize four suborders, 10–15 families, about 80 genera, and over 400 species.

Physical characteristics

Cladocerans are small animals that are often shaped like flat disks. Most range in size from 0.008 to 0.1 in (0.2–3 mm), although one species, *Leptodora kindtii* can grow as large as 0.7 in (18 mm) in length. Their bodies are not clearly segmented like those of other crustaceans; however, three parts can be distinguished—head, thorax, and abdomen. The head is typically dome-shaped with large compound eyes and five pairs of appendages. Among the appendages are two pairs of antennae—a small pair that serves a sensory function and a larger pair that is used for swimming. The other three pairs of appendages function in securing food.

A thin transparent shell called the carapace encloses the thorax and abdomen of most cladocerans. The carapace may be patterned; some species have stiff spines protruding from their shells. The thorax holds four to six pairs of leaflike legs that are used for gathering food, filtering water, or grasping mates during copulation. The cladoceran digestive tract is contained within the thorax and abdomen. A pair of claws

used for cleaning the thoracic legs may extend from the end of the carapace.

Distribution

Cladocerans are distributed worldwide.

Habitat

Most cladocerans are restricted to freshwater habitats, although eight species are found in marine waters. Cladocerans abound in lakes, ponds, slow-moving streams, and rivers. They are found from the Arctic to the Southern Ocean, from sea level to alpine ponds. Some cladocerans are benthic (found on the bottom of a lake, sea, or ocean); other species live on sediment or in vegetation virtually anywhere water is present, including swamps, puddles, ditches, and ground water. One species even lives in water trapped by mosses living on trees in a Puerto Rican cloud forest.

Behavior

Cladocerans migrate up and down in the water column on a daily cycle. They typically come to the surface at night and move to lower depths in the daytime. By seeking deeper waters during the day, cladocerans are protected by the darkness from such visual predators as fishes; they then maximize growth by moving to warmer surface waters at night. Occasionally, large swarms of cladocerans are observed at the surface during the day, typically as a response to light and the threat of predatory fishes. When cladocerans encounter predators, pelagic species often swim away rapidly. If caught by small

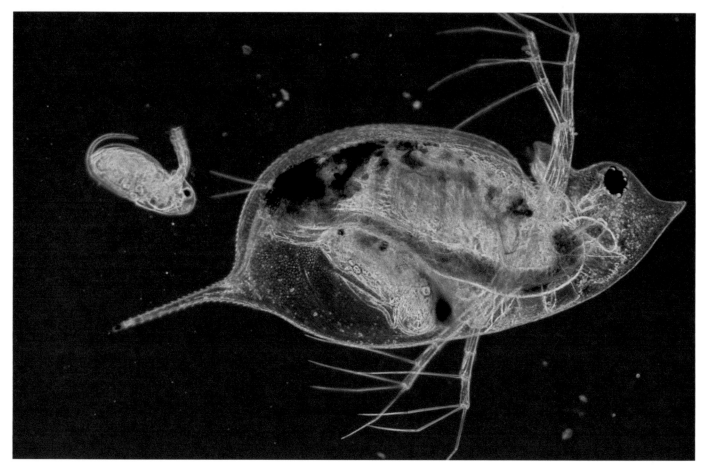

A water flea (*Daphnia magna*) giving birth. (Photo by M. I. Walker/Science Source/Photo Researchers, Inc. Reproduced by permission.)

predators, some cladocerans fight to break free; if they escape, they stop moving and sink deeper into the water column.

Feeding ecology and diet

Cladocerans use their thoracic legs to produce a constant current of water that allows them to filter food particles. The food items are collected in a groove at the base of their legs and mixed with mucus to form a bolus or mass that is moved forward towards the mouth. Because most cladocerans are filter feeders, they eat organic detritus of all kinds, including algae, protozoans, and bacteria. Although cladocerans cannot select individual food items, they can decide whether to accept or reject the mass of food collected in their mucus; they may reject the bolus if it contains toxic algae or other undesirable particles. A few genera of cladocerans are predatory and capable of seizing their prey, which consists primarily of other zooplankton. Because of their small size, cladocerans have a wide variety of predators, including fishes, amphibians, birds, predaceous zooplankton, and insects.

Reproductive biology

Cladocerans can reproduce both sexually and asexually; the mode of reproduction depends on environmental conditions, but most species reproduce asexually most of the time. Re-

production begins as water temperatures rise in the spring, but the population declines in the summer due to overcrowding and competition for food. A second peak of reproduction and population growth may occur in autumn.

Cladoceran eggs develop in about two days in the female's brood chamber, which is located between the body and the carapace. In most cladoceran species, the clutch size increases with body size, while the development time of the eggs is inversely related to temperature. Some eggs produced asexually develop into young immediately. If conditions are unfavorable, both sexually- and asexually-produced eggs may enter a resting state called diapause, during which they are resistant to heat, desiccation (drying out), and freezing. Most eggs develop into females. The production of male offspring is triggered by such environmental signals as crowding, changes in food concentrations, or decreasing day length.

During episodes of sexual reproduction, males use their first antennae to locate female mates. After finding a receptive female, the male grasps the edge of her carapace. The female continues to swim, carrying the male with her. After some time, the swimming ceases and both extend their postabdominal region, presumably to allow the male to eject his sperm as close to the female's brood pouch as possible. Mating may take anywhere between 15 minutes and several hours to complete.

Conservation status

Cladocerans are common throughout the world, and distributions of most species are stable or expanding, often due to human activities that introduce non-native species to new areas. No species are considered threatened, and none are listed by the IUCN.

Significance to humans

Although cladocerans are not recognized by most humans, they are vital for sustaining functions that humans value in many aquatic ecosystems. In particular, cladocerans are a critical link in the food chain that enables aquatic habitats to support valuable species of fish.

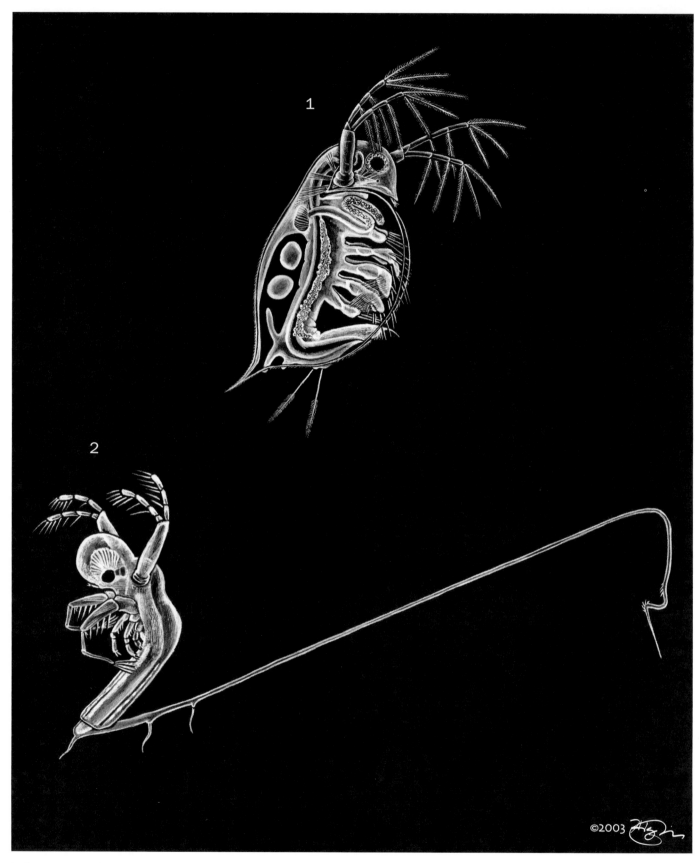

1. Common water flea (*Daphnia pulex*); 2. Fishhook water flea (*Cercopagis pengoi*). (Illustration by Jonathan Higgins)

Species accounts

Fishhook water flea
Cercopagis pengoi

FAMILY
Cercopagidae

TAXONOMY
Cercopagis pengoi Ostroumov, 1891, Caspian Sea.

OTHER COMMON NAMES
None known.

PHYSICAL CHARACTERISTICS
C. pengoi lacks the ventral (lower surface) portion of the carapace; as a result, its limbs are uncovered and its body is very mobile. The dorsal (upper surface) portion of the carapace remains and protects the brood cavity. The head of this cladoceran is rounded and contains strong mandibles (jaws) that are hardened with chitin. *C. pengoi* has only four pairs of thoracic legs, which are covered with setae (bristles) and spines. It has a long caudal (tail-like) appendage extending from its abdomen. Females are 0.05–0.08 in (1.2–2 mm) in length, and males are 0.04–0.06 in (1.1–1.4 mm) long; the caudal appendage may be 5–7 times the length of the body.

DISTRIBUTION
Native to the Caspian, Black, Azov, and Aral Seas; introduced to the Baltic Sea and the North American Great Lakes.

HABITAT
Inhabits brackish water as well as freshwater seas, lakes, and reservoirs.

BEHAVIOR
In its native range, this species migrates vertically on a daily basis; it lives at depths of 164–197 ft (50–60 m) during the day and rises to the surface layer at night. Young individuals are not found below 66–98 ft (20–30 m). Strong migration patterns are not observed in the Baltic Sea or Great Lakes.

FEEDING ECOLOGY AND DIET
C. pengoi is a predatory species; it actively catches prey and eats the tissues and soft body parts. Prey items include such other zooplankton as rotifers, copepods, protozoans, and other cladocerans. *C. pengoi* may be consumed by such larger planktivorous fish as herring and smelt.

REPRODUCTIVE BIOLOGY
C. pengoi reproduces asexually for most of the summer. Sexual reproduction takes place in late autumn.

CONSERVATION STATUS
Not listed by the IUCN.

SIGNIFICANCE TO HUMANS
The hook on the caudal appendage of *C. pengoi* often becomes tangled in fishing gear, fouling the equipment. In addition, *C.*

Cercopagis pengoi

pengoi competes with fish for food, and high populations of this species may reduce the number of fishes in a water body. Further, the fishhook water flea may intensify the eutrophication (overgrowth of algae resulting from an increased supply of nutrients) of water bodies by eating zooplankton that would normally graze on the algae. ◆

Common water flea
Daphnia pulex

FAMILY
Daphnidae

TAXONOMY
Daphnia pulex DeGeer, 1778.

OTHER COMMON NAMES
French: Puce d'eau commune; German: Wasserfloh.

PHYSICAL CHARACTERISTICS
D. pulex is a small laterally-flattened cladoceran. The body is covered by a carapace, but the head, large compound eye, and antennae remain open to view. The common water flea has two pairs of antennae and five pairs of leaf-like limbs on its body. The spine is located at the posterior end of the carapace. Individuals typically range in size from 0.008–0.1 in (0.2–3 mm).

DISTRIBUTION
Holarctic distribution across North America, Europe, and Asia; temperate zones of North America, South America, and Europe.

HABITAT
Freshwater lakes, ponds, rivers, and streams.

BEHAVIOR
Water fleas migrate vertically in the water column on a diel (regular 24-hour) basis. When predatory fishes are present in the water, *D. pulex* forms defensive structures, particularly small protrusions in the neck region called neckteeth.

FEEDING ECOLOGY AND DIET
D. pulex eats algae and such small zooplankton as protozoans or rotifers. In turn, water fleas are consumed by small fishes and predatory insects.

REPRODUCTIVE BIOLOGY
D. pulex reproduces asexually during the spring and early summer. Later in the season, some asexually-produced eggs become males, a change that allows sexual reproduction to begin. Eggs produced by sexual reproduction are covered in thick shells. Females have 3–9 young per brood, and these young mature in 6–8 days after leaving the brood pouch.

CONSERVATION STATUS
Not listed by the IUCN.

SIGNIFICANCE TO HUMANS
The common water flea is commonly maintained in laboratories and often used to test for toxic substances in water. ◆

Daphnia pulex

Resources

Books

Thorp, J. H., and A. P. Covich, eds. *Ecology and Classification of North American Freshwater Invertebrates.* 2nd ed. New York: Academic Press, 2001.

Periodicals

Benoit, H. P., O. E. Johannsson, D. M. Warner, W. G. Sprules, and L. G. Rudstam. "Assessing the Impact of a Recent Predatory Invader: The Population Dynamics, Vertical Distribution, and Potential Prey of *Cercopagis pengoi* in Lake Ontario." *Limnology and Oceanography* 47, no. 3 (2002): 626–635.

Katherine E. Mills, MS

Phyllocarida
(Leptostracans)

Subphylum Arthropoda

Class Malacostraca

Subclass Phyllocarida

Number of families 3

Thumbnail description
Small crustaceans with a laterally flattened carapace enclosing the bases of the thoracopods (legs), a hinged rostrum covering stalked eyes, and a tapering abdomen ending in a forked tail

Illustration: *Levinebelia maria*. (Illustration by Dan Erickson)

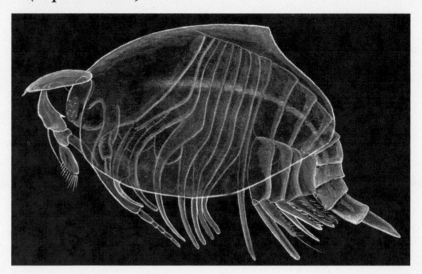

Evolution and systematics

Only one fossil leptostracan has been recognized in the literature, the Permian *Rhabdouraea bentzi*. It has a separate family status (Rhabdouraeidae; Schram and Malzahn, 1984). However, the fossil is only that of an abdomen and makes comparison to other leptostracans difficult. The regular segmentation of the body, the trunk musculature, nervous system, circulatory system, presence of a furca, and absence of uropods indicate that the Leptostraca are primitive malacostracans. Three extant families are recognized: Nebaliidae (with genera *Nebalia*, *Nebaliella*, *Dahlella*, and *Speonebalia*); Nebaliopsidae (with only one genus, *Nebaliopsis*, and only one species, *N. typica*); and Paranebaliidae, containing genera *Paranebalia* and *Levinebalia*. Almost 35 species are described. There is not a common name for the leptostracans, but names such as "leaflike-shrimps," "mud-shrimps," and "sea fleas," are used in the literature.

Physical characteristics

Leptostracans have a bivalved carapace (the two valves are fused dorsomedially but have no hinge) enclosing head and thorax, or enclosing head, thorax, and part of the abdomen, laterally compressed and smooth. The anterior end of the carapace dorsum continues as a movable articulated rostrum. The eyes are stalked, compound, with visual elements either present or absent (genus *Dahlella*). The antennules (antenna one) are usually biramous and the antennae (antenna two) are uniramous. The eight pairs of thoracic limbs (thoracopods) are leaflike turgor appendages, foliaceous (phyllopodous), and biramous. The abdomen has six pairs of appendages (pleopods): the first four pairs are birramous and large, and the last two are uniramous and reduced. The seventh abdominal segment continues as a telson having a long caudal furca.

Most leptostracans are 0.19–0.59 in (5–15 mm) long, but one species (*Nebaliopsis typica*) is a giant at nearly 1.5 in (4 cm) in length.

In life, specimens appeared mostly transparent (colorless), except for the bright red eyes. When viewed with the naked eye the thoracic region appears almost white. All leptostracans, except *Nebaliopsis typica*, show pronounced sexual dimorphism. Males also occur much more rarely than females. The carapace of the male is less deep than that of the female and the antennae are much longer.

Distribution

The genera of Leptostraca are distributed differently. *Nebalia* is cosmopolitan. *Nebaliella* is confined to cold waters, being found in Antarctica, southern Australia, and the high latitudes of the Northern Hemisphere. *Nebaliopsis* is a pelagic genus with a worldwide distribution. *Paranebalia* is found in central America, Bermuda, New Caledonia, and Australia. *Levinebalia* has been recorded only in Australia and New Zealand. *Dahlella* was collected from hydrothermal vents near the Galápagos, and *Speonebalia* has been recorded from only marine caves in the Turks and Caicos Islands.

Habitat

All leptostracans are marine, and have been recorded in waters from 3.2 ft (1 m) deep to more than 6,561 ft (2,000 m); most species occur in <656 ft (<200 m). Many leptostracans, with the exception of the pelagic *Nebaliopsis typica*, seem to prefer mud bottoms that are low in oxygen content. However, leptostracans can be found in a variety of habitats ranging from the intertidal zone to the abyssal deeps.

Behavior

Young animals and females rest in the same place for hours, their thoracic limbs beating rhythmically, moving respiratory currents of water through the carapace. The mature males may swim long distances propelled by the four anterior pairs of pleopods. Each of these pleopods is hooked to its partner so that each pair functions as a unit when swimming.

When placed in laboratory tanks or small, concave "dishes" of glassware (watch glasses) with mud, leptostracans will swim right to the bottom and burrow in. Frequently, not one part of the body is moved at all, and even the heartbeat slows down; these are functional and physiological adaptations to living in low-oxygen environments.

Feeding ecology and diet

Living in muddy environments, often where organic matter is abundant, it is assumed that leptostracans are filter feeders, but they are often taken in baited traps, which indicates that they act as at least opportunistic scavengers.

Most of the benthic leptostracans suspension feed by stirring up bottom sediments. They are also capable of grasping relatively large bits of food (detritus particles, dead animals) directly with the mandibles.

Reproductive biology

The eggs of all leptostracans, except possibly in *Nebaliopsis*, are carried by the female in the thoracopod brood chamber beneath her carapace. Development is embryonic and the young hatch as post larvae. These post-larval, or mancoid, stages differ from the adults in having a rudimentary fourth pleopod. The free-swimming, pelagic larva of *Nebaliopsis* has been found in plankton collections. In this genus, the eggs may not be carried in a brood pouch but laid directly into the water.

Water temperatures influence the length of time taken to reach maturity, the size at maturity, and the incubation time of young.

Transformations in the shape of the carapace, antennae, pleopods, and furca are gradual from molt to molt in immature and subadult males, and deviate from the female morphology, which generally remains unchanged except when reproductive.

Conservation status

Nothing is known. No species are listed by the IUCN.

Significance to humans

Nebalia bipes is considered to be desirable as a living fish food and for larval rearing because of its suitable size and tolerance to the deterioration of bottom conditions.

1. *Levinebalia maria*; 2. *Nebalia hessleri*; 3. *Dahlella caldariensis*. (Illustration by Dan Erickson)

Species accounts

No common name
Dahlella caldariensis

ORDER
Leptostraca

FAMILY
Nebaliidae

TAXONOMY
Dahlella caldariensis Hessler, 1984, Galápagos.

OTHER COMMON NAMES
None known.

PHYSICAL CHARACTERISTICS
Body approximately 0.31 in (8.1 mm) long (base of rostrum to tip of telson of largest individual); carapace 0.23 in (6 mm) long, ovoid, deepest one-third from front, 1.3–1.5 times longer than deep; posterior edge emarginate, exposing at least part of pleonite one dorsally. Rostrum 0.4 carapace length, three times longer than wide, without keel (ventral projection), but with proximal midventral device that loosely interlocks with eye structures. Eyestalk strongly curved (banana shape), surface with denticles; live at deep-sea hydrothermal vents, without visual surface (pigmentation).

DISTRIBUTION
Common at hydrothermal vents on the Galápagos, spreading center and the East Pacific Rise at 13°N and 21°N; 8,040–8,595 ft (2,450–2,620 m) depth.

HABITAT
Found in the throat of vent openings and clumps of mussels and vestimentiferans ("beard worms").

BEHAVIOR
Nothing is known.

FEEDING ECOLOGY AND DIET
Hessler (1984) suggested the toothed anterior edge of the eyestalk (pointed laterally beyond the carapace) may be used to scrape surfaces to loosen food such as bacterial encrustations.

REPRODUCTIVE BIOLOGY
Smallest individuals at all localities (0.035 in [0.9 mm] from base of rostrum to tip of telson) were considered by Hessler (1984) the first free-living instar, called mancoid stage.

CONSERVATION STATUS
Not listed by the IUCN.

SIGNIFICANCE TO HUMANS
None known. ◆

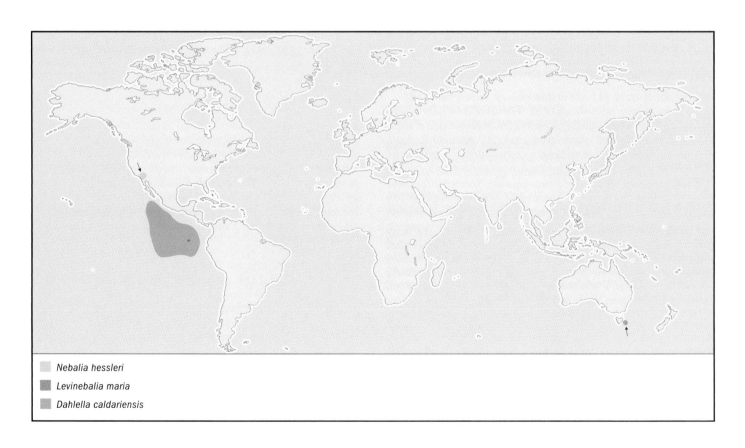

Nebalia hessleri

Levinebalia maria

Dahlella caldariensis

No common name
Nebalia hessleri

ORDER
Leptostraca

FAMILY
Nebaliidae

TAXONOMY
Nebalia hessleri Martin Vetter, and Cash-Clark, 1996, Scripps Canyon, La Jolla, San Diego, California, United States.

OTHER COMMON NAMES
None known.

PHYSICAL CHARACTERISTICS
Total length (excluding setation) up to 0.59 in (15 mm) (average total length 0.38 in [9.8 mm]). Carapace length of mature female is up to 0.26 in (6.8 mm), approximately 1.5 times longer than deep, unsculptured, with usual posterodorsal indention. Rostrum long, clearly extending beyond eye, distally rounded, length approximately 2.7 times width; with ventral projection or "keel" more or less rectangular, unpaired, but with slight medial depression gently sloping upward toward ventral surface of rostrum and with slightly protruding, blunt anterolateral corners. Compound eye is large, well developed; pigmented (visual surface), extensive, covering at least distal half. Telson short, approximately as long as wide, rectangular, or with sides slightly diverging posteriorly. Caudal furcae elongate, approximately twice length of telson, and sometimes greater than twice its length;; acute spines along posterior dorsal borders of pleonites. Males have second antenna that is curved rather sharply in an anterior direction, but considerable variation exists (from extending almost directly anteriorly up to appearing slightly cork-screwed).

DISTRIBUTION
To date, known only from the type locality.

HABITAT
Detrital mats with low oxygen levels in submarine canyons off the coast of southern California at depths of 60 ft (19 m).

BEHAVIOR
Nothing is known.

FEEDING ECOLOGY AND DIET
Within the detritus mat habitat of these animals, vertebrate and invertebrate carcasses are frequently seen and exploited by *Nebalia hessleri*. In the laboratory, this leptostracan feeds on large pieces of meat (e.g., fish, shrimp, fish-containing cat food, and squid) by scraping off chunks, using the peduncle of the first antenna, which bears a row of strong spines. The animals often remained in contact with the meat for five minutes or more, and produced a continuous fecal strand that sometimes rivaled the length of the animal. In such cases, the fecal strand was the same color as the food and thus possibly was poorly processed.

REPRODUCTIVE BIOLOGY
Eggs, brooded under the carapace of the female, are gold or cream colored.

CONSERVATION STATUS
Not listed by the IUCN.

SIGNIFICANCE TO HUMANS
None known. ◆

No common name
Levinebalia maria

ORDER
Leptostraca

FAMILY
Paranebaliidae

TAXONOMY
Levinebalia maria Walker-Smith, 2000, Tasman Sea.

OTHER COMMON NAMES
None known.

PHYSICAL CHARACTERISTICS
Female entire length 0.18 in (4.8 mm). Carapace length 0.11 in (3 mm), depth 0.07 in (2 mm), emarginate, dorsally convex, anterior and posterior margin rounded, 3.6 times length of rostrum, posterior margin reaching pleonite four, surface not sculptured. Rostrum length 2.8 times width, greatest depth 0.2 times length, subterminal spine present, keel absent. Eyestalks pigmented, dorsal margin slightly convex, without denticles or cuticular outgrowths. Caudal furca 2.4 times long as wide, 1.5 times as long as telson. Juvenile male (fully mature male has not been found) entire length 0.17 in (4.4 mm).

DISTRIBUTION
Tasman Sea, 9.3 mi (15 km) E of Maria Island, Tasmania, Australia, at 334 ft (102 m) depth.

HABITAT
Epibenthic, in relation with fine muddy bryozoan sand and coarse, shelly sand. In waters from 328 to 656 ft (100–200 m) deep.

BEHAVIOR
Nothing is known.

FEEDING ECOLOGY AND DIET
Nothing is known.

REPRODUCTIVE BIOLOGY
Nothing is known.

CONSERVATION STATUS
Not listed by the IUCN.

SIGNIFICANCE TO HUMANS
None known. ◆

Resources

Books

Brusca, R. C., and G. J. Brusca. *Invertebrates*, 2nd ed. Sunderland, MA: Sinauer Associates, Inc., 2003.

Calman, W. T. "Crustacea." In *A Treatise on Zoology*, Part 7, third fascicle, edited by R. Lankester. London: Adam and Charles Black, 1909.

Cannon, H. G. "Leptostraca." In *Klassen und Ordnungen des Tierreichs*, Vol. 5, Section 1, Book 4, no. 1, edited by H. G. Bronn. Leipzig, Germany: Akademische Verlagsgesellschaft, 1960.

Dahl, E., and J.-W. Wägele. "Sous-classe des Phyllocarides (Phyllocarida Packard, 1879)." In *Traité de Zoologie. Anatomie, Systématique, Biologie. Crustacés*, Tome VII, Fascicule II, edited by J. Forest. Paris: Masson et Cie, 1996.

Fretter, V., and A. Graham. *A Functional Anatomy of Invertebrates*. London: Academic Press, 1976.

Haney, T. A., and J. W. Martin. "Leptostraca." In *Common and Scientific Names of Aquatic Invertebrates of the United States and Canada*, edited by D. Turgeon. Bethesda, MD: American Fisheries Society Special Publication, forthcoming.

Hessler, R. R., and F. R. Schram. "Leptostraca as Living Fossils." In *Living Fossils*, edited by N. Eldredge and M. Stanley. Berlin: Springer-Verlag, 1984.

Kaestner, A. *Invertebrate Zoology*, Vol. 3, *Crustacea*. New York: Wiley-Interscience, 1970.

McLaughlin, P. A. *Comparative Morphology of Recent Crustacea*. San Francisco, CA: W. H. Freeman and Company, 1980.

Rolfe, W. D. I. "Phyllocarida." In *Treatise on Invertebrate Paleontology*, Part R, Arthropoda 4, Vol. 1, edited by R. C. Moore. Lawrence: Geological Society of America and University of Kansas Press, 1969.

Ruppert, E. E., and R. D. Barnes. *Invertebrate Zoology*, 6th ed. Fort Worth, Texas: Saunders College Publishing, 1994.

Schmitt, W. L. *Crustaceans*. Ann Arbor: University of Michigan Press, 1965.

Schram, F. R. *Crustacea*. New York: Oxford University Press, 1986.

Periodicals

Brahm, C., and S. R. Geiger. "On the Biology of the Pelagic Crustacean *Nebaliopsis typica* G. O. Sars." *Bulletin of the Southern California Academy of Sciences* 65 (1966): 41–46.

Dahl, E. "Crustacea Leptostraca, Principles of Taxonomy and A Revision of European Shelf Species." *Sarsia* 70 (1985): 135–165.

Hessler, R. R. "*Dahlella caldariensis*, New Genus, New Species: A Leptostracan (Crustacea, Malacostraca) from Deep-sea Hydrothermal Vents." *Journal of Crustacean Biology* 4, no. 4 (1984): 655–664.

Martin, J. W., and G. E. Davis. "An Updated Classification of the Recent Crustacea." *Natural History Museum of Los Angeles County Science Series* 39 (2001): 1–124.

Martin, J. W., E. W. Vetter, and C. E. Cash-Clark. "Description, External Morphology, and Natural History Observations of *Nebalia hessleri*, New Species (Phyllocarida: Leptostraca), from Southern California, with a Key to the Extant Families and Genera of the Leptostraca." *Journal of Crustacean Biology* 16, no. 2 (1996): 347–372.

Olesen, J. "A New Species of *Nebalia* (Crustacea, Leptostraca) from Unguja Island (Zanzibar), Tanzania, East Africa, with a Phylogenetic Analysis of Leptostracan Genera." *Journal of Natural History* 33, no. 12 (1999): 1789–1809.

Schram, F. R., and E. Malzahn. "The Fossil Leptostracan *Rhabdouraea bentzi*." *Transactions of the San Diego Society of Natural History* 20 (1984): 95–98.

Spears, T., and L. G. Abele. "Phylogenetic Relationships of Crustaceans with Foliaceous Limbs: An 18S rDNA Study of Branchiopoda, Cephalocarida, and Phyllocarida." *Journal of Crustacean Biology* 19 (1999): 825–843.

Walker-Smith, G. K. "*Levinebalia maria*, a New Genus and New Species of Leptostraca (Crustacea) from Australia." *Memoirs of Museum Victoria* 58, no. 1 (2000): 137–148.

Walker-Smith, G. K., and G. C. B. Poore. "A Phylogeny of the Leptostraca (Crustacea) with Keys to the Families and Genera." *Memoirs of Museum Victoria* 58, no. 2 (2001): 383–410.

Other

"Anchialine Fauna of the Bahamas (*Speonebalia cannoni*)." April 16, 2003 [July 30, 2003.] <http://www.tamug.edu/cavebiology/fauna/bahamafaunalist.html>.

"The Biology of Sea Fleas." [July 30, 2003.] <http://www.museum.vic.gov.au/crust/nebbiol.html>.

"Leptostraca." [July 30, 2003.] <http://atiniui.nhm.org/peet/leptostraca/index.html>.

"Leptostraca—Sea Fleas (Tree of Life Web Project)." 2002 [July 30, 2003.] <http://tolweb.org/tree?group=Leptostraca&contgroup=Malacostraca>.

"Leptostracan Morphology...A Look with the Scanning Electron Microscope." [July 30, 2003.] <http://atiniui.nhm.org/people/haney/morphologySEM.html>.

Lowry, J. K. "Crustacea, the Higher Taxa—Leptostraca (Malacostraca)." October 2, 1999 [July 30, 2003.] <http://www.crustacea.net/crustace/www/leptostr.htm>.

Estela C. Lopretto, PhD

Stomatopoda

(Mantis shrimps)

Phylum Anthropoda

Subphylum Crustacea

Class Malacostraca

Order Stomatopoda

Number of families 17

Thumbnail description
Predatory, shrimplike crustaceans with acute vision and complex behavior that subdue prey and defend themselves by means of specialized forelimbs modified with stabbing spines or heavily calcified "elbows" used as clubs

Photo: Orange mantis shrimps (*Lysiosquilla* sp.) live in burrows and prey on small fishes which they "spear" with their sharp claws. (Photo by David Hall/Photo Researchers, Inc. Reproduced by permission.)

Evolution and systematics

Order Stomatopoda, whose member species are commonly called mantis shrimps or mantis prawns, is included within phylum Arthropoda, subphylum Crustacea, class Malacostraca, and subclass Hoplocarida ("armed shrimp"). The Stomatopoda are sorted into five superfamilies, 17 families, 109 genera, and about 450 species. Mantis shrimps are also commonly called stomatopods, prawn killers, squilla, thumbsplitters, and split-toes; the last two names were coined by fishermen who found out the hard way about mantis shrimp weaponry.

Mantis shrimps are only distant relatives of shrimps and lobsters, despite the inclusion of "shrimp" in the common name and the superficial resemblance between mantis shrimps and shrimps. The Stomatopoda diverged from other malacostracans (which include lobsters, crabs, and shrimp) about 400 million years ago. The raptorial appendages evolved their initial forms around 200 million years ago.

Physical characteristics

A typical mantis shrimp looks, at first glance, like an elaborated, elongated shrimp, with stalked eyes and, according to species, may sport brilliant splashes or coats of garish colors, or may be more cryptically colored. Colors vary considerably among and within species, and between sexes. The body may be cylindrical or somewhat flattened dorsoventrally. The thorax is shield-shaped and the abdomen is conspicuously segmented. At the front of the head, moveable somites bear the

eyes and double antennae. The prominent, stalked eyes are often vividly colored iridescent emerald, ruby, or sapphire.

Eight pairs of limbs arise from the thorax. The front-most five limb pairs, the maxillipeds, are manipulatory, used in hunting and feeding; the second pair comprises the deadly raptorial appendages, the stomatopods' most conspicuous claim to fame. Behind the raptorials are three more pairs of maxillipeds, then three pairs of walking limbs, or pereaopods. The abdomen bears five pairs of swimming limbs, the pleopods, which also bear the gills, and ends in an enlarged, fanlike tail, made up of the shieldlike telson, flanked by a pair of uropods, or tail appendages. Stomatopoda ("stomach-mouth") alludes to the first five pairs of thoracic limbs, since these are used in feeding.

The formidably armed second pair of maxillipeds, the raptorial appendages, are kept folded and tucked away underneath the animal's head and body when not in use. The folded appendages, still partly visible, account for the "mantis" part of the common name, since they recall the praying mantis, the familiar insect, which rests its raptorial appendages similarly.

All mantis shrimp species, and both sexes, are predatory, but employ only two basic attack modes and appropriate equipment, being either spearers or smashers. The most distal joint, or dactyl, on the second pair of appendages bears, in spearers, an array of dactylar teeth, or spines, and, in smashers, a heavily calcified "elbow" used as a very potent club. The raptorial appendages look and act like jacknives, folded similarly into grooves when at rest, and whipping open and into action in a fraction of a second.

The business edge of a spearer's raptorial appendage may bear from two to 20 spines, or dactylar teeth, barbed at their tips, efficient and effective arrays of spears for impaling prey, then hauling it in for feeding.

If a stomatopod loses one or both raptorial appendages through injury, it will regrow them gradually over four successive molts. An individual will forcibly remove its own damaged raptorial appendage by use of its other maxillipeds.

Among the five superfamilies, both smasher and spearer species are found in the Gonodactyloidea and Lysiosquilloidea. Superfamilies Squilloidea, Bathysquilloidae, and Erythrosquilloidea contain only spearers.

The earliest stomatopods, before the evolution of the raptorial appendages, likely fed by grubbing in seabottom mud for resident small creatures, some of which the stomatopods grabbed in the mud while flushing out more mobile types. The fleeing prey prompted the evolution, within the early stomatopods, of elongated, barbed maxillipeds able to flash out and skewer prey on the run. The smasher type of raptorial appendages developed later, probably several different times, from the original barbed appendages.

Stomatopod vision is extraordinarily keen and the eyes are elaborately constructed, being the most highly developed visual organs among crustaceans. Stomatopod eyes are assemblies of ommatidia, or facets, like those in the eyes of other arthropods. However, those assemblies in a central midband region of specialized ommatidia are unique in stomatopods.

The eyes, in a close-up view, have the appearance of highly burnished ovoids with two dark spots and the distinctive midband region. The midband region divides the eye into separate hemispheres, each with a pupil and focal point, allowing binocular vision and depth perception in a single eye. The midbrand region also carries its own pupil and focal point, allowing trinocular vision in each eye. Since both eyes each carry three focal points, stomatopod vision is hexnocular, coordinating imagery from six focal points. The eyes are thereby capable of fine-tuned depth perception and range, enabling their owners to be aware of subtleties of speed and distance of prey to be able to attack with fiendish accuracy.

The midband region of the eye, made up of six rows of specialized ommatidia, can perceive color and polarized light. Four rows carry 16 differing sorts of photoreceptor pigments, 12 for color sensitivity, others for color filtering. The pigmented filters, arrayed within the ommatidia like sunglasses, allow fine-tuning of sensitivity to certain wavelengths of light. The eyes are capable of distinguishing up to 100,000 colors, ranging from infrared through the visible color spectrum and into the ultraviolet. The remaining two rows within the midband region are sensitive to polarized light.

The hemispheres are not equipped for color vision and are primarily sensitive to forms. Although the midband region color-scans only a small segment of the stomatopod's visual field, the animal can move the eyes with delicate precision in a wide range, scanning its environment with the sensitive midland region.

A resting but wary stomatopod will often keep its stalked eyes aloft, like periscopes, achieving 360° vision.

Stomatopods can change the spectral sensitivity of their eyes if descending into deeper waters, where the remaining light is mostly blue and highly polarized. The eyes will adjust to the low, blue-shifted light by becoming more sensitive to the blue end of the visual spectrum.

Supplementing the eyes are the antennae, or antennules, sensitive to water-borne odors and turbulence. A special feature of stomatopod antennae is a leaflike flap, or scale, fringed with hairlike setae, to amplify the sensitivity of the antennules. Studies by Caldwell have found that, at least in some species, the antennal scales and the uropods, the paddle-shaped appendages of the tail, reflect polarized light. Caldwell's findings suggest that by moving the antennae and uropods, and thereby communicating flashes of polarized reflected light, which stomatopod eyes can see, individual animals may communicate with one another. Experiments lend support to the probability that an individual mantis shrimp can detect and identify the odor of another individual of the same species.

Stomatopod sexes can be easily distinguished. Males bear a pair of long, slender penes, or sperm transferal organs, articulating at the bases of the last paraeopods, or walking limbs. The female gonopores, or sperm receptacles, are visible as a narrow slit on the sternum between the first peraeopod pair. Many species go further with sexual dimorphism, the males being bigger, with larger raptorial appendages and telsons than females.

Distribution

Mantis shrimps live offshore of most, if not all, continental landmasses that straddle the tropical and subtropical latitudes, and offshore of tropical and subtropical islands.

Habitat

Stomatopods are primarily tropical or subtropical, shallow-water marine animals, with a few species in cool temperate or subantarctic waters. Most species prefer the intertidal or subtidal zones. Exceptions are species of superfamily Bathysquilloidea, which set up and keep house on outer continental shelf habitats down to 4,920 ft (1,500 m) below sea level.

Spearers prefer soft, sandy, or muddy sea bottoms for homemaking, digging out burrows or moving into burrows abandoned by other creatures. Smashers prefer to seek out and move into suitably shaped and sized hollows in hard substrates such as coral and rock. Species of the Gonodactyloid superfamily favor the hard, rough substrates like rock and coral, and so are the dominant stomatopod species on coral reefs. Species within the superfamilies Squilloidea and Lysiosquilloidea burrow in soft, level, shallow sea bottoms of sand.

The peacock mantis shrimp (*Odontodactylus scyllarus*) is a "smasher" species with an incredibly powerful strike capability; it is a favorite among aquarium owners. (Photo ©Tony Wu/www.silent-symphony.com. Reproduced by permission.)

Behavior

All aspects of stomatopod behavior are defined or influenced by the animal's unique raptorial appendages, and that behavior is bewilderingly complex. Mating rituals and competitive interactions between rivals of a species involve elaborate visual communication cues and actions.

Mantis shrimp activity, depending on species, may be diurnal, nocturnal, or crepuscular (dawn and dusk). Some become active during moonlit nights. Males guard pregnant females. Individuals of some species probably recognize other individuals of the same species by sight and odor. So far, stomatopods are the only invertebrate type in which the ability to tell apart non-mated individuals has been supported by studies. Individual recognition among stomatopoda would be especially important in coral reef habitats, with their limited available cavities. In this high-competition, high-risk environment, it is best to be able to recognize the bullies on the block.

All mantis shrimp species live in soft-substrate burrows or hard rock crevices. Burrow-dwellers either dig out their burrows or move into one abandoned by some other creature. A burrow may be up to 33 ft (10 m) long, and may have several entrances.

When resting, mantis shrimp, depending on species, may plug the burrow entrance with a rock, or position the claws or telson at the entrance. Individuals of the species, *Echinosquilla guerini*, have telsons equipped with spines, suitable to their function as shields for the burrow entrance.

Newly mature individuals of either sex compete for living space. In substrates suitable for burrowing, most commonly sand, the competition can be fierce, but less so than in coral reefs, since malleable sand permits flexibility in where an individual establishes its burrow. Coral reefs, on the other hand, are built of hard material, always limiting the number of crevices available for colonization by mantis shrimps, so that the competition among coral reef stomatopods for living space is intense. It is among the rock-dwelling species that senses are most keen and communication between individuals is most complex.

Size difference is the usual criterion in individual competition, with the larger opponent chasing off the smaller. More similarly sized opponents, when threat displays fail, may go on to a fight. A spat between rivals is a dangerous affair, since the sharp tines of spearers' raptorials can cut and shred, while the clubbed appendages of smashers can deal a single, fatal blow. In fights, opponents use their flexible telsons as shields and fenders. The battle ends in retreat by one individual or in injury.

A burrowing stomatopod, once ensconced, will defend its home vigorously against other mantis shrimps of the same species, and drive off other marine animals that come too close to the burrow. Depending on size and strength, newcomer stomatopods may force weaker residents out of their burrows or crevices, or the established stomatopod may drive off the intruder.

Burrowing stomatopods show less in-species aggressiveness than do species that live in rock or coral crevices. Competing burrowers generally go through elaborate threat rituals until one admits defeat and leaves, without any actual fighting, since the loser can simply go elsewhere and dig out a new burrow. This sort of flexibility does not exist for stomatopods dependent on rocky crevices for shelter, since the animals cannot carve out new crevices in the tough material; the number of available crevices is always limited. Contenders for crevices have no choice but to fight in earnest. On the other hand, since these are creatures that can kill opponents with a single blow of the raptorial appendage, the rock-dwelling stomatopod species, for defense, sport the brightest and most varied colors and stage the most elaborate threat behaviors among stomatopod species.

A common but dramatic threat display among stomatopods is the meral spread, in which an individual elevates its thorax and spreads wide its antennules, antennal scales, raptorial appendages, and peraeopods. The display creates an impression of inflated size and exposes a brightly colored depression, the meral spot, on the merus, or proximal segment, of the raptorial appendage. Throughout subclass Hoplocarida, flashing the meral spot is the prelude to attacking the rival, so that displaying the meral spot has an unusually intimidating effect on a rival, even one of another stomatopod species. *Gonodactylus smithii* is quick to resort to the meral spread and display, sometimes able to chase off other, more aggressive species with its vivid purple and white meral spots.

The efficacy of the flashed meral spot in banishing rivals seems to depend on how early and how aggressively an individual displays the meral spread in an in-species confrontation. Victory in most cases goes to the individual who displays earlier and more aggressively. Flashing the meral spot early in an encounter between individuals usually ends in victory for the displayer, without a fight ensuing. Females guarding egg clutches are quick to display the meral spread to other stomatopods.

Mantis shrimp molt, or shed and replace, their exoskeletons, like other arthropods. Molting times vary with species, size, age and reproductive state. Adults of superfamily Gonodactyloidea molt every three or four months.

Learning experiments on gonodactylid mantis shrimps were carried out by M. J. Reaka, who set up an aquarium furnished with a black-painted flask and no other available cover, introduced a single stomatopod individual from a permanent aquarium, and then noted its behavior and how long it took to locate the entrance and move into the flask. The stomatopod was allowed to rest within the flask for an hour, then withdrawn and returned to its original aquarium. A few days later, the same individual was introduced to the testing aquarium and its behavior noted. Each tested individual was exposed to five such trials. The more subsequent the trial, the less time the stomatopod took to find the flask, on the average cutting its time from 52 hours on the first trial to only one hour on the fifth.

Feeding ecology and diet

Typically, a spearer lies in wait at the entrance of its burrow, body tucked within, raptorial claws tucked beneath, head and eyes out and alert for movements of passing, soft-bodied prey like fish and shrimp. When the right victim comes into a vulnerable range, the spearer will strike out and up at the prey animal in an almost invisibly fast extension, impaling and retracting the armed claws. The victim secured, the stomatopod begins feeding formalities.

Smashers are more likely to roam about, crawling or swimming, as active hunters, depending on mobility, or lack of it, among prey animals, since smashers hunt for creatures with hard exteriors, including shelled mollusks and other crustaceans. A smasher will cripple a crab with multiple punches of the heel to the crab's claws, legs, and carapace, then drag the battered creature to the mantis's burrow to be feasted upon at leisure. Smashers deal with snails and clams by toting them back to the burrow, wedging them against the burrow wall, then punching them into fragments, later casting the shell shards from the burrow.

Mantis shrimps are capable of some of the fastest movements in living nature. A stomatopod strike at a prey animal, including the unfolding motion, can pass in 2 milliseconds. The strike of a praying mantis takes 100 milliseconds, the length of time of one human eye blink. Smashers add enormous force to their strike speed. Two of the larger smasher species, *Hemisquilla ensigera* and *Odontodactylus scyllarus*, pack a punch nearly equal to the impact of a .22 caliber bullet, powerful enough to break open the double-layered safety glass of public aquariums, which in fact they have done.

Reproductive biology

Most stomatopod species are solitary, with opposite sexes staying together only during mating. But some stomatopod species among families Lysiosquillidae and Nannosquillidae are monogamous, very rare among invertebrates. A monogamous stomatopod pair may remain together as long as both are alive, which can be 15–20 years.

Monogamous stomatopod pairs share one burrow and divide up duties, the female tending the eggs while the male hunts food for himself and the female. In non-monogamous species, the sexes only associate during mating and brooding. Some somatopod species mate any time of the year, while others have well-defined, limited periods of female receptivity that occur just before a female lays eggs.

Males may leave their burrows or crevices and seek out females for mating, or both sexes may roam about, looking for partners. Males perform elaborate mating rituals in front of potential mates. The female makes the final decision as to which male she favors for mating. In non-monogamous species, the female may accept one or several males for mating during any one fertile period.

Stomatopod fertilization is internal, and mating behavior somewhat parallels the mammalian. Males carry a pair of penes, penis-like organs attached to the bases of the last pair of pareopods. During mating, the penes erect themselves much as do mammalian penises. Male and female copulate underside-to-underside, the male inserting the penes into the female gonopores, on her ventral side. The sperm is stored in a small pocket within the body, just behind and within the gonopores.

The female extrudes the fertilized eggs through the gonopores, where they are fertilized by the stored sperm just before they emerge. The female may not lay her egg mass immediately, but may wait weeks before doing so, waiting for strong ocean currents that are favorable to dispersing the young.

The female uses a gummy exhudate from glands on her ventral thorax to glue the extruded eggs into a single, portable mass that she carries with her or plasters to the wall of her burrow. Hatching may start up to three weeks after the egg mass is laid. A female will tend her eggs from 10 days to two months, depending on species. Hatchlings may remain in the burrow with the mother for a time period of a week to two months, or depart immediately upon hatching, for the pelagic stages of their development. Females of non-monogamous species do not eat while tending the egg mass.

During breeding, the male stomatopod, in both monogamous and solitary species, stands guard over the female before she spawns, and, if a solitary species, only leaves to make or find a new home when the female has spawned. Should he encounter the same female in her burrow while looking for a home, he leaves her undisturbed for up to two weeks after separation. The female, in her burrow, sensing an approaching stomatopod, shunts water currents outward from the burrow and toward the approaching individual with fanning motions of her maxillipeds. The male mate apparently recognizes the female's scent in the disturbed water, and leaves her alone.

Like many marine life forms, newly hatched stomatopod individuals pass through several distinct stages from hatching to maturity. The cycle from hatchling to adult takes about three months. The early larval stages may be benthic (sea bottom) and passed through within the mother's burrow, or pelagic (open sea), but all later stages are pelagic. At those

stages the larvae are an abundant component of plankton, and thus a significant food source for plankton-feeding fish. Stomatopod larvae are translucent, glassy, ephemeral-looking beings with wiry, skeletal bodies and huge, bulbous eyes.

A typical larva of superfamily Lysiosquilloidea hatches as an antizoea, with five pairs of biramous thoracic appendages and no abdominal appendages. The antizoea develops into an erichthus, with two or fewer intermediate denticles, or tooth-like projections, on the telson, and pleopods sprouting in sequence from front to rear.

A larva of a species within superfamily Squilloidea or Gonodactyloidea hatches as a pseudozoea, with two pairs of thoracic appendages and four or five pairs of pleopods. A squilloid pseudozoeae develops into an alima, with four or more intermediate denticles on the telson, while a gonodactyloid pseudozoeae develops into an erichthus. The development cycles of superfamilies Erythrosquilloidea and Bathysquilloidea are only poorly known.

Stomatopod postlarvae settle toward the sea bottom, resembling adults at this stage and living in the manner of adults.

Conservation status

No species are listed by the IUCN, and all species are apparently widespread and abundant. Nevertheless, stomatopod species inhabiting coral reefs are vulnerable to depletion because of the massive destruction being wrought on coral reefs through blasting and sewage.

Significance to humans

The presence of stomatopoda can enable ecologists to gauge the health of coral reef ecosystems. Reef-dwelling stomatopods are especially sensitive to adverse conditions such as the introduction of petroleum, sewage, and agricultural runoff. Censuses of stomatopod populations on coral reefs are undertaken frequently to estimate the health or contamination of the reefs.

Keeping mantis shrimps in aquaria has become a fairly popular hobby among marine pet keepers who prefer the unusual. There are books and Web sites covering aspects of maintaining stomatopods in home aquariums.

Species of the family Squilla are edible and tasty. Originally overlooked as commercial food sources because individuals of most species live solitarily, which makes harvesting difficult, a fishery industry harvesting *Squilla mantis* is underway in the Mediterranean and Adriatic.

1. *Lysiosquillina maculata*; 2. *Nannosquilla decemspinosa*; 3. Peacock mantis shrimp (*Odontodactylus scyllarus*). (Illustration by John Megahan)

Species accounts

Giant mantis shrimp
Hemisquilla ensigera

FAMILY
Hemisquillidae

TAXONOMY
Two subspecies: *Hemisquilla ensigera ensigera* (Owen, 1832) and *Hemisquilla ensigera californiensis* (Stephenson, 1977).

OTHER COMMON NAMES
English: Blueleg mantis shrimp.

PHYSICAL CHARACTERISTICS
A giant among mantis shrimps, adults often reaching a total length of 13.7 in (35 cm) or more. Individuals are brightly colored, with yellow frontal areas, deep blue appendages, and transverse orange-brown stripes across the dorsal abdomen.

DISTRIBUTION
Found off the coast of southern California to the Gulf of California.

HABITAT
Shallow sea bottoms with sand and/or mud and rubble.

BEHAVIOR
Active, aggressive hunter.

FEEDING ECOLOGY AND DIET
A smasher, actively hunts crabs, clams, and snails, which it breaks up with its clubbed raptorial appendages. Not content with hard-shelled prey, it attacks and eats fish, battering them into immobility.

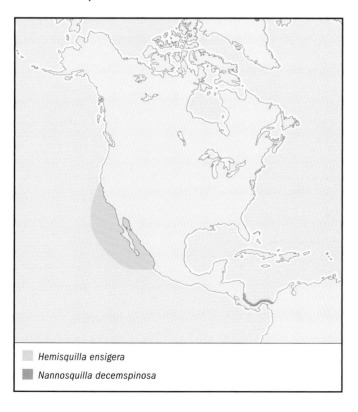

■ *Hemisquilla ensigera*
■ *Nannosquilla decemspinosa*

REPRODUCTIVE BIOLOGY
Males seek out females, displaying before females in their burrows; female makes the final choice of partner. Male fertilizes female, guards her until the larvae have left for the open sea, then departs.

CONSERVATION STATUS
Not listed by the IUCN.

SIGNIFICANCE TO HUMANS
Individuals of this species often end up as pets in public reef aquaria. Its punch is nearly equal to the impact of a .22 caliber bullet. ◆

No common name
Lysiosquillina maculata

FAMILY
Lysiosquillae

TAXONOMY
Lysiosquillina maculata Risso, 1816.

OTHER COMMON NAMES
None known.

PHYSICAL CHARACTERISTICS
One of largest mantis shrimps, adults routinely attaining lengths of 15 in (38 cm) or more. Coloration in males tends toward the cryptic, being predominantly dark-speckled gray, brown, and dull orange. Females are a more conspicuous orange color; both sexes have distinctive, alternating light and dark transverse bands on the dorsal abdomen.

DISTRIBUTION
Eastern North Atlantic and Mediterranean.

HABITAT
Muddy sea bottoms, coastal to upper-slope continental shelves.

BEHAVIOR
Spearer, digs out burrows in sea bottom mud, then sits patiently in entrance for passing prey. Adults form monogamous pairs.

FEEDING ECOLOGY AND DIET
An individual waits at entrance of burrow, its forequarters partly exposed and its eyes aloft and alert. When an edible, soft-bodied creature, often a fish, passes within range, it springs forward, whipping its raptorial appendages from beneath body, then forward and upward, impaling the prey on the spearlike projections, then hauling in the prey.

REPRODUCTIVE BIOLOGY
Form monogamous pairs that may remain together as long as both partners are alive. Male guards the brooding female and hunts food for both of them.

CONSERVATION STATUS
Not listed by the IUCN.

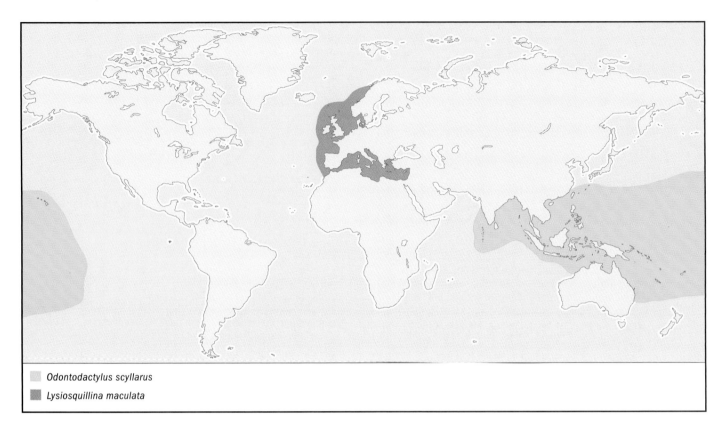

Odontodactylus scyllarus

Lysiosquillina maculata

SIGNIFICANCE TO HUMANS
None known. ◆

No common name
Nannosquilla decemspinosa

FAMILY
Nannosquillidae

TAXONOMY
Nannosquilla decemspinosa Rathbun, 1910.

OTHER COMMON NAMES
None known.

PHYSICAL CHARACTERISTICS
Only 0.98 in (2.5 cm) long, and cryptically colored.

DISTRIBUTION
Pacific coast of Panama.

HABITAT
Shallow, muddy sea bottoms close to shore.

BEHAVIOR
Has attracted special attention from biologists, because, when beached by tides or storms, it moves itself around and back into the ocean by rolling up its body and rolling like a self-propelled wheel, covering as far as 6.5 ft (2 m) at a sprint, rolling 20–40 times, at 72 revolutions per minute, or 1.5 body lengths per second. When the rolling animal slows down, it uses its entire body as a spring to propel itself upwards and forwards into the next roll. So far, it and the caterpillar of the mother-of-pearl moth (*Pleurotya ruralis*) are the only two known animal species above microscopic size that use this form of locomotion.

FEEDING ECOLOGY AND DIET
A spearer, it is a passive hunter, waiting at the entrance of its burrow, and striking out with its raptorial appendages to impale soft-bodied prey.

REPRODUCTIVE BIOLOGY
Males seek out females, displaying before females in their burrows; female makes final choice of partner. Male fertilizes female, guards her until the larvae have left for the open sea, then departs.

CONSERVATION STATUS
Not listed by the IUCN.

SIGNIFICANCE TO HUMANS
Its ability to roll like a wheel renders it an object of fascination and a subject of scientific curiosity. ◆

Peacock mantis shrimp
Odontodactylus scyllarus

FAMILY
Odontodactylidae

TAXONOMY
Odontodactylus scyllarus Linnaeus, 1758.

OTHER COMMON NAMES
None known.

PHYSICAL CHARACTERISTICS
Reaches lengths of 6.7 in (17 cm). Impact of its strike equals or exceeds that of *H. ensigera*. There is considerable sexual dichromatism. Mature males are brilliantly colored, with emerald green bodies and crimson and blue antennules and maxil-

lipeds; females are more cryptically colored with olive or brown. Juveniles are bright yellow. An odd feature is its telson, which is thin and weak, despite the strength of the predatory strike, which is so powerful that it can kill an individual in a confrontation between same-species rivals.

DISTRIBUTION
Indopacific Ocean, including Hawaii.

HABITAT
Coral reefs.

BEHAVIOR
Smasher; diurnally active, but may also go hunting nocturnally during a full moon. When building a burrow, an individual lines the interior with pieces of coral, rock, and shell.

FEEDING ECOLOGY AND DIET
Actively hunts hard-shelled animals like clams, snails, and crabs, splitting shells with repeated blows of its raptorial appendages, then dragging the prey back to the burrow.

REPRODUCTIVE BIOLOGY
Males seek out females, displaying before females in their burrows; female makes final choice of partner. Male fertilizes female, guards her until the larvae have left for the open sea, then departs.

CONSERVATION STATUS
Not listed by the IUCN.

SIGNIFICANCE TO HUMANS
Favored for keeping in private aquaria. ◆

Resources

Books

Ensminger, Peter J. *Life Under the Sun*. London and New York: Academic Press, 1997.

Hof, Cees H. J. "The Taphonomy, Fossil Record, and Phylogeny of Stomatopod Crustaceans." PhD diss. Amsterdam: University of Amsterdam, 1998.

Reaka, M. L., ed. *The Ecology of Coral Reefs*. Rockville, MD: NOAA Undersea Research Program, 1985.

Steger, R. "The Behavioral Ecology of a Panamanian Population of the Stomatopod, *Gonodactylus bredini* (Manning)." PhD diss. Berkeley: University of California, Berkeley, 1985.

Periodicals

Ahyong, S. T. "Phylogenetic Analysis of the Stomatopoda (Malacostraca)." *Journal of Crustacean Biology* 17, no. 4 (1997): 695–715.

———. "Revision of the Australian Stomatopod Crustacea." *Records of the Australian Museum* Supplement 26 (2001): 1–326.

Ahyong, S. T., and C. Harling. "The Phylogeny of the Stomatopod Crustacea." *Australian Journal of Zoology* 48 (2000): 607–642.

Caldwell, R. L. "Interspecific Chemically Mediated Recognition in Two Competing Stomatopods." *Marine Behavioral Physiology* 8 (1985): 189–197.

———. "Recognition, Signalling and Reduced Aggression between Former Mates in a Stomatopod." *Animal Behavior* 44 (1992): 1–19.

———. "A Test of Individual Recognition in the Stomatopod *Gonodactylus festae*." *Animal Behavior* 33 (1985): 101–106.

Caldwell, R. L., and H. Dingle. "Stomatopods." *Scientific American* 234 (1975): 80–89.

Cronin, T. W., N. J. Marshall, R. L. Caldwell, and N. Shashar. "Specialization of Retinal Function in the Compound Eyes of Mantis Shrimps." *Vision Research* 34 (1994): 2639–2656.

Erdmann, M. V., and R. L. Caldwell. "Stomatopod Crustaceans as Bioindicatiors of Marine Pollution Stress on Coral Reefs." *Proceedings of the 8th International Coral Reef Symposium* 33 (1997): 1521–1526.

Full, R., K. Earls, M. Wong, and R. Caldwell. "Locomotion Like a Wheel?" *Nature* 365 (1993): 495.

Hof, C. H. J. "Fossil Stomatopods (Crustacea: Malacostraca) and their Phylogenetic Impact." *Journal of Natural History* 32 (1998): 1567–1576.

Land M. F., N. J. Marshall, D. Brownless, and T. W. Cronin. "The Eye Movements of the Mantis Shrimp *Odontodactylus scyllarus* (Crustacea: Stomatopods)." *Journal of Comparative Physiology* 167 (1990): 155–166.

Manning, R. B. "Stomatopod Crustacea of Vietnam: The Legacy of Raoul Serène." *Crustacean Research*, Special Edition No. 4 (1995).

Manning R. B., R. K. Kropp, and J. Dominguez. "Biogeography of Deep-sea Stomatopod Crustacea, Family Bathysquillidae." *Progress in Oceanography* 24 (1990): 31–316.

Marshall, N. J. "A Unique Color and Polarization Vision System in Mantis Shrimps." *Nature* 333 (1988): 557–560.

Marshall N. J., M. F. Land, C. A. King, and T. W. Cronin. "The Compound Eyes of Mantis Shrimps (Crustacea, Hoplocarida, Stomatopoda). I. Compound Eye Structure: The Detection of Polarised Light." *Philosophical Transactions of the Royal Society, Series B* 334 (1991): 33–56.

Marshall N. J., M. F. Land, C. A. King, and T. W. Cronin. "The Compound Eyes of Mantis Shrimps (Crustacea, Hoplocarida, Stomatopoda). II. Color Pigments in the Eyes of Stomatopod Crustaceans: Polychromatic Vision by Serial and Lateral Filtering." *Philosophical Transactions of the Royal Society, Series B* 334 (1991): 57–84.

Reaka M. J. "On Learning and Living in Holes by Mantis Shrimp." *Animal Behavior* 28 (1980): 111–115.

Other

"Lurker's Guide to Stomatopods." July 27, 2003 [August 1, 2003]. <http://www.blueboard.com/mantis/index.htm>.

Kevin F. Fitzgerald, BS

Bathynellacea

(Bathynellaceans)

Phylum Arthropoda

Subphylum Crustacea

Class Malacostraca

Order Bathynellacea

Number of families 2

Thumbnail description
Small, worm-like, blind Crustacea with head, thorax, and abdomen. Thorax with seven pairs of walking legs and a reduced eighth one, which in the male is transformed for copulatory purposes. Abdomen with one pair of appendages on the last segment, one pair on the first, and very rarely also one pair on the second segment.

Illustration: *Notobathynella williamsi.* (Illustration by John Megahan)

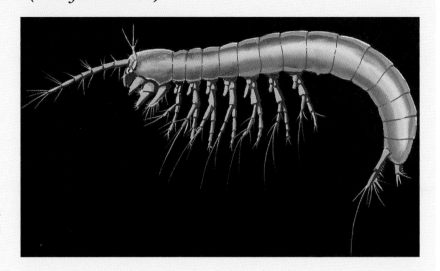

Evolution and systematics

Bathynellcea have no fossil record. Their closest relatives are the Anaspidacea in the Southern Hemisphere. Both groups are confined to fresh water, but have had marine ancestors. This is evidenced by their fossil relatives, the Palaeocaridacea, which, during the Carboniferous and Permian, inhabited the littoral of the then tropical seas in the Northern Hemisphere. Transition into fresh water was achieved independently, first by the Bathynellacea and later also by the Anaspidacea. Unlike the Anaspidacea, the Bathynellacea have disappeared almost completely from surface waters today.

When compared with adults of related Crustacea taxa, Bathynellacea appear very different, but they share a remarkable similarity with the larvae of these other Crustacea. In fact, Bathynellacea have the appearance of larvae, indicating that they have had larval-like ancestors. There is good evidence that the postembryonic development of their surface-living ancestors passed through a series of several larval stages. In the course of adaptation to life in the groundwater, abbreviation of this development caused by precocious sexual maturity at an early stage of larval development marked the beginning of the evolution of subterranean Bathynellacea. Today they reach adulthood at a stage that corresponds to the last larval stage of their ancestor. A morphological consequence of this adaptational process was that their body became continually smaller and its structure increasingly simpler. As a result, they finally fitted into the small spaces between sand particles of the groundwater-bearing strata. Ecologically they progressed from a benthic life through a colonization of coarse substrates with large interstitial spaces to an existence in ever smaller sediments with ever narrower spaces between the particles. The ancestral form of the Bathynellacea had conquered a completely new habitat and marked the beginning of an impressive radiation leading to a worldwide colonization of the groundwater.

The order Bathynellacea comprises two families, 60 genera, and about 200 species. The two families are Bathynellidae and Parabathynellidae.

Physical characteristics

Bathynellacea range in length between 0.02 in (0.5 mm) and 0.14 in (3.5 mm). As an adaptation to their subterranean existence, they do not have eyes and lack body pigment. The body is divided into head, thorax, and abdomen. The head carries the antennae with chemo- and mechanosensory structures as well as the mouthparts. The mandibles, used for biting, are symmetrical in structure. The thorax has seven pairs of biramous walking legs. The eighth pair is reduced in both sexes and in the male is transformed into a characteristic copulatory organ. The abdomen carries one pair of appendages on the first and last somite (pleotelson). There can also be a pair of appendages on the second abdominal somite. The pair on the last somite (uropods) does not form a tailfan together with the pleotelson, as is the case in several related groups of Crustacea. The basal segment (sympodite) of the uropods carries a characteristic row of spines. Bathynellidae can be identified by the second antennae being directed anteriorly, by the last abdominal somite (pleotelson) carrying two dorsal setae, by the rim of the upper lip (labrum) being smooth and not serrate, by the outer branches (exopodites) of all seven walking legs being one-segmented, by the first abdominal appendages (pleopods) being two-segmented, and by the sympodite of the uropods being relatively short.

In contrast, Parabathynellidae have the second antennae bent backwards and carry lateral setae on the pleotelson. The rim of their labrum is serrate or fringed with fine setae. The exopodites of the walking legs are composed of one or more segments and only rarely do the exopodites of all the legs have the same number of segments. The first pleopods, if present, are one-segmented or represented by only two setae. The sympodite of the uropods is more elongate than in the Bathynellidae.

Distribution

Bathynellacea are found in the groundwater nearly all around the globe. They are absent in the Antarctic and in the Northern Hemisphere in those areas that had been covered by ice during the last glaciation. Recolonization has been observed at a few places but has not progressed very far beyond the old ice frontier. Bathynellacea are not known from Central America and are also absent from volcanic islands and several islands of continental origin (e.g., New Caledonia, Fiji, Caribbean islands).

Both families have a worldwide distribution, but Bathynellidae are more common in temperate regions of the world than in the tropics. Parabathynellidae have a more even distribution but do not reach as far north as Bathynellidae in the Northern Hemisphere. All recolonizers of previously ice-covered areas belong to the Bathynellidae.

There are two hypotheses to explain the distribution of Bathynellacea. Both agree on marine ancestors as the starting point of the adaptational process and on plate tectonics as a major factor. The first hypothesis maintains that the ancestors of Bathynellacea invaded fresh surface waters and that from their larvae the Bathynellacea arose to become inhabitants exclusively of the groundwater. Subsequent spread in the groundwater itself led to today's worldwide occurrence. According to the second hypothesis, as a first step a marine ancestor became adapted to an interstitial life in littoral sands and in a second step had to switch to freshwater conditions as a result of marine shoreline regression. Multiple such invasions of the groundwater caused by repeated sea-level changes at different geological times together with plate tectonics led to the currently observed worldwide distribution.

Habitat

Bathynellacea are typical inhabitants of groundwater and can be collected in wells, in sandy banks of rivers and sandy shores of lakes, in caves, and in springs. One species is known from a hot spring in Africa at a temperature of 131°F (55°C). Bathynellacea are absent from surface waters. Only two species are known from Lake Baikal, where they occur on sandy patches down to 4,725-ft depth (1,440-m), but not in the open water. A few species have been recorded from near the seashore and tolerate brackish water conditions. Only one species is polyhaline.

Behavior

Bathynellacea are elegant, persistent crawlers and ungainly, short-winded swimmers. Their crawling movements are a combination of swimming and walking. As an escape reaction or when caught in blind alleys of the interstitial labyrinth, they engage in a specialized turning maneuver that consists of placing the abdomen beneath the rest of the body, sliding it along the ventral side towards the head, and turning at the same time about the longitudinal axis to return to a ventral position.

Feeding ecology and diet

Not much is known about the feeding and food of Bathynellacea. Plant debris (detritus), nematodes, and turbellarian worms have been observed from the guts of various specimens. Among Parabathynellidae, in particular, there is great variation in types of mouthparts, indicating specialization on different types of food. Most Bathynellacea may be omnivorous, but many are certainly specialized, feeding on detritus, protozoans, and/or bacteria. One species has mouthparts with setae suited for scraping off material from sand grains.

Reproductive biology

In Bathynellacea the sexes are separate. Mating behavior has never been observed. The eggs are shed freely and singly. The egg passes through a nauplius stage. After hatching, postembryonic development passes through two phases, a larval (parazoëal) phase with three stages, and a juvenile (bathynellid) phase with a variable number of stages.

Conservation status

Most species are known only from their type locality. In Europe, from where more is known than elsewhere about areas of distribution of single species, some are of local occurrence while others have a wider distribution. None seem to be threatened, and none are listed by the IUCN Red List.

Significance to humans

Bathynellacea are part of a complex groundwater biocoenosis made up of bacteria, protozoans, fungi, and metazoans (animals). All of these organisms act together to decompose particles washed into the groundwater from outside. These particles would otherwise clog the spaces between the sand grains, preventing groundwater from circulating freely. Bigger particles are broken up by Bathynellacea, and after they are passed through the gut, they are further degraded by protozoans and bacteria. Thus the interstitial spaces are kept open. Humans benefit from this ecological service because it helps to keep drinking water clean.

Species accounts

No common name
Antrobathynella stammeri

FAMILY
Bathynellidae

TAXONOMY
Bathynella natans stammeri Jakobi, 1954, from a well at Möhrendorf, near Erlangen, Germany.

OTHER COMMON NAMES
None known.

PHYSICAL CHARACTERISTICS
Can be distinguished from other European bathynellids by the specific structure of the eighth leg and by a hyaline, cone-shaped protrusion at the inner side of the coxa of the seventh leg of the male as well as by the basal

Antrobathynella stammeri

segment of the uropods of both sexes carrying a row of four spines of which the anterior one is bigger than the others and separated from them by a distinct gap.

DISTRIBUTION
Widely distributed in Europe from Ireland to Romania.

HABITAT
Known from all groundwater habitats.

BEHAVIOR
Agile crawler from the second postembryonic stage onwards; clumsy when just hatched from the egg, and may fall prey to cyclopoid copepods.

Antrobathynella stammeri

FEEDING ECOLOGY AND DIET
Feeds on detritus and worms of different kind.

REPRODUCTIVE BIOLOGY
Reproduces all year round, deposits eggs singly. Embryonic phase takes two months, parazoëal phase at 48°F (9°C), about three months, bathynellid phase about four months. Becomes adult after nine months of development but not sexually mature. There are further molts, four in the male and five in the female, until sexual maturity is reached. Lives several years; two years are recorded from laboratory cultures.

CONSERVATION STATUS
Not listed by IUCN. Does not appear to be threatened.

SIGNIFICANCE TO HUMANS
None known. ◆

Resources

Books

Coineau, Nicole. "Sous-classe des Eumalacostracés (Eumalacostraca Grobben, 1892) Super-ordre des Syncarides (Syncarida Packard, 1885)." In *Traité de Zoologie* 7, no. 2, *Crustacés*, edited by Jean Forest. Paris: Masson et Cie, 1996.

———. "Syncarida." In *Encyclopaedia Biospeologica 2*, edited by C. Juberthie and V. Decou. Bucarest, Romania: Société de Biospéologie, 1998.

Periodicals

Camacho, Ana I. "Historical Biogeography of *Iberobathynella* (Crustacea, Syncarida, Bathynellacea), an Aquatic Subterranean Genus of Parabathynellids, Endemic to the Iberian Peninsula." *Global Ecology & Biogeography* 10 (2001): 487–501.

Schminke, Horst Kurt. "Adaptation of Bathynellacea (Crustacea, Syncarida) to Life in the Interstitial ("Zoea Theory")." *Internationale Revue der Gesamten Hydrobiologie* 66 (1981): 575–637.

Horst Kurt Schminke, PhD

Anaspidacea

(Anaspidaceans)

Phylum Arthropoda
Subphylum Crustacea
Class Malacostraca
Order Anaspidacea
Number of families 4

Thumbnail description
These crustaceans do not have a carapace, one thoracic somite is incorporated into the head, with its appendages modified as maxillipeds, and the remaining thoracic legs have both epipods and exopods

Illustration: *Anaspides tasmaniae.* (Illustration by Michelle Meneghini)

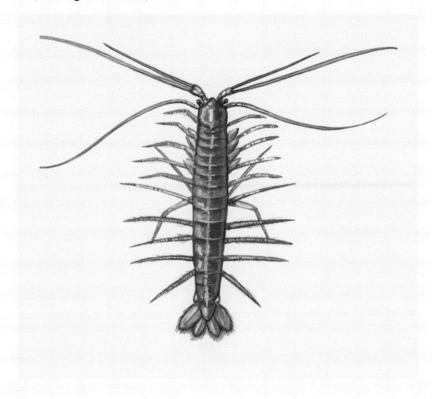

Evolution and systematics

The order Anaspidacea contains four families that have sometimes been divided into two suborders, the Anaspidinea and Stygocaridinea. The family Anaspididae has been known since the Triassic and exhibits strong relationships to the extinct order Palaeocaridacea. Anaspidaceans are generally thought to be among the most primitive of eumalacostracan crustaceans (that is, members of the subclass Eumalacostraca), especially because their thoracic limbs possess both epipods and exopods. However, details of their circulatory system, foregut morphology, and other aspects of their internal anatomy suggest that anaspidaceans are derived from ancestors leading to the eucaridan decapods. The lack of a carapace has caused debate about their exact placement among eumalacostracan crustaceans, but anaspidaceans are generally thought to be ancestral to at least the line leading to decapods, if not some of the other eumalacostracan groups as well.

Physical characteristics

The anaspidacean head does not bear a carapace. Eyes may be on stalks, sessile on the head, or absent altogether. The first thoracic somite is fused to the head while the remaining seven thoracic somites are free. The first thoracic appendage is modified as a maxilliped, that is, it has a different morphology from the remaining seven pairs of thoracic limbs, and is modified as

a feeding structure. Thoracic limbs two through eight have the endopod developed into full walking legs. On these appendages, the epipods, which emanate from the outside margin of the coxa, function as flattened gills, and the exopods, which come from the basis, function for water movement. Sometimes the exopods are not present. The pleopods may be strongly developed, reduced, or absent. In the Anaspididae, the telson and uropods form a tail fan, but in other families, the uropods are elongate and the telson short. Anaspidaceans can be up to 1.9 in (5 cm) in length, but some species are less than 0.39 in (1 cm). Color is usually a dull brown.

Distribution

Anaspidaceans show a classic Gondwana relict distribution pattern, being found only in Tasmania, southeastern Australia, New Zealand, and southern South America. The family Anaspididae are known only from a small number of localities in Tasmania and Victoria, Australia, while the Stygocarididae are more widely dispersed with species known from Victoria, Australia, New Zealand, Chile, and Argentina.

Habitat

The larger anaspidaceans are generally found in cool mountain streams, lakes, and swamps, while the smaller stygocaridineans are dwellers of the groundwater, living among the sand grains. The swamp-dwelling anaspidaceans live in the burrows of freshwater crayfish, while those found in lakes tend to live in the algal macrophyte mats on the lake bottom, and the stream-dwellers patrol over and among larger rocks of the streambed.

Behavior

Anaspidaceans are not good swimmers. Instead, they spend most of their time walking over the substrate. The exopods of the thoracic legs are in nearly constant motion, most likely circulating fresh oxygen-bearing water past the flap-like epipods. When walking, the legs move in a metachronal pattern, which continues to the pleopods. In fact, in the larger species, the pleopods have the same motion as the walking legs, so that at first glance the animal looks to have a continuous set of legs all the way to the posterior end of the body. When startled, anaspidaceans are capable of an upward jump in which the body is flexed about midway along the back. On relaxation, the animal settles to the bottom and walks about as if nothing had happened. There appears to be no territoriality in anaspidaceans. When two individuals meet, they may touch antennae, but as often as not, one merely walks over the body of the other.

Feeding ecology and diet

Anaspidaceans are generalist feeders, eating organic detritus obtained from the substratum. The larger species may also be scavengers and appear capable of scraping organic films from the surfaces of small pebbles.

Reproductive biology

Mating has so far not been observed in this group. Eggs are laid freely on vegetation or stones and are not guarded or cared for. Hatching occurs 30–60 weeks after the eggs are laid. The young emerge as juveniles. The longer developmental periods are associated with an over-wintering dormancy period.

Conservation status

Anaspidaceans are restricted in their distribution, but for the most part occur in areas where landscape development is minimal. As with many Gondwanan freshwater crustaceans, however, the introduction of trout into the rivers and streams by European colonizers has meant that some species survive only in the small tributaries where fishes cannot go. These crustaceans evolved in the absence of freshwater fishes, so they have no natural defenses against those introduced predators. At present, none of the anaspidaceans are considered to be threatened, and none are listed by the IUCN.

Significance to humans

Anaspidaceans represent an interesting evolutionary branch of crustaceans, and as such, are important in telling the history of life on Earth.

Species accounts

No common name
Anaspides tasmaniae

FAMILY
Anaspididae

TAXONOMY
Anaspides tasmaniae Thomson, 1892.

Anaspides tasmaniae

OTHER COMMON NAMES
None known.

PHYSICAL CHARACTERISTICS
Eyes stalked; body elongate, with thoracic and abdominal somites of nearly similar size; pleopod of outer ramus long and capable of touching the substrate; telson and uropods forming a tail fan. (Illustration shown in chapter introduction.)

DISTRIBUTION
Tasmania.

HABITAT
Freshwater streams and shallow pools.

BEHAVIOR
Walking involves the use of both thoracic and abdominal appendages, moving in a metachronal beat from anterior to posterior. It employs a vertical escape jump when startled, but otherwise walks constantly about in search of food.

FEEDING ECOLOGY AND DIET
Has been observed to gnaw on large plant fragments, scrape the surfaces of small pebbles with its mouth appendages, and will scavenge carcasses of small dead organisms.

REPRODUCTIVE BIOLOGY
Mating behavior is unobserved at present. Eggs are laid on plants or bark. Young hatch as juveniles.

CONSERVATION STATUS
Not listed by the IUCN.

SIGNIFICANCE TO HUMANS
An intellectual curiosity. ◆

Resources

Books
Schram, F. *Crustacea.* Oxford, U.K.: Oxford University Press, 1986.

Periodicals
Swain, R., and C. I. Reid. "Observations on the Life History and Ecology of *Anaspides tasmaniae.*" *Journal of Crustacean Biology* 3 (1983): 163–172.

Les Watling, PhD

Euphausiacea
(Krill)

Phylum Arthropoda
Subphylum Crustacea
Class Malacostraca
Order Euphausiacea
Number of families 2

Thumbnail description
Small marine crustaceans known as "krill" that are found in all the world's oceans and are critical to the marine ecosystem, providing a link between plankton and larger species in the food chain

Photo: Krill (*Euphausia superba*) cooked by thermal water on Deception Island, Antarctica. (Photo by John Shaw. Bruce Coleman, Inc. Reproduced by permission.)

Evolution and systematics

The class Malacostraca comprises three subclasses: Eumalacostraca, Hoplocarida, and Phyllocarida. The subclass Eumalacostraca comprises the superorders Eucarida, Pancarida, Peracarida, and Syncarida. The superorder Eucarida includes the order Decapoda (crabs, lobsters, crayfish, shrimp) and the order Euphausiacea.

There are two families within the order Euphausiacea: Euphausiidae and Bentheuphausiidae. Ten genera comprise the family: Euphausiidae *Euphausia, Meganyctiphanes, Nematobrachion, Nematoscelis, Nyctiphanes, Pseudeuphausia, Stylocheiron, Tessarabrachion, Thysanoessa,* and *Thysanopoda.*

There is no fossil record of the order Euphausiacea. Today, the order includes about 85 species, of which five are found in the Antarctic, including the dominant *Euphausia superba*. The cold-adapted krill has benefited from climate change in the Antarctic, completely replacing the temperate-zone pelagic fishes that inhabited the region in the late Eocene.

Physical characteristics

Krill have the same basic body plan as other crustaceans such as lobsters or shrimp. Their elongated cephalothorax bears up to 13 pairs of limbs, 6–8 of which form a net-like structure, with bristles adapted for sieving food from the water. An additional five pairs of paddle-like limbs called swimmerets or pleopods, used for propelling the krill through the water, are found on the segmented abdomen and tail. Unlike more advanced crustaceans, however, euphausiids have exposed gills, which lie below the carapace.

Krill have two pairs of antennae, prominent compound eyes, and transparent skin with red spots of pigment. Mycosporine-like amino acids in their tissues, derived from algae they consume, absorb UV light and help to prevent sun damage. The gut is visible through the skin and may appear variously colored depending on their diet. Krill species range from less than 0.5 in (1.25 cm) to several inches (centimeters) in length.

Krill are sometimes called "light-shrimp," a name deriving from bioluminescent organs on their eyestalks and body, called photophores, that produce a yellow-green or blue light. These may be used for mating displays or to confuse predators. The bioluminescent protein, luciferin, is believed to be obtained through consumption of dinoflagellates.

Distribution

Krill are found in all the world's oceans, from tropical regions to the Arctic and Antarctic.

Habitat

Krill live in coastal waters, in the open ocean, and around sea ice. Most live within reach of the surface, where they feed and spawn, although there are a few little-known species living at depths of up to 16,400 ft (5,000 m).

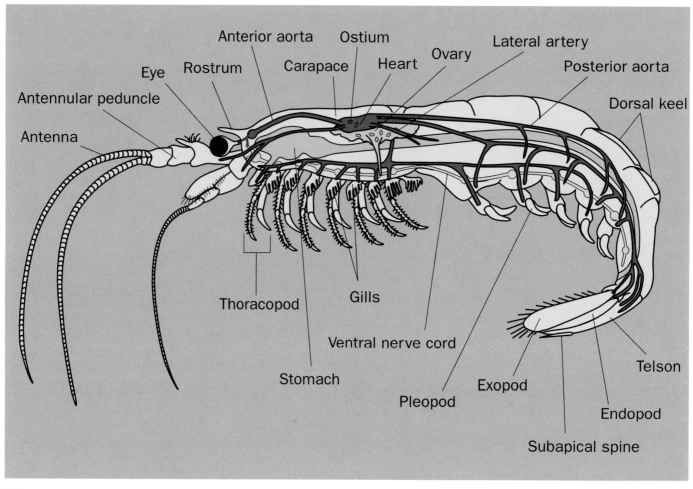

Krill anatomy. (Illustration by Christina St. Clair)

Behavior

Eighteen krill species, including the most common varieties, have been observed to congregate in large groups called swarms. Depending on the species and location, krill swarms may pack as many as one million individuals, or hundreds of pounds (kilograms) of biomass, into each cubic foot (0.03 cubic m), and can extend over several hundred square miles (square kilometers). The density of organisms is not constant; in some areas, there may only be a few individuals per ft^2 (0.09 m^2). There may also be aggregations called shoals, less dense than swarms, with 10–100 krill per 35 cubic ft (1 cubic m). Swarming behavior is generally less apparent during the winter than at other times of year.

In most krill species, the swarms tend to be found at lower depths during the day to escape predation. The cooler temperatures also allow them to conserve energy by slowing their metabolism. At night, they rise to the surface in order to feed, accomplishing these vertical migrations through control of their buoyancy.

The pleopods of euphausiids allow them to maneuver, and in fact, being heavier than water, they must swim in order to stay afloat. They do this sporadically, alternating with periods of rest. Despite their swimming ability, krill do not propel themselves over large horizontal distances, and although not strictly planktonic (drifting) during most of their lifecycle, they are to some degree at the mercy of the current. However, by adjusting their buoyancy, they may be able to take advantage of current variations at different depths.

Feeding ecology and diet

Krill are generally surface feeders, and phytoplankton is an important component of their diet; it must grow where light is available for photosynthesis. Other elements of the diet, depending on the species, may include algae, diatoms, and copepods.

Krill filter their food from the water as they swim, using a "feeding basket" formed from bristles on their thoracic legs. As water is squeezed through the basket, the food is left behind, and the krill use their legs to convey it forward to their mouth.

An unusual adaptation seen in krill is the ability to reduce their size in response to scarcity of food. Unlike most other crustaceans, krill continue to molt throughout their lifespan. Under austere conditions, they may produce a new exoskeleton of a smaller size and shrink, using some of their body protein (they do not maintain significant fat stores) for fuel.

Krill are key organisms in the ecology of the oceans, providing an important food source not only for whales but also for other marine mammals, fishes, cephalopods, and sea birds. Their concentration in large swarms provides ample nutrition even for very large animals such as whales. Krill are critical in translating the food yield of plankton further up the food chain.

Reproductive biology

Reproduction only takes place where food is abundant; eggs are rich in lipids. Male krill produce spermatophores, which they transfer to the female using the uppermost abdominal appendages. The sperm is stored by the female and later released for fertilization as the eggs pass out of the genital opening. Females may spawn several times during the season, each brood consisting of thousands of free-floating eggs. Spawning occurs near the surface. The higher temperature of the shallow water allows faster development of the eggs, thereby limiting exposure to predators, and ensures that the offspring hatch into a food-rich environment.

After hatching as larvae, krill mature through juvenile stages (called nauplius, protozoea, zoea, and cyrtopia) into the adult form over a period of a few months, with segments and appendages as well as growth added at new molts.

Adult krill shed their sexual characteristics after the summer spawning season, and return to a juvenile-like state. They mature again in the spring. The krill lifespan is between two and 10 years, depending on the species.

Conservation status

Krill are abundant, and no species are listed by the IUCN. However, ecologists are wary of the possibility of over-fishing because of the key role that krill play in ocean ecosystems. Other potential threats to the krill population include habitat destruction and climate change.

Significance to humans

About 330,000 tons (300,000 kg) of krill per year are fished commercially, mostly for fish and animal feed. The largest single use is in aquaculture. Krill pigments are responsible for the pink color of salmon flesh, and krill-based feed is used extensively for farmed salmon as well as yellowtail and other fish.

Northern krill (*Meganyctiphanes norvegica*) eat algae and zooplankton that live on ice. (Photo by D. Larsen. Bruce Coleman, Inc. Reproduced by permission.)

In some regions such as Japan, krill products are produced for human consumption. Krill provide a plentiful, high-protein food source, but processing problems are significant. The organisms deteriorate quickly after death, because of the release of powerful digestive enzymes. In addition, their exoskeletons, which have a high-fluoride content, must be removed before the meat can be processed into food products for humans.

Other potential krill products being explored include food additives (proteins, omega-3 fatty acids) and enzymes and other biochemical products for pharmaceutical and industrial use.

In addition to harvesting the krill themselves, fishermen sometimes seek flocks of seabirds feeding on krill as indicators that a swarm is present and, therefore, other krill predators of interest, such as salmon, are probably in the vicinity.

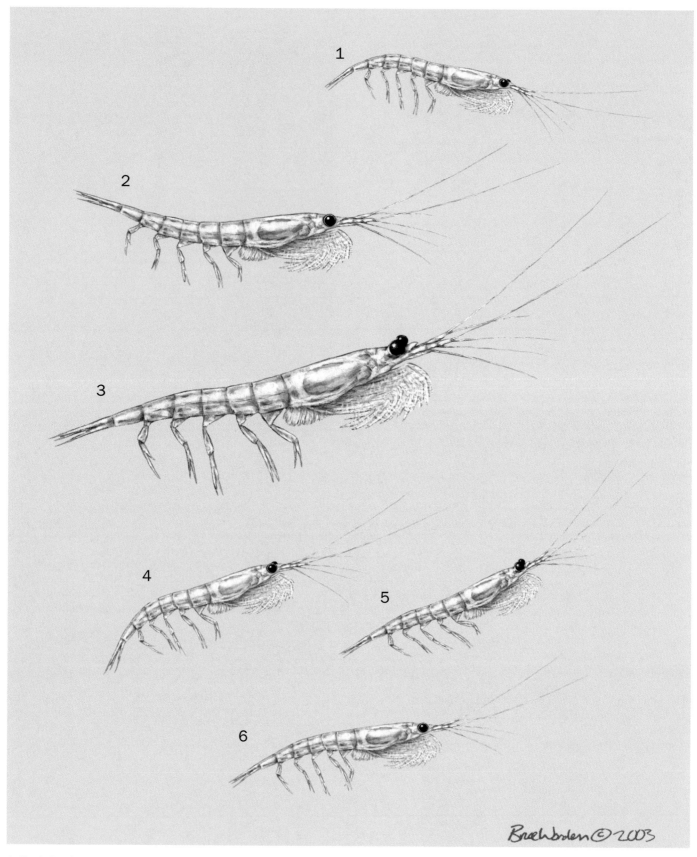

1. North Pacific krill (*Euphausia pacifica*); 2. Nordic krill (*Meganyctiphanes norvegica*); 3. Antarctic krill (*Euphausia superba*); 4. *Thysanoessa inermis*; 5. *Thysanoessa spinifera*; 6. *Thysanoessa raschii*. (Illustration by Bruce Worden)

Species accounts

North Pacific krill
Euphausia pacifica

FAMILY
Euphausiidae

TAXONOMY
Euphausia pacifica Hansen, 1911.

OTHER COMMON NAMES
Japanese: Isada.

PHYSICAL CHARACTERISTICS
About 0.75 in (1.9 cm) long, weighing about 0.003 oz (0.1 g).

DISTRIBUTION
North Pacific Ocean, from North America to Japan.

HABITAT
Open water, beyond the continental shelves or at their edges. Observed from the surface to about 984 ft (300 m) in depth.

BEHAVIOR
Migrates vertically to the food-rich sea surface in the early part of the night, feeds, and then sinks again upon satiation. Tends not to feed during the day even if food is relatively abundant below the surface.

FEEDING ECOLOGY AND DIET
Diet consists of phytoplankton and zooplankton. Predators include salmon, cod, herring, halibut, rockfish, whales, and seabirds.

REPRODUCTIVE BIOLOGY
Lifespan is about two years, somewhat longer for females. In general, females take part in two spawning seasons, while males take part in only one.

CONSERVATION STATUS
Not listed by the IUCN. One of the most common krill species along the west coast of North America.

SIGNIFICANCE TO HUMANS
Studies have been done to determine the feasibility of establishing fisheries off the California coast; this species makes up about 70% of the harvest of a controlled active krill fishery in British Columbia, products of which are mainly used in fish feed for aquaculture and aquaria. Also fished commercially off the coast of Japan. ◆

Antarctic krill
Euphausia superba

FAMILY
Euphausiidae

TAXONOMY
Euphausia superba Dana, 1850.

OTHER COMMON NAMES
English: Southern krill.

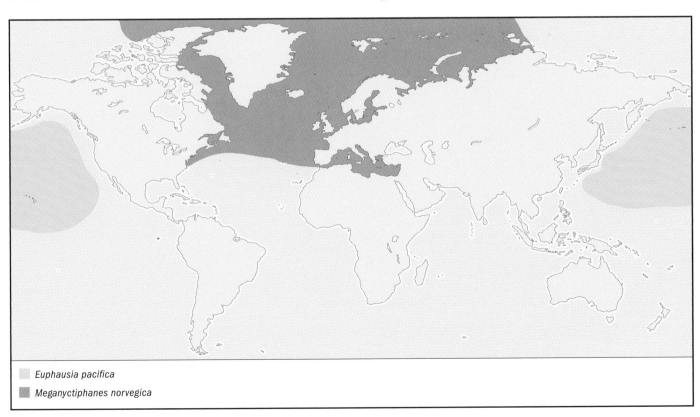

■ *Euphausia pacifica*
■ *Meganyctiphanes norvegica*

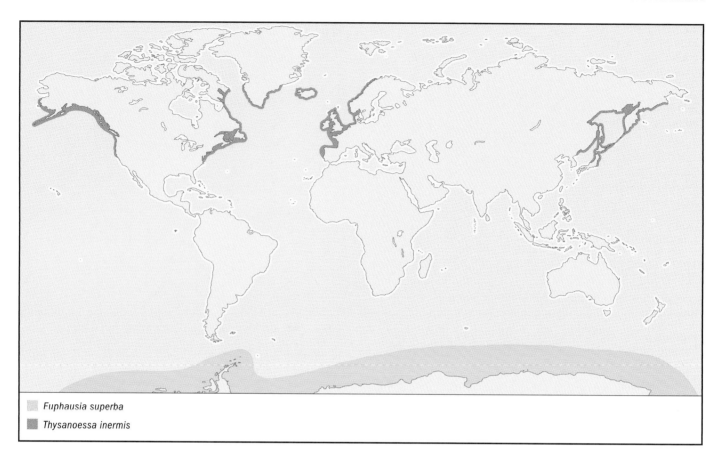

Euphausia superba

Thysanoessa inermis

PHYSICAL CHARACTERISTICS
Among the largest species of krill; up to 2.5 in (6.5 cm) in length at maturity and weighing up to 0.07 oz (2 g). Gut appears green from predominance of phytoplankton in diet.

DISTRIBUTION
Southern ocean around Antarctica, over an area of about 13,500,000 mi² (35,000,000 km²), almost five times the area of Australia.

HABITAT
Found from the surface to about 1,640 ft (500 m) depth. They congregate under and around sea ice edges to feed. Swarms may extend under the sea ice to about 8 mi (13 km) in from the edge, especially in the winter. Areas just south of sea ice edges have been found to have especially high krill densities. Juveniles in particular seek haven from predators in small crevices in the ice. Larger topographical features under the ice such as keels and pressure ridges also attract higher krill densities.

BEHAVIOR
Density in swarms may be as high as 30,000 krill per 35 cubic ft (1 cubic m). Some swarms appear to be segregated by age or sex. Usually rise to the surface at night to feed, remaining deeper during the day. Occasionally will surface during the day, for unknown reasons. An anti-predation strategy is that, upon being disturbed, some individuals will suddenly molt while fleeing, leaving behind their exoskeletons as decoys. This may be effective when fleeing predators such as seabirds that target individual krill, but offers no protection against high-volume feeders such as whales.

FEEDING ECOLOGY AND DIET
Feeds mainly on phytoplankton; also sea ice algae, and copepods. Strategies for surviving the Antarctic winter include the ability, demonstrated in laboratory studies, to withstand months of starvation; they accomplish this without significant fat reserves by reducing their size, metabolizing some of their own mass in the process and producing a smaller exoskeleton at their next molt. Absent large schools of pelagic fish such as the herring of the North Atlantic, this species occupies this niche as the keystone species of the ecosystem; they are staple food of many of the larger organisms in the Southern Ocean, including fish, squid, marine mammals, and birds.

REPRODUCTIVE BIOLOGY
Females lay as many as 10,000 eggs at a time, and may spawn several times in a five-month season. The eggs sink thousands of feet (meters) before they hatch, after which the larvae spend 10 days making their way back to the surface to feed and mature. Grows to its maximum size over three to five years; lifespan up to seven years.

CONSERVATION STATUS
Not listed by the IUCN. The Antarctic krill biomass, estimated at about 550 million tons (490,000,000 metric tons), may be among the largest of any multi-cellular animal. In response to an explosion of commercial fishing in the 1970s, a treaty called the Convention on the Conservation of Antarctic Marine Living Resources (CCAMLR) was signed in 1981 and took effect in 1982 under the auspices of the Antarctic Treaty System. It sets conservative catch limits in order to protect against overfishing of krill and encourage recovery of whale populations.

SIGNIFICANCE TO HUMANS
Fished commercially by several countries, despite logistical problems and expense associated with operating in the Antarctic. Most fisheries are in the South Atlantic, around the Antarctic Peninsula and South Georgia Island. Current catch is almost 100,000 tons (90,000 metric tons) per year. Processed Antarctic krill is made into products for human consumption, food for domestic animals and aquaculture, and sport fishing bait. ◆

Nordic krill
Meganyctiphanes norvegica

FAMILY
Euphausiidae

TAXONOMY
Meganyctiphanes norvegica M. Sars, 1857.

OTHER COMMON NAMES
English: Northern krill.

PHYSICAL CHARACTERISTICS
Length about 1.2–1.5 in (3–4 cm); relatively short legs. Weighs about 0.017 oz (0.5 g) at maturity. Gut appears red because of a diet of copepods.

DISTRIBUTION
North Atlantic, Arctic Ocean, and Mediterranean Sea.

HABITAT
Wide climactic range, from sub-Arctic zones around Greenland and Scandinavia down to North Carolina and the Mediterranean Sea. Pelagic, found from the surface to a depth of 984 ft (300 m).

BEHAVIOR
Surface swarms may contain almost 800,000 individuals per 35 cubic ft (cubic m) and extend for more than 1,075 ft² (100 m²). Swarms may be segregated by sex. Like most other krill species, it rises to the surface at night to feed. This endogenous rhythm has been found to be influenced by moonlight and synchronized to the lunar cycle; perturbations associated with the lunar eclipse have been observed.

FEEDING ECOLOGY AND DIET
Feeds mainly on copepods; surveys in the Northeast Atlantic found *C. finmarchicus* making up 85–95% of prey mass. Will also consume other foods, including phytoplankton and other euphausiids. Predators include marine mammals, fishes, and birds.

REPRODUCTIVE BIOLOGY
Spawning occurs in the winter in the Mediterranean, but in colder waters in the summer. Maturity occurs after about a year, and individuals generally participate in two breeding seasons. Lifespan is about 2.5 years.

CONSERVATION STATUS
Not listed by the IUCN.

SIGNIFICANCE TO HUMANS
Commercial fisheries in Canada, and small-scale harvesting in the Mediterranean. ◆

No common name
Thysanoessa inermis

FAMILY
Euphausiidae

TAXONOMY
Thysanoessa inermis Kroyer, 1846.

OTHER COMMON NAMES
None known.

PHYSICAL CHARACTERISTICS
About 1.2 in (3 cm) long at maturity, weighing about 0.005 oz (0.15 g).

DISTRIBUTION
North Atlantic and North Pacific.

HABITAT
Coastal waters in cool, temperate regions, surface to about 984 ft (300 m) depth. Does not breed above 65–70°N latitude.

BEHAVIOR
Tends to remain near the sea bottom during the winter. Daytime surface swarms may appear during the spring and summer, the result of diurnal vertical migration. Reproduction carried out while swarming.

FEEDING ECOLOGY AND DIET
Primarily feeds on phytoplankton; up to one-third of the diet may consist of copepods, depending on location. In the northeast Atlantic, surveys have found *C. finmarchicus* to be the dominant prey. Predators include pelagic fish such as capelin as well as larger marine animals.

REPRODUCTIVE BIOLOGY
Spawns in May and June, during the spring phytoplankton bloom. Individuals may participate in two spawning seasons. Lifespan is 2–4 years.

CONSERVATION STATUS
Not listed by the IUCN.

SIGNIFICANCE TO HUMANS
Commercial fisheries off the coast of Japan and in the Gulf of St. Lawrence, Canada. ◆

No common name
Thysanoessa raschii

FAMILY
Euphausiidae

TAXONOMY
Thysanoessa raschii M. Sars, 1864.

OTHER COMMON NAMES
None known.

PHYSICAL CHARACTERISTICS
About 0.98 in (2.5 cm) long at maturity, weighing about 0.004 oz (0.13 g).

DISTRIBUTION
Arctic Ocean, northern North Atlantic, and northern North Pacific.

Thysanoessa raschii

Thysanoessa spinifera

HABITAT
Cool temperate to Arctic regions, near shoreline, from the surface to about 984 ft (300 m) in depth.

BEHAVIOR
Tends to remain below 328 ft (100 m) depth in the winter. In the summer, resumes its diurnal vertical migration and surface swarming.

FEEDING ECOLOGY AND DIET
Diet is primarily phytoplankton; little copepod feeding observed.

REPRODUCTIVE BIOLOGY
Spawning occurs in the spring in the temperate zone, and during the summer in colder waters. Maturity reached in the second year; individuals generally participate in two spawning seasons. Lifespan is two years or more.

CONSERVATION STATUS
Not threatened.

SIGNIFICANCE TO HUMANS
Experimental fisheries in the Gulf of St. Lawrence, Canada. ◆

No common name
Thysanoessa spinifera

FAMILY
Euphausiidae

TAXONOMY
Thysanoessa spinifera Holmes, 1900.

OTHER COMMON NAMES
None known.

PHYSICAL CHARACTERISTICS
Approximately 0.75 in (1.9 cm) long.

DISTRIBUTION
North Pacific Ocean, from the Americas to Asia.

HABITAT
Cooler coastal waters over the continental shelves.

BEHAVIOR
Swarms at the surface during the day from spring through fall. This behavior increases surface predation, and it is not known what benefit outweighs this disadvantage and has led to this behavior.

FEEDING ECOLOGY AND DIET
Diet consists of phytoplankton and zooplankton. Predators include salmon, rockfish, seabirds, and whales.

REPRODUCTIVE BIOLOGY
Lifespan is about two years.

CONSERVATION STATUS
Not listed by the IUCN. One of the most common krill species along the West Coast of North America, where they provide up to 80% of the diet of some whale species.

SIGNIFICANCE TO HUMANS
Among the species harvested in a controlled commercial fishery in British Columbia. Studies have been done to determine the feasibility of establishing fisheries off the California coast. ◆

Resources

Books

Howard, Dan. "Krill." In *Beyond the Golden Gate: Oceanography, Geology, Biology and Environmental Issues in the Gulf of the Farallones,* edited by Herman A. Karl, John L. Chin, Edward Ueber, Peter H. Stauffer, and James W. Hendley II. Reston, VA: United States Geological Survey Circular 1198, 2002.

Nicol, Stephen. *Krill Fisheries of the World,* Technical Paper 167. Rome: Food and Agriculture Organization of the United Nations, 1997.

Periodicals

Bamstedt, U., and K. Karlson. "Euphausiid Predation on Copepods in Coastal Waters of the Northeast Atlantic." *Marine Ecology Progress Series* 172 (1998): 149–168.

Brierly, A. S., et al. "Antarctic Krill Under Sea Ice: Elevated Abundance in a Narrow Band Just South of Ice Edge." *Science* 295 (March 8, 2002): 1890–1892.

Dalpadado, P., and H. R. Skjoldal. "Abundance, Maturity and Growth of the Krill Species *Thysanoessa inermis* and *T.*

longicaudata in the Barents Sea." *Marine Ecology Progress Series* 144 (1996): 175–183.

Nakagawa, Y., Y. Endo, and H. Sugisaki. "Feeding Rhythm and Vertical Migration of the Euphausiid *Euphausia pacifica* in Coastal Waters of North-eastern Japan During Fall." *Journal of Plankton Research* 25, no. 6 (2003): 633–644.

Tarling, G., F. Buchholz, and J. Matthews. "The Effect of Lunar Eclipse on the Vertical Migration Behavior of *Meganyctiphanes norvegica* in the Ligurian Sea." *Journal of Plankton Research* 21, no. 8 (1999): 1475–1488.

Wheeler, D. L., et al. "Database Resources of the National Center for Biotechnology Information." *Nucleic Acids Research* 28, no. 1 (2000): 10–14.

Other

International Council of Scientific Unions, Scientific Committee on Antarctic Research. "Evolution in the Antarctic (EVOLANTA): Science Plan and Implementation Plan" [July 25, 2003]. <http://www.nioo.knaw.nl/projects/scarlsssg/docs/Evolbiosciplan.pdf>.

Sherri Chasin Calvo

Amphionidacea
(*Amphionids*)

Phylum Arthropoda

Subphylum Crustacea

Class Malacostraca

Order Amphionidacea

Number of families 1

Thumbnail description
Single-species order of small crustaceans with carapace covering head and thorax; females carry developing eggs in a brood chamber; development is multi-staged and complex

Illustration: *Amphionides reynaudii.* (Illustration by Bruce Worden)

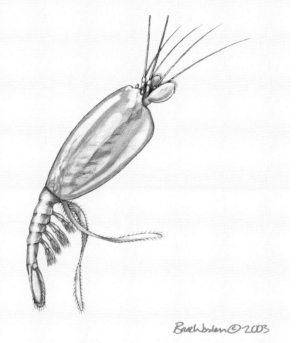

Evolution and systematics

Amphionids have been studied very little, and as a result, the amount of information about them is very limited. No adult males or brooding females have yet been found.

Until the 1970s, the amphionidacea were classified as a family among the 30-odd families of caridean shrimps of the infraorder Caridea, because of similarities between the early larval stages of amphionidacea and those found among caridean shrimp species. Larval amphionids at various stages were sometimes classed as distinct species. In 1973, Williamson proposed reclassifying the amphionidacea as a separate, single-species order within the superorder Eucarida. He based his proposal on characteristics unique to the amphionids, including the brood chamber of the female and the structure of the first pair of pleopods in the female.

For the purposes of this chapter, the taxonomic breakdown for the Amphionidacea is: phylum Arthropoda, subphylum Crustacea, class Malacostraca, and order Amphionidacea. There is one family, Amphionididae, one genus, *Amphionides*, and one species, *Amphionides reynaudii* (H. Milne Edwards, 1832).

Amphionides reynaudii is the single extant species in its order that would have once contained others all descended from an ancestor in which females shed eggs directly into the water before evolving the distinctive brood chamber.

Physical characteristics

Amphionids are marine crustaceans with a number of unique, identifying characteristics. An adult amphionid is about 1 in (2.6 cm) long. The head and thorax are covered by a thin, shell-like carapace that encloses the dorsal and ventral surfaces and about half the animal's total length, with considerable space between the ventral body surface and the bottom of the carapace. There are gaps in the front and rear of the carapace. The thorax and abdomen are distinct and differentiated.

The head bears a pair of stalked, multifaceted eyes, the stalks each carrying a structure that is probably a photophore, or light-producing organ. There are two sets of biramous (two-branched) antennae, each antenna of the second pair bearing large, leaf-shaped accessory structures, the antennal scales, or scaphocerites. There are seven pairs of thoracic appendages, the paraeopods, all but the seventh being biramous. Only the first pair, close to the mouth, is functional, the maxillipeds being used for swimming. There is a wide gap between the first and second pair of appendages. The second to sixth pairs are thin and sticklike. They are useless for swimming and probably are not remnants of true maxillipeds. In the female, the fifth thoracic paraeopods are much longer than the rest, while the coxae, or proximal segments of the sixth paraeopods, carry genital pores. The seventh thoracic segment lacks appendages in the female, but bears a pair of uniramous (unbranched) appendages in juvenile males. Pairs of feathery pleurobranchs (gills) are mounted on the third to seventh thoracic somites (segments).

The heart, placed dorsally in the rear part of the thorax, is well developed. In females, ovaries run from the heart region to oviducts that lead to the genital pores in the sixth thoracic appendages.

Two pairs of longitudinal muscles link the posterior thorax and the abdomen. The abdomen divides into six somites, of which five bear appendages, the pleopods. In the female, the pleopods on the first abdominal segment are very long, thin, and flattened, able to reach forward to about half the length of the carapace. When extended forward, these specialized appendages are able to cover the rear and part of the forward gap of the carapace, thus rendering it a brood chamber. The unusually long fifth thoracic pareaopods may serve to clean and tend the eggs. In the male, judging by the larval stages, the first pair of pleopods is stoutly constructed and capable of vigorous propulsion.

Following the first pair of pleopods are four more pairs of shorter, biramous pleopods. The last segment carries a pair of uropods, or tail appendages.

Distribution

The Amphionidacea live throughout the oceans of the world, with some concentration in the equatorial regions. (Specific distribution not known; no map available.)

Habitat

Larval and juvenile forms live in a shallow, planktonic layer at depths of 90–300 ft (30–100 m). Adult females have been found in a depth zone of 21,000–111,000 ft (700–3,700 m).

Behavior

Judging from incomplete information, free-swimming larval and juvenile forms consume microorganisms as food; the pre-adult forms reside within swarms of nutrient-rich oceanic plankton. Adult forms sink to deeper ocean levels for breeding. Newly hatched individuals rise into the shallow planktonic layer to start feeding and maturing through the larval and juvenile stages. Males are able to swim faster than females, by virtue of the males' strong first pair of abdominal segments, giving them a survival edge, yet the only purpose of the males may be to mate once.

Feeding ecology and diet

What amphionids eat and how they feed is poorly known. Larval and juvenile forms probably feed on algae and microscopic fauna. Adult females have only a degenerate mandible and maxillule (jaw parts) and reduced digestive tracts with stomachs that are still full of food ingested during the pre-adult stage. They may not feed at all as adults, but live off their stored food.

Reproductive biology

Young pass through up to 13 larval stages, then molt to a juvenile (postlarval) form, and then developed into the adult form. The number of larval stages may vary among regions and even among individuals in one region. Forms intermediate between known larval stages are often found.

The eggs form within the ovaries, pass through the oviducts to the genital aperture in the coxae of the sixth thoracic appendages, and then into a brood chamber defined by the carapace, the ventral thorax, and the first pair of pleopods, which can extend forward and over the bottom of the carapace. Although mating details are still not known, fertilization most likely takes place in the brood chamber and the developing eggs are kept there until hatching. Hatchlings are presumably released by the female pulling her first abdominal appendages loose enough to provide a gap through which the young may escape.

The photophores of the eyestalks may be used for signaling and attracting the opposite sex.

Conservation status

This species is not listed by the IUCN, but it is probably not threatened because of its wide distribution.

Significance to humans

The value of these little-studied crustaceans is mainly scientific. They are probably destined to attract more attention and study by marine biologists due to their unique array of characteristics that offer new insights into evolution and adaptation.

Resources

Books

Holthuis, L. B. *The Recent Genera of the Caridean and Stenopodidean Shrimps (Decapoda); with an Appendix on the Order Amphionidacea.* Leiden, The Netherlands: Nationaal Natuurhistorisch Museum, 1993.

Periodicals

Heegaard, P. "Larvae of Decapod Crustacea: The Amphionidae." *Dana Expedition*, Report 77 (1969): 1–67.

Lindley, J. A., and Hernández, F. "The Occurrence in Waters Around the Canary and Cape Verde Islands of *Amphionides reynaudii*, the Sole Species of the Order Amphionidacea (Crustacea: Eucarida)." *Revista de la Academia Canaria de las Ciencias* 11, no. 3–4 (1999): 11–119.

Williamson, D. I. "*Amphionides reynaudii* (H. Milne Edwards), Representative of a Proposed New Order of Eucaridan Malacostraca." *Crustaceana* 25, no. 1 (1973): 35–50.

Kevin F. Fitzgerald, BS

Decapoda

(Crabs, shrimps, and lobsters)

Phylum Arthropoda
Subphylum Crustacea
Class Malacostraca
Order Decapoda

Thumbnail description
Crustaceans with a large carapace covering the head and thorax and enveloping the gill chambers; also possess five pairs of legs

Number of families 151

Photo: The bubble coral shrimp (*Vir philippinensis*) is very widespread throughout Indo-Pacific waters. This species is always found in association with the bubble-shaped coral. They often come in pairs, and they live in the crevices among the bubbles of the coral, using the spaces for shelter. They appear not to do any harm to the coral, and may help to keep the coral clean. This particular specimen is pregnant with a load of eggs. (Photo ©Tony Wu/www.silent-symphony.com. Reproduced by permission.)

Evolution and systematics

The order Decapoda falls within the class Malacostraca, a group of crustaceans that comprises twice as many species as all other crustacean classes combined. Malacostracans are distinguished by bodies divided into 19 segments (five head, eight thoracic, and six abdominal segments) and the location of their genital openings. Many of the divisions within the Malacostraca are based in part on the number of thoracic appendages that have been integrated into the head region to function as mouthparts; these appendages are termed maxillipeds. In the decapods, the first three thoracic appendages are maxillipeds, leaving five pairs of legs or ten feet for walking. The order name Decapoda literally means "ten-footed."

The two suborders within the Decapoda are based in part on differences in the gill structure. The gills of the Dendrobranchiata consist of bundles of branching filaments, while those of the Pleocyemata are unbranched— either as filaments (trichobranchs) or more commonly as unbranched leaflike plates (phyllobranchs). The two groups also differ in their method of reproduction: female Pleocyemata brood their eggs on their abdominal appendages until they hatch, while female Dendrobranchiata spawn their eggs into the sea.

Many of the subsequent divisions within the Pleocyemata can be recognized by the number and position of appendages with claws. The classification of crustaceans remains in flux, however, reflecting the great degree of uncertainty as of 2003 about the relationships among the groups. The following list represents one of the more common classifications of the major groups within the Decapoda:

- Suborder Dendrobranchiata
- Superfamily Penaeoidea, the penaeid shrimps
- Superfamily Sergestoidea, the sergestids
- Suborder Pleocyemata
- Infraorder Caridea, the true shrimps
- Infraorder Stenopodidea, the boxer shrimps
- Infraorder Astacidea, the clawed lobsters and crayfish
- Infraorder Thalassinidea, the ghost and mud shrimps
- Infraorder Palinura, the spiny and slipper lobsters
- Infraorder Anomura, the hermit crabs, king crabs, and porcelain crabs
- Infraorder Brachyura, the true crabs

The earliest decapod to appear in the fossil record is *Palaeopalaemon newberryi*, which lived during the late Devonian period, about 360 million years ago (mya). It shares characters of both lobsters and shrimps, leaving unanswered questions about its taxonomic affiliations. The Dendrobranchiata are considered the most primitive extant decapods, but early fossils of dendrobranchs are lacking. Fossil representatives of all of the infraorders except the Stenopodidea are known from the Triassic (213 mya) and Jurassic (145 mya) periods onward, and the Brachyura have shown extensive radiation (diversification) since the Eocene epoch, 33.7 mya.

Physical characteristics

In addition to having five pairs of legs, decapods are distinguished by the carapace that covers the dorsal portion of the head and thorax, forming a single functional unit termed the cephalothorax. The sides of the carapace extend downward to envelop the gills, forming lateral branchial chambers. Although the vast majority of decapods are aquatic and breathe with gills, some terrestrial forms have developed blood vessels in the inner surface of the branchial chambers so that they function as lungs.

Decapods have two pairs of antennae, as do all crustaceans. The first pair typically bears special chemosensory structures that govern the sense of smell, while the second pair is often elongate and tactile. The foremost legs often have claws that perform several functions related to feeding, mating, and defense.

The structure and function of the abdomen varies among the different decapod taxa. Lobsters and shrimplike forms have large muscular abdomens that terminate in a flattened tail fan. To evade predators, these animals snap their tail fans rapidly beneath their abdomens, which propel them backwards. Shrimps and other decapods that don't have heavy or thick exoskeletons use their abdominal appendages to swim forward. At the other extreme, the true crabs or brachyurans have lost most of their abdominal appendages; their abdomen plays no role in locomotion. The few remaining appendages in brachyurans are used only for egg attachment in females or as copulatory structures (gonopods) in males.

Decapods exhibit tremendous diversity in shape, size, and color. They range in size from minute parasitic pea crabs to the giant Japanese spider crab *Macrocheira kaempferi*, which has legs spanning up to 12 ft (3.7 m). Many species have distinctive color patterns, while others are able to change color by expanding or contracting specialized groups of pigment cells in the epidermis underlying their exoskeleton.

The many variations of the basic decapod body plan reflect the great success of this group and their adaptation to a wide range of habitat types and ecological roles. Although the familiar terms "shrimp," "lobster," and "crab" have no formal taxonomic meaning, they do represent three distinct and recognizable body plans. Of the shrimplike forms, the Dendrobranchiata and the Caridea are the most familiar; they include all commercially harvested shrimps and prawns.

The Dendrobranchiata have perhaps the least-modified body plan of all the Decapoda and also show little diversity in form. The first three pairs of legs on penaeid shrimps have small claws that are equal in size. This is a relatively small group of approximately 350 species, and is primarily tropical or subtropical in distribution. Some Caridea superficially resemble the penaeids, but the third pair of legs never has claws, and the abdomen usually has a pronounced hump in the middle. This diverse group shows an unusually wide range of variation in body form; there are about 1800 described species.

The Stenopodidea are a very small group (25 species) of tropical shrimps. Like the penaeids, they have claws on their first three pairs of legs, but their third pair is greatly enlarged. The ghost and mud shrimps (Thalassinidea) are an especially problematic group whose taxonomic affinities remain unclear. The Thalassinidea are thin-shelled burrowing forms that often have large claws on the first pair of legs and small claws on the second.

Heavily armored crustaceans with large muscular abdomens are commonly referred to as "lobsters"; again, this category combines groups with little relationship to one another besides their desirability as food. The Astacidea (crayfish and clawed lobsters) infraorder encompasses about 800 species that have large claws on the first pair of legs and small claws on the next two. The infraorder Palinura is a small (130 species) group found primarily in tropical and subtropical waters, which contains the spiny and slipper lobsters along with some deep-water forms that should probably be classified elsewhere. Slipper and spiny lobsters lack true claws on their front appendages and have a distinctive and unusual flattened larval form known as a phyllosoma.

Crabs have the most compact decapod body form. In the true crabs or Brachyura, the abdomen is greatly reduced in size, folded beneath the body, and involved only in reproduction. Only the first pair of legs has claws. This is a very successful group, accounting for roughly half the 10,000 known species of decapods. The Anomura is composed of a wide range of unusual crablike animals, including the hermit crabs, king crabs, sand crabs, and porcelain crabs. It is an almost entirely marine group, with about 1800 species worldwide. Like the Brachyura, the Anomura usually have claws on their first pair of legs; however, anomuran crabs can be distinguished by the fact that their last pair of legs is much smaller and often tucked into the gill chamber. In addition, their second pair of antennae is positioned beside the eyes rather than between them as in the brachyurans.

Anomurans have the widest range of variation in abdominal form of any of the decapods. Hermit crabs have elongate, soft, asymmetrical abdomens that they protect with an empty snail shell. The king crabs are believed to have evolved from hermit crabs; although the abdomen is tucked underneath the body, all retain the asymmetry of hermit crabs and some even have completely soft, unprotected abdomens. At the other extreme, porcelain crabs have thin flat abdomens superficially similar to those of true crabs, but have retained a tail fan and a limited ability to swim.

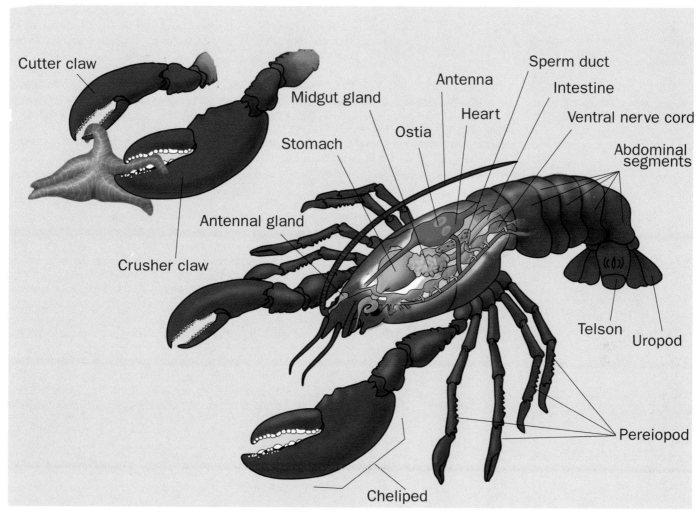

Decapod anatomy. (Illustration by Christina St. Clair)

Distribution

Decapod crustaceans occur worldwide. The marine species are most diverse in the Indo-Pacific region but relatively uncommon in extreme deep-sea and polar waters. Freshwater crabs and crayfish are most diverse in Southeast Asia and North America, respectively; crayfish occur naturally on all continents except Africa and Antarctica. Terrestrial decapods are largely limited to tropical and subtropical regions and occur on all continents except Europe and Antarctica. Approximately 90% of all species of decapods are marine; fewer than 1% are terrestrial.

Habitat

Although there are some species of midwater shrimps that seldom if ever encounter the bottom of the ocean, the vast majority of decapods are associated with benthic habitats. Marine decapods live in all types of habitats from intertidal mud flats to deep-sea hot vents. Such highly structured habitats as rock or coral reefs, seagrass beds, and mangrove forests support the greatest numbers of species, but even featureless sand

and mud bottoms support many types that are adept at burying themselves. Most decapods emerge to feed only at night when fewer predatory fishes are active.

Structures to hide in are also important for freshwater decapods; they are most abundant in vegetated areas. Many freshwater decapods construct burrows that allow them to remain in contact with ground water when their pond dries up. Terrestrial decapods are also dependent on water; while some crabs can occur as much as 9 mi (15 km) from the ocean and up to 3,280 ft (1,000 m) above sea level, they all have planktonic larvae and must return to the ocean to breed.

Behavior

Decapods exhibit many complex and even spectacular behaviors. The Caribbean spiny lobster *Panulirus argus* sometimes migrates toward deeper water in long lines or queues of as many as 65 individuals. The reason for these migrations is not entirely clear, but seems to be associated with avoidance of winter storms. Juvenile red king crabs (*Paralithodes*

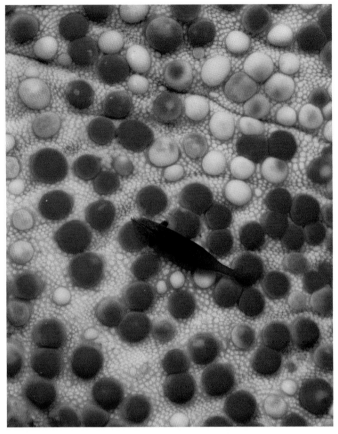

This individual of the genus *Periclimenes* is on the bottom of a pin cushion star (*Culcita* sp.). (Photo ©Tony Wu/www.silent-symphony.com. Reproduced by permission.)

camtschaticus) often gather together into mounds that may contain thousands of individuals, possibly to deter predators.

Many decapods use visual and even auditory signals to communicate with one another. Sounds are usually produced with some type of stridulating surface, and this means of communication is most common in terrestrial and semiterrestrial species. Stridulation refers to sounds produced by rubbing body parts together. Communication using pheromones appears to be common in aquatic species, especially in conjunction with mating. Pheromones are released in the urine via the antennal gland; when crayfish fight they literally blow pheromone-laden urine into the face of their opponent.

Experiments have demonstrated that crabs and lobsters are capable of such complex learning as navigating a path through a maze. Crabs that are offered novel items of prey quickly learn the most effective way to feed on them, and decapods have been taught to respond to a specific cue or to discriminate among colors.

The activity patterns of intertidal species are often synchronized with the tidal cycle. For example, a crab may emerge to feed only during nighttime high tides, and this pattern becomes set in the animal's own biological clock. When the crab is placed in captivity away from all tidal influence, its set rhythm of activity can persist for days or even weeks.

Many marine decapods form symbiotic associations with other organisms. Some shrimp set up cleaning stations where fishes line up to be picked over for parasites. Others may live full-time with a fish, building and maintaining a shared burrow while the fish watches for predators. Far more have established less formal associations with larger organisms that offer protection from predators, such as the many kinds of shrimps that associate with sea anemones.

Feeding ecology and diet

Decapods employ a wide variety of feeding techniques, ranging from filter feeding, grazing, and deposit feeding to predation. Some are specialists that use just one of these methods, while others are generalists that make use of several different techniques depending on the circumstances. One of the most common misconceptions about crabs and other decapods is that they are primarily scavengers, since many are harvested from baited pots. Most large marine crustaceans are actually very efficient predators and only scavenge when the opportunity arises. The decapod body plan allows for a great degree of specialization in feeding structures; this specialization is especially apparent in the structure of the claws. Fast, slender claws can be used to snatch elusive prey, while massive, strong claws armed with molars can exert tremendous force on mollusks and other hard-shelled prey. Much of the elaborate ornamentation and other features seen in marine snails and other mollusks is a side effect of predation by shell-crushing crabs, as these groups are engaged in what is essentially an evolutionary arms race.

In many cases decapods have evolved asymmetrical claws, thus providing more than one type of tool for subduing and extracting prey. For example, the pistol or snapping shrimp of the family Alphaeidae have one enormously enlarged claw that produces a loud noise when snapped and can stun or kill prey. Apart from its use as a weapon, however, the large claw is virtually useless for most other functions, and the much smaller claw on the other side is used to convey food to the shrimp's mouth. On the other hand, claws are not essential to predatory decapods: palinurid lobsters and many shrimps lack large claws, but are still remarkably effective predators. Spiny lobsters have exceptionally strong mandibles (jaws) that are used to crush mollusks and other prey, and many shrimp are quite skilled in using their walking legs to envelop worms or smaller crustaceans.

In many tropical areas land crabs play the role of earthworms, as the primary recyclers of fallen leaves and other plant material. They also help to turn over the soil with their burrowing activities. Many marine and freshwater species are herbivorous as well, but most will occasionally ingest animal material when it is available.

A number of anomuran crabs and caridean shrimps are exclusively filter feeders, pulling detritus and plankton out of the water with their maxillipeds, antennae, or modified legs. Many more are deposit feeders, consuming much the same type of material after it has settled out of the water column. Deposit feeding crabs often have downturned claws that are adapted for scooping up sediment so that it can be processed by the animal's mouth parts.

Reproductive biology

With the exception of one species of crayfish, decapods reproduce sexually; and in almost all cases the sexes are separate. A number of caridean shrimps are protandric hermaphrodites, which means that they mature as males first and later change into females; a few species retain male structures and become functional hermaphrodites. Some species of snapping shrimps live in colonies with only one reproducing "queen," much like social insects.

In species where there is intense competition for mates, the claws of males are often proportionately much larger than those of females. An extreme example of this disproportion is seen in the fiddler crabs (*Uca* spp.), in which one of the claws of the male is useless for feeding and functions only to attract females and duel with other males. Courtship and mating can take anywhere from seconds to weeks, depending on the species. Copulation in penaeid and caridean shrimps is often an instant affair, while mating in crabs is often a long process involving extended periods of guarding before and after mating. In these cases the female crab can only mate while in her soft-shell condition, so she must molt her exoskeleton immediately prior to mating. A female preparing to molt releases pheromones that attract males; the male will embrace and carry the female for days or even weeks preceding her molt. Male hermit crabs are often seen dragging a smaller female about by the shell in anticipation of mating.

Fertilization can be external or internal, depending on the taxon. Many female brachyuran crabs have internal receptacles for sperm storage where sperm can remain viable for years, and males have evolved a number of different strategies to help insure that it is their sperm that fertilizes the eggs. The male will often stand guard over the female while she is soft to prevent others from mating with her, or block her genital openings with a sort of "chastity belt" to prevent matings with other partners.

In all groups except the Dendrobranchiata the eggs are brooded on the female's pleopods (abdominal appendages) until they hatch. Females clean and aerate their egg masses, and a chemical cue from the eggs stimulates the female to violently shake the mass when the larvae are ready to be released. Parental care generally ends with hatching, although young crayfish often continue to associate with their mother for protection. Some tropical crabs that make use of the freshwater trapped in bromeliad plants practice true maternal care, bringing food to their developing young and protecting them from predators.

Decapods in cold and temperate regions usually release their larvae in the spring to coincide with plankton blooms, while those in the tropics often reproduce year round. In most cases development consists of several planktonic larval stages followed by a stage that makes the transition to a benthic existence; in crabs these are known as the zoea and megalops stages, respectively. The megalops has well-developed pleopods and can swim like a shrimp, but once it finds an appropriate place to settle it molts to the first juvenile stage and the pleopods are lost.

Conservation status

In the year 2002 there were 176 species of decapods listed by the IUCN. Of these, 159 were freshwater crayfish; all but two of the remaining species were shrimps or brachyuran crabs that also live in freshwater. Freshwater species often have very limited distributions, making them especially susceptible to habitat destruction or degradation. Only one marine species was listed, but virtually nothing is known about the populations of most marine decapods, especially those that are not commercially exploited. Three species of crayfish and three species of shrimp were listed as threatened in the United States.

Significance to humans

The order Decapoda encompasses nearly all of the crustaceans that are used for human consumption, and supports many large and valuable fisheries. In addition, penaeid shrimps and crayfish are extensively cultured for food in many parts of the world.

A number of human fatalities have been caused by the consumption of poisonous crabs. Several reef-dwelling species of Indo-Pacific crabs appear to acquire toxins from their food; since toxicity varies with the crab's diet and location, it can be very difficult to determine whether one of these crabs is safe to eat. In other areas, decapods are often host to such schistosome parasites as lung flukes, which can infect humans who eat raw or poorly cooked freshwater crabs or crayfish.

In some rice farming areas land crabs are considered serious pests, both because they eat the plants and because they dig burrows that drain water from the fields. The burrowing activities of thalassinid shrimps have had serious effects on oyster culture; when present in large numbers they loosen the substrate to such an extent that the oysters sink into it and are smothered.

The intentional and unintentional introduction of some decapods to new areas has caused a number of detrimental effects. For example, the accidental introduction of the European green crab (*Carcinus maenas*) to the eastern coast of the United States has resulted in serious population reductions in the clam fisheries there. Various intentional introductions of freshwater crayfish have resulted in crop damage and threats to native species of crayfish.

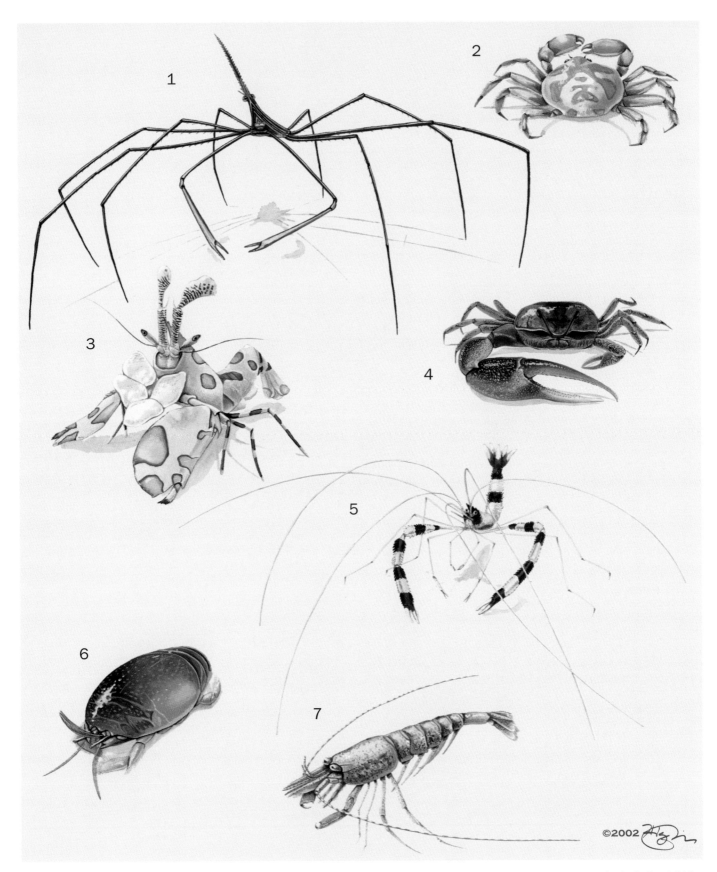

1. Yellowline arrow crab (*Stenorhynchus seticornis*); 2. Pea crab (*Pinnotheres pisum*); 3. Harlequin shrimp (*Hymenocera picta*); 4. Sand fiddler crab (*Uca pugilator*); 5. Banded coral shrimp (*Stenopus hispidus*); 6. Pacific sand crab (*Emerita analoga*); 7. Sevenspine bay shrimp (*Crangon septemspinosa*). (Illustration by Jonathan Higgins)

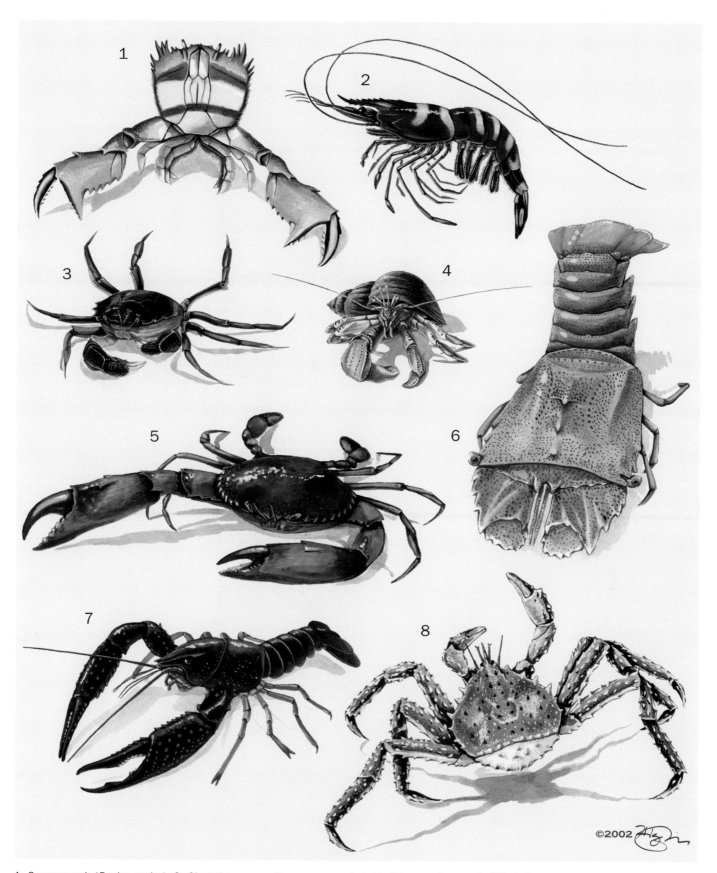

1. Spanner crab (*Ranina ranina*); 2. Giant tiger prawn (*Penaeus monodon*); 3. Chinese mitten crab (*Eriocheir sinensis*); 4. Common hermit crab (*Pagurus bernhardus*); 5. Mangrove crab (*Scylla serrata*); 6. Flathead locust lobster (*Thenus orientalis*); 7. Red swamp crayfish (*Procambarus clarkii*); 8. Red king crab (*Paralithodes camtschaticus*). (Illustration by Jonathan Higgins)

Species accounts

Red swamp crayfish

Procambarus clarkii

FAMILY
Cambaridae

TAXONOMY
Cambarus clarkii Girard, 1852, Texas, United States.

OTHER COMMON NAMES
English: Red crayfish, Louisiana crayfish; French: Écrevisse américaine, écrevisse de Louisiane, écrevisse rouge; German: Roter Amerikanischer Sumpfkrebs; Italian: Gambero americano, gambero rosso della Louisiana.

PHYSICAL CHARACTERISTICS
Body length to 4.7 in (120 mm). Claws long and narrow. Color dark red, sometimes nearly black; claws with bright red tubercles (small nodules) and red tips.

DISTRIBUTION
Native to the southern United States and northern Mexico. Widely introduced and now found in Europe, Africa, Central and South America, and Southeast Asia.

HABITAT
Sluggish streams, swamps and ponds, especially where there is aquatic vegetation and leaf litter. Most abundant in areas that are seasonally flooded. Tolerant of a wide range of conditions, including brackish water.

BEHAVIOR
Can dig burrows over 2 ft (0.6 m) deep to reach the water table during the dry season.

FEEDING ECOLOGY AND DIET
Omnivorous. Feeds on a wide variety of plants and smaller animals, including insects, snails, tadpoles, and small fishes.

REPRODUCTIVE BIOLOGY
Mature males occur in two forms. Form I is the sexually active stage, distinguished by enlarged claws, hooks at the base of some of the walking legs, and hardened gonopods. Following the reproductive season, males molt and revert to smaller-clawed Form II. Females store sperm and extrude eggs several weeks to months after mating. Eggs are brooded 2–3 weeks; there are no free-swimming larval stages, and the hatchlings are recognizable as small crayfish. The young may stay with the mother for several weeks. There are typically two generations per year. Lifespan is 12–18 months in the wild.

CONSERVATION STATUS
Not listed by the IUCN. An abundant species whose range is continually expanding due to intentional introductions and escape from aquaculture operations.

SIGNIFICANCE TO HUMANS
One of the most important aquaculture species, widely introduced throughout the world due to fast growth rate and tolerance of a wide range of conditions. Often raised in crop rotation with rice in the United States. Introduced into eastern

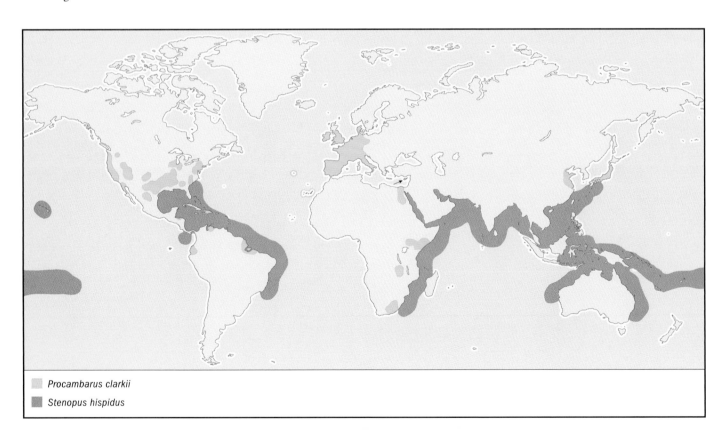

Procambarus clarkii
Stenopus hispidus

Africa to help control snails that host schistosome parasites. War emblem for the Houmas Indians. Commonly sold for home aquariums; also used as bait. ◆

Sevenspine bay shrimp

Crangon septemspinosa

FAMILY
Crangonidae

TAXONOMY
Crangon septemspinosa Say, 1818, eastern coast of United States.

OTHER COMMON NAMES
English: Salt-and-pepper shrimp, sand shrimp; French: Crevette de sable, crevette grise.

PHYSICAL CHARACTERISTICS
Length to 3 in (75 mm). First leg is subchelate. Body is somewhat dorsoventrally compressed; rostrum small. Color typically grayish with dark and light speckling, matching the sediment in its habitat.

DISTRIBUTION
Northern Gulf of Saint Lawrence to eastern Florida; Alaska.

HABITAT
Typically found from 0 to 115 ft (0–35 m), but collected from waters as deep as 1,460 ft (450 m). Occurs in a wide variety of habitats but most abundant on mud flats, sand, and in eelgrass beds; very tolerant of low salinities.

BEHAVIOR
Remains buried in the sediment in the daytime, emerging at night to seek prey on and above the bottom. Populations appear to migrate offshore to deeper water in the winter, possibly to avoid extremely cold temperatures.

FEEDING ECOLOGY AND DIET
Primarily a predator, feeding on mysids, amphipods, worms, and mollusks. May also prey on newly-settled flatfish. Food is generally swallowed whole and quickly ground up in the gastric mill of the cardiac stomach. The bay shrimp is known to scavenge opportunistically and will also eat diatoms, algae, and detritus.

REPRODUCTIVE BIOLOGY
Studies of closely related species of *Crangon* reveal that most individuals are protandric hermaphrodites (maturing first as males and then changing to female as they get larger) while some "primary females" remain the same sex throughout life. Eggs are brooded on the pleopods; large egg-bearing females migrate into estuaries in spring and release their larvae, which are carried out to sea, while smaller females release their larvae in the late fall. Larvae pass through seven zoeal stages before settling out; those hatched in the spring migrate to shallow areas where they grow rapidly.

CONSERVATION STATUS
Not listed by the IUCN. One of the most abundant shrimps in the northwestern Atlantic, and an important prey item for many species. Population trends not known.

SIGNIFICANCE TO HUMANS
Although occasionally fished in the past, this species is considered too small to be worthwhile for human consumption. Sometimes sold as fish bait. ◆

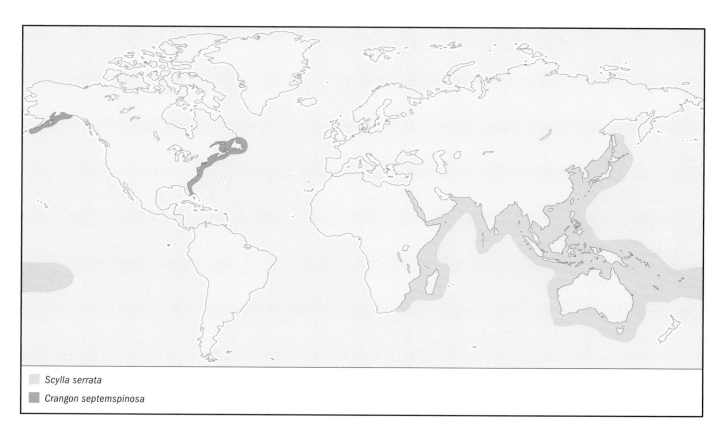

Scylla serrata

Crangon septemspinosa

Chinese mitten crab
Eriocheir sinensis

FAMILY
Grapsidae

TAXONOMY
Eriocheir sinensis H. Milne-Edwards, 1853, China.

OTHER COMMON NAMES
English: Chinese river crab; hairy-fisted crab; French: Crabe chinois; German: Wollhandkrabbe.

PHYSICAL CHARACTERISTICS
Carapace somewhat square in outline and only slightly wider than long; about 3.1 in (80 mm). Claws have a thick, dense covering of hair. Color is olive green to dark brown.

DISTRIBUTION
Native to China and Korea. Accidentally introduced to Germany in the early 1900s and spread to other parts of northern Europe; first appeared in San Francisco Bay in the early 1990s.

HABITAT
Freshwater rivers and streams; estuaries.

BEHAVIOR
Burrows into mud banks, especially in areas that are under tidal influence. Readily leaves the water during migrations to bypass dams and other obstructions.

FEEDING ECOLOGY AND DIET
Omnivorous; feeds primarily on aquatic plants, but also on insects and other aquatic invertebrates.

REPRODUCTIVE BIOLOGY
Catadromous, which means that it lives in fresh water but returns to the ocean to reproduce. Adults migrate to salt water in the fall to mate; both sexes spawn only once and then die. Females carry broods of 250,000 to one million eggs; larvae are planktonic for one to two months and pass through five zoeal stages. Megalopae settle out in estuaries; as juveniles they live initially in burrows constructed in intertidal zones before starting their upstream migration. Known to migrate as far as 800 mi (1300 km) upstream.

CONSERVATION STATUS
Not listed by the IUCN. Overfished in some areas of its native range. Populations in Europe have fluctuated dramatically since their introduction.

SIGNIFICANCE TO HUMANS
Highly prized as food in China and often cultured. When present in large numbers, migrating adults have caused considerable damage to fishing nets and their catch in Europe. Burrowing activities contribute to the erosion of stream banks and have raised concern about the integrity of dikes and similar structures. Mitten crabs are intermediate hosts for lung flukes, which can be transmitted to humans who eat raw or undercooked crabs. ◆

Pacific sand crab
Emerita analoga

FAMILY
Hippidae

TAXONOMY
Hippa analoga Stimpson, 1857, California.

Emerita analoga
Eriocheir sinensis

OTHER COMMON NAMES
English: Pacific mole crab; Spanish: Limanche; pulga de mar

PHYSICAL CHARACTERISTICS
Carapace length to 1.4 in (35 mm). Body smooth and egg-shaped; legs flattened and lacking claws. Color grayish.

DISTRIBUTION
Occasionally found as far north as Kodiak, Alaska, but more typically found between Washington State and Magdalena Bay in Baja California. Range resumes again in the southern hemisphere from Peru to the Straits of Magellan.

HABITAT
Found on surf-swept sand beaches.

BEHAVIOR
Tends to migrate up and down the beach with the tide, staying in the surf zone and rapidly burying itself with its flattened legs and uropods (part of the tailfan). Specimens are often found buried in sand at low tide.

FEEDING ECOLOGY AND DIET
Filter-feeds on detritus and phytoplankton, using its long feathery second pair of antennae.

REPRODUCTIVE BIOLOGY
Mating occurs in California in late spring and summer; most larvae are hatched in July and August. This species has five zoeal stages.

CONSERVATION STATUS
Not listed by the IUCN. Extremely abundant and important prey for fish and shorebirds. Population trends not studied.

SIGNIFICANCE TO HUMANS
Commonly used for bait in recreational fishing. Has shown potential as an indicator species for monitoring toxic algal blooms and pollution. ◆

Harlequin shrimp
Hymenocera picta

FAMILY
Hymenoceridae

TAXONOMY
Hymenocera picta Dana, 1852, Raraka, Tuamotu Archipelago.

OTHER COMMON NAMES
English: Clown shrimp, dancing shrimp, painted shrimp; German: Östliche Harlekingarnele; Italian: Gamberetto arlecchino, gamberetto dipinto.

PHYSICAL CHARACTERISTICS
Length to about 2 in (5 cm). Claws on the second pair of legs are formed into extremely large and distinctive flattened plates. Body and claws brilliantly marked with purple or red blotches on a white or cream background.

DISTRIBUTION
East Africa and the Red Sea to Indonesia, across northern Australia to Hawaii, Panama, and the Galapagos Islands.

HABITAT
Found intertidally and subtidally on coral reefs, hiding among branches of coral.

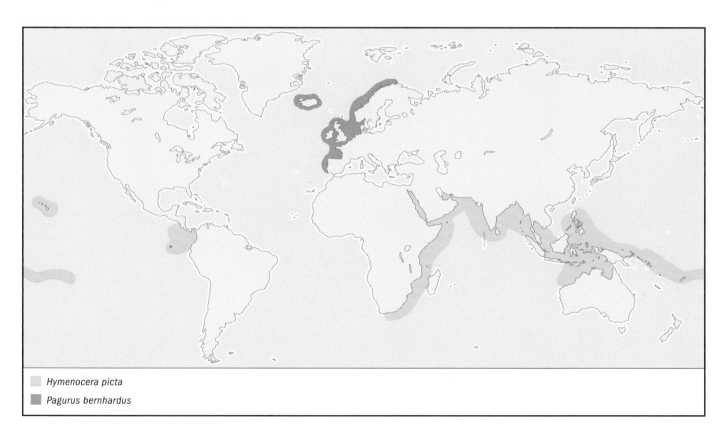

Hymenocera picta

Pagurus bernhardus

BEHAVIOR
Pairs are territorial. Single individuals are much more active than those in pairs.

FEEDING ECOLOGY AND DIET
Feeds on sea stars; detects prey by scent. The large, flattened claws are used to dislodge and turn over the sea star, which is eaten from the tips of the arms inward. It can take up to a week for a pair of these shrimp to eat a sea star.

REPRODUCTIVE BIOLOGY
Lives in male-female pairs. Females mate after molting, which occurs every 18–20 days. They produce about 1000 eggs per brood. Eggs hatch within 18 days, releasing well-developed larvae with a short period of planktonic development.

CONSERVATION STATUS
Not listed by the IUCN.

SIGNIFICANCE TO HUMANS
Popular in the aquarium trade; one of the few species of marine crustaceans that can be entirely bred and raised in captivity. Will prey on small crown-of-thorns sea stars (*Acanthaster plancki*), so may play a role in controlling this coral-eating species. ◆

Red king crab
Paralithodes camtschaticus

FAMILY
Lithodidae

TAXONOMY
Maja camtschatica Tilesius, 1815, Kamchatka.

OTHER COMMON NAMES
English: Alaska king crab; American king crab; French: Crabe royal; German: Kamtschatkakrabbe, Königskrabbe; Italian: Granchio reale; Norwegian: Kamtsjatkakrabbe.

PHYSICAL CHARACTERISTICS
An exceptionally large species, with a carapace up to 11 in (280 mm) in width and a leg span that can exceed 6 ft (1.8 m). Carapace and legs covered with numerous sharp spines; right claw larger than left and only three pairs of walking legs visible. Color reddish-brown.

DISTRIBUTION
Sea of Japan to northern British Columbia. Introduced into the Barents Sea in the 1960s; it has since spread westward as far as Norway.

HABITAT
Found at depths of 10–1,190 ft (3–366 m). Usually occurs on fairly open sand or mud bottoms.

BEHAVIOR
Two-year-old juveniles often gather by the hundreds or thousands to form spectacular mounds known as pods in shallow water. The pods are thought to discourage predators; they disperse shortly after dusk as the crabs forage for prey, and form again before dawn.

FEEDING ECOLOGY AND DIET
Predator on a variety of benthic invertebrates, particularly echinoderms (brittle stars, sea stars, sand dollars, and sea urchins) and barnacles, worms, mollusks, and sponges.

REPRODUCTIVE BIOLOGY
Mating occurs when the female crab is in her soft-shell condition. Prior to molting she is carried by a male who grasps her by the claws in a face-to-face position. Fertilization is external,

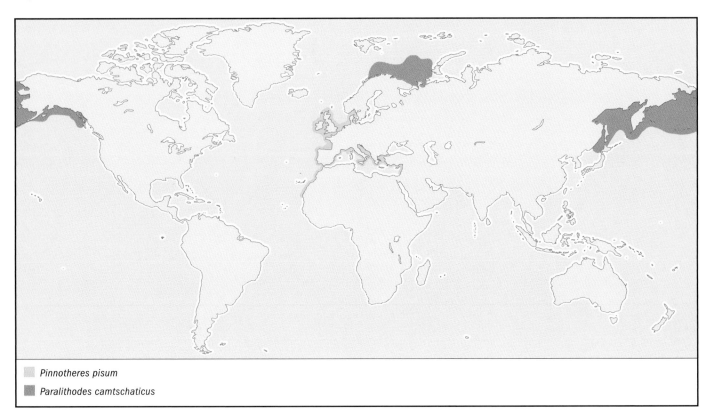

Pinnotheres pisum
Paralithodes camtschaticus

with the male crab using the reduced fifth pair of legs to spread sperm onto the female's pleopods. The eggs are extruded immediately after mating but take nearly a year to hatch; depending on the size of the female the clutches range from 150,000–400,000 eggs. Larval development consists of four zoeal stages followed by a megalops, which settles out in structurally complex habitats.

CONSERVATION STATUS
Not listed by the IUCN. Alaskan populations experienced dramatic declines in the early 1980s, prompting a closure of the fishery and a shift to other species of king crab. Most populations have since shown little recovery. Largely unregulated fishing in the western Bering Sea has resulted in greatly reduced landings from that region as well. The population introduced in the Barents Sea appears to be increasing in size.

SIGNIFICANCE TO HUMANS
Formerly the target of one of the most valuable fisheries in United States waters. Commercial harvesting is carried out with large baited pots. ◆

Yellowline arrow crab
Stenorhynchus seticornis

FAMILY
Majidae

TAXONOMY
Cancer seticornis Herbst, 1788, Curaçao.

OTHER COMMON NAMES
French: Crabe lance, crabe nez pointu; German: Gelblinien-Pfeilkrabbe, Karbische Spinnenkrabbe, Pfeil-Gespensterkrabbe; Portugucsc: Carangucjo-aranha.

PHYSICAL CHARACTERISTICS
Carapace is triangular, with a very long spiny rostrum. Legs are extremely long and slender. Color golden brown with many light and dark lines; fingers of claws blue.

DISTRIBUTION
Found in the western Atlantic from North Carolina and Bermuda to Santa Catarina, Brazil.

HABITAT
Occurs from very shallow water to 1,190 ft (366 m) on a wide variety of substrates, including coral, rocks, and calcareous algae.

BEHAVIOR
Leaves its shelter and migrates along soft corals and gorgonians at dusk, returning to the same location at dawn. Sometimes found in association with the long-spined sea urchin *Diadema antillarum*.

FEEDING ECOLOGY AND DIET
In the field, appears to feed throughout the day on material that accumulated on the surface of its body and appendages as it climbed about during the previous night. Has been observed scavenging and preying on mollusk siphons and small worms in captivity, and has also been observed attaching material to its rostrum for later consumption.

REPRODUCTIVE BIOLOGY
Like other spider crabs, females do not have to molt prior to mating. Egg-bearing females occur throughout the year; eggs hatch in 12 days. The two zoeal stages and megalops take about 20 days to develop.

CONSERVATION STATUS
Not listed by the IUCN. One of the more commonly observed crabs in the Caribbean. Population trends not known.

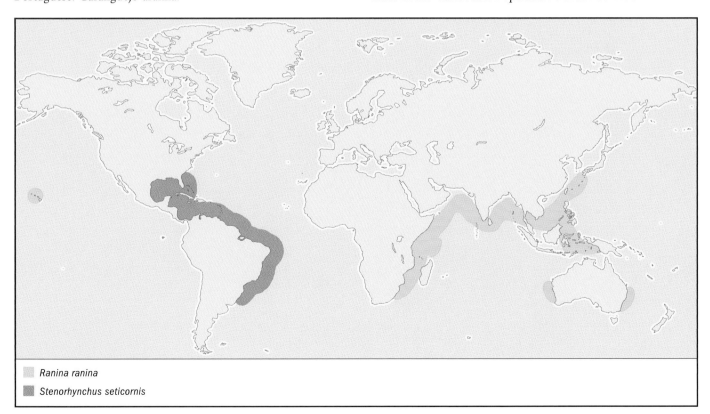

Ranina ranina
Stenorhynchus seticornis

SIGNIFICANCE TO HUMANS
A popular species in the tropical marine aquarium trade. As of 2003 it has not been commercially propagated. ◆

Sand fiddler crab
Uca pugilator

FAMILY
Ocypodidae

TAXONOMY
Ocypoda pugilator Bosc, 1802, "Caroline."

OTHER COMMON NAMES
English: Atlantic sand fiddler crab; French: Crabe violoniste; German: Winkerkrabbe.

PHYSICAL CHARACTERISTICS
Carapace width to 1 in (26 mm). Eyestalks very long and slender. Chelipeds grossly unequal in size in males, small and equal-sized in females.

DISTRIBUTION
Eastern coast of the United States from Cape Cod, Massachusetts to Pensacola, Florida.

HABITAT
Found in intertidal zones on sand and sand-mud beaches in such sheltered areas as the edges of salt marshes and the banks of tidal creeks.

BEHAVIOR
Constructs unbranched burrows in sand. Activity is limited to low tide periods; when fiddler crabs detect the tide in their burrows, they plug the entrances with sand.

FEEDING ECOLOGY AND DIET
Feeds at low tide by scraping up mud with the claws and sorting it with the mouthparts for detritus, bacteria, diatoms, and other organic material. Only the small claw is used in feeding, so males have to feed for twice as long as females.

REPRODUCTIVE BIOLOGY
Males sit near the opening of their burrow and wave their large claw, drumming it against the sediment to attract females. Mating occurs in the male's burrow, where the females remain for two weeks until the eggs are ready to hatch. Larvae released during nocturnal neap high tides and carried out of the estuary; following five zoeal stages, the megalopae return to settle out 6–8 weeks later during a flooding spring tide.

CONSERVATION STATUS
Not listed by the IUCN. Occurs in tremendous numbers. Chief threats are habitat loss and pesticides.

SIGNIFICANCE TO HUMANS
Occasionally appears in the pet trade. ◆

Common hermit crab
Pagurus bernhardus

FAMILY
Paguridae

TAXONOMY
Cancer bernhardus Linnaeus, 1758, North Sea.

OTHER COMMON NAMES
English: Bernard's hermit crab, soldier crab; French: Bernhard l'hermite; German: Einsiedlerkrebs; Norwegian: Eremittkreps; Spanish: Bruja.

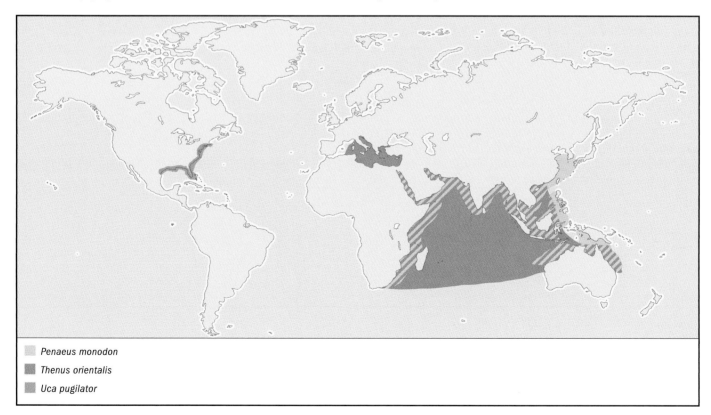

Penaeus monodon

Thenus orientalis

Uca pugilator

PHYSICAL CHARACTERISTICS
Total length to 4 in (100 mm); carapace to 1.5 in (40 mm).
Right claw much larger than left, with a uniform covering of
small granules and spines. Color yellowish-red.

DISTRIBUTION
Northern and western Europe from Norway to Portugal; Iceland.

HABITAT
Small specimens often found in tidal pools in the middle and
low intertidal zones, while larger ones are found offshore. Usually found on sandy and rocky bottoms at depths less than 100
ft (30 m) but collected as deep as 5,850 ft (1,800 m).

BEHAVIOR
Unlike most crustaceans, the hermit crab is more active in the
daytime than at night. Large specimens often use shells of the
common whelk *Buccinum undatum*; smaller individuals are
found in a wide variety of shells including moon snails (*Polin-
ices* spp.) and periwinkles (*Littorina* spp.). Shells selected based
on combination of weight, volume, and aperture size, and often
covered with the hydroid *Hydractinia echinata* or the anemone
Calliactis parasitica, whose stinging cells (nematocysts) help to
deter predators. The crab actively transplants anemones onto
its shell; it prods the anemone with its legs in a manner that
elicits its release so that the anemone can reattach to the hermit's shell.

FEEDING ECOLOGY AND DIET
Adults are deposit feeders, sifting through sediment for organic
material and small animals and also scavenging on discarded
bycatch from trawls. Juveniles often filter feed by using their
maxillipeds.

REPRODUCTIVE BIOLOGY
Females become sexually mature in their first year; those in intertidal zones breed during the winter months. The eggs are
carried on the pleopods and take about 43 days to hatch at a
water temperature of 46.4–50°F (8–10°C); females generally
produce two broods each year. Larval development consists of
four zoeal stages followed by a megalops, which seeks out a
small shell in the intertidal zone.

CONSERVATION STATUS
Not listed by the IUCN.

SIGNIFICANCE TO HUMANS
Commonly displayed in public aquaria and sometimes used as
bait. Larvae are important prey for juvenile salmon. ◆

Giant tiger prawn
Penaeus monodon

FAMILY
Penaeidae

TAXONOMY
Penaeus monodon Fabricius, 1798, Indonesia.

OTHER COMMON NAMES
English: Blue tiger prawn, leader prawn, panda prawn, black
tiger shrimp; French: Crevette géante tigrée; German: Bären-
schiffskielgarnele; Spanish: Camarón tigre gigante.

PHYSICAL CHARACTERISTICS
Largest of the penaeid shrimps, reaching a length of 13.2 in
(336 mm). Body is dark in color with several bold white and
black bands.

DISTRIBUTION
Eastern coast of Africa and the Red Sea, then east to India,
Australia, and Japan.

HABITAT
Juveniles found in nearshore areas and mangrove estuaries;
adults found on silty or sandy bottoms from shallow water to
525 ft (162 m).

BEHAVIOR
Remains buried in the sediment in the daytime. Groups of
200–300 have been observed moving about in shallow water at
dawn and dusk.

FEEDING ECOLOGY AND DIET
Omnivorous, but more predatory than other penaeids. Feeds
primarily on smaller shrimp, crabs, and mollusks; also on algae,
worms, and fishes. Is known to ingest sediment and make use
of organic material and bacteria associated with the sediment.
Feeding activity increases during ebbing tides.

REPRODUCTIVE BIOLOGY
Mating occurs at night, just above the bottom and after the female molts. Sperm is deposited into a special structure called
the thelycum on the underside of the female's thorax. Produces
250,000–800,000 eggs, which are freely released into the water
and hatch within 18 hours into nauplius larvae. Larvae pass
through six nonfeeding naupliar stages in two days; larval development is completed in 12 days. Life span is probably less
than two years.

CONSERVATION STATUS
Not listed by the IUCN. Greatest threat is loss of mangrove
habitat, much of which is being destroyed to create shrimp
farms for culturing *P. monodon*. Farm production has dropped
off dramatically in some areas due to problems with disease.

SIGNIFICANCE TO HUMANS
Most important species in shrimp aquaculture, due to its ex-
ceptionally fast rate of growth, high value, and large size.
Fished commercially with trawl nets. ◆

Pea crab
Pinnotheres pisum

FAMILY
Pinnotheridae

TAXONOMY
Cancer pisum Linnaeus, 1767, Algeria.

OTHER COMMON NAMES
English: Linnaeus's pea crab; French: Pinnothère; German:
Erbsenkrabbe; Spanish: Cangrejo de los mejillones.

PHYSICAL CHARACTERISTICS
Pea crabs are very small in size. The carapace is very round in
outline, to 0.5 in (14 mm) wide in females and 0.3 in (8 mm)
in males. Sexes differ in morphology, with males having much
stouter claws and a hard exoskeleton while females are soft-
bodied.

DISTRIBUTION
Found from southern Scandinavia to western Africa, and throughout the Mediterranean.

HABITAT
In the mantle cavity of such bivalve mollusks as mussels, and inside ascidians.

BEHAVIOR
Although researchers once thought that pea crabs did no harm to their hosts, they have found that their activities do cause damage to the delicate gills of bivalves; the crabs are now considered parasites.

FEEDING ECOLOGY AND DIET
Pea crabs feed on the detritus and phytoplankton that is filtered out by the host's gills, intercepting it before it reaches the host's mouth.

REPRODUCTIVE BIOLOGY
After developing hairlike setae (bristles) on their walking legs that allow them to swim, males leave their host to find and mate with a female. Brooding females are found from April to October in the northern part of their range. Larvae develop in the plankton through four zoeal stages before settling out as megalopae and seeking a host.

CONSERVATION STATUS
Not listed by the IUCN.

SIGNIFICANCE TO HUMANS
None known. ◆

Mangrove crab
Scylla serrata

FAMILY
Portunidae

TAXONOMY
Cancer serratus Forskål, 1775, Red Sea.

OTHER COMMON NAMES
English: Indo-Pacific swamp crab, mud crab, Samoan crab, serrated swimming crab; French: Crabe des palétuviers; Spanish: Cangrejo de manglares.

PHYSICAL CHARACTERISTICS
Large, reaching a carapace width of 11 in (280 mm). Carapace is smooth, with four teeth between eyes and nine alongside the eye. Claws are massive; last pair of legs is flattened for swimming.

DISTRIBUTION
Found widely throughout the tropical Pacific and Indian Oceans, from Africa and the Red Sea to Australia, Japan, and east to Tahiti. Introduced into Hawaii in the 1920s. The distribution and other information given here is a synthesis of information regarding a number of very closely related species of *Scylla* that were not previously distinguished from one another.

HABITAT
Typically found in such muddy and brackish-water areas as mangrove forests, estuaries, and river mouths.

BEHAVIOR
Typically nocturnal, spending the daytime in a burrow that can be up to 6.5 ft (2 m) deep. The flattened, paddle-like rear legs can be used for swimming or for rapidly burying itself in the sediment.

FEEDING ECOLOGY AND DIET
An aggressive predator that uses its large, powerful claws to break open and feed on bivalves, snails, and barnacles; also preys on crabs, shrimp, and fish.

REPRODUCTIVE BIOLOGY
Mating occurs in summer after the female molts, and she may remain in the male's burrow for several days after mating. Stored sperm can remain viable in the spermatheca for at least seven months. Clutches contain 2–8 million eggs and take 2–4 weeks to develop; brooding females may move offshore during this period. Larvae complete five zoeal stages in 2–3 weeks, and the megalops stage lasts at least six days.

CONSERVATION STATUS
Not listed by the IUCN. Severely overfished in many countries, and also affected by loss of critical mangrove habitat. Regulations protecting immature specimens and egg-bearing females (which are considered a delicacy) are often nonexistent or ignored. Although increasingly being cultured, these are essentially "grow out" operations that rely on wild-caught juveniles, so wild populations continue to be depleted. Considerable research has been conducted on large-scale production of laboratory-reared larvae, with promising results.

SIGNIFICANCE TO HUMANS
Important in both commercial and artisanal fisheries. Sometimes considered a serious pest in oyster culture operations. ◆

Spanner crab
Ranina ranina

FAMILY
Raninidae

TAXONOMY
Cancer raninus Linnaeus, 1758, "Mari Indico."

OTHER COMMON NAMES
English: Kona crab, red frog crab; German: Froschkrabbe; Hawaiian: Papa'i-kua-loa.

PHYSICAL CHARACTERISTICS
Males grow as large as 5.9 in (150 mm); females, 4.5 in (115 mm). Claws very unusual in form, with the fixed finger extending at right angles to the hand. Walking legs greatly flattened. Carapace red with a series of white dots and covered with rounded spines; abdomen visible in dorsal view.

DISTRIBUTION
Eastern coast of Africa across Indian Ocean to Indonesia, Australia, Japan, and Hawaii.

HABITAT
Usually found in coralline sand on the outer side of reefs, at depths of 10–325 ft (3–100 m).

BEHAVIOR
Spends most of the time partially buried in sand, with only the anterior part of the carapace visible. Movement is forward and

backward, unlike the typical sideways walk of other crabs. Spanner crabs will often assume a vertical position before fighting, with the claws extended straight out from the sides of the body.

FEEDING ECOLOGY AND DIET
Reportedly ambushes small fish and other organisms from its hiding place in the sand. Commonly found feeding on discarded bycatch from prawn trawlers.

REPRODUCTIVE BIOLOGY
Females mate in a hard-shell condition throughout the year. Mating is initiated by the male, who pries the female out of the sand and maneuvers her into position. After mating, the female reburies herself and the male guards her. Males appear unable to distinguish the sex of prospective mates and often pry up other males. Brood size ranges from 80,000–150,000; eggs take 39–44 days to hatch and females spend much of this time buried. There are eight zoeal stages and a megalops stage; larval development takes 1–1.5 months.

CONSERVATION STATUS
Not listed by the IUCN. Typically fished with baited tangle nets; undersized crabs are often injured when removed from the nets and suffer high mortality. Following some declines in the population of spanner crabs in the 1990s, Australia implemented monitoring and management plans for a sustainable fishery. Population trends in other areas are not known.

SIGNIFICANCE TO HUMANS
Largest commercial fishery for spanner crabs is in Australia, primarily for export to Asian markets. Annual world commercial catch is approximately 980 tons (1,000 metric tons). ◆

Flathead locust lobster
Thenus orientalis

FAMILY
Scyllaridae

TAXONOMY
Scyllarus orientalis Lund, 1793, Indonesia.

OTHER COMMON NAMES
English: Moreton Bay bug, Northern shovelnose lobster, reef bug; French: Cigale raquette; Spanish: Cigavros; cigarra chata.

PHYSICAL CHARACTERISTICS
Length to 11 in (280 mm). Second antennae formed into broad flat plates; all legs lack claws. Eyes located at anterior corners of the carapace. Rear edge of fifth segment of abdomen has median dorsal spine. Color brownish with reddish brown granules.

DISTRIBUTION
Eastern coast of Africa through the Red Sea; rarely in the Mediterranean Sea; Indian Ocean to Indonesia, southern Japan, and Australia.

HABITAT
On soft mud or fine silty sand bottoms; typically found at a depth of 80–195 ft (25–60 m) but found as deep as 2,080 ft (640 m).

BEHAVIOR
Nocturnal. Remains buried in the sediment in the daytime and emerges to feed at night. The large flattened antennae serve as rudders when the lobster flips its tail backward to evade predators. Known to migrate over 50 mi (80 km).

FEEDING ECOLOGY AND DIET
Feeds primarily on clams that it wedges open by using the sharp tips of its legs. Also known to feed on fish and smaller crustaceans.

REPRODUCTIVE BIOLOGY
Reproduction occurs year round, but peaks in spring and early summer. Females carry broods of 16,000–60,000 eggs. Larval development consists of four phyllosoma stages and a settling puerulus stage, and takes about four weeks. Both sexes mature at 4.25 in (108 mm); estimates of longevity range from 3–6 years.

CONSERVATION STATUS
Not listed by the IUCN. Population not known for whole range; however, some localized populations (e.g., Moreton Bay in Australia) have experienced declines attributed to overfishing.

SIGNIFICANCE TO HUMANS
Commercially marketed for food but not the focus of a targeted fishery; taken incidentally in trawls for prawns or scallops. ◆

Banded coral shrimp
Stenopus hispidus

FAMILY
Stenopodidae

TAXONOMY
Palaemon hispidus Olivier, 1811, Australasiatic seas.

OTHER COMMON NAMES
English: Banded boxer shrimp; barber pole shrimp; bandanna prawn; spiny prawn; French: Crevette de corail; German: Rot-Weiß gebänderte Scherengarnele; Italian: Gamberetto meccanico.

PHYSICAL CHARACTERISTICS
Body to 2 in (5 cm) in length. Third pair of legs with greatly enlarged claws; claws and body spiny and boldly marked with broad red and white bands. Antennae long, bright white and conspicuous.

DISTRIBUTION
Circumtropical, but not yet known from the eastern Atlantic. Has one of the widest distributions of any marine invertebrate.

HABITAT
Among rocks and coral reefs, from water as shallow as 10 ft (3 m) to at least 680 ft (210 m).

BEHAVIOR
Largest and most conspicuous of the cleaner shrimps, waving its white antennae to attract fish to its cleaning station. Capable of remembering and recognizing its mate by chemosensory means.

FEEDING ECOLOGY AND DIET
As a cleaner shrimp, the banded shrimp uses its small claws to remove ectoparasites from fishes as well as damaged tissue

around injuries. Also preys on smaller invertebrates and scavenges opportunistically.

REPRODUCTIVE BIOLOGY
Monogamous; found in male-female pairs that can form while they are still juveniles. Female is often the larger of the pair. Reproduces nearly year round, with females molting and mating within two days of hatching the previous brood. Larvae develop as planktonic zoeae; in captivity they take 120–210 days to complete development. Number of larval stages may vary with conditions; as many as nine zoeal stages have been ob-

served. Banded shrimp may delay settlement until an appropriate location is found.

CONSERVATION STATUS
Not listed by IUCN. Population trends unknown. Because of this shrimp's role as a cleaner, its overcollection for the aquarium trade may affect the health of reef fishes. Efforts were underway in 2002 to establish commercial methods of captive propagation.

SIGNIFICANCE TO HUMANS
Extremely popular species for tropical marine aquariums. ◆

Resources

Books

Bauer, Raymond T., and Joel W. Martin, eds. *Crustacean Sexual Biology.* New York: Columbia University Press, 1991.

Bliss, Dorothy E. *Shrimps, Lobsters and Crabs.* New York: Columbia University Press, 1989.

Burggren, Warren W., and Brian R. McMahon. *Biology of the Land Crabs.* Cambridge, U.K.: Cambridge University Press, 1988.

Crane, Jocelyn. *Fiddler Crabs of the World.* Princeton, NJ: Princeton University Press, 1975.

Factor, Jan Robert. *Biology of the Lobster.* New York: Academic Press, 1995.

McLaughlin, Patsy A. *Comparative Morphology of Recent Crustacea.* San Francisco: W. H. Freeman and Company, 1980.

Schram, Frederick R. *Crustacea.* Oxford, U.K.: Oxford University Press, 1986.

Warner, G. F. *The Biology of Crabs.* New York: Van Nostrand Reinhold Company, 1977.

Williams, Austin B. *Shrimps, Lobsters, and Crabs of the Atlantic Coast of the Eastern United States, Maine to Florida.* Washington, DC: Smithsonian Institution Press, 1984.

Periodicals

Dew, C. B. "Behavioral Ecology of Podding Red King Crab, *Paralithodes camtschatica." Canadian Journal of Fisheries and Aquatic Sciences* 47, no. 10 (1990): 1944–1958.

Diesel, R. "Maternal Care in the Bromeliad Crab, *Metopaulias depressus*: Protection of Larvae from Predation by Damselfly Nymphs." *Animal Behaviour* 43, no. 5 (1992): 803–812.

Skinner, D. G., and B. J. Hill. "Feeding and Reproductive Behaviour and their Effect on Catchability of the Spanner Crab *Ranina ranina." Marine Biology* 94, no. 2 (1987): 211–218.

Zhang, Dong, et al. "Mating Behavior and Spawning of the Banded Coral Shrimp *Stenopus hispidus* in the Laboratory." *Journal of Crustacean Biology* 18, no. 3 (1998): 511–518.

Other

Blue Crab Home Page. July 2001 [25 July 2003]. <www.blue-crab.net>.

Crayfish Home Page. 24 Nov. 2002 [25 July 2003]. <http://crayfish.byu.edu/crayhome.htm>.

Fiddler Crabs (Genus *Uca.* 31 March 2003 [25 July 2003]. <www.fiddlercrab.info/>.

Gregory C. Jensen, PhD

Mysida
(Mysids)

Phylum Arthropoda
Subphylum Crustacea
Class Malacostraca
Order Mysida
Number of families 4

Thumbnail description
Small shrimplike crustaceans with a flexible carapace enveloping the thoracic region along the sides; stalked eyes; and a well-developed tail fan

Photo: The glass shrimp *Mysis relicta* is found in freshwater. (Photo by James H. Robinson/Photo Researchers, Inc. Reproduced by permission.)

Evolution and systematics

Fossil mysids have been dated as far back as the Triassic period, about 248–213 million years ago (mya). A group of fossil crustaceans known as Pygocephalomorpha, which includes a number of Paleozoic genera from the Carboniferous and Permian periods (360–248 mya), is possibly related to the mysids.

Two suborders for the order Mysidacea, the Lophogastrida and Mysida, were recognized in 1883. This arrangement persisted for nearly a century until some researchers proposed alternative taxonomic schemes to explain the relationships among mysidaceans, including raising both suborders to order level. This proposal, which has been sustained by a number of experts, is followed here.

The order Mysida includes four families: Petalophthalmidae, with six genera; Mysidae, with six subfamilies (one, the Mysinae, comprises seven tribes) and almost 140 genera; Lepidomysidae, with only one genus, *Spelaeomysis*; and Stygiomysidae, also with only one genus, *Stygiomysis*. The order as a whole includes slightly over 1,000 described species.

Mysids are sometimes known as opossum shrimps because of the marsupium, or external pouch, formed by specially developed plates on the inner sides of the thoracic limbs of adult females.

Physical characteristics

Most mysids are fairly small, between 0.39 and 1.18 in (10 and 30 mm) long. They have a shieldlike carapace that covers the cephalon (head region) and most of the thorax. The carapace is fused with the first three (in rare cases the fourth) thoracic somites (segments). The eyes are usually stalked and movable; the cornea is generally developed with visual elements, but tends to be reduced with increasing depth of habitat. The antennules (antenna 1) and antennae (antenna 2) are biramous (forked). Male mysids typically bear a setose (bristly) lobe, the processus masculinus, on the peduncle of the antennules.

The thorax has eight pairs of pereopods or thoracic limbs, all of which are divided into two branches, an endopod and an exopod. The endopods of the first and sometimes the second pereopod are usually transformed into gnathopods (specialized appendages for feeding), which differ considerably from the remaining limbs. Female mysids have a marsupium made of fewer than seven pairs of lamellae (oostegites or brood plates), in which the embryos are kept until they grow into juveniles. The abdomen consists of six somites, generally similar in form but with the last somite longer than the others. Each of the first five abdominal segments bears a pair of pleopods. The pleopods are biramous, frequently reduced in the female and sometimes in the male. They are often sexually modified in the male. The telson, or posterior extremity of the body, has a pair of appendages known as uropods, which form a well-developed tail fan. A statocyst, which is a tiny organ related to the animal's sense of balance, is usually present in the endopod of the uropods.

Mysids themselves are often glassy or transparent; they can be seen only when the observer notices their black eyes darting about in the water. Most mysids have a body pattern formed by dark star-shaped chromatophores (clusters of pigmented cells) against a light background color. Some species turn dark when placed against a black background; others that are usually light green and found among green algae may change to dark olive. Deep-sea mysids are often red.

Close-up of the head of the species *Stygiomysis cokei*. (Photo by Dr. Jerry H. Carpenter, Northern Kentucky University. Reproduced by permission.)

Distribution

Mysids are widespread over all continents. They live in a variety of aquatic environments, including coastal and open sea waters, estuaries and other brackish water ecosystems, and continental freshwater lakes and rivers. In addition, a few species have been found in different groundwater habitats and in anchialine (from Greek words that mean "near the sea") caves, which are coastal caves formed from limestone or volcanic rock that are flooded with seawater.

Habitat

Mysids are originally marine crustaceans. They are a highly adaptive group, however, which makes them effective invaders of new habitats, including brackish water and freshwater environments. As a whole, the group is essentially pelagic, although commonly epibenthic, which means that they live on or immediately above the surface of the sediment. Some species burrow into the sediment, live just above the sandy or muddy bottom, or migrate between substrates at the bottom and the surface waters. A few are strictly pelagic species; some live in shallow water in the littoral zone among macroalgae, in crevices along rocky shores, or on sandy beaches. Many species of deep-sea mysids are found on or just above the ocean floor at various depths, including abyssal waters at depths of 18,700–23,622 ft (5,700–7,200 m).

Behavior

Most sand-burrowing and coastal mysids perform a diel (24-hour cyclical) vertical migration, rising and dispersing into the water column at night and returning to deeper water towards dawn. Many are benthic by day and pelagic at night. A few species rest on algae during the day or on stones and cliff ledges; only a few mysids bury themselves in sand. *Gastrosaccus* throws up sand grains by moving its thoracic limbs while lying on the sand. *Paramysis* digs ditches in muddy sand with its first three pereopods. This species can dig for long periods of time, producing open ditches as long as 1.9–3.9 in (5–10 cm) within an hour.

Almost all bathypelagic and bottom-dwelling mysids, even those that burrow in sand, rise in the water column at night. The stimulus for this migration may be light intensity; the time it takes the mysids to rise depends on the speed and depth of water currents. This nocturnal pattern is most noticeable during spawning periods, and may be related to the dispersal of young mysids leaving the marsupium. Littoral species sometimes move to deeper water in fall and return to the shoreline in spring or summer.

Some species of mysids form swarms. These swarms may be several miles long and three or more feet in diameter.

Mysids are primarily swimmers; all members of the class can swim up, down, forward, and backward. The females have reduced pleopods and swim with the exopod (external branch) of their pereopods. They hold the exopods out to the sides and rotate them so that the tip describes an oval. They move their limbs continuously in a slightly different phase and draw water from the sides toward their upper surface. In this way two strong currents of water, one parallel to the abdomen and the other some distance from it, drive the body forward. Many species hold their bodies in a horizontal position with the dorsum up while swimming. A few hold the anterior part of their bodies in an almost vertical position.

Some species swim in schools. School formation among mysids depends on optic signals during the day and probably sensitivity to water currents generated by the swimming movements of their neighbors at night.

Mysids that are suddenly disturbed jerk backward by flexing their abdomen and tail fan against their thorax. Bottom dwellers walk slowly on their endopods with their exopods also in constant motion. The eyes and statocysts keep the body horizontal even in the dark. If a mysid is illuminated from the side, it will turn its back toward the light. The angle of turning may be greater than 45° especially if the light is strong. The animal's optic control over body position is able to override the statocysts.

Feeding ecology and diet

Most mysids are filter feeders, removing fine detritus, rotifers, mollusk larvae, diatoms, and other planktonic organisms out of the water while swimming just above the bottom and creating a suspension feeding current. All filter-feeding mysids may also feed raptorially; that is, they may actively capture selected prey from the environment. They have species-specific feeding modes; some species can switch from one feeding mode to another according to food availability.

Members of some mysid genera, including *Neomysis* and *Siriella*, have been seen to catch small live crustaceans (copepods, cladocerans, amphipods) as well as small mollusks. Mysids use their gnathopods to seize and feed on zooplankton as well as to strain phytoplankton and particulate debris. These appendages move the food under the animal's mandibles (jaws) and press it against these cutting appendages.

On the other hand, mysids are the prey of many larger predators around the world, including invertebrates, fishes, birds, seals and whales.

Reproductive biology

Male mysids do not actively search for females during reproduction. After shedding a previous brood, the female soon molts and is ready to breed again. At that time she produces a pheromone, or chemical substance that stimulates the antennules or the antennular processus masculinus of nearby males. Mating is very quick and takes place at night. The male lies under the female either head-to-tail and belly-to-belly, or doubles up and grasps the anterior part of the female's abdomen with his antennae. The sperm are either injected into the female's brood pouch or shed between the mating individuals and swept by currents produced by the thoracic appendages into the marsupium. The copulating pair soon separate, and within half an hour the female's eggs are extruded into her brood chamber and fertilized there.

The incubation period and frequency of mating depends on the species and the water temperature, and can range from a few weeks to several months. The young are shed as juveniles with complete sets of appendages. Released mysids need about a month to reach their adult stage at a water temperature of about 68°F (20°C).

Conservation status

The IUCN has placed three species, all anchialine stygobitic mysids (*Bermudamysis speluncola*, *Platyops sterreri*, and *Stygiomysis hydruntina*) on its Red List as Critically Endangered and another one as Vulnerable.

Regression or even complete extinction of mysid populations as a result of human impact has been documented. Mysids are endangered by domestic and industrial pollution of coastal waters; dredging of canals (for land use, fisheries, and navigation); artificial redirection of waters (for river traffic and dykes); groundwater drainage; and the repeated application of pesticides. Subterranean ecosystems are variously threatened by tourism, agriculture, urbanization, and the construction of hydroelectric reservoirs.

Significance to humans

Some mysids are used as fish food in commercial aquaculture. On the Island of Jersey, mysids are compounded into a paste called "cherve," which is sold to mullet anglers for bait. In the Orient, mysids are harvested commercially for human consumption. In Japan, *Neomysis intermedia* and *N. japonica* are used for *tsukudani*, a popular dish made with soy sauce.

Mysids are also used for scientific research. They are excellent experimental organisms because they are easy to collect, relatively easy to handle, and stay healthy in the laboratory for long periods of time.

1. *Amathimysis trigibba*; 2. *Stygiomysis cokei*; 3. *Mysis relicta*; 4. *Hemimysis margalefi*; 5. *Spelaeomysis bottazzii*. (Illustration by Dan Erickson)

Species accounts

No common name
Spelaeomysis bottazzii

FAMILY
Lepidomysidae

TAXONOMY
Spelaeomysis bottazzii Caroli, 1924, caves near Castro Marina, Otranto, Italy.

OTHER COMMON NAMES
None known.

PHYSICAL CHARACTERISTICS
This mysid has a short carapace with the last two thoracic somites exposed on the dorsal surface. The rostrum is broadly rounded and the eyes reduced in size. Antennal scale (a lobe-like modification of the external branch or exopod of antenna 1) small, without apical suture, outer margin smooth, setose, without terminal spine. The second thoracic limb (pereopod) has developed as a gnathopod; the exopod is well developed. The marsupium in the adult female is composed of seven pairs of oostegites. The sixth and seventh abdominal somites are fused.

DISTRIBUTION
Southern Italy in the Salentine Peninsula around Lecce; the caves of Zinzulusa, Buco dei Diavoli and L'Abisso; artificial wells near Gallipoli and the area around Bari; groundwaters around Gargano.

HABITAT
This mysid lives in brackish underground waters. It is a euryhaline (tolerant of a wide range of salt concentrations) and eurythermal (adaptable to a broad range of temperatures) species that can live in darkness as well as full or dim light.

BEHAVIOR
Spelaeomysis bottazzii has been kept alive for more than four months in the laboratory under different lighting conditions, in fresh as well as salt water, with temperatures ranging from 50°F–64°F (10°C–18°C).

FEEDING ECOLOGY AND DIET
This species feeds on diatoms and other autotrophic (self-nourishing) aquatic microorganisms.

REPRODUCTIVE BIOLOGY
Spelaeomysis bottazzii has a long period of development in the marsupium as well as a particular reproductive strategy that is not known in other mysids. Females of this species assume an immature form after molting at the end of the incubation period, probably in order to build up nutrition reserves before starting a new reproductive cycle.

CONSERVATION STATUS
This species is known from fewer than five localities scattered over an area of less than 38.6 square miles (100 square kilometers) in southern Italy. The IUCN Red List categorizes it as Vulnerable.

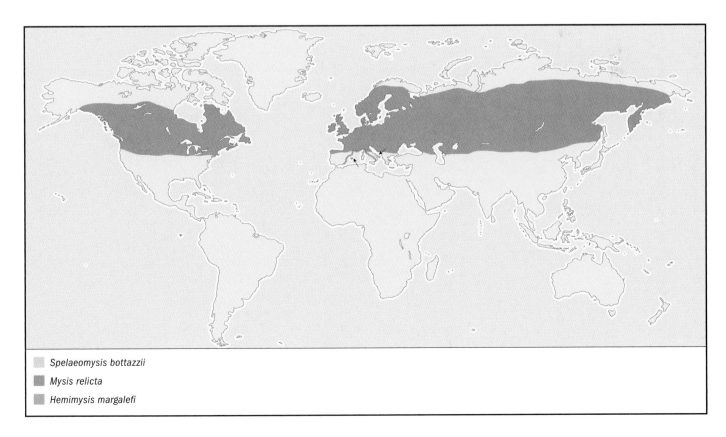

☐ *Spelaeomysis bottazzii*

■ *Mysis relicta*

▨ *Hemimysis margalefi*

SIGNIFICANCE TO HUMANS
None known. ◆

No common name
Amathimysis trigibba

FAMILY
Mysidae

TAXONOMY
Amathimysis trigibba Murano and Chess, 1987, Isthmus Reef, Catalina Island, California, United States.

OTHER COMMON NAMES
None known.

PHYSICAL CHARACTERISTICS
The size range for adult specimens is 0.1–0.13 in (2.7–3.5 mm). The general form is sturdy. The frontal margin of the carapace leads to a broadly rounded rostral plate. The side margins of the rostrum are evenly convex, partially covering the eyestalks in females. The eyestalks are exposed in males. The posterior margin of carapace leaves the last thoracic somite exposed. There are three tubercles (nodules) on the dorsal surface located between the frontal margin of the carapace and the cervical groove. The eyes are developed, and slightly longer than broad; the cornea is wider than the stalk. The antennal scale has a long apical lobe. Female marsupium has two pairs of oostegites, with the posterior pair considerably larger than the anterior pair. Telson entire, 1.4 times as long as broad; lateral margins naked, proximal two-thirds convex and distal one-third concave; distal margin transverse, one-fifth of maximum width at base, armed with 2 pairs of spines, inner pair two-fifths of telson length.

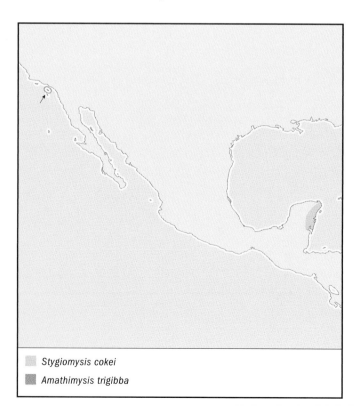

☐ *Stygiomysis cokei*
■ *Amathimysis trigibba*

DISTRIBUTION
Catalina Island, Pacific coast of California, United States.

HABITAT
Found at a depth of 16.4–75.4 ft (5–23 m). Most abundant within the suspended detritus layer just over the sandy substrate, but also found among low benthic algae growing on rocks.

BEHAVIOR
Rises at night several feet into the water column where it occurs in densities of about 8.5 per m3.

FEEDING ECOLOGY AND DIET
Nothing is known.

REPRODUCTIVE BIOLOGY
Nothing is known.

CONSERVATION STATUS
Not listed by the IUCN.

SIGNIFICANCE TO HUMANS
None known. ◆

No common name
Hemimysis margalefi

FAMILY
Mysidae

TAXONOMY
Hemimysis margalefi Alcaraz Riera, and Gili, 1986, dark submarine cave on the northeastern coast of Mallorca in the western Mediterranean, 39°45′N, 3°26′E, at a depth of 39.3 ft (12 m) depth.

OTHER COMMON NAMES
None known.

PHYSICAL CHARACTERISTICS
Living specimens are bright red in color. The total size of an adult specimen from the anterior margin of the carapace to the distal end of telson ranges from 0.16 to 0.21 in (4.12–5.49 mm). General form robust. Carapace short, emarginate posteriorly, thoracic somites 6 and 7 exposed dorsally; anterior margin forming a broad angle rostrum; cervical groove present. Eyes large, globular, slightly broader than the eyestalk, extended laterally beyond the limits of the carapace; cornea pigment black. Antennal scale lanceolate, narrow, about 5.5 times as long as broad in the broadest part; outer margin straight or slightly curved outward; the proximal 1/5 of its length naked (without a spine or a tooth). Telson short; lateral margins converging distally, armed with 8–11 spines; width at distal end less than half the broadest part; telson cleft for about 1/5 of its length, cleft armed with 20–30 small teeth; apical lobes with a long, strong spine in the distal end.

DISTRIBUTION
Mediterranean; northeastern coast of Mallorca.

HABITAT
Dark submarine caves.

BEHAVIOR
This species demonstrated adaptability to a range of temperatures and resistance to acute thermal stress during a series of laboratory experiments.

FEEDING ECOLOGY AND DIET
Nothing is known.

REPRODUCTIVE BIOLOGY
Nothing is known.

CONSERVATION STATUS
Not listed by the IUCN.

SIGNIFICANCE TO HUMANS
This species is being used to study the physiology of mysids. ◆

No common name
Mysis relicta

FAMILY
Mysidae

TAXONOMY
Mysis relicta Lovén, 1862, cold water lakes in northern Europe.

OTHER COMMON NAMES
None known.

PHYSICAL CHARACTERISTICS
The total size of mature specimens is 0.59–0.98 in (15–25 mm) long. The antennal scale is only four times as long as it is wide. The exopod on the third pleopod of mature males is well developed and has five segments while the exopod on the fourth pleopod has six. The telson has a wide bifurcated tip.

DISTRIBUTION
Mysis relicta has a circumpolar distribution. It is found throughout the northern latitudes (above 42°N) in the Great Lakes of North America; Green Lake, Trout Lake, and Lake Geneva in Wisconsin; the Finger Lakes of New York; a handful of Canadian shield lakes; Europe, Scandinavia, and Russia. Considered a relict (survivor) of the great glaciers of the Pleistocene Epoch (1.8 million years ago), its natural distribution has been increased by its introduction into continental waters as an addition to the forage base of sport fisheries.

HABITAT
Cold freshwater. During warm months almost restricted to temperatures range as low as 39.2°F (4°C). It never lives in waters warmer than 57.2°F (14°C). In northern Germany this species does not occur in water with an oxygen concentration lower than 4 cm³/liter; in Wisconsin, however, it lives in water with an oxygen concentration as low as 1 cm³/liter.

BEHAVIOR
Members of this species are swimmers, remaining in deep cold water during summer and moving in winter to shallower waters to reproduce. They remain near the bottom during daytime and rise at dusk toward the surface to forage. Smaller juveniles often lead the migration, rise higher in the water column than the adults, and are the last to descend. Adapted to living in dark environments, their eyes are easily damaged by strong light.

FEEDING ECOLOGY AND DIET
Mysis relicta is an opportunistic feeder with both filter-feeding and predatory habits. It feeds mainly on detritus stirred up from the bottom, including diatoms and unicellular algae. The remains of small crustaceans (cladocerans and copepods) have also been found in its stomach contents, however, which indicates that it also feeds heavily on zooplankton and carrion.

REPRODUCTIVE BIOLOGY
Brood size is positively related to the female's size. A female 0.51–0.59 in (13–15 mm) long carries 10–20 embryos while a female 0.66–0.82 in (17–21 mm) long carries as many as 25–40 embryos. The young are carried in the brood pouch for 1–3 months and leave the marsupium when they are (3–4 mm) long. Mature individuals are 0.51 in (13 mm) long. *M. relicta* lives for over a year and as long as two years, reproducing once or twice before it dies.

CONSERVATION STATUS
Not listed by the IUCN; often introduced as important food item for forage fish.

SIGNIFICANCE TO HUMANS
Mysis relicta is commercially important to fisheries managers as prey for a large number of coldwater fishes, including lake trout, brown trout, rainbow trout, kokanee salmon, various coregonids, burbot, smelts, and alewives. ◆

No common name
Stygiomysis cokei

FAMILY
Stygiomysidae

TAXONOMY
Stygiomysis cokei Kallmeyer and Carpenter, 1996, Temple of Doom Cave (328 ft [100 m] in from cave entrance), 3.7 mi (6 km) northwest of Tulum, Quintana Roo, Mexico, at a depth of 32.8–65.6 ft (10–20 m).

OTHER COMMON NAMES
None known.

PHYSICAL CHARACTERISTICS
The size of this mysid ranges from 0.35 to 0.86 in (9.0–22.0 mm). The length of the colorless wormlike body is 7.0–7.2 times the width. The length of the carapace is about one-fifth the length of the body. The abdominal somites are smooth and rounded on the dorsal surface. Antenna 1 is about one-half the length of the body. Four paired ventral lamellae or oostegites (typical for *Stygiomysis* females and a number unique among the Mysida) extending medially and anteriorly from the proximal part of pereiopods 3–6; each oostegite single flexible membranous flap, rounded and elongated anteriorly. The length of the telson is about 1.7–2.0 times its width, or one-sixth the length of the body. It has 15 spines arranged in five groups of three on its posterior margin.

DISTRIBUTION
Coastal inland caves, Quintana Roo, Yucatan Peninsula, Mexico; the type locality, Mayan Blue Cave; Carwash Cave; and Naharon Cave.

HABITAT
These mysids are stygobites, or cave dwellers. All caves from which this species was collected are completely underwater and entered through water-filled limestone sinkholes. The specimens were collected at depths of 32.8–65.6 ft (10–20 m) in the freshwater layer, occasionally in the upper part of the halocline (a well-defined vertical gradient of salinity). Conditions re-

mained relatively constant with temperatures around 76.1°F–77.9°F (24.5°C–25.5°C), pH 6.8–7.0, low oxygen (near 2.0 ppm), and high carbon dioxide levels (44–864 ppm).

BEHAVIOR

The behavior of *S. cokei* was observed in a laboratory setting. Open containers of cave water readily lost carbon dioxide, causing the pH to rise. The mysids kept their tails almost straight up at a right angle when the pH was comfortably low. As the pH rose, their tails gradually dropped in proportion to the increase; in extreme conditions, their tails were nearly horizontal. They also lowered their tails to a nearly horizontal position while walking. When the animals were forced off their substrate in the caves or the laboratory, they displayed frantic and ineffective swimming movements. The uropods spread away from the telson to make a wide tail fan when *S. cokei* is walking. In healthy specimens, the respiratory beating of pereiopods 1–7 occurred in sequences of 3–9 seconds, followed by rest periods that lasted 2–45 seconds. As carbon dioxide levels dropped and the pH rose, their rest periods increased to as long as 50 minutes.

FEEDING ECOLOGY AND DIET

Primarily filter feeders.

REPRODUCTIVE BIOLOGY

Nothing is known.

CONSERVATION STATUS

Not listed by the IUCN.

SIGNIFICANCE TO HUMANS

None known. ◆

Resources

Books

Bowman, T. E. "Mysidacea." In *Stygofauna Mundi: A Faunistic, Distributional, and Ecological Synthesis of the World Fauna Inhabiting Subterranean Waters*, edited by L. Botosaneanu. Leiden, The Netherlands: E. J. Brill/Dr. W. Backhuys, 1986.

Brusca, R. C., and G. J. Brusca. *Invertebrates*. 2nd ed. Sunderland, MA: Sinauer Associates, Inc., 2003.

Calman, W. T. "Crustacea." In *A Treatise on Zoology*, Part 7, third fascicle, edited by R. Lankester. London: Adam and Charles Black, 1909.

Fretter, V., and A. Graham. *A Functional Anatomy of Invertebrates*. London: Academic Press, 1976.

Kaestner, A. *Invertebrate Zoology*. Vol. 3, *Crustacea*. New York: Wiley-Interscience, 1970.

Mauchline, J. *The Biology of Mysids and Euphausiids*. Part I, *The Biology of Mysids*. London: Academic Press, 1980.

McLaughlin, P. A. *Comparative Morphology of Recent Crustacea*. San Francisco: W. H. Freeman and Company, 1980.

Morgan, M. D. *Ecology of Mysidacea* (Developments in Hydrobiology, Vol. 10). The Hague, The Netherlands: W. Junk Publishers, 1982.

Müller, H.-G. *World Catalogue and Bibliography of the Recent Mysidacea*. Wetzlar, Germany: Wissenschaftler Verlag, Laboratory for Tropical Ecosystems, Research and Information Service, 1993.

Murano, M. "Mysidacea." In *South American Zooplankton*. Vol. 2, edited by D. Boltovskoy. Leiden, The Netherlands: Backhuys Publishers, 1999.

Nesler, T. P., and E. P. Bergerson, eds. *Mysids in Fisheries: Hard Lessons from Headlong Introductions*. Bethesda, MD: American Fisheries Society, 1991.

Nouvel, H., J.-P. Casanova, and J.-P. Lagardère. "Ordre des Mysidacés (Mysidacea Boas 1883)." In *Traité de Zoologie. Anatomie, Systématique, Biologie*. Tome VII, Fascicule III A. *Crustacés Péracarides* (Mémoires de l'Institut Océanographique, 19), edited by J. Forest. Monaco: Musée Océanographique de Monaco, 1999.

Pennak, R. W. *Fresh-Water Invertebrates of the United States*. 3rd edition. New York: John Wiley and Sons, 1989.

Pesce, G. L., L. Juberthie-Jupeau, and F. Passelaigue. "Mysidacea." In *Encyclopaedia Biospeologica*, edited by C. Juberthie and V. Decu. Moulis, France and Bucharest, Romania: Société de Biospéologie, 1994.

Ruppert, E. E., and R. D. Barnes. *Invertebrate Zoology*. 6th ed. Fort Worth, TX: Saunders College Publishing, 1994.

Schmitt, W. L. *Crustaceans*. Ann Arbor: University of Michigan Press, 1965.

Schram, F. R. *Crustacea*. New York: Oxford University Press, 1965.

Wittmann, K. J. "Global Diversity in Mysidacea, with Notes on the Effects of Human Impact." In *Crustaceans and the Biodiversity Crisis. Proceedings of the Fourth International Crustacean Congress, Amsterdam, The Netherlands*. Vol. 1, edited by F. R. Schram and J. C. von Vaupel Klein. Leiden, The Netherlands: Brill, 1999.

Periodicals

Alcaraz, M., T. Riera, and J. M. Gili. "*Hemimysis margalefi* sp. nov (Mysidacea) From a Submarine Cave of Mallorca Island, Western Mediterranean." *Crustaceana* 50, no. 2 (1986): 199–203.

Ariani, A. P., and K. J. Wittmann. "Alcuni aspetti della biologia della riproduzione in *Spelaeomysis bottazzi* Caroli (Mysidacea, Lepidomysidae)." *Thalassia Salentina* 23, Suppl. (1997): 193–200.

Beeton, A. M., and J. A. Bowers. "Vertical Migration of *Mysis relicta* Loven." *Hydrobiologia* 93 (1982): 53–61.

Bowers, J. A., and H. A. Vanderploeg. "In situ Predatory Behavior of *Mysis relicta* in Lake Michigan." *Hydrobiologia* 93 (1982): 121–131.

Caroli, E. "Su di un misidaceo cavernicolo (*Spelaeomysis bottazzi* n. g., n. sp.) di Terra d'Otranto." *Atti della Accademia Nazionale dei Lincei Rendiconti, Classe di Scienze Fisiche, Matematiche e Naturali* ser. 5, 33 (1924): 512–513.

Cooper, S. D., and C. R. Goldman. "Opossum Shrimp (*Mysis relicta*) Predation on Zooplankton." *Canadian Journal of Fisheries and Aquatic Sciences* 37 (1980): 909–919.

De Jong-Moreau, L., and J.-P. Casanova. "The Foreguts of the Primitive Families of the Mysida (Crustacea, Peracarida): A

Resources

Transitional Link Between Those of the Lophogastrida (Crustacea, Mysidacea) and the Most Evolved Mysida." *Acta Zoologica* 82 (2001): 137–147.

Evans, M. S., R. W. Bathelt, and C. P. Rice. "PCBs and Other Toxicants in *Mysis relicta*." *Hydrobiologia* 93 (1982): 205–215.

Folt, C. L., J. T. Rybock, and C. R. Goldman. "The Effect of Prey Composition and Abundance on the Predation Rate and Selectivity of *Mysis relicta*." *Hydrobiologia* 93 (1982): 133–143.

Grossnickle, N. E. "Feeding Habits of *Mysis relicta*—An Overview." *Hydrobiology* 93 (1982): 101–107.

Kallmeyer, D. E., and J. H. Carpenter. "*Stygiomysis cokei*, New Species, A Troglobitic Mysid From Quintana Roo, Mexico (Mysidacea: Stygiomysidae)." *Journal of Crustacean Biology* 16, no. 2 (1996): 418–427.

Lasenby, D. C., T. G. Northcote, and M. Furst. "Theory, Practice, and Effects of *Mysis relicta* Introductions to North American and Scandinavian Lakes." *Canadian Journal of Fisheries and Aquatic Sciences* 43 (1986): 1277–1284.

Lejeusne, C., and P. Chevaldonné. "Cave-Dwelling Invertebrates of the NW Mediterranean: Silent Victims of Global Warming?" *Geophysical Research Abstracts* 5, 58–96 (2003).

Martin, J. W., and G. E. Davis. "An Updated Classification of the Recent Crustacea." *Natural History Museum of Los Angeles County Science Series* 39 (2001): 1–124.

Mauchline, J., and M. Murano. "World List of the Mysidacea, Crustacea." *Journal of the Tokyo University of Fisheries* 64, no. 1 (1977): 39–88.

Moreau, X., D. Benzid, L. De Jong, R.-M. Barthélémy, and J.-P. Casanova. "Evidence for the Presence of Serotonin in Mysidacea (Crustacea, Peracarida) as Revealed by Fluorescence Immunohistochemistry." *Cell and Tissue Research* 310, no. 3 (2002): 359–371.

Morgan, M. D., and S. T. Threlkeld. "Size Dependent Horizontal Migration of *Mysis relicta*." *Hydrobiologia* 93 (1982): 63–68.

Murano, M., and J. R. Chess. "Four New Mysids from Californian Coastal Waters." *Journal of Crustacean Biology* 7, no. 1 (1987): 182–197.

Pesce, G. L., and B. Cicolani. "Variation of Some Diagnostic Characters in *Spelaeomysis bottazzii* Caroli (Mysidacea)." *Crustaceana* 36, no. 1 (1979): 74–80.

Sell, D. W. "Size-Frequency Estimates of Secondary Production by *Mysis relicta* in Lakes Michigan and Huron." *Hydrobiologia* 93 (1982): 69–78.

Vanderploeg, H. A., J. A. Bowers, O. Chapelski, and H. K. Soo. "Measuring in situ Predation by *Mysis relicta* and Observations on Underdispersed Microdistributions of Zooplankton." *Hydrobiologia* 93 (1982): 109–119.

Other

"Biology of Opposum [sic] Shrimps." Mysidacea Gallery. [6 Aug. 2003]. <http://www.museum.vic.gov.au/crust/mysibiol.html>.

CaveBiology.com. 9 July 2003 [6 Aug. 2003]. < http://www.tamug.edu/cavebiology/sitemap.html>.

European Register of Marine Species. 29 Feb. 2000 [6 Aug. 2003]. <http://erms.biol.soton.ac.uk/lists/brief/Mysidacea.shtml>.

"Misidacei." Dispense di Zoologia. 19 June 2003 [6 Aug. 2003]. <luciopesce.interfree.it/zoologia/mysids.html>.

"Mysidacea: Families, Subfamiles, and Tribes." Crustacea.net. Oct. 2000 [6 Aug. 2003]. <http://www.crustacea.net/crustace/Mysidacea/index.htm>.

"Mysids." Pesce's Home Page. 19 Feb. 2002 [6 August 2003]. <http://www.geocities.com/mediaq/fauna/mysids.html>.

"*Mysis relicta*." Department of Fisheries and Wildlife, University of Minnesota. 8 Dec. 1997 [6 Aug. 2003]. <http://www.fw.umn.edu/nresexotics3001/mysisrelicta.html/mysispage>.

"*Mysis relicta* Lovèn." Bonito Feldberg's Home Page. [6 Aug. 2003]. <http://homepages.compuserve.de/BonitoFeldberg/mysis.html>.

"Nonindigenous Aquatic Species." Biological Resources Division, United States Geological Survey. 6 Oct. 1999 [6 Aug. 2003]. <http://nas.er.usgs.gov/crustaceans/docs/my_relic.html>.

"Opossum Shrimp (*Mysis relicta* Loven)." *NR 615, Autumn 2001.* [6 August 2003]. <http://www.ag.ohio-state.edu/exotic/nr615au01/fetzer/format.htm>.

Quintana Roo Speleological Survey. 1 Dec. 2002 [6 Aug. 2003]. < http://www.caves.org/project/qrss/biospeleo.htm>.

"World Groundwater Mysids." Pesce's Home Page. 19 Feb. 2002 [6 Aug. 2003]. <http://www.geocities.com/mediaq/mysid.html>.

Estela C. Lopretto, PhD

Lophogastrida

(Lophogastrids)

Phylum Arthropoda

Subphylum Crustacea

Class Malacostraca

Order Lophogastrida

Number of families 2

Thumbnail description

Small shrimplike marine crustaceans with elongate carapaces covering, but not fused to, all thoracic somites (segments); stalked eyes; and strongly developed swimming legs on the first five abdominal somites

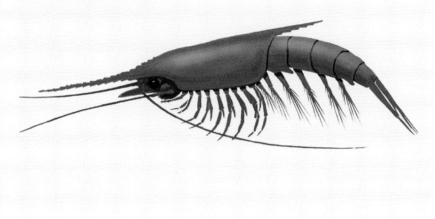

Illustration: Giant red mysid (*Gnathophausia ingens*). (Illustration by John Megahan)

Evolution and systematics

Lophogastrids appear to be most closely allied to the strictly fossil pygocephalomorphs; indeed, species recognizable as lophogastrid relatives are known from the Pennsylvanian Period, about 325–286 million years ago. In these groups the sternites (ventral shields) are wide, with the thoracic legs being located on the outer edges of the somites.

The order Lophogastrida contains two extant families, the Lophogastridae with six genera, and the Eucopiidae with the single genus *Eucopia*. Many scientists now regard lophogastrids as a suborder of the order Mysidacea. They differ from the other mysid suborder in having gills on most thoracic appendages; no statocyst (organ governing the sense of balance) in the uropod; and no modifications of the anterior pleopods in the males. Some authors have suggested that any resemblance to other mysids is purely superficial and that the lophogastrids should have their own order. The latter classification is followed here.

Physical characteristics

The body of a lophogastrid is long and shrimplike, with the head and thorax covered by a loosely fitting carapace. The carapace is extended in front as a rostrum, which is an anatomical structure resembling a bird's beak. The rostrum can be very large in one genus of lophogastrids. The rear portion of the carapace may cover from five to seven thoracic somites. It extends over the sides of the animal to the bases of the thoracic legs. Both pairs of lophogastrid antennae are biramous; that is, they have two branches. All lophogastrids have eyes on moveable stalks. The antenna (=second antenna) exopod (outer branch) is shaped like a large scale. The mandible, or lower jaw, is of the rolling crushing type, with a toothed incisor and large crushing molar. The mandible palp, which is a jointed sensory appendage, can be rather large and extends upwards in front of the head to the area between the antennular peduncles (small stemlike process at the base of an antenna). In some genera, at least, the maxillules, or first pair of maxillae, possess a backwardly-directed endopod (posterior process) in the form of a palp. On the maxilla (upper jaw), the exopod forms a large setose (bristly) lobe that resembles the scaphognathite (lateral flap on the second maxilla) of decapod crustaceans.

The thoracic limbs of lophogastrids are modified in various ways; however, all have well-developed endopods and exopods. The first pair is always modified as a maxilliped because the first thoracic somite is incorporated into the head. In essence, such modifications usually involve shortening of the endopod as well as the development of specialized lobes and setae for handling food. The remaining thoracic limbs, now called pereopods, are numbered from one through seven. The pereopods are all usually similar to one another in the structure of the endopod as well as in the strength of the swimming setae (bristles) on the exopods. In the genus *Eucopia*, however, the first three pereopods are short and stout with subchelate (pincerlike) tips; the fourth through the sixth pereopods are highly modified as elongate grasping appendages; and the seventh pereopod is reduced to a limb with bristles. Most lophogastrid pereopods bear gills at their base. Oostegites, or brood plates, are present on all seven pairs of pereopods.

All lophogastrids, both males and females, have well-developed biramous propulsive pleopods on the first five abdominal somites, and broad flattened uropods on the sixth abdominal somite. The uropods do not possess statocysts. A flattened and somewhat pointed telson terminates the body. In males of the genus *Gnathophausia*, the endopod of the second pair of pleopods is slightly modified for sperm transfer.

Distribution

Lophogastrids occur in all oceans except the Arctic. They are generally bathypelagic (found below 3,280 ft [1,000 m]), but some species may be found in waters as shallow as 164 ft (50 m) and others as deep as 13,120 ft (4,000 m). Most of the known species are found primarily in the Pacific and Indian Oceans, but a few are common in the Atlantic.

Habitat

Members of this group occur in the pelagic oceanic zone (the open sea beyond the continental shelf).

Behavior

Few observations have been published on the behavior of living lophogastrids. They are obviously difficult to observe but have been reared in captivity recently at the Monterey Bay Aquarium in California. Because lophogastrids are bathypelagic they spend almost all their time swimming, using the pleopods for propulsion and the exopods to maintain a flow of water around the gills for respiration.

Feeding ecology and diet

The lophogastrids that have been investigated to date are almost all predators. Others whose feeding habits are not known seem to lack filter setae on their mouth appendages, suggesting that they too are predators. Only *Gnathophausia* seems to use its mouthparts to filter large particles from the seawater.

Reproductive biology

Lophogastrid mating has not been observed. Eggs are extruded into the female's ventral brood pouch where they are kept until hatching. When the young hatch, they possess a full set of appendages on the thorax.

Conservation status

No species are listed by the IUCN.

Significance to humans

Lophogastrids are undoubtedly eaten by various pelagic fishes, some of which are commercially important.

Species accounts

Giant red mysid
Gnathophausia ingens

FAMILY
Lophogastridae

TAXONOMY
Gnathophausia ingens Dohrn, 1870.

OTHER COMMON NAMES
English: Deep water giant mysid; German: Rote Riesengamele.

PHYSICAL CHARACTERISTICS
The carapace possesses a very long rostrum, which extends almost to the end of the antennule, and a long posterior spine that reaches to the end of the second abdominal somite. The carapace is also folded inward along the ventral margin, creating a semi-enclosed gill chamber. The antennal scale is also very long, unsegmented, and sharply pointed at its apex. *Gnathophausia ingens* is the largest pelagic crustacean known, reaching a length of 12.2 in (31 cm). (Illustration shown in chapter introduction.)

DISTRIBUTION
Giant red mysids are commonly found in deep waters below the tropical and subtropical waters of the Atlantic, Pacific, and Indian Oceans. Although they have been sampled between

656–13,123 ft (200–4,000 m), they are most common at depths of 1,965–4,920 ft (600–1500 m).

HABITAT
Bathypelagic in the oceanic zone (open sea).

BEHAVIOR
Little is known about the behavior of this species.

FEEDING ECOLOGY AND DIET
Gnathophausia ingens apparently feeds on large particles filtered from the seawater. It also feeds on small dead aquatic organisms.

REPRODUCTIVE BIOLOGY
Gnathophausia ingens has a very long period of larval development, estimated to be about 530 days. Adult females probably have more than one brood, and live for almost 3000 days, based on growth estimate data. From hatching to adult, an individual giant red mysid passes through 13 instars (stages between molts).

CONSERVATION STATUS
Not listed by the IUCN.

SIGNIFICANCE TO HUMANS
Probably serves as food for larger fishes that are harvested commercially. ◆

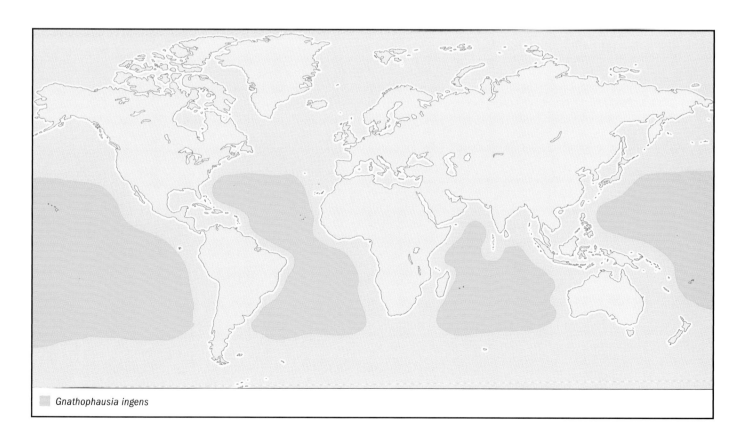

☐ *Gnathophausia ingens*

Resources

Books

Nouvel, H., J.-P. Casanova, and J. P. Lagardère. *Ordre des Mysidacés. Mémoires de l'institut océanographique.* Monaco: [n.p.], 1999.

Schram, F. R. *Crustacea.* Oxford, U.K.: Oxford University Press, 1986.

Periodicals

Childress, J. J., and M. H. Price. "Growth Rate of the Bathypelagic Crustacean *Gnathophausia ingens* (Mysidacea: Lophogastridae). I. Dimensional Growth and Population Structure." *Marine Biology* 50 (1978): 47–62.

Tattersal, O. S. "Mysidacea." *Discovery Reports* 28 (1955): 1–190.

Les Watling, PhD

Cumacea
(Cumaceans)

Phylum Arthropoda
Subphylum Crustacea
Class Malacostraca
Order Cumacea
Number of families 8

Thumbnail description
Cumaceans are quickly recognizable by their large carapace, which covers the first three thoracic somites and narrow abdomen. The first thoracic appendage is modified as a maxilliped, which has both feeding and respiratory functions.

Illustration: *Diastylis sculpta.* (Illustration by Dan Erickson)

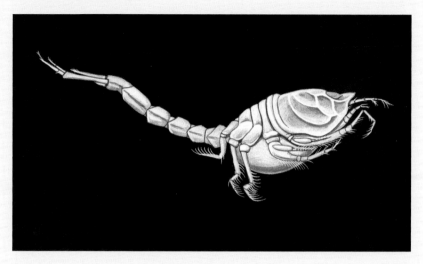

Evolution and systematics

Cumaceans are the most modified of the orders usually grouped together in the superorder Peracarida. The fossil record is sparse, but a few specimens from the Mesozoic have been found. Eight families of cumaceans are recognized. As many as 26 families once were used to divide the cumacean genera, but many contained only one or two genera. Modern phylogenetic methods are being used to revise the current scheme. Molecular and morphological evidence suggests that reduction of the cumacean telson, which is a feature of three families, has occurred only once.

Physical characteristics

Cumaceans have a large carapace that extends backward and ventrally to cover the first three thoracic somites. The anterior aspect of the head usually bears a single middorsal eye, but some species have two eyes, each member of the pair being located anterolaterally. The mandibles are usually of the basic, generalized feeding type with strong molar and incisor, but the molar occasionally is modified into a long styliform process. Because the first three thoracic somites are fused to the head, the appendages on those somites are modified as maxillipeds. The first pair of maxillipeds has a complicated structure. The endopod functions as a feeding device, and the epipod elaborates into a large and sometimes convoluted gill. The second pair of maxillipeds also is involved in feeding, but in females the coxae of these appendages have small posteriorly directed brood plates. The third pair of maxillipeds usually is leg-like but can have other functions, such as forming an opercular covering over the more anterior mouth appendages. The other five pairs of thoracic appendages function as walking legs. These appendages may have exopods, or the exopods may be reduced or absent. The first five pairs of abdominal appendages are known as pleopods. With a single exception (a species from the deep sea), pleopods are not found on females. Among males, pleopods may be absent, may occur in reduced numbers, or may be fully formed. A freely articulated telson is present in five families but is fused to the last abdominal somite in three families. The last pair of abdominal appendages is the uropods. In most cumaceans the uropods are composed of a peduncle that leads to two rami, the endopod and the exopod. The structure of the uropod often has bearing on classification at the genus level.

Distribution

All seas, from shallow bays and estuaries to the deepest trenches. Caspian, Aral, and Black Seas. Some families, such as Diastylidae, Lampropidae, and Ceratocumatidae, are diverse in the deeper and colder waters of the ocean, whereas Nannastacidae and Bodotriidae are most diverse in shallow tropical waters. Pseudocumatidae is especially diverse in the Black and Caspian Seas.

Habitat

Most cumaceans are found in marine and brackish waters, but some range into water that is fresh for at least some of the time. Cumaceans tend to live in soft sediment where there is reasonable current motion but not heavy wave action. They bury themselves just below the sediment–water interface. Members of the family Gynodiastylidae are found in algal turf on the surfaces of stones or coral rubble.

Behavior

Cumaceans generally sit with their bodies just submerged in the sediment. They need to maintain contact with the overlying water to pump oxygenated water over their gills. The respiratory current generated by the movement of the first maxilliped epipod enters at the base of the walking legs, passes under the carapace, and exits at the front of the carapace, where part of the epipod forms a siphon. Some members of several shallow-water species after sunset undergo vertical migration from the sediment into the water column.

Feeding ecology and diet

Most cumaceans are fine-particle feeders. They obtain the particles by manipulating sediment grains with the mouth appendages or scraping the surfaces of larger particles. There has been some speculation that cumaceans are filter feeders, but there is no direct evidence to support that idea. Some members of the family Nannastacidae are considered, on the basis of the structure of the mandible, to be carnivores, feeding on small organisms such as foraminiferans or meiofaunal metazoans.

Reproductive biology

In most cumacean species males are semipelagic, swimming about until they find a female. In a few species within several families, males have lost their swimming ability and have developed grasping antennae, which are used to hang on to the abdomen of females until mating has occurred. Eggs are carried in a ventral brood pouch on the female, where they are incubated until hatching as juveniles. There is no pelagic larval stage. Both males and females of most species reach terminal molts at sexual maturity.

Conservation status

No species are listed by the IUCN. No cumaceans have been studied to the extent that it is known whether they are endangered. A few have been transported to foreign shores, where they have established themselves as invasive species.

Significance to humans

One cumacean species is known to be important as food for juvenile salmon along the northwest coast of North America. Others are commonly found in the stomachs of juvenile ground fish, such as cod. Most cumaceans are very small and insignificant from the perspective of marine food webs.

1. *Cyclaspis longicaudata*; 2. *Diastylis sculpta*; 3. *Diastylis rathkei*. (Illustration by Dan Erickson)

Species accounts

No common name
Cyclaspis longicaudata

FAMILY
Bodotriidae

TAXONOMY
Cyclaspis longicaudata Sars, 1865.

OTHER COMMON NAMES
None known.

PHYSICAL CHARACTERISTICS
The carapace is nearly globose and is completely smooth. The free thoracic somites are quite small, and the abdomen is very long and narrow. The male has five pairs of pleopods.

DISTRIBUTION
Cold waters of the North Atlantic, northern Norway to northeastern United States at depths from 395 to 16,400 ft (120–5,000 m).

HABITAT
Cyclaspis longicaudata is found in sandy mud. The sand-sized particles in deep sea sediment are derived from foraminiferan and pteropod shells.

BEHAVIOR
Nothing is known. The morphological features suggest *C. longicaudata* is not likely to spend much time swimming, except in the mature male stage.

FEEDING ECOLOGY AND DIET
Nothing is known.

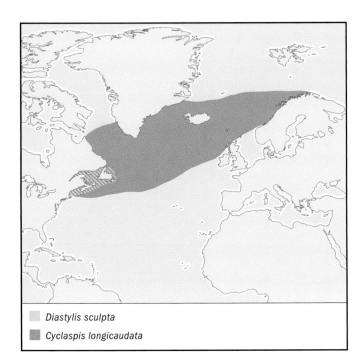

☐ *Diastylis sculpta*
▨ *Cyclaspis longicaudata*

REPRODUCTIVE BIOLOGY
Juveniles and females with young in the brood pouch have been found at almost any time of the year, so it is likely that at least some populations are reproductive throughout the year.

CONSERVATION STATUS
Not listed by the IUCN.

SIGNIFICANCE TO HUMANS
Cyclaspis longicaudata has been found in the stomachs of some deep-dwelling fishes. ◆

No common name
Diastylis sculpta

FAMILY
Diastylidae

TAXONOMY
Diastylis sculpta, Sars 1871.

OTHER COMMON NAMES
None known.

PHYSICAL CHARACTERISTICS
Diastylis sculpta gets its name from the series of large ridges on the carapace that give it a highly sculptured appearance. The carapace of males is somewhat smoother, being streamlined for hydrodynamic purposes. The male has two pairs of pleopods.

DISTRIBUTION
Coast of eastern North America from the Gulf of Saint Lawrence to Long Island Sound.

HABITAT
Muddy sand bottoms in shallows of bays and estuaries.

BEHAVIOR
Diastylis sculpta spends most of the day nestled into surface sediment. At night it makes short forays into the overlying water; the purpose of these forays is not known.

FEEDING ECOLOGY AND DIET
Although its specific food has not been investigated, *D. sculpta* spends most of its time in the sediment sifting through the particles, most likely looking for microalgae and other high-quality organic particles.

REPRODUCTIVE BIOLOGY
In the Woods Hole, Massachusetts, region, *D. sculpta* is found from July to January with young in the brood pouch. More than one generation is represented during this period.

CONSERVATION STATUS
Not listed by the IUCN.

SIGNIFICANCE TO HUMANS
Diastylis sculpta is a common food of bottom-feeding fishes, especially flatfishes. ◆

Resources

Books

Bacescu, M. *Cumacea I: Crustaceorum Catalogus.* Part 7. The Hague, The Netherlands: SPB Academic Publishing, 1988.

————. *Cumacea II: Crustaceorum Catalogus.* Part 8. The Hague, The Netherlands: SPB Academic Publishing, 1992.

Sars, G. O. *Cumacea: An Account of the Crustacea of Norway.* Christiania, Norway: Cammermeyers Forlag, 1899–1900.

Periodicals

Gamo, S. "Studies on the Cumacea (Crustacea, Malacostraca) of Japan, Part III." *Seto Marine Biological Laboratory* 16 (1968): 147–92.

Gerken, S. "The Gynodiastylidae (Crustacea: Cumacea)." *Memoirs of the Museum Victoria* 59 (2001): 1–276.

Watling, L. "Revision of the Cumacean Family Leuconidae." *Journal of Crustacean Biology* 11 (1991): 569–82.

Les Watling, PhD

Tanaidacea

(Tanaids)

Phylum Arthropoda

Subphylum Crustacea

Class Malacostraca

Order Tanaidacea

Number of families 20

Thumbnail description
Small, mostly marine-dwelling crustaceans that are among the most diverse and abundant creatures in some marine environments

Illustration: *Apseudes intermedius.* (Illustration by Dan Erickson)

Evolution and systematics

The fossil record of tanaids is better known from the Mazon Creek faunas from the Middle and Upper Pennsylvanian (Carboniferous). These faunas are distinguished from others by the numbers and types of species and by their unusually fine preservation. The suborder Anthracocaridomorpha was noted in 1980 and is known from fossils only.

The systematics and phylogeny of the Tanaidacea are poorly understood. However, there is a consensus that suggests abandoning the ancient divisions Monokonophora and Dikonophora for three new suborders proposed in 1983: Apseudomorpha, Neotanaidomorpha, and Tanaidomorpha. Regarding phylogeny, Apseudomorpha shows the most plesiomorphic features, Tanaidomorpha presents the most apomorphic or derived features, and Neotanaidomorpha is placed in an evolutionary position between the Apseudomorpha and the Tanaidomorpha. Three suborders, three superfamilies, and 20 families comprise the Tanaidacea.

Physical characteristics

There is a carapace covering the cephalothorax, which is formed from the head fused with the first two thoracic segments; the cephalothorax bears one pair of antennules and one pair of antennae. Compound eyes can be absent or present. The two first pairs of thoracic appendages (thoracopods) are additional mouthparts called maxillipeds, the second pair (called gnathopods) is chelate, which is a distinctive characteristic of Tanaidacea. Thoracopods 3–8 are ambulatory pereopods (articulate walking legs), which are usually similar in form. In some families of Tanaidacea, the distal articles of the first pereopod are flattened, possibly to use in digging or even in swimming. In Tanaidacea, the pleopods can be present or

absent; when present, they are used in swimming and to produce a ventilatory current in tube-dwelling species. In the final body region, the telson and the last one or two pleonites are fused as a pleotelson, which bears a pair of terminal biramous or uniramous appendages called uropods. Ovigerous females possess a brood pouch (marsupium), formed from flattened plates arising from an inner proximal margin of coxa of certain pereopods (oostegites), within which the eggs are developed.

Most species of Tanaidacea are small, ranging from 0.039 to 0.78 in (1 mm–2 cm) in length. The tanaidaceans display some color patterns, mainly yellowish and blue or gray tonalities.

Distribution

The Tanaidacea are distributed in all oceans of the world, from tropical to temperate regions, and even in polar waters.

Habitat

Most species of Tanaidacea are benthic, although some species have been found in plankton. The tanaidaceans can be found at a wide variety of depths, from the littoral zone to deep waters, but the better known species are those that live in shallow waters. Although the tanaidaceans are almost exclusively marine dwellers, some species have penetrated freshwater habitats. This is the case of some euryhaline species. Some euryhaline species can also tolerate a wide range of temperatures, inhabiting from tropical to temperate regions, living in rivers, lakes, and estuaries. Most of the offshore species are found in sand, mud, and kelp holdfasts. Other species can be found in intertidal regions of sandy or muddy shores. There are some species with different habitats: *Hexapleomera robusta* inhabits minute tubes in fissures between the carapace scales of marine turtles; *Pagurapseudes* spp. is adapted to fit the pleon into small gastropod shells like the hermit crabs. Many tanaids can produce a cementlike material from tegumental glands in the pereonal cavity; this cement emerges from the tips of the dactyli of the anterior pereopods and is used to construct tubes with particles of sand and detritus. Generally, the tubes are open at both ends, and tanaids use the pleopods to create water currents. During the incubatory period, ovigerous females may close the ends of their tubes. The tubes built by tanaids stabilize the sediment, favoring the colonization of other sedentary species.

Behavior

Nothing is known of the behavior of tanaids.

Feeding ecology and diet

The Tanaidacea, like many others crustaceans, have exploited several feeding strategies. Most tanaids utilize feeding mechanisms that usually involve direct manipulation of food by the mouthparts and the pereopods, especially chelate anterior legs. They are raptorial feeders, eating detritus and its associated microorganisms. The larger food particles are manipulated by the chelipeds and maxillipeds and transferred toward the mouth. Some members of the Tanaidacea possess better-developed branchial chambers, maxillae, maxillipedal epipodites, and exopods on the chelipeds, and the first pair of pereopods that are used in filter or suspension feeding. In this feeding mechanism, the simultaneous action of the thoracic limbs creates a suspension-feeding current by alternate movements of adjacent limb pairs. Thus, surrounding water is drawn into the interlimb spaces, and particles are retained by setae on the inner face of these appendages. Then, the retained particles are carried to a midventral food channel and moved anteriorly toward the mouth. Some tanaids are predators, catching the prey with chelate pereopods or even directly

with the mouth appendages. Then, with various mouthparts, mainly the mandibles, they shear, tear, or grind the prey.

Reproductive biology

As in all other peracarid crustaceans, tanaid offspring are held within a marsupium on the ventral face of an ovigerous female during their early development, and generally are released only when they have developed most of their appendages. The fertilized eggs develop into embryos, which gradually develop into the first larval stage (manca I), and then into a second, more mobile larval stage (manca II). In this second stage, the manca has partially formed the sixth pereopods; this is when the tanaid generally leaves the marsupium. In tube-dwelling tanaids, the larval stage is followed by a juvenile stage with completely formed appendages; these juveniles will develop either into preparatory males or preparatory females. However, hermaphroditism can occur in tanaids in many ways. For example, in *Apseudes spectabilis*, some specimens possess both mature eggs and male gonads filled with sperm. Hence, several researchers believe that self-fertilization is possible.

In the case of conventional reproduction of tube-dwelling species, a copulatory male enters the tube of a female, and the courtship takes place for a long period of time. Male and female lie with their ventral sides adjacent to one another, and sperm are deposited in the marsupium. After closing the marsupium, the female releases the eggs. Then, the female expels the male from the tube, closes the ends of the tube, and incubates the brood.

Conservation status

Nothing is known of the conservation status of tanaids. No species are listed by the IUCN.

Significance to humans

No significance to humans is known.

1. *Paraleiopus macrochelis*; 2. *Apseudes intermedius*. (Illustration by Dan Erickson)

Species accounts

No common name
Apseudes intermedius

FAMILY
Apseudidae

TAXONOMY
Apseudes intermedius Hansen, 1895, St. Vincent (Cape Verde Island).

OTHER COMMON NAMES
None known.

PHYSICAL CHARACTERISTICS
The last five thoracic segments have rounded lateral expansions with few bristles. Oostegites are little developed; telson is twice as long as broad; antennal flagellae with six articles. Chela has the same length as carpus; posterior region of carpus broader than anterior region; inner margin smooth. Outer branch of uropods with seven articles.

DISTRIBUTION
Cape Verde Islands, Mediterranean Sea, and Brazil (Rio de Janeiro).

HABITAT
Shallow waters, in sandy or muddy bottoms.

BEHAVIOR
Nothing is known.

FEEDING ECOLOGY AND DIET
Probably a suspension feeder, eating detritus and microorganisms.

REPRODUCTIVE BIOLOGY
Individuals are sexually dimorphic and engage in conventional reproduction with courtship.

CONSERVATION STATUS
Not listed by the IUCN.

SIGNIFICANCE TO HUMANS
None known. ◆

No common name
Paraleiopus macrochelis

FAMILY
Kalliapseudidae

TAXONOMY
Paraleiopus macrochelis Brum, 1978, Santa Cruz, Espírito Santo (Brazil).

OTHER COMMON NAMES
None known.

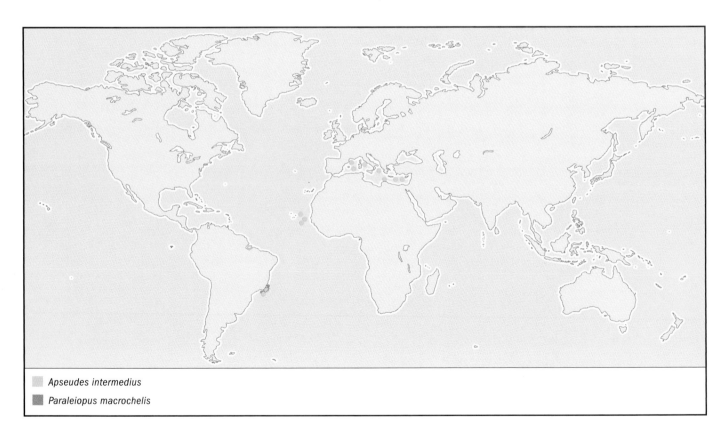

Apseudes intermedius
Paraleiopus macrochelis

PHYSICAL CHARACTERISTICS
Pereonites with rounded margins; carapace as long as broad. Pleon with five pleonites; pleotelson short, as long as broad; posterior region rounded; antennules longer than antennae.

DISTRIBUTION
Known only in Brazil (Espírito Santo and Rio de Janeiro).

HABITAT
Shallow waters, unconsolidated bottoms (sand or mud).

BEHAVIOR
Nothing is known.

FEEDING ECOLOGY AND DIET
Probably a suspension or raptorial feeder or both, eating a microorganisms and detritus.

REPRODUCTIVE BIOLOGY
Individuals are sexually dimorphic, without any characteristics of hermaphrodites.

CONSERVATION STATUS
Not listed by the IUCN.

SIGNIFICANCE TO HUMANS
None known. ◆

Resources

Books
Brusca, Richard C., and Gary J. Brusca. *Invertebrates.* Sunderland, MA: Sinauer Associates, Inc., 2003.

Holdich, D. M., and D. A. Jones. *Tanaids: Keys and Notes for the Identification of the Species of England.* Cambridge, U.K.: Cambridge University Press, 1983.

Martin, Joel W., and George E. Davis. *An Update Classification of the Recent Crustacea.* Los Angeles: Natural History Museum of Los Angeles County, Science Series 39, 2001.

Paulo Ricardo Nucci, PhD

Mictacea
(Mictaceans)

Phylum Arthropoda
Subphylum Crustacea
Class Malacostraca
Order Mictacea
Number of families 2

Thumbnail description
Very small crustaceans, generally allied with peracarids, that are constructed much like thermosbaenaceans and spelaeogriphaceans, but there is no carapace and only rudimentary appendages on the abdomen

Illustration: *Mictocaris halope.* (Illustration by John Megahan)

Evolution and systematics

The order Mictacea includes two families, three genera, and five species. This group has been included in the superorder Peracarida on the basis of a posteriorly directed flap on the walking legs that is thought to be homologous to the brood plates of other peracarids.

Physical characteristics

The body is elongate and cylindrical. There is no carapace emanating from the posterior margin of the head. In fact, the side of the head is also without pleurae, so the mandible is quite visible. Eye lobes may be present or absent, but functional eyes are absent. One thoracic somite is fused to the head and its appendage has been modified as a maxilliped. There are seven walking legs, the first six of which have exopods. On the abdomen, the first five somites have minute one-segmented pleopods. Attached to the last abdominal somite is the telson and a pair of biramous uropods.

Distribution

Of the five known species, one was described from the deep sea off the north coast of South America, one from the continental slope off southeastern Australia, and the others from marine caves in Bermuda, Bahamas, and the Cayman Islands. Only the Bermudan species, *Mictocaris halope*, is known from several specimens.

Habitat

Mictaceans are found in marine caves or rubble.

Behavior

The animals move both by walking and swimming.

Feeding ecology and diet

Because of the design of the mouthparts and the small size of the body, mictaceans are thought to be detritivores.

Reproductive biology

Copulation is presumed but has not been observed. Young hatch as manca, missing the last pair of pereopods.

Conservation status

No species is threatened or listed by the IUCN, but those species confined to cave systems are obviously dependent on the health of the cave waters being maintained.

Significance to humans

They are of intellectual interest only.

Species accounts

No common name
Mictocaris halope

FAMILY
Mictocariidae

TAXONOMY
Mictocaris halope Bowman and Iliffe, 1985, Bermuda.

OTHER COMMON NAMES
None known.

PHYSICAL CHARACTERISTICS
Body elongate and cylindrical; no functional eyes; no carapace. Most appendages are less setose in this species than in the other species. (Illustration shown in chapter introduction.)

DISTRIBUTION
Bermuda.

HABITAT
Sediments and rock of a marine cave.

BEHAVIOR
Spends most of its time swimming or crawling on the substratum. When walking, the antennae are held out to the sides and are folded backwards somewhat during swimming events. In many respects, acts like spelaeogriphacean.

FEEDING ECOLOGY AND DIET
Material in the gut suggests it is a detritivore. Scraping of small food particles from the substratum was observed in aquaria.

REPRODUCTIVE BIOLOGY
Copulation presumed but has not been observed. Young hatch as manca, missing last pair of pereopods.

Mictocaris halope

CONSERVATION STATUS
Not threatened at present, but its conservation status depends on the water quality of the cave systems of Bermuda not being degraded.

SIGNIFICANCE TO HUMANS
None known. ◆

Resources

Periodicals

Bowman, T. E., and T. M. Iliffe. "*Mictocaris halope*, a New Unusual Peracaridan Crustacean from Marine Caves on Bermuda." *Journal of Crustacean Biology* 5 (1985): 58–73.

Hessler, R. R. 1999. "Ordre des Mictacés. Traité de Zoologie, Tome VIII, Fasc. 3A." *Mémoires de l'Institut Océanographique* 19 (1999): 87–91.

Sanders, H. L., R. R. Hessler, and S. P. Garner. "*Hirsutia bathyalis*, a New Unusual Deep-sea Benthic Peracaridan Crustacean from the Tropical Atlantic." *Journal of Crustacean Biology* 5 (1985): 30–57.

Les Watling, PhD

Spelaeogriphacea
(Spelaeogriphaceans)

Phylum Arthropoda
Subphylum Crustacea
Class Malacostraca
Order Spelaeogriphacea
Number of families 1

Thumbnail description
Small, cave-dwelling, freshwater crustaceans having a short carapace, eye lobes without eyes, and appendages on all abdominal somites

Illustration: *Spelaeogriphus lepidops.* (Illustration by Dan Erickson)

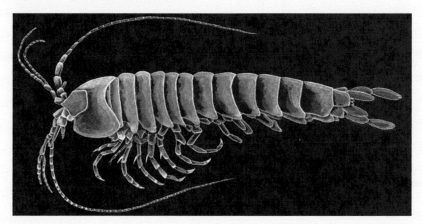

Evolution and systematics

Spelaeogriphaceans are known from one fossil species, found in New Brunswick, Canada, and three recent species. The first living species was found in a stream at the bottom of Bat Cave in Table Mountain outside Cape Town, South Africa. The two additional species are known from caves in Brazil and Western Australia. Spelaeogriphaceans are members of the short, branchial carapace clade within the superorder Peracarida. There is only one family, Spelaeogriphidae.

Physical characteristics

Spelaeogriphaceans have elongate, cylindrical bodies. A short carapace extends posteriorly from the back of the head. Both pairs of antennae are elongate and biramous. On the head are eye lobes, but no trace of the eye can be found. The first thoracic somite is fused to the head and its appendage is a maxilliped, which has a branchial epipod that extends into the branchial chamber formed by the short carapace. The mouthparts, including the maxilliped, are armed with long, stiff setae, suggesting they are used in a sweeping motion during feeding. The remaining thoracopods, known as pereopods, have endopods developed for walking, and exopods used to circulate water past the body. It is possible that the anterior exopods are also used for respiratory gas exchange. In females, the first five pereopods bear small brood plates. The appendages of the abdominal somites, known as pleopods, are moderately strongly developed, except for the last pair. At the end of the abdomen is a large fleshy telson and a pair of flattened uropods.

Distribution

Spelaeogriphaceans are known only from caves in South Africa, Brazil, and Western Australia.

Habitat

They are found in freshwater streams or pools in caves.

Behavior

Little is known about the behavior of these animals. They are unable to burrow into the sand or to swim. They walk about on the surface of the sand in the cave streams or pools and are easily swept about if the current is moderately strong.

Feeding ecology and diet

Spelaeogriphaceans appear to feed on plant detritus washed into the caves. They sweep their mouth appendages over the substrate in order to pick up small particles.

Reproductive biology

Nothing is known about mating or development. Females carry 10–12 eggs in the brood pouch.

Conservation status

Although no species are listed by the IUCN, the South African species, *Spelaeogriphus lepidops*, is protected locally and can be collected only by permit. Numbers seem to be stable.

Significance to humans

The species in this order are important as relics of past life.

Species accounts

No common name
Spelaeogriphus lepidops

FAMILY
Spelaeogriphidae

TAXONOMY
Spelaeogriphus lepidops Gordon, 1957, Table Mountain, Cape Town, South Africa.

OTHER COMMON NAMES
None known.

PHYSICAL CHARACTERISTICS
First spelaeogriphacean species to be described. It is pure white to transparent and often is detected by the dark material in gut. The carapace is short and covers the branchial epipod of the maxilliped. (Illustration shown in chapter introduction.)

DISTRIBUTION
Known only from one stream and a pool within the Bat Cave system in Table Mountain, Cape Town, South Africa.

HABITAT
Lives in freshwater of very low pH, about 4.2.

FEEDING ECOLOGY AND DIET
Appears to eat plant detritus that washes into the cave system from the bogs on the mountain top. It uses the long setae on its mouthparts to sweep detrital particles into its mouth.

BEHAVIOR
Because it lives at such low pH, it does not appear to become fouled. In experiments where pH the water was raised to 7, animals became covered with stalked ciliates. Lowering of pH eliminated fouling. Only when animals became fouled by fungal hyphae in culture was any grooming behavior observed.

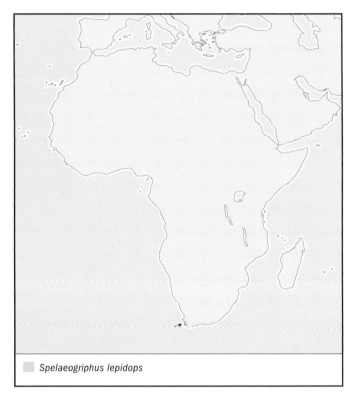

◻ *Spelaeogriphus lepidops*

REPRODUCTIVE BIOLOGY
Nothing is known about mating or development. Unpublished observations suggest animal hatches with all thoracic appendages intact.

CONSERVATION STATUS
The cave system is protected. Collection for research is by permit only.

SIGNIFICANCE TO HUMANS
An interesting relict of past crustacean evolution. ◆

Resources

Books
Schram, F. R. *Crustacea*. Oxford, U.K.: Oxford University Press, 1986.

Periodicals
Gordon, I. "On Spelaeogriphus, a New Cavernicolous Crustacean from South Africa." *Zoology* 5 (1957): 31–47.

Poore, G. B., and W. F. Humphreys. "First Record of Spelaeogriphacea (Crustacea) from Australia—A New Genus and Species from an Aquifer in the Arid Pilbara of Western Australia." *Crustaceana* 71 (1998): 721–742.

Les Watling, PhD

Thermosbaenacea

(Thermosbaenaceans)

Phylum Arthropoda
Subphylum Crustacea
Class Malacostraca
Order Thermosbaenacea
Number of families 4

Thumbnail description
Small crustaceans with a short carapace (shell) and seven free thoracic somites (segments)

Illustration: *Thermosbaena mirabilis.* (Illustration by Dan Erickson)

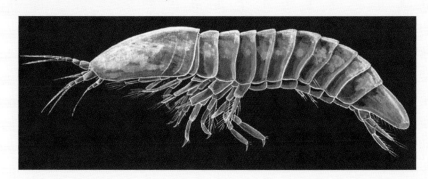

Evolution and systematics

Thermosbaenacea is an order comprising seven genera arranged in four families. A total of 34 species are known as of 2003. Thermosbaenaceans were first classified within their own superorder but are now generally regarded as members of the superorder Peracarida. They and other peracarid orders that possess a short branchial carapace have sometimes been grouped into a superorder known as Brachycarida. A recent study of thermosbaenacean evolution called attention to the close relationship of thermosbaenaceans to mictaceans, an order known only since 1985, and spelaeogriphaceans, very rare crustaceans that live in fresh groundwaters.

Physical characteristics

Thermosbaenaceans have a small, elongate body. The head is short and possesses a short carapace that extends backwards over the first few thoracic somites. The first thoracic somite is fused to the head; the appendage of this somite is modified as a maxilliped, or feeding appendage, located behind the jaw. There are seven free thoracic somites, each with a biramous walking leg. Each walking leg consists of a coxa (basal segment attached to the body) and basis (segment attached to the coxa); an exopod with one or two segments; and an endopod with five segments. There are six abdominal somites known as pleonites; only the first two have short appendages called pleopods. A broad and flattened telson and a pair of flattened uropods form the tail fan, located posterior to the last pleonite. The head appendages, especially the maxillae and maxillipeds, are usually broad and carry a variety of setae (bristles). The maxilliped often has a branchial epipod that extends backwards under the short carapace. The carapace of mature females develops a large inflated area that shelters the eggs.

Distribution

The distributional pattern of thermosbaenaceans lies within the limits of the ancient Tethys Sea, an ancient body of wa-

ter that existed during the Mesozoic Era, 248–65 million years ago. It extended from what is now Mexico across the Atlantic Ocean and Mediterranean Sea into central Asia. All Holocene species are found in the zone once covered by the Tethys Sea or along its former coastlines. All extant species in this order are known from ground waters.

Habitat

The first thermosbaenacean discovered, *Thermosbaena mirabilis*, was found living in hot springs near El Hamma, Tunisia, where the Romans built bathing houses. The name of the animal derives from the fact that these are thermal springs with water temperatures of 111–118°F (44–48°C). Other thermosbaenaceans were subsequently found in underground springs; most species, however, do not live in thermal springs. In the Caribbean, several species were found in wells drilled for drinking water. Others have been found in caves with water running through them. The water in some of these caves is oligohaline (low salinity) rather than fresh. In addition, a few species in the genus *Halosbaena* are known from interstitial marine waters or marine caves.

Behavior

Thermosbaenaceans move primarily by walking, but can also swim by using their thoracic limbs for propulsion.

Feeding ecology and diet

Thermosbaenaceans living in hot springs feed on blue-green algae, diatoms, and other microalgae lining the rocks. Little is known about the diet of species living in other habitats. The mouth appendages of these organisms, however, are well festooned with setae located on the distal edges of the appendage segments, suggesting they might be used to sweep small particles from the crustacean's substrate. One species is known to feed on plant detritus.

Reproductive biology

Mating has not been observed. The eggs are incubated in a dorsal brood pouch formed by the swollen carapace of the mature female, and are bathed in water passing past the respiratory epipod. The young resemble miniature adults when they hatch. In one species, the young hatch before the sixth and seventh legs are fully formed, and so must undergo some development outside the brood pouch.

Conservation status

The type species of the order seems to have disappeared from the type locality. Most thermosbaenaceans are protected because their habitats are protected. None are listed by the IUCN.

Significance to humans

Thermosbaenaceans have no known significance to humans.

Species accounts

No common name
Thermosbaena mirabilis

FAMILY
Thermosbaenidae

TAXONOMY
Thermosbaena mirabilis Monod, 1924, El Hamma, Tunisia.

OTHER COMMON NAMES
None known.

PHYSICAL CHARACTERISTICS
The body is more or less cylindrical in shape, with the thorax somewhat shorter than the abdomen. The thorax bears only five pairs of legs, each with well-formed exopods. The maxilliped is modified and lacks an endopod. The abdomen has two small pairs of pleopods on its first and second segments. The telson is quite large, as long as the last three abdominal somites. Eyes are not present. (Illustration shown in chapter introduction.)

DISTRIBUTION
Known only from a limited number of thermal springs in Tunisia.

HABITAT
Thermal springs with temperatures above 111°F (44°C), generally with highly mineralized water.

BEHAVIOR
Crawls on the surfaces of rocks in search of food.

FEEDING ECOLOGY AND DIET
Thermosbaena mirabilis has been found to feed on several species of blue-green algae in the thermal springs.

REPRODUCTIVE BIOLOGY
The testes are located in posterior part of the head and the first somite of the thorax. Long vasa deferentia lead from the thorax to the end of the abdomen and then back to the eighth thoracic somite where the male gonopore is located. Ovaries occupy the entire thorax in females, with the gono-

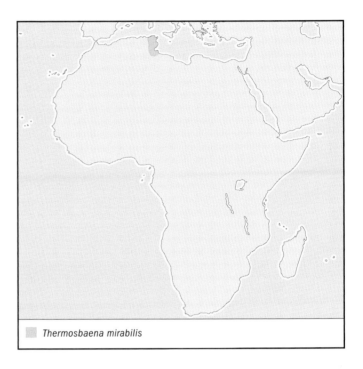

Thermosbaena mirabilis

pore on the sixth thoracic somite. When mature the ovary also extends into the abdomen. Eggs are carried in a brood pouch formed by a lobe of the female carapace, although deposition of the eggs in this location has never been observed. When shed from the brood pouch, the young have five pairs of thoracic limbs beyond the maxilliped and look like miniature adults.

CONSERVATION STATUS
The animal disappears when the baths are cleaned, but apparently repopulates afterward, perhaps from a larger population living deep underground. Not listed by IUCN.

SIGNIFICANCE TO HUMANS
None known. ◆

Resources

Books
Schram, F. R. *Crustacea.* Oxford, U.K.: Oxford University Press, 1986.

Periodicals
Wagner, H. P. "A Monographic Review of the Thermosbaenacea (Crustacea: Peracarida)." *Zoologische Verhandelingen* 291 (1994): 1–338.

Les Watling, PhD

Isopoda

(Pillbugs, slaters, and woodlice)

Phylum Arthropoda
Subphylum Crustacea
Class Malacostraca
Order Isopoda
Number of families Approximately 120

Thumbnail description
Small, generally gray, usually flat, marine, freshwater, or terrestrial animals with numerous legs; some species are parasitic

Photo: Pill woodlouse (*Armadillidium vulgare*) rolled up in defensive mode. (Photo by Nigel Cattlin/ Holt Studios Int'l/Photo Researchers, Inc. Reproduced by permission.)

Evolution and systematics

With approximately 10,000 known species in 10 suborders, the order Isopoda falls under the class Malacostraca, subphylum Crustacea, phylum Arthropoda. Five of the predominant suborders are as follows:

- Asellota, marine and freshwater isopods

- Epicaridea, parasitic isopods that live on or in other crustaceans

- Flabellifera, marine or estuarine species, including a few parasitic taxa

- Oniscidea, mainly terrestrial isopods, including the familiar pillbugs, sowbugs, and woodlice

- Valvifera, marine species

Isopods are perhaps most intriguing as models of the evolutionary transition from marine to terrestrial habitats. The terrestrial suborder Oniscidea is believed to have arisen either from the marine suborder Flabellifera, specifically from the family Cirolanidae, or from the marine suborder Asellota. The ancestors of the oniscideans were likely similar to the genus *Ligia*, which inhabits the ocean shoreline and has both aquatic and terrestrial characteristics. These characteristics include a primitive water-conducting system and the ability to swim, a mode of locomotion that terrestrial isopods no longer possess.

The order Isopoda, which dates to at least 300 million years ago, once was thought to be monophyletic. Recent findings, however, suggest that the suborder Flabellifera has a separate phylogeny. Early isopods were shallow marine inhabitants, then spread to freshwater, deep marine, and terrestrial areas, where they live today.

Physical characteristics

A diverse order of crustaceans, the isopods are mainly small, at least slightly dorsoventrally flattened, gray or brown organisms with numerous legs, called pereopods. Their bodies are divided into the head, which includes fused maxillipeds that form the so-called cephalon; the leg-bearing thorax or pereon; and the abdomen, or pleon. The length of isopods typically ranges from approximately 0.2 to 0.6 in (5–15 mm), but these animals can be smaller (0.02 in [0.5 mm]) or much larger. The largest species, at 19.7 in (50 cm) long, is *Bathynomus giganteus*. Isopods lack the carapace of many other crustaceans, instead having a cephalic shield. All but the parasitic forms have at least 14 walking legs—two on each of the seven somites that make up the pereon. The legs usually are short, but they can be quite long and spider-like in species such as *Munna armoricana*, which has legs as long as or longer than the body. Unlike many other crustaceans, isopods have unstalked rather than stalked compound eyes. They typically have two, well-developed antennae equipped with stiff sensory setae. Another vestigial pair of antennae is present. In addition, the ventral plates of the thorax, specifically those of the second through fifth thoracic segments, form the brood pouch used by females to house their developing young.

Two pairs of white, oval structures are apparent on the first two abdominal segments of terrestrial isopods. Desert forms have five pairs. These structures are called pleopods and are appendages modified for respiration. Pleopods contain pseudotracheae, which trap air and give the pleopods their white appearance. In addition to using this source of oxygen, terrestrial isopods breathe by diffusion of gas directly through the cuticle.

The appearance of parasitic forms of isopods differs somewhat from that of other isopods. Females commonly have asymmetric bodies—rather like a pillbug viewed in a funhouse mirror. If they are present at all, legs often are developed only on one side of the body. The mandibles are sharp, piercing devices. The male looks more like a typical pillbug or sowbug, having a symmetrical, oval body. The male is much smaller than the female and often is found attached to her abdomen. Male and female parasitic isopods have two pairs of antennae, but both are vestigial at best.

Distribution

Worldwide. Many isopod species have extended their ranges with the help of humans. Marine species have moved across the ocean in the bilge waters of sea-faring tankers, and terrestrial forms find welcome hiding places in the dark, damp storage areas of various transportation vehicles. *Armadillidium vulgare* is an example of a species that has successfully invaded habitats in the New World from its original distribution around the Mediterranean Sea.

Habitat

Isopods can be found in a wide range of habitats from marine or freshwater areas to deserts but are best known from their terrestrial haunts under logs, in or beneath rotting wood, or in other damp areas. Isopods are gill breathers, and even the terrestrial species, which number approximately 4,000, need wet habitats.

Among the terrestrial isopods, rock slaters of the family Ligiidae, inhabit littoral (seashore) areas. Pillbugs are species of the families Armadillididae and Armadillidae and are typically found in grasslands and arid habitats. Sowbug is the common name given to species of the families Oniscidae and Porcellionidae. These isopods favor forests and semiarid areas, respectively. A few species live in deserts. *Hemilepistus reaumuri*, for example, lives in deserts of northern Africa and the Middle East.

Marine and estuarine species, which number approximately 4,500, often live in shallow coastline waters, but numerous species, particularly those in the suborder Asellota, have successfully invaded the deep sea. They frequently inhabit burrows they make in the sediment or in vegetation. The freshwater species, numbering approximately 500, similarly are sediment burrowers. A wood-boring isopod, *Sphaeroma terebrans*, burrows into the aerial roots of mangrove trees, which periodically flood. A few, including *Ligia* species, are transitional between terrestrial and marine habitats and exist in the semiterrestrial, rocky coastline along the ocean.

Numerous species, particularly those in the suborder Epicaridea, are parasitic, blood-sucking forms that live on or in various animals, including barnacles, crabs, and shrimps.

Behavior

Terrestrial isopods are most often found in dark nooks and crannies of decaying logs and underneath rocks and leaf litter. Many do, however, venture into the daylight. Pillbugs even have a positive phototactic response, particularly toward sunset. When temperatures during the day become too high, terrestrial isopods generally move quickly to underground hiding places, where higher humidity can help the animals avoid desiccation. As another defense against desiccation, when temperatures rise to 68°F–86°F (20°C–30°C), pillbugs apparently become attracted to the odors of conspecifics and group together. The bunching behavior decreases the exposed surface area of each individual. If necessary, pillbugs are able to drink water droplets by taking up water through tail projections and diverting it along lateral, exterior grooves (collectively called a water transport system) to the mouth.

By spending a good deal of time hidden in or under logs and leaf litter, terrestrial isopods gain considerable protection from many of their predators, but they have additional defenses. One is camouflage. Isopods are generally brownish-gray or gray, colors that conceal them well against the ground, a log, or a rock. Another predator deterrent comes from repugnatorial glands on the thorax. These glands release a secretion that is unpleasant to predators and is enough to ward off most attacks. The European pillbug (*Armadillidium klugii*) is unusual in that it has aposematic coloration that mimics the orange, hourglass marking typical of a European black widow spider (*Latrodectus mactans tredecimguttatus*). The spider is a venomous species that predators avoid. The copycat coloration allows these isopods to reap the rewards of the spider's warning pattern.

A strange example of another species that affects the behavior of pillbugs is an acanthocephalan worm that parasitizes these isopods. The life cycle of the worm begins when its eggs are passed in the feces of birds, specifically starlings (*Sturnus vulgaris*). Pillbugs eat the feces and ingest the worms. The worms hatch inside the pillbug, where they grow to approximately 0.1 in (2–3 mm) long (approximately one third of the total length of the pillbug). In addition to crowding pillbugs' internal organs and rendering females sterile, the worms alter pillbugs' behavior. Infected pillbugs move from their normal damp, dark areas to wide-open spaces. Starlings feed on pillbugs and easily find the now exposed individuals. Infected starlings are the final host for the worms, which lay eggs that are passed through feces to repeat the cycle.

Isopods as a group are perhaps best known for the ability of some to roll into a ball. With this posture, called conglobation, isopods effectively use their armor-like dorsal surface to shield the softer body parts from predators and from water loss. Not all isopods can conglobate, but the behavior is common among terrestrial species. Even some intertidal and littoral species, such as *Campecopea hirsuta* and *Tylos*, respectively, can roll up. When conglomated, many species enclose their antennae in the ball, but some, such as *Armadillidium*, leave the antennae outside. Despite the acrobatics involved in conglobation, many species of isopods cannot right themselves if they are turned on their backs. This is particularly true if the species is especially rounded dorsally. Sowbugs, the

Pillbug bunching behavior. (Illustration by Christina St. Clair)

flatter, terrestrial isopods, cannot conglobate but can right themselves easily.

Among marine and coast-living isopods, level of activity is frequently related to the tides. As water rises and lowers along the coast, some dune-dwelling species, such as *Tylos punctatus*, move up or down the beach slope to station themselves just beyond the water. Other marine organisms, such as *Eurydice pulchra*, are inactive during neap tide but become active just after a high tide and heighten their activity approximately three or four days after a new or full moon. Swimming when the water is high helps these isopods follow the water's rise and fall up and down the beach slope.

Littoral and seashore isopods may be either diurnal or nocturnal. *Ligia* species are an example of the former, and *Tylos* of the latter.

Desert isopods eke out a living through various behaviors. *Hemilepistus reaumuri*, for example, retains the family unit with parents tending juveniles in the burrow throughout the summer.

Feeding ecology and diet

As a group, isopods are omnivores, eating everything from living and dead vegetation to fungi and from living and dead animals to fecal matter. The terrestrial forms, commonly called pillbugs or sowbugs, are mostly detritus feeders, scouring the forest floor for decaying organic matter. Their diet is wide ranging and may include fruits and tender shoots, dead and dying vegetative matter, fungi, and their own as well as other organisms' feces. Scientists have studied the isopod habit of eating feces, called coprophagy. Research indicates that nearly one tenth of the isopod diet may be the animals' own waste products, which are believed to replenish the digestive microorganisms the system requires and to provide some nutrition. When deprived of this dietary source, isopods grow more slowly than normal. Some species, such as those in the genus *Platyarthrus*, eat feces or regurgitated pellets of ants. These isopods, which are blind and white, live in ant nests.

Predators of terrestrial isopods include various spiders, such as those in the family Dysderidae, that can penetrate the isopod's hard coat. Amphibians and birds also take a toll, as do centipedes. Terrestrial isopods are particularly vulnerable to predators when they molt and temporarily lose their hard, protective covering.

Marine isopods feed primarily on algae, diatoms, and other vegetation in addition to wood and vegetative detritus. A few, such as *Cirolana* species, eat the decaying flesh of dead animals, especially fishes. Predators of these marine species are

A coney (*Cephalopholis fulva*) with the parasitic isopod *Anilocra laticaudata* attached to it. (Photo by Andrew Martinez/Photo Researchers, Inc. Reproduced by permission.)

primarily fishes. Parasitic isopods also exist. Some, such as *Lironeca* and *Aega*, attach to fishes. Others, including *Stegias clibanarii*, are parasitic during only one stage of their lives. They live on yolk during the first developmental stage, go through a second, parasitic stage, and become free living in the third stage.

Reproductive biology

The sexes are typically separate. Male isopods transfer sperm through the second, or the first and second, pair of pleopods. Mating generally occurs by the male climbing onto the back of the female and bending his abdomen to her ventral gonopores for sperm transfer. The female is fertile only as she goes through maturation molt, but male-female pairs can form a day or two early. During maturation molt, females shed the posterior half of the exoskeleton and two or three days later shed the anterior half. Soon after mating, the female sheds her eggs, which range in number from half a dozen to several hundred, into a brood pouch, or marsupium. The ovaria and marsupium are attached by thin tubes. Egg size varies between species and among females of the same species.

Generally, larger females have larger eggs. In general, young develop in the brood pouch for the next 8–12 weeks. One or two broods per year are common, and females of many species can store sperm for as long as a year. Juveniles that leave the brood pouch are called mancas. Mancas are almost identical to adults but are missing the last pair of thoracic legs. Among burrow-dwelling species, mancas may find and live in tunnels individually or remain with the mother in a family burrow. In the burrows, the juveniles harden and darken through successive molts.

A few isopods, including the parasitic *Lironeca* and *Aega* species, are protandric hermaphrodites that switch from male to female as they develop. *Pseudione* and other parasitic forms, however, have separate sexes.

Isopods typically live one or two years, but some survive five years. The longest-lived species known is *Armadillo officinalis*, which can live nine years.

Conservation status

The IUCN lists 39 species of isopods as threatened. They include 22 species categorized as Vulnerable, seven as Critically Endangered, nine as Endangered, and one as Extinct in the Wild. The extinct species is the Socorro isopod, *Thermosphaeroma thermophilum*, a member of the family Sphaeromatidae. Found only in Sedillo Spring, Socorro County, New Mexico, the population became extinct in 1988 when a valve control system failed and cut off flow to the area. The problem has since been repaired, and previously obtained individuals have been bred in captivity and reintroduced to the spring. Additional captive populations are held by three organizations, including the Santa Fe, New Mexico, Department of Game and Fish.

Significance to humans

The feeding and burrowing behavior of some marine and estuarine isopods, such as *Sphaeroma* and *Limnoria*, can cause considerable damage to wooden pilings, docks, and other underwater structures. Terrestrial isopods are generally harmless, although large numbers can cause vegetation damage, particularly in gardens and greenhouses.

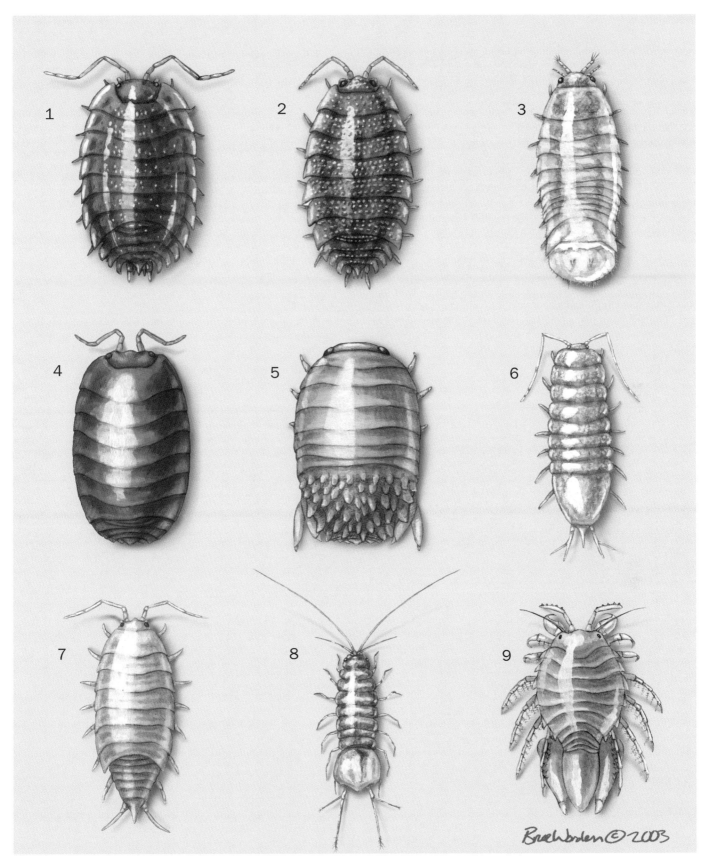

1. Common shiny woodlouse (*Oniscus asellus*); 2. Common rough woodlouse (*Porcellio scaber*); 3. Gribble (*Limnoria quadripunctata*); 4. Common pill woodlouse (*Armadillidium vulgare*); 5. *Sphaeroma terebrans*; 6. *Lirceus fontinalis*; 7. Common pygmy woodlouse (*Trichoniscus pusillus*); 8. Water louse (*Asellus aquaticus*); 9. Sand isopod (*Chiridotea caeca*). (Illustration by Bruce Worden)

Species accounts

Common pill woodlouse
Armadillidium vulgare

FAMILY
Armadillidiidae

TAXONOMY
Armadillidium vulgare Latreille, 1804, cosmopolitan.

OTHER COMMON NAMES
French: Cloporte vulgaire; German: Kugelassel.

PHYSICAL CHARACTERISTICS
The common pill woodlouse usually is dark gray and often has distinct rows of spots. The color sometimes varies to brown or red. The antennae but not the legs are visible from above. The oval body is approximately twice as long as it is wide and can reach a length of 0.7 in (18 mm).

DISTRIBUTION
Originally the Mediterranean periphery; now nearly ubiquitous in temperate climates.

HABITAT
The common pill woodlouse, once having the genus name *Armadillo*, is found in wide-ranging habitats, including forests, grasslands, and even sand dunes. This woodlouse also is common in cultivated areas and in greenhouses.

BEHAVIOR
The common pill woodlouse can conglobate (roll into a ball). The woodlice move about in the open but prefer areas of high humidity and thus frequently are found under leaf litter or logs. Under normal conditions, the common pill woodlouse moves slowly through its habitat and becomes more active in dry air when it actively seeks a more humid area. Studies of daily, foraging movements reveal that individuals travel an average of 43 ft (13 m) each day in the summer but only 22 ft (6.6 m) a day in the winter. In cold months, individuals have been found as deep as 10 in (25 cm) beneath the soil surface.

FEEDING ECOLOGY AND DIET
These herbivorous and detrivorous creatures feed on tender plant shoots as well as dead and decaying plant matter. Their dietary habits shift during periods of drought, when these isopods switch from being mainly vegetarians to being scavengers.

REPRODUCTIVE BIOLOGY
Reproduction is cyclical and triggered by environmental factors, such as rising temperatures and longer days. In addition, females experience an acceleration of ovarian maturation in the presence of males, although this effect is lessened in northern climates. Mating occurs immediately before the parturitional molt, usually from late spring to early summer, but sometimes the period extends from late winter to early fall. Females in northern populations have one brood per year, and those in

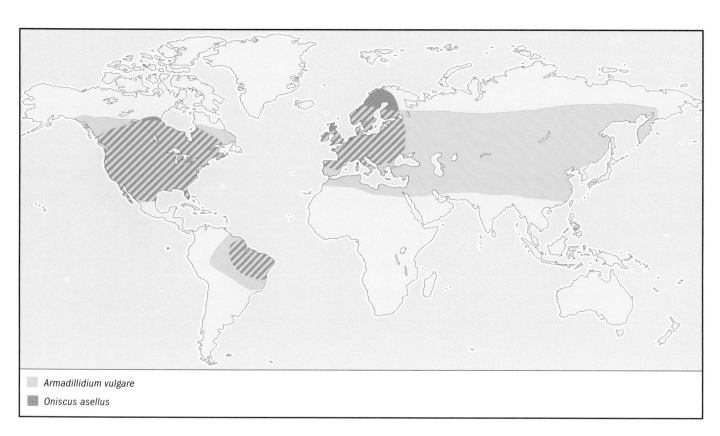

Armadillidium vulgare

Oniscus asellus

southern populations typically have two, sometimes three. Broods can include more than 100 eggs, but typically approximately one half of eggs develop into mancas. Eggs move into a brooding pouch, or marsupium. These woodlice can store sperm, which remains viable for approximately one year.

CONSERVATION STATUS
Not listed by the IUCN.

SIGNIFICANCE TO HUMANS
Large numbers in a garden or greenhouse can take a toll on tender, young, plant shoots but normally do not pose a considerable threat. ◆

Water louse
Asellus aquaticus

FAMILY
Asellidae

TAXONOMY
Asellus aquaticus Linnaeus, 1758, type locality not specified.

OTHER COMMON NAMES
English: Hog slater, hoglouse; French: Aselle, cloporte d'eau; German: Wasserassel.

PHYSICAL CHARACTERISTICS
Individuals are brownish and have long, oval, dorsoventrally flattened bodies. Some have two yellow, fluorescent stripes down their backs. Males are typically larger than females, reaching approximately 0.8 in (20 mm). Individuals that live in caves and other dark environments are unpigmented and have no visible eyes.

DISTRIBUTION
Europe.

HABITAT
A freshwater isopod, the water louse lives primarily in surface-water areas, including ponds and slow-moving creeks. Researchers have found well- and cave-dwelling populations of this species.

BEHAVIOR
The male guards its intended mate for several days before fertilization by hoisting up the female from the substrate and carrying her underneath him. Unpaired males sometimes are successful in attempts to take the place of mating males by struggling with and separating united pairs.

FEEDING ECOLOGY AND DIET
The diet includes coarse, particulate, vegetative matter taken from the sediment.

REPRODUCTIVE BIOLOGY
Water lice reproduce from early spring to mid autumn. Observations differ on whether males prefer larger females. Heftier females produce more and faster-hatching eggs but are a heavier load for males to carry during the precopulatory, mate-guarding period. Males that choose larger females may cut back on precopulatory mate guarding and carry the female nearer to her fertile period, which occurs during her final molting.

CONSERVATION STATUS
Not listed by the IUCN.

Asellus aquaticus
Chiridotea caeca
Lirceus fontinalis

SIGNIFICANCE TO HUMANS
None known. ◆

No common name
Lirceus fontinalis

FAMILY
Asellidae

TAXONOMY
Lirceus fontinalis Rafinesque-Schmaltz, 1820, Kentucky, United States.

OTHER COMMON NAMES
None known.

PHYSICAL CHARACTERISTICS
Lirceus fontinalis is somewhat teardrop-shaped and has a rounded head and a body that tapers posteriorly. It has two long and apparent antennae and long and noticeable legs. Males are slightly larger than females, length ranging from 0.4 to 0.6 in (10–15 mm). The length of females averages approximately 0.2–0.4 in (6–10 mm).

DISTRIBUTION
Eastern United States.

HABITAT
Streams, particularly among mats of filamentous algae, but also along bare silt, sand, and gravel substrates and in submerged leaf litter.

BEHAVIOR
During courtship, males of *L. fontinalis* first struggle with, then dominate, and finally defend their intended partners in precopulatory mate guarding. Defense is necessary to prevent a single male from dislodging the paired male and taking his place. During precopulatory mate guarding, the male carries the female beneath him for one to three days. Females are receptive for only a short time, so the behavior appears to assure that the male is in the correct location when the female begins her maturation molt and is ready to be fertilized. Males seem to prefer females who are close to parturitional molt.

FEEDING ECOLOGY AND DIET
Lirceus fontinalis is a detritus feeder, sustaining itself on organic particulates in the water and algal mats. Predators include various species of stream-dwelling fishes, such as green sunfish (*Lepomis cyanellus*) and sculpins (*Cottus* species). Results of laboratory experiments have suggested that these isopods detect sunfish chemicals and use these cues to avoid the fish by hiding in vegetation, remaining motionless, or burying themselves in the substrate. *Lirceus fontinalis* also uses avoidance tactics when it detects changes in current caused by approaching swimming fishes.

REPRODUCTIVE BIOLOGY
Males inseminate females during the maturation molt. Females do not store sperm. Females may produce 20–120 eggs, which are deposited into a ventral marsupium, or brood pouch. Larger females produce more eggs. Every 0.0007 oz (2 mg) in body weight correlates to approximately 10 eggs. After hatching, the juveniles spend approximately three weeks in the mar-

supium, where they proceed through several molts. Upon leaving the marsupium, the juveniles live independently.

CONSERVATION STATUS
Not listed by the IUCN.

SIGNIFICANCE TO HUMANS
None known. ◆

Sand isopod
Chiridotea caeca

FAMILY
Idoteidae

TAXONOMY
Chiridotea caeca Say, 1818, northwest Atlantic Ocean.

OTHER COMMON NAMES
None known.

PHYSICAL CHARACTERISTICS
Adults are dorsoventrally flattened with a thorax that is almost round when viewed from above and have long, robust legs equipped with conspicuous, plumose setae. A long, pointed abdomen follows. Individuals reach approximately 0.6 in (15 mm) in length and 0.3 in (7 mm) in width.

DISTRIBUTION
Western Atlantic Ocean, from Nova Scotia to Florida.

HABITAT
This benthic, marine species populates the coarse, sandy bottom of the intertidal, sometimes subtidal, zone along the ocean shoreline.

BEHAVIOR
A burrowing species of the suborder Valvifera, the sand isopod uses its hindmost pereopods to dig tunnels in the sand. If they are removed from their tunnels, perhaps by wave action, sand isopods swim—often upside down—to the substrate, where they again seek underground protection. Mating occurs while the pair is buried in the substrate.

FEEDING ECOLOGY AND DIET
Apparently carnivorous. Little is known about the diet and feeding ecology of this burrowing species. Predators include a variety of fishes, including flounders and puffers, and shore birds.

REPRODUCTIVE BIOLOGY
Eggs, which are oval shaped, number two to three dozen, occasionally reaching six dozen. Juvenile instars approximately 0.1 in (2.5 mm) long and 0.05 in (1.25 mm) wide appear in the spring. Juveniles go through approximately six instars and mature at a length of approximately 0.4 in (10 mm). Females become ovigerous in December. Mating can take several days. Females undergo the last maturation molt while in amplexus with males. Sand isopods produce one brood per year.

CONSERVATION STATUS
Not listed by the IUCN.

SIGNIFICANCE TO HUMANS
None known. ◆

Gribble
Limnoria quadripunctata

FAMILY
Limnoriidae

TAXONOMY
Limnoria quadripunctata Holthuis, 1949, Netherlands.

OTHER COMMON NAMES
German: Bohrassel.

PHYSICAL CHARACTERISTICS
The gribble, a 14-legged isopod, is more cigar shaped than oval and reaches a length of approximately 0.1–0.2 in (3–6 mm). Gribbles have two pairs of short antennae and are generally light grayish brown.

DISTRIBUTION
Northern Hemisphere.

HABITAT
The gribble, a coastal, marine organism, makes and inhabits tunnels and grooves that it makes at or just below the surface of water-exposed wood.

BEHAVIOR
Gribbles appear able to orient themselves and swim toward wood. Experimental investigation indicates that these isopods are attracted to marine fungi and vegetation, as well as to chemicals from other members of the species on wood, and orient themselves to these markers with their antennae. One pair appears to detect chemicals and the other seemingly picks up odors. Gribbles are able to conglobate.

FEEDING ECOLOGY AND DIET
Gribbles consume wood, fungi on the wood, and waterborne detritus.

REPRODUCTIVE BIOLOGY
Young gribbles develop in their mother's tunnel, many digging their own tunnels into the wood. Although males remain with females after mating, it is unclear whether males participate in parental care of offspring.

CONSERVATION STATUS
Not listed by the IUCN.

SIGNIFICANCE TO HUMANS
The wood-boring activities of gribbles damage wood pilings, docks, and other wooden structures. In large infestations, wood depth can decrease approximately 0.8 in (2 cm) per year. ◆

Common shiny woodlouse
Oniscus asellus

FAMILY
Oniscidae

TAXONOMY
Oniscus asellus Linnaeus, 1758, Europe.

OTHER COMMON NAMES
German: Mauerassel.

PHYSICAL CHARACTERISTICS
The common shiny woodlouse, which has a somewhat shiny dorsal surface, has the typical isopod appearance of an oval,

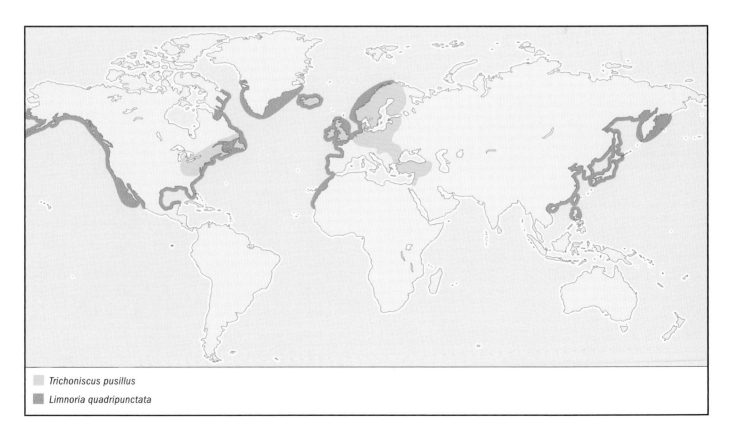

☐ *Trichoniscus pusillus*
■ *Limnoria quadripunctata*

grayish brown body and a pair of long antennae. It reaches a length of approximately 0.6 in (16 mm).

DISTRIBUTION
Western and northern Europe originally; eastern Europe and North America currently.

HABITAT
This woodlouse lives in almost any damp, terrestrial area, preferring forests, and often living under rocks and deadfall.

BEHAVIOR
The common shiny woodlouse and others of its family cannot roll up (conglobate). They shun the light and seek warm areas of high humidity. A characteristic behavior is the tendency to firmly latch onto rocks and other surfaces.

FEEDING ECOLOGY AND DIET
Eats vegetative matter, primarily lime, ash, and alder leaves. Research has indicated that growth and fecundity increase when food intake includes dicotyledonous rather than mono-cotyledonous leaves. Growth also improves with a copper-fortified diet.

REPRODUCTIVE BIOLOGY
The common shiny woodlouse typically has two broods and produces 27–33 eggs. Juveniles are dull, dark gray and have a pale marking on the pleon.

CONSERVATION STATUS
Not listed by the IUCN.

SIGNIFICANCE TO HUMANS
Although common in gardens and often in greenhouses, the common shiny woodlouse does little damage to vegetation. ◆

Common rough woodlouse
Porcellio scaber

FAMILY
Porcellionidae

TAXONOMY
Porcellio scaber Latreille, 1804, western Europe.

OTHER COMMON NAMES
German: Kellerassel.

PHYSICAL CHARACTERISTICS
The common rough woodlouse has the typical oval shape of terrestrial isopods but has a rough rather than shiny dorsal surface and frequently is orange at the base of the antennae. The coloration is gray, brown, or orangish brown, often with gray blotches. This woodlouse reaches a length of approximately 0.7 in (17 mm).

DISTRIBUTION
Native to western Europe; north to Iceland, south to South Africa and South America.

HABITAT
The common rough woodlouse lives in damp, dark locations, often beneath logs or rocks, in forests or along waterways, sometimes in meadows, and often extends into cultivated areas, such as gardens and greenhouses. This woodlouse is abundant in splash zones along ocean shorelines.

BEHAVIOR
Lengthening daylight is a trigger for reproduction. Mainly active at night, the common rough woodlouse cannot roll into a ball, as can many other terrestrial isopods. Surveys of this

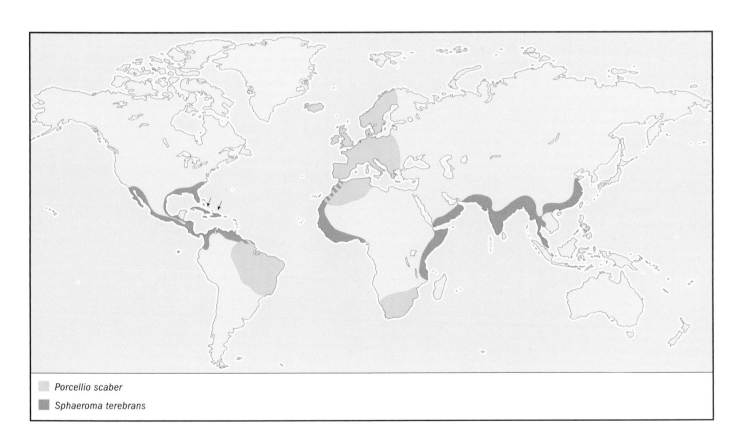

Porcellio scaber
Sphaeroma terebrans

species in Europe show that individuals often travel into trees in the summer and back to the soil in the fall. This behavior is common to many other terrestrial isopods.

FEEDING ECOLOGY AND DIET
The common rough woodlouse is herbivorous, detrivorous, and coprophagous. It seems to prefer poplar leaves and leaves with fungal growth, and it eats decaying pine needles. Results of experiments indicate that ingestion of feces is important in building up copper reserves and in invigorating gut bacteria that are important in digestion.

REPRODUCTIVE BIOLOGY
Females have one to three broods per year. The eggs, numbering one to three dozen, are approximately 0.03 in (0.7 mm) in diameter, and the young develop in the female's marsupium. Studies of broods indicate than more than 80% have multiple paternity.

CONSERVATION STATUS
Not listed by the IUCN.

SIGNIFICANCE TO HUMANS
The common rough woodlouse sometimes becomes a pest in gardens or greenhouses, where they feed on new plant growth. ◆

No common name
Sphaeroma terebrans

FAMILY
Sphaeromatidae

TAXONOMY
Sphaeroma terebrans Bate, 1866, India.

OTHER COMMON NAMES
None known.

PHYSICAL CHARACTERISTICS
The rough-surfaced pleotelson of *Sphaeroma terebrans* is slightly pointed. Females are larger than males, averaging a length of 0.3–0.4 in (8–10 mm) compared with the male length of 0.26–0.33 in (6.5–8.5 mm).

DISTRIBUTION
Mangrove forests worldwide, including Florida and Kenya.

HABITAT
Sphaeroma terebrans, a wood-boring species, lives chiefly in the aerial roots of red mangroves (*Rhizophora mangle*), although it also inhabits fallen trees and the roots of other plants.

BEHAVIOR
Juveniles live in burrows, either their own or often those of the parent. These "family burrows," which lie at the end of the mothers' burrows (fathers do not participate in parental care), also often house one to eight uninvited juveniles from the related species *Sphaeroma quadridentatum. Sphaeroma quadridentatum* does not burrow but seeks refuge in crevices or other small hideaways. *Sphaeroma terebrans* females either do not distinguish the *S. quadridentatum* juveniles from their own or simply tolerate them. A survey of the burrows of reproductive *S. terebrans* females placed the proportion of those harboring *S. quadridentatum* juveniles at 30% and showed that young *S. terebrans* receive a shorter period of parental care in the shared burrows. Parental care may include general housekeeping duties, such as removal of waste, and ventilation and excavation of burrows.

FEEDING ECOLOGY AND DIET
The diet of this apparent filter feeder likely includes waterborne planktonic algae. During high tides, water rushes over the mangrove roots and into the burrows, delivering algae to the isopods.

REPRODUCTIVE BIOLOGY
Reproductive activity is cyclical, peaking in the fall and in the late spring to early summer. A female has one or two broods per year, which develop from eggs in the marsupium. The young emerge at the manca stage approximately 0.1 in (2.5 mm) long. Larger females have the largest broods, and typically host 5–20 juveniles in their burrows, although some females host as many as five dozen juveniles. Some juveniles do not use family burrows, instead finding and living in their own tunnels. Juveniles do not grow much while in the family burrow, remaining approximately 0.1 in (2.5 mm) long. The lifespan is approximately 10 months.

CONSERVATION STATUS
Not listed by the IUCN.

SIGNIFICANCE TO HUMANS
S. terebrans causes considerable damage to red mangroves. ◆

Common pygmy woodlouse
Trichoniscus pusillus

FAMILY
Trichoniscidae

TAXONOMY
Trichoniscus pusillus Brandt, 1833, Europe.

OTHER COMMON NAMES
None known.

PHYSICAL CHARACTERISTICS
A small, reddish brown (rarely violet) woodlouse with a shiny dorsal surface that is slightly mottled with white. The legs noticeably extend beyond the body. The common pygmy woodlouse reaches a length of only 0.2 in (5 mm). Males are slightly larger than females and have additional white markings where the genital apparatus attaches to the pleon segments.

DISTRIBUTION
Europe and North America.

HABITAT
The common pygmy woodlouse typically inhabits damp soil and leaf litter in forested areas but also is found in nearly every other temperate habitat, including grasslands and sparsely vegetated fields.

BEHAVIOR
The common pygmy woodlouse does not seen to mind being close to conspecifics. Reports of several thousand per 11 ft² (1 m²) are not uncommon.

FEEDING ECOLOGY AND DIET
Diet is mainly vegetative detritus.

REPRODUCTIVE BIOLOGY

Bisexual, parthenogenetic, and mixed populations of common pygmy woodlice exist. Parthenogenetic populations in Britain are composed almost entirely of females. Two to three broods are typical, breeding occurring between May and September. Eggs number between 4 and 17 and are approximately 0.01 in (0.3 mm) in diameter. An average female produces five to seven juveniles per brood.

CONSERVATION STATUS

Not listed by the IUCN.

SIGNIFICANCE TO HUMANS

None known. ◆

Resources

Books

Alikhan, M. A., ed. *Crustacean Issues 9: Terrestrial Isopod Biology.* Rotterdam, The Netherlands: A. A. Balkema, 1995.

Raham, R. G. "Pill Bug Strategies." In *Dinosaurs in the Garden: An Evolutionary Guide to Backyard Biology.* Medford, NJ: Plexus Publishing, 1988.

Warburg, M. R. *Evolutionary Biology of Land Isopods.* Berlin: Springer-Verlag, 1993.

Periodicals

Brooks, R. A. "Colonization of a Dynamic Substrate: Factors Influencing Recruitment of the Wood-Boring Isopod, *Sphaeroma terebrans* onto Red Mangrove (*Rhizophora mangle*)." *Oecologia* 127 (2001): 522–532.

Hassall, M., and S. P. Rushton. "The Role of Coprophagy in the Feeding Strategies of Terestrial Isopods." *Oecologia* 53 (1982): 374–381.

Holomuzki, J. R., and T. M. Short. "Habitat Use and Fish Avoidance Behaviors by the Stream-Dwelling Isopod *Lirceus fontinalis.*" *Oikos* 52 (1988): 79–86

McDermott, J. J. "Biology of *Chiridotea caeca* (Say, 1818) (Isopoda: Idoteidae) in the Surf Zone of Exposed Sandy Beaches along the Coast of Southern New Jersey, U.S.A." *Ophelia* 55 (2001): 123–135.

Sparkes, T. C., D. P. Keogh, and R. A. Pary. "Energetic Costs of Mate Guarding Behavior in Male Stream-Dwelling Isopods." *Oecologia* 106 (1996): 166–171.

Thiel, M. "Reproductive Biology of a Wood-Boring Isopod, *Sphaeroma terebrans*, with Extended Parental Care." *Marine Biology* 135 (1999): 321–333.

Zimmer, M., S. Geisler, S. Walter, and H. Brendelberger. "Fluorescence in *Asellus aquaticus* (Isopoda: Asellota): A First Approach." *Evolutionary Ecology Research* 4 (2002): 181–187.

Organizations

British Myriapod and Isopod Group. E-mail: steve.gregory@northmoortrust.co.uk Web site: <http://www.salticus.demon.co.uk>

Inland Water Crustacean Specialist Group. Denton Belk, 840 E. Mulberry Ave., San Antonio, TX 78212-3194 USA. Phone: (210) 732-8809. Fax: (210) 732-3943. Web site: <http://www.iucn.org/themes/ssc/pubs/sgnewsl.htm>

International Isopod Research Group. Web site: <http://www .uni-kiel.de/zoologie/institut/limnologie/IIRG.htm>

Other

"World List of Marine, Freshwater and Terrestrial Isopod Crustaceans." Department of Systematic Biology, Invertebrate Zoology, Smithsonian National Museum of Natural History. [1 Aug. 2003]. <http://www.nmnh.si.edu/iz/isopod/>

Leslie Ann Mertz, PhD

Amphipoda
(*Amphipods*)

Phylum Arthropoda

Subphylum Crustacea

Class Malacostraca

Order Amphipoda

Number of families 155

Thumbnail description
Diverse group of crustacean arthropods ranging in size from 0.2 in (5 mm) to 9.8 in (25 cm) in length

Photo: Skeleton shrimps (*Caprella* sp.) have raptor like claws, presumably used to capture prey (copepods, crustacea larvae, worms, amphipods, etc.) floating by in the current. (Photo ©Tony Wu/www.silent-symphony.com. Reproduced by permission.)

Evolution and systematics

The order Amphipoda is made up of three suborders (some scientists recognize the Ingolfiellidea as a suborder rather than a family of the Gammaridea), 155 families, and more than 6,000 species. The three suborders are Gammaridea, with 126 families; Caprellidea, with 8 families; and Hyperiidea, with 21 families.

The fossil record of crustacean arthropods is patchy, and fossils of amphipods are almost non-existent. The few that have been found can be traced back to the Cambrian period. The Gammarids appear to be the most primitive of the amphipods, with Hyperiids and Caprellids showing more specialization in body form, behavior, and ecological relationships.

Physical characteristics

Amphipods tend to have laterally compressed bodies that curve to form a "C." Although there is wide variation in body form, the general body type is made up of a head, thorax, and abdomen. The head has compound eyes of varying sizes and well-developed pairs of first and second antennae. The seven multisegmented thoracic appendages are made up of two pairs of claw-like gnathopods used for grasping and five pairs used for crawling, jumping, and burrowing. Gills are found on the thorax. The abdomen has three pairs of appendages (pleopods) used for swimming and moving water through a burrow, and three appendages (uropods) are used for jumping, burrowing, or swimming. Most amphipods are small, 0.2–0.6 in (5–15 mm) long, but deep sea benthic forms can reach over 9.8 in (25 cm) in length.

Distribution

Amphipods are a diverse group of crustacean arthropods found in virtually all habitats of the world. Most are marine

but 1,200 species are known to inhabit fresh water, and almost 100 species are terrestrial.

Habitat

Most amphipods are benthic, living in burrows of mud or among detritus. Some live in fresh water among decaying leaves. Others live among sand grains on beaches. Oceanic forms are found in the water column, living the majority of their lives associated with gelatinous zooplankton (jellies, ctenophores, and thalicean tunicates).

Behavior

Gammarids live under decaying leaves or can make burrows in sand or mud. Hyperiids live at least part of their lives associated with gelatinous zooplankton. Caprellids attach themselves to algae, hydroids, and other small structures. Cyamids live as ectoparasites on marine mammals in species specific relationships.

Feeding ecology and diet

Amphipods can be herbivores, carnivores, or scavengers. In many instances, amphipods help breakdown decaying animals and plants. Hyperiid amphipods live most of their lives attached to gelatinous zooplankton, and *Phronima* eats the inside of thalicean tunicates, fashioning the remaining tunic into a barrel that it uses as a brood chamber. Cyamid amphipods eat the skin of the marine mammals they live on.

Reproductive biology

In many amphipods fertilization takes place when the male attaches to a female, transferring sperm to her genital duct.

Scuds mating. (Photo by Tom Branch/Photo Researchers, Inc. Reproduced by permission.)

Fertilized eggs are incubated in the female's ventral brood chamber formed by modified thoracic appendages. Development is direct so the newly hatched amphipods look much like their parents.

Conservation status

As a group, no amphipods are known to be in danger of extinction, and none are listed by the IUCN. Those that are ectoparasites in species-specific relationships with endangered marine mammals are at risk.

Significance to humans

In many habitats amphipods are important in breaking down decaying matter. They are an important part of the food chain for some commercially harvested species.

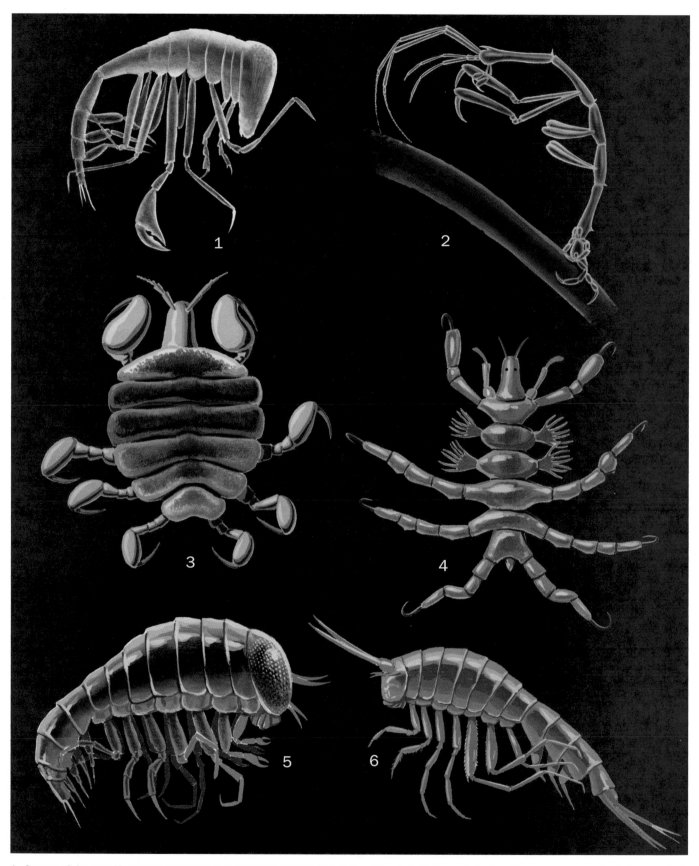

1. Cooper of the sea (*Phronima sedentaria*); 2. Skeleton shrimp (*Caprella californica*); 3. Gray whale lice (*Cyamus scammoni*); 4. Sperm whale lice (*Neocyamus physeteris*); 5. *Hyperia galba*; 6. *Scina borealis*. (Illustration by John Megahan)

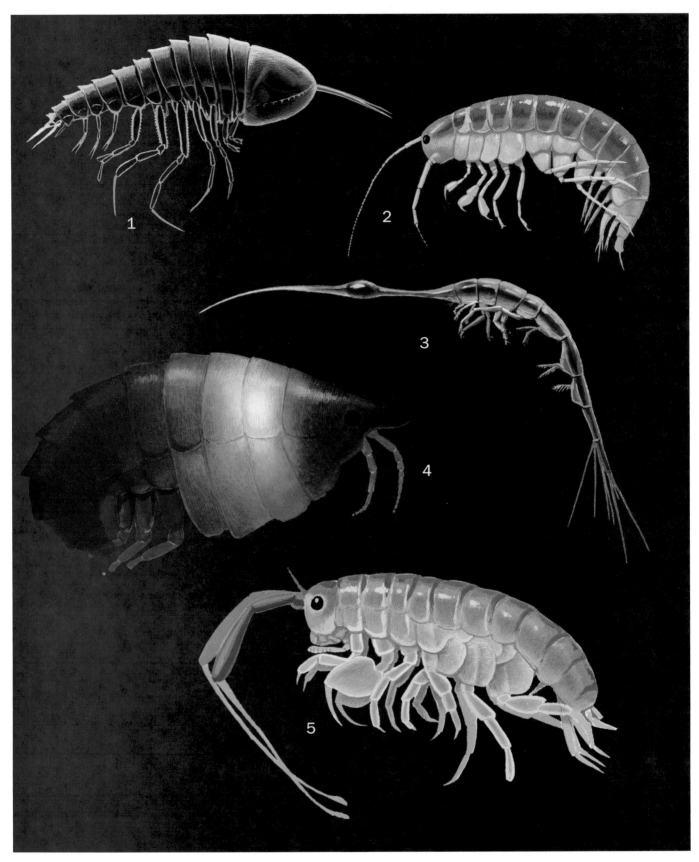

1. *Cystisoma fabricii*; 2. *Gammarus lacustris*; 3. *Rhabdosoma brevicaudatum*; 4. *Pleustes platypa*; 5. Beach hopper (*Orchestoidea californiana*).
(Illustration by John Megahan)

Species accounts

Skeleton shrimp
Caprella californica

FAMILY
Caprellidae

TAXONOMY
Caprella californica Stimpson, 1857.

OTHER COMMON NAMES
None known.

PHYSICAL CHARACTERISTICS
Caprellids are up to 1.38 in (35 mm) in length and have long slender bodies with small abdomens. The tips of the appendages have prehensile claws that show their adaptation for grasping and climbing.

DISTRIBUTION
Caprellids are exclusively marine and are found in shallow waters in all oceans of the world. *Caprella californica* can be found coastally from central to southern California.

HABITAT
Commonly found subtidally grasping onto hydroids, algae, and bryozoans.

BEHAVIOR
They are known to bow and scrape the substrate they are attached to as they gather food, prompting some to call them the praying mantis of the sea.

FEEDING ECOLOGY AND DIET
They are omnivores feeding on diatoms, detritus, protozoans, smaller amphipods, crustacean larvae, and other tiny attached or floating food items.

REPRODUCTIVE BIOLOGY
Fertilized eggs are kept within a brood pouch made by broad, leaf-like thoracic appendages. Development is direct in that the newly hatched juveniles look very much like their parents.

CONSERVATION STATUS
Not listed by the IUCN.

SIGNIFICANCE TO HUMANS
None known. ◆

Gray whale lice
Cyamus scammoni

FAMILY
Caprellidae

TAXONOMY
Cyamus scammoni Dall, 1872.

OTHER COMMON NAMES
English: Whale flea.

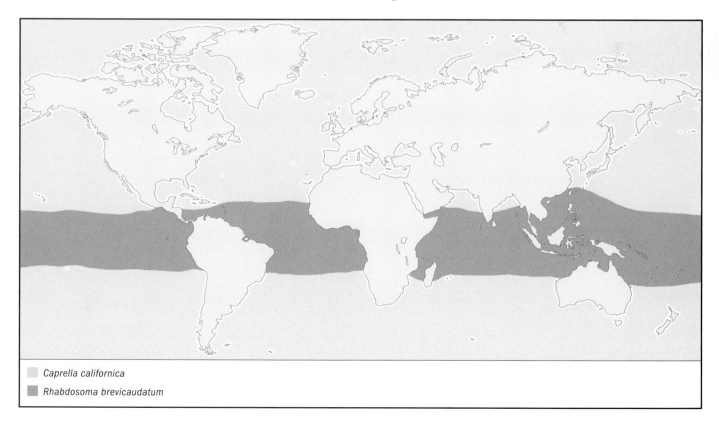

Caprella californica

Rhabdosoma brevicaudatum

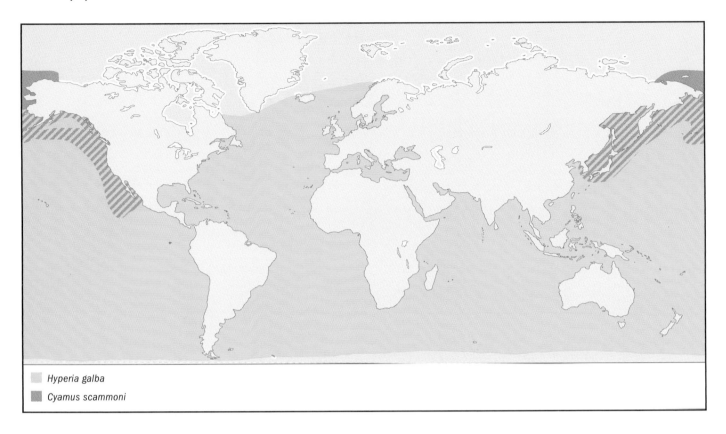

Hyperia galba

Cyamus scammoni

PHYSICAL CHARACTERISTICS
Body up to 1.2 in (30 mm) in length and, unlike other amphipods, is dorsoventrally flattened. Whale lice have hooks at the tips of their appendages that they use to sink into the skin of gray whales (*Eschrichtius robustus*). Their gills are exposed in the thoracic region and are highly coiled. Body is orange.

DISTRIBUTION
These lice are found exclusively on gray whales, and their range matches that of the whales, ranging all along the west coast of the eastern North Pacific Ocean.

HABITAT
Cyamus scammoni are species specific ectoparasites on gray whales.

BEHAVIOR
Commonly found in large groups around the barnacle *Cryptolepas rachianecti*, feeding on dead and dying whale skin. There is a report that the lice benefit the whale by eating the skin around the barnacles, eventually causing the barnacle to fall off the whale.

FEEDING ECOLOGY AND DIET
Gray whale lice are scavengers. Their diet consists mainly of dead and dying skin of the whales on which they live. They also eat detritus attached to the whale's skin.

REPRODUCTIVE BIOLOGY
Fertilized eggs are kept within a brood pouch made by broad, leaf-like thoracic appendages. Newly hatched juveniles look very much like their parents. Baby whales are infested by rubbing up against their mothers.

CONSERVATION STATUS
Not listed by the IUCN.

SIGNIFICANCE TO HUMANS
None known. ◆

Sperm whale lice
Neocyamus physeteris

FAMILY
Caprellidae

TAXONOMY
Neocyamus physeteris Pouchet, 1888.

OTHER COMMON NAMES
English: Whale flea.

PHYSICAL CHARACTERISTICS
Body up to 0.4 in (10 mm) in length and, unlike other caprellid amphipods, is dorsoventrally flattened. Whale lice have hooks at the tips of their appendages that they use to grasp onto the skin of the sperm whale *Physeter macrocephalus*. Their gills are exposed in the thoracic region and appear as clumps of finger-like projections. Body is yellow to orange in color.

DISTRIBUTION
Neocyamus physeteris are only found on sperm whales. Sperm whales are found in all major oceans.

HABITAT
Found only on sperm whales.

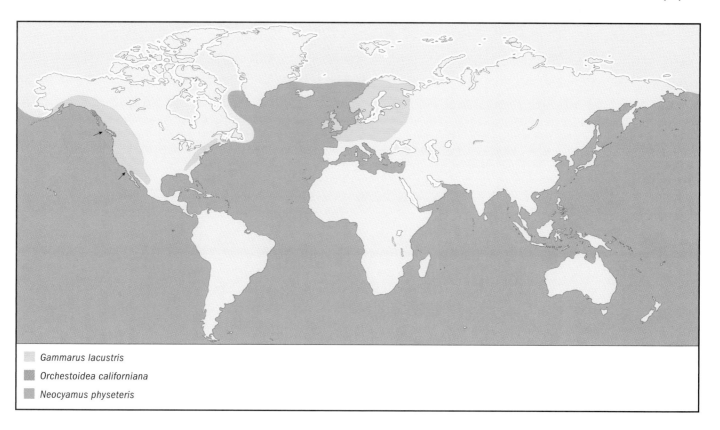

Gammarus lacustris

Orchestoidea californiana

Neocyamus physeteris

BEHAVIOR
Their hooked appendages enable them to hang onto the skin of the whale.

FEEDING ECOLOGY AND DIET
Scavengers that feed on the whale's skin, diatoms, and other items attached to the skin.

REPRODUCTIVE BIOLOGY
Fertilized eggs are kept within a brood pouch made by broad, leaf-like thoracic appendages. Newly hatched juveniles look very much like their parents. Whale lice infest other whales by jumping onto them when whales rub up against each other. Baby whales are infested by rubbing up against their mothers.

CONSERVATION STATUS
Not listed by the IUCN. Sperm whale lice are as endangered as the sperm whales they exclusively inhabit.

SIGNIFICANCE TO HUMANS
None known. ◆

No common name
Gammarus lacustris

FAMILY
Gammaridae

TAXONOMY
Gammarus lacustris Sars, 1864.

OTHER COMMON NAMES
None known.

PHYSICAL CHARACTERISTICS
Gammarus lacustris has the typical gammarid body plan. Males reach 0.87 in (22 mm), and females can be up to 0.71 in (18 mm) in length.

DISTRIBUTION
Originally described from northwestern Europe, this species can be found across North America, with populations found in mountain freshwater bodies of water.

HABITAT
Found among detritus in shallow, cold small bodies of fresh water, including lakes, ponds, streams, swamps, and springs.

BEHAVIOR
Can be very abundant and can be seen scurrying around, under, and among detritus on the shores of lakes and streams.

FEEDING ECOLOGY AND DIET
Scavenges on decaying matter and single-celled algae called diatoms.

REPRODUCTIVE BIOLOGY
Fertilized eggs are kept within a brood pouch made by broad, leaf-like thoracic appendages. Newly hatched juveniles look very much like their parents. Egg-carrying females occur from March to September, but this varies upon latitude and water temperature.

CONSERVATION STATUS
Not listed by the IUCN.

SIGNIFICANCE TO HUMANS
None known. ◆

Beach hopper
Orchestoidea californiana

FAMILY
Gammaridae

TAXONOMY
Orchestoidea californiana Brandt, 1851.

OTHER COMMON NAMES
English: Beach flea, sand hopper, sandflea, long-horned beach hopper.

PHYSICAL CHARACTERISTICS
Orchestoidea californiana reaches a length of 1.1 in (28 mm). This species has the typical amphipod body form. Eyes very small. Possesses a long pair of slender, second antennae that are orange or rosy red.

DISTRIBUTION
Found from Vancouver Island, British Columbia, to Laguna Beach, California.

HABITAT
Found on exposed beaches of fine sand backed by dunes.

BEHAVIOR
Hoppers can jump along on the sand using the posterior part of the abdomen and the terminal uropods as a spring. Mature individuals make a burrow in the sand that can be 12 in (30 cm) deep. When storm waves pound the beach, these hoppers are known to take refuge in areas higher up on the shore. They tend to hide in their burrows most of the day. At sunset they hop along the shoreline looking for piles of seaweed or other matter washed up by the waves.

FEEDING ECOLOGY AND DIET
Orchestoidea californiana can be observed feeding on seaweeds at night to avoid high daylight temperatures and predators such as shorebirds and racoons.

REPRODUCTIVE BIOLOGY
Mating occurs in the burrows from June until November. Males deposit sperm in a gelatinous mass on the underside of the female. After the male leaves the burrow, the female fertilizes the eggs, which are dark blue and kept within a brood pouch made by broad, leaf-like thoracic appendages. Newly hatched juveniles look very much like their parents.

CONSERVATION STATUS
Not listed by the IUCN.

SIGNIFICANCE TO HUMANS
None known. ◆

No common name
Pleustes platypa

FAMILY
Gammaridae

TAXONOMY
Pleustes platypa Barnard and Given, 1960.

OTHER COMMON NAMES
None known.

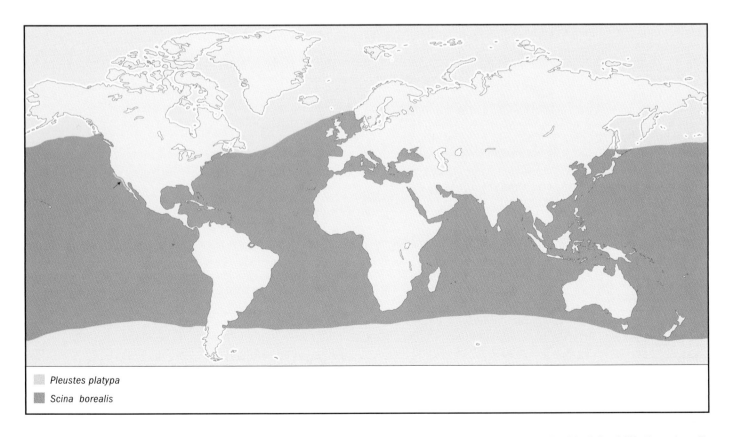

Pleustes platypa
Scina borealis

PHYSICAL CHARACTERISTICS
About 0.35 in (9 mm) in length with a long, broad rostrum. Normally the body has yellow or brown bands.

DISTRIBUTION
Found in kelp beds in southern California.

HABITAT
Found living on giant kelp *Macrocystis pyrifera*, among individuals of the small snail *Mitrella carinata*, which the amphipod closely resembles.

BEHAVIOR
They are known to mimic the snail *Mitrella carinata*, but the reasons are unclear. It has been suggested that fish may not eat them due to their resemblance to this snail. This is one of a very few examples of amphipod-molluscan mimicry.

FEEDING ECOLOGY AND DIET
Pleustes platypa is an omnivore that feeds on diatom films and encrusting organisms like bryozoans growing on kelp blades.

REPRODUCTIVE BIOLOGY
Nothing is known.

CONSERVATION STATUS
Not listed by the IUCN.

SIGNIFICANCE TO HUMANS
None known. ◆

No common name
Cystisoma fabricii

FAMILY
Hyperiidae

TAXONOMY
Cystisoma fabricii Stebbing, 1888.

OTHER COMMON NAMES
None known.

PHYSICAL CHARACTERISTICS
Females to 3.6 in (92 mm), males to 2 in (50 mm) in length. The large dorsally rounded head is made up of two large eyes. The body is completely transparent, soft and delicate. Musculature is very weak, and thus they are weak swimmers.

DISTRIBUTION
Found circumoceanic in midwater at 655–3,280 ft (200–1,000 m) depths.

HABITAT
Due to their weak swimming abilities, it is believed that these amphipods cling to gelatinous zooplankton most, if not all, of their lives.

BEHAVIOR
Very little is known about behavior, other than the fact that they are found associated with gelatinous zooplankton.

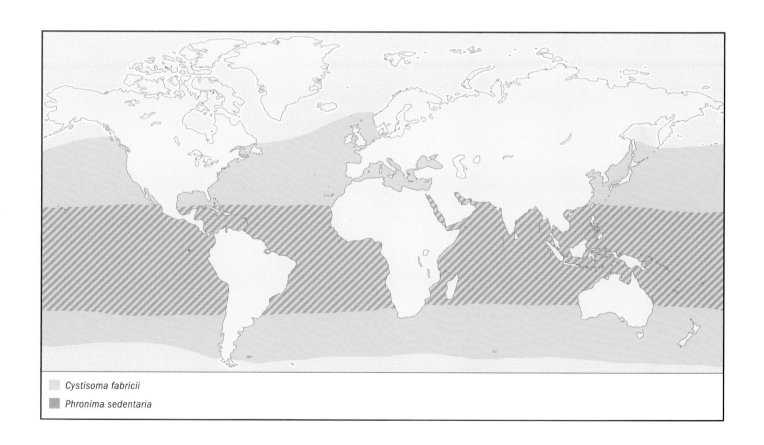

Cystisoma fabricii

Phronima sedentaria

FEEDING ECOLOGY AND DIET
Nothing is known.

REPRODUCTIVE BIOLOGY
Fertilized eggs are kept within a brood pouch made by broad, leaf-like thoracic appendages. Newly hatched juveniles look very much like their parents.

CONSERVATION STATUS
Not listed by the IUCN.

SIGNIFICANCE TO HUMANS
None known. ◆

No common name
Hyperia galba

FAMILY
Hyperiidea

TAXONOMY
Hyperia galba Montagu, 1813.

OTHER COMMON NAMES
None known.

PHYSICAL CHARACTERISTICS
Females are 0.08–0.18 in (2–4.5 mm) long; males are 0.06–0.24 in (1.5–6.0 mm) long. Due to their massive eyes compared to their small body, these amphipods are placed in the infraorder Physocephalata ("large head"). Males have elongated first and second antennae. Both sexes have large eyes that take up most of the head. Body is a pale brown, with maroon eyes.

DISTRIBUTION
Found circumglobal at depths of 1,640–3,610 ft (500–1,100 m).

HABITAT
Commonly found on midwater gelatinous zooplankton (hydromedusae, scyphomedusae, siphonophores, ctenophores, and salps).

BEHAVIOR
They are known to make vertical daily migrations from 3,280 ft (1,000 m) during the day to less than 1,640 ft (500 m) at night.

FEEDING ECOLOGY AND DIET
Feed on the gelatinous zooplankton they are frequently found living upon. They are also found in food pouches of jellyfish, where they feed on food collected by the jellies.

REPRODUCTIVE BIOLOGY
Females with eggs are most common in the spring, and young hatch in the summer. Newly hatched juveniles look very much like their parents. No one knows where the released young spend their time until they enter the adult population.

CONSERVATION STATUS
Not listed by the IUCN.

SIGNIFICANCE TO HUMANS
None known. ◆

Cooper of the sea
Phronima sedentaria

FAMILY
Hyperiidae

TAXONOMY
Phronima sedentaria Forskal, 1775.

OTHER COMMON NAMES
French: Tonnelier de la Mer.

PHYSICAL CHARACTERISTICS
Females to 1.65 in (42 mm), males to 0.6 in (15 mm) in length. Males have an elongated basal segment of the first antenna. The body is transparent except for four small retinal masses and occasional pale purple chromatophores on the thorax and abdomen. The eyes make up about one-quarter of the body, and therefore this species is placed in the infraorder Physocephalata ("large head"). Each of the eyes is split into a medial and lateral eye. The medial eyes are large with long cones angled to small retinal masses. The lateral eyes are small and located near the mouth.

DISTRIBUTION
Coopers of the sea are found circumtropically in midwater at 0–3,610 ft (0–1,100 m) depths.

HABITAT
Commonly found in midwater pelagic habitats but can be found migrating all the way to the surface in higher latitudes and midwater in the deep sea.

BEHAVIOR
The abdominal pleopods are strong, allowing the female to swiftly move the barrel (of her salp home) through the water. The female wedges itself into the barrel and, by beating the pleopods, can swiftly move the barrel through the water. The amphipod can change directions by quickly somersaulting in the barrel. The medial eyes are effective in searching out the open end of the barrel, and the lateral eyes are helpful in visualizing the brood.

FEEDING ECOLOGY AND DIET
Phronima sedentaria is carnivorous on gelatinous zooplankton, chaetognaths, and euphausiids.

REPRODUCTIVE BIOLOGY
Fertilized eggs are kept within a brood pouch made by broad, leaf-like thoracic appendages. Newly hatched juveniles look very much like their parents. Females catch gelatinous zooplankton like salps and pyrosomes and eat out the insides, climb inside, fashion the tunic into a barrel, and use it as a brood chamber. The female grabs newly hatched young and puts them into the inner wall of the barrel. While attached as a group, the female catches prey and shares it with her brood. The young also eat the barrel material. The barrel slowly is eaten and decomposes so that the young swim away to live solitary lives. No one knows how or when the males and females get together for fertilization of the eggs. Males are not known to make barrels.

CONSERVATION STATUS
Not listed by the IUCN.

SIGNIFICANCE TO HUMANS
None known. ◆

No common name
Rhabdosoma brevicaudatum

FAMILY
Hyperiidae

TAXONOMY
Rhabdosoma brevicaudatum Stebbing, 1888.

OTHER COMMON NAMES
None known.

PHYSICAL CHARACTERISTICS
Elongate, needle-like body to 1.2 in (30 mm) in length. Large bulbous, fused eyes behind an elongate rostrum and in front of a long thin neck.

DISTRIBUTION
Found equatorially in Atlantic and Pacific oceans in midwater from 165–1,640 ft (50–500 m) deep.

HABITAT
Found on gelatinous zooplankton.

BEHAVIOR
Little is known.

FEEDING ECOLOGY AND DIET
Very little is known about what this amphipod eats. It is believed to feed on the gelatinous zooplankton it inhabits.

REPRODUCTIVE BIOLOGY
Fertilized eggs are kept within a brood pouch made by broad, leaf-like thoracic appendages. Newly hatched juveniles look very much like their parents.

CONSERVATION STATUS
Not listed by the IUCN.

SIGNIFICANCE TO HUMANS
None known. ◆

No common name
Scina borealis

FAMILY
Hyperiidae

TAXONOMY
Scina borealis G. O. Sars, 1883.

OTHER COMMON NAMES
None known.

PHYSICAL CHARACTERISTICS
Body length of 0.28–0.32 in (7–8 mm). Body color is red. Due to the small eyes compared to the size of the large body, these amphipods are put into the infraorder Physosomata. First antennae large, extending to one-third of the body length. Sixth and seventh thoracic legs have greatly elongated segments.

DISTRIBUTION
Found circumoceanic in midwater at 165–9,845 ft (50–3,000 m) depths.

HABITAT
Found associated with gelatinous zooplankton, especially siphonophores.

BEHAVIOR
When disturbed this amphipod protrudes its large antennae and sixth and seventh thoracic legs out, looking very much like a child's jack. The enlarged first antennae can bioluminesce when disturbed. Commonly found associated with gelatinous zooplankton.

FEEDING ECOLOGY AND DIET
Believed to feed on the gelatinous zooplankton they are found inhabiting.

REPRODUCTIVE BIOLOGY
Fertilized eggs are kept within a brood pouch made by broad, leaf-like thoracic appendages. Newly hatched juveniles look very much like their parents.

CONSERVATION STATUS
Not listed by the IUCN.

SIGNIFICANCE TO HUMANS
None known. ◆

Resources

Books

Brusca, R. C., and G. L. Brusca. *Invertebrates.* Sunderland, MA: Sinauer Associates, Inc., 1990.

Dhermain, F., L. Soulier, and J.-M. Bompar. "Natural Mortality Factors Affecting Cetaceans in the Mediterranean Sea." In *Cetaceans of the Mediterranean and Black Seas: State of Knowledge and Conservation Strategies*, edited by G. Notarbartolo di Sciara. Monaco: Report to the ACCOBAMS Secretariat, 2002.

Martin, J. W., and G. E. Davis. "An Updated Classification of the Recent Crustacea." Science Series No. 39. Los Angeles: Natural History Museum of Los Angeles, 2001.

Morris, R. H., D. P. Abbott, and E. C. Haderlie. *Intertidal Invertebrates of California.* Stanford, CA: Stanford University Press, 1980.

Parks, P. *The World You Never See: Underwater Life.* Skokie, IL: Rand McNally, 1976.

Smith, R. I., and J. T. Carlton. *Light's Manual: Intertidal Invertebrates of the Central California Coast.* Berkeley: University of California Press, 1976.

Vinogradov, G. "Amphipoda." In *South Atlantic Zooplankton*, Vol. 2, edited by D. Boltovskoy. Leiden, The Netherlands: Backhuys, Leiden, 1999.

Resources

Yamaji, I. *Illustrations of the Marine Plankton of Japan.* Osaka, Japan: Hoikusha Publishing Co., 1976.

Periodicals

Bousfield, E. L. "An Updated Commentary on Phlyetic Classification of the Amphipod Crustacea and Its Applicability to the North American Fauna." *Amphipacifica* 3 (2001): 49–120.

Bousfield, E. L., and E. A. Hendrycks. "A Revision of the Family Pleustidae (Crustacea: Amphipoda: Leucothoidea). Systematics and Biogeography of Component Subfamilies. Part I." *Amphipacifica* 1 (1994): 17–58.

Brusca, G. J. "The Ecology of Pelagic Amphipoda, I. Species Accounts, Vertical Zonation and Migration of Amphipoda from the Waters off Southern California." *Pacific Science* 21 (1967): 382–393.

———. "The Ecology of Pelagic Amphipoda, II. Observations on the Reproductive Cycles of Several Pelagic Amphipods from the Waters off Southern California." *Pacific Science* 21 (1967): 449–456.

Harbison, G. R., D. C. Biggs, and L. P. Madin. "The Associations of *Amphipoda hyperiidea* with Gelatinous Zooplankton—II. Associations with Cnidaria, Ctenophora and Radiolaria." *Deep Sea Research* 24 (1977): 465–488.

Holsinger, J. R. "The Freshwater Amphipod Crustaceans (Gammaridae) of North America." *Biota of Freshwater Ecosystems, Identification Manual No. 5, Environmental Protection Agency* (1972): 17–24.

Laval, P. "Hyperiid Amphipods as Crustacean Parasitoids Associated with Gelatinous Zooplankton." *Oceanography Marine Biology Annual Review* 18 (1980): 11–56.

Samaras, W. F., and F. E. Durham. "Feeding Relationship of Two Species of Epizoic Amphipods and the Gray Whale, *Eschrichtius robustus.*" *Bulletin of the Southern California Academy of Sciences* 84 (1984): 113–126.

Schell, D. M., V. J. Rowntree, and C. J. Pfeiffer. "Stable-Isotope and Electron-Microscopic Evidence That Cyamids (Crustacea: Amphipoda) Feed on Whale Skin." *Canadian Journal of Zoology* 78 (2000): 721–727.

Other

The Amphipod Homepage. [17 July 2003]. <www.web.odu .edu/sci/biology/amphome/>.

The Biology of Amphipods. [17 July 2003]. <www.museum.vic .gov.au/crust/amphbiol.html>.

Michael S. Schaadt, MS

Thecostraca
(Cirripedes and relatives)

Phylum Arthropoda
Subphylum Crustacea
Class Maxillopoda
Subclass Thecostraca
Number of families 48

Thumbnail description
Highly modified, mainly free-living sessile and parasitic crustaceans, usually enclosed within a calcareous carapace or forming a chitinous saclike body

Photo: Species of the genus *Synagoga*, found on the species *Antipathella wollastoni* at a depth of 131 ft (40 m) in waters off of the Azores. (Photo by Peter Wirtz. Reproduced by permission.)

Evolution and systematics

The calcareous carapaced forms of thecostracans have provided the richest fossils. The oldest known thecostracan fossil is dated from the Middle Cambrian; traces of the parasitic forms (without carapaces) have been dated from Cretaceous. The great diversity of the species began to occur in the Upper Cretaceous and was essentially completed by the end of the Miocene.

Until 1834 barnacles were classified as mollusks because of their calcareous shells. Only in 1829, when thecostracan larval stages were first discovered, were their affinities with other crustaceans fully recognized.

Thecostracans are included among maxillopodan groups. The group Maxillopoda is not generally accepted as a natural group, and there are doubts over its monophyly and component groups. The thecostracan lineage is founded on morphological, ontogenetic, molecular, and fossil data. Some ascothoracicans show some similarities with other maxillopodans, and are now considered the primitive group of thecostracans. Cirripedes are a notable exception among maxillopodans because of their adaptations to a sessile way of life. The facetotectans are one of the biggest remaining mysteries of crustacean diversity; the latter group's affinity with the tantulocarids is still under investigation.

Two-thirds of thecostracan species are free-living or commensals, and one-third have different degrees of parasitism. There are about 1,400 species divided into three infraclasses: Cirripedia (barnacles), with three suborders (Thoracica, Acrothoracica, and Rhizocephala) and 41 families with free-living, commensal, and parasitic species; Ascothoracica, with six families, includes ecto- and endoparasites of cnidarians and echinoderms; Facetotecta is composed of one genus, *Hansenocaris*, which is microscopic, Y-shaped, and free-living.

Physical characteristics

As in some extant ascothoracicans, the body is primitively composed of a head containing some cephalic structures, a thorax with six segments and appendages, and a segmented abdomen. In all groups the head is reduced, and the abdomen has no limbs; the second antennae is absent in some groups. Telson and compound eyes are absent in adults. The mouth appendages have some reductions and modifications. Most adult cirripedes are modified for a life attached to an object, or as a parasite.

Cirripedes are unique among crustaceans because they are sessile. A number of barnacle peculiarities may be correlated with sessility. During the larval stage almost all cirripedes find

Common goose barnacles (*Lepas anatifera*), with legs extended for feeding. (Photo by A. Flowers & L. Newman. Reproduced by permission.)

an object to attach to by using their first antennae (antennules), after which the preoral region becomes fixed to the object. Later, during metamorphosis, the body, mouth, eyes, and the adductor muscles separate from the antennules. All openings of the body are located on the side of the body that is opposite from the object they are attached to, that is, the free side of their body. Barnacles are known as animals that sit on their heads and kick food into their mouths.

A bivalved and chitinous carapace (mantle) encloses the body in ascothoracicans. In cirripedes, the mantle forms a sac. In the thoracicans, the mantle secretes calcareous plates, which form an outer wall or shell that is permanent. Thoracian species of barnacles such as goose barnacles are pedunculate (stalked), while sessile species are non-pedunculate. The pedunculate species have a fleshy peduncle that hangs along the head and entire body. Facetotectans, ascothoracicans, acrothoracicans, and rhizocephalans have no calcareous plates and are all chitinous.

The mantle cavity is a spacious chamber where the mouth, anus, and sexual organs are located and from which the larvae are freed into seawater.

Most thecostracans are recognized as crustaceans due to their paired, chitinous, and jointed thoracic appendages, the cirri, which can be uni- or biramous as in other crustaceans, and are sometimes heavily fringed with bristles. Cirri are mainly used during feeding and as a respiratory organ. Only rhizocephalans in their larval stages have appendages; adults have a ramified (branching off) structure that penetrates the tissues of the host. Rhizocephalans are the most highly modified of all thecostracans.

Facetotectans are less than 0.039 in (1 mm) long. Most parasites and simple forms are only a few millimeters in length. Stalked barnacles range from a few millimeters to more than 27.5 in (70 cm) in length. The majority of sessile species are a few inches (centimeters) in length and can reach 9 in (23 cm) in height and up to 3.1 in (8 cm) in diameter. Free-living barnacles are white, pink, red, purple, orange, violet, or brown.

Distribution

Thecostracans are exclusively marine and/or estuarine. Most species are intertidal or subtidal. Some thoracican species live in the high tide and others are found near abyssal hydrothermal vents. The group occurs worldwide, but barnacles are less conspicuous in tropical rocky shores. A number of species have commensal relationships with some pelagic animals and their distribution is limited only by the range of their host.

Habitat

Primitive ascothoracicans, facetotectans, and males of some species are free-swimming. Some ascothoracicans attach to their hosts using a prehensile first antenna, which has glands that secrete cement for the attachment. The cement is produced throughout their lifespan, and repairing partial detachment is possible. The attachment of most thecostracans is done by cyprid larvae after settlement. These animals can live on almost any hard object in the seawater. Most free-living sessile barnacles attach to rock. Common pedunculate and sessile barnacles attach to inanimate objects such as wood, floating logs, bottoms of ships, wooden pilings, and empty bottles. Commensals and parasites attach to living organisms, including pelagic animals such as corals, sea anemones, jellyfishes, mussels, crabs, shrimps, lobsters, copepods, other barnacles, echinoderms, tunicates, sea turtles, and the skin of whales and sharks. Thecostracans attach to their host or object by the cypnid larval stage.

Ascothoracicans bore on calcareous substrates such as mollusk shells, dead corals, or carapaces of sea urchins. They bore using chitinous teeth, as well as by excreting chemicals that lead to dissolution. As larvae, parasites, attach to some part of the host's body, making a perforation on the tegument.

Behavior

Most thecostracans can move about freely only as larvae. In some species, adults retain the ability to swim throughout their lives, attaching only temporarily for feeding.

Barnacles are very resistant to abiotic factors. Many species of sessile barnacles, common in rocky shores, live in the intertidal zone on the coast. During low tide, these animals are exposed to the air. They hermetically close the valves present in the carapace to avoid desiccation, high temperatures, and freshwater rain. They can form bands for miles along the coast, with high population densities of 1,000–2,500 individuals in 15.5 in^2 (100 cm^2). Some species have a high growth rate in areas with high wave rates, because turbulance and strong currents promote the movement of plankton toward the coast line; plankton is a major food source for barnacles.

Cyprid larvae settle in dense numbers in areas where other living or dead barnacles occur. A protein present in the exoeskeleton of older attached individuals has been shown to attract larvae. This behavior ensures that individuals will be close enough for cross-fertilization and settlement to take place, as these animals are sessile.

Feeding ecology and diet

Most thoracicans are filter feeders. They feed actively by extending their long, feathery, birramous cirri out of the carapace, in a fan-like manner, to filter feed on suspended material from the surrounding water. The bristles of the cirri overlap to form an effective filtering net. The water is filtered and the food is passed to the mouthparts.

Northern rock barnacles (*Balanus balanoides*). (Photo by Animals Animals ©E. R. Degginger. Reproduced by permission.)

Food particles range from 0.0000787 to 0.039 in (2 µm to 1 mm) in size, and includes detritus, bacteria, algae, and zooplankton. Food is detritus for those species that are found within estuaries and bays.

Cirripedes can be predators. Stalked barnacles are capable of preying upon larger planktonic animals by coiling a single cirrus around the prey.

Ectoparasitic thecostracans send roots into the tissues of their pelagic hosts in order to feed. Some parasites have modified mouthparts that form suctorial cones for piercing tissues and sucking out the body fluids of their hosts. In rhizocephalans, the ramified body lacks an alimentary system, and nutrients are absorbed directly from the host's tissues.

Starfishes, snails, fishes, worms, and birds feed on barnacles.

Reproductive biology

Most species are hermaphrodites, but some are accompanied by additional males, and are called complementals. Ascothoracicans, acrothoracicans, and rhizocephalans have separate sexes. Males of these groups are called dwarves. Complemental and dwarf males are greatly reduced in size and do not feed frequently. They attach to females.

Free-living barnacles generally cross-fertilize, because a suitable substrate almost always contains a large number of adjacent individuals. The penis of the free-living barnacle can be extended out of the body and into the mantle cavity of another individual in order to deposit sperm.

In all cirripedes, the eggs develop within the ovisac, present in the mantle cavity. In most species, free-swimming nauplii larvae hatch from the eggs. Other naupliar instars occur before the transformation to cyprid larvae. The entire body is enclosed within a bivalved carapace; one pair of sessile compound eyes and six pairs of thoracic appendages are also present.

Conservation status

No species are listed by the IUCN.

Significance to humans

Barnacles often attach to the bottom of ships, where they grow to such a degree that they can reduce its speed by as much as 35%. Significant effort and money have been expended toward the development of special paints that will prevent barnacles from attaching.

The barnacle species *Balanus nubilis* is eaten by Native Americans in the United States. The rock barnacle, *Balanus psittacus*, can reach 9 in (23 cm) in height and 3.1 in (8 cm) in diameter, and is a popular local seafood in South America.

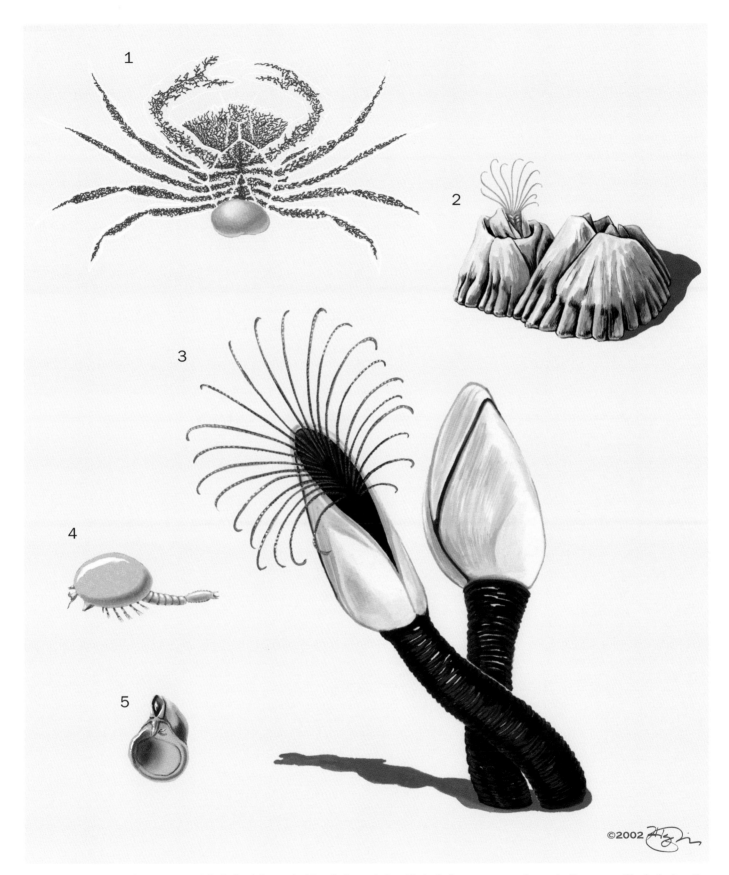

1. Root-like barnacle (*Sacculina carcini*); 2. Rock barnacle (*Semibalanus balanoides*); 3. Common goose barnacle (*Lepas anatifera*); 4. *Ascothorax ophioctenis*; 5. *Trypetesa lampas*. (Illustration by Jonathan Higgins)

Species accounts

No common name
Trypetesa lampas

ORDER
Apygophora

FAMILY
Trypetesidae

TAXONOMY
Alcippe lampas Hancock, 1849.

OTHER COMMON NAMES
None known.

PHYSICAL CHARACTERISTICS
Species does not have a calcareous carapace. Females reach 0.78 in (2 cm) in length; body is colorless or yellowish and covered by a large mantle. Cavity is exposed to water by the narrow fissure-like orifice. The preoral region has a wide and flat disc plate that larvae use to form attachments to a host's shell. Thorax is bent by segmentation. Three pairs of appendages within the mouth; one pair of reduced cirri near the mouth. Only three uniramous pairs of cirri located at the end of the thorax. Females have an incomplete gut so ceca are able to reach into various parts of the body; anus is absent. Dwarf males are 0.047 in (1.2 mm) long, bottle-shaped, legless, and attach to females. Antennae are the only appendages present and there are no internal organs other than reproductive ones.

DISTRIBUTION
Shores of the Pacific and Atlantic Oceans in the Northern Hemisphere.

HABITAT
Adult females spend their lives boring into the shells of living or dead gastropods and hermit crabs; they then live within the aperture or the columela. Adults bore with the non-mineralized spines present on the mantle, and by secreting chemicals.

BEHAVIOR
When molting, the female does not shed the non-mineralized layers present on the attachment disc.

FEEDING ECOLOGY AND DIET
Females expand their mantle and bend their body away from the mantle slit so that water is forced in and out of the mantle cavity. Three posterior cirri then collect particles of food from the water. Cirri present in the mouth draw up against the posterior cirri to sweep particles into the mouth.

REPRODUCTIVE BIOLOGY
Sexes are separate; males attach to females as cyprid larvae and undergo metamorphosis into dwarf males. Testes open into extensible penis to fertilize eggs. Nauplii hatch from eggs and after four naupliar stages, cyprid larvae with six pairs of appendages and compound eyes search for a suitable substrate in which to burrow.

CONSERVATION STATUS
Not listed by the IUCN.

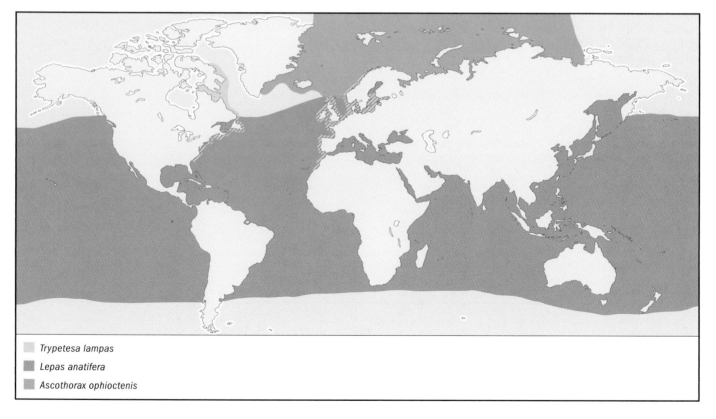

Trypetesa lampas

Lepas anatifera

Ascothorax ophioctenis

SIGNIFICANCE TO HUMANS
None known. ◆

No common name
Ascothorax ophioctenis

ORDER
Dendrogastrida

FAMILY
Ascothoracidae

TAXONOMY
Ascothorax ophioctenis, Djakonov, 1914, Novaya Zemlya, Russia, at a depth of 328 ft (110 m).

OTHER COMMON NAMES
None known.

PHYSICAL CHARACTERISTICS
Females 1.1–1.5 in (3–4 cm) in length; males one-third to one-sixth female's size. Body consists of head, six thoracic segments, and five abdominal segments; last segment bears the caudal rami. The first antenna is long, thick, and subchelate; cement glands are absent. The body is often orange in color. Males have bivalved, oval, laterally compressed, and uncalcified carapaces surrounding the mantle cavity; the carapace is attached to the head, enclosing the head and thorax. The adductor muscle can close the carapace.

Females are heart-shaped with a bivalved carapace that is much larger than the trunk. First and sixth thoracic cirri are uniramous, the others are biramous. Thorax and abdomen are both short, and the thorax is sharply bent at the fourth segment.

DISTRIBUTION
Found from Greenland to Norway, and east to Novaya Zemlya and Franz Josef Land; including the Barents, Greenland, Norwegian, Kara, and Laptev Seas.

HABITAT
Parasitic; lives in the bursa present between two arms of its host, the serpent stars *Ophiocten sericeums* and *O. gracilis*. Once it has attached and is fixed, the host's bursa becomes inflated so that the parasite's presence can be recognized from the outside.

BEHAVIOR
Males attach to female's body using the subchelate first antennae. In females, one or both antennal chelae are extended and attach to bursa wall of the host. Slit present between carapace valves toward the bursal opening permits limbs to circulate a current of water that allows breathing to occur. Prevents its host from developing gonads.

FEEDING ECOLOGY AND DIET
Mouthparts of males and females are stylet-like structures enclosed in a conical labrum modified for piercing and sucking. Labrum is pressed to bursal wall where penetration of host occurs. Gutter-like first maxillae forms a tube through which cell debris and fluids are drawn from host.

REPRODUCTIVE BIOLOGY
First abdominal segment bears male intromittent organ. Fertilization is external, taking place in the brood chamber of the female. Nauplii hatch from eggs and develop within the brood chamber where they metamorphosize into cyprid-like larvae that are capable of feeding and have a bivalved carapace, but no second antennae. During the larval stage—called the ascothoracid stage—young leave the brooding chamber present in females and begin to search for their own host.

CONSERVATION STATUS
Not listed by the IUCN.

SIGNIFICANCE TO HUMANS
None known. ◆

Root-like barnacle
Sacculina carcini

ORDER
Kentrogonida

FAMILY
Sacculinidae

TAXONOMY
Sacculina carcini Thompson, 1836.

OTHER COMMON NAMES
Dutch: Krabbezakje.

PHYSICAL CHARACTERISTICS
Body of adult females is made of tissue and is covered with a thin cuticle devoid of calcareous plates, traces of segmentation, distinct regions, appendages, or an alimentary tract. The body consists of a central nucleus, which attaches itself onto the mid-gut of the host from where it sends root-like appendages (the interna) into every part of the host's body. This nucleus connects to the exterior of the host with the external knob sac (the externa). Males are dwarves, microscopic cyprids that attach to females.

DISTRIBUTION
Occurs in European seas; predator of the green crab, *Carcinus maenas* and other crabs.

HABITAT
Uses its root-like extensions to attach to the host's visceral organs, nervous system, gonads, and appendages, but do not penetrate inside the host.

BEHAVIOR
A host's gonads are castrated by this parasite; in adult males, the abdomen becomes feminized. Young crabs do not form gonads when infested.

FEEDING ECOLOGY AND DIET
Nourishment is accomplished by diffusing nutritive substances from host's blood. It may attack organs with enzymes it secretes. Males receive nourishment from the female's body.

REPRODUCTIVE BIOLOGY
External sac has branched ovaries and a pair of male sperm receptacles. A male cyprid swims into the female's body. After attachment, male extrudes its whole cell content into the female. Male takes up residence and its function is to produce sperm. Fertilization and brooding occur in the externa; nauplii hatch out of the eggs, pass through four naupliar stages. One parasite can release many broods during one season, but the peak of reproduction is during summer; there is a rest period in the winter. The period of incubation is 12–35 days; after 10 days the

parasite settles. Female cyprids attach to the base of their host's bristles. Females abandon their limbs, thorax, mantle, and carapace, and develop a curved stylet through which their cells pass into the host's body. After an internal phase, which can last up to 3 years, females produce an externa, attracting male cyprids.

CONSERVATION STATUS
Not listed by the IUCN.

SIGNIFICANCE TO HUMANS
Suggested as a biological control agent against the invasive green crab, *Carcinus maenas*, which is negatively affecting coastal ecosystems worldwide. ◆

Common goose barnacle
Lepas anatifera

ORDER
Pedunculata

FAMILY
Lepadidae

TAXONOMY
Lepas anatifera Linnaeus, 1758.

OTHER COMMON NAMES
English: Goose barnacle, gooseneck barnacle; French: Anatife; Portuguese: Anatifas, conchas marrecas, lepas.

PHYSICAL CHARACTERISTICS
Average body length is 9.8 in (25 cm); largest specimen found was 29.5 in (75 cm). Color of body is dark brown and calcareous plates are whitish. Body divided into two regions: the peduncle (stalk), and the capitulum. The peduncle is fleshy, large, and long, and it attaches to the substrate using the first antennae. The body is compressed laterally, covered by two folds of mantle, where five thin calcareous plates are attached. The carina is a dorsal unpaired plate, which forms a central keel. Paired scuta are large, and are located at the anterior region of the body. Paired terga are short and are located at the posterior-most region of the body. Six pairs of thoracic, biramous cirri bordered with chaetae are visible through an aperture present in the mantle cavity. The adductor muscle closes the mantle cavity. In the mantle cavity, there is a short head, a thorax with six thoracic, biramous limbs, a mouth, and a long, setose penis.

DISTRIBUTION
Cosmopolitan in tropical and temperate seas.

HABITAT
Small colonies attach to floating objects such as logs and ships.

BEHAVIOR
Cirri, stalk, and mantle are covered by sensory bristles (chaetae). Stimulation of these chaetae causes the withdrawal of the body and the mantle aperture to close.

FEEDING ECOLOGY AND DIET
Plankton is collected with movements by the thoracic cirri. From between plates, cirri are extended and spread out like a fan-shaped net; cirri are then withdrawn and the aperture is closed. Filtering of food from water is done rhythmically through combined movements of the cirri and the closure of the aperture; food is then transferred to mouthparts.

REPRODUCTIVE BIOLOGY
Hermaphroditic, but without complemental males. The long, extensible penis is inserted into the mantle cavity of adjacent individuals. Eggs brood within the mantle cavity and nauplii hatch from eggs. After several molts, larvae reach the cyprid stage. The cyprid attaches to the substrate using its antennae and begins metamorphosis; during this stage it can grow almost 0.039 in (1 mm) per day, attaining a length of 3.9 in (10 cm) within 113 days.

CONSERVATION STATUS
Not listed by the IUCN.

SIGNIFICANCE TO HUMANS
Scientific name means "goose carrier" because they resemble goose eggs. They attach to ships. ◆

Rock barnacle
Semibalanus balanoides

ORDER
Sessilia

FAMILY
Archaeobalanidae

TAXONOMY
Balanus balanoides Linnaeus, 1767.

OTHER COMMON NAMES
English: Acorn shell, common barnacle, acorn barnacles, stalk-less barnacle; French: Balane, gland-de-mer; Portuguese: Bálanos, bolotas-do-mar, caracas, glandes-do-mar.

PHYSICAL CHARACTERISTICS
Reaches 0.19–0.59 in (5–15 mm) in diameter. Carapace is conical, with six gray or white fused calcareous plates forming a lateral wall. The dorsal anterior plate is the largest. The carine is posterior and two pairs of plates are located laterally. On top, two pairs of short calcareous and articulated plates, the terga and scuta (one pair on each side) form the operculum in a diamond-shaped arrangement; it encloses the mantle cavity and can be opened and closed by muscles. When rotated laterally, they form an aperture through which leads to the mantle cavity and the six pairs of birramous cirri bordered by chaetae, the first three being shorter than other three. Tissue inside opercular aperture is usually white or pinkish. Attaches to rock using the flattened and broad membranous attachment disc.

DISTRIBUTION
Found in intertidal rocky shores in North America, Europe, and the Arctic.

HABITAT
Common and abundant in rocky shores. It settles on a wide variety of solid objects including pilings, rocks, and shell crabs. Prefers low-tide areas occasionally splashed by water.

BEHAVIOR
The calcareous carapace protects the animal, and the operculum hermetically seals the carapace; when the tide goes out, they close the carapace. Survives freezing weather during the winter in tide zones of the Arctic Ocean, as well as the daily dry 6–9 hours between high tides during summer. Eggs and nauplii in

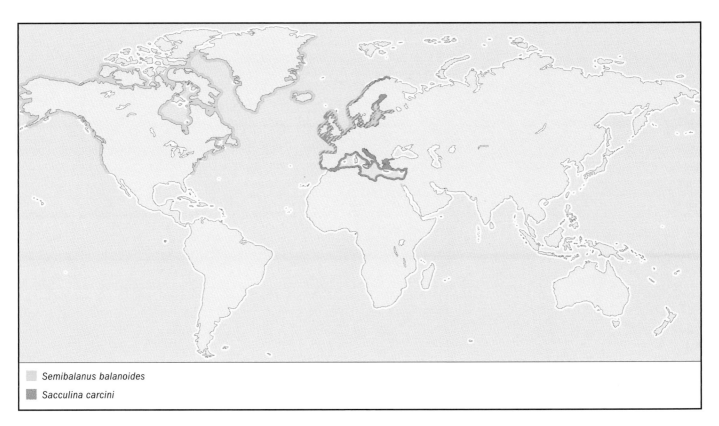

Semibalanus balanoides
Sacculina carcini

the mantle cavity also survive. In very dense colonies, young do not find enough space to grow, so the wall plates become elongate, resembling stalks, elevating the feeding position as in pedunculate barnacles; on rocky coasts, they form a wide band that can stretch for miles. Turbulence and waves are of importance since the current brings plankton for feeding. Fewer individuals settle in areas without strong waves.

FEEDING ECOLOGY AND DIET
Feeds only during high tide when the carapace is covered by water. Food is small and collected by filtering; animal opens operculum and extends the last three long pairs of cirri, which are spread and curved outward resembling a fan-shaped net. It rhythmically extends and retracts the cirri and collects material, transferring it to mouthparts. Cirri can do 33–37 beats per minute to collect food. Preyed on extensively by the dog whelk, *Nucella lapillus*, and the shanny, *Lipophrys pholis*.

REPRODUCTIVE BIOLOGY
Hermaphrodite; has an extensible penis that fertilizes neighboring individuals. Eggs are fertilized in the mantle cavity; gravid individuals each contain 6,000–13,000 nauplii in mantle cavity. After some stages as a nauplius, a cyprid larva begins to seek a place of attachment. The antennae and cement glands make the attachment onto objects. During metamorphosis animals can grow to a diameter of 0.23–0.27 in (6–7 mm) in 58 days. Lives for 3–8 years. In some localities, all young molt synchronously. Individuals in same place hatch nauplii at same time; a good number of larvae settle in same place, assuring cross-fertilization.

CONSERVATION STATUS
Not listed by the IUCN.

SIGNIFICANCE TO HUMANS
Used in research. ◆

Resources

Books
Anderson, D. T. *Barnacles. Structure, Function, Development and Evolution.* Melbourne, Australia: Chapman and Hall, 1994.

Brusca, R. C., and G. J. Brusca. *Invertebrates.* 2nd ed. Sunderland, MA: Sinauer Associates Inc., 2003.

Kaestner, A. *Invertebrate Zoology.* New York: Interscience Publishers, 1970.

Schram, F. *Crustacea.* New York: Oxford University Press, 1986.

Other
Cato, Paisley, and Patricia Beller. "Marine Invertebrates—Barnacles." [August 21, 2003]. <http://www.sdnhm.org/research/marine-inverts/localshells.html>.

Davey, Keith. "Life on Australian Seashores." [August 21, 2003]. <http://www.mesa.edu.au/friends/seashores/barnacles.html>.

"Introduction to the Cirripedia—Barnacles and Their Relatives." [August 21, 2003]. <http://www.ucmp.berkeley.edu/arthropoda/crustacea/maxillopoda/cirripedia.html>.

Tatiana Menchini Steiner, PhD

Tantulocarida
(Tantulocaridans)

Phylum Arthropoda
Subphylum Crustacea
Class Maxillopoda
Subclass Tantulocarida
Number of families 4

Thumbnail description
Tiny parasitic crustaceans that spend most of their lives attached to the external body surface of their hosts, a wide range of other marine crustaceans

Photo: The tantulus larva and globular adult female of *Microdajus langi*, a tantulocaridan parasite, attached to its host, a tanaidacean crustacean. (Photo by G. A. Boxshall. Reproduced by permission.)

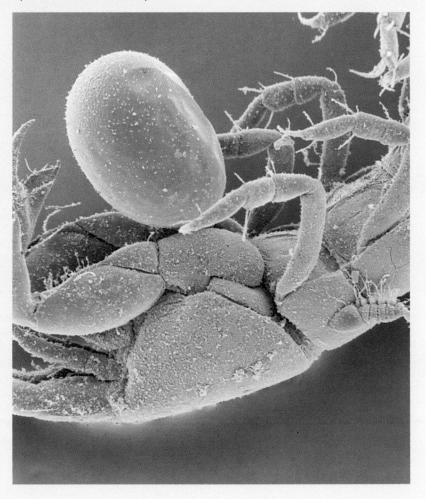

Evolution and systematics

The subclass Tantulocarida is currently classified in the class Maxillopoda, and it is regarded as most closely related to the barnacles (the Thecostraca), with which it shares a similar body plan and a similar position of genital openings in both sexes. The Tantulocarida comprises four families, 22 genera, and about 30 species. No orders have been established for this group.

Physical characteristics

The adult asexual female consists of a minute head, a neck of varying length, and a sac-like trunk full of eggs or developing tantulus larvae. This is the largest stage in the tantulocaridan life cycle and may attain lengths of up to 0.08 in (2 mm). It is attached to the exoskeleton of its host by a tiny oral sucker, about 0.000472–0.000591 in (12–15µm) in diameter. This stage has no limbs at all and no genital apertures, and it appears to release mature larvae by rupturing the trunk wall. The sexual female is less than 0.02 in (0.5 mm) in length and

consists of a large cephalothorax and a five-segmented post-cephalic trunk. The cephalothorax carries a pair of sensory antennules but no mouthparts. A small number of large eggs lie within the cephalothorax, and it also carries a conspicuous median genital opening, interpreted as a copulatory pore. The first two of the trunk segments each carry a pair of biramous thoracic legs, which appear to be used for grasping, and the fifth segment bears the elongate caudal rami. The adult male resembles the sexual female in size and basic body plan, with a large cephalothorax and a six-segmented trunk, but it has more pairs of limbs: vestigial sensory antennules, six pairs of biramous swimming legs, a well-developed median penis, and the caudal rami. Adults of both sexes develop within posteriorly located, sac-like expansions of the trunk of the attached tantulus larva.

Distribution

Knowledge of tantulocaridan distributions is incomplete, in part because they are often overlooked due to their minute body size. Species have been reported from the North and South Pacific and North and South Atlantic oceans, as well as from both Arctic and Antarctic waters.

Habitat

Tantulocaridans spend most of their lives attached to their hosts, which include isopod, tanaid, amphipod, cumacean, ostracod, and copepod crustaceans. The dispersal and infective stage in the life cycle, the tantulus larva, has also been found living free in marine sediments. The sexual adults have never been collected away from the host, but probably inhabit the hyperbenthic zone, just above the sea bed.

Behavior

Little is known of tantulocaridan behavior. After release from the mother, infective larvae spend some time in the sediment before encountering a suitable benthic or hyperbenthic host. Host location and attachment mechanisms are poorly understood in these forms, which lack eyes and antennules, the main sensory interfaces of other crustaceans.

Feeding ecology and diet

Tantulocaridans are ectoparasitic and do not appear to feed away from their hosts. They attach to the external skeleton of their hosts by means of an adhesive oral disc. The host surface is punctured by a stylet, which is protruded out through a minute pore in the center of this disc. Nutrients are obtained via the puncture into the host. There is evidence of an absorptive rootlet system extending from the oral disc of the tantulocaridan and penetrating through the tissues of the host. Tantulocaridans exhibit varying degrees of host specificity: for example, members of the family Deoterthridae occur on cumacean, isopod, tanaid, amphipod, ostracod, and copepod hosts, members of the Microdajidae on tanaid hosts only, members of the Doryphallophoridae on isopods only, and members of the Basipodellidae on copepods only.

Reproductive biology

Tantulocaridans have a bizarre double life cycle, involving a sexual phase and an asexual phase. The asexual phase is en-

countered much more frequently than the sexual phase. The sac-like asexual female releases fully formed tantulus larvae, which are capable of infecting a new host and developing directly into another asexual female, without mating and without even molting. The tantulus larva is minute, ranging from 0.00335 in (85µm) to about 0.00709 in (180 µm) in length. It comprises a head, which has an oral disc but lacks any limbs, and a trunk of six leg-bearing segments and a maximum of two limb-less abdominal segments. The swimming legs are biramous and have reduced endites. After successfully infecting a host, the tantulus larva develops into the asexual female, and the postcephalic trunk of the larva is shed, so the female remains attached to the host by the adhesive oral disc of the preceding larval stage. The trunk of the female expands to accommodate the growing larvae until they are released.

In the sexual phase, the cycle again begins with the infective tantulus larva attaching to its host by the oral disc. A sac-like expansion begins to grow near the back of the trunk, within which either an adult male or an adult female then develops. The precise location of this expansion varies according to family. Both are supplied with nutrients from the host, transported via an umbilical cord originating in the still-attached larval head. Fully formed adults develop within the sac, which remains attached to the host by the oral disc of the larva. On reaching maturity these sexual adults are released by rupturing of the sac wall. These sexual stages have never been observed alive, but it is assumed that the male, which has well-developed swimming legs and a large cluster of antennulary chemosensors, actively searches out and locates the receptive female. The male carries a large penis and presumably inseminates the female by the single ventral copulatory pore. The fertilized eggs are presumed to develop within the expandable cephalothorax of the female until ready to hatch, as a fully formed tantulus.

Conservation status

Information on the abundance and distribution of tantulocaridans is fragmentary. No species are listed by the IUCN.

Significance to humans

Tantulocaridans have no known significance to humans.

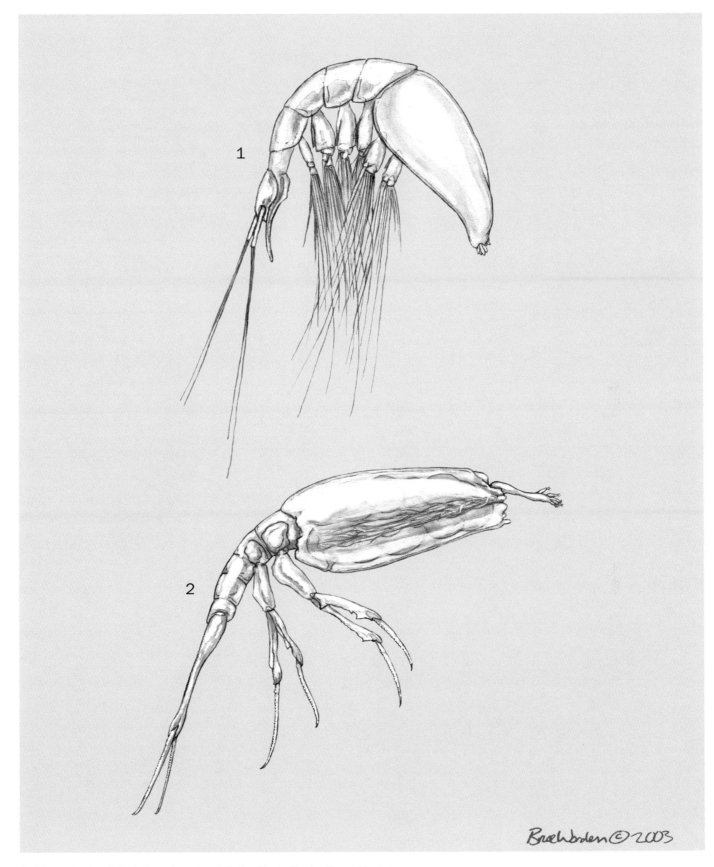

1. *Microdajus langi*; 2. *Itoitantulus misophricola*. (Illustration by Bruce Worden)

Species accounts

No common name
Itoitantulus misophricola

FAMILY
Deoterthridae

TAXONOMY
Itoitantulus misophricola Huys, Ohtsuka, and Boxshall, 1992, Okinawa, Japan.

OTHER COMMON NAMES
None known.

PHYSICAL CHARACTERISTICS
Tantulus larva have well-developed rami, with endites on second to fifth pairs of swimming legs; and sixth leg with recurved apical spine. Trunk sac containing developing sexual male formed posterior to sixth thoracic tergite. Adult male with recurved penis, unsegmented abdomen, and distinct caudal rami.

DISTRIBUTION
Okinawa in southern Japan to the Philippines.

HABITAT
Parasitic on benthic or hyperbenthic copepods living at depths of 550–6,725 ft (167–2,050 m).

BEHAVIOR
Nothing is known.

FEEDING ECOLOGY AND DIET
Parasitizing copepods of the orders Harpacticoida (family Styracothoracidae) and Misophrioida (family Misophriidae).

REPRODUCTIVE BIOLOGY
This was the first tantulocaridan for which the complete double life cycle was elucidated.

CONSERVATION STATUS
Not listed by the IUCN.

SIGNIFICANCE TO HUMANS
None known. ◆

No common name
Microdajus langi

FAMILY
Microdajidae

TAXONOMY
Microdajus langi Greve, 1965, Norway.

OTHER COMMON NAMES
None known.

PHYSICAL CHARACTERISTICS
Tantulus larva have reduced rami and lacking endites on swimming legs. Trunk sac containing developing sexual male formed posterior to sixth thoracic tergite. Adult male with slender stylet-like penis, with unsegmented abdomen and with distinct caudal rami.

DISTRIBUTION
Southern Norway to the west coast of Scotland.

HABITAT
Parasitic on tanaid crustaceans in marine sediments, at depths of 70–430 ft (22–130 m).

BEHAVIOR
Nothing is known.

FEEDING ECOLOGY AND DIET
Specific to tanaid hosts of the family Leptognathiidae.

REPRODUCTIVE BIOLOGY
Sexual female stage unknown.

CONSERVATION STATUS
Not listed by the IUCN.

SIGNIFICANCE TO HUMANS
None known. ◆

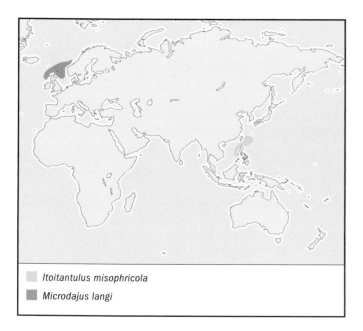

☐ *Itoitantulus misophricola*
■ *Microdajus langi*

Resources

Periodicals

Boxshall, G. A., and R. J. Lincoln. "The Life Cycle of the Tantulocarida (Crustacea)." *Philosophical Transactions of the Royal Society of London* B315 (1987): 267–303.

Huys, R., G. A. Boxshall, and R. J. Lincoln. "The Tantulocaridan Life Cycle: The Circle Closed?" *Journal of Crustacean Biology* 13 (1993): 432–442.

Geoffrey Allan Boxshall, PhD

Branchiura

(Fish lice)

Phylum Anthropoda
Subphylum Crustacea
Class Maxillopoda
Subclass Branchiura
Number of families 1

Thumbnail description
Branchiurans are ectoparasites of fishes, mainly living in freshwater habitats; they have flattened bodies comprised of five limb-bearing segments; the head has well-developed carapace lobes

Photo: *Argulus* attached to its goldfish host. The branching gut lobes contain partly digested blood. Photo by G. A. Boxshall. Reproduced by permission.)

Evolution and systematics

The subclass Branchiura comprises just four genera placed in a single family, the Argulidae, and about 175 species.

Physical characteristics

Branchiuran fish lice have flattened bodies, which have a low profile when attached to their hosts. The body comprises a head of five limb-bearing segments and a short trunk divided into a thoracic region, carrying four pairs of strong swimming legs and a short, unsegmented abdomen. The head has well-developed carapace lobes, which are posterior extensions of the dorsal head shield that mostly cover the legs on both sides of the body and may extend further to cover the abdomen. These carapace lobes contain the highly branched gut caecae and have two specialized areas ventrally, which are traditionally referred to as "respiratory areas," but appear to be involved in regulating the internal body fluids. Anteriorly, on the ventral surface of the head lie the short antennules and antennae. Both are provided with claws and are important organs of attachment to the host. The distal segments of the antennules are sensory and carry arrays of short setae.

Branchiurans have a tubular sucking mouth equipped with rasping mandibles located at the tip of the mouth tube. In *Argulus* there is a retractable poison stylet located just in front of the mouth. This stylet is absent in *Chonopeltis* and *Dipteropeltis*. The maxillules are developed into powerful muscular suckers in the adults of all genera, except *Dolops*, which retains long clawed maxillules into the adult phase. The maxillae are uniramous (one branched) limbs with spinous (spine-like) processes on the basal segments and small claws at the tip. The four pairs of thoracic swimming legs are biramous and directed laterally. The first and second legs commonly carry an additional process, the flagellum, originating

near the base of the exopod. The third and fourth legs are usually modified in the male and are used for transferring sperm to the female during mating. The abdomen contains the paired testes in the male and, in the female, the paired seminal receptacles, where sperm are stored until needed to fertilize eggs. The abdomen terminates in paired abdominal lobes separated by the median anal cleft, in which lies the anus and the minute paired caudal rami.

Distribution

Argulus species occur in freshwater habitats on all continental land masses. Species of *Dolops* exhibit a southern distribution, occurring in freshwater in southern Africa, South America, and Australasia (Tasmania). *Chonopeltis* species occur only in African freshwaters, while the sole species of *Dipteropeltis* is restricted to South America.

Habitat

Branchiurans are ectoparasites of fishes, but are occasionally reported from the tadpoles of amphibians. They live mainly in freshwater habitats, both running and static water, and may occur at high density in artificial water bodies such as reservoirs, ornamental fishponds, and fish farms. A few species of *Argulus* infest estuarine and coastal marine fishes, but they do not occur in oceanic waters.

Behavior

Only the behavior of *Argulus* is well known; little is known of the other genera. After taking a meal, a mature female *Argulus* will leave its host and begin to lay eggs in rows on any hard, submerged surface. The eggs are cemented to the substrate and abandoned. These eggs hatch into free-swimming larvae equipped with setose (bristly) swimming antennae and mandibles and rudiments of the maxillules, maxillae, and first two pairs of swimming legs. These larvae function as a dispersal phase and molt into the second stage, in which strong claws have replaced the setae on the antenna and the setose palp of the mandible is lost. Branchiurans are parasitic from the second stage onwards, but appear to leave the host and then find a new host at intervals throughout development. Changes during the larval phase are gradual, mainly involving the development of the thoracic legs and reproductive organs, except for the maxillule, which undergoes a metamorphosis around stage five, changing from a long limb bearing a powerful distal claw to a powerful circular sucker. This is one of the most remarkable transformations known for any arthropod limb.

Feeding ecology and diet

Branchiurans attach to the skin of their fish host and feed on its blood and external tissues. They have rasping mandibles, which scrape tissues into the opening at the tip of the tubular sucking mouth. In *Argulus*, the poison stylet is used to inject a secretion into the host. The secretion may contain digestive enzymes to begin to break up host tissues before ingestion. Paired labial stylets, lying within the opening of the mouth tube, are also secretory and may produce secretions with a similar pre-digestive function. Host blood is also taken and is digested within paired, lobate gut caecae that lie within the carapace lobes.

Reproductive biology

The sexes are separate and, in most branchiurans, males transfer sperm directly to the females using a variety of structures on the third and fourth thoracic legs. In *Dolops*, however, sperm are transferred in chitinous packages called spermatophores.

Conservation status

No species are listed by the IUCN.

Significance to humans

Branchiurans are important pests in fish culture facilities, mainly in freshwater facilities, but occasionally in marine fish farms.

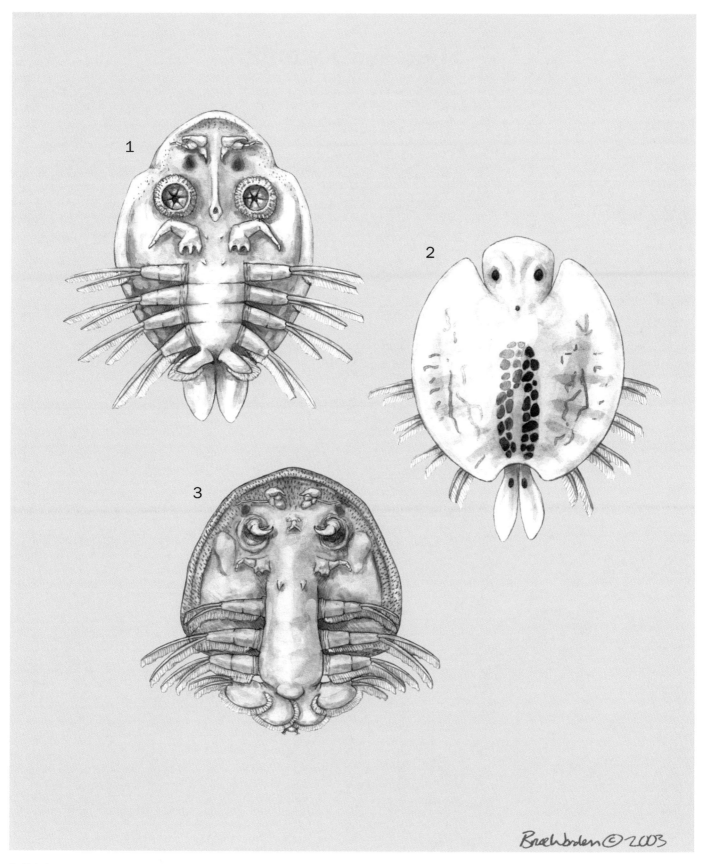

1. Fish louse (*Argulus foliaceus*); 2. *Argulus japonicus*; 3. *Dolops ranarum*. (Illustration by Bruce Worden)

Species accounts

Fish louse
Argulus foliaceus

ORDER
Arguloida

FAMILY
Argulidae

TAXONOMY
Argulus foliaceus Linnaeus, 1758, Europe.

OTHER COMMON NAMES
German: Karpfenläuse.

PHYSICAL CHARACTERISTICS
Abdominal lobes broadly rounded at tip; anal cleft less than half the length of abdomen. First to third legs, with darkly pigmented patches near base. Male with triangular process on posterior surface of leg two, directed towards base. Body length to 0.39 in (10 mm) for female and 0.35 in (9 mm) for male.

DISTRIBUTION
Europe, through to central Asia and Siberia.

HABITAT
Ectoparasitic; attaching to wide variety of freshwater fishes.

BEHAVIOR
On hatching, the larvae swim actively in the water column for about 2–3 days, after which their infectivity decreases.

FEEDING ECOLOGY AND DIET
Feeding externally on host, will take host epidermis and blood.

REPRODUCTIVE BIOLOGY
Eggs are laid in strings, of between two and six rows, containing up to 400 eggs; hatch after 25 days (at 59°F [15°C]), and development time is very dependent on temperature. Eggs within a string tend to hatch within 4–6 days. First larval stage lasts about six days, and molts occur at intervals of about 4–6 days until maturity.

CONSERVATION STATUS
Not listed by the IUCN.

SIGNIFICANCE TO HUMANS
May occur as epizootic infestation in fish hatcheries and other facilities. Can cause severe mortality in cultured fish stocks and can transmit viral diseases such as spring viraemia between fishes. ◆

Argulus foliaceus
Argulus japonicus
Dolops ranarum

No common name
Argulus japonicus

ORDER
Arguloida

FAMILY
Argulidae

TAXONOMY
Argulus japonicus Thiele, 1900, China.

OTHER COMMON NAMES
None known.

PHYSICAL CHARACTERISTICS
Abdominal lobes acutely rounded at tip; anal cleft more than half the length of abdomen. First to third legs, without any darkly pigmented patches near base. Male leg two with rounded processes at either end of posterior margin. Body length to 0.35 in (9 mm) for female and 0.31 in (8 mm) for male.

DISTRIBUTION
Originally described from China, this species has spread into Europe, North America, Australasia, and Africa by the movement of fish stocks, particularly of koi carp, in the ornamental fish trade.

HABITAT
Ectoparasitic; attaching to freshwater fishes, especially on koi carp, goldfish, and other ornamental fishes.

BEHAVIOR
Newly hatched larvae swim strongly towards light. May live 3–5 days off the host.

FEEDING ECOLOGY AND DIET
Feeding externally on host, will take host epidermis and blood.

REPRODUCTIVE BIOLOGY
Eggs laid in multiple rows, in batches of up to 200, hatching after 15–30 days, depending on temperature. Larval development is similar to *Argulus foliaceus*, but fewer larval stages are reported.

CONSERVATION STATUS
Not listed by the IUCN.

SIGNIFICANCE TO HUMANS
A pest species, particularly of ornamental fishes, that has become established on several continents after accidental introduction. ◆

No common name
Dolops ranarum

ORDER
Arguloida

FAMILY
Argulidae

TAXONOMY
Dolops ranarum Stuhlman, 1891, Africa.

OTHER COMMON NAMES
None known.

PHYSICAL CHARACTERISTICS
Maxillules forming large paired claws. Lacking preoral poison spine and lacking hooks on the antennules. Body length 0.23–0.27 in (6–7 mm).

DISTRIBUTION
Sub-Saharan Africa.

HABITAT
Permanent freshwater bodies and rivers.

BEHAVIOR
Larvae resemble miniature adults, swim actively, and appear to be infective immediately on hatching.

FEEDING ECOLOGY AND DIET
Ectoparasitic; attaching to wide range of freshwater fishes and, rarely, to amphibian tadpoles.

REPRODUCTIVE BIOLOGY
Males transfer sperm in spherical spermatophores, which are placed on the female during mating. Eggs are laid in clusters and those on the periphery tend to hatch first. Development takes about 30 days at 73.4°F (23°C).

CONSERVATION STATUS
Not listed by the IUCN.

SIGNIFICANCE TO HUMANS
None known. ◆

Resources

Books

Overstreet, R. M., I. Dyková, and W. E. Hawkins. "Branchiura." In *Microscopic Anatomy of Invertebrates*. Vol. IX, *Crustacea*, edited by F. W. Harrison. New York: J. Wiley and Sons, 1992.

Periodicals

Gresty, K. A., G. A. Boxshall, and K. Nagasawa. "The Fine Structure and Function of the Cephalic Appendages of the Branchiuran Parasite, *Argulus japonicus* Thiele." *Philosophical Transactions of the Royal Society of London* B339 (1993): 119–135.

Geoffrey Allan Boxshall, PhD

Mystacocarida
(*Mystacocarids*)

Phylum Arthropoda
Subphylum Crustacea
Class Maxillopoda
Subclass Mystacocarida
Number of families 1

Thumbnail description
Very small, vermiform crustaceans with appendages behind those of the head strongly reduced or absent

Illustration: *Derocheilocaris typicus.* (Illustration by Jonathan Higgins)

Evolution and systematics

Mystacocarids have recently been grouped with, among others, Copepoda, Ostracoda, and Cirripedia, and in the class Maxillopoda. These groups all have in common shortened bodies with at most 11 trunk segments behind the head, of which six are thoracic segments. There are 13 known species of mystacocarids, which are grouped into a single family containing two genera. Some scientists now classify this group at the order level.

Physical characteristics

Mystacocarids have shortened, vermiform bodies, with about one-third of the total body length being taken up by the head. The trunk consists of a very short maxillipedal somite, followed by nine more-or-less similar somites, the first four of which bear reduced appendages. The last five trunk somites (abdomen) do not have appendages. The trunk ends in a large telson (=anal) somite, which has a pair of simple, terminal, furca-like appendages. All trunk somites and the posterior section of the head have dorsolateral toothed furrows of unknown function. The head appendages are the most conspicuous aspect of mystacocarids. The antennules are very long, about half the length of the body. The antennae, mandibles, and both pairs of maxillae are composed of large circular segments whose shape is maintained by fluid pressure. The antennae and mandibles are biramous, whereas both pairs of maxillae are uniramous. The maxillae are armed with long setae on their inner margin.

Distribution

Members of the genus *Derocheilocaris* are known from the coastlines of eastern North America, including the Gulf of Mexico, the Atlantic coast of southern Europe to the tip of southern Africa, as well as the Mediterranean. Species in the genus *Ctenocheilocaris* are known from the southern coasts of eastern and western South America, as well as Western Australia.

Habitat

All mystacocarids live among the sand grains of outer coastal beaches. Their distribution within one beach may be quite patchy, so they are often not found in areas where they were previously known to occur.

Behavior

Mystacocarids move among sand grains. They use the antenna and mandibles as well as the surface of the trunk to push against the sand grain surfaces. In order to move efficiently, they require sand grains both above and below the body.

Feeding ecology and diet

Mystacocarids most likely graze on microalgae and bacteria living on the surfaces of sand grains. The movements of the mouth appendages may also be responsible for capturing particles in the interstitial spaces.

Reproductive biology

Sexes are separate. Copulation has not been observed, but it is known that fertilized eggs are shed freely into the interstitial habitat. Development is direct, proceeding though a series of molt stages in which body somites and appendages are added in a gradual manner. Beyond the first three head appendages, body somites are always added, then limb primordia, and finally the definitive appendage.

Conservation status

No mystacocarid species are known to be threatened, even though some species are known from single localities. None are listed by the IUCN.

Significance to humans

These small animals are most likely only of intellectual interest, not being part of any food web leading directly to fish or other consumable marine organisms.

Species accounts

No common name
Derocheilocaris typicus

ORDER
Mystacocarida

FAMILY
Derocheilocaridae

TAXONOMY
Derocheilocaris typicus Pennak and Zinn, 1943.

OTHER COMMON NAMES
None known.

PHYSICAL CHARACTERISTICS
Very conservative body plan, with differences among them being in the details of the appendages and the sizes of trunk structures such as the toothed furrows. Has a large and robust maxillule with slight divisions of the precoxa, coax, and basis. The endites on the mouth appendages bear robust setulose setae. Caudal furca short, with terminal seta almost as long as basal article. (Illustration shown in chapter introduction.)

DISTRIBUTION
Found along the Atlantic coast of the United States from Cape Cod to southern Florida.

HABITAT
Lives deep in the beach, often several feet (meters) inland from the low-tide line where the seawater penetrates at high tide, but often individuals are above the water table at low tide.

BEHAVIOR
Crawls with its antennae among the sand grains.

FEEDING ECOLOGY AND DIET
Feeds on small particles in the interstitial spaces on or microalgae and bacteria scraped from the surfaces of sand grains.

REPRODUCTIVE BIOLOGY
Eggs are laid freely in the beach and development is direct, proceeding from a metanauplius with four post-cephalic

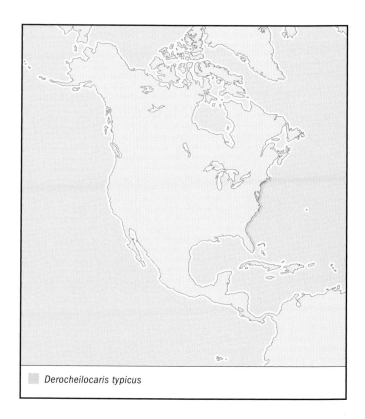

Derocheilocaris typicus

somites. Develops through six metanaupliar, and one juvenile stage before reaching adult size. One additional molt as an adult occurs.

CONSERVATION STATUS
Not listed by the IUCN.

SIGNIFICANCE TO HUMANS
First mystacocarid species to be discovered; a new crustacean order was created to house it. ◆

Resources

Periodicals
Lombardi, J., and E. E. Ruppert. "Functional Morphology of Locomotion in *Derocheilocaris typica* (Crustacea: Mystacocarida)." *Zoomorphology* 100 (1982): 1–10.

Pennak, R. W., and D. J. Zinn. "Mystacocarida, a New Order of Crustacea from Intertidal Beaches in Massachusetts and Connecticut." *Smithsonian Miscellaneous Collections* 103 (1943): 1–11.

Les Watling, PhD

Copepoda
(Copepods)

Phylum Arthropoda
Subphylum Crustacea
Class Maxillopoda
Subclass Copepoda
Number of families 220

Thumbnail description
Copepods are characterized by a body plan that consists of five head segments, seven thoracic segments, and a limbless abdomen of four segments; body lengths range from 0.019–7.8 in (0.5 mm to 20 cm); occur virtually everywhere there is water

Photo: Light micrograph showing egg sacs on both sides of the abdomen of a female of the species *Cyclops viridis*, a tiny, free-swimming, freshwater crustacean. (Photo by Laguna Design/Science Photo Library/Photo Researchers, Inc. Reproduced by permission.)

Evolution and systematics

The Copepoda is classified within the class Maxillopoda, on the basis of the body plan, which consists of five head segments, seven thoracic segments (of which the last bears the genital openings in both sexes), and a limb-less abdomen of four segments. The Copepoda is a large group, currently comprising nine orders, about 220 families, and about 13,500 species. The orders include:

- Platycopioida: a small order comprising just one family with three genera and about 12 species found in the near-bottom plankton community in coastal waters and in the plankton of flooded marine caves.

- Calanoida: the dominant order of planktonic copepods, comprising 42 families found from the surface to the greatest depths of the ocean, as well as in the plankton of freshwater lakes and ponds.

- Misophrioida: a small order comprising three families and a total of 32 species found in the near-bottom plankton community in marine waters and in the plankton of flooded marine caves.

- Mormonilloida: this order consists of two species in one family. Both are widely distributed in the deep-water plankton of the world's oceans.

- Harpacticoida: a large and diverse order of mainly bottom-living forms. A few of the 52 families are members of the marine plankton community, and some are even parasitic on animal hosts, but the great majority are benthic (bottom living), including forms that are interstitial (living in the spaces between sediment particles) and forms that live on the surface of the sediment or on macroalgae.

- Gelyelloida: this order consists of just two species in one family. Both are found only in groundwater in the karstic regions of southern Europe.

- Cyclopoida: a large and diverse group, comprising about 83 families exhibiting a wide range of lifestyles from planktonic to benthic, as well as symbiotic. The benthic forms include members of the family Cyclopidae that have invaded groundwater habitats. The symbiotic forms include many that are parasitic on marine and freshwater fishes, as well as on a whole range of marine invertebrate hosts. In addition, some cyclopoid families are extremely abundant in marine zooplankton communities.

- Siphonostomatoida: all 37 families of this order have symbiotic lifestyles. They use as hosts virtually every phylum of multicellular animals from sponges up to vertebrates, including mammals (whales). Most are marine, but a few species from three families live in freshwater.

- Monstrilloida: this order consists of a single family containing about 80 species, all of which occur in marine zooplankton as adults, but have larvae living as endoparasites of worms and mollusks.

Physical characteristics

Copepods are typically small, with a body length in the range of 0.019–0.78 in (0.5–2 mm), although some free-living forms attain lengths of up to 0.7 in (18 mm), and some highly modified parasites can reach 7.8 in (20 cm) in length. The copepod body plan consists of two regions, the anterior prosome and the posterior urosome. The prosome comprises

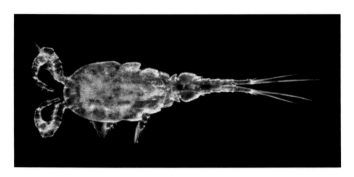

Light micrograph of a freshwater copepod (*Cyclops* sp.). This tiny plank-tonic crustacean swims by generating hopping movements with its ap-pendages. (Photo by Laguna Design/Science Photo Library/Photo Researchers, Inc. Reproduced by permission.)

the cephalosome, plus either four or five leg-bearing seg-ments, according to group. The cephalosome is composed of the segments that carry the five pairs of head limbs typical of crustaceans, including antennules, antennae, mandibles, max-illules and maxillae, as well as the first thoracic segment car-rying the maxillipeds. The six segments are covered by the single, dorsal head shield, and the median nauplius eye lies frontally, beneath the shield. Behind the cephalosome are the free prosomal segments with swimming leg pairs, either four or five of them. In the gymnoplean body plan, as found in Platycopioida and Calanoida, there are five leg pairs, whereas in the podoplean body plan, as found in all the other seven orders, there are only four pairs. The boundary between the prosome and the urosome is less distinct in the Harpacticoida, many of which have slender cylindrical bodies. The urosome is primitively four- or five-segmented depending on the group, and the last segment bears the anus and caudal rami. The most characteristic feature of copepods is the form of the swimming legs: members of each leg pair are fused to a me-dian intercoxal sclerite, which ensures that left and right legs always beat together. The legs are two-branched (biramous) and each branch (endopod and exopod) consists of a maxi-mum of three segments. Antennules can express up to 28 ar-ticulating segments in a single axis, but segment numbers are always less than this because of the failure of expression of one or more joints. The female typically carries the eggs in paired egg sacs, although many marine calanoids broadcast their eggs instead.

Distribution

Copepods occur virtually everywhere there is water—in all oceans and on all continental land masses, including Antarctica.

Habitat

In the oceans, free-living copepods dominate the zoo-plankton community in all temperature regimes from polar to tropical, and on land they occur in freshwater and inland saline waters on all continents. They occur in the plankton that inhabits the water column from the surface to the great-est depths. Copepods also occur on and in sediments, and in the extensive freshwater, groundwater realm.

Behavior

The basic lifecycle comprises six naupliar and five cope-podid stages preceding the adult. Many parasites, however, have large yolk-filled eggs and an abbreviated developmental pattern. The commonest abbreviation is the reduction or even the complete loss of the naupliar phase. Sealice, for example, have only two non-feeding naupliar stages, instead of the usual six. The nauplius phase is typically a dispersal phase.

Feeding ecology and diet

Copepods exploit an enormous variety of food sources and have diverse feeding behaviors. In the plankton, many feed on small particles such as unicellular phytoplankton, protists, and other microorganisms. They capture these particles by gener-ating water currents, using the slow swimming movements of antennae and maxillipeds. At the small scale of physics where copepods operate, water behaves in a highly viscous manner, so the currents created (known as laminar flow fields) act like a conveyor belt drawing food particles in towards the other mouthparts. Lateral movements of mandibular palps and max-illules then draw water as well as particles into range of the maxillae, which capture the particles in a raptorial manner. Other oceanic copepods are scavengers and specialist detritus feeders, which have acute chemosensory systems associated with their feeding systems. Benthic copepods feed on organic matter of all kinds, both living and dead. Some families are specialist associates of macroalgae. About half of all copepod species are symbionts living in association with multicellular animal hosts, including sponges, polychaetes, echinoderms, mollusks, crustaceans, and chordates, particularly tunicates and fishes. Their lifestyles are often poorly known, but many are parasitic, living inside or on the outer surface of their hosts.

Reproductive biology

Copepods have separate sexes, and mating behavior typi-cally involves chemical signalling between the sexes. Females release pheromones, which provide information concerning her species identity as well as her state of sexual receptivity. Males home in on these pheromonal trails in a variety of ways. Copulation results in the transfer of a single or a pair of chitinous packets of sperm (spermatophores) onto the fe-male genital region, and these usually discharge their con-tents into storage organs (seminal receptacles) inside the female.

Conservation status

A few species are listed by the IUCN, but most copepods are too small or too poorly documented for their conserva-tion status to have been assessed. A few species known from restricted and possibly threatened sites such as caves should be regarded as vulnerable or endangered, but these categories have rarely been applied to copepods.

Significance to humans

Copepods are hyper-abundant. They dominate the largest habitat on Earth, the open pelagic biome, and it is estimated that there are more individual copepods on the planet (1.37×10^{21}) than there are insects. They play a vital role in the economy of the oceans, forming the middle link in food chains leading from phytoplankton up to commercially important fish species. Consideration of global carbon fluxes suggests that copepods play an equally important role in the global carbon cycle, since a major proportion of fixed carbon dioxide passes through the oceanic food web. Some freshwater copepods act as vectors of human parasites such as guineaworm, while others are important predators of mosquito larvae and are actively used in biological control of mosquitoes in malarial areas. Sealice and other fish parasitic copepods are major pests in aquaculture of both fishes and mollusks.

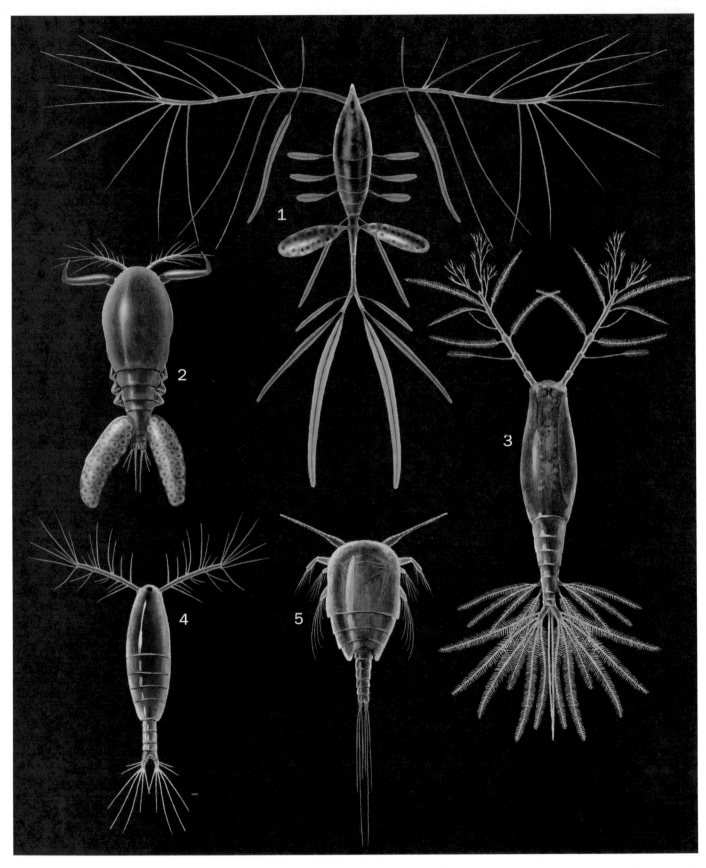

1. *Oithona plumifera*; 2. *Ergasilus sieboldi*; 3. *Monstrilla grandis*; 4. *Onychodiaptomus sanguineus*; 5. *Benthomisophria palliata*. (Illustration by John Megahan)

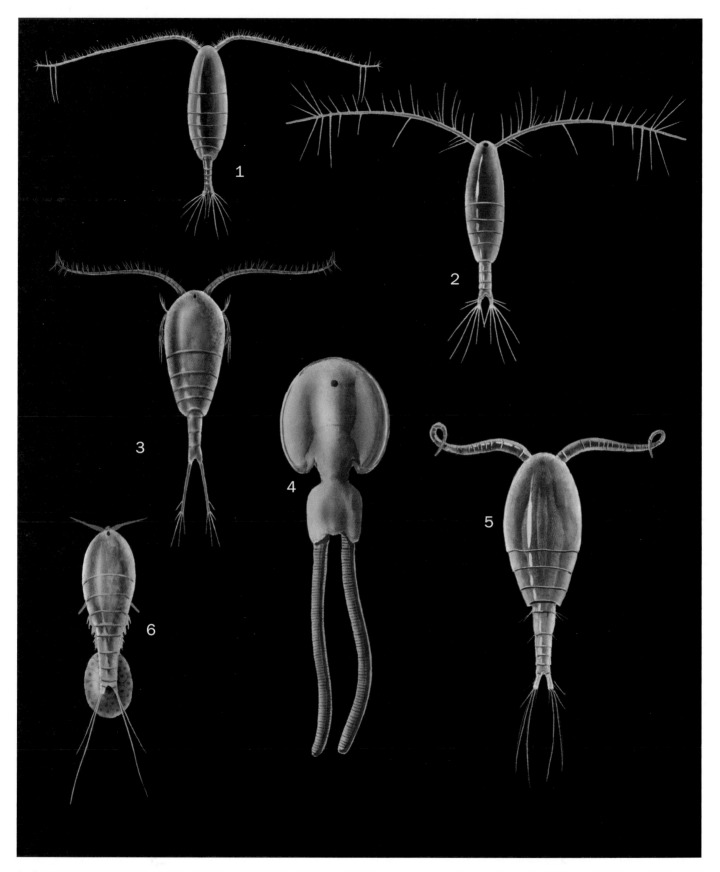

1. *Calanus finmarchicus*; 2. *Acartia clausi*; 3. *Temora longicornis*; 4. Salmon louse (*Lepeophtheirus salmonis*); 5. *Mesocyclops leuckarti*; 6. *Tigriopus californicus*. (Illustration by John Megahan)

Species accounts

No common name
Acartia clausi

ORDER
Calanoida

FAMILY
Acartiidae

TAXONOMY
Acartia clausi Giesbrecht, 1889, Mediterranean Sea.

OTHER COMMON NAMES
None known.

PHYSICAL CHARACTERISTICS
Body length 0.045–0.048 in (1.15–1.22 mm) for female and 0.039–0.042 in (1–1.07 mm) for male. Body gymnoplean, with somewhat slender prosome and short, slender urosome of three segments. Antennules long, indistinctly 18-segmented in female; male antennule geniculate on right side only. Antenna with four-segmented exopod and with 6 to 8 supernumerary setae on allobasal segment. Maxillipeds reduced distally. Swimming legs 1–4 all biramous, with three-segmented exopods and two-segmented endopods; fifth legs reduced in female, uniramous ending in tapering spinous process. Male fifth legs asymmetrical, specialized for grasping female and transferring spermatophore during mating.

DISTRIBUTION
Cosmopolitan in temperate and subtropical waters. (Specific distribution map not available.)

HABITAT
Found in near surface depths 0–164 ft (0–50 m), in shallow coastal waters and embayments; rarely in open oceanic waters.

BEHAVIOR
Typically 4–5 generations per year, but up to seven generations reported. Lifecycle consists of six naupliar stages, followed by five copepodid larval stages before final molt into adult. Exhibits daily vertical migration, but only over short vertical range. Feeds near surface waters at night, returning to deeper water during daytime.

FEEDING ECOLOGY AND DIET
A small particle feeder that feeds in near-surface coastal waters; generates water currents by the slow swimming movements of antennae and first legs and captures food particles in laminar flow fields; first legs are involved instead of maxillipeds. Mostly herbivorous; diet includes wide range of minute unicellular phytoplankton such as diatoms, coccolithophores, and dinoflagellates, and ciliate protists. Selective in choice of food particle type.

REPRODUCTIVE BIOLOGY
Mating takes place in water column. Mating behavior is not known, but probably involves male detection of pheromone trail laid down by female. Male grasps female using left fifth leg, and transfers single spermatophore with tip of right leg.

Spermatophore discharges into seminal receptacle in female genital region. Sperm stored in seminal receptacle; stored sperm from single mating is probably sufficient for lifetime egg production by female; remating is rare. Female can produce numerous batches of eggs, broadcast into water column at night. Two types of eggs produced, subitaneous eggs, which hatch after about one day, and resting eggs, which have a spiny external coat and sink to lie on sediment. Resting eggs typically act as overwintering stage, hatching in spring.

CONSERVATION STATUS
Not listed by the IUCN.

SIGNIFICANCE TO HUMANS
Plays vital role in economy of coastal seas; forms middle link in food chain leading from phytoplankton up to commercially important fish species in estuarine fish spawning grounds and in coastal waters. ◆

No common name
Calanus finmarchicus

ORDER
Calanoida

FAMILY
Calanidae

TAXONOMY
Calanus finmarchicus Gunnerus, 1765, Norway.

OTHER COMMON NAMES
None known.

PHYSICAL CHARACTERISTICS
Body length up to 0.21 in (5.4 mm) for female and 0.14 in (3.6 mm) for male. Body gymnoplean, comprising large prosome incorporating five leg-bearing segments and short, slender urosome of four segments; antennules long, 25-segmented in both sexes; male antennules non-geniculate. Swimming legs 1–5 all biramous, with three-segmented rami; fifth legs with toothed inner coxal margin.

DISTRIBUTION
Northern part of temperate North Atlantic Ocean, up to and including Arctic Ocean. (Specific distribution map not available.)

HABITAT
Most of the population found in upper 1,640 ft (500 m) of oceanic water column, but precise depth of greatest population density varies with time of day and with season.

BEHAVIOR
Typically only one generation per year, but up to three in some regions. Lifecycle consists of six naupliar stages, followed by five copepodid larval stages before final molt into adult. All stages exhibit daily vertical migration, but this is most pronounced in later stages. Population rises to near surface waters

0–98 ft (0–30 m), commencing just before sunset, feeds at surface during night, then begins to either sink or swim downwards by dawn, returning to daytime depths 98–1,640 ft
(30–500 m). Can overwinter in a resting (diapause) phase at
fourth or fifth copepodid stage, descending to 1,640 ft (500 m)
and spending winter there, sustained by stored lipids.

FEEDING ECOLOGY AND DIET
A small particle feeder that feeds at night in near-surface
oceanic waters. Generates water currents by the slow swimming movements of antennae and maxillipeds. Lateral movements of mandibular palps and maxillules then draw water and
food particles into range of the maxillae, which capture the
particles. Diet includes wide range of minute unicellular phytoplankton and protists such as diatoms, coccolithophores, dinoflagellates, silicoflagellates, and tintinnids.

REPRODUCTIVE BIOLOGY
Mating takes place in water column. Sexually receptive female
releases sexually attractant chemicals (pheromones), leaving
trail up to 3.2 ft (1 m) long as she gradually sinks or slowly
swims. Males detect chemicals in trail and begin pursuit. Once
in close proximity, male detects hydromechanical disturbance
caused by female swimming motions. Male grasps female
using maxillipeds, then transfers single spermatophore with
tip of fifth leg. Spermatophore discharges its sperm contents
into copulatory pore, leading to seminal receptacle in female
genital region. Using stored sperm, single female can produce
numerous batches of eggs over period of 60–80 days. Eggs
broadcast into water column, at rate of one about every 30
minutes, typically during afternoon or at night. Eggs hatch
after about one day.

CONSERVATION STATUS
Not listed by the IUCN.

SIGNIFICANCE TO HUMANS
The dominant copepod in the northern North Atlantic, it plays
vital role in economy of the oceans, forming middle link in
food chain leading from phytoplankton up to commercially important fish species, many of which feed on this species either
as larvae or as adults. Plays equally important role in global
carbon cycle, since large proportion of fixed carbon dioxide
passes through oceanic food web as phytoplankton consumed
by *C. finmarchicus.* ◆

No common name
Onychodiaptomus sanguineus

ORDER
Calanoida

FAMILY
Diaptomidae

TAXONOMY
Onychodiaptomus sanguineus Forbes, 1876, North America.

OTHER COMMON NAMES
None known.

PHYSICAL CHARACTERISTICS
Body length up to 0.08 in (2.1 mm) for both sexes. Body
gymnoplean; with posterolateral corners of prosome produced

into spinous processes. Antennules long, 25-segmented in female; male antennule geniculate on right side only; antenna
with eight-segmented exopod. Swimming legs 1–4 all biramous, with three-segmented rami; middle exopodal segment
without spine on outer margin. Fifth legs biramous
in female. Male fifth legs asymmetrical, right leg with claw
for grasping female and left leg for transferring spermatophore during mating.

DISTRIBUTION
Canada and United States. (Specific distribution map not available.)

HABITAT
Shallow ponds, including temporary waters and small lakes.

BEHAVIOR
Typically only one generation per year. Life cycle consists of
six naupliar stages, followed by five copepodid larval stages before final molt into adult. Produce resting (diapause) eggs that
can lie in the sediment for considerable periods before hatching; eggs were still viable after up to 300 years in the sediment
bank of a New England (United States) pond.

FEEDING ECOLOGY AND DIET
A small particle feeder that generates water currents by the
slow swimming movements of antennae and maxillipeds and
captures food particles in laminar flow fields. Herbivorous,
with diet including wide range of minute unicellular phytoplankton such as diatoms.

REPRODUCTIVE BIOLOGY
Mating takes place in water column. Sexually receptive female
releases pheromones as she swims, leaving trail that is followed
by male. Once in close proximity, male grasps female using
strongly geniculate right antennule. Single spermatophore
transferred during copulation, using tip of fifth leg. Diaptomids lack seminal receptacles, so spermatophore discharges its
contents over female's genital area to form an attached spermatophoral mass. Attached spermatophoral mass effectively
prevents further inseminations and is displaced when the egg
sacs are extruded. After hatching of eggs and release of sacs,
another copulation is required before another batch of eggs
can be produced.

CONSERVATION STATUS
Not listed by the IUCN.

SIGNIFICANCE TO HUMANS
Diaptomids form middle link in food chain leading from phytoplankton up to commercially important fish species in freshwater habitats. ◆

No common name
Temora longicornis

ORDER
Calanoida

FAMILY
Temoridae

TAXONOMY
Temora longicornis O. F. Müller, 1792, North Sea.

OTHER COMMON NAMES
None known.

PHYSICAL CHARACTERISTICS
Body length 0.039–0.045 in (1–1.15 mm) for female and
0.039–0.053 in (1–1.35 mm) for male. Body gymnoplean comprising broad, somewhat triangular prosome and short, slender
urosome of three segments ending in elongate caudal rami,
about nine times longer than wide. Antennules long, 24-seg-
mented in female; male antennule geniculate on right side
only. Swimming legs 1–4 all biramous, with three-segmented
exopods and two-segmented endopods; fifth legs of female uniramous and three-segmented. Male fifth legs asymmetrical,
specialized for grasping female and transferring spermatophore
during mating.

DISTRIBUTION
Temperate waters of North Atlantic and North Pacific Oceans.
(Specific distribution map not available.)

HABITAT
Found in near-surface depths (0–164 ft [0–50 m]) in coastal
waters.

BEHAVIOR
Typically 3–5 generations per year, but up to six generations
reported. Life cycle consisting of six naupliar stages, followed
by five copepodid larval stages before final molt into adult. Exhibit daily vertical migration, but only over short vertical
range. Feeds in near-surface waters during night, returning to
deeper water at daytime.

FEEDING ECOLOGY AND DIET
A small particle feeder that feeds in near-surface coastal waters,
generates water currents by the slow swimming movements of
antennae and maxillipeds and captures food particles in laminar
flow fields. Omnivorous, with diet including wide range of
minute unicellular phytoplankton such as diatoms, dinoflagellates, ciliates, and tintinnids. Exhibits selectivity in choice of
food particle size and type.

REPRODUCTIVE BIOLOGY
Mating takes place in water column. Sexually receptive female
releases pheromones, leaving trail. Within 0.039–0.078 in (1–2
mm) of trail, male can detect chemicals in trail up to 10 seconds after female passed, and begins pursuit. Once in close
proximity, male detects hydromechanical disturbance caused by
female swimming motions, and grasps female using fifth legs.
Single spermatophore transferred during copulation, using tip
of fifth leg. Spermatophore discharges its sperm contents into
copulatory pore, leading to seminal receptacle in female genital
region. Using stored sperm, single female can produce numerous batches of eggs, which are broadcast into water column at
rate of 16–32 per day. Two types of eggs produced, subitaneous eggs, which hatch after about one day, and resting eggs,
which sink to sediment. Resting eggs typically provide over-
wintering stage, hatching in spring.

CONSERVATION STATUS
Not listed by the IUCN.

SIGNIFICANCE TO HUMANS
Plays vital role in economy of coastal seas; forms middle link in
food chain leading from phytoplankton up to commercially important fish species in coastal waters. ◆

No common name
Mesocyclops leuckarti

ORDER
Cyclopoida

FAMILY
Cyclopidae

TAXONOMY
Mesocyclops leuckarti Claus, 1857, Germany.

OTHER COMMON NAMES
None known.

PHYSICAL CHARACTERISTICS
Body length 0.02–0.05 in (0.5–1.3 mm) for female, 0.03–0.04 in
(0.8–1 mm) for male. Body cyclopiform, with podoplean division
into prosome and uro-some. Female antennules 17-segmented;
male antennules 16-segmented, geniculate on both sides. Antenna
with single outer seta on basis, representing exopod. Mandible with
palp reduced to three setae on small papilla. Maxillule with large
and well-armed arthrite, palp small. Maxilla powerful, with strong
claws on allo-basis; outer margin of syncoxa wrinkled. Maxilliped
smaller than maxilla. Legs 1–4 biramous, with three-segmented
rami; third ex-opodal segment of legs with only two outer spines.
Fifth legs with single free exopodal segment bearing two long se-
tal elements, one originating apically, the other on mid-medial
margin. Caudal rami 3–4 times longer than wide.

DISTRIBUTION
Europe and Asia; numerous previous records from Africa
shown to be misidentifications of closely related species. (Specific distribution map not available.)

HABITAT
The water column of freshwater bodies, especially ponds and lakes.

BEHAVIOR
Lifecycle consists of six nauplius stages, followed by five copepodid larval stages before final molt into adult.

FEEDING ECOLOGY AND DIET
Carnivorous; will take variety of small invertebrates, including
chironomid larvae and cladocerans (water fleas) caught in water
column. Prey are captured by the powerful, clawed maxillae
and broken up by action of maxillules and mandibles.

REPRODUCTIVE BIOLOGY
Mating probably takes place in the water column. Males guard
copepodid five-stage females, until they molt into adult and become sexually receptive. Male guards juvenile female by grasping onto her dorsal side. During copulation, male changes
position to grasp female around her fourth legs while transferring paired spermatophores directly onto ventral surface of
genital region with tips of fourth legs. Spermatophores discharge into paired copulatory pores located ventrally. Sperm
are stored in voluminous seminal receptacle prior to fertilization. Females produce eggs in paired egg sacs extruded from
genital openings on dorsal surface of genital region.

CONSERVATION STATUS
Not listed by the IUCN.

SIGNIFICANCE TO HUMANS
Acts as vector for the human parasite, the guineaworm (*Dra-
cunculus*), in Indian subcontinent. Previously thought to be vector in Africa also, but now known that African populations

represent closely related, but different species. It is also an important predator of mosquito larvae and has been used for biological control of mosquitoes in malarial areas. ◆

No common name
Ergasilus sieboldi

ORDER
Cyclopoida

FAMILY
Ergasilidae

TAXONOMY
Ergasilus sieboldi von Nordmann, 1832, Germany.

OTHER COMMON NAMES
None known.

PHYSICAL CHARACTERISTICS
Body length up to 0.078 in (2 mm) for female, 0.031–0.035 in (0.8–0.9 mm) for male. Body cyclopiform, consisting of ovoid prosome and slender five-segmented urosome. Caudal rami with only four caudal setae. Antennules six-segmented in both sexes, non-geniculate in males. Antenna terminating in single curved claw in both sexes; no trace of exopod present. Mandible with three spinulate blades, lacking palp. Maxillule reduced to lobe bearing three setae. Maxillae comprising triangular proximal segment (syncoxa) and spinulate distal process (basis). Maxillipeds absent in female, four-segmented with terminal claw in male. Swimming legs biramous with three-segmented rami, except exopod of fourth leg two-segmented. Fifth legs with single free exopodal segment bearing three setae.

DISTRIBUTION
All of Europe, extending eastward into Asia. (Specific distribution map not available.)

HABITAT
Adult females are external parasites, infesting the gills of a wide range of freshwater fishes and attaching to the gill filaments. Developmental stages are planktonic and occur in the water column of freshwater bodies.

BEHAVIOR
Ergasilids have unique lifecycle: eggs hatch into free-swimming naupliar phase, which comprises six stages, followed by first to fifth copepodids preceding the adult. First nauplius 0.003 in (0.08 mm) in length, sixth nauplius 0.01 in (0.29 mm). Only adult female is parasitic, seeking and attaching to a wide variety of freshwater fishes. Female attaches to gill filament of host, using clawed antennae that penetrate host gill epithelium. One, sometimes two, generations during the spring and summer; the population survives over winter as adult females on fishes.

FEEDING ECOLOGY AND DIET
Nauplii are planktotrophic, feeding on unicellular phytoplankton such as diatoms. Feeding ecology of copepodid stages still unknown, although mouthparts are adapted for surface feeding, as in parasitic adult female. Adult female feeds by rasping at surface epithelium of gill filaments of host fishes, using spinulate mandible blades.

REPRODUCTIVE BIOLOGY
Adults mate in water column. Male transfers pair of spermatophores that are placed over slit-like, paired openings on dorsal surface of female. After mating, female only seeks and attaches to a fish host on which she produces a series of egg sacs over course of lifetime.

CONSERVATION STATUS
Not listed by the IUCN.

SIGNIFICANCE TO HUMANS
Serious pest of cultured fishes, especially in Eastern European freshwaters; heavy infestations, with up to 13,400 individual parasites on a single host fish, have been recorded. Such outbreaks cause mass mortality, reduced growth, and render fishes susceptible to secondary bacterial and fungal infections. ◆

No common name
Oithona plumifera

ORDER
Cyclopoida

FAMILY
Oithonidae

TAXONOMY
Oithona plumifera Baird, 1843, North Sea.

OTHER COMMON NAMES
None known.

PHYSICAL CHARACTERISTICS
Body length 0.039–0.059 in (1–1.5 mm) for female, 0.029–0.039 in (0.75–1 mm) for male. Body cyclopiform, ovoid prosome extended into pointed rostrum frontally; slender urosome nearly as long as prosome. Female antennule with very long setae; male antennules geniculate on both sides. Antenna with exopod represented by two setae on outer margin. Mandible with two spinulate setae on distal margin of basis; maxilliped well developed. Legs 1–4 biramous, with three-segmented rami; outer basal setae long, plumose. Fifth leg reduced to outer basal seta as well as long plumose seta representing exopod.

DISTRIBUTION
Cosmopolitan in world's oceans. (Specific distribution map not available.)

HABITAT
Found in near-surface depths (0–328 ft [0–100 m]) in coastal and open oceanic waters.

BEHAVIOR
Life cycle consists of six naupliar stages and five copepodid stages preceding the adult stage.

FEEDING ECOLOGY AND DIET
Omnivorous small particle feeder, with diet including wide range of minute unicellular phytoplankton such as diatoms, dinoflagellates, ciliates, and tintinnids. Little known of food capture mechanisms.

REPRODUCTIVE BIOLOGY
Mating probably takes place in swarms, which form in discontinuities in the water column. Male probably grasps female around her fourth legs during copulation and transfers paired

spermatophores directly onto her genital region. Spermatophores discharged into paired copulatory pores located on ventral surface of female genital region. Sperm are stored in seminal receptacles prior to fertilization. Females produce eggs in paired egg sacs extruded from genital openings on dorsal surface of genital region.

CONSERVATION STATUS
Not listed by the IUCN.

SIGNIFICANCE TO HUMANS
Forms middle link in oceanic food web leading up to commercially important fish species. Plays equally important role in global carbon cycle. ◆

No common name
Tigriopus californicus

ORDER
Harpacticoida

FAMILY
Harpacticidae

TAXONOMY
Tigriopus californicus Baker, 1912, California, United States.

OTHER COMMON NAMES
None known.

PHYSICAL CHARACTERISTICS
Body length 0.047–0.055 in (1.2–1.4 mm). Body slender, cylindrical, with inconspicuous boundary between prosome and urosome. Female antennules short, nine-segmented; male antennules geniculate on both sides. Antenna with three-segmented exopod. Mandible with well-developed biramous palp. First swimming legs modified, biramous with both rami prehensile and armed with array of apical claws. Legs 2–4 biramous, with three-segmented rami. Fifth legs with large endopodal lobe and single free exopodal segment in both sexes; bearing five setae in male.

DISTRIBUTION
West coast of North and South America. (Specific distribution map not available.)

HABITAT
Rock pools in intertidal zone on seashore. Rock pools are an extreme habitat, which can undergo wild fluctuations in salinity and temperature over course of tidal cycle.

BEHAVIOR
Life cycle comprising six naupliar stages and five copepodid stages preceding adult. All stages are bottom-living; even nauplii are adapted for creeping over surfaces, not for swimming.

FEEDING ECOLOGY AND DIET
Omnivorous surface feeders; will take variety of microorganisms, algae, and organic material contained in surface film in rock pools; will also eat tissues of various metazoans, including sponges.

REPRODUCTIVE BIOLOGY
Males exhibit pre-copulatory mate guarding, in which adult male will grasp onto juvenile (copepodid stage three onwards) females and hold them for several days, until they undergo final molt into sexually receptive adult. Males can distinguish between developmental stages of female and between closely related species by surface chemical properties. During copulation, male transfers single spermatophore onto ventral genital region of female. Spermatophore discharges into seminal receptacle where sperm stored until used. Fertilized eggs retained in single ventral egg mass until ready to hatch.

CONSERVATION STATUS
Not listed by the IUCN.

SIGNIFICANCE TO HUMANS
Very hardy, surviving well in laboratory conditions, so has been used as a model animal for scientific study of genetics of marine copepods. ◆

No common name
Benthomisophria palliata

ORDER
Misophrioida

FAMILY
Misophriidae

TAXONOMY
Benthomisophria palliata Sars, 1920, North Atlantic Ocean.

OTHER COMMON NAMES
None known.

PHYSICAL CHARACTERISTICS
Body length 0.149 in (3.8 mm) in male, 0.157 in (4 mm) in female. Body podoplean. First leg bearing segment free, but entirely concealed beneath carapace-like extension of dorsal head shield. Antennules 18-segmented in female; male antennule geniculate on both sides and used to grasp female during mating. Swimming legs 1–4 biramous, with three-segmented rami; fifth legs of female two-segmented with broad triangular protopodal segment bearing single exopodal segment, endopod represented by inner seta.

DISTRIBUTION
Cosmopolitan in deep areas of major oceans. (Specific distribution map not available.)

HABITAT
Found in deep water plankton below 8,200 ft (2,500 m), but concentrated in near-bottom layer (hyperbenthic zone) at depths of 9,840–13,120 ft (3,000–4,000 m).

BEHAVIOR
Lacks eyes, so feeding and mating behavior dominated by chemosensory and hydromechanical sensory systems. Lateral margins of dorsal cephalic shield carry fields of cone organs interpreted as producing anti-fouling secretion important in grooming behavior.

FEEDING ECOLOGY AND DIET
An opportunistic gorger, possibly a scavenger; will take relatively large food items, including other copepods, chaetog-

naths, and cnidarians, which can be accommodated in highly distensible midgut. Presumably, swimming motions of mouthparts create flow fields that it samples for chemical traces originating from potential food items. When gut is full, the entire body swells.

REPRODUCTIVE BIOLOGY
Little known about early development. Shallow water relatives have single non-feeding, yolk-like (= lecithotrophic) naupliar stage that hatches from large eggs and molts after one day into first copepodid. Has five copepodid stages preceding the adult stage. Mating probably takes place in near-bottom zone, with male grasping female around her fourth legs while transferring paired spermatophores directly onto her genital region. Spermatophores discharge into seminal receptacles. Female produces large yolky eggs that are probably loosely attached in an egg mass to her urosome, rather than contained in an egg sac.

CONSERVATION STATUS
Not listed by the IUCN.

SIGNIFICANCE TO HUMANS
None known. ◆

No common name
Monstrilla grandis

ORDER
Monstrilloida

FAMILY
Monstrillidae

TAXONOMY
Monstrilla grandis Giesbrecht, 1891, North Atlantic.

OTHER COMMON NAMES
None known.

PHYSICAL CHARACTERISTICS
Body length to 0.147 in (3.75 mm) for female, 0.074 in (1.9 mm) for male. Body consisting of elongate prosome and short urosome of five segments and bearing large caudal rami. Antennules short and five-segmented in both sexes; geniculate on both sides in male. Antennae, mandibles, labrum, maxillules, paragnaths, maxillae, and maxillipeds all missing in adults of both sexes. Legs 1–4 biramous, with indistinctly three-segmented rami. Fifth legs reduced, bilobed, with three setae on outer lobe and two on inner. Female with large median genital opening and long paired ovigerous (egg-bearing) spines.

DISTRIBUTION
Northeastern Atlantic Ocean. (Specific distribution map not available.)

HABITAT
Adults of both sexes are members of coastal zooplankton community, living mainly in upper (epipelagic) layers. Larval development takes place within host, which may be either a gastropod mollusk or a polychaete worm.

BEHAVIOR
Free-swimming naupliar larva is tiny, with body length of about 0.002 in (0.05 mm) in closely related species. Nauplius is infective stage, seeking out potential host, attaching by means

of clawed antennae and mandibles, and burrowing through body surface of host. Once inside the host's blood system, nauplius transforms into sac-like body and larval limbs grow into absorptive rootlets. Development proceeds inside host until fifth copepodid stage, which exits host through body wall, undertakes single molt into adult, and begins short, non-feeding planktonic phase.

FEEDING ECOLOGY AND DIET
Free-swimming adults are non-feeding, lacking any mouthparts or food-gathering apparatus. Naupliar stage is internal parasite of polychaete worm, absorbing nutrients from the host via a paired rootlet system derived from modified naupliar appendages.

REPRODUCTIVE BIOLOGY
Females carry eggs on long, ovigerous spines; eggs hatch into infective nauplii that locate a host and burrow into its tissues. Little is known about mating behavior: male presumably grasps female using geniculate antennules and transfers spermatophores with tips of swimming legs.

CONSERVATION STATUS
Not listed by the IUCN.

SIGNIFICANCE TO HUMANS
None known. ◆

Salmon louse
Lepeophtheirus salmonis

ORDER
Siphonostomatoida

FAMILY
Caligidae

TAXONOMY
Lepeophtheirus salmonis Krøyer, 1837, Scandinavian waters.

OTHER COMMON NAMES
Norwegian: Laxlus.

PHYSICAL CHARACTERISTICS
Body length 0.27–0.49 in (7–12.5 mm) in female, 0.17–0.26 in (4.5–6.7 mm) in male. Body dorsoventrally flattened, giving low profile when attached to its host; comprising cephalothorax, incorporating first to third leg-bearing segments, free segment bearing leg four, genital complex consisting of fused fifth leg-bearing and genital somites, and a long, free abdomen. Frontal plates lacking lunules. Antennules two-segmented in both sexes; antennae and maxillipeds with strong apical claws used for attachment to surface of host. Oral cone consisting of upper lip (labrum) and lower lip (labium); mandibles stylet-like, passing into oral cone at base, via lateral slits. Third legs forming apron-like structure and completing posterior margin of cephalothoracic sucker. Fourth leg uniramous. Genital complex of female with paired genital openings from which long egg strings are extruded, plus paired copulatory pores into which spermatophores discharge their sperm.

DISTRIBUTION
North Atlantic and North Pacific Oceans, including Arctic waters. (Specific distribution map not available.)

HABITAT
Ectoparasitic on salmonid fishes, especially Atlantic salmon, while in marine waters, but abandon salmon when they return to rivers to spawn. Presence of sealice on salmon in rivers is interpreted as sign that they are freshly run from sea.

BEHAVIOR
Life cycle consists of two naupliar stages, followed by the infective copepodid stage, then four chalimus stages and two pre-adult stages on the host, before final molt into adult. Naupliar stages are non-feeding and depend solely on yolk; each lasts about 24 hours. Free-swimming copepodid stage can survive up to 10 days, but is most successful in finding a host in first 48 hours. On host, molts into first of four chalimus stages, characterized by possession of chitinous frontal filament, used to attach developing sealouse to host via anchor at tip that embeds in host skin. Fourth chalimus molts into first pre-adult, which is temporarily attached via frontal filament but later moves freely over host surface.

FEEDING ECOLOGY AND DIET
Sealice feed by rasping at skin of salmon host with tips of stylet-like mandibles and dislodging pieces of tissue. When feeding at one site on host, sealice can erode skin and begin to feed on blood. Feeding activity on host can cause pathological lesions and blood loss, which weaken the fish and render it more susceptible to secondary infections.

REPRODUCTIVE BIOLOGY
Mating occurs on surface of fish. Only adults are mature and mate, but adult male exhibits mate guarding behavior in which he finds and remains attached to pre-adult female until her final molt into sexually receptive state. Male transfers pair of spermatophores and female stores sperm in paired seminal receptacles until used to fertilize series of egg strings. Each female can produce several pairs of egg strings, each containing a single stack of up to 700 disc-shaped eggs. Eggs hatch serially.

CONSERVATION STATUS
Not listed by the IUCN.

SIGNIFICANCE TO HUMANS
Sealice are the greatest health hazard to farmed salmonids in northern Europe and North America, causing estimated losses of $30 million per year in Europe alone, because of direct cost of salmonid mortality as well as indirect costs of chemotherapy, reduced growth rate, and reduced market value of infected fish. ◆

Resources

Books

Boxshall, G. A. "Copepoda." *Microscopic Anatomy of Invertebrates.* Vol. IX. *Crustacea,* edited by F. W. Harrison. Chichester, U.K.: John Wiley and Sons, 1992.

Boxshall, G. A., and Defaye, D., eds. *Pathogens of Wild and Farmed Fish: Sea Lice.* Chichester, U.K.: Ellis Horwood, 1993.

Huys, R., and G.A. Boxshall. *Copepod Evolution.* London: The Ray Society, 1991.

Mauchline, J. *The Biology of Calanoid Copepods. Advances in Marine Biology.* New York and London: Academic Press, 1998.

Geoffrey Allan Boxshall, PhD

Ostracoda
(Mussel shrimps)

Phylum Arthropoda
Subphylum Crustacea
Class Maxillopoda
Sublass Ostracoda
Number of families 46

Thumbnail description
Generally small crustaceans with reduced body entirely enclosed within an often-calcified bivalved carapace

---

Photo: Members of *Herpetocypris reptans*, like all ostracods, have bilaterally symmetrical bodies. (Photo by Hubert Kranemann/OKAPIA/Photo Researchers, Inc. Reproduced by permission.)

Evolution and systematics

Ostracods have a long fossil history, being known from the Cambrian, and have undergone extensive radiations, especially since the early Mesozoic. Including fossil forms, the subclass Ostracoda can be divided into six orders, of which three, the Palaeocopida, Myodocopida, and Podocopida, contain recent species. The Palaeocopida contains one modern family, the Punciidae, known from the seas around New Zealand. Myodocopids are predominantly swimmers; one suborder lives in the benthic boundary layer just above the seabed, while the other suborder has radiated into the pelagic zone of the sea. Podocopids are generally smaller than other ostracods and, for the most part, live as epibenthos. Within the Podocopida, there is a tendency for reduction of appendage segments or rami, and from turgor appendages to ones with more exoskeletal integrity and strength. Podocopid carapaces also tend to be more highly ornamented with ridges and spines than those seen in the other orders. The origin of the Ostracoda has long been a mystery. They are unlike all other crustacean groups in their body organization, having retained some larval features such as feeding with antennae as well as with other mouthparts. In addition, the nauplius of ostracods is unique in possessing carapace folds from the earliest stage. The paucity of appendages on the ostracod body has resulted in considerable confusion regarding the homologies of some of the post-cephalic legs. A recent study by Tsukagoshi and Parker (2000) suggests that ostracods have the 5-6-5 or 5-7-4 trunk segmentation pattern of other maxillopodans.

Physical characteristics

Ostracods are most notable in having their entire body enclosed within the carapace folds. As a result, there are many reductions in body segments and appendages. The carapace fold originates at the posterior margin of the larval head shield and extends anteriorly, laterally, and posteriorly to cover the body. Dorsally, the carapace fold is divided by a hinge and flexible cuticle. A system of adductor muscles is used to close the "valves," as the two parts of the carapace fold are called, and hydrostatic pressure applied at the top of the mandible serves to open the valves. The two valves are asymmetrical, with one valve fitting inside the other. Because the valves are often heavily calcified, ostracod valves are common features of the fossil record, have been studied extensively, and a specialized terminology has been developed to describe the fine details. Ostracods have typical crustacean head appendages: antennules, antennae, mandibles, maxillules, and maxillae. In contrast to many other crustacean groups, however, ostracods use their antennae, and sometimes their antennules, for locomotion. As a consequence, these appendages are short and robust. The antennules are uniramous and composed of five to eight segments. The antennae are biramous, with one ramus, the exopod, often reduced in the podocopids. Ostracod mandibles have a coxal gnathobase, with the remainder of the appendage developed into a palp, which is often biramous. Posterior to the mandible is the maxillule, which is quite variable in morphology. The protopod usually has a set of setose endites, and from the protopod arises a short endopod used in manipulating food particles and a dorsally directed setose vibratory plate. It is not

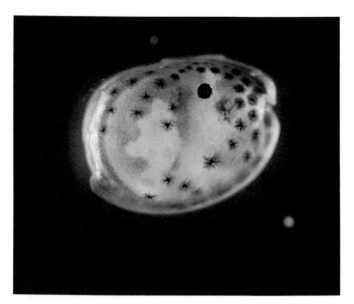

These small zooplanktonic marine ostracods or mussel shrimps are protected from predators by their bivalve carapace. Many other species are benthic and are well-known for their ability to glow at night. (Photo by A. Flowers & L. Newman. Reproduced by permission.)

known with certainty whether this vibratory plate is an exopod or an epipod. The trunk limbs have had various names applied to them. Posterior to the maxillule is the maxilla, but this limb has usually been called a maxilliped, or first trunk limb, in response to the degree to which the limb looks like the maxillule or the next two pairs of trunk limbs, and is used for feeding or locomotion. If used for feeding, the endopod is shortened and forwardly directed and the vibratory plate is a smaller version of that on the maxillule. If used for locomotion, the maxillae are changed considerably in shape, with the endopod directed posteriorly and the vibratory plate reduced to a simple seta. Of course there are many examples of intermediate forms for this appendage. The next two pairs of trunk limbs are the thoracopods, and they are also quite variable in design. Sometimes there is also variation between right and left limbs as well as sexual dimorphism. In some cases, these limbs are absent.

Myodocopids have transformed the second thoracic limb into a long multiarticulate vermiform appendage capable of extending into the dorsal space between the body and the hinge area of the valves, presumably as a cleaning device. In the most recently evolved podocopids, these trunk limbs look like typical crustacean walking legs and have often been termed pereopods. According to recent research, there are several limbless trunk segments following the two pairs of thoracopods, and the body typically ends in a pair of caudal rami. In most ostracods, the penes of the male are quite large relative to the remainder of the body. In higher podocopids, the penial lobes can be as much as one-third of the body length and are very complicated, highly chitinous structures.

Distribution

Ostracods are found in nearly all marine and freshwater environments, as well as some moist terrestrial habitats.

Habitat

Most ostracod species are benthic, or epibenthic, living on the substrate or on other organisms. One group, the myodocopids, has developed exclusively pelagic species, which can be found throughout the oceanic waters of the world.

Behavior

When benthic ostracods walk, they open the carapace valves, push out the walking legs and antennae, and amble across the substrate with a rocking motion. Swimming in pelagic species involves pushing the terminal section of the antennae out the antennal notches in the carapace and moving the limbs in a rowing motion. Shallow-water ostracods seem to spend most of the day in motion, probably foraging for food particles. The few observations done at night indicate that they usually do not move about at night.

Feeding ecology and diet

Ostracods were long thought to be filter feeders, but recent evidence shows that they are quite capable of grazing on diatoms and other fractions of marine detritus. Most ostracods are apparently detritus feeders, with a few being capable of scavenging or predation.

Reproductive biology

Ostracods mate by putting the opened ventral side of the valves together, or the male may mount the female dorsally from behind and insert the large penes into the open valves of the female. Eggs may be laid freely in the environment or, as in some podocopids, are incubated inside the valves, usually above the abdomen. The first larval stage that hatches is a nauplius, with the carapace fold covering the body and appendages. At this stage, the animal can walk or swim by means of the antennae and mandibles. At each successive molt, new legs or leg primordia are usually added. Adult stages are reached in 5–8 molts, depending on order and family. Lifespan is usually a year or less, but little is known about this for deep-sea-dwelling species.

Conservation status

Most ostracod species occur in sufficient numbers and are not threatened, and none are listed by the IUCN. A few, however, such as the entocytherids, are symbionts of freshwater crayfish and isopods, and so may become rare as their hosts disappear.

Significance to humans

The fossilized valves of ostracods have long been used by paleontologists as indicators of past habitat and climate conditions. Because no species of ostracod extends over a long geological time period, they are useful indicators of ages of deposits.

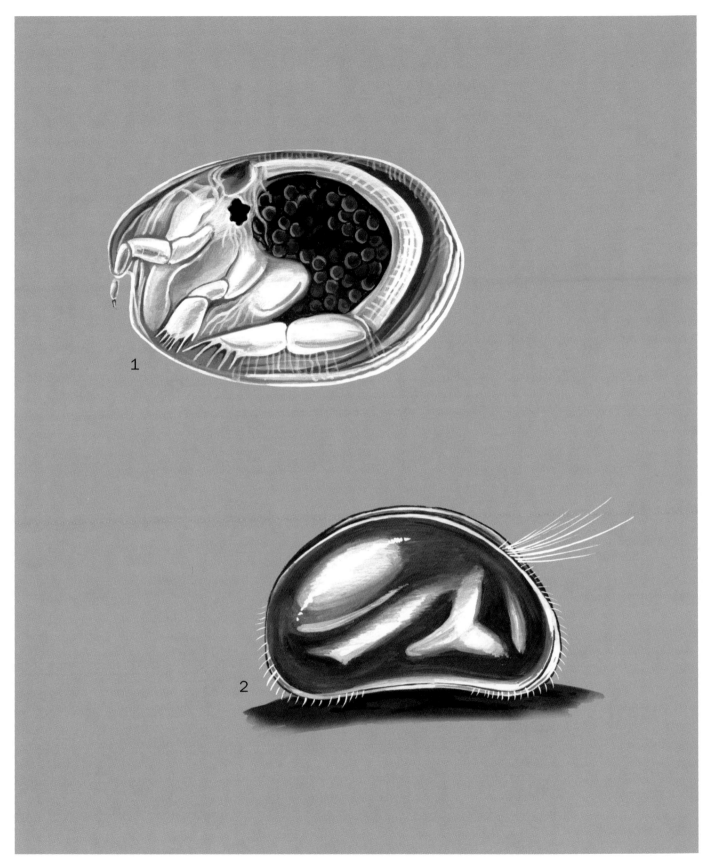

1. *Vargula hilgendorfii*; 2. *Heterocypris salinus*. (Illustration by Patricia Ferrer)

Species accounts

No common name
Vargula hilgendorfii

ORDER
Myodocopida

FAMILY
Cypridinidae

TAXONOMY
Vargula hilgendorfii (Muller, 1890), Japan.

OTHER COMMON NAMES
None known.

PHYSICAL CHARACTERISTICS
Typical myodocopid, with rounded, smooth, unpigmented, bivalved carapace marked anteriorly with a rostrum and large antennal notch. Pair of black lateral eyes are visible through the carapace. Caudal furca is extremely large and visible extending through the ventral valve opening. At 0.12 in (3 mm), is relatively large for ostracod.

DISTRIBUTION
Pacific coast of central Japan.

HABITAT
Very common and easy to collect on sandy bottoms in moderately shallow water.

BEHAVIOR
Stays in sediment in daytime and is most active at night, either hopping across the substratum or swimming in lower part of water column. Swimming and digging in the sediment are accomplished by synchronous strokes of the antennae. In probable escape response to predators, it used its rather large furca to push itself into sand. Escape from sand was also accomplished by a sudden push from the caudal furca, launching it into overlying water. In general, it burrowed only a few millimeters into the sediment.

FEEDING ECOLOGY AND DIET
In laboratory, it has been observed to attack living prey such as polychaete annelids, to scavenge on fish carcasses, or to eat artificial aquarium foods. The mandibular palps are used primarily to hold the food while it uses its fourth limbs (maxillules), caudal furcae to open the integument or body wall. Endites transfer food items to opening of the esophagus.

REPRODUCTIVE BIOLOGY
Observations are few, but it appears to spend 30–60 minutes off the bottom in a precopulatory mode with the male grasping the female. Mating occurs on the bottom with the pair lying in reverse ventral position. Male transfers spermatophore to female. Eggs are brooded under carapace of female. After hatching, there are five instars leading to the adult. Youngest juveniles were able to crawl, swim, and dig into the bottom sediment.

CONSERVATION STATUS
Not listed by the IUCN.

SIGNIFICANCE TO HUMANS
As with most myodocopids, it is preyed upon by fish. ◆

No common name
Heterocypris salinus

ORDER
Podocopida

FAMILY
Cyprididae

TAXONOMY
Heterocypris salinus (Brady).

OTHER COMMON NAMES
None known.

PHYSICAL CHARACTERISTICS
Valves are conspicuously pigmented with vertical brown streaks on either side of the muscles' scars.

DISTRIBUTION
Known from Europe, the Azores, North Africa, and western Asia. (Specific distribution unknown; no map available.)

HABITAT
Brackish water pools along the coast as well as inland saline waters. Tolerance experiments showed that they tolerate 10 parts per thousand salinity the best, especially at lower temperatures. In the field, tolerance to wide ranges of salinity have been seen.

Vargula hilgendorfii

BEHAVIOR
Active only during the day. At low light levels, they creep towards deeper parts of the pools in which they are found. They walk about the pool constantly with no resting period observed during daylight hours.

FEEDING ECOLOGY AND DIET
Feeds exclusively on small green alga cells, especially desmids and *Chlorella*-like algae.

REPRODUCTIVE BIOLOGY
Thirty to 40 red to orange oval eggs are generally attached to the substratum or to thalli of green algae. Three genera-tions are produced during the summer in the Swedish Baltic, with the third generation being an over-wintering generation. The eggs of the third generation are always laid in the bottom sediments so as to be protected from the winter ice.

CONSERVATION STATUS
Not listed by the IUCN.

SIGNIFICANCE TO HUMANS
None known. ◆

Resources

Books
Benson, R. H., et al. *Treatise on Invertebrate Paleontology, Part Q, Arthropoda 3*. Lawrence, KS: Geological Society of America and University of Kansas Press, 1961.

Hartmann, G., and M.-C. Guillaume. 1996. *Classe des Ostracodes. Traité de Zoologie*, VII. *Crustaces*, Fasc. 2. Paris: Masson et Cie, 1996.

Periodicals
Ganning, B. "On the Ecology of *Heterocypris salinus*, *H. incongruens* and *Cypridopsis aculeate* (Crustacea: Ostracoda) from Baltic Brackish-water Rockpools." *Marine Biology* 8 (1971): 271–279.

Vannier, J., and K. Abe. "Functional Morphology and Behavior of *Vargula hilgendorfii* (Ostracoda: Myodocopida) from Japan, and Discussion of Its Crustacean Ectoparasites: Preliminary Results from Video Recordings." *Journal of Crustacean Biology* 13 (1993): 51–76.

Vannier, J., K. Abe, and K. Ikuta. "Feeding in Myodocopid Ostracods: Functional Morphology and Laboratory Observations from Videos." *Marine Biology* 132 (1998): 391–408.

Les Watling, PhD

Pentastomida
(Tongue worms)

Phylum Arthropoda

Subphylum Crustacea

Class Pentastomida

Number of families 8

Thumbnail description
Parasitic tongue worms inhabit the respiratory systems of terrestrial vertebrates

Photo: Adult females of the species *Porocephalus crotali* in the lung of a freshly killed rattlesnake. The coils of the uterus are plainly visible through the cuticle (scale in centimeters). (Photo by John Riley. Reproduced by permission.)

Evolution and systematics

Pentastomida once was classified as a minor phylum, a fact reflected in most modern textbooks of parasitology. The classification is being changed, however. The evolutionary history of tongue worms is unique among parasites. The fossil record apparently extends to the late Cambrian period (500 million years ago [mya]), exceeding that of the next oldest parasites, certain copepods, by some 370 million years. Tiny fossil tongue worms have been etched from ancient fine-grained, deep-water limestone when the rock has been dissolved with dilute acid. In most respects these 0.04 to 0.08 in long (1–2 mm) larvae are indistinguishable from modern tongue worms. Therein lies a conundrum, because extant adult tongue worms are parasitic in terrestrial vertebrates, which do not appear in the fossil record until the late Devonian, approximately 350 mya. Nonetheless, fossil agnathan fishes are known from the Upper Cambrian, and it is noteworthy that all of the fossil tongue worms discovered so far have been collected from limestone containing diverse conodont fauna. Conodonts are primitive fish-like chordates. Although a marine ancestry for tongue worms seems assured, the exact nature of the ancestral host may never be known. In addition to having anterior pairs of claws, some of these ancient tongue worms have two pairs of vestigial nonsegmented trunk limbs. These structures are lacking in other fossil specimens and in all modern forms. A compounding problem is that the larvae of modern tongue worms are highly modified for tissue migration. Clawed limbs, together with other limb-like outgrowths evident in early larval development within the egg, have so far proved impossible to homologize with euarthropod head and trunk appendages.

A relation between tongue worms and branchiuran crustaceans (class Maxillopoda, subclass Branchiura) is supported by results of studies of sperm of other species and by comparison of ribosomal 18S recombinant RNA sequences in the two groups. In numerous extant parasitic maxillopods representative of several subclasses, the degree of cephalization together with the development of the trunk and anchoring devices is highly variable, to the extent that many adults bear no resemblance to their free-living counterparts. Always, however, the first larval instar, the nauplius, is easily recognized. The highly modified first instar of tongue worms, the so-called primary larva, is very different from its putative naupliar forebear because it has evolved to penetrate and traverse tissues in (mostly) terrestrial hosts. These larvae hatch with only two limb-bearing head somites and three trunk somites. Subsequent growth of the trunk is by a form of pseudometamerism, without the addition of further somites. The development of some copepods, relatives of branchiurans, also fits this pattern.

The present, still unresolved debate hinges heavily on the relative merits of the fossil evidence versus that of ribosomal RNA and sperm morphology. In this entry, tongue worms are considered a class of crustaceans. The class Pentastomida contains eight families and approximately 110 species assorted between two orders—the primitive Cephalobaenida and the advanced Porocephalida.

Cephalobaenida
- Cephalobaenidae. Three genera. *Cephalobaena*, one species found in snakes; *Raillietiella*, more than 35 species found in amphibians, snakes, lizards, amphisbaenans, and birds; *Rileyiella*, one species found in mammals.
- Reighardiidae. One genus. *Reighardia*, two species found in marine birds.

Porocephalida
- Sebekidae. Seven genera. *Alofia*, five species; *Leiperia*, two species; *Selfia*, one species; *Agema*, one species, all found in crocodiles; *Sebekia*, 12 species, found in

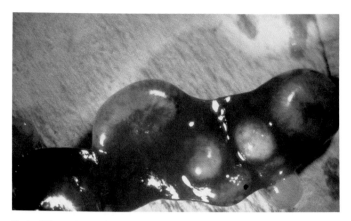

Part of the spleen of an infected mouse showing at least four encysted nymphs of *Porocephalus crotali*. These nymphs are infective to the snake definitive host, and each one is enclosed in the cuticle of the previous instar, which constrains the nymph to a C-shape. Photo by John Riley. Reproduced by permission.)

crocodiles and chelonians (one species); *Pelonia*, one species; *Diesingia*, one species, found in chelonians.

- Subtriquetridae. One genus. *Subtriquetra*, three species, found in crocodilians.

- Sambonidae. Four genera. *Sambonia*, four species; *Elenia*, two species, found in monitor lizards; *Waddycephalus*, 10 species; *Parasambonia*, two species, found in snakes.

- Porocephalidae. Two genera. *Porocephalus*, eight species; *Kiricephalus*, five species, found in snakes.

- Armilliferidae. Three genera. *Armillifer*, seven species; *Cubirea*, two species; *Gigliolella*, one species, found in snakes.

- Linguatulidae. One genus. *Linguatula*, more than six species, found in mammals.

Physical characteristics

All tongue worms inhabit the respiratory systems of terrestrial vertebrates. As a consequence, all aspects of their structure and function have adapted to life in this unusual habitat. All are aerodynamic, possessing an elongated, worm-like, mostly cylindrical body, which is rounded both anteriorly and posteriorly. The body is differentiated into an anterior head region, bearing a small ventral mouth flanked by two pairs of retractile hooks, and a long posterior trunk, which carries numerous raised annuli that are not true segments. Sexual dimorphism is pronounced, females are invariably larger than males. Females may be as small as 0.06 × 0.01 in (1.5 × 0.3 mm) (*Rileyiella*) to 4.7 × 0.4 in (12 × 1 cm) (*Armillifer*), but males are much shorter and proportionately more slender. The chitinous cuticle is thin, flexible, and translucent, so that in living specimens the body organs, suspended in an extensive fluid-filled haemocoel, are clearly visible. Peristaltic contractions of the body wall musculature affect locomotion, which is comparable with that of soft-skinned

dipteran maggots. The cuticle is shed periodically during growth, and simple metamorphosis occurs (developing nymphs resemble adults). Numerous, exceedingly fine, chitin-lined ducts erupt over the entire surface of the cuticle, each connected to an extensive system of subparietal gland cells that abound in the haemocoel immediately beneath the tegumental epidermis. Secretion from these glands, composed largely of membranous secretory droplets, is critical to prolonged survival of these parasites in the delicate environment of the lung. Large numbers of distinctive, flask-shaped cells, analogous to those involved in ion transport in other invertebrates, are embedded in the cuticle. Additional gland systems, located mainly in the head (both orders), but also flanking the intestine in porocephalids, discharge copious enzymic secretions over the head and into hook pits. There are no respiratory or excretory systems.

The sucking mouth leads into a short esophagus, which is separated by a valve from a simple undifferentiated tubular intestine. The intestine terminates at a short rectum. The simple nervous system forms initially as separate ganglia that fuse progressively during development to varying degrees within the two orders. In Porocephalida, all ganglia fuse to form a compact "brain." In cephalobaenids only the most anterior ganglia do so. A variety of small sense organs, often visible as distinct papillae, are arranged over the head, but most are concentrated on the ventral surface around the mouth.

The reproductive system differs between the two orders. The uterus of cephalobaenids is saccate, and the vagina opens anteriorly close to the junction of the head and trunk. In porocephalids, the uterus is elongate and tubular. Because it is many times longer than the body, the uterus is irregularly coiled to occupy most of the available haemocoel. The vagina opens near the anus. In both orders, the dorsal ovary leads into a paired oviduct, which passes around the gut. Close to the junction between the oviduct and the uterus are paired spermathecae, which are responsible for long-term storage of sperm.

Part of the surface of a rat that was killed, soaked in saline and then enclosed in a polyethylene bag that was left at room temperature for 24 hours. The rat had been infected with eggs of *Armillifer armillatus* four months earlier. Upon the death of this intermediate rodent host, nymphs excyst and burrow out to the surface using enzymes to digest a route through tissues. Under natural circumstances these nymphs would anticipate liberation into the stomach of a snake! (Photo by John Riley. Reproduced by permission.)

A culture of (mainly) eighth instar nymphs of *Armillifer armillatus*. These have been sieved and washed from a blood based culture medium; normally they would be resident in the lung of an African gaboon viper. The raised annuli are bands of circular muscles and gland cells and do not represent true segments. (Photo by John Riley. Reproduced by permission.)

The lower reproductive tract of males comprises paired, elongate penises—basically thin, coiled tubes of chitin—close to elaborate chitinous spicules called *dilators*. Dilators may be extruded through the anterior genital pore by muscles and thereby carry the tip of the penis either to the entrance of the spermathecal ducts in cephalobaenids or into the vagina/uterus in porocephalids. Peristalsis within the uterus of porocephalids pulls the ornamented heads of the paired penises toward the spermathecal ducts. The testis is dorsal, and the paired vasa deferentia empty into a seminal vesicle, which functions as a sperm storage depot before intromission.

Distribution

Most tongue worms, approximately 96%, live in tetrapod definitive hosts that are widespread in the tropics and subtropics. Only a few species are found outside these regions. *Porocephalus crotali* is unique in that it is cold-adapted, infecting North American rattlesnakes, which hibernate each winter. *Reighardia sternae* is a cosmopolitan species in gulls and terns, which are widespread in both hemispheres. Another cosmopolitan species, *Linguatula serrata*, lives in the nasal sinuses of canines (dogs, foxes, wolves), whereas the reindeer host of *Linguatula arctica*, inhabits the arctic tundra.

Habitat

Most tongue worms live in the lungs of their definitive hosts, although two genera, *Leiperia* and *Subtriquetra*, inhabit the trachea and nasal sinuses respectively. At least one member of the genus *Elenia* infests the throats of monitor lizards. All *Linguatula* species live in the nasal sinuses of mammals, and *Reighardia* species are located in the body cavity and air sacs of their avian hosts. An intermediate host is usual in the life cycle of tongue worms. In these cases infective larvae encyst in the tissues of arthropods or vertebrates, depending on the species.

Behavior

Because tongue worms are endoparasites of the respiratory tracts of tetrapods, what little is known about behavior has been inferred mostly from findings at autopsy. In the case of direct life cycles, eggs containing primary larvae gain entry to the definitive host through the alimentary tract as contaminants of food or water. When there are intermediate hosts in the life cycle, larvae acquired by the same means escape from the egg but invade the viscera, where they molt several times to form infective nymphs. The nymphs excyst when intermediate hosts are eaten, and larvae penetrate the stomach or intestinal wall of the final host. This stage is followed by a period of growth in the body cavity before larvae penetrate the lung through the pleura. It is possible to culture in vitro the lung-stage worms of certain species in a blood-based medium under sterile conditions. For example, developing nymphs of *Porocephalus crotali*, normally resident in the lung of their rattlesnake host, ingest ad libitum the medium in which they are suspended and molt normally through several instars to the adult stage. Thus lung-dwelling species appear not to require specific cues for successful development.

Feeding ecology and diet

With the exception of species belonging to the genus *Linguatula*, which browse on cells and mucus lining the nasal sinuses, all tongue worms feed on blood. In most cases the blood is pumped from capillaries lining the respiratory surface of the lung by the sucking action of an oral papilla. In some genera (*Alofia, Leiperia, Elenia, Waddycephalus, Kiricephalus*, and *Cuberia*) the head of females is separated from the trunk by a distinct neck, which is permanently encapsulated by inflammatory tissue and thereby anchored to the lung wall. In the lung females may also feed on inflammatory cells, as is known to occur during the development of *Porocephalus* nymphs in rodent intermediate hosts.

Reproductive biology

As far as is known, all tongue worm females become sexually mature precociously. Copulation occurs when the uterus is undeveloped and when males and females are of similar size. Copulation is a lengthy and complex process, entailing docking of the paired penises in the narrow spermathecal duct and concomitant transfer of millions of filiform sperm. Stored sperm fertilize ova released continuously from the ovaries of mature females. Fertilized eggs mature as they descend the uterus of porocephalids, and gravid females of *Armillifer* and *Linguatula* species may contain millions of eggs. In contrast, the eggs of cephalobaenids are stored temporarily in a saccate uterus until 30–50% become infectious (i.e., they contain a fully mature primary larva). Then egg deposition begins. The vagina is equipped with a sieve-like mechanism that retains small, undeveloped eggs but allows mature eggs to escape. Thus in both orders, continuous egg production is usual. Eggs shed into lungs are wafted out by the hosts' ciliation and swallowed.

Conservation status

No species are listed by the IUCN.

Significance to humans

The eggs of five tongue worm species are infective to humans, and in four of these species (*Armillifer grandis, A. armillatus, A. moniliformis*, and *A. agkistrodontis*), humans are merely an accidental intermediate host. Nearly always, nymphal infections are acquired when eggs in undercooked meat, derived from tongue worm–infected snakes, are consumed. The epidemiology of the remaining species, *Linguatula serrata* from the nasal sinuses of dogs, is complicated, because both eggs and infective larvae can become established in humans. Ingested eggs hatch to produce infective nymphs. In contrast, if ingested in contaminated offal from sheep or goats, nymphs attempt to migrate from the stomach to the nasal passages, producing acute symptoms of nasopharyngeal linguatulosis.

Resources

Books

Kabata, Z. *Parasitic Copepoda of British Fishes*. London: Ray Society, 1979.

Mehlhorn, H. *Parasitology in Focus*. Berlin: Springer Verlag, 1988.

Riley, J. "Pentastomids." In *Reproductive Biology of Invertebrates*. Vol. VI, edited by K. G. Adiyodi and R. G. Adiyodi. Oxford: IBH Publishing, 1994.

Periodicals

Almeida, W. O., and M. L. Christoffersson. "A Cladistic Approach to Relationships in Pentastomida." *Journal of Parasitology* 85 (1999): 695–704.

Böckeler, W. "Embryogenese und ZNS-Differenzierung bei Reighardia sternae, Licht- und electronenmikroskopishe Untersuchungen zur Tagmosis und systematischen Stellung der Pentastomiden." *Zoologische Jahrbucher (Anatomie)* 11 (1984): 297–342.

Buckle, A. C., J. Riley, and G. F. Hill. "The *in vitro* Development of the Pentastomid *Porocephalus crotali* from the Infective Instar to the Adult Stage." *Parasitology* 115 (1997): 503–512.

Martin, J. W., and G. E. Davis. "An Updated Classification of the Recent Crustacea." *Science Series, Natural History Museum of Los Angeles County* 39 (2001): 1–124.

Riley, J. "The Biology of Pentastomids." *Advances in Parasitology* 25 (1986): 46–128.

Riley, J., and R. J. Henderson. "Pentastomids and the Tetrapod Lung." *Parasitology* 119, supplement (1999): S89–105.

Storch, V., and B.G.M. Jamieson. "Further Spermatological Evidence for Including the Pentastomida (Tongue Worms) in the Crustacea." *International Journal of Parasitology* 22 (1992): 95–108.

Walossek, D., and K. J. Müller. "Pentastomid Parasites from the Lower Palaeozoic of Sweden." *Transactions of the Royal Society of Edinburgh, Earth Sciences* 85 (1994): 1–37.

Wingstrand K. G. "Comparative Spermatology of a Pentastomid *Raillietiella hemidactyli* and a Branchiuran Crustacean *Argulus foliaceus* with a Discussion of Pentastomid Relationships. *Biologiske Skrifter* 19 (1972): 1–72.

John Riley, PhD

Pycnogonida
(Sea spiders)

Phylum Arthropoda

Class Pycnogonida

Number of families 8

Thumbnail description
Spiderlike marine creatures that both live on and eat seaweed and some marine invertebrates

Photo: A sea spider (class Pycnogonida) on kelp. (Photo by David Hall/Photo Researchers, Inc. Reproduced by permission.)

Evolution and systematics

Pycnogonids have almost no fossil record, though three genera have been found in the Devonian in the Hunsruck Slate in Germany. Despite the lack of fossil evidence, scientists have deduced from morphological and embryonic studies that sea spiders are an old lineage of animals.

Pycnogonids are an unusual group that has been difficult to place relative to other arthropod groups. Pycnogonids are thought to represent an early divergence from the evolutionary line leading to other Chelicerates. Both pycnogonids and chelicerates have claws on the first appendages and a tubercle with simple eyes, and both lack antennae. Pycnogonids, however, differ by possessing features such as the proboscis, reduced abdomen, and ovigers.

About eight families and 1,000 species of sea spiders are known.

Physical characteristics

Pycnogonids are seldom seen because they are pale or cryptically colored. Most pycnogonids are about 0.39 in (1 cm) or less in size, but some deep-sea forms reach up to 27.5 in (70 cm) across between leg tips. They have extremely reduced bodies in which the abdomen has almost disappeared, while the legs are long and clawed. The head has a long proboscis with a terminal mouth and a single four-part eye on a central stalked tubercle.

The surface area of the thin body and legs allows diffusion of gases and wastes, since pycnogonids lack specialized respiratory or excretory structures. Digestion occurs in the gut,

which sends branches into the long legs. Most species have four pairs of walking legs, but some have five or six pairs, with the reproductive organs located in the joints of the legs. The males, and often the females, have ovigerous legs in addition to the walking legs. These legs are used for holding and carrying eggs during the breeding season and, in females, for cleaning and grooming outside of the breeding season. The body is contained in and supported by a non-calcareous exoskeleton.

Distribution

All pycnogonids are found in the marine environment. They are distributed worldwide, from tropical waters to the poles.

Habitat

Pycnogonids are benthic organisms with a habitat ranging from intertidal zones to depths over 19,865 ft (6,000 m). While most live on seaweed, corals, and sponges, one species has been discovered in the hydrothermal vent communities at the great depths of the Galápagos Rift.

Behavior

Adult sea spiders are solitary but often live in close association with invertebrate food hosts or seaweeds. Species with smaller body size move slowly around the substrate, while the larger deep-sea pycnogonids tend to be more active; some even swim, using leg motions similar to walking.

A sea spider (*Pycnogonum littorale*) feeding on anemone in the Gulf of Maine. (Photo by Andrew J. Martinez/Photo Researchers, Inc. Reproduced by permission.)

Feeding ecology and diet

Sea spiders feed on seaweeds and soft-bodied invertebrates, including hydroids, soft corals, bryozoans, anemones, and sponges. Certain species also prey on opisthobranchs, small polychaetes, and other mobile soft-bodied invertebrates. Some pycnogonids pierce the skin of their hosts with teeth at the tip of the proboscis and suck the juices out of their prey. Others tear their prey apart and pass the pieces into a proboscis for feeding.

Reproductive biology

Male sea spiders often mate with more than one female. During courtship, male pycnogonids use their ovigerous legs to induce the females to release eggs. Once the female begins to lay eggs, the male fertilizes them as the female holds them on her ovigerous legs. In some genera, the eggs are relatively large and the female only releases 4–5. In other genera, the eggs are small and the female releases more than 100 eggs. Depending on the leg type, after fertilization, the male will either gather the eggs one by one onto his legs or hooks his ovigerous legs into the egg mass and gathers most of the eggs into a single mass onto his legs.

Male sea spiders carry cemented egg clutches gathered from females until they hatch; the actual amount of time the male carries the clutches depends on species and geographic location. Upon hatching, larvae have only three pairs of legs and develop an additional pair (or pairs) upon subsequent moltings. The most common larval type is a free-swimming naulpius-like larva. However, in some species, the males continue to provide care of the larvae until after several moltings. In this case, the females produce eggs with a large amount of yolk and the young continue to live off of the yolk while attached to the male's ovigerous legs.

Conservation status

No species are listed by the IUCN.

Significance to humans

Sea spiders have no known significance to humans. They are usually not even seen, or are ignored, by humans.

1. *Colossendeis megalonyx*; 2. *Anoplodactylus evansi*. (Illustration by Joseph E. Trumpey)

Species accounts

No common name
Colossendeis megalonyx

ORDER
No order designation

FAMILY
Colossendeidae

TAXONOMY
Colossendeis megalonyx Hoek, 1881.

OTHER COMMON NAMES
None known.

PHYSICAL CHARACTERISTICS
Body is approximately 0.78 in (20 mm) long and proboscis is 1.2–1.6 times the length of the body, with a broad rounded tip. As members of the genus with the longest leg span, it has legs spanning up to 27.5 in (70 cm), with each leg bearing a long slender claw.

DISTRIBUTION
From depths of 10–16,400 ft (3–5,000 m) throughout circumpolar Antarctica with northern extension into the south Atlantic Ocean, south Indian Ocean, and South Pacific Ocean, including the Antipodes Islands off of New Zealand.

HABITAT
Nothing is known.

BEHAVIOR
Nothing is known.

FEEDING ECOLOGY AND DIET
Eats soft corals and small hydroids attached to sponges.

REPRODUCTIVE BIOLOGY
Nothing is known.

CONSERVATION STATUS
Not listed by the IUCN.

SIGNIFICANCE TO HUMANS
None known. ◆

No common name
Anoplodactylus evansi

ORDER
Pantopoda

FAMILY
Phoxichilidiidae

TAXONOMY
Anoplodactylus evansi Clark, 1963.

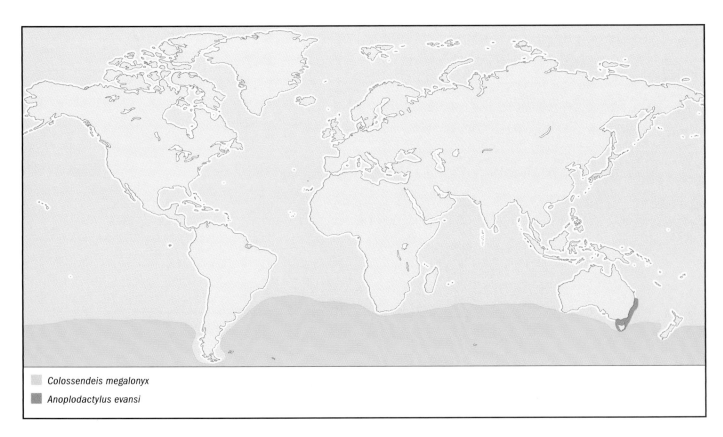

Colossendeis megalonyx
Anoplodactylus evansi

OTHER COMMON NAMES
None known.

PHYSICAL CHARACTERISTICS
Leg span up to 1.2 in (30 mm) across, with four pairs of legs.

DISTRIBUTION
Found along the New South Wales coast of Australia and as far south as Tasmania.

HABITAT
Nothing is known.

BEHAVIOR
Nocturnally active, it seeks shelter in algae during the day. It hunts opisthobranchs on benthic algae and immobilizes them with immoveable claws on the front four legs while the back four legs remain secured to the substrate. It then consumes all of the soft tissue of the prey. While solitary in the wild, it will feed in groups on the same item when captured in a holding tank. It broods its young in seaweed. This behavior limits

dispersal and can lead to resident populations at particular locations.

FEEDING ECOLOGY AND DIET
A generalist predator of small opisthobranchs, including nudibranchs and other soft-bodied invertebrates. Attacks prey up to 5–6 times its body weight. Some of the prey species secrete toxins, which it can tolerate at low levels, including compounds found in sponges, cnidarians, bryozoans, and opisthobranchs. One of its main food sources, juvenile sea hares, secretes ink that contains toxins when attacked. If the sea spider comes into contact with the toxic deterrent, it will vigorously wave the affected limb.

REPRODUCTIVE BIOLOGY
Nothing is known.

CONSERVATION STATUS
Not listed by the IUCN.

SIGNIFICANCE TO HUMANS
None known. ◆

Resources

Books

Child, C. A. *Marine Fauna of New Zealand: Pycnogonida (Sea Spiders)*. Wellington: National Institute of Water and Atmospheric Research, 1998.

Hedgpeth, J. W. "Pycnogonida." In *Treatise on Invertebrate Paleontology*. Part P, *Arthropoda 2: Chelicerata*, edited by R. C. Moore. Lawrence, KS: Geological Society of America and University of Kansas Press, 1960.

Periodicals

Arnaud, Francoise, and Margo L. Branch. "The Pycnogonida of Subantarctic Marion and Prince Edward Islands:

Illustrated Keys to the Species." *South African Journal of Antarctic Research* 21, no. 1 (1991): 65–71.

Rogers, C. N., R. de Nys, and P. D. Steinberg. "Predation on Juvenile *Aplysia parvula* and Other Small Anaspidean, Ascoglossan, and Nudibranch Gastropods by Pycnogonids." *Veliger* 43, no. 4 (2000): 330–337.

Other

Underwater Field Guide to Ross Island and McMurdo Sound, Antarctica. Scripps Institution of Oceanography, University of California, San Diego. [29 July 2003]. <http://scilib.ucsd.edu/sio/nsf/fguide/>.

Elizabeth Mills, MS

Merostomata
(Horseshoe crabs)

Phylum Arthropoda

Class Chelicerata

Subclass Merostomata

Number of families 1

Thumbnail description
Marine creatures distinguished by a large, hard exoskeleton that includes an arched, horseshoe-shaped shield in front (prosoma), a middle portion (opisthosoma), and a thin tail (telson); they are among the oldest living organisms

Photo: A horseshoe crab (*Limulus polyphemus*) trying to turn itself right side up. (Photo by Gilbert S. Grant/Photo Researchers, Inc. Reproduced by permission.)

Evolution and systematics

The subclass Merostomata is one of three branches of the chelicerate line of arthropods; the other two branches include sea spiders and terrestrial spiders. Thus horseshoe crabs are more closely related to spiders and scorpions than to other crabs. Horseshoe crabs date to the Carboniferous period (350 million years ago [mya]). Ancestral relatives from the Cambrian period (550 mya) have been found. Horseshoe crabs are classified into a single order (Xiphosura) and family (Limulidae). Four species are recognized. Many scientists now categorize Merostomata as a class rather than a subclass.

Physical characteristics

The body of a horseshoe crab is covered by a smooth greenish to dark brown exoskeleton. The exoskeleton con-

sists of three major parts: an arched, horseshoe-shaped shield in the front, the prosoma; a middle portion, the opisthosoma; and a thin tail, the telson. The prosoma bears two pairs of simple eyes on the top and a pair of compound eyes on ridges laterally along the outside. Under the exoskeleton, eight pairs of appendages are aligned along the lengthwise axis of the prosoma. The first seven pairs function in feeding. The eighth pair is fused and covers five pairs of book gills in the opisthosoma. The book gills maintain water flow for respiration, movement, and reproduction. Spines protrude from the outer edge of the opisthosoma; the number of spines varies by species. The long, thin telson extends from the back of the body.

Horseshoe crabs must shed their exoskeleton, or molt, to grow. Individuals molt 16 or 17 times during their lives. Six of these molts occur within the first year. As adults, females

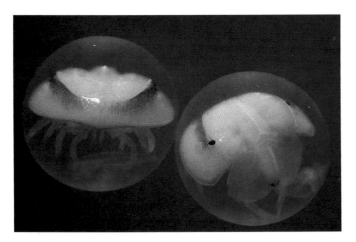

The eggs of a horseshoe crab (*Limulus polyphemus*). (Photo by Kon. Sasaki/Photo Researchers, Inc. Reproduced by permission.)

are larger than males. In the smallest species, *Carcinoscorpius rotundicauda*, females reach 15 in (38 cm) in length and 5 in (12.5 cm) in width. In *Tachypleus tridentatus*, the largest species, females attain a length of 33.5 in (85 cm) and a width of 15.5 in (39.3 cm).

Distribution

Western Atlantic coast and regions of the Indian and Pacific oceans.

Habitat

Horseshoe crabs inhabit saline portions of estuaries or near-shore coastal areas. They often live in coves, bays, or wetlands protected from strong wave action. They remain in sandy or muddy sublittoral areas except when they move onto beaches for spawning.

Behavior

As larvae, horseshoe crabs swim vigorously for hours, but they adopt diurnal activity patterns as juveniles and adults. When resting, horseshoe crabs often bury themselves in shallow burrows. Crawling along the substrate is the primary means of locomotion, but horseshoe crabs sometimes swim upside-down by using the book gills for propulsion. As adults, horseshoe crabs migrate annually from deeper near-shore waters to beaches for spawning. Individuals that are flipped onto their backs use the telson to arch the body and roll over.

Feeding ecology and diet

Larval horseshoe crabs do not feed. Feeding begins after the first juvenile stage is attained. Horseshoe crabs do not have jaws, so they use their legs to grasp and crush prey. Horseshoe crabs scavenge on almost any food items they encounter in the sediment, such as mollusks and worms. They also scrape algae off rocks. Adults are eaten by opportunistic

predators, including sharks, sea turtles, sea gulls, and terrestrial mammals. Most predation occurs on young horseshoe crabs, the larvae and eggs being eaten by fish. The eggs provide an important food source for many shorebirds during spring migration from South America to the Arctic.

Reproductive biology

Horseshoe crabs are long-lived and mature later than other invertebrates. Males mature between 9 and 11 years of age and females, between 10 and 12 years. Horseshoe crabs spawn during the spring and summer. Spawning occurs at high tide on low-energy beaches of estuaries, bays, and coves. One species (*Carcinoscorpius rotundicanda*) moves upstream into rivers to spawn.

During mating, the male grasps the edge of the female's opisthosoma. The female uses her legs and prosoma to dig a nest, into which she deposits a cluster of eggs. The eggs are fertilized by the male, and the pair moves 4–8 in (10–20 cm) farther in the sand and repeats the process. As the female digs the second nest, the excavated sand is pushed backward to cover the previous nest. Individual horseshoe crabs are capable of spawning more than once per season. The eggs hatch into trilobite larvae; after molting into juveniles, horseshoe crabs settle to the seafloor.

Conservation status

No species is listed by the IUCN. However, horseshoe crab populations have declined as the result of harvesting and habitat destruction.

Significance to humans

Horseshoe crabs have been harvested for food and bait. They also have been processed into fertilizer. Perhaps most important, horseshoe crabs have enabled numerous human health advances. Studies of the eyes of horseshoe crabs have led to therapies for human eye disorders. The blood of horse-

Horseshoe crabs (*Limulus polyphemus*) spawning. (Photo by Vanessa Vick/Photo Researchers, Inc. Reproduced by permission.)

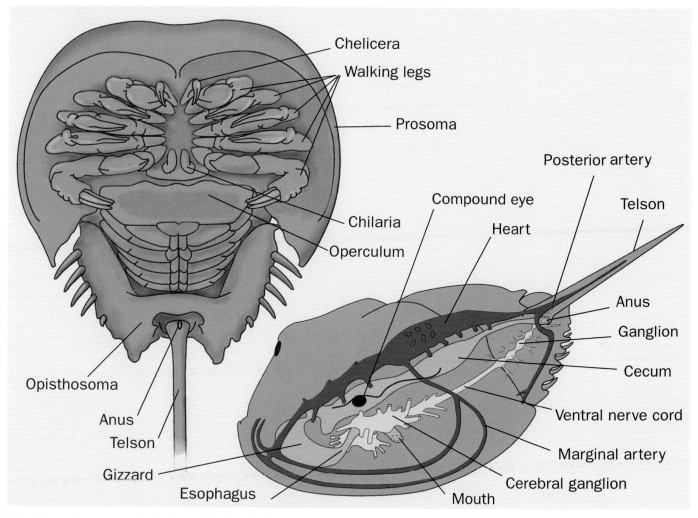

Horseshoe crab anatomy. (Illustration by Christina St. Clair)

shoe crabs forms a substance, Limulus Amebocyte Lysate (LAL), that is used to identify gram-negative bacteria in medical fluids and drugs and on surgical devices. Nontoxic and biodegradable chitin from horseshoe crabs is used in prod-

ucts such as contact lenses, surgical sutures, and skin lotion. The chitin forms a chemical that removes metals and toxins from water, and its fat-absorbing properties help remove fat and cholesterol from the human body.

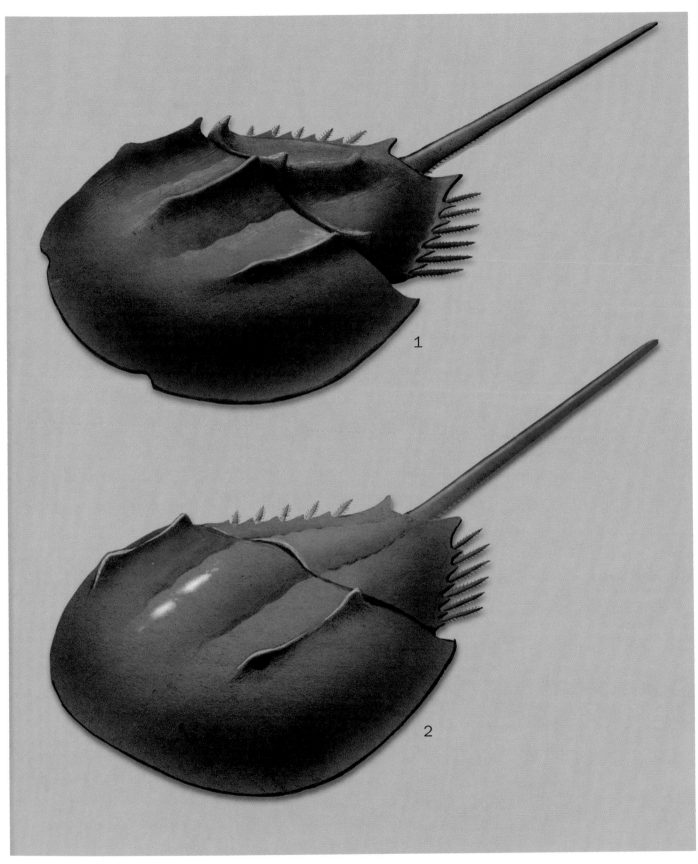

1. Japanese horseshoe crab (*Tachypleus tridentatus*); 2. American horseshoe crab (*Limulus polyphemus*). (Illustration by John Megahan)

Species accounts

American horseshoe crab
Limulus polyphemus

ORDER
Xiphosura

FAMILY
Limulidae

TAXONOMY
Limulus polyphemus Linnaeus, 1758.

OTHER COMMON NAMES
English: Atlantic horseshoe crab, horseshoe crab, king crab.

PHYSICAL CHARACTERISTICS
The American horseshoe crab is a large species. Males are smaller than females. Mean body length is approximately 14 in (35.7 cm) for males and 17 in (43.8 cm) for females. The edge of the carapace is domed. The prosomatic carapace forms a circular arch, whereas the opisthosomatic carapace is hexagonal and elongated toward the posterior of the body. This species has large compound eyes. The American horseshoe crab is greenish brown to blackish brown.

DISTRIBUTION
Atlantic coast of North America, from Long Island to the Yucatan Peninsula.

HABITAT
To a depth of more than 200 ft (60 m) in coastal areas; sandy beaches for spawning.

BEHAVIOR
The American horseshoe crab lives in deeper offshore waters during the winter and migrates into shallow coastal waters as the spawning season approaches. Adults range a maximum of 22–25 mi (35–40 km). Juveniles move into marine waters offshore of the natal beach at the end of the first summer.

FEEDING ECOLOGY AND DIET
The diet consists largely of bivalve mollusks, such as clams, and polychaete worms. Gulls feed on adult horseshoe crabs that become stranded on beaches during spawning, and many types of migratory shorebirds consume the eggs.

REPRODUCTIVE BIOLOGY
Mature individuals migrate to sandy beaches to spawn in the spring. As the tide rises, males approach the beaches in large groups. Females follow and couple with the males. Spawning occurs mostly at night and near the high-tide line. Females bury approximately 20,000 eggs in a series of clusters that are fertilized by males. The eggs hatch into trilobite larvae after 13–15 days.

CONSERVATION STATUS
Not listed by the IUCN.

Limulus polyphemus

Tachypleus tridentatus

SIGNIFICANCE TO HUMANS
In the United States, the American horseshoe crab is harvested as bait for the conch and eel fisheries. From 1850 until the 1970s, the horseshoe crab was processed for fertilizer. The blood is used to make LAL for detecting gram-negative bacteria. ◆

Japanese horseshoe crab
Tachypleus tridentatus

ORDER
Xiphosura

FAMILY
Limulidae

TAXONOMY
Tachypleus tridentatus Leach, 1891.

OTHER COMMON NAMES
English: Eastern horseshoe crab, three-spine horseshoe crab

PHYSICAL CHARACTERISTICS
The Japanese horseshoe crab has a large body and relatively small compound eyes. Males average approximately 20 in (51 cm) in length, and females are approximately 23.6 in (60 cm) long. The prosomatic carapace is domed. The opisthosomatic carapace is hexagonal but not strongly elongated, and spines protrude from its margins. This horseshoe crab is greenish gray.

DISTRIBUTION
Discrete coastal areas in the Indian and Pacific oceans—Inland Sea and North Kyushu, Japan; south of the Yangtze River in China; southwest Vietnam; Philippines; Borneo; North Celebes; northern Sulawesi; northeast Java and southwest Sumatra.

HABITAT
Deep water in winter, shallow coves in spring; muddy or sandy substrate.

BEHAVIOR
Larvae spend the first winter in the sand of the natal beach. Young then migrate to nearby mud flats. After becoming approximately 6 in (15 cm) long, individuals move into the ocean. They migrate from deeper offshore waters during the winter into shallow coastal waters in preparation for the summer spawning season.

FEEDING ECOLOGY AND DIET
The Japanese horseshoe crab eats invertebrates it encounters as it moves along the substrate. Predation on adults is minimal; fish and birds may consume eggs.

REPRODUCTIVE BIOLOGY
The Japanese horseshoe crab spawns near the high-tide line on sandy and gravel beaches during evenings in July and August. Males and females form pairs before approaching the beach. The pair deposits and fertilizes approximately 20,000 eggs divided among as many as 10 nests. Eggs hatch after approximately five weeks, and trilobite larvae emerge.

CONSERVATION STATUS
Not listed by the IUCN.

SIGNIFICANCE TO HUMANS
The blood is used to detect gram-negative bacteria. The Japanese horseshoe crab is sold for human consumption in several Asian countries, where it is considered a delicacy. ◆

Resources

Books
Tanacredi, John T., ed. Limulus *in the Limelight: A Species 350 Million Years in the Making and in Peril?* New York: Kluwer Academic/Plenum, 2001.

Other
Smithsonian Marine Station at Fort Pierce. "*Limulus polyphemus* (Horseshoe crab)." 25 July 2001 [12 Aug. 2003]. <http://www.sms.si.edu/IRLSpec/Limulu_polyph.htm>.

Katherine E. Mills, MS

⬠ Arachnida

(Spiders, scorpions, mites, and ticks)

Phylum Arthropida

Class Chelicerata

Subclass Arachnida

Number of families 648

Thumbnail description
Highly recognizable and populous eight-legged invertebrates with two body parts (a prosoma and an abdomen), pedipalps, book lungs or tracheae, sometimes poisonous fangs, and generally the ability to produce silk; they are terrestrial chelicerates (invertebrates with pincer-shaped mouthparts)

Photo: A smooth-headed scorpion (*Opisthopthal-mus*) in defensive posture. (Photo by Ann & Steve Toon Wildlife Photography. Reproduced by permission.)

Evolution and systematics

Fossil records suggest that arachnids were among the first animals to live on land, switching from water- to air-breathing. The oldest known arachnid fossils date from the Silurian Period, more than 417 million years ago. It is during this period of time that scorpions (order Scorpionida) appear to have left the water for life on land. Many paleontology experts presume that scorpions were the first animals to make the transition from water to land. In fact, the histological resemblance between the gills of king crabs and the lungs of scorpions help to support this hypothesis. However, the subphylum Chelicerafformes, as a whole, spent many millions of years in the water before it became terrestrial. More than 60,000 species of arachnids are described, although many species, especially mites, remain undiscovered or discovered-but-not-yet-described. Spiders, mites, and ticks constitute the largest and most diverse orders of arachnids. Among the extant species, scorpions are known to have had a long maritime history that continued well after some of them switched to living on land. The marine-living scorpions, at that time, were very large, some up to 3.3 ft (1 m) in length. The harvestmen (daddy longlegs) are also believed to have had a pre-terrestrial history in the sea.

Currently, arachnids constitute the subclass Arachnida, in the phylum Arthropoda. The subclass is divided into 11 distinct orders: Acari (mites, chiggers, and ticks), Amblypygi (tailless whip scorpions), Araneae (spiders), Opiliones (daddy longlegs), Palpigradi (palpigrades), Pseudoscorpiones (false scorpions), Ricinulei (ricinuleids), Schizomida (micro whip scorpions or schizomids), Scorpionida (scorpions), Solpugida (wind scorpions or solifugids), and Uropygi (whip scorpions and vinegaroons). Many scientists now categorize Arachnida at the class level.

Physical characteristics

There are at least 10 features of arachnids that are often used to describe the group, including:

- carapace may be uniform or in part segmented

- pedicel may be absent or present

- sternum may be uniform or segmented

- opisthosoma may be uniform or segmented

- chelicerae may contain two or three segments (podomeres)

- pedipalpi may be pincer-like or leg-like

- coxae of legs or pedipalpi may or may not contain gnathobases (plate-like anterior expansions)

- first leg may be used as a leg or like an antenna

- legs may be of seven segments (podomeres) or may be sub-segmented anywhere

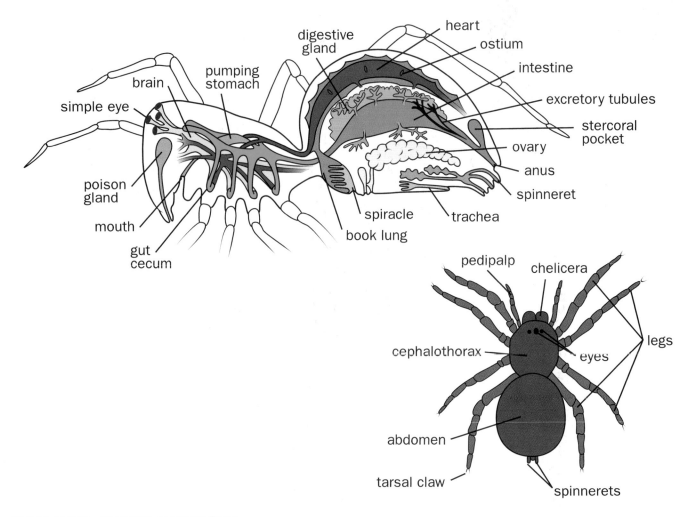

Arachnid anatomy. (Illustration by Patricia Ferrer)

- coxae may meet and hide the sternum or may be separated

Anatomical features such as two pairs of limbs, the pedipalps and chelicerae, are distinctively present but greatly modified for different uses in various arachnid species.

The 18-segment arachnid body is often protected by sternites below and tergites above, connected by a soft pleural membrane, and is divided into two tagmata: anterior and posterior. The anterior (front) part, called the cephalothorax (or prosoma), contains sense organs, mouthparts, and limbs or appendages in pairs. The cephalothorax is composed of an anterior, unsegmented region called the acron, and six true segments (each bearing a pair of appendages). It accommodates both the head and limbs. A carapace-like shield completely or partially covers the cephalothorax of arachnids. The first pair of limbs (chelicerae) attached in front of the mouth may form pincers or poison fangs, and the second pair (pedipalps) behind the chelicerae may serve as pincers, feelers, or additional legs. The other limb pairs are used for walking. The 12-segment posterior (rear) part of the body, the abdomen (or opisthosoma), contains the genital opening and other structures. The abdomen may be segmented (as in scorpions) or unsegmented (as in most ticks and spiders). Abdominal appendages are either lacking, or modified into special organs such as the spinnerets of spiders and the pectines of scorpions.

Arachnids breathe by means of tracheae (windpipes), book lungs (modified gills), or both. The mouth of arachnids is not readily noticeable from the external surface. They do not possess jaws (mandibles), but instead have cutting or piercing appendages called chelicerae. The open circulatory system distributes blood from the heart to an enlarged blood space by the use of arteries. The heart is a tubular organ located dorsal to the mid-gut, containing various openings so that blood can be returned to the heart. The central nervous system consists of two cerebral ganglia connected to a pair of sub-esophageal ganglia by means of a circum-esophageal linkage (commisure). Arachnids possess a number of sense organs, many related with the outer body covering (cuticle). The most common of these sense organs is the hair-like setae that are sensitive to various stimuli; they are generally located throughout the surface of the body.

Distribution

Arachnids are found throughout the world from equatorial to polar regions, but reach their most abundant numbers and diversities in very warm to hot, arid and tropical/subtropical regions.

Habitat

Arachnids are essentially terrestrial animals that are found in nearly every habitat around the world.

Behavior

Arachnids are terrestrial, except for some mites and a few spiders that can still be found in water. Most arachnids are solitary creatures, other than during mating periods. Even normally sedentary species will roam when in search of a mate. A courtship ritual usually precedes reproduction. A large proportion of their lives are spent in long periods of inactivity, often waiting for prey to stumble upon them. When disturbed by possible danger, they often fall motionless, acting dead, and try to appear nearly invisible to approaching enemies. Regular activities are instinctive by nature, geared primarily toward perpetuating the particular species and activated by external circumstances (such as the general environment and light intensity) and internal adaptations that have been modifying over thousands of years. Some species ambush their prey, while others chase them down. They feed on specialized prey or on many different types of food, depending on species. Arachnids also feed in various ways: as herbivores, scavengers, parasites, cannibals, and carnivores.

Feeding ecology and diet

Arachnids are predaceous, either actively hunting or patiently lying in wait for small animals such as insects. They have various structures that are geared to capturing prey. Some of these features are the segmented, stinging tail of scorpions and the abdominal spinnerets (that allow for the construction of insect traps, or webs) of spiders. Since they do not have the ability to masticate (chew) their food with their mouthparts, they are generally able only to feed on the fluids of their prey. After piercing the prey's body wall with their chelicerae, arachnids will either ingest the fluid contents or digest the tissues externally with enzyme-containing secretions that are ejected from the mid-gut (as with spiders) or the salivary glands (as with ticks and mites). A powerful suctorial pharynx draws the fluid up through the pre-oral food canal and delivers it into the mid-gut. Gaseous exchange occurs in a variety of ways. Respiratory gases may enter and leave the body through specialized structures (either lung-books or spiracles) or may diffuse through the cuticle (as in some mites and larval ticks).

Reproductive biology

During mating, a variety of complex behavior patterns are normally observed. Generally, the reproductive organs are

Neon green opiliones seen in Kodagu, Karnataka, India, on a cardamom plantation. Opiliones are also called "daddy longlegs," or "harvestmen," but this also refers to a group of spiders, so this common name is misleading. (Photo by A. Captain/R. Kulkarni/S. Thakur. Reproduced by permission.)

contained in the abdomen and open ventrally on the second abdominal somite. Male sex organs may consist of one diffuse testis or one or two compact testes. The spermatozoa produced are conveyed to a median gonopore through one or two excretory ducts (vasa deferentia). Insemination into the female may come from the male gonopore in a liquid medium (as in spiders) or may be contained in packages called spermatophores (as in ticks and scorpions). An intermittent organ or penis may or may not be present to direct the spermatozoa into the female during mating. Females possess a single or paired ovary, which may be either compact or diffuse and one or two oviducts may lead to the median gonopore. Eggs may be laid underground, in the shelter of a stone, under tree bark, enclosed in a cocoon, or other variations of these methods and structures. Females usually guard eggs or young, which are often born live and as miniatures of the adult with regard to appearance. Eggs may number from one to more than 1,000 in a single brood.

Conservation status

As a group, arachnids are considered abundant all over the world. Some species are diminished in numbers, even considered rare or endangered, because of internal circumstances (such as limitations of habitat) or external circumstances (such as human activities). The 2002 IUCN Red List includes 18 arachnid species: one as Endangered; nine as Vulnerable; one as Lower Risk/Near Threatened; and seven as Data Deficient.

Significance to humans

Most arachnids are harmless and contribute to the give and take of nature by controlling the populations of the insects they prey on or the plants, reptiles, birds, or mammals that serve as their hosts. A few species are serious agricultural pests. The bites of some spiders, such as the black widow spider and the brown recluse spider, and the stings of a few species of scorpions are dangerously poisonous to humans.

1. *Phrynus parvulus*; 2. Striped scorpion (*Centruroides vittatus*); 3. Demodicid (*Demodox folliculorum*); 4. Rocky Mountain wood tick (*Dermacentor andersoni*); 5. *Ricinoides afzelii*; 6. Book scorpion (*Chelifer cancroides*); 7. Giant whip scorpion (*Mastigoproctus giganteus*); 8. Emperor scorpion (*Pandinus imperator*). (Illustration by Bruce Worden)

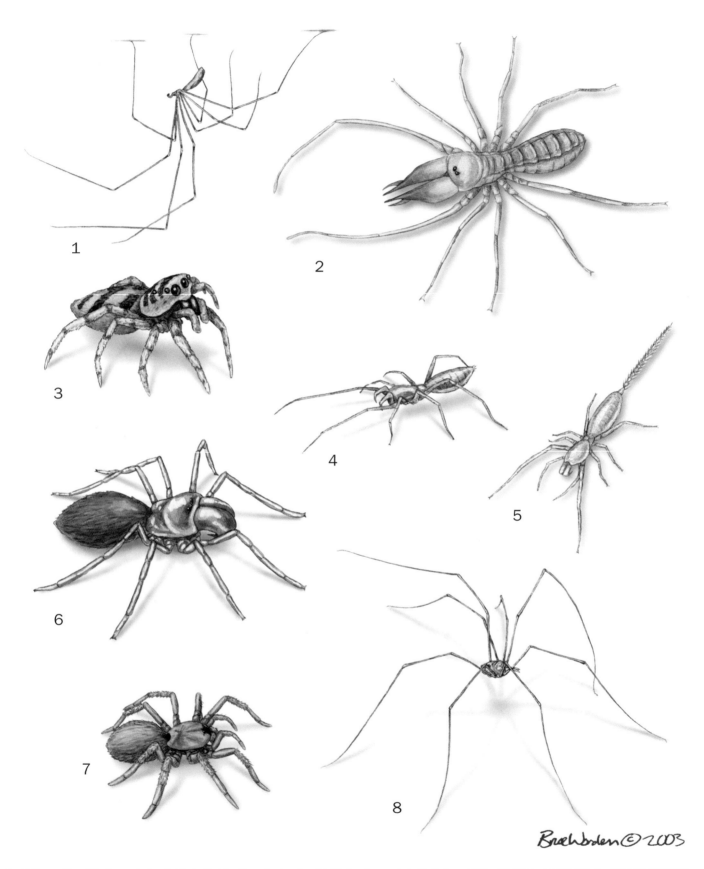

1. Cellar spider (*Pholcus phalangioides*); 2. Egyptian giant solpugid (*Galeodes arabs*); 3. Zebra spider (*Salticus scenicus*); 4. *Agastoschizomus lucifer*; 5. *Eukoenenia draco*; 6. *Atypus affinis*; 7. Spruce-fir moss spider (*Microhexura montivaga*); 8. Harvestman (*Phalangium opilio*). (Illustration by Bruce Worden)

Species accounts

Demodicid
Demodex folliculorum

ORDER
Acari

FAMILY
Demodicidae

TAXONOMY
Demodex folliculorum Berger, 1841.

OTHER COMMON NAMES
English: Face mite, hair follicle mite.

PHYSICAL CHARACTERISTICS
Microscopic, elongated parasitic mite that is 0.00394–0.0178 in (0.1–0.4 mm) in length; worm-like appearance, with distinct head-neck part and body-tail part; long, tapering annulated abdomen. Adults possess four pairs of short legs (basically stumps) on head-neck part; legs contain tiny, but strong claws. Body is mostly semi-transparent. Needle-like mouthparts are used for eating skin cells; it is covered by cuticle surface that shows numerous striations. Bodies are layered with scales, which keeps them secured to follicles of hosts. Does not have excretory opening because its digestive system produces very little waste. Peritremes are absent, and palps (pedipalpi) are reduced in number and size.

DISTRIBUTION
Worldwide.

HABITAT
Lives in hair follicles of humans; primarily on pores of facial skin and sebaceous glands of forehead, nose, and chin; often in roots of eyelashes. However, it may inhabit follicles, with or without hair, anywhere on body. More common on humans with oily skin or those who use excessive amounts of cosmetics and fail to cleanse skin properly.

BEHAVIOR
Lives in hair follicles and eyelashes with heads buried first into root. Migrates onto skin during nighttime at rate of 0.4 in (1 cm) per hour.

FEEDING ECOLOGY AND DIET
Parasitic, eating skin cells of humans.

REPRODUCTIVE BIOLOGY
Females may lay up to 20–25 oval eggs on one hair follicle. Larvae (protonymph) and nymphs (deutonymph), with physical features similar to adults, are swept by sebaceous flow to mouth of follicle. First stage larvae emerge without legs. Larvae in later stages have six legs, as opposed to eight for adults. As immature mites grow, they become tightly packed. When mature, mites leave follicle, mate, and find new follicle in which to lay eggs. Entire life cycle spent on host: 14–18 days.

■ *Atypus affinis*
■ *Demodex folliculorum*

CONSERVATION STATUS
Not listed by the IUCN.

SIGNIFICANCE TO HUMANS
Acute and chronic inflammation and infection often result when large numbers congregate in single follicle. If too many mites are buried in same follicle, the eyelash may fall out. It is basically harmless and is not known to transmit diseases, but large numbers may cause itching and skin disorders, referred to as demodicosis. The incidence of demodicosis occurs worldwide, being more prevalent in older humans. When humans are infested, they may show no symptoms. ◆

Rocky Mountain wood tick
Dermacentor andersoni

ORDER
Acari

FAMILY
Ixodidae

TAXONOMY
Dermacentor andersoni Stiles, 1908.

OTHER COMMON NAMES
English: Paralysis tick, Rocky Mountain spotted fever tick.

PHYSICAL CHARACTERISTICS
Includes a large collection of diverse types of mites, chiggers, and ticks; carries parasites and disease; has adult length of 0.08–0.65 in (2.1–16.5 mm). Specifically, unengorged females have length of 0.11–0.21 in (2.8–5.4 mm), adult males a length

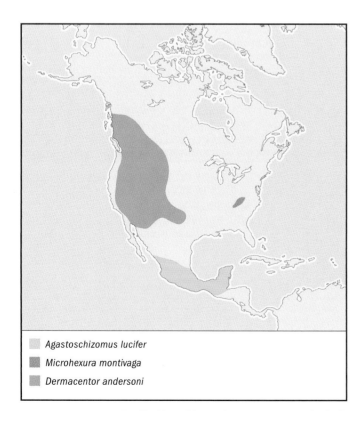

Agastoschizomus lucifer

Microhexura montivaga

Dermacentor andersoni

of 0.08–0.24 in (2.1–6.1 mm), and engorged females a length of up to 0.65 in (16.5 mm) and a width of up to 0.45 in (11.4 mm). The body, covered with hard protective covering, is pear shaped, and is dorsoventrally flattened (top to bottom). Immature instars and adult females possess a strong pro-dorsal sclerite, with the opisthosoma being covered with soft cuticle to permit engorgement. Holodorsal shield is found on adult males; posterior idiosoma is flattened, anterior gnathosoma is articulated. Tectum is well developed. Coxal glands are absent, and water control is performed through salivary glands.

Color of adult females is reddish brown with grayish white dorsal shield (scutum) near front of body, which changes to grayish color when engorged; adult males are spotted with brown and gray, and do not have distinctive white shield; they possess simple eyes, located on the margin of scutum. Capitulum is apparent from above; basis capitulum is rectangular in shape with sides not laterally produced and approximately length of mouthparts. There are 11 abdominal festoons. Anal groove is located posterior to anus. Broad spiracular plates, located on underside of body, possess blunt process that usually reaches dorsum; goblets, located within spiracular plates, are moderate in size and number.

DISTRIBUTION
Widely distributed in North America, primarily throughout Rocky Mountain states and into southwestern Canada. Specific areas are central British Columbia through southern Alberta into southwestern Saskatchewan; south through eastern Washington, Oregon, and California; all of Montana, Idaho, Wyoming, Nevada, and Arizona; western Oklahoma to northern New Mexico and Texas.

HABITAT
Mostly woods and meadows; those arid, brushy areas that provide food and protection for its usual hosts such as livestock, wild mammals, and humans. Over-winters in ground debris.

BEHAVIOR
Usually attached to its host, but when without host it will hide in cracks and crevices or in soil. If without a host at the beginning of winter, it will over-winter under groundcover to resume seeking host in spring. It will also stop looking for hosts during hottest summer months. Generally, adults will climb to top of grasses and low shrubs to attach to hosts that brush against them. Host attachment is accomplished by secreting cement-like substance around mouthparts and inserting it into host.

FEEDING ECOLOGY AND DIET
Adults generally feed for less than one hour at a time; parasitic, feeding primarily off of terrestrial birds, reptiles, and mammals and typically feeding only from late February until mid-July. All three stages (larva, nymph, and adult) can survive for more than one year without feeding. Engorged larvae, nymphs, and unfed adults normally spend cold months in grasses and leaf litter. Larvae feed throughout the summer, with nymphs continuing possibly to late summer. Males feed for about five days without engorging, become sexually mature and ready to mate, and then will resume feeding. Females feed for up to seven days (until fully engorged), during which time they mate. Fully engorged female will increase body weight from about 0.000176 oz (5 mg) to more than 0.0247 oz (700 mg). Each stage feeds on a unique host individual. Larvae, nymphs, and adults climb grass stems and bushes while searching for host. Can detect the presence of chemicals such as carbon dioxide associated with mammals, which indicates animal's presence.

REPRODUCTIVE BIOLOGY

Semelparous (reproduces once during lifetime, after which it dies). Requires blood meal before developing into its next life stage and for egg development. Mating takes place primarily on host, with female usually on top of male. Males do not become engorged. After feeding for 4–17 days, mated female descends from host and seeks protected area to lay eggs. In spring, after a preoviposition period of usually 3–11 days, she lays single cluster of usually 3,000–5,500 (but possibly 2,500–7,400) yellowish brown ellipsoidal eggs over period of 10–33 days. Female then dies within 1–14 days. During next 7–38 days, eggs hatch if temperature is 72–90°F (22–32°C).

Young six-legged larvae begin crawling in search for small rodent host (such as mice, voles, and chipmunks), dying within 30 days if unsuccessful. Unengorged first instar larva is about 0.0236 in (0.6 mm) in length. Usually feeds for 2–8 days (usually three) to engorgement, and then drops to ground to molt within 6–21 days. May survive for more than 300 days if unfed during this time. After finding suitable small- to medium-sized host (such as rabbits, ground squirrels, marmots, and skunks), nymphs reach engorgement in 3–11 days. Second instar eight-legged nymphs are 0.0551–0.0591 in (1.4–1.5 mm) in length. After completing engorgement, they drop off again and molt into adults usually in 14–15 days (possibly in 12–120 days). Adults can survive more than a year (usually about 600 days) unfed, but after finding a suitable medium- to large-sized host (such as dogs, deer, and humans), they mate on host after partial feeding. Life cycle is 1–3 years (typically 20 months), depending primarily on host availability and various environmental stresses and conditions.

CONSERVATION STATUS

Not listed by the IUCN.

SIGNIFICANCE TO HUMANS

Females may carry and transmit several diseases to humans, including Rocky Mountain spotted fever, tularemia, and Colorado tick fever in the United States, with only rare occurrences in Canada. ◆

No common name

Phrynus parvulus

ORDER
Amblypygi

FAMILY
Tarantulidae (= Phrynidae)

TAXONOMY
Phrynus parvulus Pocock, 1902.

OTHER COMMON NAMES
None known.

PHYSICAL CHARACTERISTICS

Whip spiders or tailless whip scorpions; a typical individual grows to 1.2 in (3 cm) in length, and does not have spinnerets or poison (venom) glands. Possesses one spine in between two longer spines on dorsal surface of pedipalpal tibia. Pedipalpi (palps) take on basket-like shape for capturing prey. Young usually have reddish palps with banded legs, while adults are colored more uniformly. First pair of elongated, whip-like legs has evolved from walking appendages to sensory defensive/tactical appendages. Antenniform first legs are covered with multiporous hairs used as olfactory sensilla. Legs lack pulvilli on

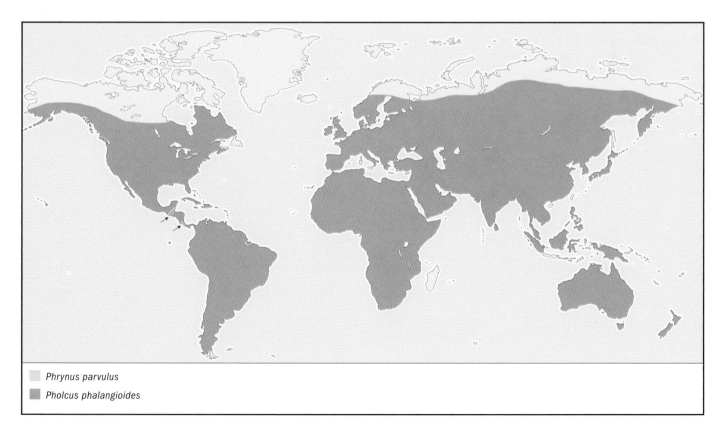

☐ *Phrynus parvulus*
■ *Pholcus phalangioides*

tips. Prosoma is wider than long, and covered by carapace. Opisthosoma is segmented and lacks telson and defensive glands. Prosoma and opisthosoma are connected by narrow pedical. Chelicera has basal article and fangs that are spider-like. Pedipalpal coxal endites are not fused, while raptorial distal articles possess strong spines and fold against spiny femur. There is a labium. Abdominal ganglia are moved to prosoma and fused with subesophageal ganglion. There is a pumping pharynx and a pumping stomach, and two pairs of book lungs.

DISTRIBUTION
Belize, Guatemala, and Costa Rica.

HABITAT
Found in cracks and crevices between rocks; under loose bark, logs, and litter; sometimes in caves, usually in forest environments. Humid tropics and subtropics are usually preferred locations.

BEHAVIOR
Nocturnal; during day, it generally lives in wide variety of places that are usually associated with base of tree trunks. Some places can be within buttressing near tree bases, deep in crevices on trunks, in burrows (dug by other animals) or other holes at base of large trunks, and sometimes behind bark. Typically, only one animal resides in a crevice; however, one male and one female may live together. One tree trunk may house many individuals. Males change resting location more often than females; it is assumed that females may be more sedentary and males may wander more in search of mates; moves with a sideways motion. Generally solitary, but majority of interactions occurs between two aggressive males or between males and females residing in same crevice or home spot. Rather docile creature that is easily frightened. When responding to threat, will use speed for escape, rarely remaining to fight enemy. If necessary, it will use palps for pinching when disturbed or under attack by enemies.

FEEDING ECOLOGY AND DIET
Predatory, eating such small creatures as crickets, moths, and millipedes. Generally avoids eating scorpions, centipedes, large ctenids (wandering spiders), and almost all ants. Usually begins to feed at dusk, not moving very far from home, and returning at dawn. Often sits quietly on vertical trunks of trees waiting for insects to pass by. Pedipalps are used for capturing prey.

REPRODUCTIVE BIOLOGY
Males often perform a ritualized posture when females are not present; involves opening one palp, and raising and holding bodies in air. When females are present, mating males engage in combat with other males by locking their chelicerae and palps together to fight, sometimes for periods of over one hour. Males who are accepted by females will deposit spermatophore after courting and then guide female to it. Apparently has a continuous breeding season, with females being observed with eggs throughout year. Once mated, females may wait a few months before laying eggs. Once laid, eggs number 20–40. Two to three broods occur each year. Female carries eggs underneath abdomen for 3.0–3.5 months. After hatching, young (prenymphs) crawl onto mother's back (abdomen) and stay there for 6–8 days; then they molt and leave. Young will molt 5–8 times in first year before reaching maturity. Thereafter, molting occurs 1–2 times each year, and growth continues throughout life, which often lasts many years.

CONSERVATION STATUS
Not listed by the IUCN.

SIGNIFICANCE TO HUMANS
It is quite harmless to humans, not even possessing any poison (venom) glands. ◆

Purse web spider
Atypus affinis

ORDER
Araneae

FAMILY
Atypidae

TAXONOMY
Atypus affinis Eichwald, 1830.

OTHER COMMON NAMES
English: Trap-door spider.

PHYSICAL CHARACTERISTICS
Has a poisonous bite, is skilled in silk manufacturing, is highly intelligent and adaptive, and is significant as insect predator; has large jaw. Length is 0.28–0.59 in (7–15 mm), with female length of 0.39–0.59 in (10–15 mm) and male length of 0.28–0.35 in (7–9 mm). Adults have glossy olive-gray legs and a reddish brown abdomen; massive abdomens, with short legs and pickaxe-type fangs. Has three tarsal claws, and no claw tufts. Possesses pedipalpal coxae, with well-developed endites (median lobes). Labium is fused with sternum, eyes are closely grouped, and there are six spinnerets.

DISTRIBUTION
Northwest Europe, especially England.

HABITAT
Lives in temperate grasslands, shrub-lands, meadows, and pastures, often in sandy or chalky areas. Usually found in densely woven, silken tubes (or "purses") sealed at both ends. The subterranean tubes, about 10 in (25 cm) long and thickness of small finger, are used as shelter and as tool to capture prey. Main part of tube is in the ground about 5.9–7.9 in (15–20 cm) in length, but it can reach up to about 20 in (50 cm). Part that sticks up above the surface is usually about 2 in (5 cm) in length, and normally located near a tree trunk. Tubes are well camouflaged with sand and various debris.

BEHAVIOR
Digs hole in ground, up to 19.7 in (50 cm) deep, and lines it with silk. Aboveground, tube extends for several inches (centimeters). Tube is covered with sand and debris, which makes it difficult for predators to see, but easy for spider to capture prey. Often lies at bottom of tube, folded up in a compact shape. Because it hides so deep in ground, it is difficult to find. It is fearful, often unable to move when disturbed by unforeseen circumstances. When in fear, it often sticks out fangs. It often lives in colonies.

FEEDING ECOLOGY AND DIET
Detects flies and other small insects walking on their silken tubes. Climbs up tube and underneath insect, sticks its fangs through tube wall, and drags prey inside and down into vertical tube to be eaten. Returns later to surface, throws out remains of prey, and repairs hole.

REPRODUCTIVE BIOLOGY

Mating takes place in tube and they stay together for several months. Then male dies and is eaten by female. Females lay 80–150 eggs in a cocoon resembling a small white bag. Spiderlings take one year to become full-grown and four years to reach maturity. Females may live for more than eight years, with usual range of 2–8 years, slightly less for males.

CONSERVATION STATUS

Not listed by the IUCN.

SIGNIFICANCE TO HUMANS

None known. ◆

Spruce-fir moss spider

Microhexura montivaga

ORDER
Araneae

FAMILY
Dipluridae

TAXONOMY

Microhexura montivaga Crosby and Bishop, 1925, Mount Mitchell, North Carolina, United States.

OTHER COMMON NAMES
None known.

PHYSICAL CHARACTERISTICS

One of smallest spiders, with adults of length 0.10–0.15 in (0.25–0.38 cm). Ranges in color from light brown to darker reddish brown. No markings on abdomen. Carapace is mostly a yellowish brown. Chelicerae are projected forward beyond anterior edge of carapace. Possesses pair of extremely long posterior spinnerets. Second pair of book lungs, which appear as light areas, is posterior to genital furrow.

DISTRIBUTION

Found only at the highest mountain peaks, at and above 5,400 ft (1,645 m) in elevation, in the southern Appalachian Mountains of western North Carolina and eastern Tennessee (Untied States). Recorded from Mount Mitchell, Yancey County, North Carolina; Grandfather Mountain, Watauga, Avery, and Caldwell Counties, North Carolina; Mount Collins, Swain County, North Carolina; Clingmans Dome, Swain County, North Carolina; Roan Mountain, Avery and Mitchell Counties, North Carolina, and Carter County, Tennessee; Mount Buckley, Sevier County, Tennessee; and Mount LeConte, Sevier County, Tennessee. Experts believe that the Mount Mitchell population has been killed off.

Ongoing surveys show that reproducing populations still survive on Grandfather Mountain in North Carolina, but are restricted to small areas of microhabitat. Both the Mount Collins and Clingmans Dome populations, if still present, are extremely small. On Roan Mountain, scattered occurrences have been found at small rock outcrop sites. At Mount Buckley, population is restricted to scattered areas of microhabitat on separate rock outcrop sites within an area of 0.5 acres (0.2 ha) in size. At Mount LeConte, research indicates that the healthiest of the surviving populations occur in four small, separate areas of rock outcrop sites.

HABITAT

Inhabit damp but well-drained moss and liverwort mats that grow on completely shaded rocks or boulders in mature, high-elevation coniferous (red spruce and Fraser fir) forests. Cannot tolerate extremes of moisture, and excessive gain or loss of moisture within body. The mats cannot be too dry (it is very sensitive to desiccation) or too wet (large drops of water can also pose a threat to it). As a result, it builds tube-shaped webs in interface between mat and rock surface (although sometimes extends into interior of mat) to control amount of moisture within surroundings. Tubes are thin-walled and typically broad and flattened with short side branches.

BEHAVIOR

Little information is known on its behavior.

FEEDING ECOLOGY AND DIET

Little information has been collected on feeding habits. No record of prey having been found in webs, nor has it been observed taking prey in the wild, but abundant springtails (tiny, wingless insects) in moss mats provide most likely source of food.

REPRODUCTIVE BIOLOGY

Little is known about its breeding habits, lifecycle, or life span.

CONSERVATION STATUS

Not listed by the IUCN. It is considered Endangered in its entire range by the U. S. Fish and Wildlife Service. It was listed as Endangered under the U. S. Endangered Species Act in February 1995, after research showed that its population size and distribution was limited to only four sites, with only one stable site left. Its populations are believed to be diminishing because of rapid decline of damp, high-elevation old-growth forest habitats (especially the Fraser fir); decline brought about by infestation of exotic insect (balsam wooly adelgid) that has been killing off fir and spruce trees, air pollution brought about by acid rain, and past land use.

SIGNIFICANCE TO HUMANS

Not known to be commercially valuable; however, because of its rarity, it is believed that collectors may seek it out. ◆

Zebra spider

Salticus scenicus

ORDER
Araneae

FAMILY
Salticidae

TAXONOMY

Salticus scenicus Clerck, 1757.

OTHER COMMON NAMES
English: Zebra jumping spider.

PHYSICAL CHARACTERISTICS

Relatively small- to medium-sized spider, with adult female length of 0.20–0.28 in (5–7 mm) and adult male length of 0.20–0.32 in (5–8 mm). Considered one of most common and well-known salticid spiders. Has very acute vision with distinctive eye arrangement of eight simple eyes (three rows of 4, 2, and 2) that enable it to focus in all directions. First median pair of eyes is largest, located on front of cephalothorax, look forward, and called "headlight" eyes. Posterior eyes are smallest in size, located on top of cehalothorax and look upward. Eyes can move in or out for focusing, and can turn up and

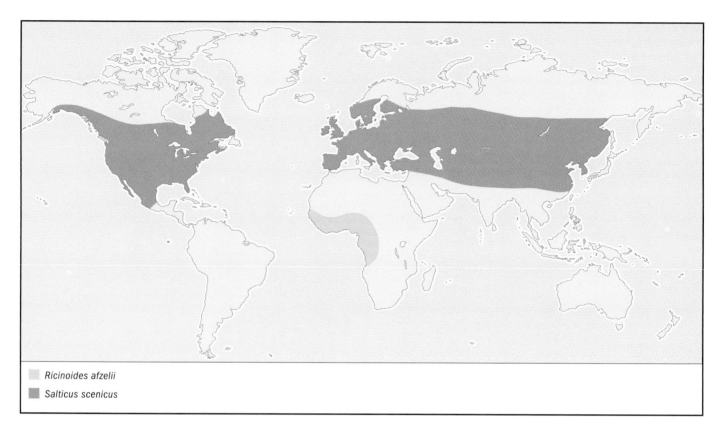

Ricinoides afzelii

Salticus scenicus

down, and left and right for 360° eyesight (called "integral binocular vision"). Nuclei of retinal cells of anterior eyes have evolved to side, out of the path of light. Can also turn its carapace more than 45° to look around. Considered to have best vision of any arthropod, especially where anterior median eyes are concerned. When eyes become dirty, they are cleaned with front two legs.

Most distinguishing feature is black body that contains white hairs, which form stripes on abdomen. Male is similar to female but with larger chelicerae, darker body color, and brightly colored brushes on appendages. Cephalothorax contains brilliant hairs, stout body, and rather short legs; eight legs are hairy and covered by sensory hairs (trichobothria). Tracheal system extends into cephalothorax. Abdomen contains digestive system, breathing apparatus, and silk-producing organs. Huge chelicerae are usually hidden behind pedipalps.

DISTRIBUTION
Northern Hemisphere, but mostly in northern Europe (and widely distributed throughout England).

HABITAT
Commonly found anywhere outside where sun is shining; especially in gardens, on rocks, stones, flowers, plant foliage, and grass, and occasionally on trees. Often found on vertical surfaces such as walls, fences, decks, patios, and doorways. At night or during rainfall, it hides in dry spots.

BEHAVIOR
Diurnal; most active during hottest days of year, mostly in early to late summer. Often attacks and kills much larger adult hobo spiders, which are competitors for food. Jumps more than it walks. Able to jump from standing start; can also jump backward and sideway with equal abilities. This type of motion is used both to capture prey and to avoid capture by predators.

Uses third and/or fourth pair of legs for jumping. Whenever it jumps, it will release thick, white, slightly viscid silky line to use as anchor to crawl back to original position. Silk is produced from special organs (spinnerets) at rear of abdomen. Also produces silken bag ("retreat") in such places as crevices, under stones, under bark, and on foliage and plants. Bags used for protection and shelter at night, resting, molting, feeding, protecting young, and during winter to hibernate.

Prey can be noticed from distance of about 12–16 in (30–40 cm), although it is reported that it can see prey up to 8 ft (2.4 m) away. At distance of about 7.9 in (20 cm), it turns its body so that front eyes point to victim; eye muscles focus on prey and the eyes move around optical axis. Able to distinguish between prey and predators, and also capable of distinguishing color. After object is recognized as eatable, it carefully moves toward victim.

FEEDING ECOLOGY AND DIET
Eats primarily insects, but also eats spiders the same size or smaller. It avoids ants. Reported to feed on mosquitoes with lengths almost twice its own. Active hunter, able to catch larger prey primarily because of its excellent eyesight during day (especially in direct sunlight) and excellent ability to jump from a stationary position. Slowly stalks potential prey by creeping very close, usually to within 2.8–5.9 in (7–15 cm). When at reachable distance, it attaches silky thread to substrate, and then jumps on prey and paralyzing it with its venomous jaws. Powerful chelicerae are then used for chewing up prey prior to sucking up liquid contents. Does not make webs for catching prey.

REPRODUCTIVE BIOLOGY
Males court females by dancing and displaying brilliant colors, distinctive marks, and bright appendages. Males deposit sperm on small web to be used as special reservoir within pedipalp to

carry seed around. They will then try to mate with females. Mating is dangerous for males, having to convince females that they are prospective mates and not prey. This activity involves various motions with front legs and moving abdomen up and down. (The more they move, the more likely they will be noticed and accepted by female.) During this time, males try to reach reproductive organ of female (epygine), located under abdomen. When sperm is successfully transferred to female, she will carry it in special compartment and use it when she is ready to fertilize eggs. Females lay their eggs in small silky bags mostly in spring and summer for the purpose of being able to protect spiderlings from predators. Females will guard young until they are ready to leave, normally after second molting period. Young usually mature in late spring and summer. Lifecycle is about one year.

CONSERVATION STATUS
Not listed by the IUCN.

SIGNIFICANCE TO HUMANS
Often considered a pest to humans, but it is actually harmless. ◆

Common harvestman
Phalangium opilio

ORDER
Opiliones

FAMILY
Phalangiidae

TAXONOMY
Phalangium opilio Linnaeus, 1758.

OTHER COMMON NAMES
English: Harvest spider.

PHYSICAL CHARACTERISTICS
Small, globular (rounded) body and very long thin legs. One major body section, eight legs, no antennae, no web spinning (silk-producing) organs, and no poison (venom) glands. Adult length is 0.14–0.35 in (3.5–9.0 mm), with males generally smaller than females. Upper body surface colored with indistinct and variable light gray to brown pattern, and lower body surface is usually light cream. Possesses two eyes located in middle of body, forcing it to look outward from sides.

DISTRIBUTION
North America, Europe, and temperate Asia.

HABITAT
Commonly found in relatively disturbed temperate habitats such as crops of alfalfa, cabbage, corn, grains, potatoes, and strawberries. Also found in wide variety of undisturbed temperate habitats, including forests, brushy areas, and open grasslands.

BEHAVIOR
Generally takes prey with its long legs as it flies by. It is active throughout summer, but most active in later summer and fall. Nocturnal, often gathering in large groups on tree trunks and interlace legs together. It uses its scent glands located on either side of body to produce peculiar smelling (but non-poisonous) fluid when it is disturbed, probably acting as a repellant to some predators. Will also intentionally detach a leg, leaving it twitching, to distract predator; this is only done in desperate situations, because detached leg is permanently lost.

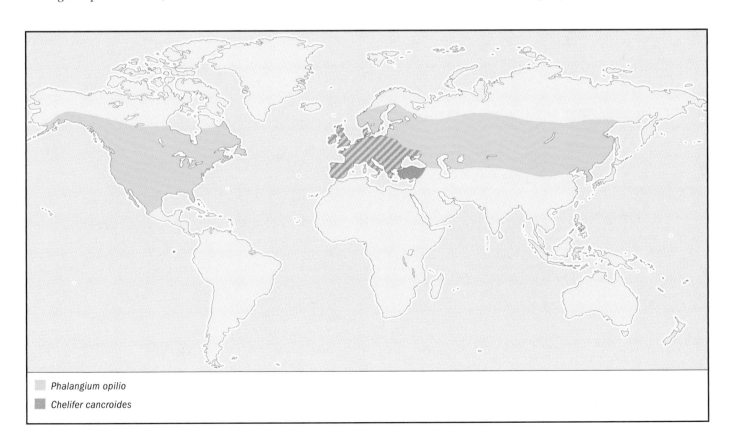

Phalangium opilio
Chelifer cancroides

FEEDING ECOLOGY AND DIET
Omnivorous, feeding on many soft-bodied pest arthropods found in crops, such as aphids, caterpillars, leafhoppers, beetle larvae, mites, and slugs. Also feeds on dead insects and other decaying plant material like rotted fruit, as well as earthworms, harvestmen, spiders, and other beneficial invertebrates. Can be cannibalistic.

REPRODUCTIVE BIOLOGY
Females lay clusters of eggs in moist areas on the ground, often under rocks, in cracks in soil, or between soil and crowns or recumbent leaves of plants. Number of eggs laid from 10 to several hundred. Eggs are generally laid in fall where they over-winter, and then hatch following spring. (In Europe, they reproduce once each year, with eggs over-wintering. In parts of North America, two or more generations may occur, with eggs, immature young, and adults often over-wintering.) The eggs hatch in 3–20 weeks or more, depending primarily on temperature. Eggs are spherical, about 0.0158 in (0.4 mm) in diameter, with a smooth surface and color changing from off-white to dark gray-brown as they mature. After hatching, immature young are similar to adults, only smaller and with legs shorter relative to body size. Immature young undergo several molts (usually seven) and reach maturity in 2–3 months, again depending on temperature.

CONSERVATION STATUS
Not listed by the IUCN.

SIGNIFICANCE TO HUMANS
It helps to control pests that feed on cultivated crops, helping to keep pest densities low. It also feeds on dead insects and other decaying material. It is medically harmless to humans. ◆

Long-bodied cellar spider
Pholcus phalangioides

ORDER
Opiliones

FAMILY
Pholcidae

TAXONOMY
Pholcus phalangioides Fuesslin, 1775.

OTHER COMMON NAMES
English: Cobweb spider, daddy longlegs spider, long-legged cellar spider, shepherd spiders, harvest spiders.

PHYSICAL CHARACTERISTICS
Characterized by eight long legs and spider-like appearance; length of 0.25–0.50 in (6.4–12.7 mm) with front legs about 1.75–1.94 in (45–50 mm) long, small bodies that are grayish brown or tan in color, and very long, somewhat translucent, skinny legs. Males are slightly shorter than females; fragile with a rectangular, elongated abdomen (oval-shaped body). Head, thorax, and abdomen are fused, with elongated, cylindrical abdomen about three times longer than wide. Chelicerae are fang-like.

DISTRIBUTION
Throughout world, but found primarily in the United States and Europe.

HABITAT
Often found in dark, damp areas such as crawl spaces, basements, closets, sink cabinets, ceilings, cellars, warehouses, garages, attics, and sheds. Also found near open doors and other similar entrances that allow flying insects to enter. Often lives upside down in stringy, irregular webs that are not cleaned, but are continually added to with new webbing, resulting in extensive web networks.

BEHAVIOR
Web (also called "net") is large, irregularly arched construction that looks similar to canopy. The spider resides on lower side hanging upside down within web. When disturbed, it will shake web violently. Anything that touches net is attacked and taken for prey, if it is not too large. Uses net as means of camouflage by whirling its body around with legs firmly attached to net. If it accidentally falls from web, it runs in a wobbly fashion, so as not to be seen easily. Once web becomes old and unusable, it constructs additional webbing attached to old web. Enemies are birds, wasps, and humans.

FEEDING ECOLOGY AND DIET
Predator and carnivore, eating almost any kind of insect or bug. Eats moths, mosquitoes, flies, gnats, and beetles that have become entangled in silky web. Will invade spiders' webs, attack resident spider, and take over web. If prey is already enwrapped in web, it eats it. New prey trapped in web will be swiftly wrapped with new silk, spun as many as several feet (meters) around victim by fourth pair of legs. The prey is then bitten, with digesting fluid injected into body. It will then proceed to suck body empty over the course of a day or two. When it is finished, it will cut remains loose, letting them fall to ground in pile of dead bodies.

REPRODUCTIVE BIOLOGY
Often lives close to mate; commonly living next door to each other. Mating can take up to several hours to complete. During process, male sperm is transferred to female by way of special cavity at beginning of uterus. Resulting spermatozoa remain in cavity until used to fertilize eggs. Time of fertilization depends largely on availability of food sources in area. Female will wrap eggs in clear sac that she keeps in her mouth (between chelicerae) for safety. Up to three sacs may be produced, each containing 13–60 eggs. Female will guard eggs until prenymphs hatch from eggs after several weeks. After about nine days, prenymphs shed old skins and tiny spiders appear; they soon leave maternal net to look for new place to build their own net. They will continue to shed their skin as they grow, with five molts needed to reach maturity. Lives for about two years.

CONSERVATION STATUS
Not listed by the IUCN.

SIGNIFICANCE TO HUMANS
Not dangerous and actually is quite beneficial in that it captures and eats many types of insects that are considered pests and poisonous spiders, including black widow and brown recluse spiders. ◆

No common name
Eukoenenia draco

ORDER
Palpigradi

FAMILY
Eukoeneniidae

TAXONOMY
Eukoenenia draco Peyerimhoff, 1906.

OTHER COMMON NAMES
None known.

PHYSICAL CHARACTERISTICS
Small (less than 0.12 in [3.0 mm] in length), can grow up to 0.11 in (2.8 mm) long, but usually are 0.0394–0.0787 in (1.0–2.0 mm) in length. It is light yellow to white, being almost colorless; has many-segmented, whip-like post-abdomen and a wide pre-abdomen. Chelicerae are thin, long, and pincer-like (chelate), with three articles and lateral moveable finger. Mouth is located at tipped end of prominent protuberance. A labium is present.

It does not have eyes, a respiratory system, or a circulatory system, but does have innervated setae that detect vibrations. Receives oxygen through very thin and colorless cuticles. No Malpighian tubules but, instead, pair of excretory coxal glands. It has carapace (propeltidium) divided in three pieces, with 11 defined segments in abdomen and four parts in sternum. The prosoma is covered by two free tergites and is attached to opisthosoma by pedicel; opisthosoma has anterior mesosoma and short posterior metasoma possessing a flagellum with many articles. Other appendages are leg-like. Pedipalpal coxae are not part of preoral cavity, and are similar to leg coxae. Several sternites are present. The first leg, used as a feeler, is positioned distally, and possesses many articles.

DISTRIBUTION
Scattered throughout the world, primarily in warmer climates. (Specific distribution unknown; no map available.)

HABITAT
Generally located in tropical regions, commonly found under rocks, half-buried stones, and sometimes in caves. During drier times, it sometimes digs under soil. Specifically, little about its habitat is known.

BEHAVIOR
It walks with first legs stretched forward in front. During drier seasons, it sometimes descends into soil. Little of its specific behavior is known.

FEEDING ECOLOGY AND DIET
There is some research evidence that it feeds on eggs of other small animals in vicinity. However, few specifics are known about its ecology and diet.

REPRODUCTIVE BIOLOGY
Males produce a spermatophore. Little of its specific reproductive biology is known.

CONSERVATION STATUS
Not listed by the IUCN.

SIGNIFICANCE TO HUMANS
None known. ◆

Book scorpion
Chelifer cancroides

ORDER
Pseudoscorpionida

FAMILY
Cheliferidae

TAXONOMY
Chelifer cancroides Linnaeus, 1758.

OTHER COMMON NAMES
English: House scorpion.

PHYSICAL CHARACTERISTICS
Slightly smaller than scorpions, with a length of 0.10–0.18 in (2.6–4.5 mm). Does not possess stinging tail, has enlarged pedipalps, and transmits spermatophore in complex courtships. Cephalothorax (also called scutum) contains six pairs of appendages: chelicerae, palpal chelae (two well-developed claws), and four pairs of legs. Femora of legs one and two are different in general structure, especially with regards to joints, from femora of legs three and four. Cephalothorax is olive-brown to dark red. Palpal chelae (reduced claws near mouth) consist of large bulbous hand, with one fixed finger and one moveable finger. Moveable finger does not have edge but instead has a subapical lobe. Accessory teeth are not contained on chelel fingers. Opisthosomal tergites are pale brown to olive-green, with darker spots. Pedipalps are tawny brown to reddish brown, with some olive coloration. Venom apparatus is well developed in both fingers of palpal chela. Cheliceral flagellum consists of three long, straight setae. Complex internal genitalia of males are heavily sclerotized. Spermathecae of females are short, rounded sacs, with sclerotic plates.

DISTRIBUTION
Throughout most of Europe.

HABITAT
Usually found under stones, beneath bark of trees, or in vegetable debris, but can also be found in human habitations and outbuildings such as stables, barns, grain stores, factories, and houses, and seems to move wherever humans locate. Often found in old books (thus, its common name). Prefers warmer regions of the world.

BEHAVIOR
Often found in groups of several dozens. Females often use pedipalps to hold onto flying insects such as houseflies to be carried to where it wants to go. Males do not commonly use this method of transportation (called phoresy). Palpal chelae are mainly used for defense/fighting, acquisition of prey, and moving small objects (like sand grains) to make nests. A small, spherical silken chamber (or cocoon) is built for hibernation during winter months, and for molting.

FEEDING ECOLOGY AND DIET
Carnivore and insectivore, eating animal tissue and arthropods such as small insects, mites, and lice. Often secures itself underneath wings of large tropical beetles to feed on parasitic mites. It may also do similar actions to legs of houseflies and other two-winged insects. Generally will grab prey with pedipalps, immobilize prey with poison glands, rip it up with chelicerae, and suck fluids from body.

REPRODUCTIVE BIOLOGY
Complex courtship and mating behaviors from both males and females, including extension of ramshorn organs of male and dance-like behavior by pair. Reproduction is through spermatophore: adult males use modified first legs to expel mass of spermatozoa. Females are oviparous; they will build nest by secreting a "brood sac" attached to body, where she will nourish young. Generally, 16–30 offspring are produced from each reproductive cycle; they will depart the mother's protection soon after hatching, after reaching a fixed, definite shape. Sexual maturity is reached in 1–2 years.

CONSERVATION STATUS
Not listed by the IUCN.

SIGNIFICANCE TO HUMANS
Often eats lice that have infected human hair. Also feeds on other creatures seen as pests for humans, such as mites and ants. It seems to offer little direct benefit to humans, although little is really known about its contribution to human life. ◆

No common name
Ricinoides afzelii

ORDER
Ricinulei

FAMILY
Ricinoididae

TAXONOMY
Ricinoides afzelii Thorell, 1892.

OTHER COMMON NAMES
None known.

PHYSICAL CHARACTERISTICS
Heavy-bodied, usually 0.12–0.20 in (3–5 mm) in length, but can reach up to 0.39 in (10 mm) long. It possesses thick exoskeleton along with cucullus (transverse hood-like flap) that covers mouthparts, and can be raised and lowered and is located at anterior edge of carapace. Prosoma is covered by carapace and is widely connected to opisthosoma, which narrows at front, forming pedicel (waist), where it is attached to prosoma (or cephalothorax). At the end of abdomen, tubucles stick out, forming anus. Chelicerae terminate into two articles that form scorpion-like pincers. Smallish pedipalpi are leg-like and also end in medium-sized pincers. Ventral and coxal endites are fused to form trough, posterior wall of postoral cavity. Leg coxae cover venter of prosoma. Short and heavy legs have no modifications except in third pair of males, which are modified to form copulatory organs. It has no eyes, and is poorly equipped with any type of sense organs. Male organs are found on metatarsus and tarsus of third pair of legs; metatarsus and first two segments of tarsus are modified by cavities and by fixed and moveable processes. Excretory organs consist of Malpighian tubules and pair of coxal glands. Circulatory system is degenerate; it has no lungs, and gas exchange takes place through trachea.

DISTRIBUTION
Scattered in tropical West Africa.

HABITAT
Dwells within soil, usually in tropical leaf litter.

BEHAVIOR
Slow-moving creature that requires dampness to survive. Little about its behavior is known.

FEEDING ECOLOGY AND DIET
Predator that feeds on tiny invertebrates and other arthropods in caves and leaf liter. Specific information on its feeding ecology and diet is sparse.

REPRODUCTIVE BIOLOGY
Males mount the backs of females, fourth legs grasping her opisthosoma. Both face same direction during reproduction. Male uses third legs to transfer sperm (possibly through spermatophore) to female's genital opening (via insemination). There is assumed to be no courtship. Normally, 1–2 eggs are laid. Eggs are carried under mother's hood, until young hatch into six-legged larvae, with subsequent molts being protonymph, deutonymph, tritonymph, and adult. About 1–2 years is taken to reach maturity, with lifespan of 5–10 years.

CONSERVATION STATUS
Not listed by the IUCN.

SIGNIFICANCE TO HUMANS
None known. ◆

No common name
Agastoschizomus lucifer

ORDER
Schizomida

FAMILY
Protoschizomidae

TAXONOMY
Agastoschizomus lucifer Brusca and Brusca, 1990.

OTHER COMMON NAMES
None known.

PHYSICAL CHARACTERISTICS
Very small, less than 0.39 in (10 mm) in length, but usually averages only 0.12 in (3.0 mm) in length. Does not have eyes. Has divided prosoma covered anteriorly by propeltidium and by two free tergites: the mesopeltidium and metapeltidium. One-third to one-tenth its greatest dimension separates large mesopeltidia. Short abdominal flagellum is located on last ab-

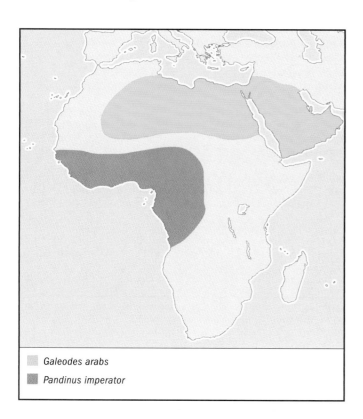

Galeodes arabs

Pandinus imperator

dominal segment. Flagellum is segmented in females. Abdomen has eight pairs of dorsoventral muscles, which are not flattened. Chelicerae lack a serrula, but have two blunt hemispherical knobs. Has fixed digit with two teeth, and no setae at base. Basitarsal spurs are symmetrically located, about one-third to one-half dorsal length of basitarsus. Feeler-like first legs are used for sensory purposes. Lacks patellae. Tarsi have eight pieces and no claws. Trochanter of fourth leg is about 2.2 times longer than wide; fourth femur is 3–5 times longer than wide. Anal glands are present. One pair of book lungs.

DISTRIBUTION
Mexico.

HABITAT
Usually caves in the tropics, living in leaf litter or under stones. Digs tunnels in soil.

BEHAVIOR
Can move backward rapidly with use of enlarged femora. Believed to be able to produce defensive chemical smell by means of repugnatorial glands.

FEEDING ECOLOGY AND DIET
Predator, but little information is known about specific prey.

REPRODUCTIVE BIOLOGY
Female hooks chelicerae into dilations of male's flagellum. Then, male deposits spermatophore to substrate and pulls female over it. Afterwards, female lays 6–30 eggs, which are attached to ventral abdomen (gonopore) until they hatch. They have one embryonic and five postembryonic instars. Mature in 2–3 years.

CONSERVATION STATUS
Not listed by the IUCN.

SIGNIFICANCE TO HUMANS
None known. ◆

Striped scorpion
Centruroides vittatus

ORDER
Scorpiones

FAMILY
Buthidae

TAXONOMY
Centrurus vittatus (Say, 1821).

OTHER COMMON NAMES
English: Common striped scorpion, striped bark scorpion.

PHYSICAL CHARACTERISTICS
Maximum length of 3 in (7.6 cm), but average about 2.4 in (6.1 cm). Characterized by dark triangular mark on front part of head region in area over median and lateral eyes, and pair of blackish, parallel, longitudinal stripes on upper surface of abdomen. Long, slender tail is longer in males than in females. Adult body color varies from yellowish to tan. Young are usually lighter in color, with last body segment and bases of pedipalps being dark brown to black. Has slender pedipalps and a long, slender tail.

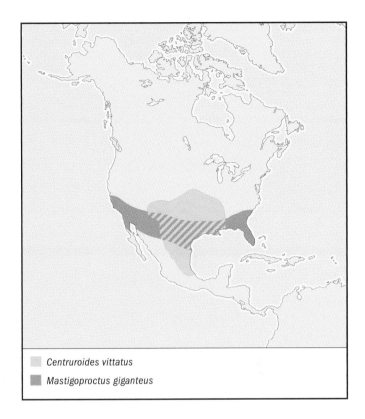

☐ *Centruroides vittatus*
▨ *Mastigoproctus giganteus*

DISTRIBUTION
Found in southwestern United States, being most heavily concentrated within the state of Texas and outward to Oklahoma, Louisiana, Arkansas, western Mississippi, extreme western Tennessee, southern half of Missouri, southern tip of Illinois, eastern tip of Kentucky, Kansas (except northwestern part), tip of south-central Nebraska, northeastern part of Arizona, southeastern part of Colorado, and New Mexico (except southwestern part); also found in northern Mexico, within the states of Tamaulipas, Coahuila, Nuevo León, Chihuahua, and Durango. Considered the most common scorpion species in the United States.

HABITAT
Found in diverse environments, being very adaptive. Mostly found in areas containing many cracks and crevices such as diverse regions of rocky areas, forests, and buildings; also found in relatively open areas such as grasslands and sand dunes. Often found in close association with humans, being commonly found indoors; often found in damp, cool areas around dead vegetation, boards, rocks, and fallen logs; often found climbing trees and walls, and are common in house attics. During the day, it usually hides under loose rocks, bark, and leaves.

BEHAVIOR
Ecomorphotype, meaning it actively moves while foraging; moves easily over vertical surfaces and clings to undersides of objects; but does not burrow. It remains in shelters during day, becoming active at night. This schedule helps it to maintain water and body temperature balance. It also dwells in crevices, willing to use any kind of crevice for protective retreat. Is able to tolerate many different climates such as very low and very high temperatures. Its venom toxicity is low.

FEEDING ECOLOGY AND DIET
Primarily insectivorous, eating small arthropods such as beetles, centipedes, crickets flies, spiders, and other small insects.

It normally hunts at night by depending on its acute sense of touch and smell. Feathery, comb-like chemical receptor organs (called pectines) located on underside between last pair of legs touch ground as it walks, helping it to track prey. When finding prey, it grabs it and crushes it with powerful pinchers. Tail is then brought over body to sting victim. This paralyzes prey, which soon dies. It will then chew prey into semi-liquid state, which can then be sucked up into tiny mouth.

REPRODUCTIVE BIOLOGY
Mating occurs primarily in spring and early summer. Has elaborate courtship ritual, which can last for hours. Female and male will grasp each other's pincers and jaws, and dance back and forth. Eventually, male deposits spermatophore on the ground and pulls female to it. She picks up the sac with a special organ on abdomen to fertilize it. Its gestation period and maturity period varies, depending on climatic and environmental conditions, with shorter periods for each in warmer parts of habitat. Embryos are nourished in female's body through placental connection. Probably has gestation period of 6–12 months, with maturity period probably of 1–2 years. Females give birth to up to 50 young at one time, but averages about 30. After birth, young climb on mother's back and soon molt. After first molt, they disperse and lead independent lives. They molt an average of six times before maturity.

CONSERVATION STATUS
Not listed by the IUCN.

SIGNIFICANCE TO HUMANS
Can inflict a very painful sting and causes extreme discomfort if touched or grabbed by a human, but it is not considered as dangerous as some of its relatives. It usually causes only minor medical problems such as local swelling and discoloration to healthy humans. If death occurs, it is because of anaphylactic shock, not from the direct toxic effects of venom. It helps control local insect populations. ◆

Emperor scorpion
Pandinus imperator

ORDER
Scorpiones

FAMILY
Scorpionidae

TAXONOMY
Pandinus imperator Koch, 1841.

OTHER COMMON NAMES
None known.

PHYSICAL CHARACTERISTICS
Has large, well-developed pincer-like pedipalps, is uniformly covered with hard cephalic shield or carapace, and has prehensile tail armed with a stinging apparatus; usually attains lengths (including tail) of 5–7 in (12.7–17.8 cm) and weighs about 1.1 oz (35 g), but can reach length of 8 in (20.3 cm) and weight of 2 oz (57 g); pregnant females usually weigh over 1.4 oz (40 g) (considered one of the largest scorpions, but not among the heaviest). As adult, males and females act and look similar; however, males are usually narrower or smaller. Has exoskeleton color of glossy dark blue or black, but some may be dark brown and occasionally even greenish; dark color acts as camouflage. Two pedipalp chela (pedipalps) have reddish brown color, and are very granular in texture. There are numerous, clearly visible sensory hairs on the pedipalps, metasoma (tail), and telson (stinger). Tail is long and made up of six segments, ending in large telson, which contains venom glands. Telson terminates in sharp curve, which serves as stinger, and is reddish brown in adults and yellowish in young. Telson of second instar is white, but soon becomes darker after each molt. Four-sectioned thorax contains pair of legs on each section, specifically on undersurface, making total of eight legs (four pairs). Behind fourth pair of legs are ventral comb-like structures known as pectines; males can be also distinguished from females by their longer pectines.

DISTRIBUTION
Western Africa, primarily in Senegal, Guinea-Bissau, Guinea, Côte d'Ivoire, Sierra Leone, Ghana, Togo, Democratic Republic of Congo, Nigeria, Gabon, and Chad.

HABITAT
Lives in tropical forests, rainforests, and savannas, preferring hot, humid environments. Lives in empty or self-made burrows up to 12 in (30 cm) in length. Often found beneath rocks, logs, tree roots, or vegetation debris.

BEHAVIOR
Sensitive to light, so is primarily nocturnal. It is unusually docile and very slow to sting. Although young use stingers in normal fashion, adults rarely use stinger to subdue prey. They prefer to kill prey with massive claws. Even when stinging in defense, adults may not inject venom. Mothers and young/siblings often live together. Mothers are occasionally cannibalistic, being known to eat a few of their young when necessary. It likes to burrow beneath soil.

FEEDING ECOLOGY AND DIET
Feeds on almost anything that is smaller in size, including arachnids, crickets, insects, small lizards, mealworms, millipedes, and small mice. Young eat pinhead crickets and other small insects. It does not generally pursue prey, but waits for unsuspecting insects and other small animals to pass by. Its eyes, which cannot form sharp images, are of little use in detecting prey. Air and ground vibrations are used primarily in determining the position of prey. When hungry, however, it moves slowly forward supported by its hind legs, with claws open and extended, and tail raised and pointed forward. It quickly strikes with stinger or grasps victim. Larger individuals rarely use stinger to capture prey; instead, they crush it with claws. Smaller and younger ones rely on stinger to subdue prey. They must predigest their food before they consume it. Once prey is subdued, they secrete digestive enzymes onto prey, which liquefies the food and readies it for consumption.

REPRODUCTIVE BIOLOGY
Males spend the majority of time looking for mates. Mating can occur year-round, but warm temperatures are required. When mating, male holds female in grasp, holding and pushing her around until finding suitable place to deposit spermatophore onto a solid substrate. He then pulls female into position over spermatophore, and she accepts it into her genital aperture. Male leaves quickly, to avoid being eaten. It is viviparous (embryos develop within mother, gaining nutrients for growth directly within specialized sacs on female's overiuterus). A highly specialized structure connects embryo's mouth to female's digestive system. Gestation period is 7–9 months. Very tiny young are born alive, with a litter of 9–35. Parental care seems important. Young stay on mother's back, as she protects and cares for them, with increased survival probabilities while

in family groups. Young are white at first, but become darker after each molt. They grow and shed entire exoskeleton several times before they are full-grown. They reach sexual maturity at around four years after seven molts. They have a lifespan of about eight years.

CONSERVATION STATUS
Not listed by the IUCN. Because of years of potential over-collection, it has been placed on the CITES Appendix II list (as Threatened) to monitor populations. Primary enemy is humans, who may have one as a pet.

SIGNIFICANCE TO HUMANS
Although large in size, it is not considered dangerous to healthy humans. Its venom is mildly venomous, with a painful sting. It has very strong pedipalps, which can give very painful pinches. Adult males will rarely sting, but young individuals and females with young can be more likely to sting. They are the most common scorpion in captivity, with many exported from western Africa each year. ◆

Egyptian giant solpugid
Galeodes arabs

ORDER
Sopugida

FAMILY
Galeodidae

TAXONOMY
Galeodes arabs Koch, 1842.

OTHER COMMON NAMES
English: Desert camel spider.

PHYSICAL CHARACTERISTICS
Measures up to 4.7 in (12 cm) in length with legs and only 2 in (5 cm) with body. Males are smaller and lighter than females, but have longer legs. Both sexes are yellowish in color. Has eight legs (four pairs), and body is in two parts: prosoma and opisthosoma (which has no pedicel, and consists of 11 somites). Has very powerful chelicerae; pedipalpi do not end in a claw but rather in a suctorial organ, and first pair of legs is long and thin, and not used for walking but as feelers. Bodies and legs are hairy. Respiratory system contains well-developed tracheal system. Has very long legs, terminal anus, and exterior lobes of propeltidium are fused posteriorly. Male cheliceral flagellum is a single, capitate (terminally enlarged), paraxially moveable seta located on mesial surface. Female operculae are not differentiated from other abdominal sternites and are not variable. Two eyes are placed on small projection near fore-edge of propeltidium.

DISTRIBUTION
Northern Africa and the Middle East, especially in the Sinai Desert.

HABITAT
Lives in sandy arid and desert environments.

BEHAVIOR
Generally nocturnal creature, however still active during day (but avoiding direct sunlight). During day, spends most time hiding in burrows or under objects looking for shade. Able to run very fast, up to 20 in (50 cm) per second for short periods

of time, though unable to sustain this pace for long periods. Considered one of the fastest arthropods. Uses rear three pairs of legs to run, and uses first pair for sensory purposes (as "feelers" to detect and pull prey into its large oversized jaws). Constructs extensive, shallow burrows under bushes and buildings by utilizing chelicerae, pedipalpi, and metatarsal and tibial rakes of second and third pairs of legs. Burrows are used for mating, defense, and shelter. Can stridulate (make a noise) by rubbing together pair of horny ridges on insides of chelicerae.

FEEDING ECOLOGY AND DIET
Exclusive carnivore, aggressively feeding on small mice, lizards, amphibians, spiders, scorpions, some small birds, and other similar animals. Hunts and feeds at night, using ground-level vibrations to detect prey. Pedipalpi are most active organs, used to pick up prey (often termites and other prey). Uses its large chelicerae not only to crush often larger prey, but also to scoop water into its mouth when drinking.

REPRODUCTIVE BIOLOGY
It mates with quick but strenuous action. Male seizes female with his legs and chelicerae, grasping her body but rarely hurting her. The caresses of his legs then so strongly affect her that she falls into a motionless hypnotic state. He then picks her up in his chelicerae and carries her to a suitable location. Next, he lays her on her side and awakens her sexually by stroking underside of her abdomen. Males do this because females will often eat male partners if they are too slow in leaving after mating. Reproduction is through spermatophore, which male produces during courtship, deposits on ground, and transfers into female gonopore with his chelicerae. She digs a deep burrow to deposit egg masses that may contain 5–164 eggs. Females may lay 1–5 egg masses. Eggs take 1–2 days to hatch. Once hatched, young remain in burrow for first two molts. First instar larvae (first "stadium") is non-moving, embryo-like creature that later molts into more active animal that looks like small adult.

CONSERVATION STATUS
Not listed by the IUCN.

SIGNIFICANCE TO HUMANS
Does not possess venom and prefers to stay away from humans, so presents little or no concern to humans. ◆

Giant whip scorpion
Mastigoproctus giganteus

ORDER
Uropygi

FAMILY
Thelyphonidae

TAXONOMY
Mastigoproctus giganteus Lucas, 1835.

OTHER COMMON NAMES
English: Desert whip scorpion, giant vinegaroon, grampus.

PHYSICAL CHARACTERISTICS
Possesses long, thin, whip-like tail instead of stinger, and has large anal glands that discharge strong defensive acids. Length about 1.0–3.2 in (25–80 mm), not including tail, and is reddish or brownish black. Both males and females are similar in appearance, with heavy pedipalps that are formed into pincers.

Carapace covers body. Has one pair of eyes, located in front of cephalothorax, and six more eyes, three off each side of head. Even though it has eight eyes, it has poor eyesight; compensates by being able to detect ground vibrations, especially those by prey. Two chelicerae (normally turned forward) are used to grasp, tear, and transfer food into mouth. Has four pairs of legs, with front-most (first) pair longer and thinner than other three pairs; first pair of legs is used to detect prey and evaluate environment (like sensory feelers), while other three pairs of legs are used for walking. Inside abdomen are two pairs of layered lungs with a whip-like telson that is usually held curled toward back; telson is used for defense, and is capable of spraying acetic and caprylic acids that originate from repugnatorial glands near anus.

DISTRIBUTION
Found throughout southernmost states in the United States; however, is more likely found in southwestern United States, especially in the Trans-Pecos region of Texas but also as far north as the panhandle of Texas and in south Texas, and in northern Mexico.

HABITAT
Commonly found in chaparral and deserts, but has also been found in grassland, scrub, pine forests, and mountains. Usually uses underground burrows for home. Also is found in burrows under logs, rotting wood, rocks, and other natural debris. Prefers humid, dark places and avoids sunlight whenever possible.

BEHAVIOR
Nocturnal, with little action seen during day. At night, it is active predator, finding prey by detecting ground vibrations to trap prey. Although basically passive, it has its own defense mechanisms that are very effective. If disturbed or in danger, it will squirt acid capable of eating through exoskeleton of invertebrates, enabling it to escape enemies or to capture prey. Uses chelicerae to pinch.

FEEDING ECOLOGY AND DIET
Feed offs many different types of insects, including crickets and roaches. Uses ground vibrations to detect movements of prey, and then uses foremost pair of walking legs, the feelers, to find prey. Uses feelers to make sure prey is surrounded and still, and then uses chelicerae, its pinchers, to pinch and capture prey. Some acid may be squirted out to kill prey. Mouthparts are used for chewing.

REPRODUCTIVE BIOLOGY
During reproduction, sperm is transferred indirectly when male deposits spermotophore, or sperm sac, on the ground. He then gently guides female over it by grasping her sensory pedipalps. He may then assist her in taking spermatophore into her genital opening by using his pedipalps. Afterwards, female will take to a sheltered spot, carrying eggs in a silken sac until they hatch. Colorless young then climb and ride on her back until they molt. They then become independent.

CONSERVATION STATUS
Not listed by the IUCN.

SIGNIFICANCE TO HUMANS
Kept as pet and is helpful in controlling roach and cricket populations. Non-poisonous but can pinch and is capable of spraying a mist of concentrated acetic acid when disturbed by humans or other creatures. ◆

Resources

Books

Adis, J., and M. S. Harvey. "How Many Arachnida and Myriapoda Are There Worldwide and in Amazonia?" In *Encyclopedia of Biodiversity*. New York: Academic Press, 2000.

Beacham, Walton, Frank V. Castronova, and Suzanne Sessine, eds. *Beacham's Guide to the Endangered Species of North America*. Detroit, MI: Gale Group, 2001.

Bosik, J. J. *Common Names of Insects and Related Organisms*. Lanham, MD: Entomological Society of America, 1997.

Cloudsley-Thompson, J. L. *Spiders, Scorpions, Centipedes and Mites*. Oxford, U.K.: Pergamon Press, 1968.

Foelix, Rainer F. *Biology of Spiders*. Cambridge, MA: Harvard University Press, 1982.

Levin, Simon Asher, ed. *Encyclopedia of Biodiversity*. San Diego, CA: Academic Press, 2001.

McDaniel, Burruss. *How to Know the Mites and Ticks*. Dubuque, IA: William C. Brown Company Publishers, 1979.

Milne, Lorus, Lorus J. Milne, and Susan Rayfield. *The Audubon Society Field Guide to North American Insects and Spiders*. New York: Knopf, 1980.

Parker, Sybil P., ed. *Synopsis and Classification of Living Organisms*. New York: McGraw-Hill Book Company, 1982.

Polis, Gary A., ed. *The Biology of Scorpions*. Stanford, CA: Stanford University Press, 1990.

Savory, Theodore. *Arachnida*, 2nd ed. London and New York: Academic Press, 1977.

Tudge, Colin. *The Variety of Life: A Survey and a Celebration of All the Creatures That Have Ever Lived*. Oxford and New York: Oxford University Press, 2000.

Woolley, Tyler A. *Acarology: Mites and Human Welfare*. New York: John Wiley and Sons, 1988.

William Arthur Atkins

Chilopoda
(Centipedes)

Phylum Arthropoda

Class Chilopoda

Number of families 21

Thumbnail description
Multi-legged predatory arthropods, mostly solitary and nocturnal, found in leaf litter and other cryptic terrestrial biotopes

Photo: A Thai centipede feeding on a house mouse. (Photo by Tom McHugh/Photo Researchers, Inc. Reproduced by permission.)

Evolution and systematics

The class Chilopoda includes five orders, 21 families, and 3,200 known species. Chilopoda belongs to the subphylum Myriapoda, which also includes millipedes (class Diplopoda) and two less diverse groups, the Pauropoda and Symphyla. The structure and function of the head endoskeleton and the mandible suggest that Myriapoda is a natural group, but some zoologists consider millipedes, pauropods, and symphylans to be more closely related to insects than to centipedes.

Four of the five living orders of centipedes share flattened heads and other adaptations for living in confined spaces, and the openings of the tracheal respiratory system are located above the legs on each side of the body. These features indicate that there is a more recent shared ancestry for the orders Lithobiomorpha (1,500 species), Craterostigmomophora (one or two species), Scolopendromorpha (550 species), and Geophilomorpha (1,100 species) than is shared with the order Scutigeromorpha (80 species). The latter group, also known as Notostigmophora, has a domed head, large multi-faceted eyes, and the openings of the tracheae are located on the upper side of the body at the back of each tergal plate. Scutigeromorphs have special tracheal lungs. It is only in the orders Scolopendromorpha and Geophilomorpha that hatchlings emerge from the egg with their full adult number of segments. This so-called epimorphic development, the distinctive structure of the testes, and the tracheae having connections between the segments all indicate that these two orders are most closely related.

The earliest fossil centipedes are from the late Silurian (418 million years ago) of Britain, and belong to the order Scutigeromorpha. Other Paleozoic centipedes include the ex-

tinct order Devonobiomorpha (one species: Devonian, New York State), and late Carboniferous members of Scolopendromorpha and Scutigeromorpha. Fossils preserved in Tertiary amber are essentially modern.

Physical characteristics

Centipede adult length ranges from 0.15 to 11.8 in (4–300 mm). The head has one pair of slender antennae, composed of 14 to more than 100 articles. Eyes are either faceted (Scutigeromorpha), or composed of one ocellus or a cluster of ocelli on each side of the head (most Lithobiomorpha and all large Scolopendromorpha), or completely lacking eyes (all Geophilomorpha, many smaller Scolopendromorpha). The mouthparts include a pair of mandibles and two pairs of maxillae. The first trunk legs are modified as mouthparts (maxillipeds) that become a functional part of the head. The maxillipeds contain a poison gland, with the venom injected through an opening near the end of the fang. The trunk has 15–191 pairs of legs, with one pair per segment, of which the last pair is usually the only one that is significantly modified: the last pair of legs has a sensory, grasping, or defensive function. The legs have six main segments, including coxa, trochanter, prefemur, femur, tibia, and a one- or two-part tarsus, and a terminal claw. Respiration is by tracheae, which are usually finely branched. The genital opening in both sexes is at the posterior end of the trunk.

As in other arthropods, the heart is dorsal and tubular, extending into the head as the aorta. The ventral nerve cord has paired ganglia in all leg-bearing segments. The brain is tripartite, as in insects. The gut is divided into an esophagus, midgut, and hindgut. The main excretory organs are a pair

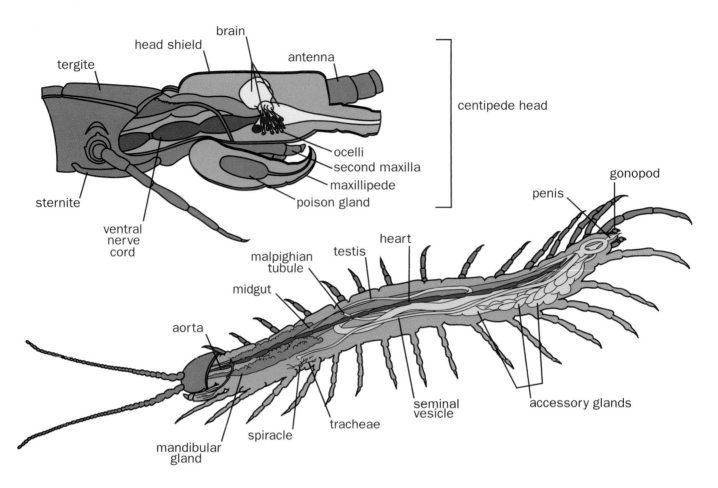

Centipede anatomy. (Illustration by Laura Pabst)

of Malpighian tubules that originate at the junction of the midgut and hindgut. An elongate ovary or testes run through much of the trunk. In both sexes, paired accessory glands originate at the genital atrium, at the rear end of the body.

Segments are of uniform length along the trunk on the underside of all centipedes. In all orders except Geophilomorpha, the tergal plates alternate between long and short along the trunk, except for between the seventh and eighth leg-bearing segment, where two long tergites occur in sequence. The tracheae open to spiracles that are confined to segments with long tergites, but are present on all trunk segments, except the last in Geophilomorpha. The number of leg pairs in centipedes totals to an odd number.

Color is highly variable. Most centipedes are drab, with the head and tergal plates yellow or brown (most Geophilomorpha and Lithobiomorpha). Large Scolopendromorpha are often brightly colored, often with a dark band across each tergal plate; in this order, the head and legs may be a different color than the trunk tergites.

Distribution

They occur worldwide, except Antarctica. Some species have become more widespread as a result of commerce and

plant introductions, carried in soil or with plants. Some species have disappeared from islands with the introduction of exotic mammals and snakes.

Habitat

Centipedes are common in wet forest and woodland, but many species inhabit dry forest, and some live in grassland or deserts. Some Geophilomorpha hunt in seaweed clusters in the littoral zone. Other species show a preference for caves; a few are confined to caves. Centipedes tolerate an elevational range from sea level to high mountain peaks, even for a single species.

Some species have quite specific microhabitats (such as rotting logs), but most thrive in a range of microhabitats, e.g., logs, bark, litter, or under stones.

Behavior

Centipedes are solitary, except for when brooding eggs or young. Contacts between members of the same species are often aggressive (sometimes even cannibalistic) or involve avoidance rituals. One genus of seashore geophilomorph is

A centipede (*Scolopendra*) eating an individual of the species *H. multifasciata*. (Photo by Danté Fenolio/Photo Researchers, Inc. Reproduced by permission.)

seen to hunt in packs, with numerous individuals feeding on the same barnacle or amphipod crustacean.

Fertilization is external, involving the transfer of a sperm packet that the female picks up with the back of her body and inserts in her genital atrium. The sexes are distinguished in Scutigeromorpha, Lithobiomorpha, and Geophilomorpha by differences in their gonopods (leg-derived structures at the back of the body). Many species have secondary sexual characteristics in the last pair of legs.

Some rituals are primarily defensive such as scolopendrids displaying the last leg pair outspread. Luminescence in some Geophilomorpha is produced by secretions from the sternal glands, which contain noxious chemicals that deter predators.

Few species are seen aboveground by day, and most are more active at night. Scutigeromorphs are inactive for long periods of time while waiting for prey. Other species show bursts in activity (e.g., captive *Scolopendra* is active for 1–2 hours on average each eighth night).

Species may inhabit deeper levels of the soil or litter during drier seasons. Some species migrate from litter to logs seasonally; seasonal migration between different forest types may occur over a small spatial scale. Apart from short-term occupation of a burrow, territoriality is unknown.

Feeding ecology and diet

Prey, which are typically other soft-bodied arthropods (including other centipedes) or worms are mostly taken alive. Large Scolopendridae can take mice, toads, birds, lizards, geckos, and small snakes as prey. Some geophilomorphs accept plants as food if denied animal prey for long enough.

Prey are often detected by the antennae, which are covered with dense mechanosensory and chemosensory hairs. The eyes do not seem to play a major role in prey detection. In some species, the last pair of legs is also used to detect or grab prey, and may be modified as pincers.

Prey are immobilized by venom injected from the maxillipede fang. The prey are held by the maxillipedes and sometimes also anterior walking legs, passed to the mouth by the first and second maxillae, and then cut up by the mandibles, which have a row of paired teeth in all centipedes, except Geophilomorpha. Geophilomorph mandibles sweep and rasp food. Salivary glands make secretions that break down the prey.

Reproductive biology

Most species have separate sexes, but some are parthenogenetic (female clones) throughout parts of their geographic range. Males have courtship rituals to entice the female to pick up a spermatophore, which is deposited on a web spun by the male in all centipedes, except Scutigeromorpha. A male initiates courtship, tapping the female's posterior legs with his antennae; this tapping ritual may last many hours. The female touches the web with the posterior end of her body so that the spermatophore lies against her genital opening or she picks up the sperm with her gonopods and deposits them in her genital atrium.

Single eggs are laid in Scutigeromorpha and Lithobiomorpha. Craterostimomorpha, Scolopendromorpha, and Geophilomorpha lay a group of 3–86 eggs that are protected by the mother, often in a hollow of a rotting log. Mothers hump their body around the egg cluster and the early hatchling instars in these three orders, ceasing feeding while brooding. Her grooming of the eggs seems to function to remove fungi. Eggs are camouflaged in soil, then abandoned in Scutigeromorpha and Lithobiomorpha.

Hatchlings have four pairs of legs in Scutigeromorpha, and six or seven pairs of legs in Lithobiomorpha; they are active from birth in those orders, and changes between subsequent instars are gradual. Hatchlings of Scolopendromorpha and Geophilomorpha have the adult number of legs. The first post-embryonic stages are incapable of hunting, and are brooded by the mother.

Breeding seasons vary for different species.

Conservation status

In general, centipede species have quite broad geographic distributions, and some are recorded from multiple continents. Many, however, are confined to narrower ranges, and some are known from single localities. A scolopendrid formerly collected in the Galápagos Islands may now be extinct. Introduced mammals and snakes on islands have decimated populations of some centipede species. Only one species (*Scolopendra abnormis*) is listed on the 2002 IUCN Red List; it is classified as Vulnerable.

Significance to humans

Centipedes have few uses to humans. Large Scolopendridae are used in the pet trade. Nearly all species are harmless to food crops and human goods (one species of geophilomorph is thought to feed on root crops). They have no role in causing or spreading diseases.

All centipedes are venomous, but most small species are incapable of piercing human skin or their bites are no worse than a bee sting. Bites by large Scolopendridae are painful, but pain and swelling pass after hours to days. There have been very few human deaths from centipede bites.

1. Scolopender (*Scolopendra morsitans*); 2. Stone centipede (*Lithobius forficatus*); 3. House centipede (*Scutigera coleoptrata*); 4. Earth centipede (*Pachymerium ferrugineum*); 5. Tasmanian remarkable (*Craterostigmus tasmanianus*); 6. Blind scolopender (*Cryptops hortensis*). (Illustration by Barbara Duperron)

Species accounts

Blind scolopender
Cryptops hortensis

ORDER
Scolopendromorpha

FAMILY
Cryptopidae

TAXONOMY
Scolopendra hortensis Donovan, 1810, a garden in England.

OTHER COMMON NAMES
None known.

PHYSICAL CHARACTERISTICS
Length of head and body usually about 0.7 in (20 mm), but up to 1.2 in (30 mm); color pale brown to orange-brown; usually 17 antennal articles; eyes lacking; 21 pairs of trunk legs; last pair of legs with a saw-like row of teeth along the last two segments (5–8 on the tibia and 2–4 on the tarsus) before the claw; prefemur of last legs with a groove along ventral side.

DISTRIBUTION
Native to Europe and North Africa; widely distributed in Britain and Ireland; introduced to Scandinavia, Iceland, and the United States, where it has a wide distribution.

HABITAT
Woodland under bark and stones; common in gardens and hothouses in introduced parts of its range.

BEHAVIOR
Quickly buries itself when disturbed; the anterior appendages push away soil, and the body wedged through loosened soil.

FEEDING ECOLOGY AND DIET
The last pair of legs has a prehensorial role. Their joints can strongly flex so that the saw-like row of spines grasps prey. The head is curved around and the maxillipede fangs are buried in the prey. Observed to kill flies, young caterpillars, and small harvestmen. Gut contents include spiders; other species of *Cryptops* have abundant earthworm remains in their guts.

REPRODUCTIVE BIOLOGY
Courtship and sperm transfer have not been seen, but spermatophores are bean-shaped with a tough, multilayered wall as in *Scolopendra*. Another species, *C. hyalinus*, has a brood size of 7–9 eggs.

CONSERVATION STATUS
Not listed by the IUCN.

SIGNIFICANCE TO HUMANS
None known. ◆

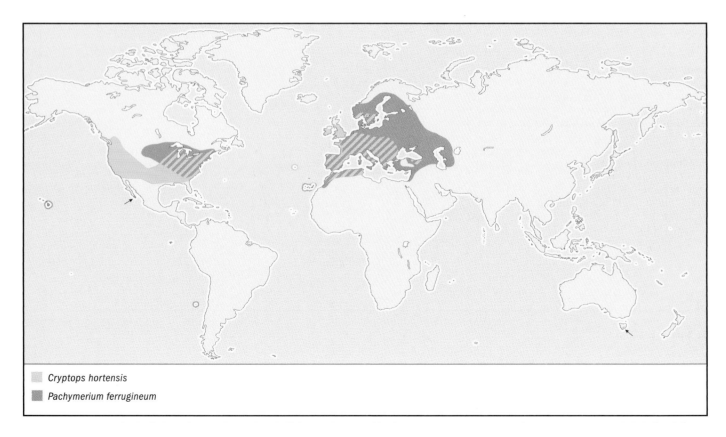

Cryptops hortensis
Pachymerium ferrugineum

Earth centipede
Pachymerium ferrugineum

ORDER
Geophilomorpha

FAMILY
Geophilidae

TAXONOMY
Pachymerium ferrugineum Koch, 1835, Europe.

OTHER COMMON NAMES
German: Erdläufer.

PHYSICAL CHARACTERISTICS
Length of head and trunk usually 1.2–1.3 in (30–35 mm), up to 1.9 in (50 mm); color reddish yellow, with head and maxillipede segment darker; eyes lacking; 43 trunk leg pairs in males, 43 or 45 in females; coxae of last leg pair not swollen, with many small coxal pores on its dorsal and ventral sides; last legs slender in female, swollen in male.

DISTRIBUTION
Throughout Europe, from Scandinavia to the Mediterranean, North Africa, Asia Minor and the Caucasus to Turkestan. Widely distributed in North America, where it is possibly introduced; also introduced to Japan, Taiwan, Hawaii, Juan Fernandez Island, and Mexico.

HABITAT
Common in coastal regions through most of its range. Inland occurrences are in grassland rather than woodland.

BEHAVIOR
Like all geophilomorphs, adapted to burrowing in the soil by elongating and contracting the body. Can tolerate long periods of submergence in water.

FEEDING ECOLOGY AND DIET
Related Geophilomorpha have been observed to feed on worms, and in the laboratory, they accept small earthworms, soft insect larvae, adult fruitflies, and collembolans. Structure of the mandible suggests that semi-fluid food is swept into the mouth.

REPRODUCTIVE BIOLOGY
Sexes distinguished externally by differences in the gonopods (two-segmented in male, stouter and single-segmented in female) and swollen last legs of male. Females lay between 20 and 55 eggs in a brood cavity (seen in moss or sand, likely also in soil); mother stops feeding for 40–50 days while guarding brood. Brooding occurs in summer. Mother usually eats brood when disturbed. Lifespan may be three years or more.

CONSERVATION STATUS
Not listed by the IUCN.

SIGNIFICANCE TO HUMANS
None known. ◆

Tasmanian remarkable
Craterostigmus tasmanianus

ORDER
Craterostigmomorpha

FAMILY
Craterostigmidae

TAXONOMY
Craterostigmus tasmanianus Pocock, 1902, near Hobart, Tasmania.

OTHER COMMON NAMES
None known.

PHYSICAL CHARACTERISTICS
Length of head and body up to 1.9 in (50 mm); color usually greenish brown with red-brown head; tapering antennae with 17–18 articles; one ocellus on each side of head shield; maxillipedes project in front of head shield; 15 pairs of trunk legs; long tergal plates subdivided in two (with a short pretergite), such that trunk appears to have 21 tergites; anogenital region enclosed in an elongate, ventrally-opening capsule with a mesh of openings for the coxal organs.

DISTRIBUTION
Throughout Tasmania. Records in New Zealand have been assigned to this species, but may be a distinct species.

HABITAT
Native forest and woodland, including rainforest, wet and dry eucalypt forest, subalpine woodland, and riparian and swamp forest; elevations from sea level to 4,265 ft (1,300 m). Favored microhabitats are in rotting logs, deep humus, and wet leaf litter.

BEHAVIOR
Relatively slow moving. Like Scolopendromorpha and Geophilomorpha, it shows maternal care of the egg cluster and hatchlings, brooding in a hollow in a rotting log. Through most of its range, individuals are active year-round.

FEEDING ECOLOGY AND DIET
Natural prey unknown, but potential prey are amphipods, isopods, millipedes, fly larvae, beetles, collembolans, and mites. In captivity, has been observed to use the maxillipedes to dig termites out of crevices in wood, but not found in association with termite mounds in the wild.

REPRODUCTIVE BIOLOGY
Sexes difficult to distinguish without examining testes or ovaries. Females found guarding egg clusters in September and guarding hatchlings in April. Number of eggs ranges from 44–77. Hatchlings emerge from the egg with 12 pairs of legs, and add the final three legs in the subsequent instar.

CONSERVATION STATUS
Not listed by the IUCN; common through much of range, but vulnerable to forest clearing and burning in drier, eastern parts of Tasmania.

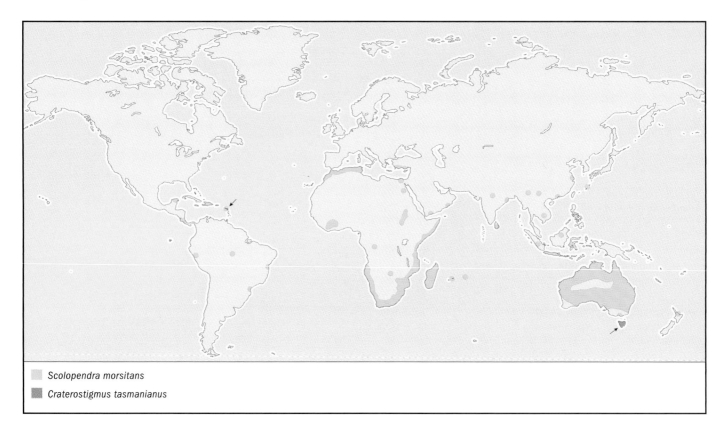

☐ *Scolopendra morsitans*
■ *Craterostigmus tasmanianus*

SIGNIFICANCE TO HUMANS
Isolated position of this species in centipede systematics gives it special scientific importance. ◆

Stone centipede
Lithobius forficatus

ORDER
Lithobiomorpha

FAMILY
Lithobiidea

TAXONOMY
Scolopendra forficata Linnaeus, 1758, Europe.

OTHER COMMON NAMES
German: Braune Steinläufer, Gemeiner Steinkriecher.

PHYSICAL CHARACTERISTICS
Length of head and body up to 1.2 in (30 mm); color chestnut brown; antenna with about 40 articles in mature specimens; cluster of ocelli (up to 40) on each side of head shield, with the posterior ocellus much larger than the others; 15 pairs of trunk legs; spiracles opening above legs on trunk segments 3, 5, 8, 10, 12, and 14; short trunk tergites 9, 11, and 13 with triangular projections at posterolateral corners; 5–9 elliptical coxal pores on each of legs 12–15.

DISTRIBUTION
Throughout Europe east to the Volga, Caucasus, Turkey, and North Africa. Widely distributed in North America, especially in the eastern United States, where it is likely an introduced species.

HABITAT
Usually found under stones, in rotting wood, or in moist leaf litter; common in woodland, grassland, and moorland as well as in gardens and greenhouses; broad elevational tolerance, from seashore to mountaintops. Moves from leaf litter in spring to logs and deep soil in summer and winter in parts of its range.

BEHAVIOR
Has been observed to climb tree trunks and wander in the open at night. Cannibalism has been observed in the laboratory and field; gut contents include other *L. forficatus*. Stimulus for taking prey is tactile or chemical.

FEEDING ECOLOGY AND DIET
Carnivorous. Gut contents include earthworms and small arthropods such as collembolans, spiders, and mites. Has been seen in the field to catch woodlice, collembolans, worms, and slugs; in captivity, feeds on flies, beetle larvae, small moths, small spiders, spider eggs, and other small arthropods. Juveniles may be specialist collembolan feeders.

REPRODUCTIVE BIOLOGY
Experiments suggest that the coxal organs on legs 12–15 emit sex-specific pheromones. Courtship ritual lasts a few hours, with tapping of last legs with antennae by both sexes, and male rocking body up and down. Male deposits spermatophore onto a 0.39 in (1 cm) wide web, then female moves over top of male and picks it up with her gonopods. Females lay single eggs nearly 0.039 in (1 mm) wide that are camouflaged with soil and abandoned; the eggs are held between a pair of stout spurs on each gonopod and the curved terminal claw of the gonopods.

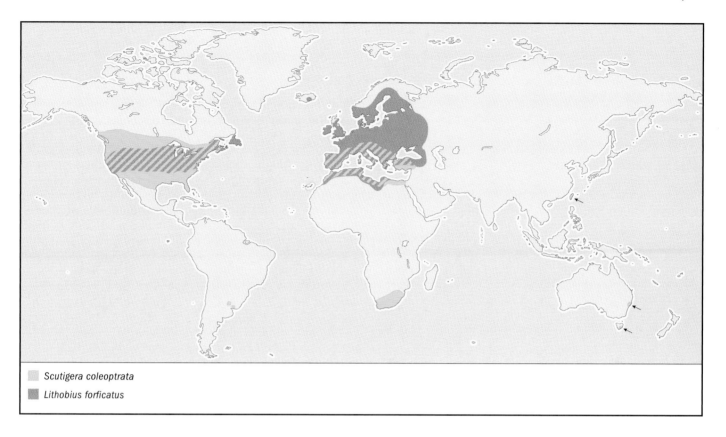

Scutigera coleoptrata

Lithobius forficatus

In captivity, female seen to lay eggs 21 times over four months. Offspring hatch with seven leg pairs, with subsequent stages have 7, 8, 10, and 12 pairs, then post-larval (15-legged) stages. Life span may be 5–6 years.

CONSERVATION STATUS
Not listed by the IUCN.

SIGNIFICANCE TO HUMANS
May be a predator on apple maggot pupae in apple orchards; preys on the symphylan *Scutigerella immaculata*, a greenhouse pest. ◆

Scolopender
Scolopendra morsitans

ORDER
Scolopendromorpha

FAMILY
Scolopendridae

TAXONOMY
Scolopendra morsitans Linnaeus, 1758, southern Europe.

OTHER COMMON NAMES
German: Skolopender, Riesenläufer.

PHYSICAL CHARACTERISTICS
Length of head and body up to 5.1 in (130 mm); color of head shield variable, trunk tergites yellow or brown with a dark band along posterior margin; cluster of four ocelli on each side of head shield; 17–23 (usually 18–21) antennal articles; maxilli-

pedes with a pair of tooth plates, each usually with five teeth; 21 pairs of trunk legs; tracheae open to triangular spiracles above legs on segments 3, 5, 8, 10, 12, 14, 16, 18, and 20; pair of median sulci on trunk tergites 2–20; prefemur of last leg with five rows of small spines (two rows on dorsal side, three rows of usually three spines on ventral side); prefemur extended as a process with 2–6 spines on its apex.

DISTRIBUTION
Very widely distributed through the tropics and warm parts of the temperate zone, including Central America and the Caribbean, much of Africa, Madagascar, South and East Asia, and Australia; a few records in tropical South America.

HABITAT
Variable, from desert to rainforest.

BEHAVIOR
Relatively fast runner, able to quickly penetrate litter when disturbed. The related *S. cingulata* burrows under stones and stays in a series of linked chambers for a few days. Contacts between individuals involve a ritual in which the animals grip each other with their last leg pair. Active at the surface throughout the year in at least parts of range.

FEEDING ECOLOGY AND DIET
Like most large scolopendrids, hunting is mostly nocturnal, the day spent in leaf litter, or under logs or bark. Gut contents show arthropod prey, including spiders, mites, centipedes, flies, beetles, ants, and termites. In the laboratory, reported to take frogs, small toads, and cockroaches.

REPRODUCTIVE BIOLOGY
As in all Scolopendromorpha, gonopods are lacking but sexes are distinguished externally by conical penis in male. Males

have flattened dorsal side of some segments on last legs. Females tend to be larger. Males deposit a bean-shaped spermatophore 0.01 in (2.5 mm) long through the penis onto a web. The mother's brood chamber is hollowed out in the soil under a stone, with 28–86 elliptical, greenish yellow eggs being laid. In parts of range (Nigeria), reaches maturity within a year, with two generations per year.

CONSERVATION STATUS
Not listed by the IUCN.

SIGNIFICANCE TO HUMANS
None known. ◆

House centipede
Scutigera coleoptrata

ORDER
Scutigeromorpha

FAMILY
Scutigeridae

TAXONOMY
Scolopendra coleoptrata Linnaeus, 1758, Spain.

OTHER COMMON NAMES
German: Spinnenassel.

PHYSICAL CHARACTERISTICS
Length of head and trunk up to 1.2 in (30 mm); color yellow or brown with three purple or blue bands along the length of the tergal plates; large compound eyes on each side of head; antenna divided into 500–600 annulations; spiracles opening as a slit on the rear margin of each of seven large tergal plates; 15 pairs of trunk legs, with tarsal segment of each flagellum divided into 250–300 annulations.

DISTRIBUTION
Native to the Mediterranean region (southern Europe and North Africa) and the Near East; introduced and widely distributed in North America and South Africa; limited distribution where introduced in Britain, northern Europe, Australia, Argentina, Uruguay, tropical Africa, and Taiwan.

HABITAT
Woodland, under pieces of wood and in litter; also found in caves; in introduced parts of its range, usually found in damp places in houses and woodpiles.

BEHAVIOR
Cannibalism has been observed. In the laboratory, specimens remain at the litter surface day and night, quickly scurrying for cover when disturbed. They are immobile while awaiting prey; contact seems necessary to recognize prey, which is grabbed immediately. Can run at 16 in (40 cm) per second.

FEEDING ECOLOGY AND DIET
In the lab, only live animals are taken. Prey includes many kinds of arthropods (flies, cockroaches, moths, spiders); generally eating only the soft parts. The anterior legs snare prey, keep it from escaping, and can hold on to captured flies while another fly is eaten.

REPRODUCTIVE BIOLOGY
Both sexes are distinguished externally by their gonopods; unique in having two pairs of male gonopods. Courtship involves partners forming a circle, tapping each other with their antennae; male rocks body up and down, and deposits a lemon-shaped spermatophore; male guides female to spermatophore, from which she removes sperm. Eggs oval, 0.05 in (1.25 mm) long. Female holds egg between gonopods, wipes it with soil, and deposits it in a soil crevice. Usually about four eggs laid per day over a breeding season of about two months (May–June in southern France). Eggs hatch in 30–38 days. Hatchlings have four pairs of legs, with subsequent stages having 5, 7, 9, 11, and 13 pairs, then five post-larval (15-legged) stages. Individuals have lived nearly three years in captivity.

CONSERVATION STATUS
Not threatened.

SIGNIFICANCE TO HUMANS
Preys on domestic flies and cockroaches. ◆

Resources

Books
Lewis, J. G. E. *The Biology of Centipedes.* Cambridge, U.K.: Cambridge University Press, 1981.

Minelli, Alessandro. "Chilopoda." In *Microscopic Anatomy of Invertebrates.* Volume 12, *Onychophora, Chilopoda, and Lesser Protostomata.* New York: Wiley-Liss, 1993.

Periodicals
Shelley, R. M. "Centipedes and Millipedes with Emphasis on North American Fauna." *The Kansas School Naturalist* 45, no. 3 (1999).

Other
International Centre of Myriapodology. June 2003 [July 24, 2003]. <http://www.mnhn.fr/assoc/myriapoda/INDEX.HTM>.

Gregory D. Edgecombe, PhD

Diplopoda
(Millipedes)

Phylum Arthropoda

Class Diplopoda

Number of families 148

Thumbnail description
Many-legged, often long-bodied, segmented animals with antennae; also possessing two pairs of legs on each body segment. The segments, which are actually two somites fused together, are called diplosegments.

Photo: A millipede (*Sigmoria aberrans*) displaying warning colors. (Photo by Gilbert S. Grant/Photo Researchers, Inc. Reproduced by permission.)

Evolution and systematics

The class Diplopoda contains about 10,000 described species in 15 orders and 148 families. Scientists believe that as many as 70,000 additional species have yet to be identified. The millipedes were once classified as a subclass of the class Myriapoda, which also contained the centipedes (now assigned to class Chilopoda). Since then, all four major myriapod groups have been given class status. The other two classes are Pauropoda and Symphyla.

Many researchers think that the millipedes may have developed during the Carboniferous period (360–286 million years ago) from the genus *Arthropleura*, a possibly diplosegmented myriapod that grew to an impressive 5.9 ft (1.8 m) long and 1.5 ft (0.45 m) wide. The largest extant millipedes, *Graphidostreptus gigas* and *Scaphistostreptus seychellarum* reach 11 in (28 cm) in length. Although the evolutionary history of the diplopods is still a disputed subject, systematists have generally agreed that diplopods and pauropods are the most closely related of the myriapods, followed by the symphylids. Some biologists have even suggested that these three groups of myriapods may be more closely related to insects than to the fourth myriapod group—the centipedes—but this view is hotly contested.

Physical characteristics

Millipedes differ from all other myriapods in having two pairs of legs on each body segment. A few segments typically have no legs or have only one pair. The first three segments form the thorax of the animal; the first of these has no legs while the second and third have one pair each. The fourth segment, which begins the abdomen, also usually has just one pair. The legs, which are uniramous (unbranched), may number from about two dozen to several hundred depending on the species. The species *Siphonophora millepeda* and *Illacme plenipes* hold the record with about 750 legs. The adult size of millipedes ranges from 0.08 in (2 mm) long in the subclass Penicillata to 11.8 in (30 cm) in *Triaenostreptus*. In general, millipedes are described as either typical (subclass Chilognatha) or bristly (subclass Penicillata).

Typical millipedes have a calcified exoskeleton and are usually long and thin. Species within the subclass Chilognatha, however, vary widely in appearance. For instance, members of the order Glomerida not only look superficially like pillbug isopods, but they also share their ability to conglobate, or roll themselves up into a ball. One morphological difference between the two groups is that the balled diplopods have dorsal plates that are similar in size from front to rear, while the posterior dorsal plates in isopods are much smaller than the anterior plates. Many chilognaths are almost round in cross section, but some, like those in the order Glomeridesmida, are flattened.

Bristly millipedes at first glance look more like hairy caterpillars than typical millipedes. They commonly have numerous transverse rows of setae (bristles) across their dorsal surface. Unlike typical millipedes, they have an uncalcified exoskeleton, so their bodies are soft. Bristly millipedes are small, reaching only about 0.16 in (4 mm), and have at most about a dozen segments.

A species of the genus *Polydesmus* mating. (Photo by David T. Roberts/Photo Researchers, Inc. Reproduced by permission.)

Overall, male and female millipedes are similar. The most outwardly noticeable difference between the two is leg length; males generally have longer legs than females. This characteristic likely assists the males in grasping females during mating.

Distribution

Millipedes do not travel far on their own, which helps to explain the vast number of species that may fill similar habitats just a few hundred miles apart. Humans, however, are efficient transporters of these animals, and have introduced species to new areas the world over. According to *The Biology of Millipedes*, "Just to give a few examples, 59 percent of the species recorded in Hordaland, Norway, by Meidell (1979) were considered to have been introduced there by man. Kime (1990b) recorded that half the species found in Britain have been introduced into North America."

Habitat

Millipedes are generally found in dark damp places, often under leaf litter, wood piles, and rocks, or in the top inch or two of soil. Most burrow by pushing their heads through the

dirt; but some, like *Polyzonium* species, also use their bodies to widen their tunnels.

A few species are arboreal (found in trees), including some bristly millipedes. These diplopods forgo burrowing and instead live in tiny cracks in tree bark. A number of spirostreptids and spirobolids are also arboreal. By contrast, some bristly millipedes are known for their preference for dry habitats. A few species, like *Archispirostreptus syriacus* and *Orthoporus ornatus*, live in deserts.

Some millipedes, like *Glomeris marginata*, are commonly seen in the open and in broad daylight.

Behavior

Perhaps the most well-known millipede behavior is their defense mechanism of conglobation—rolling up into a ball if they are short or into a spiral if they are longer. Most diplopods conglobate except the bristly millipedes. By conglobating, millipedes protect their softer and more vulnerable undersides from predators, leaving only their hard dorsal surfaces exposed. Another behavior of millipedes that protects them from both predators and desiccation (drying out) is their general preference for such damp dark places as burrows or crevices within tree bark. A few diplopods break this pattern and spend considerable time in the open.

Millipedes also defend themselves against predation via poisonous—or at least noxious-smelling—secretions they emit through pores on their sides. A few of the larger tropical species can actually squirt their secretions. The secretions of polydesmids contain cyanide. Some, like the bristly millipedes, don't produce defensive secretions.

Millipedes exhibit some unusual behaviors. For example, when *Diopsiulus regressus* feels threatened, it heaves itself off the ground and jumps 0.8–0.12 in (2–3 cm). Upon landing, it runs forward, then leaps again. The males of a few members of the order Sphaerotheriida can stridulate, making sounds by brushing their legs against the sides of their bodies. A few nocturnal millipedes glow in the dark, including members of the genus *Motyxia* (= *Luminodesmus*). It is thought that their bioluminescence warns potential predators of their noxious qualities. In addition, some millipedes, like *Calymmodesmus montanus*, form mutually beneficial relationships with ants or termites. The insects protect the diplopods from predators, and the millipedes perform housekeeping duties by eating fungi and detritus in the insects' nests and bivouacs.

Feeding ecology and diet

Most diplopods are detritus feeders, chewing up and digesting decaying leaves or other vegetation. Many, including members of the family Siphonophoridae, will also eat tender shoots or roots; a few derive nourishment from organic matter in ingested soil. Some, including those of the order Callipodida and possibly a few other species, will eat animal detritus; a number of polydesmids eat fungi.

Although many species of millipedes eat their own feces, a habit called coprophagy, the practice is not universal. Some researchers suggest that millipedes don't obtain much nutritional benefit from the fecal matter itself, instead drawing food value from the fungi growing within it.

Predators of millipedes include amphibians, reptiles, birds, carnivorous invertebrates, and some insect-eating mammals.

Reproductive biology

The sexes are separate in diplopods; most millipedes reproduce sexually. Some species are parthenogenetic, especially those in the order Polydesmida; polydesmid females produce daughters without the contribution of male sperm. Some diplopod species perform courtship rituals. In *Julus scandinavius*, for example, the male presents a gift secretion to the female. *Loboglomeris pyrenaica* males stridulate in order to entice females.

The chilognaths and the bristly millipedes differ in their methods of insemination, with the former employing direct sperm transfer and the latter indirect transfer. A chilognath

male generates a spermatophore (sperm jelled together to form a packet) and moves it from the genital opening to the gonopod, an intermittent organ. He uses this gonopod to place the spermatophore directly into the female's genital opening. A male bristly millipede instead spins a web and ejects his sperm into it. The female approaches the web and inserts the sperm into her genital opening herself. Females of all millipedes store sperm and fertilize the eggs as they deposit them.

Female diplopods lay eggs in nests in the soil, sometimes making capsules to help protect the eggs. *Narceus* species, for example, mold individual egg capsules from masticated leaves. In some species, the females guard the eggs; in a few, like some platydesmids, the males take over the sentry role. The young undergo two or three molts inside the nest, emerging in most cases with three pairs of legs. They gain segments, or "rings," and legs with each successive molt through a process called anamorphosis. Millipedes generally molt in such protected spaces as underground burrows, crevices, or even slight depressions in the soil. A few, like *Narceus americanus* and *Orthoporus ornatus*, seal themselves into chambers during this particularly vulnerable stage of their lives.

In a year or two, sometimes longer, the young diplopods molt into sexually mature individuals. Many male julids, especially *Ommatoiulus* and *Tachypodoiulus* species, are unusual in their ability to molt "backwards," so to speak, reverting from sexually mature adults to nonsexual stadia (known as intercalary or Schalt stadia). The purpose of this reverse molt, called periodomorphosis, is unknown.

The life span of millipedes varies among species and can range from one to 11 years, possibly longer.

Conservation status

No diplopods are listed by the IUCN.

Significance to humans

As detrivores, millipedes' major contribution lies in promoting overall plant decomposition. One study estimated that they add two tons of manure to each acre of the forest floor each year.

Occasionally, large numbers of millipedes are reported to damage gardens and crops. In Japan, outbreaks of the species *Parafontaria laminata* have been known to cause transportation problems. The diplopods are run over by trains and their flattened bodies stick to the rails. Large numbers of these crushed millipedes on the rail surface have actually caused railroad cars to lose traction.

In addition, humans who come in contact with some species of millipedes may have severe allergic reactions. Some species, like those in the genus *Spirobolus*, secrete a defensive chemical that irritates human skin.

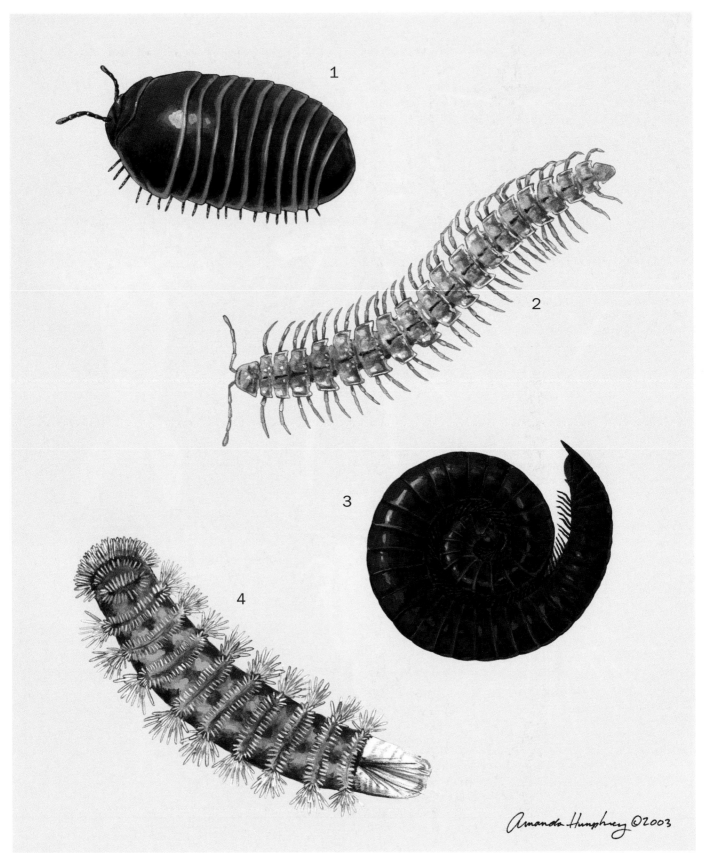

1. Pill millipede (*Glomeris marginata*); 2. Flat-backed millipede (*Polydesmus angustus*); 3. Snake millipede (*Julus scandinavius*); 4. Bristly millipede (*Polyxenus lagurus*). (Illustration by Amanda Humphrey)

Species accounts

No common name
Underwoodia iuloides

ORDER
Chordeumatida

FAMILY
Caseyidae (Underwoodiidae)

TAXONOMY
Underwoodia iuloides Harger, 1872, Simon's Harbor, Pukaskwa National Park, Ontario.

OTHER COMMON NAMES
None known.

PHYSICAL CHARACTERISTICS
This shiny, dark brown millipede is lighter on its ventral surface and has somewhat banded antennae. The adults average about 0.4 in (10 mm) in length. Like other members of this genus, these millipedes have three projections extending from the anterior gonopod.

DISTRIBUTION
Northeastern quarter of North America, stretching as far south as North Dakota, Michigan, and upstate New York, and west as far as Saskatchewan, Canada. A population has also been found in northeastern New Mexico.

HABITAT
Damp wooded areas with thick leaf litter or coastal barrens that experience frequent fog and rain.

BEHAVIOR
Little is known of this diplopod's behavior, except that it prefers to remain under stones or logs. It apparently requires habitats with high moisture content.

FEEDING ECOLOGY AND DIET
U. iuloides is thought to feed on leaf detritus.

REPRODUCTIVE BIOLOGY
Only a few males have ever been found, which indicates that the species may be parthenogenetic over much of its range. Little is known about its reproductive biology.

CONSERVATION STATUS
Not listed by the IUCN.

SIGNIFICANCE TO HUMANS
None known. ◆

Glomeris marginata

Julus scandinavius

Underwoodia iuloides

Pill millipede
Glomeris marginata

ORDER
Glomerida

FAMILY
Glomeridae

TAXONOMY
Glomeris marginata Villers, France, 1789.

OTHER COMMON NAMES
French: Gloméris marginé; German: Gemeine Saftkugler;
Finnish: Mustapallotuhatjalkainen.

PHYSICAL CHARACTERISTICS
Short dark brown or black millipedes with 12 dorsal diplosegments lined posteriorly in light brown or light gray. One segment is actually two diplosegments fused into one large plate.
Pill millipedes are dome-shaped in cross section, with 17–19
pairs of legs. They reach 0.8 in (20 mm) long and 0.3 in (8
mm) wide.

DISTRIBUTION
British Isles; western and northwestern Europe.

HABITAT
Unlike most other millipedes, *G. marginata* is less prone to
desiccation and is often seen in the open, even on sunny days.
It is usually found under leaf litter in forests, fields, and gardens.

BEHAVIOR
When threatened, *G. marginata* rolls into a ball that resembles
an isopod. Another defense mechanism is a secretion that dis-
courages most predators. Some males stridulate, and all are capable of producing a pheromone that attracts females for
mating.

FEEDING ECOLOGY AND DIET
Pill millipedes eat decomposing leaves. One study in France
showed that *G. marginata* was responsible for eating about one
of every 10 leaves that fell to the forest floor every autumn. Its
predators include hedgehogs.

REPRODUCTIVE BIOLOGY
Reproduction is cyclical with eggs produced in spring and early
summer. Brood size varies according to the size of the female,
but six to seven dozen is common. Eggs are about 0.04 in (1
mm) in diameter and deposited inside a capsule. The young
emerge as second stadia about two months after the eggs are
laid. Temperature can affect the timing of development and
delay hatching by several months. The young may take several
years to mature. Females have an active reproductive life of
several years; half a dozen broods in a lifetime is not unusual.
G. marginata can live as long as 11 years.

CONSERVATION STATUS
Not listed by the IUCN.

SIGNIFICANCE TO HUMANS
This species plays an important role in recycling dead leaves
and similar vegetable matter. ◆

Snake millipede
Julus scandinavius

ORDER
Julida

FAMILY
Julidae

TAXONOMY
Julus scandinavius Latzel, 1884, from several localities in Austria
including Kirchdorf and Wien (lectotypification is required to
establish a type locality).

OTHER COMMON NAMES
German: Schnurfüsser.

PHYSICAL CHARACTERISTICS
These black to brownish black millipedes are long and cylindrical in shape with a heavily armored dorsal surface. Adult
males are identified by mesial (located toward the middle) ventrally directed processes on the first joint (coxa) of the second
pair of legs. Adult females range from 0.6–1.5 in (16–38 mm)
in length and 0.06–0.11 in (1.5–2.7 mm) in diameter, and
males from 0.5–1.1 in (13–29 mm) long and 0.06–0.07 in
(1.4–1.9 mm) in diameter. Immature stadia are light brown.

DISTRIBUTION
British Isles and western Europe, with an introduced population in Massachusetts, northeastern United States.

HABITAT
Prefer sandy soils in forested areas; often found under leaf litter.

BEHAVIOR
Snake millipedes, like other julids, are sometimes referred to as
"bulldozers" for the way they push with their legs and force
their heads through the soil. Mating among julids occurs when
the male locates a female and climbs onto her back. He twists
himself around her body until the gonopores are adjacent. He
then produces a glandular secretion and presents it to the female before copulation.

FEEDING ECOLOGY AND DIET
Eats decaying leaf matter, preferring ash to oak leaves.

REPRODUCTIVE BIOLOGY
After the female lays her eggs in April, the young emerge in
about four to six weeks as stadium-III individuals, having undergone the first two molts in the egg. The young molt three
or four more times by the end of their first winter, and twice
again during each of the next two winters. The females, now
almost three years old, are finally ready to mate. They die
shortly after laying their eggs.

CONSERVATION STATUS
Not listed by the IUCN.

SIGNIFICANCE TO HUMANS
None known. ◆

Flat-backed millipede

Polydesmus angustus

ORDER
Polydesmida

FAMILY
Polydesmidae

TAXONOMY
Polydesmus angustus Latzel, 1884, Normandy, France.

OTHER COMMON NAMES
French: Polydesme; German: Bandfüsser; Dutch: Grote Platrug; Norwegian: Flattusenbeinet.

PHYSICAL CHARACTERISTICS
Flat-backed millipedes resemble typical centipedes in form; they are flattened dorsoventrally and have long legs. They also possess long segmented antennae and sculptured dorsal segments. The adults are dark brown, have about 20 segments, and range from 0.6–1.0 in (14–25 mm) in length and about 0.16 in (4 mm) wide.

DISTRIBUTION
Northwestern Europe, now also introduced to the southeastern United States.

HABITAT
Prefers compost piles, hiding places under tree bark, crevices in decomposing trees, and loose soil packed with organic matter.

BEHAVIOR
Like other polydesmids, this species has paranota or keels on its dorsal surface. The paranota help them to burrow into the soil by providing wedges that they can lift and lower to open up the soil in front of them and move forward.

FEEDING ECOLOGY AND DIET
Flat-backed millipedes feed on roots, dead leaves, and other vegetative detritus, as well as on strawberries and other fruits.

REPRODUCTIVE BIOLOGY
Breeding season generally runs from late spring through the summer or from late summer through mid-fall. The females store sperm and may produce several broods, although the males typically breed only once. The young become sexually mature in 1–2 years, with those hatched in the earlier breeding season maturing in one year, and those hatched in the second season in two years. According to research published in 2003, the second group is sensitive to light, which affects both development and reproduction. This study was the first to show the influence of photoperiods among diplopods.

CONSERVATION STATUS
Not listed by the IUCN.

SIGNIFICANCE TO HUMANS
None known. ◆

Bristly millipede

Polyxenus lagurus

ORDER
Polyxenida

FAMILY
Polyxenidae

TAXONOMY
Polyxenus lagurus Linnaeus, 1758, Sweden.

OTHER COMMON NAMES
German: Pinselfüsser; Dutch: Penseelpoot; Swedish: Penselfoting.

PHYSICAL CHARACTERISTICS
These light brown millipedes have an uncalcified exoskeleton, which gives them a soft body. They also possess characteristic bristles that look like transverse rows of fringe separating their diplosegments. Splotches of dark brown are visible on the dorsal surface. Bristly millipedes reach only 0.08–0.12 in (2–3 mm) in length, and have 13–17 pairs of legs (12 pairs on average) on their 11–13 segments.

DISTRIBUTION
Found in Europe, northern Africa, and western Asia, as well as northeastern North America and the area around Vancouver on the western coast of Canada.

HABITAT
Polyxenus lagurus does not burrow; instead it seeks refuge beneath or between pieces of loose bark.

BEHAVIOR
Unlike the chilognath millipedes, *P. lagurus* employs indirect sperm transfer. In this method, the male spins a small web onto which he deposits his sperm. He finishes the web with a string of so-called "signal threads" that attract the female and lead her to the sperm. The female then approaches the web and gathers the sperm into her genital opening.

FEEDING ECOLOGY AND DIET
This species finds its meals on tree bark, where it feeds on the algae growing there.

REPRODUCTIVE BIOLOGY
Although *P. lagurus* does reproduce sexually, some populations are parthenogenetic. Females in these groups produce offspring from unfertilized eggs. In sexual reproduction, however, males and females mate to generate fertilized eggs. The female lays her sticky eggs in a loose mass and then uses her tail brush to fashion a protective sheath around them.

CONSERVATION STATUS
Not listed by the IUCN.

SIGNIFICANCE TO HUMANS
None known. ◆

Polyxenus lagurus

Polydesmus augustus

Resources

Books

Hopkin, S. P., and H. J. Read. *The Biology of Millipedes.* Oxford, U.K.: Oxford University Press, 1992.

Minelli, A., ed. *Proceedings of the Seventh International Congress of Myriapodology.* Leiden, The Netherlands: E. J. Brill, 1990.

Periodicals

David, J. F., J. J. Geoffroy, and M. L. Célérier. "First Evidence for Photoperiodic Regulations of the Life Cycle in a Millipede Species, *Polydesmus angustus* (Diplopoda: Polydesmidae)." *Journal of Zoology (London)* 260 (2003): 111–116.

Dove, H., and A. Stollewerk. "Comparative Analysis of Neurogenesis in the Myriapod *Glomeris marginata* (Diplopoda) Suggests More Similarities to Chelicerates Than to Insects." *Development* 130 (2003): 2161–2171.

Niijima, K., and K. Shinohara. "Outbreaks of the *Parafontaria laminata* Group (Diplopoda: Xystodesmidae)." *Japanese Journal of Ecology* 38 (1988): 257–268.

Shear, W. A. "Millipedes." *American Scientist* 87 (1999): 232–239.

Shelley, R. M. "The Millipede Genus *Underwoodia* (Chordeumatida: Caseyidae)." *Canadian Journal of Zoology* 71 (1993): 168–176.

Organizations

British Myriapod and Isopod Group. Web site: <http://www.bmig.org.uk/>

Other

"Diplopoda." *Studies in Arthropod Morphology and Evolution.* [4 Aug. 2003]. <http://www.life.umd.edu/entm/shultzlab/vtab/diplopoda.htm>.

"The Diplopoda (Millipedes)." *Earth-Life Web.* 20 June 2003 [4 Aug. 2003]. <http://www.earthlife.net/insects/diplopoda.html>.

"*Glomeris marginata* (The Pill Millipede)." *Casual Intruders.* 2000 [4 Aug. 2003]. <http://www.the-piedpiper.co.uk/th11b(2).htm>.

"Myriapods." *Nature of the Northcoast: Millipedes.* 27 Nov. 2001 [4 Aug. 2003]. <http://www.humboldt.edu/natmus/NorthcoastNature/Miriapoda/myriapoda.html>.

"Orders of Millipedes." *Herper.com.* [4 Aug. 2003]. <http://www.herper.com/myriapods/orders.html>.

"Species Spotlight: Millipedes." *Illinois Natural History Survey.* 1 Aug. 2003 [4 Aug. 2003]. <http://www.inhs.uiuc.edu/chf/pub/surveyreports/may-jun95/milli.html>.

"Systematics: Myriapods." 11 Feb. 2003 [4 Aug. 2003]. <http://www.biols.susx.ac.uk/ugteach/cws/syst/myriapods.htm>.

Leslie Ann Mertz, PhD

Symphyla
(Symphylans)

Phylum Arthropoda

Class Myriapoda

Subclass Symphyla

Number of families 2

Thumbnail description
Small, whitish, and weakly sclerotized animals that dwell in soil

Photo: A member of the family scutigerellidae (above), a member of the family Scolopendrellidae (below). (Photo by Ernest C. Bernard, University of Tennessee. Reproduced by permission.)

Evolution and systematics

The Symphyla seem to be a very old and homogenous group, probably monophyletic. It is known from both Dominican and Baltic amber. Contrary to Diplopoda, Chilopoda, and Pauropoda (other subclasses within the Myriapoda), the Symphyla have a remarkably uniform anatomy and outer morphology. Only two families have been distinguished: Scutigerellidae, with five genera and about 125 swift-moving species, generally 0.15–0.31 in (4–8 mm) long; and Scolopendrellidae, with eight genera and about 75 generally slow-moving species, length 0.078–0.15 in (2–4 mm). Numerous papers have been published over more than 100 years, but the general knowledge of the group is still very incomplete. This is because research has been restricted to investigations based on questions posed by an early interest in the affinities of the group, and later, on sporadic studies on the composition of the fauna. Many reports have also been published on different aspects of the destructiveness, control, and population dynamics of the garden symphylan (*Scutigerella*). Many scientists now categorize Symphyla as a class rather than a subclass.

Physical characteristics

The trunk of symphylans is whitish, 0.078–0.31 in (2–8 mm) long, and has 14 segments, the same number of segments as some insects have. The gonopore is unpaired and situated in the anterior part of the body, and may be secondarily developed. However, the mouthparts and the locomotory habit show more connections with the other myriapods than with insects.

The head is heart shaped, well demarcated from the trunk, and has one pair of simple moniliform antennae, three pairs of mouthparts, and one pair of postantennal organs. Eyes are lacking. All but the two most-posterior trunk segments are

subsimilar, and have one pair of legs and an entire or subdivided tergite. The tergites are weakly sclerotized and their number is 15–24, always greater than the number of trunk segments and legs. There are 12 pairs of legs, and at the bases of most of them are short styli and coxal sacs. The latter are probably important for water and salt balance. The preanal segment has two large subconical and posteriorly directed cerci connected with spinning glands in the last trunk segment, and the anal segment is provided with a pair of long sensory hairs.

Distribution

Symphylans are subcosmopolitan. Because the taxonomy is poorly developed, many more species will be described when better identification characteristics have been discovered.

Habitat

Symphylans occur both in natural and agricultural habitats, but seldom in heavy, peaty, or very wet soils. Sometimes they penetrate to a depth of at least 3.2 ft (1 m). Moisture seems to be the most important factor determining their vertical distribution.

Behavior

Symphylans are usually present in large numbers and sometimes distinctly aggregated. They are negatively phototropic, but this response is not very strongly developed. When individuals are in motion, the antennae are kept in constant movement; when feeding, the antennae are held backwards. Symphylans may be very swift runners, which feature rapidly disappears when they are disturbed. The touching of the posterior spinnerets with a small brush will cause them to start spinning a thread from which they can be hanging in the air. The sexes are separate, and the males deposit stalked sperm-packets, which the females pick up. Nothing is known about other types of social behavior and communication. Their display and territoriality are unknown. Vertical and horizontal migrations occur when soil conditions change.

Feeding ecology and diet

Most species are probably omnivores, but the main food sources are fungal hyphae and fresh root material. Some species cause damage to growing crops both in fields and hothouses. There are more than 800 papers dealing with symphylans, and many of them describe injuries caused by the garden symphylan to crops of pineapple, beet, potato, bean, and many others. Population densities of several thousands specimens per square meter are not unusual.

Reproductive biology

The sexes are separate, and the unpaired gonopore opens out at the fourth pair of legs. Two kidney-shaped plates around the gonopore identify adult males. Fertilization is in-direct; the partners do not come in contact with one another. The pearly white eggs have a diameter of about 0.011 in (0.3 mm) and are deposited in masses of 4–25. The first larval instar has six or seven pairs of legs and is very inactive. The second larval instar has eight pairs of legs, then stages follow with nine, 10, and 11 pairs of legs before the adult stage with 12 pairs of legs is reached.

Conservation status

Most papers dealing with symphylans have focused on the destructiveness to growing crops by the garden centipede. However, endemism probably often occurs, but has been ignored in nature conservation because of the partly undeveloped taxonomy and the lack of specialists. No species are listed by the IUCN.

Significance to humans

Like the pauropods, the symphylans are largely unknown to the public. Because they have caused severe damage to growing crops in both green- and hothouses and in the field, they are well known to many growers, particularly in the United States. They are not dangerous to humans.

Species accounts

Garden symphylan
Scutigerella immaculata

ORDER
Symphyla

FAMILY
Scutigerellidae

TAXONOMY
Scolopendrella immaculata Newport, 1845, near London (United Kingdom).

OTHER COMMON NAMES
English: Garden centipede; French: Scutigerelle.

PHYSICAL CHARACTERISTICS
Large, generally 0.19–0.31 in (5–8 mm) long, whitish or light brownish. Not possible to identify in the field.

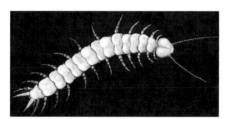
Scutigerella immaculata

DISTRIBUTION
The species most often recorded in the literature; however, the taxonomy of the genus *Scutigerella* is problematic, and the species delimitation of *S. immaculata* is uncertain. The distribution is considered to be unknown since several other species have been confused with it. (Specific distribution map not available.)

HABITAT
Lives in litter and soil in both natural and agricultural hothouse habitats.

BEHAVIOR
Swift moving, especially adults.

FEEDING ECOLOGY AND DIET
Feeds on vegetable material, most often fungal hyphae and thin roots. Can be serious pest in fields, gardens, and green- and hothouses.

REPRODUCTIVE BIOLOGY
Sexes separate; unpaired gonopore opens out at the fourth pair of legs. Two kidney-shaped plates around gonopore distinguish adult males; fertilization indirect. Pearly white eggs have diameter 0.011 in (0.3 mm); deposited in masses of 4–25. First instar larva has six or seven pairs of legs; is very inactive. Second instar larva has eight pairs of legs, then follow stages with nine, 10, and 11 pairs of legs before the adult stage with 12 pairs of legs.

CONSERVATION STATUS
Not listed by the IUCN.

SIGNIFICANCE TO HUMANS
Always has to be regarded as a possible, sometimes severe, pest of growing crops. ◆

Resources

Books

Edwards, Clive A. "Symphyla." In *Soil Biology Guide*, edited by Daniel L. Dindal. New York: John Wiley and Sons, 1990.

Scheller, Ulf. "Symphyla." In *Biodiversidad, Taxonomía y Biogeografía de Artrópodos de México: Hacia una Síntesis su Conocimiento*, edited by Jorge Llorente Bousquets and Juan J. Morrone. Tlalpan, Mexico: Comisión Nacional para el Conocimiento y Uso de la Biodiversidad (CONABIO), 2002.

Scheller, Ulf, and Joachim Adis. "Symphyla." In *Amazonian Arachnida and Myriapoda. Identification Keys to All Classes, Orders, Families, Some Genera, and Lists of Known Terrestrial Species*, edited by Joachim Adis. Sofia/Moscow: Pensoft Publishers, 2002.

Periodicals

Scheller, Ulf. "Symphyla from the United States and Mexico." *Texas Memorial Museum, Speleological Monographs* 1 (1986): 87–125.

Ulf Scheller, PhD

Pauropoda

(Pauropods)

Phylum Arthropoda

Class Myriapoda

Subclass Pauropoda

Number of families 5

Thumbnail description
Very small myriapods that live hidden in litter and soil; highly specialized to live in the edaphic (soil) environment

Photo: *Allopauropus carolinensis.* (Photo by Ernest C. Bernard, University of Tennessee. Reproduced by permission.)

Evolution and systematics

Though no fossil pauropods have been found from before the time of the Baltic amber, they seem to be an old group closely related to the Diplopoda. Their head capsules show great similarities to diplopods: both have three pairs of mouthparts and the genital openings occur in the anterior part of the body. Moreover, both groups have a pupoid phase at the end of the embryonic development. The two groups probably have a common origin.

There are two orders: Hexamerocerata and Tetramerocerata. Hexamerocerata has a 6-segmented and strongly telescopic antennal stalk and a 12-segmented trunk with 12 tergites and 11 pairs of legs. The representatives are white and proportionately long and large. The one family in this order, Millotauropodidae, has one genus and a few species. Tetramerocerata has a 4-segmented and scarcely telescopic antennal stalk, 6 tergites, and 8–10 pairs of legs. Representatives of this order are often small (sometimes very small), and white or brownish. Most species have nine pairs of legs as adults. The four families include Pauropodidae, Afrauropodidae, Brachypauropodidae, and Eurypauropodidae. Most genera and species belong to the family Pauropodidae. Many scientists now categorize Pauropoda at the class level.

Physical characteristics

These white or brownish animals are relatively short, 0.019–0.078 in (0.5–2.0 mm), with a very flexible trunk. They are blind and have many partly unusual sensory organs. There are five pairs of long sensory hairs on the trunk. The antennae are complicated, biramous, with three flagellae. They are provided with different organs to help the animal analyze the

surroundings, among them the end organs of the flagellae and the globulus at the distal part of the lower antennal branch.

There are 12 trunk segments, but the number of tergites is always less. The head is small, directed downwards, without eyes but with eyelike sensory organs. Behind the last trunk segment, there is an anal segment, the pygidium, which is horizontally divided. It has a most peculiar structure posteriorly, the anal plate (diameter 0.00039–0.00079 in [0.01–0.02 mm]), which is the greatest factor for identification purposes. Almost every species has a unique plate that is characterized by its shape (form, size, cuticular structure) and, if any, its appendages (number, length, thickness, direction, surface structure, insertion points). Each species can be identified from this plate, even at the first larval stage.

Distribution

Hexamerocerata has a purely tropical range, while in Tetramerocerata, most genera are subcosmopolitan.

Habitat

In most environments, their occurrence is patchy and the population sparse. However, sometimes they have been reported to have several thousand specimens per square meter, even in agricultural habitats. They are easy to find under stones and molting tree branches that have good contact with the underlying moist soil. They are also under moss carpets. They inhabit many plant communities and soil types and are most abundant in a zone about 3.9–7.8 in (10–20 cm) deep. However, though they cannot burrow, they can follow root canals and crevices in much deeper levels, down to the groundwater surface.

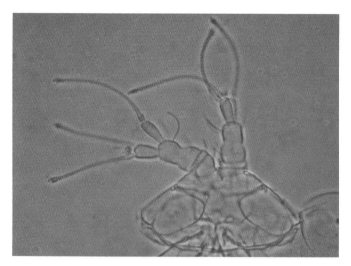

Allopauropus gracilis (Hansen), showing the 3-branched antenna that is the most prominent feature of the subclass. (Photo by Ernest C. Bernard, University of Tennessee. Reproduced by permission.)

Behavior

Pauropods have a patchy occurrence and can often be found aggregated on the underside of stones and tree branches. They are most often swift runners, with mouse-like intermittent rushes. The species most often observed can turn its body almost 180°. Pauropods are shy of light and try to disappear in crevices and soil clumps as soon as possible. The antennae rotate constantly with an unusually high rapidity to examine the environment. The sexes are separate, and males deposit small spherical sperm-packets in the soil, which the females seek and pick up. Nothing is known about other types of social behavior and communication. Their display and territoriality are unknown. Vertical migration occurs when there are changes in soil moisture.

Feeding ecology and diet

The food habits of most species are unknown, but some can eat mold or suck out fungal hyphae; at least one species even eats root hairs. Most species move swiftly with intermittent rushes very much like a miniature mouse. Reduced agility occurs in the families Brachypauropodidae and Eurypauropodidae.

Reproductive biology

Pauropods are bisexual and progoneate. The egg develops in a short pupoid phase before the first larval instar appears. In Tetramerocerata, the first larval instar has three pairs of legs, and is then followed by instars with five, six, and eight pairs of legs; the adults have eight, nine, or 10 pairs of legs. In Hexamerocerata, the first larval instar has six pairs of legs. Parthenogenetic reproduction seems to occur, particularly in areas with unfavorable environmental conditions.

Conservation status

The anticipated low number of species and the insufficient knowledge of their distribution have resulted in little consideration for pauropods among conservationists. Some species have subcosmopolitan ranges, but most of the species probably have very restricted ranges. Endemism ought to be common, at least in the tropics. No species are listed by the IUCN.

Significance to humans

Pauropods are so small that they have often been overlooked, even by many trained soil zoologists. Consequently, pauropods are little known by the general public. A single species has recently been found feeding on and damaging *Saintpaulia* cuttings in a greenhouse in the Netherlands. They are not dangerous to humans.

Species accounts

No common name
Allopauropus carolinensis

ORDER
Tetramerocerata

FAMILY
Pauropodidae

TAXONOMY
Pauropus carolinensis Starling, 1943, Duke University Forest at Durham, North Carolina, United States.

OTHER COMMON NAMES
None known.

PHYSICAL CHARACTERISTICS
About 0.039 in (1 mm) long, white or whitish in color. Can be distinguished from other species in the genus only by a high magnification microscope.

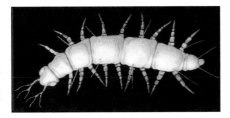

Allopauropus carolinensis

DISTRIBUTION
Known only from the eastern United States.

HABITAT
Deciduous forests.

BEHAVIOR
An active species, not yet well studied.

FEEDING ECOLOGY AND DIET
Nothing is known, probably fungicolous.

REPRODUCTIVE BIOLOGY
Nothing is known.

CONSERVATION STATUS
Not listed by the IUCN.

SIGNIFICANCE TO HUMANS
None known. ◆

Resources

Books
Scheller, Ulf. "Pauropoda." In *Biodiversidad, Taxonomía y Biogeografía de Artrópodos de México: Hacia una Síntesis su Conocimiento,* edited by Jorge Llorente Bousquets and Juan J. Morrone. Tlalpan, Mexico: Comisión Nacional para el Conocimiento y Uso de la Biodiversidad (CONABIO), 2002.

————. "Pauropoda." In *Soil Biology Guide,* edited by Daniel L. Dindal. New York: John Wiley and Sons, 1990.

Periodicals
Scheller, Ulf. "The Pauropoda (Myriapoda) of the Savannah River Plant, Aiken, South Carolina." *Savannah River Plant and National Environmental Research Park Program* 17 (1988): 1–99.

Ulf Scheller, PhD

Aplacophora

(Aplacophorans)

Phylum Mollusca

Class Aplacophora

Number of families 30

Thumbnail description
Vermiform (worm-shaped) marine mollusks lacking shells and living in the zone between the seashore and the edge of the continental shelf

---

Photo: A caudofoveatan (species unknown) seen in the Antarctic. (Photo by S. Piraino [University of Lecce]. Reproduced by permission.)

Evolution and systematics

The class Aplacophora contains two subclasses: Neomeniomorpha (also called Solenogastres) and Chaetodermomorpha (also called Caudofoveata). Most neomenioids creep by means of a narrow foot with a ventral groove that begins as a pedal pit toward the front of the animal. They have a sensory vestibule above the mouth; a single midgut organ combining stomach and digestive gland; and serial sets of muscle bands running along their sides and lower surface. Neomenioids are simultaneous hermaphrodites; they also lack ctenidia in their mantle cavities. A ctenidium is a finger-shaped or comblike structure that functions in respiration. The subclass of Neomeniomorpha comprises three orders (Pholidoskepia, Neomeniomorpha, and Cavibelonia); 24 families; 75 genera; and fewer than 250 species. The subclass Chaetodermomorpha contains six families and 15 genera. Aplacophorans in this subclass have a midgut separated into a stomach and a digestive organ, and one pair of ctenidia in their mantle cavities.

It is uncertain to what extent aplacophorans are specialized and to what extent they are primitive mollusks, but there is no evidence that they ever had shells. Their specialized features include the reduction or loss of the foot; the absence of a shell; sometimes the lack of a radula (a specialized organ unique to mollusks that allows them to scrape food from the ocean floor); and modifications of the nephridia (simple organs for excreting wastes) in certain genera to form accessory sexual organs. The possession of well-defined cerebral and pleural ganglia (groups of nerve cells) indicates that the Aplacophora are more advanced than the Polyplacophora mollusks in this respect at least. It seems probable that aplacophorans represent a secondary simplification of an ancestral form. If this hypothesis is accurate, however, the form and location of the mantle cavity in present-day aplacophorans tells researchers little about the condition of the remote ancestor. Aplacophorans have little in common with chitons (small armor-plated mollusks), although in the past the two groups were placed together in the class Amphineura.

Physical characteristics

Aplacophorans, which are also called solenogasters, are worm-shaped mollusks covered with spicules or sharp needle-like projections. The body shape varies from almost spherical to elongated and slender. These mollusks are usually less than 2 in (5 cm) in length, but adult individuals may vary from 0.039–0.078 in (1–2 mm) to 3.9 in (10 cm) or more in length.

The exterior of an aplacophoran may be spiny, smooth, or rough. The head is poorly developed, and the typical mollusk shell and foot are absent. The exoskeleton is represented only by a cuticular (horny secreted) layer that bears spicules in a variety of forms. The spicules and integument (covering) together form a character that links genera or families in this class to one another. Most aplacophorans have some specialized spicules at the entrance to the mantle cavity; these are presumably used in copulation. The cuticle and epidermis may be either thick or thin relative to the size of the species: a thick cuticle may occur together with a thin epidermis; a thin cuticle may occur with a thick epidermis; or they may be the same thickness. Glandular cells on the epidermis known as papillae may have either long stalks or no stalks at all.

Aplacophorans have a midventral longitudinal groove containing one or more ridges, which are similar in structure to the foot of other mollusks. The mantle covers the upper surface, the sides, and the greater part of the lower surface of the animal. A large gland that secretes mucus opens into the groove toward the front.

The mouth at the front of the animal opens into a muscular pharynx lined by a thick cuticle. The pharynx typically receives the products of one or two pairs of salivary glands and the radula sac. Some genera lack salivary glands. Neomenioid species creep by ciliary action of the "foot" along a sticky track of mucus produced from the ciliated, eversible pedal pit at the anterior end of the pedal groove. Both the pedal groove and the pedal pit are supplied by many mucus-secreting glands. The radula is highly variable in form. It is situated where the pharynx joins the midgut unless an esophagus is present; it may have two teeth per row, one tooth per row, or many teeth per row. The radula is lacking in 20% of known species.

The posterior end of the body contains a cavity into which one or two gametopores open as well as the anus, the copulatory spicule sacs, and the folds of respiratory tissue or papillae. The posterior cavity is believed to represent a mantle cavity. Burrowing species have a pair of gills. In *Neomenia* and in several other genera there is a circlet of laminar gills in the mantle cavity; in other genera, however, there are no gills.

Distribution

Aplacophorans are found in all oceans of the world; some genera have worldwide distribution. Although they have been sampled at depths ranging from 16 to 17,390 ft (5–5,300 m), the greatest diversity of species occurs at depths greater than 656 ft (200 m).

Habitat

Neomenioids live on hydroids, corals, or surface sediment. Caudofoveatans construct burrows in marine sediments, which they inhabit head downward.

Behavior

Nothing is known about the behavior of aplacophorans.

Feeding ecology and diet

Neomenioids feed on cnidarians—stony and soft corals, hydrozoans, zooantharians, or gorgonians. Some species prey only on specific cnidarians. Caudofoveatans ingest sediment or may be selective carnivores or scavengers. They feed mostly on faraminifera.

Reproductive biology

Aplacophorans are hermaphrodites and have paired gonads. Copulation probably occurs in those with the former condition, and spawning in the latter. Researchers have inferred from the presence of seminal receptacles, the structure of introsperm (sperm that never contact the water), and observation of living specimens of *Epimenia australis* that fertilization takes place internally.

Conservation status

No species are listed by the IUCN.

Significance to humans

Aplacophorans are used in scientific research, especially research into the evolutionary origins of mollusks.

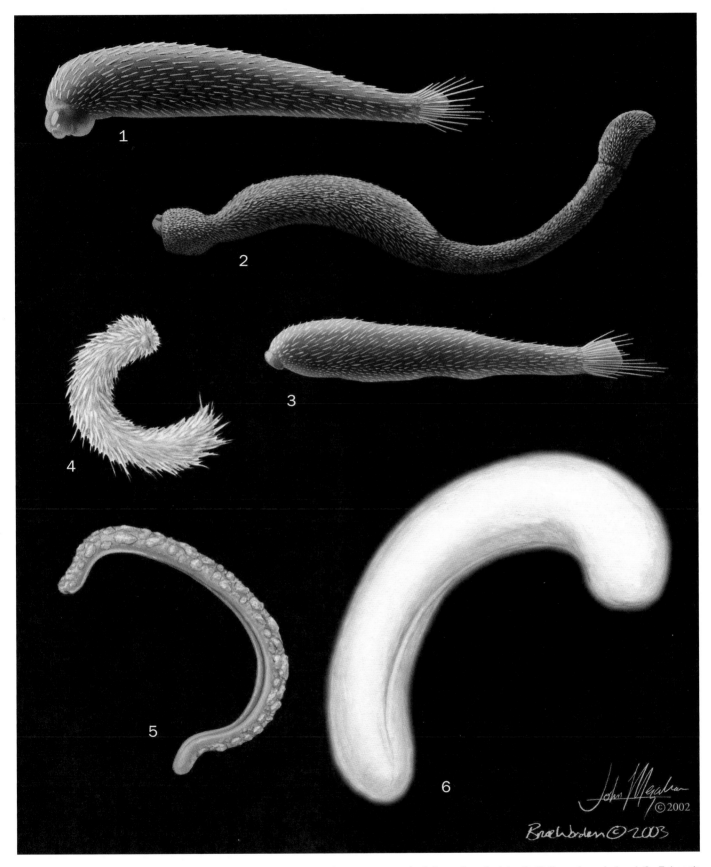

1. *Prochaetoderma yongei*; 2. *Chaetoderma argenteum* 3. *Chevroderma turnerae*; 4. *Spiomenia spiculata*; 5. *Helicoradomeria juani*; 6. *Epimenia australis*. (Illustration by Bruce Worden and John Megahan)

Species accounts

No common name
Epimenia australis

ORDER
Cavibelonia

FAMILY
Epimeniidae

TAXONOMY
Epimenia australis Thiele, 1897, Timor Sea, at a depth of 590 ft (180 m).

OTHER COMMON NAMES
None known.

PHYSICAL CHARACTERISTICS
Epimenia australis may grow as long as 4.33 in (11 cm) and as wide as 0.19–0.23 in (5–6 mm). There are irregular iridescent bright green or blue patches against a reddish brown background along the upper surface and sides of the organism, with raised reticulate (netlike) ridges and verrucae (wartlike projections) between the patches. *Epimenia australis* has spicules lying parallel to the body; the spicules cross each other running from the front of the lower surface to the rear of the upper surface and from the rear of the lower surface to the front of the upper surface. The cuticle is thicker within the ridges and verrucae, and the spicules stand entirely upright, extending in thick clumps beyond the cuticle. The head end is narrower than the posterior end when the organism is viewed from above; the posterior end is somewhat flattened dorsoventrally and is usually bent downward into a hook. The part of the head in front of the somewhat protruded pedal pit is lifted off the substrate, with the vestibule held open. The animals are stiff to handle. *Epimenia australis* has a radula with about 200 rows of teeth on a distinct pedestal; a pedal pit with two joined lobes; and several footfolds toward the front, decreasing to three and then to one toward the rear of the animal.

DISTRIBUTION
Found throughout the Indo-Pacific tropics and subtropics.

HABITAT
Epimenia australis is found close to shore with its cnidarian prey on the undersurface of sheltered coral slabs. It usually forms tightly tangled masses of as many as 11 individuals.

BEHAVIOR
Nothing is known.

FEEDING ECOLOGY AND DIET
Epimenia australis feeds on cnidarians.

REPRODUCTIVE BIOLOGY
Individual members of this species are hermaphrodites. Fertilization takes place internally.

CONSERVATION STATUS
Not listed by the IUCN.

Epimenia australis

Helicoradomenia juani

Spiomenia spiculata

SIGNIFICANCE TO HUMANS
Useful in scientific research. ◆

No common name
Helicoradomenia juani

ORDER
Cavibelonia

TAXONOMY
Helicoradomenia juani Scheltema and Kuzirian, 1991, Endeavour Segment, Juan de Fuca Ridge, 47°57′N, 129°04′W, at a depth of 7,382 ft (2,250 m).

OTHER COMMON NAMES
None known.

PHYSICAL CHARACTERISTICS
The body shape is somewhat elongated, with a fuzzy appearance. The organism may grow as long as 0.2 in (5 mm). It is narrowest toward its front end, with a sensory pit on the upper surface; its spicules are widest at the base and point backward, varying in shape from short, wide, and recurved, to long, slender and curved. There are two spicules in the pocket of each copulatory spicule. The mouth lies at the closer end of the vestibule; the pedal pit is large and often protrudes. The cuticle is thin. *H. juani* has a large radula that makes a single turn into its ventral pockets. The radula has 34–35 rows of teeth with five or six denticles (small toothlike projections).

DISTRIBUTION
Explorer, Juan de Fuca, and Gorda Ridges, at a depth of 5,906–10,732 ft (1,800–3,271 m).

HABITAT
Hydrothermal vents.

BEHAVIOR
Nothing is known.

FEEDING ECOLOGY AND DIET
Helicoradomenia juani differs from most neomenioids in that it does not eat cnidarians; it feeds instead on organic matter.

REPRODUCTIVE BIOLOGY
Adult individuals of *H. juani* are hermaphrodites.

CONSERVATION STATUS
Not listed by the IUCN.

SIGNIFICANCE TO HUMANS
Scientific research. ◆

No common name
Spiomenia spiculata

ORDER
Cavibelonia

FAMILY
Simrothiellidae

TAXONOMY
Spiomenia spiculata Arnofsky, 2000, West European Basin, 55°7.7 min N, 12°52.5 min W, at a depth of 9,505 ft (2,897 m).

OTHER COMMON NAMES
None known.

PHYSICAL CHARACTERISTICS
Spiomenia spiculata has a curved body that tapers slightly toward the posterior end; it is usually widest at its midsection. It has a dorsofrontal sensory pit, dorsoterminal sense organ, and a mantle cavity opening partly closed off by spicules. The longest epidermal spicules on *S. spiculata* are found toward the rear of the organism near the opening to the mantle cavity. There are nine different types of spicules, including two types of copulatory spicules that occur in paired groups protruding through the mantle cavity. Several accessory copulatory spicules are grouped near the opening to the mantle cavity. The pedal groove contains three types of solid spicules. The radula of *S. spiculata* makes a single turn into paired anteroventral radular pockets; there are 22–25 teeth with 22–23 denticles per tooth. In addition, this aplacophoran has unusually large paired salivary glands on its upper surface that empty into the esophagus where it joins the radular sac.

DISTRIBUTION
West European Basin at a depth of 6,560–13,120 ft (2,000–4,000 m).

HABITAT
Nothing is known.

BEHAVIOR
Nothing is known.

FEEDING ECOLOGY AND DIET
Spiomenia spiculata does not appear to depend on cnidarians as a food source. It feeds on diatoms and unidentified spicules that resemble those found in sponges.

REPRODUCTIVE BIOLOGY
Members of this species are hermaphrodites.

CONSERVATION STATUS
Not listed by the IUCN.

SIGNIFICANCE TO HUMANS
Spiomenia spiculata is used in scientific research. ◆

No common name
Chaetoderma argenteum

ORDER
Chaetodermatida

FAMILY
Chaetodermatidae

TAXONOMY
Chaetoderma argenteum Heath, 1911, southwestern Vancouver Island, British Columbia, 328–656 ft (100–200 m).

OTHER COMMON NAMES
None known.

PHYSICAL CHARACTERISTICS
This species has four distinct body regions: the anterium (with the oral shield), the anterior trunk, the posterior trunk, and the

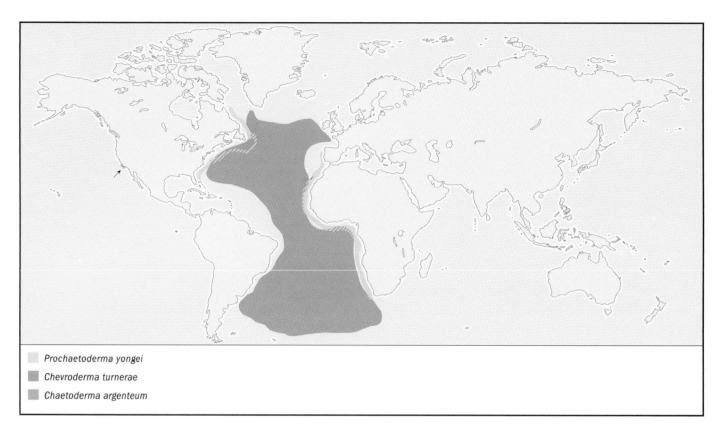

Prochaetoderma yongei

Chevroderma turnerae

Chaetoderma argenteum

posterium. This species typically has an anterior trunk either equal in diameter to, or narrower than, the neck and longer than the posterior trunk. Dense erect spicules on anterior trunk; more sparse flat spicules against posterior trunk. The oral shield is unpaired and cleft dorsally. Posteriorly the dorsoterminal sense organ is obvious and about 0.04 in (1 mm) in length in large specimens. The spicules of the posterium do not form a terminal ring. Greatest length to more than 1.6 in (40 mm). Radula with a large, cone-shaped cuticular piece and a single pair of denticles. Denticles are rather small. The cuticular dome extends proximally one-half the length of cone.

DISTRIBUTION
Found in the northeast Pacific, south of Monterey Bay in the Santa Maria Basin.

HABITAT
In sediment of fine silt and clay at depths between 328 ft (100 m) and 1,969 ft (600 m) in the sea.

BEHAVIOR
Nothing is known.

FEEDING ECOLOGY AND DIET
Nothing is known.

REPRODUCTIVE BIOLOGY
Nothing is known.

CONSERVATION STATUS
Not listed by the IUCN.

SIGNIFICANCE TO HUMANS
None known. ◆

No common name
Chevroderma turnerae

ORDER
Chaetodermatida

FAMILY
Prochaetodermatidae

TAXONOMY
Chevroderma turnerae Scheltema, 1985, North American Basin, 35°50.0′N, 64°57.5′W, 15,857 ft (4,833 m).

OTHER COMMON NAMES
None known.

PHYSICAL CHARACTERISTICS
Chevroderma turnerae is large for a prochaetodermatid species, with total body length of 0.11–0.20 in (2.8–5 mm). Body has three distinct regions reflecting internal anatomy: the anterium, the trunk, and the posterium. Trunk diameter averages 0.02 in (0.5 mm), with greatest diameter of 0.03 in (0.8 mm). The posterium is long, from two-fifth to nearly one-half the body length; it averages 0.04–0.06 in (1.1–1.6 mm) in length, with greatest length of 0.11 in (2.9 mm), and 0.01 in (0.3 mm) in diameter, with greatest diameter of 0.02 in (0.5 mm). The spicules of the trunk meet at a distinct angle along the dorsal midline. The long spicule blades of the posterium extend out from the body. There are two rows of prominent oral shield spicules; the oral shield is distinctly large and is divided into two lateral parts. The margin of the cloaca in lateral view is slanted. Radula and jaws large, teeth up to 0.00551 in (140 μm) long, jaws up to 0.0276 in (700 μm) long, and central plate long, up to 0.00197 in (50 μm), narrow and curved.

DISTRIBUTION
Chevroderma turnerae is a cosmopolitan abyssal species of the Atlantic basins, absent only in Guyana and Iberian Basins. Its depth range is also great, from a little over 6,890 ft (2,100 m) to 17,090 ft (5,208 m). It is most commonly found at depths greater than 9,840 ft (3,000 m) except in the Argentine basin, where it was taken in largest numbers at depths less than 9,840 ft (3,000 m).

HABITAT
Muddy surfaces at great depths in the sea.

BEHAVIOR
Nothing is known.

FEEDING ECOLOGY AND DIET
Thought to be omnivores, feeding on a wide variety of organic material.

REPRODUCTIVE BIOLOGY
Nothing is known.

CONSERVATION STATUS
Not listed by the IUCN.

SIGNIFICANCE TO HUMANS
None known. ◆

No common name
Prochaetoderma yongei

ORDER
Chaetodermatida

FAMILY
Prochaetodermatidae

TAXONOMY
Prochaetoderma yongei Scheltema, 1985, North American Basin, 39°46.5′N, 70°43.3′W, 4,364–4,823 ft (1,330–1,470 m).

OTHER COMMON NAMES
None known.

PHYSICAL CHARACTERISTICS
Prochaetoderma yongei is a small, slender, translucent species with flat-lying spicules oriented anterior-posterior except where they diverge along the ventral midline. Body has three distinct regions reflecting the internal anatomy: the anterium (with the oral shield), the trunk, and the posterium. Total body length averages 0.06–0.11 in (1.5–2.8 mm). Trunk diameter averages 0.01–0.02 in (0.3–0.4 mm), with greatest diameter of 0.02 in (0.6 mm). The posterium is about one-quarter the total length. An opaque, thickened patch of cuticle at the ventral junction of the trunk and posterium is characteristic of the species. Oral shield is small, and is divided into two lateral shields with indistinct spicules. The cloaca is rounded. Solid spicules are flat with the base shorter than the blade; blade is broad and triangular with median keel and sharp distal point. There is a pair of large, cutilar jaws and a small distichous radula. The central radula plate is short.

DISTRIBUTION
Very widely distributed on the continental slope between 2,625 ft (800 m) and 6,560 ft (2,000 m) in the northwestern and eastern Atlantic.

HABITAT
Muddy surfaces at great depths in the sea.

BEHAVIOR
Nothing is known.

FEEDING ECOLOGY AND DIET
Thought to be omnivores, feeding on a wide variety of organic material.

REPRODUCTIVE BIOLOGY
Nothing is known.

CONSERVATION STATUS
Not listed by the IUCN.

SIGNIFICANCE TO HUMANS
None known. ◆

Resources

Books

Barnes, Robert D. *Invertebrate Zoology*. 4th ed. Philadelphia: Saunders College Publishing, 1980.

Beesley, Pamela L., Graham J. B. Ross, and Alice Wells. *Mollusca—The Southern Synthesis, Part A*. Canberra: Australian Biological Resources Study, 1998.

Geise, Arthur C., and John S. Pearse. *Reproduction of Marine Invertebrates*. Vol. V, *Mollusks: Pelecypods and Lesser Classes*. London: Academic Press, 1975.

Purchon, R. D. *The Biology of Mollusca*. Oxford: Pergamon Press, Ltd., 1968.

Periodicals

Arnofsky, Pamela. "*Spiomenia spiculata*, Gen. et sp. nov (Aplacophora: Neomeniomorpha) Collected from the Deep Water of the West European Basin." *The Veliger* 43, no. 2 (2000): 110–117.

Scheltema, Amélie H. "The Aplacophoran Family Procahetodermatidae in the North American Basin, including *Chevroderma n.g.* and *Spathoderma n. g.* (Mollusca: Chaetodermomorpha)." *Biology Bulletin* 169, no. 2 (1985): 484–529.

Scheltema, Amélia H., John Buckland-Nicks, and Fu-Chiang Chia. "*Chaetoderma argenteum* Heath, a Northeastern Pacific Aplacophoran Mollusc Redescribed (Chaetodermomorpha: Chaetodermatidae)." *The Veliger* 34, no. 2 (1998): 204–213.

Scheltema, Amélie H., and M. Jebb. "Natural History of a Solenogaster Mollusc from Papua New Guinea, *Epimenia australis* (Thiele) (Aplacophora: Neomeniomorpha)." *Journal of Natural History* 28 (1994): 1297–1318.

Scheltema, Amélie H., and Alan M. Kuzirian. "*Helicoradomenia juanigen*, Gen. et sp. nov., a Pacific Hydrothermal Vent Aplacophoran (Mollusca: Neomeniomorpha)." *The Veliger* 34, no. 2 (1991): 195–203.

Scheltema, Amélie H., and C. Schander. "Discrimination and Phylogeny of Solenogaster Species Through the Morphology of Hard Parts (Mollusca, Aplacophora, Neomeniomorpha)." *Biology Bulletin* 198 (2000): 121–151.

Treece, Granvil D. "Four New Records of Aplacophorous Mollusks from the Gulf of Mexico." *Bulletin of Marine Science* 29, no. 3 (1979): 344–364.

Tatiana Amabile de Campos, MSc

Monoplacophora

(Monoplacophorans)

Phylum Mollusca

Class Monoplacophora

Number of families 1

Thumbnail description
Molluscans with a single symmetrical shell

Illustration: *Laevipilina antarctica.* (Illustration by Bruce Worden)

Evolution and systematics

Monoplacophorans are a small group of Cambro-Silurian fossils in the family Tryblidiidae. The single modern genus is *Neopilina*. In 1952, the *Galathea* expedition dredged 10 living specimens of *Neopilina galathea* from a deep trench of the Pacific Ocean. The specimens were found at a depth of 11,700 ft (3,570 m) in a bottom of dark, muddy clay off the west coast of Mexico. Later, four specimens of *Neopilina ewingi* were collected from the Peru-Chile Trench at a depth of slightly more than 9,840 ft (3,000 m). The monoplacophorans had been classified with the chitons or the gastropods, and it was only on examination of the soft parts of living *Neopilina* specimens that it was recognized that a new class was needed for this genus and for the fossil genera *Pilina*, *Scenella*, *Stenothecoides*, *Tryblidium*, *Archaeophiala*, *Drahomira*, *Proplina*, and *Bipulvina*. All these fossil genera share a peculiar and distinctive feature: the undersurface of the shell has three to eight muscle scars that must be interpreted as homologous, segmentally repeated structures also present in the living genus *Neopilina*. This segmental internal structure is thought to show a relation between mollusks and annelids.

The survival of *Neopilina* species undoubtedly correlates with adaptation to life at great depths, and they are perhaps considerably more specialized than other members of the class. Fossil species appear to have evolved along two lines. In one group (subclass Cyclomya), there was an increase in the dorsoventral axis of the body, leading to a planospiral shell and a reduction of gills and retractor muscles. Although they disappeared from the fossil record in the Devonian, this group may have been ancestral to the gastropods. The other line (subclass Tergomya) retained a flattened shell with five to eight retractor muscles. Although this group also disappeared from the fossil record in the Devonian, it is believed to have survived in the genus *Neopilina*.

Monoplacophorans comprise one order (Tryblidioidea), one family (Neopilininae), four genera, and 17 recent species.

Physical characteristics

Monoplacophorans have a single symmetrical shell, which varies in shape from a flattened shieldlike plate to a short cone. The shell is large and bilaterally symmetric and has a single depressed, limpet-shaped valve. The apex of the shell lies almost vertically above the anterior margin in the median line. The margin of the shell is almost circular, being slightly elongated in the sagittal plane. The shell is relatively thin, has three layers, and is slightly thicker toward the margin than in the center.

A pallial groove (the mantle cavity) separates the edge of the foot from the mantle on each side. The mantle cavity forms a shallow gutter that entirely surrounds the animal. The cavity is delimited internally by the walls of the foot and laterally by the pallial fold that underlies the margin of the shell. Anteriorly, the mantle cavity contains the mouth, which is surrounded by the anterior velar ridge of a lateral fold and a pair of tentacle ridges. Laterally, the mantle cavity contains five or six pairs of gills, each of which is suspended from the roof of the mantle cavity by a slender base. The foot is a short circular column. The flat ventral wall of the foot is thin, transparent, relatively lacking in muscular support, and covered by ciliated epithelium, and it forms a creeping sole. A pedal gland, which lies along the anterior borders of the foot, may aid creeping movement by supplying mucus. Monoplacophorans are 0.25 in (3 mm) to a little more than 1.25 in (3 cm) long and externally resemble a combination of gastropod and chiton. The head is small. The mouth is in front of the foot, and the anus is in the pallial groove at the posterior end of the body, behind the foot. In front of the mouth is a preoral fold, or velum, which extends on each side as a rather large ciliated palplike structure. Another fold lies behind the mouth and projects to each side as a mass of postoral tentacles.

Distribution

Deep seas (624–22,980 ft [190–7,000 m]) in various parts of the world, including the South Atlantic Ocean, Gulf of Aden, and a number of localities in the eastern Pacific Ocean

Habitat

Monoplacophorans live at great depths in the sea, crawling on radiolarians, attached to rocks, and in debris collected from the bottom.

Behavior

Nothing is known.

Feeding ecology and diet

Monoplacophorans feed on detritus.

Reproductive biology

The sexes are separate. There are no copulatory organs, no indication of sexual dimorphism, and no traces of sperm in the female reproductive system. It has been suggested that genital products (eggs and sperm) are discharged into the water column and that fertilization occurs externally.

Conservation status

No species are listed by the IUCN.

Significance to humans

Species in this taxonomic group are used for scientific research.

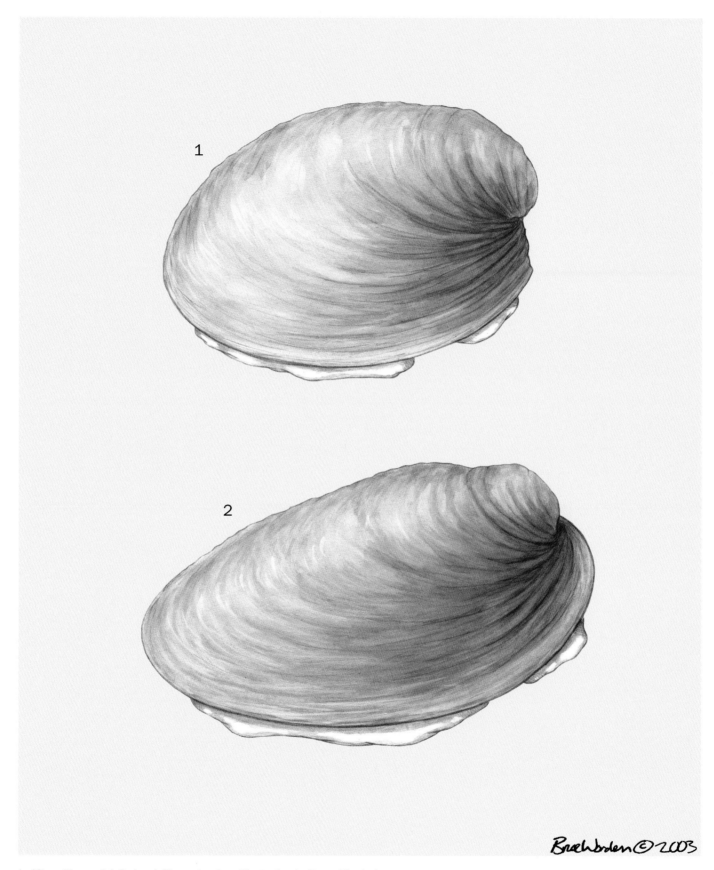

1. *Micropilina arntzi*; 2. *Laevipilina antarctica*. (Illustration by Bruce Worden)

Species accounts

No common name
Laevipilina antarctica

ORDER
Tryblidioidea

FAMILY
Neopilininae

TAXONOMY
Laevipilina antarctica Warén and Hain, 1992, *Polarstern* expedition ANT VII/4, station 245, 75°40.4'S, 029°37.2'W, 1,560 ft (480 m).

OTHER COMMON NAMES
None known.

PHYSICAL CHARACTERISTICS
The shell is small, fragile, depressed, and transparent with a flat peristome. The apex is slightly mamillate, forms an angle of approximately 60° with the basal plane, and is situated behind the anterior margin. The apical area has no distinct sculpture but has only regularly shaped impressions. Outside this area begins a uniform, concentric sculpture of low, raised ridges formed by the concentric arrangement of the prisms of the prismatic layer, which also form indistinct and fragmentary radial ridges. The shell is low. A convex curve forms gradually toward the posterior aspect, the highest point of the curve being somewhat behind the apex. The velar lobes are well developed and strongly ciliated. The anterior lip is conspicuous and rather thinly cuticularized. The postoral tentacles are short and claviforme and equipped with approximately seven short, stumpy distal appendages. There are five pairs of gills. The foot (contracted) measures 0.06 by 0.04 in (1.5 by 0.9 mm). The gonads are visible as a large, lobate, dorsal sac along each side of the animal. The anus is a simple opening in the pallial furrow.

DISTRIBUTION
Weddell Sea, Lazarev Sea, Antarctica.

HABITAT
Stones or old shells in the sea at a depth of 690–2,100 ft (210–644 m).

FEEDING ECOLOGY AND DIET
All feed by scraping off the thin layer of sediment and eating mineral particles, unidentified organic material, scattered sponge spicules, radiolarian fragments, small nematodes, and polychaete bristles.

BEHAVIOR
Nothing is known.

REPRODUCTIVE BIOLOGY
Nothing is known.

CONSERVATION STATUS
Not listed by the IUCN.

SIGNIFICANCE TO HUMANS
Research use. ◆

No common name
Micropilina arntzi

ORDER
Tryblidioidea

FAMILY
Neopilininae

TAXONOMY
Micropilina arntzi Warén and Hain, 1992, *Polarstern* expedition, station 173, 70°00.5'S, 03°16.8'W, 624–672 ft (191–204 m).

OTHER COMMON NAMES
None known.

PHYSICAL CHARACTERISTICS
The shell is very small, fragile, inflated, and almost semiglobular and has a large, bulbous apex and flat peristome. The apex is mamillate and forms an angle of approximately 45° with the basal plane. Apart from occasional small pits, there is no distinct sculpture on the slightly worn apical area. Outside this area is a fine, irregularly concentric striation. In the furrows between the ridges are numerous small pits. The shell is unusually convex with the apex well in front of the anterior edge. The highest point of the shell is slightly anterior of center. No muscle scars can be discerned on the interior. The exterior layer of the shell does not contain defined prisms. The head is

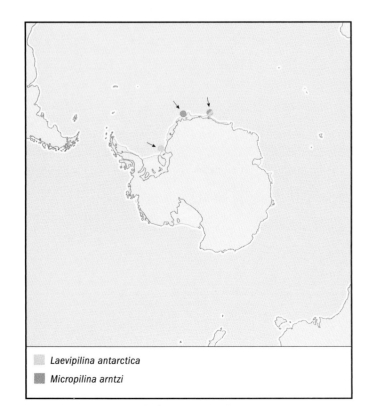

Laevipilina antarctica

Micropilina arntzi

unusually large and bulging and has short, tapering, strongly ciliated velar lappets at the sides. The anterior lip seems to be very solid and cuticularized. Postoral tentacles are not present. The foot is round with a thickened rim. Three pairs of small, simple, tubercular gills are situated in the pallial groove and lack appendages. Five small, close-set muscle bundles are situated along the central third of the body halfway between the midline and the lateral margin.

DISTRIBUTION
Lazarev Sea, Antarctica

HABITAT
Sediment bottoms with stones and shells at 624–2,500 ft (191–765 m).

FEEDING ECOLOGY AND DIET
Nothing is known.

BEHAVIOR
Nothing is known.

REPRODUCTIVE BIOLOGY
Nothing is known.

CONSERVATION STATUS
Not listed by the IUCN.

SIGNIFICANCE TO HUMANS
Used in scientific research. ◆

Resources

Books

Barnes, Robert D. *Invertebrate Zoology*. Philadelphia: Saunders College, 1980.

Geise, Arthur C., and John S. Pearse. *Reproduction of Marine Invertebrates*. Vol. V, *Mollusks: Pelecypods and Lesser Classes*. London: Academic Press, 1975.

Purchon, R. D. *The Biology of Mollusca*. Oxford, U.K.: Pergamon Press, 1968.

Periodicals

Warén, Anders, and Stefan Hain. "*Laevipilina antarctica* and *Micropilina arntzi*: Two New Monoplacophorans from the Antarctic." *The Veliger* 35 (1992): 165–176.

Tatiana Amabile de Campos, MSc

Polyplacophora
(Chitons)

Phylum Mollusca

Class Polyplacophora

Number of families 10

Thumbnail description
Mollusks with a flattened, ovoid shape, broad ventral foot, and eight (sometimes seven) dorsal shell plates that overlap one another and allow the animal to bend and mold itself onto a rock to avoid wave dislodgement

Photo: The lined chiton (*Tonicella lineata*) is a slow moving mollusk with a shell composed of multiple plates. (Photo by David Hall/Photo Researchers, Inc. Reproduced by permission.)

Evolution and systematics

The earliest fossil chitons occur in the Upper Cambrian, dating the group back nearly half a billion years. The fossil record of this group of mollusks is relatively sparse, with approximately 350 described fossil species. Chitons diversified more rapidly in recent (Cenozoic) times, and today there are approximately 1,000 living species worldwide. One-fifth of the species are found on the Pacific coast of North America, distributed from Alaska to Southern California, more than on any coast of comparable length in the world. Roughly half of all living species live in the intertidal or shallow subtidal zones.

Within the phylum Mollusca, the chitons are unique in possessing eight shell valves. However, the class Monoplacophora, with a single shell, shares several characteristics with the chitons, including eight pairs of dorsoventral pedal retractor muscles. Repeated pairs of many organs, including one to three pairs of gonoducts, three to seven pairs of excretory nephridiopores, three to six pairs of gills, and two paired atria in monoplacophorans may suggest that mollusks as a whole evolved from a segmented ancestor not unlike the chitons. Larval aplacophorans, larval polyplacophorans, and adult polyplacophorans possess seven or eight transverse dorsal rows of spicules, further strengthening the link between these two classes of mollusks. Finally, recent analyses of 18S rDNA gene sequences suggest that the mollusks are united with other eutrochozoans that possess a trochophore larva, including the annelids. The analyses also support the theory that mollusks arose from a segmented ancestor.

Nevertheless, other theories currently lean toward a non-segmented ancestor for the phylum Mollusca, based on the unsegmented coelom (unlike that of annelids), the lack of agreement in the number of paired organs in basal mollusks (e.g., aplacophorans versus polyplacophorans), and the lack of evidence of segmentation in many other classes within the phylum. These theories support the placement of the aplacophorans, with a vermiform body that lacks a shell (and often lacks a foot, radula, and most gills), as near the base of the phylogenetic tree of mollusks.

In addition, apparent differences between the shell valves of polyplacophorans and the so-called dorsal plates found in aplacophorans suggest that chitons stand alone as a uniquely derived group that arose early in the evolution of mollusks.

In conclusion, it is unclear where, within the spiralian protostomes (including numerous phyla such as the Platyhelminthes, Nemertea, Sipuncula, Echiura, Annelida, Onycophora, and Arthropoda), the mollusks arose, although the presence of a reduced coelomic area surrounding the heart (the pericardial space) suggests a coelomate (rather than acoelomate) ancestor.

Physical characteristics

Chitons are distinct in possessing eight (sometimes seven) overlapping transverse shell plates (hence, the name Polyplacophora, which means "bearer of many plates") that permit

A juvenile chiton on a cone shell. (Photo by Bill Wood. Bruce Coleman, Inc. Reproduced by permission.)

the ovoid, dorsoventrally reduced body to conform to the irregular, rocky shores on which they are most often found. Strong, paired pedal retractor muscles extend from the foot to each shell valve, which is often wing-like or butterfly-like in shape, with two lateral, anterior extensions where these muscles attach. The shell plates are composed of four layers: an outer, organic periostracum; an inner tegmentum composed of calcium carbonate and proteinaceous material (conchiolin); an inner articulamentum below the tegmentum, comprised of pure calcium carbonate (aragonite) that extends laterally, free of the tegmentum layer to form the insertion plates of each valve; and the innermost hypostracum, lying against the mantle.

The mantle, a thick, stiff tissue layer that secretes the shell, extends beyond the shell plates (and sometimes over them, such as in *Cryptochiton stelleri*). This tissue layer secretes a thin glycoprotein cuticle on the dorsal surface of the body. This cuticle may bear scales, bristles, or calcareous spicules similar to those in the class Aplacophora.

Ventrally, a broad muscular foot is bordered laterally on each side of the body by a pallial groove between the edge of the foot (medially) and the edge of the mantle (laterally), forming a chamber in which the gills (or ctenidia) are located. Within these mantle cavities, anywhere from six to 88 pairs of ciliated, bipectinate gills are located. A current of water en-

ters alongside the anterior end of the body, ventrally, on both sides of the body near the head, and travels through each of these grooves posteriorly, exiting out of the body beyond the tail end of the digestive track (the anus). This water current carries oxygenated seawater, and rids the animal of egested feces (released from the anus) as well as urine released out of a pair of nephridiopores that open laterally in the posterior mantle cavity. The nephridiopores represent the exit points of two large nephridia that filter out wastes within the coelomic cavity surrounding the heart, which is filled with blood (and often termed a heomocoel).

Most chitons feed on microalgae, scraping the surface of the rocks on which they sit with a long radular belt of 17 recurved teeth, arranged in transverse rows, that are capped with magnetite (an iron-containing hardening material) in some species. Some variation in feeding exists. Species such as *Katharina tunicata* feed on large macroalgae, including kelps (*Hedophyllum*). Other species possess spectacular modifications of the anterior portion of the girdle to trap small crustacean prey, allowing the evolution of carnivory in an otherwise completely herbivorous (or omnivorous) group.

The chiton nervous system consists of a circumenteric nerve ring around the gut, leading to ladder-like nerve cords that radiate posteriorly towards the end of the body along four lines: two paired pedal cords and two paired visceral cords. These four nerve cords are connected by a series of transverse rung-like commissures, yielding the ladder-like form of the overall system.

Reproduction in chitons involves a single gonad, formed in the dorsal hemocoel, which empties gametes (either eggs or sperm, depending on the sex of the individual) by way of two gonoducts into the mantle cavity just anterior to the openings of the nephridiopores on both sides of the foot. Eggs (which are coated with a spiny envelope) are released either singly or in strings, and are fertilized externally in the water column. Development usually leads to a lecithotrophic (yolk-filled) trochophore larva. There is no veliger stage. Some species have larger, direct-developing eggs that are brooded in the female's mantle cavity, from which a juvenile chiton is formed. The shell gland develops with seven regions on the dorsum of the juvenile; the seventh band divides later to form an eighth. Thus, eight shell plates are formed.

Distribution

Chitons are common rocky intertidal inhabitants, occurring particularly in the temperate zone.

Habitat

Chitons are found primarily on hard substrates, molding their body to the contours of the rock. They are found from the high intertidal zone to depths of more than 13,123 ft (4,000 m), and occur in tropical, temperate, and cold polar seas. Most abundant on hard substrates, especially rocks, chitons graze on surface microalgae and encrusting organisms.

Behavior

Cryptochitons and other chitons roll up when dislodged from a rock, about the only defensive trick these animals have. Several species, such as the mossy chiton (*Mopalia muscosa*), have "home scars", or areas on a rock that they return to following excursions for feeding; these place are often particularly well situated for the chiton to grasp onto to avoid dislodgement by waves as the tide comes in. Except for the predatory chiton, *Placiphorella velata*, which can quickly trap prey with its "head-flap", most chitons are highly sedentary

animals that remain stationary when the tide is out, mostly feeding at night when low tides or full water submergence keeps them moist.

Feeding ecology and diet

The radula, a chitonous ribbon covered with many rows of hard, recurved teeth, is used for feeding, most frequently on the microalgae that coat the rocks on which chitons are found. In some species, this radula contains iron, part of it as magnetite, the only known example of biological production of this common mineral. Magnetite greatly strengthens the radula, allowing many chitons to feed on hard, encrusted coralline algae.

Besides the microscopic and macroscopic algae, chitons are also known to feed on other encrusting organisms (such as bryozoans). One genus (*Placiphorella*) has evolved the remarkable ability to capture live prey such as worms and small crustaceans by using an expanded "head-flap" created by an anterior extension of the girdle, which is held above the substrate until unwary victims wander in, at which time the head-flap is rapidly closed down over the prey.

Reproductive biology

The reproductive system of chitons consists of a gonad located in front of the heart dorsally, with a pair of ducts that open into the mantle cavity at the posterior end of the body. The sexes are always separate, with sperm shed into the sea. Eggs are either shed into the sea or retained in the female's mantle cavity, where sperm that enter with the respiratory currents fertilize them. The eggs are then brooded until embryos become well-developed young chitons.

Conservation status

No chitons are listed by the IUCN.

Significance to humans

Native Americans of the Pacific Coast of North America used to eat the giant chiton, *Cryptochiton stelleri*; shell valves of this species can be found in prehistoric kitchen middens.

1. *Tonicella lineata*; 2. *Placiphorella velata*; 3. *Mopalia muscosa*; 4. *Katharina tunicata*; 5. *Cryptochiton stelleri*. (Illustration by Joseph E. Trumpey)

Species accounts

Gumboot chiton
Cryptochiton stelleri

ORDER
Neoloricata

FAMILY
Acanthochitonidae

TAXONOMY
Cryptochiton stelleri Middendorff, 1846, Kamtschatka, Russia.

OTHER COMMON NAMES
English: Gumshoe chiton.

PHYSICAL CHARACTERISTICS
Largest chiton in the world, reaching 13 in (33 cm) in length and 5 in (13 cm) in width. Has the general appearance of a wandering "meatloaf." It is distinguished both by its size and its brick-red colored, leathery mantle, which extends up and over the shell valves, obscuring them from sight. The mantle is covered with closely spaced fascicles of very short, spreading spines, or spicules. The white, butterfly-shaped shell valves are hard and frequently wash up on beaches intact.

DISTRIBUTION
From the Aleutian Islands (Alaska) south to San Miguel Island and San Nicolas Island of the Channel Islands National Park in California; northern Hokkaido Island, Japan; and Kurile Islands, Kamchatka.

HABITAT
Found on rocky shores as well as soft bottoms, in relatively protected sites near deep channels, from the low intertidal zone down to a depth of roughly 70 ft (21.3 m) in kelp beds.

BEHAVIOR
It has a relatively weak foothold on the rocks at low tide, and individuals can frequently be found lying near the base of a rock from which they have fallen at low tide. Individuals are not gregarious, and a study on the Oregon coast found that marked individuals remain within 65.6 ft (20 m) of the point of release even after two years of time. They often harbor a commensal polychaete worm, the scaleworm (*Arctonoe vittata*), ventrally in the pallial groove (mantle cavity) on one side of the foot. The commensal feeds on plankton and detritus brought in by the respiratory currents of the host.

FEEDING ECOLOGY AND DIET
Uses its many transverse rows of 17 teeth capped with magnetite, with a central tooth flanked by eight marginal teeth on either side, to feed on red algae, including *Gigartina*, *Iridaea*, *Plocamium*, and various corallines. Individuals will also eat sea lettuce (*Ulva*), kelp (*Macrocystis*), and small *Laminaria*. They grow slowly and may live 20 years or more. They have few enemies: the predaceous snail, *Ocenebra lurida*, is the sole exception; it rasps pits 0.4 in (1 cm) in diameter and 0.11–0.15 in (3–4 mm) deep into the dorsal surface of a chiton's body, exposing the yellow flesh over the valves.

☐ *Placiphorella velata*
■ *Cryptochiton stelleri*

REPRODUCTIVE BIOLOGY
Spawning in California occurs between March and May, may last over a week, and results in individuals losing up to 5% of their body mass. The cinnamon-red eggs are laid in gelatinous spiral strings up to 3.3 ft (1 m) long, which do not stick to the substratum and are broken up by waves. Release of eggs by females triggers the release of sperm by males. Trochophore larvae are liberated from the egg roughly five days post-fertilization, following a free-swimming period of up to 20 hours, then settle and begin metamorphosis.

CONSERVATION STATUS
Not listed by the IUCN.

SIGNIFICANCE TO HUMANS
Amerindians of the Pacific Coast of North America used to eat this species; shell valves are frequently found in prehistoric kitchen middens. ◆

Lined chiton
Tonicella lineata

ORDER
Neoloricata

FAMILY
Ischnochitonidae

TAXONOMY
Tonicella lineata Wood, 1815, Sitka, Alaska, United States.

OTHER COMMON NAMES
None known.

PHYSICAL CHARACTERISTICS
Body low, elongate-oval, usually less than 1.2 in (3 cm) long. Possesses shiny shell valves distinctly demarcated with zigzag lines of alternating dark and light red, light (or dark) blue and red, or whitish and red colors. Mantle girdle naked and leathery, usually yellow to green in color, sometimes banded.

DISTRIBUTION
From Alaska (in the Aleutian Islands) south to San Miguel Island in the Channel Islands National Park in California, in the Sea of Okhotsk (Russia) to northern Japan.

HABITAT
Found on temperate rocky shores, on rocks covered with erect or crustose coralline algae, in the mid to low intertidal zone down into the subtidal to depths of 180 ft (54.8 m). On the Monterey Peninsula in California, subtidal individuals show consistent color and size differences, being smaller (0.39–0.78 in [1–2 cm] long) with purple lines on the girdle. On the Oregon coast, it is frequently found living under purple sea urchins (*Strongylocentrotus purpuratus*) in the burrows that the urchins dig out of rock. The color pattern frequently matches, to some degree, the coralline algal substrate, lending chiton some degree of protective camouflage.

BEHAVIOR
Activity patterns vary with habitat: individuals near Monterey, California, remain stationary in the intertidal zone when exposed at low tide, while subtidal individuals follow a diurnal

Tonicella lineata

rhythm characterized by more twice as much movement at night than during the day.

FEEDING ECOLOGY AND DIET
Feeds on crustose coralline algae (*Lithothamnium*), which it scrapes with its strong radula, consuming the more superficial layers and removing the film of diatoms and other small organisms coating the surface of this alga. Species that feed on the lined chiton include the sea stars (*Pisaster ochraceus* and *Leptasterias hexactis*). However, in Monterey Bay, California, sea stars have rarely been observed to feed on chitons except for those that are removed from the rocks.

REPRODUCTIVE BIOLOGY
Females release eggs into the water column in April along the coasts of California and central Oregon, whereas populations on San Juan Island (off the Washington coast) release eggs in May and June. Cleavage divisions and gastrulation lead to the formation of a trochophore larvae 16–24 hours post-fertilization (depending on temperature), and they hatch roughly 43–44 hours after fertilization. Development of the trochophore larva stops sometime within 150–160 hours post-fertilization, and further development depends on contact with crustose coralline algae (or an extract thereof). Larvae undergo metamorphosis to the adult within 12 hours of settlement, becoming juvenile chitons that begin to feed around 30 days after settlement.

CONSERVATION STATUS
Not listed by the IUCN.

SIGNIFICANCE TO HUMANS
None known. ◆

Black katy chiton
Katharina tunicata

ORDER
Neoloricata

FAMILY
Mopaliidae

TAXONOMY
Katharina tunicata Wood, 1815, west coast of North America.

OTHER COMMON NAMES
None known.

PHYSICAL CHARACTERISTICS
Up to 4.7 in (12 cm) long, with a highly elongate-oval shape. Girdle thick, shiny, black, and leathery; white shell valves are deeply embedded in the girdle, exposed only in the mid-dorsal area.

DISTRIBUTION
Found in the Aleutian Islands (Alaska) to the faunal break at Point Conception (Santa Barbara County in California), and in Kamchatka, Russia.

HABITAT
Common on mid to low intertidal rocky shores exposed to strong wave action, often in direct sunlight, as well as inland waters associated with swift currents in the Pacific Northwest.

BEHAVIOR
Remains at a particular level in the tide zone, and is one of the tougher chitons, being highly resistant to wave splash and exposure to sun.

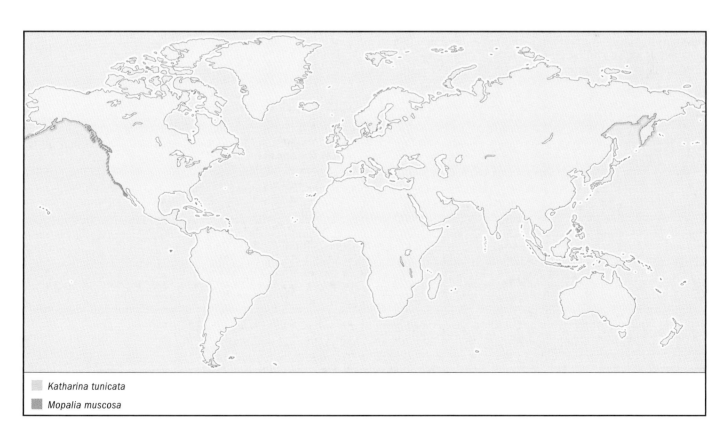

Katharina tunicata

Mopalia muscosa

FEEDING ECOLOGY AND DIET
Uses hard, magnetite-capped radular teeth to consume diatoms and red and brown algae, including kelp (*Hedophyllum*).

REPRODUCTIVE BIOLOGY
Reaches sexual maturity at a mass of 0.14 oz (4 g) and length of 1.3–1.4 in (33–36 mm). Depending on latitude, individuals spawn between March and July; later spawning occurs in colder regions. The eggs are green. Lifespan is roughly three years.

CONSERVATION STATUS
Not listed by the IUCN.

SIGNIFICANCE TO HUMANS
None known. ◆

Mossy chiton
Mopalia muscosa

ORDER
Neoloricata

FAMILY
Mopaliidae

TAXONOMY
Mopalia muscosa Gould, 1846, Puget Sound, United States.

OTHER COMMON NAMES
None known.

PHYSICAL CHARACTERISTICS
Oval to oblong body up to 3.5 in (9 cm) long with a moderately developed dorsal ridge down the midline; brown to olive or gray valve surface is lusterless, sometimes sculptured with wavy, crenulated riblets, but often eroded or overgrown by algae or worm tubes. Tan-colored girdle covered by dense assemblage of stiff, mossy, brownish red hairs or bristles with a slight notch carved at the posterior end of the girdle.

DISTRIBUTION
Found in the Queen Charlotte Islands (British Columbia, Canada) to Isla Cedros (Baja California).

HABITAT
Does well in estuaries because it tolerates a wide range of environmental conditions. Found on rocks protected from heavy surf and in tidepools in the mid to low intertidal zone.

BEHAVIOR
More tolerant of light than many other chitons, and can be seen out in the open on foggy days on the Pacific Coast of North America. It spends most of the daylight hours under rocks, however, and moves primarily at night when the animal is wet or submerged. Individuals that are exposed at low tide show homing behavior, moving within a radius of roughly 19.6 in (50 cm) from home base, frequently returning by a set pathway to original resting place when immersed by tide. Individual home ranges do not overlap, but chitons living in tidepools (which are never exposed) do not appear to maintain home ranges.

FEEDING ECOLOGY AND DIET
Feeds on algae, primarily the reds (*Gigartina papillata* and *Endocladia muricata*) as well as green alga (*Cladophora*) when available, but the gut may contain up to 15% animal matter.

REPRODUCTIVE BIOLOGY
In Central and Northern California, spawning has been noted from July to September, whereas a winter spawn has been recorded in Santa Monica Bay (Los Angeles County, California) and a spring spawn has been observed in Monterey Bay, with eggs and sperm shed into large tidepools. The green- or golden brown-colored eggs are 0.011 in (0.29 mm) in diameter and, following fertilization, develop through three cleavage divisions in 2.5 hours. Hatching occurs in 20 hours, and the larvae swim freely for several days, with settlement occurring about 11.5 days post-fertilization, provided an appropriate substratum is present. The first seven shell plates are visible within 13.5 days, with the last (eighth) shell plate appearing only about six weeks after fertilization. Sexual maturity is reached in approximately two years.

CONSERVATION STATUS
Not listed by the IUCN.

SIGNIFICANCE TO HUMANS
None known. ◆

Veiled chiton
Placiphorella velata

ORDER
Neoloricata

FAMILY
Mopaliidae

TAXONOMY
Placiphorella velata Dall, 1879, Humboldt Bay, California; Monterey Bay, California; Todos Santos Bay, California, United States.

OTHER COMMON NAMES
None known.

PHYSICAL CHARACTERISTICS
Distinguished by the girdle around the margins of shell valves, which is greatly expanded anteriorly to form a "head flap," ventrally pigmented with red and green. The body is up to 1.9 in (5 cm) long, with brown to red shell valves that are short and wide and variously mottled and streaked with green, beige, white, and brown.

DISTRIBUTION
Although uncommon, found from Forrester Island (Alaska) to Isla Cedros (Baja California) and the upper Gulf of California.

HABITAT
Found associated with coralline algae very low in the intertidal zone, in shaded depressions, or in crevices on or under rocks from the low tide line down to a depth of 50 ft (15.2 m).

BEHAVIOR
Only chiton genus to have adapted a means for capturing live prey. Other than the fast movement of the girdle clamping down on unsuspecting victims that wander into the cavity below "head flap," adults are highly sedentary.

FEEDING ECOLOGY AND DIET
May browse on encrusting sponges and algae, scraping the rocks with radula like other chitons, or acts as a predator, an unusual feeding mode among chitons. Captures worms and small crustaceans by clamping down on them quickly (within one second) with the extended "head flap," which is a very

large anterior expansion of the mantle that is lifted to form a trap for unsuspecting prey. Flap is pigmented on both sides and fringed with bristles, and thus resembles a blade of red algae. The rest of the body is often overgrown with bryozoans and algae, making it highly camouflaged as well.

REPRODUCTIVE BIOLOGY
In California, it spawns in September.

CONSERVATION STATUS
Not listed by the IUCN.

SIGNIFICANCE TO HUMANS
None known.

Resources

Books

Brusca, G. J., and Brusca, R. C. *A Naturalist's Seashore Guide: Common Marine Life of the Northern California Coast and Adjacent Shores.* Eureka, CA: Mad River Press, 1978.

Gotshall, D. W. *Guide to Marine Invertebrates: Alaska to Baja California.* Monterey, CA: Sea Challengers, 1994.

Kozloff, E. N. *Seashore Life of the Northern Pacific Coast: An Illustrated Guide to Northern California, Oregon, Washington, and British Columbia.* Seattle: University of Washington Press, 1983.

Meinkoth, N. A. *The Audubon Society Field Guide to North American Seashore Creatures.* New York: Alfred A. Knopf, 1992.

Morris, R. H., Abbott, D. P., and Haderlie, E. C. *Intertidal Invertebrates of California.* Stanford, CA: Stanford University Press, 1980.

Pearse, V., Pearse, J., Buchsbaum, M., and Buchsbaum, R. *Living Invertebrates.* Boston, MA: Blackwell Scientific Publications, 1997.

Sean F. Craig, PhD

Opisthobranchia
(Sea slugs)

Phylum Mollusca

Class Gastropoda

Subclass Opisthobranchia

Number of families 110

Thumbnail description
Marine snails found in oceans throughout the world; the diversity encompasses species that range in habit from pelagic to burrowing, in color from transparent to vividly colored, and in structure from those with protective shells to soft-bodied forms with flowing movements

Photo: This small nudibranch (*Glossodoris stellatus*) is laying eggs. The "swirl" pattern of egg laying is a common one; the color of the eggs is dependent upon the species. (Photo ©Tony Wu/www.silent-symphony.com. Reproduced by permission.)

Evolution and systematics

Opisthobranchia is a subclass of gastropod mollusks whose evolutionary ancestors date back to the Paleozoic period (543–248 million years ago). The earliest ancestors were probably all benthic, burrowing animals with rigid shells covering their soft body tissues. As members of this group evolved, a major trend was towards reduction, internalization, or loss of the shell. Numerous lineages have arisen independently that have exhibited this same evolutionary trend. In place of the shell, opisthobranchs have evolved other defensive mechanisms, including cryptic or warning coloration, secretion of toxic chemicals, and the ability to fire stinging cells obtained from animals on which they feed. The subclass, Opisthobranchia, includes a great diversity of species. More than 3,000 species are recognized, and these are classified into more than 110 families and eight orders.

Physical characteristics

Physical appearance varies greatly among members of the Opisthobranchia. Generally, their bodies are divided into three sections: head, foot, and visceral mass. In some species, these components are easily visible; in others, the sections may have merged in various ways. The head is flattened in some species, but in others, it may bear up to four types of sensory tentacles: oral tentacles, which project from the sides of the mouth; rhinophores, which are located on top of the head; propodial tentacles, which occur on the front of the foot; or posterior cephalic tentacles, which project from the posterior portion of the head.

The foot has a flattened sole that is used for creeping over substrates. Some species have wide lateral extensions of the foot that appear winglike; these structures are called parapodial lobes and are used for swimming. The visceral mass con-

tains the digestive and reproductive organs. It is covered by the mantle, which generates a shell in some species and hangs like a skirt around the body in others. The shell has become reduced, internalized, or lost in some of the more evolved species. The more primitive species have a gill located between the mantle and visceral mass near the head. In place of the gill, some of the more advanced species have cerata, small projections from their bodies, or a ruffled mantle edge to facilitate gas exchange.

Species range in size from some that are so minute that they can move between grains of sand to others that reach lengths of more than 17.7 in (45 cm). They vary in shape from rounded, shell-covered forms to elongate, often ornate, bodies. Some species are cryptically colored to blend in with their natural surroundings, while others are brightly colored, perhaps as a warning signal to predators; some of the pelagic species are nearly transparent.

Distribution

Opisthobranchs occur worldwide. The greatest diversity is in the tropical seas, but several species have been reported from the polar waters of Antarctic.

Habitat

Sea slugs are found in marine habitats such as reefs, intertidal areas, and the deep ocean. Some species live on the substrate, while others remain in the water column.

Behavior

Most species of sea slugs are benthic and live on some type of substrate as adults; a portion of these species can swim for

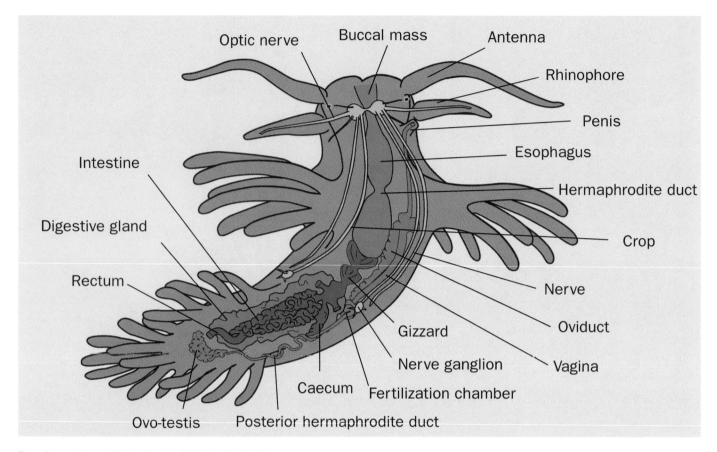

Sea slug anatomy. (Illustration by Christina St. Clair)

short periods of time as well. Benthic sea slugs are slow-moving organisms. They live relatively sedentary lives, and some species spend their entire life on one prey organism such as a sponge or coral reef. Almost all dispersal to new areas occurs during the veliger larval stage; veligers will settle out of the water column only when suitable substrate is present. Some larvae have crossed entire ocean basins because of lack of suitable settlement substrate.

A smaller portion of species remain planktonic and float in the water column throughout their lives. Most pelagic species migrate up and down the water column on a daily cycle. They move closer to the surface at night and to deeper waters during the day; however, in some species, this pattern is reversed.

Sea slugs exhibit several defensive behaviors. Some species secrete toxic chemicals or retain stinging cells from animals they eat; these defenses can be used to ward off or harm potential predators. Other species, typically benthic, swim to escape predators, while pelagic species sink in the water column to avoid predation.

Feeding ecology and diet

Sea slugs feed on a wide variety of organisms. Some are herbivores that eat algae, while others are filter feeders that take in particles from the water. Most are carnivores that eat

many types of animals, including hydra, sponges, corals, barnacles, worms, other mollusks, and even the eggs of other sea slugs, cephalopods, or fishes. Despite the diversity of organisms consumed by sea slugs as a group, many species are highly specialized feeders that consume only a single type of food item; some will only prey upon a single genus or species.

Most sea slugs have a pair of jaws and a radula, a rasping, tonguelike organ with rows of small teeth, that are used for feeding. The radula is capable of scraping, piercing, tearing, or cutting food particles. Species of sea slugs that lack a radula may suck in whole prey like a vacuum, pry tissues off of their prey, or have special adaptations for trapping food items. To overcome defensive mechanisms of their prey, such as spines, exoskeletons, or stinging cells, many sea slugs cover their food with mucus.

Few organisms prey on sea slugs; some sea slugs use chemical defenses to keep predation to a minimum. The sea slugs often take up distasteful or toxic chemicals from their own prey and, in turn, use them to deter predators; some nudibranchs even eat the stinging cells off of jellyfish and Portuguese man-of-war and use them to ward off potential predators. One of the most frequently reported acts of predation occurs between two different types of opisthobranch mollusks. *Navanax inermis*, a species of the order Cephalaspidea, feeds on nudibranchs as it crawls along the substrate; it follows the trail of slime left by a nudibranch, sneaks up onto the prey, and sucks

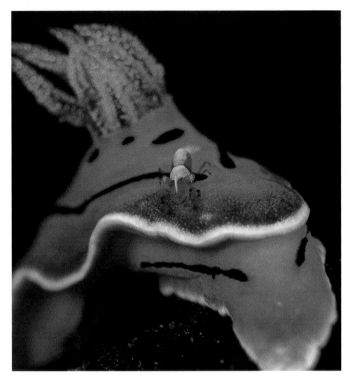

This individual of the *Chromodoris willani* species has a tiny emperor shrimp (*Periclimenes imperator*) riding on it. The emperor shrimp does not appear to harm its host, and may help to keep the host clean. (Photo ©Tony Wu/www.silent-symphony.com. Reproduced by permission.)

This species *Hypselodoris bullocki* is very common on tropical reefs throughout Asia; its body has a wide range of color variation, though all have the same basic body shape. These four specimens are mating. Nudibranchs are hermaphroditic, and often congregate like this for mass breeding. (Photo ©Tony Wu/www.silent-symphony.com. Reproduced by permission.)

the organisms in with a sort of suction tube from inside its body.

Reproductive biology

Sea slugs are hermaphrodites; each individual has both male and female sexual organs and produces sperm and eggs. Some simultaneous hermaphrodites possess male and female organs at the same time, while others begin life as one sex and then transform to the other gender. Individuals do not exchange their own sperm and eggs; the reproductive system of sea slugs keeps these products separate within an individual.

Mating behaviors are highly variable in sea slugs. Some species mate as pairs; others form long chains of individuals. In some species, particularly those that are simultaneous hermaphrodites, both sperm and eggs are reciprocally exchanged between partners at the same time. Individuals in other species behave as either male or female during a mating session. Mating may last from minutes to days, depending on the species. Fertilization may not take place immediately; an organ called the seminal receptacle can store sperm for several months until the eggs they will fertilize have reached maturity.

Eggs are laid in masses ranging from hundreds to millions of individual eggs; the masses are sometimes formed into chains or globules on the substrate or in the water column, and some species protect their egg masses with a mucous cov-

ering. When eggs hatch, planktonic larvae emerge; these larvae, called veligers, soon develop a shell around them. A few species retain their eggs inside their bodies, and young emerge as juveniles, thereby eliminating the planktonic larval stage.

Conservation status

Although little is known about their populations, sea slugs are not considered threatened or endangered. No species are listed by the IUCN.

Significance to humans

Many sea slugs are admired by humans who are fortunate enough to view them during snorkeling or diving activities. Beyond their aesthetic appeal, they are of little significance to humans.

1. *Glossodoris atromarginata*; 2. *Pupa solidula*; 3. *Aeolidiella sanguinea*; 4. *Tylodina corticalis*; 5. *Corolla spectabilis*; 6. *Elysia viridis*. (Illustration by Joseph E. Trumpey)

Species accounts

No common name
Pupa solidula

ORDER
Cephalaspidea

FAMILY
Acteonidae

TAXONOMY
Pupa solidula (Linnaeus, 1758).

OTHER COMMON NAMES
None known.

PHYSICAL CHARACTERISTICS
Among the most primitive of the opisthobranchs, it has a hard calcified shell that is adorned with a spiral pattern of black dots. A relatively small, colorless animal lives inside the shell. Flaps on the head overlap the front of the shell and protects the mantle cavity from sand as it crawls along or burrows into the substrate. Most adults are approximately 0.39–0.59 in (10–15 mm) in shell length.

DISTRIBUTION
Tropical IndoPacific.

HABITAT
Sandy substrates of marine waters.

BEHAVIOR
None known.

FEEDING ECOLOGY AND DIET
Feeds on polychaete worms; other prey items and its predators have not been reported.

REPRODUCTIVE BIOLOGY
Hermaphroditic and reproduces on an annual cycle. Eggs are laid in the spring, and veliger larvae settle to the substrate a few months later. Young produced in one year reach maturity and reproduce the following year. Most individuals die after spawning, but a few survive into their second year of life.

CONSERVATION STATUS
Not listed by the IUCN.

SIGNIFICANCE TO HUMANS
None known. ◆

No common name
Tylodina corticalis

ORDER
Notaspidea

FAMILY
Tylodinidae

TAXONOMY
Tylodina corticalis (Tate, 1889).

OTHER COMMON NAMES
None known.

PHYSICAL CHARACTERISTICS
One of the more primitive members of its order. Retains a large heavily calcified shell of circular shape on the exterior of its body. When crawling, the body becomes elongate and extends beyond the shell. Has a large head with prominent feeding tentacles and rhinophores. The body is bright yellow, and individuals may reach size of 2 in (5 cm).

DISTRIBUTION
In Australia from southern Queensland to southwestern Australia around the southern coast.

HABITAT
From the intertidal zone to depths of around 328 ft (100 m), typically on its food sponge.

BEHAVIOR
None known.

FEEDING ECOLOGY AND DIET
Feeds on sponges in the family Aplysinellidae. Predators have not been reported.

REPRODUCTIVE BIOLOGY
Mating behavior has not been described, but it is known to lay ribbons of eggs on its food sponge.

CONSERVATION STATUS
Not listed by the IUCN.

SIGNIFICANCE TO HUMANS
None known. ◆

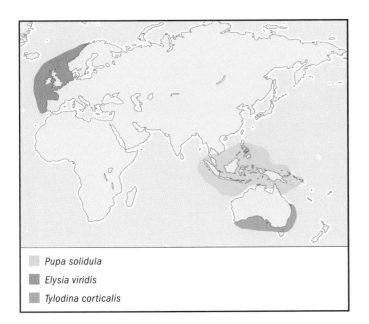

■ *Pupa solidula*
■ *Elysia viridis*
■ *Tylodina corticalis*

No common name
Aeolidiella sanguinea

ORDER
Nudibranchia

FAMILY
Aeolidiidae

TAXONOMY
Aeolidiella sanguinea (Norman, 1877).

OTHER COMMON NAMES
None known.

PHYSICAL CHARACTERISTICS
An elongate body that is covered in cerata, thin, fleshy projections that arise in regular rows from the digestive tract. Long oral tentacles and prominent rhinophores are present on the head. The body, cerata, and tentacles are pale yellow, orange, or red in color, depending on the individual's diet. The sole of the foot and tips of the cerata are white.

DISTRIBUTION
Atlantic coasts of Ireland, France, and Scotland.

HABITAT
Found in the intertidal zone and sublittoral areas; often in muddy inlets or on rocky coasts.

BEHAVIOR
Has been described as slow and aggressive. Is more active at night than during the day.

FEEDING ECOLOGY AND DIET
Feeds on sea anemones. Predators have not been reported.

REPRODUCTIVE BIOLOGY
Mating behavior has not been described. Eggs are laid in a spiral thread with scalloped sections.

CONSERVATION STATUS
Rare, but not considered threatened. Not listed by the IUCN.

SIGNIFICANCE TO HUMANS
None known. ◆

No common name
Glossodoris atromarginata

ORDER
Nudibranchia

FAMILY
Chromodorididae

TAXONOMY
Glossodoris atromarginata (Cuvier, 1804).

OTHER COMMON NAMES
None known.

PHYSICAL CHARACTERISTICS
Characterized by an elogate body with a highly sinuous mantle edge. Has two rhinophores on the head that are similar in appearance to small horns; gills that undulate like coral polyps are found on the dorsal portion of the body. Although the body may range in color from cream to pale brown, the mantle edge, rhinophores, and gill edges are always black. Adults reach lengths up to 3.15 in (8 cm).

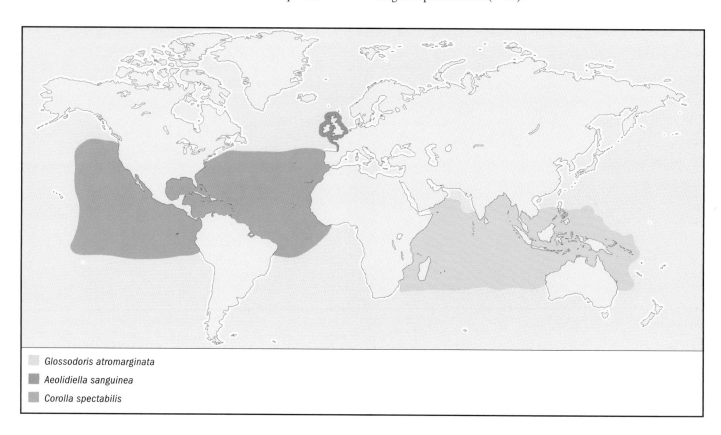

Glossodoris atromarginata
Aeolidiella sanguinea
Corolla spectabilis

DISTRIBUTION
Throughout the tropical and subtropical Pacific and Indian Oceans.

HABITAT
Inhabits reefs, intertidal, and subtidal areas of coastal waters; ranges from low intertidal to depths of 92 ft (28 m).

BEHAVIOR
Nothing is known.

FEEDING ECOLOGY AND DIET
Feeds on several species of siliceous sponges. Predators have not been reported.

REPRODUCTIVE BIOLOGY
Simultaneous hermaphrodite; it typically mates as pairs. The individuals position themselves alongside one another and exchange both sperm and eggs. Eggs are laid in ribbon-like masses with a mucous sheath surrounding them.

CONSERVATION STATUS
Not listed by the IUCN.

SIGNIFICANCE TO HUMANS
None known. ◆

No common name
Elysia viridis

ORDER
Sacoglossa

FAMILY
Elysiidae

TAXONOMY
Elysia viridis (Montagu, 1804).

OTHER COMMON NAMES
None known.

PHYSICAL CHARACTERISTICS
A flat, elongated, leaflike body that may grow to 1.7 in (45 mm) in length; the parapodia may stretch alongside the body or be held over the dorsal mid-line; body color varies from green to red depending on the types of algae that have been the recent diet of the organism; red, blue, and green specks of color are typical on the body; white patches may occur near the edge of the parapodia, and black markings sometimes occur on the head. Feeding tentacles and rhinophores are prominent on the front of the head.

DISTRIBUTION
North Atlantic from Norway to the Mediterranean. Also observed in North Africa and China.

HABITAT
Found on a variety of shallow-water algae.

BEHAVIOR
Nothing is known.

FEEDING ECOLOGY AND DIET
Eats filimentous (*Cladophora* and *Chaetomorpha*) and coenocytic (*Codium* and *Bryopsis*) algae. Retains plastids, often chloroplasts,

from the algae it eats and uses these cells to continue photosynthesis within its own body. Predators have not been reported.

REPRODUCTIVE BIOLOGY
Hermaphroditic; spawning occurs from May to October. Egg masses contain as many as 10,000 eggs. Embryos develop for 5–12 days before veliger larvae emerge. Probably has an annual life cycle.

CONSERVATION STATUS
Not listed by the IUCN.

SIGNIFICANCE TO HUMANS
None known. ◆

No common name
Corolla spectabilis

ORDER
Thecosomata

FAMILY
Cymbulidae

TAXONOMY
Corolla spectabilis (Dall, 1981).

OTHER COMMON NAMES
None known.

PHYSICAL CHARACTERISTICS
A small gelatinous species; its body is nearly transparent, with the dark gut being the most visible part. Lacks an external shell, but a gelatinous internal structure provides skeletal support. Has large oval wing plates that extend laterally from the body; the lobes of the foot have become fused to form a proboscis. Large individuals can reach lengths of 3 in (8 cm), with wing plates spanning 6 in (16 cm).

DISTRIBUTION
Atlantic and eastern Pacific Oceans between approximately 40°N and 5°S latitude.

HABITAT
Near the surface of marine waters.

BEHAVIOR
Flaps its large lateral wing plates to maintain position or move in the water column. Swims away rapidly when disturbed (up to speeds of 18 in/sec [45 cm/sec]), typically shedding the mucous sheet used for feeding. May form large aggregations at the water's surface.

FEEDING ECOLOGY AND DIET
Feeds on zooplankton that it collects with a large, delicate mucous sheet that is produced by glands along the edge of the wing plate. Traps prey by slowly sinking in the water column and entangling planktonic particles as it descends. Mucous web and entangled food items ingested by the proboscis. Predators have not been reported.

REPRODUCTIVE BIOLOGY
Protandric hermaphrodite; it matures and functions first as a male and then as a female. Mating behavior has not been

reported. Eggs are spawned in mucous strings that may extend to 1.6 ft (0.5 m) in length.

CONSERVATION STATUS
Not listed by the IUCN.

SIGNIFICANCE TO HUMANS
None known. ◆

Resources

Books

Lalli, Carol M., and Ronald W. Gilmore. *Pelagic Snails: The Biology of Holoplanktonic Gastropod Mollusks.* Stanford, CA: Stanford University Press, 1989.

Other

"Sea Slug Forum." July 10, 2003 [July 27, 2003]. <http://www.seaslugforum.net>.

Katherine E. Mills, MS

Pulmonata

(Lung-bearing snails and slugs)

Phylum Mollusca

Class Gastropoda

Subclass Pulmonata

Number of families Approximately 120

Thumbnail description
Mainly terrestrial but also freshwater and marine slugs and snails, almost all having an enclosed lung

---

Photo: A garden snail (*Helix aspersa*) issuing defensive "foam." (Photo by Holt Studios, Int./Photo Researchers, Inc. Reproduced by permission.)

Evolution and systematics

The earliest fossil pulmonate land snails date back over 300 million years to the Carboniferous period of North America and Europe, although there is some controversy as to whether these belong to primitive pulmonate families such as the Ellobiidae, or to more advanced families of the major terrestrial pulmonate group Stylommatophora. The first certain stylommatophoran fossils date from the Lower Cretaceous (about 140 million years ago [mya]), and become more common during the Upper Cretaceous, with several genera attributable to modern families such as the Plectopylidae, Streptaxidae, and Camaenidae present. Most extant families appear in the fossil record for the first time during the Cenozoic era (from 65 mya).

The subclass Pulmonata evolved from opisthobranch (marine gastropod mollusks of the subclass Opisthobranchia) ancestors, although it is uncertain to which opisthobranch group they are most closely related. The pulmonates include a number of marine and freshwater families within several orders, but are predominantly terrestrial. By far the largest group of land pulmonates is the order Stylommatophora, but there are also several primitive families of slugs (Veronicellidae, Rathousiidae, and Onchidiidae) and snails (Ellobiidae), which are wholly or partly terrestrial. This entire group, excluding

the ellobiids, is sometimes united as the Geophila, but it unclear whether it is monophyletic. Molecular studies show that the Stylommatophora are monophyletic, and may have evolved from a terrestrial ellobiid ancestor.

Classification within the Stylommatophora has traditionally been based on the morphology of the pallial system (which includes the lung, kidney, and ureter). Four primary groups, or suborders, were recognized: the Orthurethra, which were thought to be the most primitive, and from which the other three evolved; the Sigmurethra, which includes most of the stylommatophoran families; the Mesurethra, which is a small group with a reduced ureter; and the Heterurethra, which contains only the amphibious family Succineidae.

However, a recent molecular study by Wade, Mordan, and Clarke (2001) shows a very different pattern of phylogenetic relationships within the Stylommatophora. It suggests that there is a primary dichotomy between a smaller, "achatinoid" clade, which includes the "sigmurethran" families Achatinidae, Subulinidae, Freussaciidae and Streptaxidae, and a larger "nonachatinoid" clade, comprising all the remaining sigmurethran families, plus the Orthurethra, Heterurethra, and Mesurethra. The Orthurethra are shown to be a relatively advanced group equivalent to other large superfamilies. The

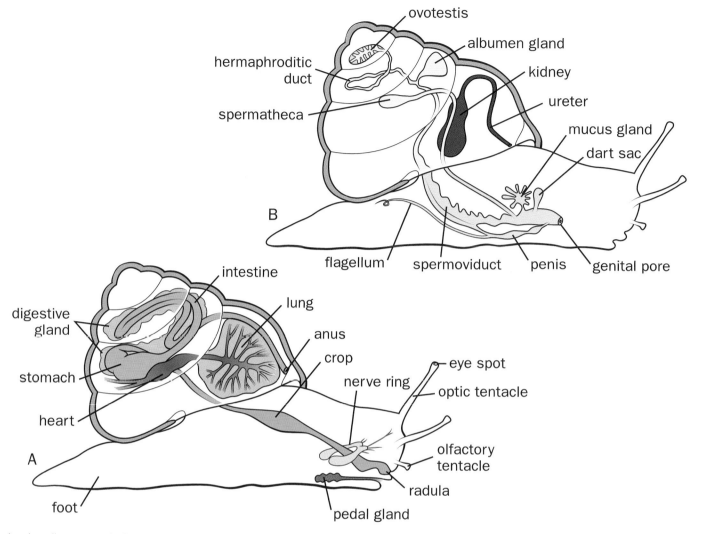

Land snail anatomy. A. General anatomy; B. Urogenital system. (Illustration by Patricia Ferrer)

succineids (Heterurethra) are the sister group of a family of tropical slugs, the Athoracophoridae (together called the Elasmognatha), and are also included within this larger "nonachatinoid" clade, as are the various families which comprised the Mesurethra. The various families of both slugs and predatory pulmonates are shown to have evolved independently numerous times.

There is no general agreement on the classification of terrestrial pulmonates, but that of Vaught (1989) represents a good recent compromise. Vaught's Stylommatophora contains 81 families arranged in 26 superfamilies. There are probably around 25,000 species worldwide. Additionally there are four families of "primitive" pulmonates with terrestrial representatives, which have relatively few species.

The four orders of Pulmonata, based on Burch 1982 and Jeffrey 2001, are the Stylommatophora (terrestrial snails and slugs), Basommatophora (mostly freshwater, also some terrestrial and marine snails and slugs); Systellommatophora (sluglike); and Actophila (marine, intertidal, and brackish-water snails).

The order Basommatophora (15 families, 50 or more genera, and at least 4,000 species) contains a number of relatively primitive families. The order as a whole inhabits a wide range of habitats. The majority of species are freshwater, but there are some terrestrial and intertidal marine families. Some of the better-known families include the Lymnaeidae, dextrally coiled pond snails, some of them amphibious, and some having limpetlike forms.

Members of the family Physidae, sinistrally coiled pond snails, commonly called tadpole snails or pouch snails, are widespread and can tolerate some pollution of their waters. The family Planorbidae, the wheel snails, orb snails or ramshorn snails, is the largest family, in terms of the number of genera and species, of freshwater pulmonates, and is found on all continents and most islands. Many Physidae species are intermediate hosts of the larvae of parasitic worms.

The order Systellommatophora (five families, 11 genera, 50 or more species) comprises sluglike animals with the anus located at the posterior end of the body, in contrast to the normal state of affairs for gastropods in having the anus close to the front of the body. Species of the family Onchidiidae have a posterior pulmonary sac, and those in the family Veronicellidae have discarded the pulmonate lung altogether and instead respirate through their skin.

The order Actophila (five families, nine or more genera, 25 or more species) comprises snails inhabiting salt marshes and mangrove stands. Most species are tropical. In most species, the inner shell is a single cavity, no longer compartmentalized.

Physical characteristics

The Stylommatophora conform to the typical pulmonate body plan. Their main diagnostic characters are the two pairs of retractile tentacles on the head. The upper, longer pair have eyes at the tips, and are concerned with sight and distant olfaction; the smaller, lower pair sense the immediate substrate and food, and are also used in trail following. Additionally, stylommatophorans have a long pedal gland which lies beneath a membrane on the floor of the visceral cavity, and an excretory system with a well-developed secondary ureter.

The shell encloses most of the body organs and is usually dextral (coiling clockwise), but is sinistral (anticlockwise) in a few groups; in some species both coiling directions are represented. The shell is normally cryptically colored, but may be brightly patterned in arboreal species. Some 16 stylommatophoran families are slugs or semi-slugs, in which the shell has been drastically reduced or even lost entirely. In these groups the lung cavity also is greatly reduced and accessory respiratory structures such as the mantle folds are developed. The male and female systems open together as a single genital pore.

Among the primitive pulmonate groups with terrestrial representatives, the tropical veronicellid and rathousiid slugs have two pairs of tentacles, the upper pair with the eyes at the tips; the veronicellids have no lung. Onchidiid slugs, most of which are littoral, have a single pair of tentacles with eyes at the tips. All three of these families lack any form of shell. Ellobiids are snails possessing a single pair of cephalic tentacles, with eyes at the base. The genital openings of the male and female systems are separate in all four of these families.

The largest snails are the giant African land snails of the family Achatinidae, which have been recorded with a shell height of 8 in (20 cm), whereas the tiny European snail *Punctum pygmaeum* has an adult shell diameter of around 0.04–0.06 in (1–1.5 mm). The foot of the North American banana slug can reach a length of 10 in (25 cm) when crawling.

Pulmonates within the orders Basommatophora, Systellommatophora, and Actophila have diversified into an impressively wide range of forms, with an extensive and often odd variation of shell shapes and details of lung anatomy. There are species with dextrally and sinistrally coiled shells, planispiral shells, limpetlike forms with rounded or conical shells, and sluglike forms. All but the family Veronicellidae bear a functional lung derived from the mantle. The lung is a hollow space within the mantle cavity, lined with tissue liberally supplied with blood vessels for gas exchange. The lung communicates with the outside via a small passage and opening called the *pneumostome*. The lung is rendered active by muscular movements within the mantle that alternately compress and expand the lung cavity. Some aquatic and marine families have evolved secondary gill-like structures within their lungs, while others have evolved gill-like organs on the outer body. Limpetlike forms have evolved from more snail-like ancestors several times within the Basommatophora.

An adult gray garden slug (*Succinea* spp.) with eggs. (Photo by Kenneth H. Thomas/Photo Researchers, Inc. Reproduced by permission.)

Basommatophora bear a single pair of tentacles, the eyes placed at the bases of the tentacles. This category includes more primitive members of the subclass than the land snails or Stylommatophora. Some breathe air, others take oxygen from the water, and some are amphibious, either using a secondary gill when in the water or coming to the surface periodically to breathe.

Among the nonstylommatophorans, two rather primitive families, the Amphibolidae and the Glacidorbidae, still carry an operculum—a moveable, doorlike, chitinous lid that the animals use to close and protect the open end of the shell.

Distribution

Land pulmonates occur in most parts of the globe, extending from hot deserts to beyond the Arctic Circle. They have found their way to even the most remote oceanic islands of the Atlantic and Indo-Pacific.

The nonstylommatophoran pulmonates have spread themselves geographically about as widely as the Stylommatophora, including freshwater bodies and tropical to temperate intertidal zones, on all major landmasses and many islands.

Habitat

Pulmonates occur in most terrestrial habitats, except for the extreme polar and desert regions, although the highest species diversity is found in the moist tropics. They typically live on or near the ground surface, although some families of slugs spend most of their lives burrowing in the soil, and many snails (and a few slugs) are arboreal.

Pulmonates outside the Stylommatophora inhabit tropical and temperate regions. Basommatophorans are mostly freshwater, living in ponds, lakes, streams, and rivers. Systellommatophorans are terrestrial and intertidal-marine, and Actophila prefer tropical salt marshes and mangrove stands.

Behavior

Other than during mating, there is little communication between land pulmonates. "Trail following" is used by some species in locating a mate, or in the case of mollusk-feeding predators, their prey. There is little evidence of territoriality, but crowding can induce aggressive behavior and cannibalism.

Pulmonates are hermaphrodites, and cross-fertilization resulting from reciprocal and simultaneous copulation is the norm. However, several groups are capable of self-fertilization. Sperm exchange can either be internal or external, and if the former, is often mediated by a spermatophore.

The daily activity pattern of most species can be described as crepuscular, the animals being active at dusk and dawn, with varying degrees of activity during the night. Weather is also an important factor, with moisture, but not heavy rainfall, inducing foraging even during the day. In dry conditions snails estivate, withdrawing into the shell and secreting an epiphragm to cover the shell aperture. Typical estivation sites are well above ground level, attached to shaded rock faces or on the stems of plants. During the winter many hibernate, buried in the soil or leaf litter. Homing behavior is quite well developed in some species.

As a whole, nonstylommatophorans lead slow, solitary existences foraging. A few species are carnivorous. Some freshwater pulmonates can regulate buoyancy—sinking, floating, and rising within the water by regulating the air content of the lung. Some can even navigate upside-down across the air-water surface.

Feeding ecology and diet

Most pulmonates are herbivorous, living on fresh and/or dead plant material. Some species, especially arboreal ones, graze on algal films and have specially modified spadelike radular teeth. Slugs and snails also commonly feed on fungi. Several families of slugs and snails are predators, and are specialized for feeding on, for example, earthworms (testacellid slugs) or other snails (spiraxid snails). Most diurnal activity is concerned with foraging, and is crepuscular or nocturnal.

Many insect groups feed on snails and slugs: carabid beetles eat large numbers of slugs, and the larva of the glowworm *Lampyris* feeds almost exclusively on snails. Flatworms also attack slugs and snails. Birds may use a stone like an anvil to break open the snail's shell. Mammals, such as badgers, rats, and hedgehogs, amphibians, and some reptiles are also important predators.

The vast majority of nonstylommatophoran pulmonates leisurely graze on living and dead plant material, algal-bacterial film on intertidal rocks, and, in some cases, carrion. A few species are carnivorous, among them species of the family Glacidorbidae (Basommatophora), snagging and consuming worms and other small aquatic or marine animal life.

Reproductive biology

Courtship is often elaborate in pulmonates and is known to last up to 36 hours in some slugs. It can involve the partners biting each other with their radulae, and the exchange of calcareous darts that pierce the skin and carry chemicals that stimulate sexual activity.

Most species lay eggs in batches in shallow cavities in the soil. Often these are calcified, either containing granules of calcium carbonate or having a calcareous shell. Most eggs are a just few millimeters in diameter, but the shelled eggs of *Strophocheilus* can reach a maximum diameter of 2 in (5 cm). The brooding of eggs by some Pacific endodontoid snails in the specially enlarged shell umbilicus is a form of maternal behavior, and some arboreal snails construct brood chambers of leaves in which the eggs are laid.

In representatives of several families, eggs are retained in the uterus until they hatch; true viviparity involving a form of placenta appears to be rare. The number of eggs laid varies enormously. Some species lay just one or two eggs each year, whereas the giant African land snail *Achatina* can lay several thousand in a lifetime. Incubation time varies with species and temperature, sometimes taking as little as three to four weeks, but extending to well over one year in the slug *Testacella*.

All species in the orders Basommatophora, Systellommatophora, and Actophila are hermaphroditic. Individuals are capable of fertilizing or being fertilized, but only a few species are able to do so simultaneously, since in all but the exceptions, sperm and unfertilized eggs mature at different times. Most species can self-fertilize. Some species can reproduce parthenogenetically, laying unfertilized eggs that develop into adults. Species in the family Siphonariidae still retain a veliger larva during development.

Conservation status

Over 1,000 species of terrestrial pulmonates are listed by the IUCN, of which almost 200 are listed as Extinct, and a further 116 as Critically Endangered. These are mostly endemic snails in the Pacific Islands, where the main causes of extinction are the destruction of the native forest and the introduction of predators for biological control.

The IUCN also includes about 1,000 freshwater mollusks on the 2002 Red List. Problems besetting nonstylommatophoran pulmonates include habitat destruction and pollution of freshwater bodies and intertidal marine areas.

Significance to humans

Many slugs, and fewer species of snails, are major agricultural pests, attacking crops both above and below ground. They also act as the intermediate hosts of important nematode and fluke parasites of humans and domesticated animals. Helicellid land snails carry the fluke *Dicrocoelium dendriticum*, which infects cattle, and the giant African land snail harbors the larvae of the nematode *Angiostrongylus cantonensis*, which can cause meningitis in humans. Several of the larger species of snails are eaten as food.

Basommatophoran species, particularly in the family Physidae, are intermediate hosts for parasitic worms that go on to enter the systems of humans and domesticated animals. Some species are used as food by humans.

1. Great gray slug (*Limax maximus*); 2. Spanish slug (*Arion lusitanicus*); 3. Roman snail (*Helix pomatia*); 4. Rosy wolfsnail (*Euglandina rosea*); 5. Giant African land snail (*Lissachatina fulica*); 6. Agate snail (*Achatinella mustelina*); 7. *Onchidium verruculatum*; 8. *Glacidorbis hedleyi*; 9. New Zealand freshwater limpet (*Latia neritoides*). (Illustration by Patricia Ferrer)

Species accounts

No common name
Glacidorbis hedleyi

ORDER
Basommatophora

FAMILY
Glacidorbidae

TAXONOMY
Glacidorbis hedleyi Iredale, 1943, Blue Lake, New South Wales, Australia.

OTHER COMMON NAMES
None known.

PHYSICAL CHARACTERISTICS
Snails with flat-coiled, nearly symmetrical shells ranging from 0.08–0.11 in (2.0–2.8 mm) diameter.

DISTRIBUTION
Southeastern and southwestern Australia, and Tasmania.

HABITAT
Lakes, ponds, streams, and rivers, usually in acidic waters.

FEEDING ECOLOGY AND DIET
A scavenger, feeding on carrion, and a carnivore, snagging and eating small freshwater fauna.

BEHAVIOR
Individuals spend their days hunting and scavenging in their aquatic habitats.

REPRODUCTIVE BIOLOGY
Only partly understood. Females lay eggs containing already well-developed embryos.

CONSERVATION STATUS
Vulnerable.

SIGNIFICANCE TO HUMANS
None known. ◆

New Zealand freshwater limpet
Latia neritoides

ORDER
Basommatophora

FAMILY
Latiidae

TAXONOMY
Latia neritoides Gray, 1850.

OTHER COMMON NAMES
None known.

PHYSICAL CHARACTERISTICS
Limpetlike form bearing a black shell with an oval shape, the shell up to 0.4 in (11 mm) long, 0.3 in (8 mm) wide, and 0.1 in (4.5 mm) high. A strong, muscular foot equipped with mucous glands and cilia enables the animal to cling to and move about on rocks and logs in rushing streams.

DISTRIBUTION
North Island, New Zealand.

HABITAT
Cool, rushing streams and rivers.

FEEDING ECOLOGY AND DIET
Consumes the surface films of algae and bacteria covering underwater rocks.

BEHAVIOR
Navigates rushing streams underwater by its powerful, muscular foot, using it as both a suction cup and an organ of motility. If disturbed, individuals release a bioluminescent substance based on the chemical interactions of luciferin and luciferase, as in a majority of bioluminescent animals.

Most likely employs its bioluminescence as a defense. The luminous substance, when released by the animal, forms into tiny, individual globules or droplets that could act as lures, the bright, speeding blobs stimulating a predatory fish into chasing the droplets and not the limpet. Or, the bioluminescence may serve to induce a "startle response" in a predator, halting and confusing it while the limpet crawls to safety.

REPRODUCTIVE BIOLOGY
Mating takes place all year long but least often in July. Individuals each carry an ovitestis, spermoval duct, and uterus. Individuals often mate in triplets, the middle partner both donating and receiving sperm. A fertilized individual will lay several egg clusters or capsules within a few days after mating. A typical egg capsule is oval, transparent and gelatinous, 0.2 in (5 mm) long and 0.08 in (2 mm) wide, with a tough outer membrane, and holding about 30 eggs. These capsules appear to be able to repel growth of fungi and other pathogens. The eggs hatch in about 30 days. The embryos are able to produce the luminescent substance mix 10 days before hatching. Newly hatched latia feed on the eggshells and capsule before moving on to adult fare. The hatchlings have planispiral shells that soon change to the limpetlike form. Average life span is about three years.

CONSERVATION STATUS
Not listed by the IUCN.

SIGNIFICANCE TO HUMANS
Since these limpetlike snails are sensitive to pollutants in running water, they serve as indicators of the health of stream and river systems where they live. ◆

Euglandina rosea

Arion lusitanicus

Achatinella mustelina

Agate snail

Achatinella mustelina

ORDER
Stylommatophora

FAMILY
Achatinellidae

TAXONOMY
Achatinella mustelina Mighels, 1845, Waianae, Oahu, Hawaii,
United States.

OTHER COMMON NAMES
English: Oahu tree snail.

PHYSICAL CHARACTERISTICS
Medium-sized snail. Shell height up to 1 in (2.5 cm) and width
up to 0.5 in (1.2 cm). Shell glossy with light brown or dark
brown and cream spiral stripes, and small columellar tooth in
shell aperture.

DISTRIBUTION
Restricted to mountains of Waianae Range, Oahu, Hawaii,
United States.

HABITAT
Arboreal.

BEHAVIOR
Nocturnal, but can become active during day in rain. Sedentary; individuals may never move from single tree or shrub.

FEEDING ECOLOGY AND DIET
Grazes on algal film found on leaves and branches.

REPRODUCTIVE BIOLOGY
Grows slowly, reaching full size around seven years and living
for up to 10 years. Lays no more than one egg per year.

CONSERVATION STATUS
Listed as Critically Endangered by the IUCN. Main threats
are from habitat destruction, introduced predatory snail *Euglandina rosea*, and brown rat. Fewer than 10,000 individuals remain.

SIGNIFICANCE TO HUMANS
Provides valuable source of material for study of evolutionary
processes. Early Hawaiians used the colorful shells for barter
and ornamentation. ◆

Giant African land snail

Lissachatina fulica

ORDER
Stylommatophora

FAMILY
Achatinidae

TAXONOMY
Achatina fulica Bowdich, 1822, Mauritius.

OTHER COMMON NAMES
French: Achatine.

Helix pomatia

Limax maximus

Lissachatina fulica

PHYSICAL CHARACTERISTICS
Large; shell up to 6 in (15 cm), brown with darker zigzag markings; body dark gray or black.

DISTRIBUTION
Native to East Africa and possibly Madagascar; has been introduced by humans throughout much of tropics.

HABITAT
Lives near human habitation, in gardens, fields, plantations, and secondary woodland.

BEHAVIOR
Active at dusk, dawn, at night, or even during daytime when it is raining or overcast. Estivates during extreme drought.

FEEDING ECOLOGY AND DIET
Voracious herbivore, feeds primarily on living plant material.

REPRODUCTIVE BIOLOGY
Very high reproductive potential, maturing in as little as five months, and laying several batches of 100–200 eggs at a time.

CONSERVATION STATUS
Not listed by the IUCN.

SIGNIFICANCE TO HUMANS
Extremely serious agricultural pest, attacking economically important crops and seedlings. ◆

Spanish slug
Arion lusitanicus

ORDER
Stylommatophora

FAMILY
Arionidae

TAXONOMY
Arion lusitanicus Mabille, 1868, Sierra d'Arabida, Portugal.

OTHER COMMON NAMES
German: Spanische wegschnecke.

PHYSICAL CHARACTERISTICS
Large slug up to 5.5 in (14 cm) long with strong tubercles on body, brick red to dark brown in color, with distinct foot fringe. Lacks dorsal keel on tail.

DISTRIBUTION
Eastern and central Europe, northward to southern England and Scandinavia. Recently range has extended rapidly.

HABITAT
Tends to live near sites of human activity.

BEHAVIOR
Daily activity periods typically around dusk and dawn, but strongly influenced by weather.

FEEDING ECOLOGY AND DIET
Feeds around dusk and dawn; mainly on herbaceous plants, with preference for cultivated plants.

REPRODUCTIVE BIOLOGY
Produces single brood each year, with hatchlings found in late autumn and early spring, and adults from early summer onward.

CONSERVATION STATUS
Not listed by the IUCN.

SIGNIFICANCE TO HUMANS
Serious agricultural pest throughout much of range. ◆

Roman snail
Helix pomatia

ORDER
Stylommatophora

FAMILY
Helicidae

TAXONOMY
Helix pomatia Linnaeus, 1758, France.

OTHER COMMON NAMES
French: Escargot de Bourgogne; German: Weinbergschnecke; Spanish: Cargol de Borgonya.

PHYSICAL CHARACTERISTICS
Largest European snail, with a globular shell reaching a diameter of 2 in (5 cm), color creamy white with pale brown spiral bands. Body gray with paler tubercles.

DISTRIBUTION
Widespread in Central and Southeast Europe; introduced into United Kingdom, Scandinavia, and Spain.

HABITAT
Woods, hedges, and herbs, in calcium-rich areas. Extends up to 6,500 ft (2,000 m) in the Alps.

FEEDING ECOLOGY AND DIET
Herbivorous, feeds principally on living vegetation.

BEHAVIOR
Hibernates during winter, digging shallow hole and forming chalky epiphragm over shell aperture. Shows strong homing ability over 150–300 ft (50–100 m).

REPRODUCTIVE BIOLOGY
Courtship elaborate, taking several hours and involving exchange of love darts. Batches of about 40 eggs laid in ground from late spring to summer, hatching three to five weeks later. Matures in three to four years and can live up to 10 years.

CONSERVATION STATUS
Not listed by the IUCN, but protected under the terms of the Bern Convention on the conservation of European wildlife and natural habitats.

SIGNIFICANCE TO HUMANS
Prized as food, especially in France, and was farmed by the Romans. A vineyard pest. ◆

Great gray slug
Limax maximus

ORDER
Stylommatophora

FAMILY
Limacidae

TAXONOMY
Limax maximus Linnaeus, 1758, Sweden.

OTHER COMMON NAMES
German: Tigerschnegel.

PHYSICAL CHARACTERISTICS
Large slug, reaching 8 in (20 cm) extended length. Pale brown or gray, with darker longitudinal bands on side and spotting on mantle. Short keel on back of tail.

DISTRIBUTION
Southern and western European species, extending into southern Scandinavia; introduced into the United States, South Africa, New Zealand, and Australia.

HABITAT
Common in woods, hedges, and gardens.

FEEDING ECOLOGY AND DIET
Herbivorous, feeds on fresh leaves and fungi.

BEHAVIOR
Crepuscular in habit.

REPRODUCTIVE BIOLOGY
Courtship elaborate, involving circling for up to one hour. Slugs then ascend a vertical surface, then hang entwined, suspended from mucus thread. They evert the penes, which hang down, and exchange sperm masses using special claspers at tips. Penes are then withdrawn, carrying partner's sperm mass. Eggs laid in soil.

CONSERVATION STATUS
Not listed by the IUCN.

SIGNIFICANCE TO HUMANS
Occasional pest of ornamental and food plants, especially fruit. ◆

Rosy wolfsnail
Euglandina rosea

ORDER
Stylommatophora

FAMILY
Spiraxidae

TAXONOMY
Euglandina rosea Férussac, 1821, Florida.

OTHER COMMON NAMES
German: Rosige wolfsschnecke.

PHYSICAL CHARACTERISTICS
Up to 3 in (8 cm) in length. Body long and slender; mouth with hornlike sensory "lips." Glossy, rose-colored, translucent shell.

DISTRIBUTION
Native to the Southeastern United States, from North Carolina to Louisiana; introduced into numerous islands in Indo-Pacific, as well as to Bermuda, Japan, and China.

HABITAT
Cosmopolitan, lives in wide variety of habitats.

FEEDING ECOLOGY AND DIET
Voracious carnivore, preys on other snails and slugs.

BEHAVIOR
Follows slime trails of prey, which it senses with special "lips."

REPRODUCTIVE BIOLOGY
Lays 25–40 eggs per year.

CONSERVATION STATUS
Not threatened, but its introduction onto Pacific Islands in an attempt to control the giant African land snail represents probably the most serious threat to native snail fauna. Has caused the extinction of numerous species of tree snails, such as *Achatinella* on Hawaii, and *Partula* on Society Islands.

SIGNIFICANCE TO HUMANS
Used unsuccessfully as control agent against introduced snails of agricultural importance. ◆

No common name
Onchidium verruculatum

ORDER
Systellommatophora

FAMILY
Onchiidae

TAXONOMY
Onchidium verruculatum Cuvier, 1830.

OTHER COMMON NAMES
None known.

PHYSICAL CHARACTERISTICS
Sluglike marine mollusks with no trace of a shell, ranging in size from 0.4–3.0 in (10–70 mm) long. The dorsal surface is covered by a tough, leathery mantle with a bumpy, rocky-looking surface. The mantle contains an air-breathing lung for use out of water, and has a pneumostome in the posterior region, near the anus, which is also posteriorly placed. When underwater, the animal relies for respiration on its skin and on gill-like protrusions on its dorsal surface. A most interesting and little-understood feature is an array of light-sensitive papillae on the dorsal surface.

DISTRIBUTION
Intertidal zones of continents and islands surrounding the Indian Ocean.

HABITAT
Intertidal zones, mud flats, and mangrove stands.

FEEDING ECOLOGY AND DIET
Forages both in and out of the ocean in its intertidal environment, scraping off and eating the films of algae and bacteria that flourish on the surfaces of rocks and mud in this sort of habitat.

BEHAVIOR
While foraging, an individual leaves a trail of mucus charged with identifying biochemicals, enabling the animal to follow the trail back to its home, usually a sheltered space between rocks.

REPRODUCTIVE BIOLOGY
When mating, reciprocal fertilization occurs. The fertile eggs are laid in clutches of 60–100, each egg in a tubular capsule, the eggs connected to one another by strands, and the entire egg mass enclosed in a jellylike blob. The parents attach an egg mass to the rocky wall within their shelter.

CONSERVATION STATUS
Not listed by the IUCN.

SIGNIFICANCE TO HUMANS
None known. ◆

Resources

Books

Abbott, R. Tucker. *A Compendium of Landshells.* Melbourne, FL: American Malacologists, Inc., 1989.

Barker, G. M., ed. *The Biology of Terrestrial Mollusks.* Oxford, U.K.: CABI Publishing, 2001.

———. *Mollusks as Crop Pests.* Oxford, U.K.: CABI Publishing, 2002.

Burch, J. B. *Freshwater Snails (Mollusca: Gastropoda) of North America.* Cincinnati, OH: U.S. Environmental Protection Agency, 1982.

Mead, Albert R. *The Giant African Land Snail: A Problem in Economic Malacology.* Chicago: University of Chicago Press, 1961.

Fretter, Vera & J. Peake, eds. *Pulmonates: Functional Anatomy and Physiology.* Vol. 1. London: Academic Press. 1975.

———. *Pulmonates: Systematics, Evolution, and Ecology.* Vol. 2A. London: Academic Press, 1978.

———. *Pulmonates: Economic Malacology.* Vol. 2B. London: Academic Press, 1979.

Solem, Alan. *The Shell Makers.* New York: John Wiley and Sons, 1974.

South, A. *Terrestrial Slugs: Biology, Ecology, and Control.* London: Chapman and Hall, 1992.

Vaught, Kay Cunningham. *A Classification of the Living Mollusca.* Melbourne, FL: American Malacologists Inc., 1989.

Periodicals

Meyer-Rochow V. B., and S. Moore. "Biology of *Latia neritoides* Gray 1850 (Gastropoda, Pulmonata,

Resources

Basommatophora): The Only Light-producing Freshwater Snail in the World." *Internationale Revue der gesamten Hydrobiologie*. 73, no. 1 (1988): 21–43.

Ponder, W. F. "Glacidorbidae (Glacidorbidacea: Basommatophora), New Family and Superfamily of Operculate Freshwater Gastropods." *Zoological Journal of the Linnean Society, London* 87 (1986): 53–83.

Wade, Christopher M., Peter B. Mordan, and Bryan Clarke. "A Phylogeny of the Land Snails (Gastropoda: Pulmonata)." *Proceedings of the Royal Society, Series B* 268 (2001): 413–422.

Other

Jeffery, Paul. "GastroClass." Natural History Museum, London. 21 August 2003 [21 August 2003]. <http://www.nhm.ac.uk/palaeontology/i&p/gastroclass/gastroclass.htm>.

Peter B. Mordan, PhD
Kevin F. Fitzgerald, BS

Patellogastropoda

(True limpets)

Phylum Mollusca

Class Gastropoda

Order Patellogastropoda

Number of families 5

Thumbnail description
Cap-shaped snails typically found on wave-exposed rocky shores

Photo: Individuals of the genus *Scurria* on the rocky shores of Chile. (Photo by S. Piraino [University of Leece]. Reproduced by permission.)

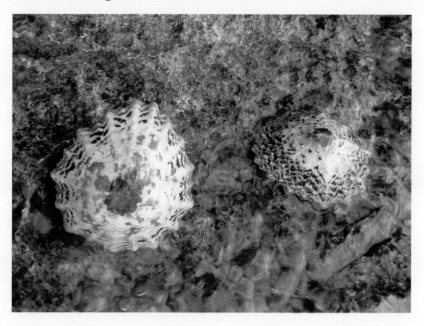

Evolution and systematics

There are numerous limpet taxa in the Paleozoic fossil record, however none possesses unique characteristics that convincingly place these taxa within the order Patellogastropoda. The earliest patellogastropod verified by shell microstructure is from the Triassic of Italy, but it is in the Late Cretaceous and Tertiary periods that many of the living higher taxa have their first occurrences in the fossil record.

Phylogenetic analysis places the patellogastropods at the base of the gastropod tree as the sister taxon to all other gastropods. Although all living patellogastropods are limpets with cap-shaped shells, it is likely that like other living limpets they are descended from coiled snails. However, possible ancestors have not yet been identified in the fossil record.

The patellogastropods are divisible into two major suborders, the Patellina and Acmaeina. The Patellina includes the families Patellidae and Nacellidae. The Acmaeina includes the families Lepetidae, Acmaeiadae, and Lottiidae. Because of their simple shell morphology and anatomy, classifications of the patellogastropods have tended to underestimate their diversity. The application of molecular techniques is providing help in resolving the evolutionary history of this group, and systematic and nomenclatural revisions are certain.

Physical characteristics

All living patellogastropods have cap-shaped shells. The apex of the shell is typically situated at the center of the shell or slightly towards the anterior. All shells are sculpted with concentric growth lines; in many species, additional radial ribs extend from the apex to the shell margin. These ribs may be very fine like growth lines or broad and strongly raised off the surface. The shell aperture is typically oval. The inner surface of the shell bears a horse-shaped muscle scar that opens anteriorly where the head is located. The head has one pair of tentacles and the mouth opens ventrally for feeding on the substrate. Inside the mouth is the radula. The patellogastropod radula has very few robust "teeth," which are brown in color because of the presence of iron compounds. Patellogastropods have two gill configurations: in the Patellina, the gill is located around the edge of the foot and extends around the aperture, while in the Acmaeina, the gill is located over the head, as it is in other gastropod species.

Patellogastropods range in size from about 0.19–7.8 in (5–200 mm) in length. The smallest species and the largest are typically found in the lowest intertidal zone or subtidally. Most species in the intertidal zone average between 0.78–1.5 in (20–40 mm) in length. Subtidal species are typically white or pink in color and intertidal species are typically drab brown or gray with white spots and radial rays. Coloration in patellogastropods is closely associated with their diet, and often the shell is similar in color to the substrate on which the limpet occurs because of the incorporation of plant compounds into the shell.

Distribution

The highest diversity of patellid species in the world occurs along the rocky shores of southern Africa. *Patella* faunas also occur in Western Europe and Australia. The highest diversity of Acmaeina species occurs around the Pacific Rim, while the Nacellidae reach their highest diversity in the tropical and warm temperate Pacific Ocean. Often, patellogastropod

The ventral side of a member of the genus *Patella*. (Photo by S. Piraino [University of Leece]. Reproduced by permission.)

faunas represent an aggregation of lineages accumulated through geological time and are subject to the vagaries of colonization, origination, radiation, and extinction. Australia, like southern Africa, is the most diverse, with representatives of patellid, nacellid, and lottiid lineages. In Chile, only nacellid and lottiid lineages are present; in Europe, there are only lottiid and patellid taxa; and only lottiid species are found in the western United States.

Habitat

Patellogastropods typically occur on intertidal rock substrates. In near-shore subtidal habitats, they are commonly associated with calcareous substrates. Numerous species utilize, as adults, specific types of host plants as substrates, including the Laminariales and Fucales of the Phaeophyta (brown algae), the Cryptonemiales (corallines) of the Rhodophyta (red algae), and marine grasses (Zosteraceae, Angiospermae). The biogeographic distribution of recent marine plant limpets is cosmopolitan; they occur in all major oceans, except the Arctic Ocean. In the deep sea, patellogastropods are found at both cold seep and hydrothermal vent sites.

Behavior

Most patellogastropods are dioecious, although both simultaneous and protandric hermaphroditism are present in the taxon. Protandry is often correlated with territorial species. Because most patellogastropods discharge their gametes directly into the sea where fertilization and development take place, there is no courtship or mating between individual limpets.

Territoriality has independently evolved at least three times in the Patellogastropoda; the most famous example being the territorial Patellidae of South Africa. Territoriality is typically associated with specific food reserves, and limpets become highly aggressive by using their shells as battering rams to drive both conspecifics and other herbivorous species

from their territory. Territorial species are often larger than related non-territorial species, and many territorial species are also protandric hermaphrodites—beginning life as males before becoming females—often upon the acquisition of a feeding territory.

Activity patterns of patellogastropods are often complicated. At high tide, moving limpets are susceptible to aquatic predators such as fish and crabs. At low tide, species are especially vulnerable to shore birds and foraging mammals. Moreover, low tide also places intertidal patellogastropods under physiological stress due to the effects of drying. Many species are most active at night during low tide when visual predation is less effective. Migratory movements of patellogastropods are limited to a general up-shore pattern, with recruitment ocurring in the lower intertidal and later movement leading to life in higher intertidal zones.

Feeding ecology and diet

Patellogastropods are grazers, and feed by removing diatoms, algal spores, and small bits of plant material from the substrate. Very few species are able to feed directly on large algae, although there are several species that live on algae or marine angiosperms and feed directly on the host plant. Often the shape of the radular teeth reflects the limpet's dietary preference—equal-sized blunt teeth are often seen in species that feed on coralline algae, while species that graze on rock substrates have unequal-sized, pointed teeth, and species that feed on marine angiosperms have broad, flat-topped teeth. Some tropical patellogastropods have small "gardens" of specific algal species that are maintained directly adjacent to the limpets' home site on the substrate.

Predators of patellogastropods include predatory gastropods, sea stars, nemertean worms, fishes, lizards, small mammals, and shore birds. Shore birds such as oystercatchers are especially voracious predators and can remove limpets in such numbers that entire expanses of the intertidal can become green from algal growth due to the lack of limpet feeding.

Reproductive biology

The eggs are typically small, about 0.0035 in (90 μm) in diameter, and contain sufficient yolk reserves to support the developing limpet through settling and metamorphosis. Larger species produce millions of eggs per reproductive season and typically have yearly cycles. Smaller species produce far fewer eggs, but can be gravid and capable of spawning gametes year-round. Patellogastropod larvae pass through a trophophore stage and a veliger stage before settling and undergoing metamorphosis. Parental brood protection has evolved at least twice in the patellogastropods. In some species, the eggs are retained in the mantle cavity over the head, where they are fertilized, and develop into crawl-away young. Other taxa have internal brood chambers and copulatory structures. In some Antarctic species, spawning occurs when individuals stack one on top of another, thereby reducing the distance between individuals and increasing the probability of fertilization.

Conservation status

Most patellogastropods are relatively common and broadly distributed. One exception is a group of *Patella* species endemic to the Azores that have been subject in recent years to over harvesting. There are also other species groups that are restricted to the Hawaiian Islands, and a unique brackish-water species that is known only from estuaries in India and Burma and has not been seen alive in more than 100 years.

One species is listed as Extinct by the IUCN: *Lottia alveus*. The extinction of this species does not appear to have been related to human activities. However, current local declines of patellogastropod species (such as in the Azores and the Hawaiian Islands) are almost always associated with over harvesting by humans for food.

Significance to humans

Because of their large size and intertidal habitat, patellogastropod species have been important components of aboriginal diets for more than 150,000 years, and there is evidence that human predation has (and continues to) reduced both maximum and mean limpet size at some localities. Patellid species are also finding use in biological-monitoring studies of the health of rocky-shore communities because of their ubiquitous presence and role in rocky-shore systems. There appears to be little economic interest in most patellogastropod species, although *Cellana* species are a significant fishery in Hawaii and patellids have been over harvested at the Azores in recent years.

Some of the flatter limpet species are often used as shell jewelry because of their medallion-like shape and nacreous shell surfaces. The most famous depiction of a limpet in art is the English painter J. M. W. Turner's *War: The Exile and the Rock Limpet* (1842), which figures a specimen of *Patella* being contemplated by Napoleon Bonaparte during his exile on St. Helena in the Atlantic Ocean.

Species accounts

Shield limpet
Lottia pelta

FAMILY
Lottiidae

TAXONOMY
Lottia pelta Rathke, 1833, Sitka, Alaska, United States.

OTHER COMMON NAMES
None known.

PHYSICAL CHARACTERISTICS
Shell of medium height with slightly anterior apex, sculptured with heavy radial ribs, fine riblets, and/or concentric growth lines. Shell color ranges from blue-black to light brown with or without white markings; white markings typically drawn out into radial rays and blotches; interior

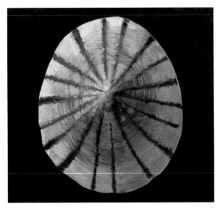

Lottia pelta

shell surface blue-white with brown stained interior muscle scar. Up to 2.3 in (60 mm) in length.

DISTRIBUTION
North Pacific: northern Honshu, Japan, to Baja California, Mexico.

HABITAT
Mid-intertidal zone to immediate subtidal, found on rock substrates as well as large brown algae, and other mollusks such as mussels.

BEHAVIOR
Some specimens change habitats during life, and ribbing and color patterns change as the individual changes substrates; records of these changes accumulate in the shell. Active at night.

FEEDING ECOLOGY AND DIET
Grazes on diatoms, blue-green algae, sporelings, and encrusting algae. When occurring on brown algae, excavates a feeding depression directly into the stipe.

REPRODUCTIVE BIOLOGY
Sexes are separate; larger intertidal individuals >0.39 in (>10 mm) spawn annually in the spring. Smaller and subtidal individuals can be reproductively active throughout year.

CONSERVATION STATUS
Not listed by the IUCN.

Patella vulgata
Lottia pelta

SIGNIFICANCE TO HUMANS
An important food source to early coastal occupants. ◆

Common limpet
Patella vulgata

FAMILY
Patellidae

TAXONOMY
Patella vulgata Linne, 1758, northern Europe.

OTHER COMMON NAMES
None known.

PHYSICAL CHARACTERISTICS
Conical shell of medium height with apex slightly anterior; shell color gray to dirty white, often with yellow to red markings; shell sculptured with strong radiating ribs and growth lines; interior of shell gray-green with prominent muscle scar surrounding

Patella vulgata

darker apical area. In living animals, bottom of foot ranges from yellow to brown in color and is surrounded by a pallial gill that completely encircles the animal. Up to 2.3 in (60 mm) in length.

DISTRIBUTION
Northeastern Atlantic; from Norway to Portugal.

HABITAT
Common across intertidal belt from barnacle zone to immediate subtidal. Habitats range from exposed coastlines to quiet embayments.

BEHAVIOR
A homing species that often returns to a home depression on the rock surface. Moderate territoriality with some aggression towards conspecifics.

FEEDING ECOLOGY AND DIET
A grazer on diatoms, blue-green algae, sporelings, and encrusting algae; can feed on detritus-covered surfaces in embayments.

REPRODUCTIVE BIOLOGY
Protandric hermaphrodite (begins life as male and later changes to female); spawns once a year during the fall and winter, depending on the location within the distribution; typically early in the north and later in the south.

CONSERVATION STATUS
Not listed by the IUCN.

SIGNIFICANCE TO HUMANS
Likely an important food source to early coastal occupants; still gathered and eaten today in many locales. ◆

Resources

Books
Branch, G. M. "Limpets: Evolution and Adaptation." In *The Mollusca*. Volume 10, *Evolution*, edited by E. R. Trueman and M. R. Clarke. Orlando, FL: Academic Press, 1985.

Periodicals
Branch, G. M. "The Biology of Limpets: Physical Factors, Energy Flow, and Ecological Interactions." *Oceanography and Marine Biology Annual Review* 19 (1981): 235–380.

Lindberg, D. R. "The Patellogastropoda." *Prosobranch Phylogeny. Malacological Review, Supplement* 4 (1988): 35–63.

David Lindberg, PhD

Vetigastropoda

(Slit and top shells)

Phylum Mollusca
Class Gastropoda
Superorder Vetigastropoda
Number of families Approximately 12

Thumbnail description
A large and diverse group of marine gastropods found from the intertidal zone to the deep sea. Most maintain some bilateral asymmetry in their organ systems; many have shells with slits or other secondary openings.

Photo: An individual of the species *Haliotis tuberculata lamellosa*, showing its epipodial tentacles, a characteristic only of the vetigastropods. (Photo by S. Piraino [University of Leece]. Reproduced by permission.)

Evolution and systematics

The origin of the Vetigastropoda in the Paleozoic era (544–248 million years ago) is difficult to interpret because the fossil record contains such large and diverse gastropod clades (groups of organisms with features reflecting a common ancestor) as the Bellerophonta and Euomphalina. These clades may or may not be related to living vetigastropods. Morphological analysis of the fossil shells suggests that their coiling parameters (i.e., the way the shells were built) are different from those of extant vetigastropods. In addition, there is no consensus as to whether the Bellerophonta were gastropods at all or whether they were untorted (without twists) Monoplacophora. Traditionally, shells with slits or emarginations (notched margins) are common features of Palaeozoic gastropods and are often regarded as diagnostic of the most "primitive" vetigastropods. The earliest putative gastropod (*Aldanella*) completely lacks these structures, however, and it is possible that slits and emarginations were derived more than once from ancestors without slits. In addition, living representatives of some of the most basal vetigastropods have poorly developed shell elaborations or lack them altogether.

The vetigastropods represent a large subset of the snails that were once commonly known as the Archaeogastropoda. Around 1950 some researchers began to recognize that Archaeogastropoda was actually a collection of several distinct lineages, many of which were subsequently removed and given a higher taxonomic rank. Ultimately, most of the remaining "archaeogastropods" were found to possess a small sensory organ called a busicle at the base of their gills, and were grouped together as vetigastropods on the basis of this character. After the reclassification, researchers found that the species in this group shared other characters, many of them sensory structures.

Today the vetigastropods form a well demarcated group placed near the base of the gastropod tree. There are several major branches within the Vetigastropoda. One branch includes the Lepetelloidea or Pseudococculina, which were formerly groped together with the Cocculoidea in the Cocculiniformia. Another branch includes such vetigastropods with shell slits and emarginations as the Pleurotomarioidea, Fissurelloidea, and Haliotoidea. A third branch includes groups that have shells without secondary shell openings, such as the Trochoidea. The Sequenzoidea are also members of the Vetigastropoda. Neomphalida, a group of organisms that lives near deep-sea hot vents and has a unique combination of characters, may also belong to the Vetigastropoda.

Physical characteristics

Vetigastropod shells range from squat and globe-shaped to elongate turreted structures. Limpet morphology has evolved at least six times in the vetigastropods; in many examples a semi-coiled component is still present in the early shell. Shell sculpture varies widely from simple concentric growth lines, which may be barely visible on the shell surface, to heavy radial and axial ribbing as well as everything in between. The shell aperture, or opening, is typically oval and often tangential to the coiling axis. Most species have an operculum (small lidlike organ) that is used to cover the aperture after the head and foot have been pulled back into the shell. The animals are supple and have a single pair of cephalic tentacles as well as a distinct snout containing the mouth. The lateral sides of the animal typically bear sensory

The species *Bolma rugosa* is relatively common in the Mediterrarean Sea. Its hard operculum is used to make beautiful gold jewelry since they keep their brilliant orange color for many years. (Photo by S. Piraino [University of Leece]. Reproduced by permission.)

epipodial tentacles. When present, copulatory organs are typically part of the right cephalic tentacle. The vetigastropod radula, or toothed ribbon that aids in feeding, is rhipdoglossate as in the Cocculiniformia and Neritopsina. The gills, kidneys, and hearts of many vetigastropods are bilaterally asymmetrical.

Vetigastropods range in size from the minute Scissurelloidea and Skeneoidea, which may be less than 0.08 in (2 mm) long, to members of the Haliotoidea, which may be more than 11.8 in (300 mm) in length. External color patterns are typically drab, but such groups as the Tricolioidea as well as some Trochoidea and Pleurotomarioidea have bright color markings and glossy shells. Reddish shades are the most common. Many vetigastropod shells are iridescent because of the presence of nacre or mother-of-pearl on their inner surfaces.

Distribution

Vetigastropods are distributed throughout all oceans of the world—in the tropics, temperate regions and even under polar ice.

Habitat

Vetigastropods are found in most marine habitats from the intertidal zone to the deep sea. They occur on rocky substrates; on and in soft sediments; on such exotic deep sea habitats as waterlogged food, the egg cases of sharks and skates, and plant debris; and in close association with other marine invertebrates. Some impressive radiations of vetigastropods have occurred at deep-sea hydrothermal vents and cold seeps. Many of the smallest taxa are found in interstitial (spaces between sand grains) habitats.

Behavior

Most vetigastropods are dioecious, although some deep-sea members of the group are hermaphrodites. Vetigastropods usually discharge their gametes directly into the sea for fertilization and development; thus there is no courtship or mating between individuals in the majority of species. Several vetigastropods have been shown to make escape responses in reaction to other predatory snails as well as sea stars. Escape responses include swaying and tilting the shell to avoid the sea star's tube feet, as well as short spurts of rapid movement away from potential predators after tissue contact.

Intertidal species are seldom active at low tide, to avoid physiological stress due to drying. Most subtidal and intertidal species are active at night when predators who rely on sight are less effective. Migratory movements of patellogastropods are limited to a general up shore pattern, from recruitment in the lower intertidal to life in higher intertidal zones.

Feeding ecology and diet

Most vetigastropods feed on such encrusting invertebrate organisms as sponges, bryozoans, and tunicates. Feeding directly on such plant material as algae and marine angiosperms has evolved in several groups, most notably the Haliotoidea and Trochoidea. Deep-sea vetigastropods typically ingest sediment; those associated with exotic substrates likely feed on microbes that are actively decomposing the substrate. Filter feeding has evolved in several vetigastropod species, including *Umbonium* in the Trochoidea.

Predators of vetigastropods include other predatory gastropods, sea stars, fishes, small mammals, and shore birds. Such shore birds as oystercatchers are especially voracious predators.

Reproductive biology

Vetigastropods generally have small eggs that produce lecithotrophic or nonfeeding larvae. Direct development has evolved in several species. Some brooding species use such features of the shell as the umbilicus or surface sculpture to hold the developing young. Unlike the patellogastropods, many vetigastropods secrete egg envelopes and have glandular pallial structures that produce masses of jelly-coated eggs on the sea bed. Early development may occur in some species within these masses of jelly containing the spawn. Copulatory organs in vetigastropods are often derived from cephalic and tentacular structures. Internal fertilization has evolved in several groups and is especially common in species living near deep-sea vents.

Larger species produce millions of eggs per reproductive season and typically have yearly cycles of spawning. Smaller species produce fewer eggs, but can be gravid and capable of spawning year round. Vetigastropod larvae pass through trochophore and veliger stages before settling and undergoing metamorphosis.

Conservation status

Most vetigastropods are relatively common, as one would expect of pelagic dispersing species. Endemic taxa are found among such usual biogeographical features as isolated islands in large oceanic basins and in such smaller seas as the Mediterranean. The only vetigastropods on the IUCN Red List are two members of the Skeneoidea, *Teinsotoma fernandesi* and *Teinsotoma funiculatum* from São Tomé and Principe in the Gulf of Guinea off western Africa. Both are Data Deficient. At more regional levels, several abalone species of the genus *Haliotis* are threatened both by overharvesting and by a pathogen in western North America that leads to a fatal wasting disease called withering syndrome. The overharvesting of large *Fissurella* species or *lapas* has also been problematic in Chile. In the tropical Pacific, the commercial harvest of *Trochus* species for the button industry began in the early twentieth century and has significantly reduced populations in some areas. Lastly, the shell trade puts tremendous value on some species of Pleurotomarioidea; collectors pay hundreds of dollars for exquisite specimens.

Shells of the species *Haliotis tuberculata lamellosa* are commonly seen in the Mediterranean Sea. (Photo by S. Piraino [University of Leece]. Reproduced by permission.)

Significance to humans

The significance of vetigastropods to humans includes the use of their shells by inhabitants of the northern Pacific islands, southern Africa, and Australia and New Zealand to make fishhooks, buttons and beads (especially nacreous groups), or bowls (large abalone species). The most common use of vetigastropods was and continues to be as food—for subsistence as well as *haute cuisine* (e.g., abalone steaks).

1. *Lepetodrilus elevatus*; 2. Red abalone (*Haliotis rufescens*); 3. Top shell (*Trochus niloticus*). (Illustration by Emily Damstra)

Species accounts

Red abalone
Haliotis rufescens

ORDER
Haliotoidea

FAMILY
Haliotidae

TAXONOMY
Haliotis rufescens Swainson, 1822.

OTHER COMMON NAMES
French: Ormeau du Pacifique; German: Rote Seeohr; Italian: Abalone rosso.

PHYSICAL CHARACTERISTICS
Shell is ear-shaped with a reduced posterior coil producing a flat bowl-shaped shell. Outer shell surface is red in color and sculpted with broad radial ribs. Interior of shell is iridescent mother-of-pearl. Grows as long as 11.8 in (300+ mm).

DISTRIBUTION
Western United States.

HABITAT
Intertidal and shallow subtidal zones, among or close to kelp forests.

BEHAVIOR
H. rufescens larvae are sensitive to specific chemicals released by coralline algae and use these cues to begin settlement on the substrate.

FEEDING ECOLOGY AND DIET
Feeds primarily on large brown and red algae.

REPRODUCTIVE BIOLOGY
Broadcast spawner; fertilization takes place externally. Has relatively short phase of pelagic development.

CONSERVATION STATUS
Not listed by the IUCN. Overfished in much of its southern range; northern sport fisheries are regulated with size and catch limits.

SIGNIFICANCE TO HUMANS
Has been an important shoreline food item in California for over 13,000 years. Shell provides mother-of-pearl for jewelry and other forms of ornamentation; larger specimens serve as natural bowls. ◆

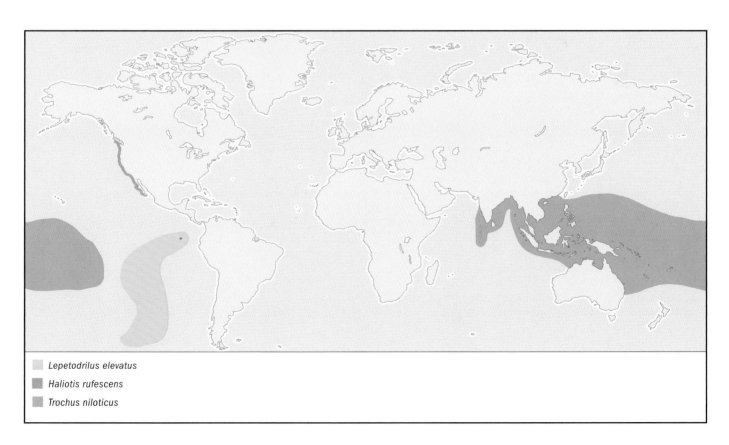

Lepetodrilus elevatus
Haliotis rufescens
Trochus niloticus

No common name
Lepetodrilus elevatus

ORDER
Lepetodriloidea

FAMILY
Lepetodrilidae

TAXONOMY
Lepetodrilus elevatus McLean, 1988, East Pacific Rise, Pacific Basin.

OTHER COMMON NAMES
None known.

PHYSICAL CHARACTERISTICS
Shell aperture is oval, with apex at posterior end and a deeply convex dorsal shell surface. Shell sculpture has concentric growth lines. Shell is whitish gray in color and covered by a greenish brown periostracum (exterior membrane). Interior of shell is whitish with a horseshoe-shaped muscle scar. Reaches 0.16–0.31 in (4–8 mm) in length.

DISTRIBUTION
East Pacific Rise and Galápagos Rift regions, Pacific Basin.

HABITAT
Deep-sea hydrothermal vents at a depth of 7,874–8,858 ft (2,400–2,700 m) associated with the tubeworm *Riftia*.

BEHAVIOR
Nothing is known.

FEEDING ECOLOGY AND DIET
Grazes on *Riftia* tubes, ingesting bacteria and detritus.

REPRODUCTIVE BIOLOGY
Dioecious; male penis is located near base of right cephalic tentacle.

CONSERVATION STATUS
Not listed by the IUCN.

SIGNIFICANCE TO HUMANS
None known. ◆

Top shell
Trochus niloticus

ORDER
Vetigastropoda

FAMILY
Trochidae

TAXONOMY
Trochus niloticus Linnaeus, 1758.

OTHER COMMON NAMES
French: Troca; German: Kreiselschnecke; Local dialect: Lala, ammót, alileng, troka, susu bundar.

PHYSICAL CHARACTERISTICS
A large conical shell with a broad expansive base. Shell marked with purple-pink wavy lines, often obscured by coralline algae. Aperture is ovoid with small open umbilicus. Up to 1.6–3.9 in (40–100 mm) in height.

DISTRIBUTION
Tropical Indo-Pacific waters.

HABITAT
Coral reefs and lagoons.

BEHAVIOR
Primarily nocturnal; rests on coral heads and basalt substrates during daylight hours.

FEEDING ECOLOGY AND DIET
Feeds on filamentous algae and diatoms.

REPRODUCTIVE BIOLOGY
Broadcast spawner; external fertilization and relatively short period of pelagic development.

CONSERVATION STATUS
Not listed by the IUCN. Heavily overfished throughout its range for button blanks and food. Harvesting is now limited with size and bag limits; other regions are under full to partial closure. Aquaculture and restoration studies are being conducted to "reseed" reef flats.

SIGNIFICANCE TO HUMANS
Important source of food and decorative material from its nacreous shell. ◆

Resources

Periodicals

Haszprunar, G. "Sententia: The Archaeogastropoda: A Clade, a Grade, or What Else?" *American Malacological Union Bulletin* 10 (1993): 165–177.

Hickman, C. S. "Archaeogastropod Evolution, Phylogeny and Systematics: A Re-Evaluation." *Malacological Review Supplement* 4 (1988): 17–34.

Salvini-Plawen, L., and G. Haszprunar. "The Vetigastropoda and the Systematics of Streptoneurous Gastropoda (Mollusca)." *Journal of Zoology (London)* 211 (1987): 747–770.

David Lindberg, PhD

Cocculiniformia
(Deep-sea limpets)

Phylum Mollusca
Class Gastropoda
Order Cocculiniformia
Number of families 2

Thumbnail description
Small white limpets that occur on waterlogged wood, whale bone, and cephalopod beaks in the deep sea

Photo: An individual of the species *Addisonia excentrica* inside the egg capsule of the species *Scyliorhinus canicula*, in waters off Acitrezza, Sicily, Italy. (Photo by Danilo Scuderi. Reproduced by permission.)

Evolution and systematics

All living Cocculiniformia have cap-shaped shells. The earliest cocculiniform limpets are found in Tertiary sediments of New Zealand where they are associated with fossilized wood. They are likely descended from coiled snails, like other living limpets, but like the patellogastropods, possible ancestors have not yet been identified in the fossil record.

Living cocculiniform limpets were unknown until the advent of deep-sea exploration in the seventeenth century. At the time, a variety of substrates were recovered by bottom dredges, including cephalopod beaks, waterlogged wood, fish and whale bones, and the egg cases of sharks and skates; all of these substrates were found to have small white limpets living on their surfaces. Almost every species was placed in different genera and families based on their gill morphology and digestive systems, which were also loosely correlated with the different substrates from which these limpets came. In the late twentieth century, these families were grouped into the Cocuiliniformia. However, subsequent phylogenetic analyses based on morphology and molecules showed that there were actually two convergent groups represented. This resulted in the Cocculiniformia being restricted to the Cocculinoidea, while the Lepetelloidea were transferred to the Vetigastropoda, where they represent an early branch in that taxon. However, the placement of the Cocculinoidea among other gastropods remains problematic. Some workers would place it near the base of the gastropod tree, thereby representing an early offshoot in gastropod evolution, while other placements include a possible relationship to the Neritopsina.

The order Cocculiniformia includes two families: the Cocculinidae and the Bathysciadiidae.

Physical characteristics

All living cocculiniforms have cap-shaped shells. The apex of the shell is typically situated at the center or nearer the posterior end of the shell. Shells are sculpted with concentric growth lines, and in some species, additional fine radial threads or beads extend from the apex to the shell margin, while in other species, the surface appears cancellate. The shell aperture is typically oval. The inner surface of the shell bears a horseshoe-shaped muscle scar that opens anteriorly where the head is located. The head is not very flexible and the mouth opens ventrally for grazing on the substrate. Copulatory organs are present and are typically part of the right cephalic tentacle. The Cocculiniform radula is rhipdoglossate, as in the Vetigastropoda and Neritopsina. The Cocculiniform gill consists of a vestigial pseudoplicate structure that does not resemble the typical molluscan gill or ctendium.

Cocculiniformia range in size from about 0.2–0.6 in (5–15 mm) in length, and are white in color and covered with a periostracum.

Distribution

Members of the Cocculiniformia are distributed throughout the world's oceans at bathyal and abyssal depths.

Habitat

Cocculiniformia are associated with three substrates in the deep sea: cephalopod beaks, waterlogged wood, and whale bone.

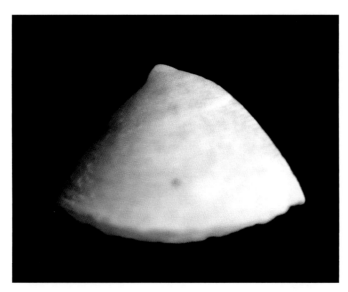

A member of the species *Lepetella laterocompressa,* found in the Sicily Channel, Italy. (Photo by Danilo Scuderi. Reproduced by permission.)

Behavior

Because of the deep-sea habitat of members of this group, little is known of their behavior. The presence of copulatory structures in these simultaneous hermaphrodites suggests a mating system for reciprocal cross-fertilization, which in other gastropods (such as the Heterobranchia) occurs in conjunction with ritualized courting behavior.

Feeding ecology and diet

Cocculiniformia that live on cephalopod beaks are unlikely to be receiving nutrition directly from the beak's chitin substrate. It is more likely that they are feeding and utilizing nutrients from microbes and fungi that are breaking down the chitin on the sea bottom. A similar situation is found in the wood-dwelling cocculiniforms that are found on decaying rather than fresh wood, which is also rich in microbes. Cocculiniforms associated with whale bones are feeding on the sulfate-reducing bacteria that oxidized the lipids found in the bones.

Predators of cocculiniform limpets are not known.

Reproductive biology

All members of the Cocculinoidea studied thus far have been found to be simultaneous hermaphrodites with distinct regions of the gonad producing eggs and sperm. The eggs are rich in yolk and development is thought to be lecithotrophic as in other basal gastropods such as the Patellogastropoda and the Vetigastropoda. Because of the ephemeral nature of the substrates used by members of the Cocculinifomia, larval duration must be sufficient for the larvae to mature and locate suitable habitat on the deep-sea bottom. Such a strategy often incorporates relatively rapid maturation, followed by an extended period of larval competence for locating suitable substrates. This strategy is also often associated with year-round reproduction to maximize the probability of locating patchy habitats. Moreover, in the deep sea, seasonal clues such as day length, water temperature, storm periods, food availability, etc. are nonexistent.

Conservation status

Conservation status of members of the Cocculinifomia is not known. While their habitat at first estimation would appear to be incredibly sparse and rare given the vastness of the deep-sea bottom, the fact that another group of gastropods (the Lepetelloidea in the Vetigastropoda) has also evolved to utilize similar substrates in the deep sea suggests that this first impression is incorrect. However, human interactions with the substrates may pose a serious threat as well as enhancement to these species. For example, increased logging and near-shore construction have undoubtedly increased the amount of wood in the ocean, and the end of commercial whaling will provide an increase in the amount of whale falls in the deep sea. Such an increase has likely increased the amount of substrate available to the Cocculinoidea. In contrast, fisheries' over-exploitation of squid stocks could jeopardize the habitat of the Bathysciadiidae by reducing the number of squid beaks that accumulate on the bottom.

No Cocculiniformia species are listed on the IUCN Red List.

Significance to humans

Because of their occurrence in the deep sea, the Cocculinoidea have no significant interaction with humans.

Species accounts

Japanese deep-sea limpet
Cocculina japonica

FAMILY
Cocculliidae

TAXONOMY
Cocculina japonica Dall, 1907, Sea of Japan.

OTHER COMMON NAMES
None known.

PHYSICAL CHARACTERISTICS
Shell aperture oval with apex close to center; shell sculpture of concentric growth lines with radial punctuations extending from apex to shell margin. Shell white-gray covered with thin periostracum. Interior of shell whitish with inconspicuous muscle scar. Length 0.15–0.31 in (4–8 mm).

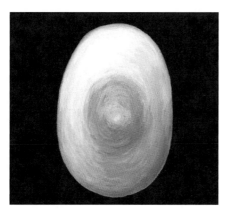

Cocculina japonica

DISTRIBUTION
Found at 164–2,460 ft (50–750 m). Northwestern Pacific Ocean and Sea of Japan.

HABITAT
Waterlogged wood.

BEHAVIOR
Nothing is known.

FEEDING ECOLOGY AND DIET
Feeds on microbes associated with the breaking down of wood on the deep-sea bottom.

REPRODUCTIVE BIOLOGY
Simultaneous hermaphrodite with copulatory structure adjacent to right cephalic tentacle.

CONSERVATION STATUS
Not listed by the IUCN.

SIGNIFICANCE TO HUMANS
None known. ◆

▢ *Cocculina japonica*

Resources

Books
Haszprunar, G. "Comparative Anatomy of Cocculiniform Gastropods and Its Bearing on Archaeogastropod Systematics." In *Prosobranch Phylogeny. Malacacogical Review, Supplement 4*, edited by W. F. Ponder. New York: Academic Press, 1988.

Periodicals
Marshall, B. A. "Recent and Tertiary Cocculinidae and Pseudococculinidae (Mollusca: Gastropoda) from New Zealand and New South Wales." *New Zealand Journal of Zoology* 12 (1985): 505–546.

Strong, E. A., M. G. Harasewych, and G. Haszprunar. "Phylogeny of the Coccilinoidea (Mollusca, Gastropoda)." *Invertebrate Biology* 122 (2003): 114–125.

David Lindberg, PhD

Neritopsina

(Nerites and relatives)

Phylum Mollusca
Class Gastropoda
Order Neritopsina
Number of families 6

Thumbnail description
Widely diverse gastropods, generally small to medium-sized, that coil their shells differently than other gastropods and lack a central shell axis

Photo: A member of the species *Smaragdia viridis*, found in water off of Lampedusa Island, Sicily, Italy. (Photo by Danilo Scuderi. Reproduced by permission.)

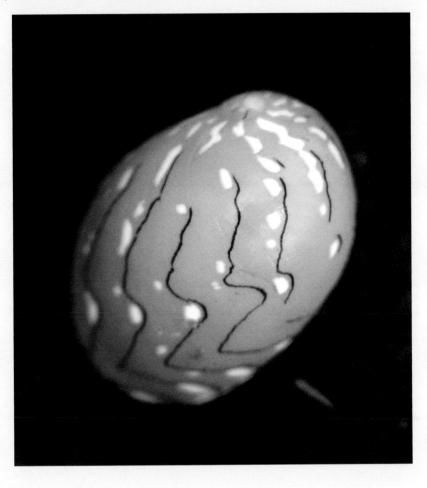

Evolution and systematics

The earliest unequivocal record of the order Neritopsina is Late Sularian-Devonian (428–374 million years ago). Earlier Ordovician records are based on protoconch and adult shell similarities. The Neritopsina were noted as being very distinct from other "archaeogastropods" in the early twentieth century, but only since the 1970s has that placement been broadly accepted.

Neritopsina coil their shells differently than other coiled gastropods and therefore lack a central shell axis, the columella, and most species absorb the internal partitions of the shell as they grow, permitting the snail's body to be more limpet-like rather than coiled, irrespective of its shell.

The Neritopsina are the first clade in the gastropod lineage that has undergone the extensive evolutionary radiations observed across the Gastropoda. The group has shell morphologies that range from coiled conical snails (Hydrocenidae) to limpets (Phenacolepadidae), and even slugs (Titiscaniidae). While conical shells are seen in Hydrocenidae, none appear to have developed very high-spired shells. Multiple terrestrial (Helicinidae, Hydrocenidae) and freshwater invasions have oc-

curred (Neritidae), with some freshwater taxa still having an estuarine or marine larval phase.

The order Neritopsina contains the families Neritidae, Phenacolepadidae, Neritopsidae, Helicinidae, Hydrocenidae, and Titiscaniidae.

Physical characteristics

Neritopsina shell morphology covers most forms seen in the Gastropoda: coiled to limpet, and even shell-less slugs. The Neritopsina have a distinctive juvenile shell, or protoconch, and most species absorb the internal partitions of the shell as they grow, permitting the snail's body to remain more limpet-like rather than coiled. *Neritopsis radula* does not absorb the internal partitions of its shell and, based on this and other anatomical characters, is thought to represent the most basal taxon. The most typical coiled morphologies are seen in the terrestrial groups Helicinidae and Hydrocenidae, and are similar to those seen in coiled vetigastropod groups such as the Turbinidae. All limpet morphologies are marine groups and include both conical and coiled forms, again similar to that seen

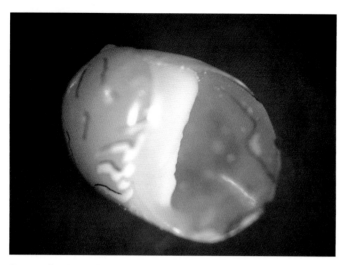

Smaragdia viridis. (Photo by Danilo Scuderi. Reproduced by permission.)

in vetigastropods groups (e.g., Fissurelloidea and Haliotoidea). The most common marine form is a coiled limpet-like morphology with a flat shelf posterior to the aperture. This porcelanous shelf is often ridged, forming tooth-like structures. Shell sculpture varies widely from simple concentric growth lines to taxa with heavy radial ribbing. Several freshwater species have spines that extend from the penultimate whorl, and are often covered by a dark, thick periostracum. Most species have a calcium carbonate operculum that is used to cover the aperture, and the operculum has an apophysis, or spur, on the internal surface. The animals are supple and they have a single pair of cephalic tentacles (hydrocenid species lack tentacles, but the eyes are stalked). Some deepwater and limpet-like taxa bear sensory epipodial tentacles.

The neritopsine female reproductive system is complex, with multiple openings, fertilization is internal, and a muscular copulatory structure is typically found on the right side of the head; a penis is lacking in the terrestrial groups. Spermatophores to facilitate sperm transfer are present and sperm morphology is also distinct in the group. Gelatinous egg capsules are produced and these may be further packaged into calcium carbonate-impregnated egg masses. Free-living veliger larvae are planktotrophic, but direct development is present in some freshwater species and in all terrestrial taxa. The neritopsine radula is rhipdoglossate, as in the Cocculiniformia and Vetigastropoda.

Neritopsines are generally small- to medium-sized gastropods, and range between 0.07–1.5 in (2–40 mm), a relatively small size range within the Gastropoda. External color patterns are variable from solid dark and light colors to bright greens. Shell markings are often geometrical in zigzag shapes. Terrestrial neritopsines often have bright color markings and glossy shells, while others are more subdued, mottled, and blend into their habitats well.

Distribution

Neritopsines are distributed worldwide and have radiated and diversified throughout all tropical and subtropical oceans of the world. Terrestrial species are also distributed worldwide, primarily in the tropics. More temperate taxa occur in the freshwater systems of Europe and Asia, and terrestrially in both the Old and New Worlds.

Habitat

In marine and freshwater realms, neritopsines are typically associated with hard substrates such as coral, beach rock, basalt outcrops and cobbles, and others. Several taxa (Smaragdiinae) are also associated with marine angiosperms in shallow near-shore marine habitats. Two terrestrial invasions, the Helicinidae and the Hydrocenidae, are both associated with moist forest floors and the trunks of trees. The Helicinidae are also found in xeric habitats, and may be partially or completely aboreal. Both terrestrial groups are often associated with limestone geologies. Multiple freshwater invasions have also occurred within the Neritidae, including the *Septaria* of freshwater streams on tropical Pacific islands and the radiation of *Theodoxus* species in the river systems of Europe and central Asia. In the tropical Pacific, some freshwater taxa still have an estuarine or marine period in their larval phase before returning to freshwater streams and rivers. Neritopsines also occur in the deep sea, and are associated with dysoxyic habitats and vent and seep communities.

Behavior

Mating behavior of neritopsines is poorly known. Because of the presence of separate sexes and copulatory structures, some mating behavior is likely present in this group. Clustering is one of the most common behaviors seen in intertidal neritopsines and the clusters are thought to reduce risks of both predation and desiccation. In these multilayered clusters, water loss is reduced and evaporative cooling helps regulate body temperature. Nerites are also known to have trail-following behavior; they can follow both their own and conspecifics' mucus trails on the substrate. These trails may be important in relocating microhabitats for both feeding and resting, and may play a role in building aggregations.

Display behaviors are unknown in the neritopsines, but the anti-predator responses to other predatory gastropods include shell elevation, rotations, flailing of tentacles, and rapid movement. Territorial behavior, or defense of a home range, is also unknown. Most activity patterns in intertidal species are mediated by light and tidal patterns. Up-shore migration is common in intertidal species, with the largest individuals often found at the highest intertidal levels. Freshwater taxa with an estuarine or marine larval phase have behaviors that cue the larvae to settle at the mouths of rivers and streams, followed by movement against the flow, literally crawling upstream into the freshwater habitat.

Feeding ecology and diet

Neritopsines are grazers on algal spores, diatoms, and detritus. Some freshwater species such as *Theodoxus* are also carnivorous, feeding on aquatic insect larvae. Terrestrial representatives feed on detritus, algal spores, moss, and lichens. Deep sea and vent taxa are likely detritivores as well.

Foraging behavior for marine intertidal forms is closely tied to both light and tidal cycles. Many intertidal species move only when awash, and not when full exposed to aerial conditions or when completely submerged. Movement at low tide at night is often common. Many intertidal species also rest at a higher level in the intertidal zone and feed at a lower one.

Predators of neritopsines include crabs, fishes, and other predatory gastropods. Several tropical crab species have large, heavy claws that can effortlessly crush the shells of intertidal nerite species. Tropical reef fishes are also well equipped to both remove and crush nerites. Predatory gastropods are also common predators on neritopsine species, and several species have escape responses to the approach of a carnivorious species.

Reproductive biology

Little is known of the courtship and mating behaviors of neritopsine species. What is known is that the sexes are separate; the female reproductive system is quite complex with as many as three distinct openings into the mantle cavity; and a penis is present in all but the terrestrial species. Eggs are laid in gelatinous capsules, either singly or as multiples in an egg mass. In most marine species, the embryos pass through a trochophore stage within the capsule before hatching out as feeding veliger larvae. Some freshwater species and all terrestrial species have direct development, and hatch as juvenile snails. There is no record of parental care or brooding in this group, and seasonality in broadcast spawning individuals appears correlated with food availability both for the adults and the larvae. In the freshwater *Theodoxus* species, most of the eggs function as food for the single surviving juvenile.

Conservation status

Most nerite marine species are quite common on open shores. However, freshwater and terrestrial species are often more restricted in their distribution and more susceptible to disturbance; this is certainly the case in the Neritopsina. Ten freshwater members of the Neritidae and 23 terrestrial snails (17 Helicinidae and six Hydrocenidae) are listed on the IUCN Red List. Habitat destruction is the largest threat to these species, with deforestation directly affecting terrestrial species and increased sedimentation load impacting freshwater species. Freshwater species are also affected by water diversion and reservoir projects, as well as pollution.

Significance to humans

The Neritopsina, especially the smaller species, have been used by humans as decorative "beads." Larger marine species have been used as food items by subsistence gatherers since prehistoric time. Freshwater, and some marine, species often serve as intermediate hosts for trematode parasites, and terrestrial species in North America are sometimes considered as agriculture pests.

Species accounts

Plicate nerite
Nerita plicata

FAMILY
Neritidae

TAXONOMY
Nerita plicata Linne, 1758.

OTHER COMMON NAMES
None known.

PHYSICAL CHARACTERISTICS
Globose, but relatively high-spired shell for the group; strong, spiral ribs, with two large denticles inside aperture. Colors range from almost solid dark to cream colored, with sparse dark markings. Length 0.39–0.59 in (10–15 mm).

Nerita plicata

DISTRIBUTION
Indo-Pacific.

HABITAT
Intertidal zone on beach rock and coral.

BEHAVIOR
Nocturnal feeding excursions during low tide periods.

FEEDING ECOLOGY AND DIET
Grazer on algal spores, diatoms, and crusts.

REPRODUCTIVE BIOLOGY
Internal fertilization (spermatophore); intertidal egg mass impregnated with calcium carbonate and attached to substrata.

CONSERVATION STATUS
Not listed by the IUCN.

SIGNIFICANCE TO HUMANS
Common in shell craft, and used as experimental organism in tropical zonation studies. ◆

Nerita plicata

Theodoxus fluviatilis

River nerite
Theodoxus fluviatilis

FAMILY
Neritidae

TAXONOMY
Theodoxus fluviatilis Linne, 1758.

OTHER COMMON NAMES
German: Gemeine Kahnschnecke.

PHYSICAL CHARACTERISTICS
Globose, with large penultimate whorl; aperture is oval with raised edge. Exterior shell surface primarily smooth with weak spiral grooves, black to dark brown in color with white blotches and markings. Length 0.23–0.47 in (6–12 mm).

Theodoxus fluviatilis

DISTRIBUTION
Europe.

HABITAT
Freshwater river systems and estuaries of Western Europe and Great Britain, and drainages into Baltic and Black Seas.

BEHAVIOR
Uses escape behavior from predatory leeches.

FEEDING ECOLOGY AND DIET
Grazer on algae, diatoms, insect larvae, and detritus.

REPRODUCTIVE BIOLOGY
Internal fertilization; benthic egg masses attach to hard substrata.

CONSERVATION STATUS
Historical distribution substantially reduced by pollution, but currently returning to rivers and tributaries. Not listed by the IUCN.

SIGNIFICANCE TO HUMANS
Host for trematode larvae, and a freshwater pollution indicator species. ◆

Resources

Books
Ponder, W. F. "Superorder Neritopsina." In *Mollusca: The Southern Synthesis*. Part B, *Fauna of Australia*, vol. 5, edited by P. L. Beesley, G. J. B. Ross, and A. Wells. Melbourne, Australia: CSIRO Publishing, 1998.

Periodicals
Bourne, G. C. "Contributions to the Morphology of the Group Neritiacea of Aspidobranch Gastropods. Part I. The Neritidae." *Proceedings of the Zoological Society of London* (1908): 810–887.

———. "Contributions to the Morphology of the Group Neritiacea of Aspidobranch Gastropods. Part II. The Helcinidae." *Proceedings of the Zoological Society of London* (1911): 759–809.

David Lindberg, PhD

Caenogastropoda

(Caenogastropods)

Phylum Mollusca

Class Gastropoda

Order Caenogastropoda

Number of families Approximately 100

Thumbnail description
The most diverse gastropod group living today; they include almost every gastropod shell form and are successful in every major habitat on Earth

Photo: Macrograph of the species *Xenophora pallidula*, a member of the carrier shell family, named for cementing broken shells and stones to its own shell, a decorative variation on the theme of camouflage. (Photo by Science Photo Library/Photo Researchers, Inc. Reproduced by permission.)

Evolution and systematics

The earliest caenogastropods appear in late Silurian and early Devonian rocks from about 400 million years ago. From this initial appearance, the caenogastropods have radiated into the most diverse gastropod group living today. This diversity spans every aspect of biodiversity and includes morphology, habitat, behavior, and reproductive mode, among others. Partly because of this tremendous diversity, they were mistakenly divided into two groups: the Mesogastropoda and the Neogastropoda. However, today the term "mesogastropod" is understood to be just a level or grade of gastropod evolution, not an evolutionary lineage. In contrast, the Neogastropoda do represent a monophyletic group within the Caenogastropoda and first appear during the Cretaceous, although the apex of their diversification is in the Upper Cretaceous and during the Tertiary. Another recognized clade within the Caenogastropoda is the Architaenoglossa. Members of the Architaenoglossa were previously considered to be "lower mesogastropods," but are now recognized as the sister taxa of the Sorbeoconcha (which also includes the Neogastropoda). Thus, the Caenogastropoda are recognized as containing the Architaenoglossa and the Sorbeoconcha. The Architaenoglossa contain the freshwater group Ampullarioidea and the terrestrial group Cyclophoroidea. The Sorbeoconcha consist of all remaining caenogastropods. Finer level relationships within the Caenogastropoda remain problematic despite ongoing morphological and molecular studies. This lack of resolution based on morphological characters occurs in part because the earlier variation seen in or-

gan systems and structures of other groups (e.g., the kidney/heart complex, radula) has become relatively fixed in caenogastropods (and especially in the Sorbeoconcha). Where novelty has evolved, it is often similar across taxa and confounds convergent evolution. In molecular studies, appropriate molecular markers for the divergence times within this taxon have yet to be identified.

The Caenogastropods are composed of the Architaenoglossa and the Sorbeoconcha. The Architaenoglossa include the Ampullarioidea and Cyclophoroidea, and the Sorbeoconcha consist of the Cerithiomorpha, Littorinimorpha, Ptenoglossa, and Neogastropoda. There are about 100 families within these smaller clades.

Physical characteristics

Caenogastropod shells are typically coiled and almost every shell form is found within this group, from flat squat shells to globose ones and even long, narrow, tightly coiled ones. A few are limpet-like, and one group, the Vermetidae, has shells that uncoil and look like worm-tubes; in a few caenogastropods, the shell is reduced to an internal remnant in the snail's body. The most pronounced morphological change is in groups such as the Entoconchidae, which are shell-less, worm-like internal parasites of echinoderms.

A major feature of the caenogastropods is modification to the pallial cavity, which contains the gill (or ctenidium) and associated sense organs and openings from the kidneys,

The red volute species (*Xenophora pallidula*) is a member of the carrier shell family. (Photo by Michael McCoy/Photo Researchers, Inc. Reproduced by permission.)

gonad, and intestine. In the Architaenoglossa, the terrestrial Cyclophoroidea have lost the ctenidium and osphradium, and the pallial cavity has been modified as a lung. In contrast, the freshwater Ampullarioidea have a single left ctenidium in addition to a new structure located on the right side of the pallial cavity that serves as a lung. In the Sorbeoconcha, pallial cavity water flow is driven by the large, ciliated osphradium, as well as by the ctenidium. There is also the formation of an anterior (or inhalant) notch, or siphon, in the shell and the mantle in many taxa. Some taxa also have a posterior notch in the shell that is associated with the exhalant current, and in some this resembles a shell-like tube.

The shell is never nacreous, and an operculum is present in most adults. Apart from members of the Neogastropoda, the radula usually has only seven teeth in each row. The radula of neogastropods has five teeth to one tooth in each row; it is altogether absent in some species.

The caenogastropods include some of the largest and smallest gastropods, ranging in adult size from 0.019–15.7 in (0.5–400 mm). The caenogastropods are also the most colorful of the gastropods and have the most elaborate markings and patterns.

Distribution

Caenogastropods occur worldwide in all marine, estuarine, freshwater, and terrestrial habitats.

Habitat

Caenogastropods occur in almost every habitat found on Earth. Most are found in the marine environment, where they extend from the high tide mark to the deepest oceans; several groups live in freshwater or terrestrial habitats. In the terrestrial realm, caneogastropods can be found in the wettest en-

vironments of tropical rainforests and in the driest deserts, where annual activity patterns may be measured in hours. Some of the smallest caenogastropods live below ground in the lightless world of aquifers and caves, and others interstitially in groundwater. Marine diversity is highest near shore and becomes reduced as depth increases beyond the shelf slope. Like many other organisms, caenogastropods reach their highest diversity in the tropical western Pacific and decrease in diversity toward the poles.

Behavior

Caenogastropod behavior can be characterized as feeding, fighting, fleeing, and mating, and most of these behaviors are primarily driven by chemoreception. Behavioral responses to visual cues are primarily limited to shadow responses and phototaxis, and because of the supposed lack of visual acuity, display behaviors are not known. However, salt marsh *Littorina* species have been shown to use vision to assess both shade intensity and shape orientation. In addition, in some Strombidae species, males sequester and guard females; fighting between males has been noted.

Like other gastropods that occur in the intertidal, caenogastropod activity and feeding behaviors vary with the tidal cycle: snails are inactive at low tide (except at night) and become active as the tide rises. In the subtidal, diurnal/nocturnal behaviors are important in avoiding predation; in the open ocean, vertical migrations of pteropods have been documented, with the snails moving to deeper water during daylight, but coming within 328 ft (100 m) of the surface at night. Migrations of upper and lower shore *Littorina* species have also been documented and may be associated with assortative mating.

Feeding ecology and diet

The majority of caenogastropods are carnivores, although herbivory has evolved in several groups, including the Littorinidae, Columbellidae, and Cypraeidae; many of these taxa also are capable of feeding on detritus as well as encrusting organisms such as tunicates, sponges, and bryozoans. Other caenogastropods have evolved filter-feeding behavior, such as the Clyptraeidae, and some occur as either endo- or ectoparasites in and on echinoderms (Endoconidae). Several large groups are scavengers (Nassariidae), feeding on dead fishes, crustaceans, and other organisms that come to rest on the bottom. Some neogastropods such as cone shells (Conidae) have poison glands and harpoon-like radular teeth that are used to actively hunt worms, other mollusks, and fishes. Off the bottom and up in the water column, members of the Janthinidae are drifting, pelagic carnivores feeding on jellyfishes, whereas the pelagic Heteropoda are active swimmers in search of other mollusks (including small cephalopods), crustaceans, and fishes.

Because of the diversity and abundance of caenogastropods in such a broad range of habitats, predators of caenogastropods include most marine, estuarine, freshwater, and terrestrial carnivores. These include fishes, crustaceans, sea stars, other mollusks, birds, and mammals.

Reproductive biology

Most caenogastropods have a penis to copulate or to exchange spermatophores. The caneogastropod reproductive systems are complex. The male system typically consists of the testes, a prostate gland, and a vas deferens that leads to the penis, which is typically located on the right side of the head. The female system typically has two glands following the ovary: the albumen gland and the capsule gland. Also, there is typically a bursa copulatrix that receives sperm during mating and a seminal receptacle where sperm is stored prior to being used to fertilize the eggs. The addition of the albumen gland and the capsule gland is of special significance because nutritional resources can now be encapsulated with the eggs in the capsule. Egg size is reflected in the initial size of the juvenile shell, or protoconch, and this feature has been useful in distinguishing feeding and non-feeding larvae in both recent and fossil taxa—non-feeding larvae tend to have larger eggs than feeding taxa. The first gastropod larval stage is typically a trochophore that transforms into a veliger and then settles and undergoes metamorphosis to form a juvenile snail. In some caenogastropods, the young hatch as trochophores, grow into veligers, feed on maternally provided albumen, grow, and ultimately break through the capsule and crawl away. In other taxa, they may be released from the capsule as veliger larvae and feed in the plankton before settling and metamorphosing into juvenile snails. Both simultaneous and protandric hermaphrodites are present in the caenogastropods, although the majority of species are dioecious. In freshwater and terrestrial caenogastropods, direct development with crawl-away larvae is the norm.

Conservation status

Marine species tend to be more common than freshwater and terrestrial species, and are often less restricted in their distribution and less susceptible to disturbance. Ten freshwater members of the Neritidae and 23 terrestrial snails (17 Helicinidae and six Hydrocenidae) are listed on the IUCN Red List. Many species are over-harvested for food, such as *Strombus* (queen conch in the Bahamas). Habitat destruction is the biggest threat to these species, with deforestation directly affecting terrestrial species and increased sedimentation load impacting freshwater species. Freshwater species are also affected by water diversion and reservoir projects, as well as pollution.

Significance to humans

Because of their tremendous diversity in form and color, caenogastropods have figured more prominently in human cultural than most other gastropod groups. The Phoenicians were famous throughout the ancient Mediterranean world for producing a royal purple dye made from organs found in members of the family Muricidae. Cowry shells were first used as money in China in 1200 B.C. So many cultures have used shells for money that it is the most widely and longest used currency in human history. Shell trumpets (constructed primarily of caenogastropod species such as *Tonna*, *Bursa*, *Trophon*, and *Cassis*) are found in many parts of the world (Polynesia, Japan, South America, and the Mediterranean), where they were used as both signaling and musical instruments. In India, the *shanka* has been used Hindu rituals for more than 1,000 years. In art, caenogastropod shells often decorate still-life paintings because of their brilliant colors and forms; one of the most famous shell paintings is undoubtedly Rembrandt's 1650 Cone Shell (*Conus marmoreus*).

Approximately 20 species of cone shells are known to be dangerous to humans, and stings from three species (*Conus geographus*, *Conus textile*, and *Conus tulipus*) have resulted in fatalities. However, cone shell venom is also being used to produce drugs for the control of pain. Lastly, shell collecting has been popular since the early nineteenth century and certain caenogastropod groups such as cone shells, cowries, and *Murex* have been among some of the most valuable and sought-after species.

1. Apple snail (*Ampullaria canaliculata*); 2. Geography cone shell (*Conus geographus*). (Illustration by Emily Damstra)

Species accounts

Apple snail
Ampullaria canaliculata

FAMILY
Ampullariidae

TAXONOMY
Ampullaria canaliculata Lamarck, 1819.

OTHER COMMON NAMES
None known.

PHYSICAL CHARACTERISTICS
Large globose shell, with whorls separated by deep channel; large, deep umbilicus. Shell smooth and yellow-green to brown in color, sometimes with darker spiral bands. Height 1.7–2.9 in (45–75 mm).

DISTRIBUTION
South America, but introduced throughout Asia for food. (Specific distribution map not available.)

HABITAT
Freshwater.

BEHAVIOR
Upon sensing chemical cues of potential predators (e.g., turtles) or of crushed snails, it releases, drops to the bottom, and buries itself.

FEEDING ECOLOGY AND DIET
Omnivorous; feeds on algae, terrestrial plants, detritus, and animal matter.

REPRODUCTIVE BIOLOGY
Internal fertilization; pink to white egg masses deposited above water line in aerial conditions.

CONSERVATION STATUS
Not listed by the IUCN.

SIGNIFICANCE TO HUMANS
Intermediate host for trematodes, significant rice pest in Asia, and common in the aquarium trade. ◆

Geography cone shell
Conus geographus

FAMILY
Conidae

TAXONOMY
Conus geographus Linne, 1758.

OTHER COMMON NAMES
None known.

PHYSICAL CHARACTERISTICS
Flat-spired shell, with knobby whorls; aperture elongate and slightly flared at anterior end. Shell marked with diffuse gold-brown tent markings, some densely grouped into bands. Length 2.3–4.7 in (60–120 mm).

DISTRIBUTION
Tropical Indo-Pacific. (Specific distribution map not available.)

HABITAT
Coral reef flats and fore-reef, in sand, and among coral rubble.

BEHAVIOR
Feeds at night by capturing and anesthetizing small schools of fish within its rostrum before individually stinging them.

FEEDING ECOLOGY AND DIET
Feeds on worms, other snails, and fishes.

REPRODUCTIVE BIOLOGY
Internal fertilization.

CONSERVATION STATUS
Not listed by the IUCN.

SIGNIFICANCE TO HUMANS
Highly toxic venom; bites have been fatal. ◆

Resources

Books

Graham, A. "Evolution within the Gastropoda: Prosobranchia." In *The Mollusca*. Vol. 10, *Evolution*, edited by E. R. Trueman and M. R. Clark. New York: Academic Press, 1985.

Ponder, W. F. "Superorder Caenogastropoda." In *Mollusca: The Southern Synthesis*. Part B. *Fauna of Australia*, vol. 5, edited by P. L. Beesley, G. J. B. Ross, and A. Wells. Melbourne, Australia: CSIRO Publishing, 1998.

Periodicals

Ponder, W. F., and Lindberg, D. R. "Towards a Phylogeny of Gastropod Mollusks: An Analysis Using Morphological Characters." *Zoological Journal of the Linnaean Society*, 119 (1997): 83–265.

David Lindberg, PhD

Bivalvia

(Bivalves)

Phylum Mollusca
Class Bivalvia
Number of families 105

Thumbnail description
Bilaterally symmetrical mollusks, with a reduced
head and typically two external shell valves,
many of which are commercially important for
human consumption and pearl production; some
have major impacts on the world economy and
environment

Photo: A coquina clam (*Donax variabilis*) underwater, with its siphons extended. (Photo by Harry
Rogers/Photo Researchers, Inc. Reproduced by
permission.)

Evolution and systematics

The evolutionary history of bivalves is represented by an
extensive fossil record. It begins in the Cambrian period
(544–505 million years ago [mya]) with a laterally compressed stenothecid monoplacophoran (primitive single-
shelled marine mollusk) as the most likely immediate
ancestor. By the Middle Ordovician period (about 460 mya),
recognizable members of all modern subclasses had appeared. Bivalves formed important components of marine
communities since they first diversified, especially in shallow-
water marine sediments, but also in the intertidal zone, the
deep sea, and freshwater habitats. In the Cretaceous period
(145–65 mya), epibenthic rudist bivalves formed tropical
reef-like structures similar to modern coral reefs. Rudists
were massive extinct cemented bivalves that had two different-
sized shell valves.

Most authors think that scaphopods and bivalves are
closely related taxa, based on the configuration of the nervous
system, lateral expansion of the mantle, and elongation of the
foot for burrowing. Within the class of bivalves itself, the
widely variable shell characteristics (shape, sculpture, color,
hinge teeth) have been historically and consistently used in
identification and classification at all levels. Various other evolutionary schemes have been based primarily on single organ
systems, especially the ligament, stomach, digestive tract, and
gills. Elements of each of these continue to be important in
modern comprehensive phylogenetic analyses.

Current classification schemes recognize five subclasses of
bivalves. The Protobranchia (e.g., families Nuculidae, Solemyidae) are presumably the most primitive, using simple gills
solely for respiration and palp proboscides (enlarged labial
palps, normally used for sorting food particles) for collecting
food from the sediment surface. The Pteriomorphia (e.g.,
Mytilidae, Pteriidae) include many of the most familiar bivalves, all of which share an epibenthic habitat (byssate or cemented), an unfused mantle edge, and a reduced foot. Byssate
bivalves are attached to their substrate by a byssus (bundle of
elastic collagen-rich threads) secreted by the foot. The Pale-
oheterodonta (e.g., Unionidae, Margaritiferidae) include the
freshwater mussels, with their specialized glochidia larvae.
Heterodonta (e.g., Veneridae, Donacidae) is the most species-
rich and most widely distributed subclass, containing the classic burrowing clams with well-developed hinges, siphons, and
active feet. The Anomalodesmata (e.g., Pandoridae, Clavagellidae) include the most unusual and most specialized bivalves, featuring modified ctenidia (gills), an edentulous hinge,
fused mantle margins, and hermaphroditism. The evolutionary relationships, and thus the accepted classification, among
and within bivalve subgroups continue to be revised through
the application of phylogenetic analysis. These analyses makes
use of a wide range of morphological (especially anatomical)
and molecular characters.

Physical characteristics

Most bivalved mollusks have laterally compressed bodies
and a shell consisting of two calcareous valves hinged dorsally
by interlocking teeth and an elastic ligament. The shell valves
are usually similar in size, sculpture, and color, and retain a

permanent record of shell growth in their concentric layers. Growth can be traced from the first stage (prodissoconch) at the umbo (rounded or pointed extremity) to the latest stage at the ventral margin. The shells of many bivalve species are rather bland, although some groups (e.g., Pectinidae, Spondylidae) show characteristic colors or color patterns. Lining each valve is mantle tissue that secretes the aragonitic or calcitic shell, including its organic outer layer, or periostracum, and the interior aragonitic nacre (mother-of-pearl) present in many species. Between the two shell-mantle layers, an internal mantle (or pallial) cavity contains the ctenidia and visceral mass (containing digestive and reproductive organs), the latter ending in a muscular extensible foot. The foot is usually equipped with a gland that secretes the byssus for attachment to hard substrates.

The posterior mantle edge of many bivalves is fused into incurrent and excurrent siphons that direct water in and out of the mantle cavity for respiration, feeding, and discharge of waste and reproductive products. In most bivalves, water flows in and out of the posterior end; some, however, are secondarily modified for an anterior-posterior flow. Paired adductor muscles connect the inner valve surfaces, enabling the bivalve to close; relaxation of these muscles allows the ligament to open the valves. The head of bivalves is reduced, so that the cephalic eyes, tentacles, and radular teeth typical of most mollusks are absent. The stomach is one of the most complex organs of bivalves, comprising various ciliated sorting areas and ridges as well as the crystalline style, which is an enzymatic rod that rotates against a hardened gastric shield lining the dorsal stomach surface to facilitate extracellular digestion. The bivalve nervous system consists simply of three pairs of ganglia connected by nerve connectives. The circulatory system is equipped with a three-chambered heart, simple vessels, and open hemocoels. The hemostatic pressure of the fluid in these blood cavities is responsible for the expansion and retraction of many bivalve structures, particularly the foot, siphons, and tentacles.

Distribution

Bivalves are found worldwide, including aquatic habitats at high and low latitudes. All are aquatic, requiring water for reproductive processes, respiration, and typically for feeding. One supratidal species (*Enigmonia*) lives in the tidal spray on mangrove leaves or seawalls in Australia, achieving the most nearly "terrestrial" life mode of any bivalve. Several independent lineages of bivalves have invaded freshwater habitats, where they have diversified to produce one of the most endemic faunas as well as some of the most important biofoulers among invertebrates. Many species have been introduced to non-native locations accidentally or intentionally, the latter often in commercial aquaculture for human use or consumption.

Habitat

All bivalves are aquatic, requiring water for reproductive processes, respiration, and typically for feeding. They range in depth from the intertidal zone to the deep sea; one supratidal species (*Enigmonia*) lives in the tidal spray on mangrove

leaves or seawalls in Australia, achieving the most "terrestrial" life-mode of any bivalve. Several independent lineages of bivalves have invaded freshwater habitats, where they have diversified to produce one of the most endemic faunas as well as some of the most important biofoulers among invertebrates. Most bivalves are free-living, either epibenthic (e.g., Pectinidae) or infaunal, burrowing into sand or mud with the muscular foot (e.g., Veneridae, Donacidae). Others are cemented by one valve (e.g., Ostreidae) or permanently attached by byssal threads (e.g., Mytilidae, Dreissenidae). Specialized members of the class burrow into rock or wood (e.g., Pholadidae, Teredinidae), using one or a combination of chemical and mechanical methods. Commensal and parasitic forms (e.g., Galeommatoidea) live associated with, attached to, or within the bodies of other invertebrates.

The habitat of a bivalve is often reflected in the form of its shell. Nestlers and cementing bivalves frequently take the shape of their substrates. Individuals in calm waters often have more delicate or leaflike shell sculpture than their counterparts in fast-flowing currents.

Behavior

Most bivalves are relatively sedentary organisms; however, many are capable of considerable levels of activity. The former class name Pelecypoda means "hatchet-foot," referring to the laterally compressed foot typically used for burrowing in sand or mud. Bivalves move downward into the substrate by extending the foot into the sediment, anchoring the foot by expanding its tip, and pulling the shell downward toward the anchor by muscular action. Byssally attached bivalves (e.g., Mytilidae, Dreissenidae) can break their byssal threads to relocate, and use the foot to move across a hard substrate in a sequence similar to that used for burrowing. They then produce a new byssus for reattachment. Other bivalves can actively swim by waving the foot or tentacles (e.g., Solenidae, Limidae) or by jet-propelling themselves by rapidly clapping the shell valves together (e.g., Pectinidae).

Although the reduction of the bivalve head has eliminated cephalic eyes and other sense organs, many bivalves (e.g., Galeommatidae, Pectinidae) have tentacles and/or photoreceptors along the mantle margins or in the vicinity of the siphons. These structures allow bivalves to respond to changes in light intensity by retracting the siphons and closing the valves. More sophisticated eyes, equipped with retina and lens, are found in several families of epibenthic bivalves (e.g., Cardiidae, Pectinidae).

Feeding ecology and diet

Most bivalves are suspension feeders, filtering food particles from the water column. The expansive ctenidia, in addition to functioning in gas exchange, are the main feeding organs. Their cilia-covered surface collects and sorts particles from currents flowing through the mantle cavity, conveying them to marginal food grooves, then anteriorly toward the labial palps flanking the mouth.

The most primitive bivalves were probably deposit feeders, collecting detritus from the sediment surface. This method is

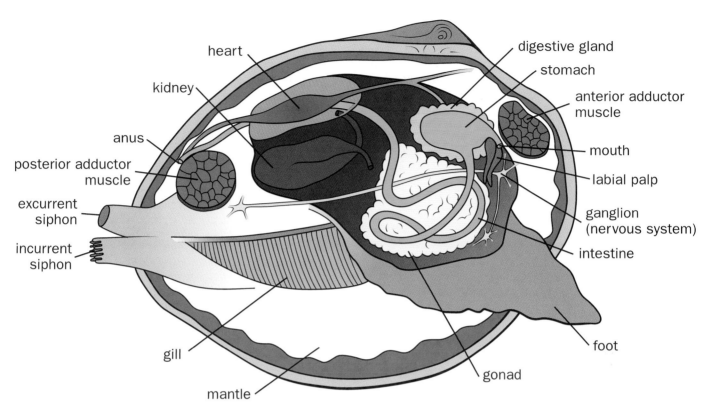

Bivalve anatomy. (Illustration by Patricia Ferrer)

still used by living Nuculoida, using specialized structures known as palp proboscides. Other specialists feed by direct absorption of dissolved organic matter, or DOM (e.g., Solemyidae), or by active capture of small crustaceans and worms through use of a raptorial incurrent siphon (e.g., Cuspidariidae). Others possess symbiotic organisms, supplementing their energy reserves with by-products from their inhabitants. Examples of symbiotic relationships include chemoautotrophic bacteria in Solemyidae and Lucinidae that facilitate habitation of anoxic muds, and zooxanthellae in Cardiidae that provide photosynthetic products in shallow eutrophic waters. The wood-eating Teredinidae are enabled by symbiotic cellulolytic (cellulose-digesting) bacteria that are stored in pouches along the bivalve's esophagus.

Reproductive biology

Bivalves are usually dioecious, with eggs and sperm shed into the water column where external fertilization occurs. Some species are consecutive or simultaneous hermaphrodites, with protandry (male phase preceding female phase) most common. Internal fertilization has been recorded for a few groups (Galeommatoidea, Teredinidae), using tentacles or siphons as copulatory organs. External sexual dimorphism is evident in only a few bivalves (Carditidae, Unionoidea).

Larval development is plesiomorphically planktotrophic, with free-swimming veliger larvae that feed in the plankton for a few weeks. Some bivalves brood their larvae in the suprabranchial chamber or in specialized brood pouches, releasing late-stage veligers or direct-developed juveniles through the excurrent opening. Settlement of larvae is time-dependent but is often delayed in the absence of suitable habitat. Freshwater mussels (Unionoidea) are characterized by specialized glochidia larvae that require attachment to the gills or fins of fish to complete their life cycles. Many of these larvae have specialized hooks for attachment, and some bivalve-fish relationships are species-specific. Many unionoideans possess specialized flaps on the mantle edge that they wave in the water column to attract the attention of the required fish; some of these "lures" mimic small fish or the invertebrate prey of the fish host.

Conservation status

Freshwater pearl mussels (Unionoida) are among the world's most gravely threatened fauna. In eastern North America, the group's center of evolutionary diversification, 35% of the 297 native species are presumed extinct, with another 69% formally listed as endangered or threatened. Human-introduced pollution, especially from agriculture and industry, as well as other forms of habitat alteration (dredging, damming) have been blamed for much of the decline. Such factors can adversely impact not only the mussels themselves, but also the obligate fish hosts of their larvae, potentially resulting in population declines. Introduced species, especially the Asian clam (*Corbicula*) and two species of zebra mussels (*Dreissena*) have further impacted unionoid populations through competition for space and food resources. One-hundred ninety-five species of bivalves have been placed on the 2002 IUCN Red List; all but 10 of these are freshwater pearl mussels. Twenty-nine species of freshwater pearl mussels are protected under CITES.

Marine bivalves are much less affected by human activities. There are no known recent extinctions in this group, and none are currently listed as threatened or endangered. Many species are partially protected by local and national laws regulating the commercial and private harvesting of shellfish. The giant clams (*Tridacna*, *Hippopus*; Tridacnidae) are the single marine group regulated internationally, as a result of overcollecting; eight species are included on the 2002 IUCN Red List, and the entire family is protected under CITES.

Significance to humans

Many kinds of bivalves, especially clams, cockles, mussels, oysters, and scallops, have served as important food sources for fish, vertebrates, other invertebrates, and humans. Aboriginal populations of many cultures have left evidence of eating bivalves in their kitchen middens (mounds or deposits of refuse from meals). Recent practices rely both on harvesting wild populations and on aquaculture in either open or closed aquatic systems. Members of the marine Pteriidae and freshwater Unionoidea have been sources of natural pearls and mother-of-pearl shell for centuries. Since the 1950s, cultured pearls have increased the quantity and quality of this biological gem through aquaculture and husbandry. The Japanese perfected the process of culturing pearls using pearl oysters of the species *Pinctada fucata* (Gould, 1850).

Bivalves have also had negative impacts on human activities. Because most bivalves are filter feeders, they are frequent vectors of human disease related to the concentration of bacteria, viruses, pesticides, industrial wastes, toxic metals, and petroleum derivatives from the water column. Shipworms (Teredinidae) have a long historical record of bioerosion of such human-made wooden structures as ships and docks. Species introduced in freshwaters of the United States, such as the biofouling zebra mussel *Dreissena*, have required millions of dollars to repair clogged water treatment plants and irrigation systems. Damage caused by such species to the environment, in terms of altered habitat and impact on native species, is irreversible; their spread has been largely unstoppable.

1. Eastern American oyster (*Crassostrea virginica*); 2. Queen scallop (*Chlamys opercularis*); 3. Yo-yo clam (*Divariscintilla yoyo*); 4. Common blue mussel (*Mytilus edulis*); 5. European pearly mussel (*Margaritifera margaritifera*); 6. Noble pen shell (*Pinna nobilis*); 7. Black-lipped pearl oyster (*Pinctada margaritifera*); 8. *P. margaritifera* internal view. (Illustration by Barbara Duperron)

1. Zebra mussel (*Dreissena polymorpha*); 2. Shipworm (*Teredo navalis*); 3. Giant vent clam (*Calyptogena magnifica*); 4. Giant clam (*Tridacna gigas*); 5. Northern quahog (*Mercenaria mercenaria*); 6. Watering pot shell (*Brechites vaginiferus*); 7. Coquina clam (*Donax variabilis*); 8. Coquina clam color morph 1; 9. Coquina clam color morph 2. (Illustration by Barbara Duperron)

Species accounts

Common shipworm
Teredo navalis

ORDER
Myoida

FAMILY
Teredinidae

TAXONOMY
Teredo navalis Linnaeus, 1758, The Netherlands.

OTHER COMMON NAMES
English: Gribble, pileworm, ship's worm; French: Taret commun; German: Schiffsbohrwurm.

PHYSICAL CHARACTERISTICS
Shell is white, triangular in shape, inflated, and coarsely sculptured with ridges used for burrowing; reduced in size and permanently gaping. It covers only the anterior end of a much larger worm-like soft body. Generally 4 in (10 cm) in total length, but may grow as long as 2 ft (60 cm). Mantle secretes a calcareous lining for wood burrow. The posterior end has short siphons and unsegmented shovel-shaped calcareous pallets used to close the burrow. Stomach has a wood-storing caecum or blind pouch.

DISTRIBUTION
Worldwide in temperate seas, spread by wooden vessels and ballast water. Probably native to northeastern Atlantic; definitely introduced to San Francisco Bay, but additional differentiation of native versus introduced range is unclear. Reported from such far-flung locations as southern Brazil, Zaire, South Australia, the Black Sea, New England, and British Columbia.

HABITAT
Burrows into floating or stationary untreated wood in seawater. Tolerates salinities ranging from normal seawater to 4 ppt.

BEHAVIOR
Specialized for boring in wood by using ridged shell valves to rasp into wood surface. During burrowing, the animal's disk-like foot acts as suction cup to hold shell tightly against end of burrow. When disturbed, it withdraws into burrow and seals opening with specially shaped pallets.

FEEDING ECOLOGY AND DIET
Specialized for boring into and digesting wood with the assistance of symbiotic cellulolytic bacteria and specialized wood-storing caecum on stomach. Ctenidia are well-developed and equipped with food groove, indicating retained ability to filter-feed.

REPRODUCTIVE BIOLOGY
Protandric hermaphrodite. Extended excurrent siphon probably (as known in other Teredinidae) used as copulatory organ for sperm transfer to adjacent individual. Larvae are brooded in gills to veliger stage. Juveniles grow to maturity in eight weeks. Several generations produced per year. Capable of invading new wood only at time of larval settlement.

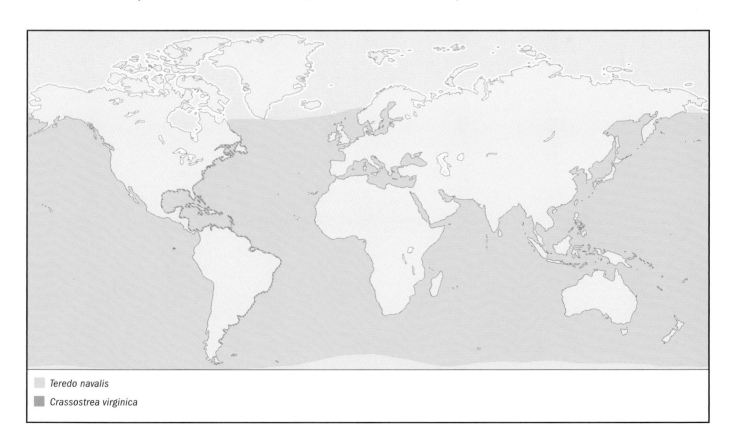

☐ *Teredo navalis*
■ *Crassostrea virginica*

CONSERVATION STATUS
Not listed by the IUCN. Considered a nuisance species due to wood-destroying capabilities.

SIGNIFICANCE TO HUMANS
Called "termites of the sea." Damage to wooden ships and piers, especially in warm tropical waters, recorded as early as Roman times (A.D. 2). Countered by coating wooden surfaces with tar and pitch in the fifteenth and sixteenth centuries, later by copper plating. Responsible for more than $900 million in damages to wooden piers and quays in San Francisco, California, between 1819 and 1821. ◆

Common blue mussel
Mytilus edulis

ORDER
Mytiloida

FAMILY
Mytilidae

TAXONOMY
Mytilus edulis Linnaeus, 1758, European oceans.

OTHER COMMON NAMES
English: Blue mussel, common mussel; French: Moule bleue; German: Miesmuschel; Japanese: Murasakii-gai; Norwegian: Blåskjell; Spanish: Mejillón; Swedish: Blåmussla.

PHYSICAL CHARACTERISTICS
The shell is narrowly elongated and triangular in shape, with the umbo at the pointed anterior end. It is rounded at the posterior end, moderately inflated, smooth, not gaping, with an adherent periostracum varying from dark blue or purple to brown. Generally 2–4 in (5–10 cm) in length, the largest reaching 8 in (20 cm). Shell shape and size influenced by environment, especially tidal level, habitat type, and population density. Interior is faintly nacreous (pearly). Soft body secretes strong stalk of byssal threads exiting shell at ventral margin; mantle edge is yellowish brown; anterior adductor muscle narrow and elongated.

DISTRIBUTION
Northern temperate coastlines from Scandinavia to France; Iceland; maritime Canada to North Carolina; both coasts of southern tip of South America. Raised in aquaculture in China, many European countries, Canada, and United States.

HABITAT
Epibenthic, attached by byssus to hard substrates from high intertidal to shallow subtidal waters. Prefers full oceanic salinity, although dwarfed individuals have been recorded at 4–5 ppt. Lower depth usually limited by presence of such predators as seastars, gastropods, and crabs. Forms large populations (mussel beds) on rocky shores that support extensive associated fauna and flora.

BEHAVIOR
Sessile (permanently attached).

FEEDING ECOLOGY AND DIET
Filter feeder.

REPRODUCTIVE BIOLOGY
Dioecious; broadcast spawner, with females releasing eggs in short yellow-orange rod-shaped masses that quickly break down in the water column. The veligers are initially

Pinna nobilis
Mytilus edulis
Brechites vaginiferus

lecithotrophic (nourished by stored yolk), but are later plank-totrophic, spending 1–4 weeks in the water column. Settling is triggered by the presence of a filamentous substrate (bry-ozoans, hydroids, algae). Growth rate in culture may be as much as 3 in (80 mm) per year. Life span may be as long as 24 years. Hybridizes with the Mediterranean *Mytilus galloprovin-cialis* in northern Europe.

CONSERVATION STATUS
Not listed by the IUCN. Collection for human consumption is regulated by fisheries and public health agencies in most areas.

SIGNIFICANCE TO HUMANS
Human food source worldwide. Recorded as vector of human gastroenteritis, hepatitis A, and viral hepatitis. Used in world-wide "Mussel Watch" programs as an indicator organism for monitoring coastal water quality. Extensively used as model organism in scientific studies. ◆

Eastern American oyster

Crassostrea virginica

ORDER
Ostreoida

FAMILY
Ostreidae

TAXONOMY
Crassostrea virginica (Gmelin, 1791), American and [West] In-dian Oceans.

OTHER COMMON NAMES
English: American oyster, Atlantic oyster, cove oyster, eastern oyster; French: Huître de Virginie; German: Amerikanische Auster. Also has commercial varietal names: Blue points (Long Island, New York), Lynnhavens (Virginia).

PHYSICAL CHARACTERISTICS
Shell is elongated and ovate-triangular in shape, irregular and variable in outline, roughly sculptured with undulating concen-tric ridges. The left (cemented) valve more deeply cupped, right valve flattened, not gaping. Color is whitish to gray. Hinge teeth reduced, ligament strong. Interior is porcelain white with purple-stained muscle scar. Generally grows about 3 in (8 cm) in length, but specimens have been found as long as 14 in (36 cm). Soft body with central adductor muscle and unfused mantle margins.

DISTRIBUTION
Atlantic coast of North America from Canada to Florida, the Gulf of Mexico, and the West Indies to Brazil and Argentina. Introduced to Europe, the western coast of United States, and Hawaii; populations established only in British Columbia and Hawaii. Raised in aquaculture in Australia, New Zealand, Eu-rope, Chile, Japan, Mexico, and eastern United States.

HABITAT
Epibenthic, forming oyster reefs or beds on hard surfaces in shallow coastal estuaries of reduced salinity. *Crassostrea virginica* tolerates a wide range of salinity (5–35 ppt) and temperature. Complex physical structure of oyster beds provides refuge for many other organisms.

BEHAVIOR
Tight shell closure allows individuals to survive many days of aerial exposure. The acidity of the shell liquor (commercial term for water within mantle cavity) is mediated by the buffer-ing action of calcium carbonate dissolved from shell.

FEEDING ECOLOGY AND DIET
Filter feeder.

REPRODUCTIVE BIOLOGY
Protandric hermaphrodite; broadcast spawner, triggered by warming water temperatures. Veligers are planktonic for about three weeks, settling in response to adult oyster shells. Newly attached juveniles are called "spat." Hermaphrodites are rare; instances are recorded of secondary sex reversal (female revert-ing to male). Three to seven years in culture are required to attain market size of 3 in (8 cm).

CONSERVATION STATUS
Not listed by the IUCN. Collection for human consumption is regulated by fisheries and public health agencies in most areas. Many historically important harvesting areas, such as Prince Edward Island, are severely depleted as of 2003.

SIGNIFICANCE TO HUMANS
Human food source, including many famous local varieties (Chesapeake Bay, Apalachicola Bay, Prince Edward Island), both through wild harvest and aquaculture. Oyster reefs stabi-lize shorelines against erosion and filter estuarine water. Recorded as vector of human disease caused by uptake and concentration of toxic metals and lipophilic organic con-taminants. ◆

Queen scallop

Aequipecten opercularis

ORDER
Pterioida

FAMILY
Pectinidae

TAXONOMY
Aequipecten opercularis (Linnaeus, 1758), Mediterranean Sea. Numerous named color forms and varieties.

OTHER COMMON NAMES
French: Vanneau.

PHYSICAL CHARACTERISTICS
Shell is round in outline and compressed, with subequal ante-rior and posterior auricles (ears) strongly delimited, gaping be-low each auricle, with about 20 finely sculptured radial ribs. Shell color is highly variable; may be white, red, orange, mot-tled or solid, with the right valve lighter in color than the left. Interior is white, with grooves reflecting external ribs. May grow as large as 3 in (80 mm) in diameter. Soft body with sin-gle central adductor muscle; mantle margin equipped with nu-merous sensory tentacles and eyes.

DISTRIBUTION
Mediterranean Sea and eastern Atlantic coast from Norway to the Cape Verde Islands, the Azores and the North Sea. Under experimental aquaculture in Spain, France, and United Kingdom.

HABITAT

Epibenthic on all substrates except rocky bottoms. Found in depths from the intertidal zone to 1,312 ft (400 m); most common at about 130 ft (40 m).

BEHAVIOR

Actively swims in response to threat by clapping shell valves together, forcing water to exit mantle cavity in a manner resembling jet propulsion.

FEEDING ECOLOGY AND DIET

Filter feeder.

REPRODUCTIVE BIOLOGY

Simultaneous hermaphrodite, broadcast spawner.

CONSERVATION STATUS

Not listed by the IUCN, and not protected except by local fishery regulations. Fished extensively until 1970s, when populations declined and queen scallops became less important in the commercial market.

SIGNIFICANCE TO HUMANS

Human food source (adductor muscle or whole). The queen scallop symbol was originally worn on heraldic insignia to signify that the wearer had made a pilgrimage to the Christian shrine of St. James in Santiago de Compostela, Spain. The symbol later identified its bearer or ancestor as a Crusader or other type of pilgrim. ◆

Watering pot shell

Brechites vaginiferus

ORDER

Pholadomyoida

FAMILY

Clavagellidae

TAXONOMY

Brechites vaginiferus (Lamarck, 1818), Aden, Red Sea. One recognized subspecies or variety, *B. vaginiferus australis* (Chenu, 1843), Australia.

OTHER COMMON NAMES

English: Australian watering pot, vaginal watering pot.

PHYSICAL CHARACTERISTICS

Shell is narrowly tube-shaped and irregularly sculptured, with the remnant of its 0.2-in (4 mm) larval shell in a lateral "saddle" near the anterior end. The anterior end is flared, comprising a perforated plate (reflected in common name) fringed by open tubules; posterior end has "plaited ruffles." Color is white and chalky. Grows as long as 12 in (295 mm). Soft body with large bottle-green siphons fringed with white tentacles and ringed by a periostracal band. The apex is coated with sand grains. Adults lack adductor muscles.

DISTRIBUTION

Red Sea and Gulf of Aden southward down the eastern coast of Africa to Zanzibar, northeast to the mouth of Gulf of Oman. The variety *australis* from western Australia is found from the Kimberly Archipelago to the Abrolhos Islands.

HABITAT

Infaunal, buried vertically in soft sediment with only the open tube at the posterior end protruding into water. Found at

depths of 0–66 ft (0–20 m). Cements itself to rock crevices in some habitats. Shell form reflects the habitat—long and thin in sand, wider and shorter in rock crevices.

BEHAVIOR

Tube is adventitious (has a single layer), thought to be produced all at once by the adult animal, analogous to the calcareous burrow lining in the Teredinidae. Shell material is subsequently added at posterior end to produce ruffles and lengthen tube. Animal can withdraw deeply into tube as protection against most predators. Cannot burrow further into sediment or reburrow if dislodged.

FEEDING ECOLOGY AND DIET

Filter feeder; pedal disk (foot) against perforated anterior plate contracts and relaxes to pump interstitial water into mantle cavity.

REPRODUCTIVE BIOLOGY

Simultaneous hermaphrodite; probably broadcast spawner, with planktonic veligers. Prodissoconch phase, pre-tube juveniles, and life span unknown.

CONSERVATION STATUS

Not listed by the IUCN; generally considered rare throughout its range.

SIGNIFICANCE TO HUMANS

None known. ◆

Noble pen shell

Pinna nobilis

ORDER

Pterioida

FAMILY

Pinnidae

TAXONOMY

Pinna nobilis Linnaeus, 1758, Mediterranean Sea.

OTHER COMMON NAMES

English: Fan shell, Mediterranean pen shell; French: La Grande Nacre, jambonneau; German: Grosse Steckmuschel; Spanish: Gran nacra.

PHYSICAL CHARACTERISTICS

Shell is narrowly elongated, compressed, and triangular in shape; umbo at pointed anterior end, bluntly flattened or rounded at gaping posterior end. Surface is fragile, sculptured with dense erect scales in younger specimens that erode to smooth areas in older parts of large adults. Exterior is light tan to orange in color; interior lightly nacreous in anterior portion. *P. nobilis* is the second largest living bivalve, growing to 35.5 in (90 cm) in length. Soft body with pallial organ; secretes long strong stalk of gold or brown byssal threads exiting shell ventrally near umbo.

DISTRIBUTION

Mediterranean Sea.

HABITAT

Infaunal, half-buried vertically in sand, mud, gravel, or seagrass, in waters up to 197 ft (60 m) in depth. Shells form an important hard substrate for biocoenotic aggregations in soft-bottom habitats, especially mollusks and sponges, thereby increasing overall local species richness.

BEHAVIOR

The pallial organ actively clears broken shell and other debris from mantle cavity. Commensal crustaceans live outside the valves, retreating into mantle cavity when threatened; crab's entry stimulates pen shell to close, thereby protecting the crustacean.

FEEDING ECOLOGY AND DIET

Filter feeder.

REPRODUCTIVE BIOLOGY

Protandric hermaphrodite; broadcast spawner. Life span about 20 years.

CONSERVATION STATUS

Not listed by the IUCN. Overcollected by divers; considered endangered throughout the Mediterranean. Protected by national laws in Spain, France, and other coastal European countries.

SIGNIFICANCE TO HUMANS

Human food source (adductor muscle). The historical use of natural pearls, both nacreous and porcelaneous (orange to black), is documented as far back as the third century B.C. Gold-colored byssus threads as long as 2 ft (61 cm) were used in weaving textiles since Roman times, including cloth for gloves, shawls, stockings and cloaks. The mythological Golden Fleece was probably spun from byssus threads of this bivalve. ◆

Black-lipped pearl oyster

Pinctada margaritifera

ORDER

Pterioida

FAMILY

Pteriidae

TAXONOMY

Pinctada margaritifera (Linnaeus, 1758), Indian Ocean. Recognized local varieties include *P. m.* var *galtsoffi*; Bartsch, 1931, of Hawaii, and *P. m.* var *cumingii* (Saville-Kent, 1890) of French Polynesia.

OTHER COMMON NAMES

English: Tahitian pearl oyster; French: Huître à lèvres noires, nacre; German: Schwarzlippigen Perlenauster; Arabic: Sadaf; Italian: Ostrica dalle labbra nere.

PHYSICAL CHARACTERISTICS

Shell is round in outline, compressed, with anterior and posterior auricles faintly delimited. It is ornamented with concentric, radially aligned rows of overlapping fragile lamellae, not gaping. Color is overall blackish with white to green mottling. Generally 6–10 in (15–25 cm) in diameter. Interior is thickly nacreous in a wide range of colors including gray, blue, pink, yellow, and green, but seldom black as common name implies. Soft body with a gray or black foot and expansive gills. The orange mantle margin is unfused. The pearl oyster secretes strong byssal threads that exit the shell through a byssal notch below the anterior auricle.

☐ *Divariscintilla yoyo*

■ *Pinctada margaritifera*

■ *Mercenaria mercenaria*

DISTRIBUTION
Widespread in Indian Ocean and western to central Pacific, including Hawaiian Islands. Raised in pearl cultivation ventures in French Polynesia, Cook Islands, Gilbert Islands, Marshall Islands, Solomon Islands, southern China, northern and western Australia, Seychelles, and the Sudan. Variety *galtsoffi* (Bartsch, 1931), is cultured in Hawaii for potential restocking of natural habitat.

HABITAT
Epibenthic, byssally attached to hard substrates in atoll lagoons and coral reefs with calm clear (nutrient-poor) waters at depths of 3–130 ft (1–40 m).

BEHAVIOR
Water currents associated with normal physiological processes (respiration, excretion) eliminate most nonconsumable particles and microfauna. Natural pearl formation often results when these processes and muscular movements cannot dislodge foreign particles. Such particles lodge between shell and mantle or within mantle tissue, and are subsequently coated with nacre. More often, natural pearls form around internal parasites.

FEEDING ECOLOGY AND DIET
Filter feeder.

REPRODUCTIVE BIOLOGY
Protandric hermaphrodite, broadcast spawner. Cultured pearl industries in South Pacific collect wild free-swimming larvae on spat collectors, subsequently raising juveniles to adult size for nucleation (insertion of a piece of mantle tissue and a pearl bead). Life span in wild is more than 30 years.

CONSERVATION STATUS
Not listed by the IUCN. Collection by diving has been illegal in French Polynesia since the collapse of the mother-of-pearl industry and the concurrent rise of the cultured pearl industry in the 1960s. Collecting has been illegal in Hawaii since 1930, where *P. margaritifera* was fished to near extinction for natural pearls at Pearl and Hermes Reef in a period of only three years (1927–1930). The population had still not recovered by the early 1990s.

SIGNIFICANCE TO HUMANS
Pinctada margaritifera is historically important as a primary source of mother-of-pearl for carving and inlay. The earliest cases from the Red Sea and Persian Gulf are from sites dating from ancient Sumeria (2300 B.C.) in Iraq and from the Greco-Roman era in Cyprus and Jerusalem. At present *P. margaritifera* is the source of black Tahitian pearls, both natural and cultured. The latter have been produced commercially and most extensively in French Polynesia and the Cook Islands since the 1960s. ◆

European pearly mussel
Margaritifera margaritifera

ORDER
Unionioida

FAMILY
Margaritiferidae

TAXONOMY
Margaritifera margaritifera (Linnaeus, 1758), boreal (northern) Europe.

OTHER COMMON NAMES
English: Eastern pearl shell, freshwater pearl mussel, river pearl mussel; French: Moule perlière; German: Flußperl-

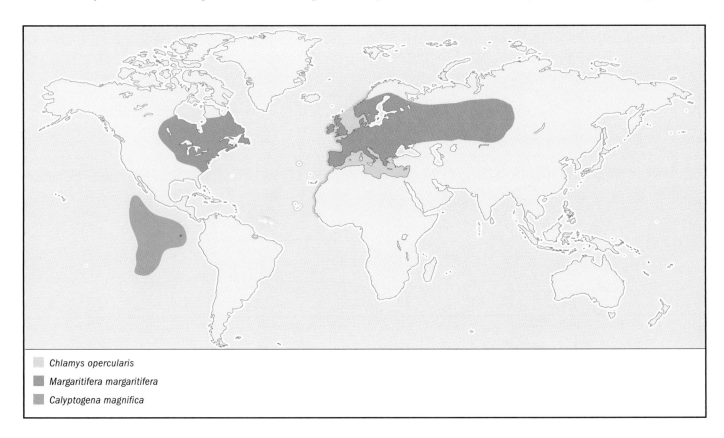

Chlamys opercularis
Margaritifera margaritifera
Calyptogena magnifica

muschel; Danish: Flodperlemusling; Japanese: Kawashinju-gai; Norwegian: Elvemusling; Swedish: Flodpärlmussla.

PHYSICAL CHARACTERISTICS
Shell is elongated and oval in shape with a subcentral umbo; moderately inflated, not gaping, with brown adherent periostracum. Grows to 4–5 in (10–13 cm) in length. Interior is thickly nacreous. Soft body with accessory muscles attaching mantle to shell, with unfused mantle margin forming functional siphons; has no "lure" on mantle margin.

DISTRIBUTION
Freshwater streams in Europe, northwestern Asia and northeastern North America. Raised in aquaculture in Europe to restock diminished populations.

HABITAT
Infaunal; prefers cool, calcium-poor, fast-flowing freshwater streams lined with cobble and gravel. Lives partly buried in substrate in a near-vertical position with posterior end uppermost.

BEHAVIOR
Very sedentary. Can sense presence of host fish in vicinity, releasing clouds of glochidia larvae.

FEEDING ECOLOGY AND DIET
Filter feeder.

REPRODUCTIVE BIOLOGY
Dioecious; males release sperm into water column; females collect sperm by normal filtering process; fertilization occurs within female. Larvae are brooded in all four demibranchs (half gills) and released as small hooked glochidia. "Parasitic" larval phase of *M. margaritifera* is dependent on salmonid fish (trout, salmon) as hosts; the glochidia attach to the host's gills.

Reaches reproductive maturity at 12–15 yrs; maximum life span 130–200 years.

CONSERVATION STATUS
Harvesting for pearls in Europe was reserved for the church and aristocrats as early as the medieval period. Although early laws were prompted by the commercial value of pearls, they served to protect this long-lived species, most effectively in Saxony (Germany). Less efficiently managed areas (Scotland, Ireland, Norway, Sweden) experienced declines and extirpation. Additional declines have been attributed to poor water quality, habitat alteration, and declines in host fish populations. Classified as Endangered on the 2002 IUCN Red List. Listed on Appendix II of the 1979 Convention on the Conservation of European Wildlife and Natural Habitats.

SIGNIFICANCE TO HUMANS
Primary source of natural Eurasian freshwater pearls, especially in Germany, Scotland, and Russia; largely restricted to royal and religious use. Pearl growth requires 20 years or more in cold waters of native habitat. ◆

Giant clam
Tridacna gigas

ORDER
Veneroida

FAMILY
Cardiidae

TAXONOMY
Tridacna gigas (Linnaeus, 1758), Amboina, Indonesia.

Tridacna gigas

Donax variabilis

Dreissena polymorpha

OTHER COMMON NAMES
English: Gigas clam; French: Bénitier géant, tridacne géant.

PHYSICAL CHARACTERISTICS
Shell is heavy, fluted with 4–6 rather smooth folds; has central umbo and relatively small byssal notch. Color is whitish. Largest extant bivalve species; the largest existing specimen, in the American Museum of Natural History, is 4 ft, 5.9 in (136.9 cm) long and weighs 579.5 lbs (262.8 kg). Interior is porcelain white. Mantle is brightly colored, yellowish brown to olive-green with iridescent blue-green spots, fused except for the incurrent and excurrent siphons.

DISTRIBUTION
Southwestern Pacific from Philippines to Micronesia. Raised in aquaculture on various Pacific Islands for the aquarium trade.

HABITAT
Epibenthic; found on coral reefs, partly embedded in sand or rubble, at depths of 6–66 ft (2–20 m).

BEHAVIOR
Tridacha gigas is permanently sessile as adult, positioned hinge-down in reef, exposing open edge and mantle to sunlight. Valves remain widely gaping unless disturbed. Reaction to stimuli is rapid, although closure is slow due to great amount of water requiring discharge from mantle cavity. Larger specimens cannot completely close shells. Although little documented evidence exists, folklore claims that divers have drowned after catching a foot or hand in the closing "maws" of giant "killer" clams. Closing speed is slow and the bivalves certainly unaggressive, but the strength of closure and the sharpness of the valve edges can inflict serious injuries on unwary divers.

FEEDING ECOLOGY AND DIET
Mantle tissue exposed to sunlight supports photosynthesis of obligate commensal zooxanthellae (dinoflagellate algae, *Symbiodinium* sp.), analogous to the condition in reef corals. The algae provide 90% of the host's metabolized nutrients. *T. gigas* supplements nutrition derived from algae by filter feeding and uptake of dissolved organic matter (DOM).

REPRODUCTIVE BIOLOGY
Protandric hermaphrodite, reaching sexual maturity in 5–6 years; broadcast spawner, probably triggered by water temperature. Growth rate is estimated at 2 in (5 cm) per year in young individuals. Life span is uncertain; estimates vary wildely from several decades to 100 years.

CONSERVATION STATUS
Classified as Vulnerable by the IUCN. Populations have been reduced by overharvesting for food, the aquarium trade, and the curio trade. Listed on CITES Appendix II.

SIGNIFICANCE TO HUMANS
Human food source (adductor muscle), harvested by native populations on Pacific Islands. Shells historically used for making tools (mallets, hoes, scrapers); also used intact as water basins (and in churches worldwide as baptismal fonts, suggested by the French vernacular name). Non-nacreous pearls have little commercial value, although the largest pearl on record is the oblong "Pearl of Allah," 9 in (22.9 cm) long and 14 lbs (6.35 kg) in weight, from a *T. gigas* specimen collected in the Philippines in 1934. ◆

Coquina clam
Donax variabilis

ORDER
Veneroida

FAMILY
Donacidae

TAXONOMY
Donax variabilis (Say, 1822), Georgia and eastern Florida, United States.

OTHER COMMON NAMES
English: Bean clam, butterfly clam, donax clam, southern coquina, variable coquina; French: Donax de Floride; Japanese: Kocyo-naminoko.

PHYSICAL CHARACTERISTICS
Shell is unequally triangular, with a subcentral umbo. The shorter anterior end is radially sculptured; the shell is otherwise smooth, not gaping. Polychromic, in wide variety of colors including white, yellow, orange, pink, purple, and blue; frequently radially striped. Interior is non-nacreous, often deep purple, with a denticulate (finely toothed) margin. Grows as long as 1 in (20 mm).

DISTRIBUTION
Eastern coast of North America from Chesapeake Bay to Florida; Gulf of Mexico to Yucatan.

HABITAT
Infaunal, on intertidal sandy beaches with wave action, sometimes numbering in thousands per square meter.

BEHAVIOR
Intertidal migration behavior is well documented. Coquina clams use their muscular foot to repeatedly rebury themselves after being washed from sand by incoming waves. Migration is both vertical (between tide levels) and horizontal (along the beach).

FEEDING ECOLOGY AND DIET
Filter feeder.

REPRODUCTIVE BIOLOGY
Dioecious, broadcast spawner. Life span is about 1–2 years.

CONSERVATION STATUS
Not listed by the IUCN.

SIGNIFICANCE TO HUMANS
Human food source, locally in "coquina broth." Coquina rock, a subfossil conglomerate of *Donax* shells and sand, was used as a building material by early Spanish settlers in North America. More recently, it has been used in ornamental landscaping. ◆

Zebra mussel
Dreissena polymorpha

ORDER
Veneroida

FAMILY
Dreissenidae

TAXONOMY
Dreissena polymorpha (Pallas, 1771), Caspian Sea and Ural River, Russia.

OTHER COMMON NAMES
French: Moule zébrée; German: Dreikantmuschel; Italian: Mitilo zebrato.

PHYSICAL CHARACTERISTICS
Shell is narrowly elongated and triangular in shape, as the animal's German name suggests. The umbo lies at the pointed anterior end. The shell is rounded at the posterior end, inflated, smooth, narrowly gaping ventrally, with a sharp midvalve keel producing a flattened ventral surface. The shell is variably banded with dark-brown-and-cream "zebra" markings. Grows to about 2 in (5 cm) in length. Has an internal shelf-like septum at the umbo for muscle attachment. Soft body with short, separated siphons; produces byssal threads that exit the shell at a ventral byssal gape.

DISTRIBUTION
Native to the Black and Caspian Seas; introduced to western Europe through canals and inland waterways during the nineteenth century; currently established in most European countries. Introduced to North America in 1985 through ship ballast water released in the Great Lakes; has spread throughout the Mississippi River and other major U.S. drainages.

HABITAT
Epibenthic, in lakes, streams and estuaries of all sizes, byssally attached to any hard surface—including rocks, wood, boat hulls, waste materials, other zebra mussels and, importantly, native freshwater mussels, which are subsequently smothered. Found in waters 0–195 ft (0–60 m) in depth. Concentrated beds of zebra mussels can contain as many as 100,000 individuals per square yard (meter).

BEHAVIOR
Sessile; smaller individuals are mobile, frequently detaching byssus threads, moving to new locations and reattaching. Dispersal is facilitated by human-mediated mechanisms, especially the movement of vessels and release of ship ballast water containing larvae.

FEEDING ECOLOGY AND DIET
Efficient filter feeder, substantially reducing density of suspended matter in the water, which may have a detrimental effect on other filter-feeding organisms.

REPRODUCTIVE BIOLOGY
Dioecious, prolific broadcast spawner, with females releasing 40,000 to one million eggs per season. Free-swimming veliger (unusual for freshwater invertebrates) is initially lecithotrophic, later planktotrophic; larval life 18–28 days; settlement cues nonspecific. Life span about two years.

CONSERVATION STATUS
Not listed by the IUCN. Considered a nuisance species in Europe and North America.

SIGNIFICANCE TO HUMANS
Introduction and spread in Europe and North America have inflicted serious economic and environmental damage due to clogging of power plant intakes and competition with native pearl mussel populations. Attachment to boat motors, docks, buoys, and pipes also has negative impact on recreation industries. Wide-ranging scientific studies are focusing on the biology and ecology of zebra mussels in order to control their spread and proliferation in North America. ◆

Yo-yo clam
Divariscintilla yoyo

ORDER
Veneroida

FAMILY
Galeommatidae

TAXONOMY
Divariscintilla yoyo Mikkelsen and Bieler, 1989, Indian River Lagoon, Fort Pierce Inlet, Florida, United States.

OTHER COMMON NAMES
None known.

PHYSICAL CHARACTERISTICS
Animal is globular, about 0.5 in (10–15 mm) in length, translucent white, with a small wedge-shaped, fragile, permanently gaping internal shell covered nearly entirely by external mantle folds. Anterior cowl is wide and flaring; there are two long retractable anterodorsal "cephalic" tentacles and a single short pallial tentacle next to the excurrent siphon on the posterodorsal midline. The foot is large and muscular with an elongated narrow posterior portion used in byssal attachment and "hanging" behavior. There are three to seven sensory flower-like organs on the visceral mass near the mouth.

DISTRIBUTION
Restricted to type locality, Indian River Lagoon in eastern Florida, United States.

HABITAT
Epibenthic; bysally attached to smooth walls of shallow-water sand burrows constructed by mantis shrimp *Lysiosquilla scabricauda* (Lamarck, 1818), in a commensal relationship.

BEHAVIOR
Yo-yo clams crawl in a snail-like fashion on a muscular foot, seeking a suitable vertical surface for attaching a short byssus thread secreted by a gland in the anterior portion of the foot. Once attached, the thread is picked up by a gland at the posterior tip of the foot, from which the animal then "hangs." Periodic contractions of mantle muscles for clearing the mantle cavity of waste products jerk the entire animal in the manner of a yo-yo.

FEEDING ECOLOGY AND DIET
Filter feeder; presumably benefits from currents generated by host mantis shrimp in its burrow as well as feeding on particles from the shrimp's predatory activities.

REPRODUCTIVE BIOLOGY
Simultaneous hermaphrodite; broods larvae in outer demibranch and suprabranchial chamber; releases shelled, swimming veligers. Life span unknown.

CONSERVATION STATUS
Not listed by the IUCN, although restricted range near commercial harbor makes it particularly vulnerable to habitat alteration.

SIGNIFICANCE TO HUMANS
None known. ◆

Giant vent clam
Calyptogena magnifica

ORDER
Veneroida

FAMILY
Kelliellidae

TAXONOMY
Calyptogena magnifica Boss and Turner, 1980, Galápagos Rift.

OTHER COMMON NAMES
English: Magnificent Calypto clam, vesicomyid clam.

PHYSICAL CHARACTERISTICS
Shell is oval with a subcentral umbo; moderately inflated with irregular concentric growth lines, slightly gaping. Color is white and chalky, with a yellow or brown periostracum persistent at margins only. Grows as long as 10 in (26 cm). Interior is porcelain white. Mantle and foot are iridescent pink; siphons short; visceral mass is red due to presence of hemoglobin in blood; byssus present only in juveniles.

DISTRIBUTION
Deep sea, on the Eastern Pacific Rise from 21°N to 22°S and the Galápagos Rift.

HABITAT
Endemic to hydrothermal vent systems of the tropical Pacific Ocean. Epibenthic; restricted to the sulfide-rich hydrothermal vent environment, nestled in crevices and among other mollusks around the vents at depths about 8,200 ft (2,500 m).

BEHAVIOR
Observed at hydrothermal vents maintaining nearly vertical position with posterior end upward; not byssally attached. Adults do not respond to crabs or shrimp crawling over shells, but respond experimentally to the manipulator arm of a submersible, suggesting they do not perceive crustaceans as predators. Octopods are probable predators.

FEEDING ECOLOGY AND DIET
Dependent upon energy produced by sulfur-oxidizing bacteria in gills. Alimentary tract reduced; ability to filter feed is questionable.

REPRODUCTIVE BIOLOGY
Reproductive mode incompletely known; probably broadcast spawner. Produces large yolky eggs, suggesting the larvae are lecithotrophic with limited dispersal capabilities (but carrying own food reserves), probably utilizing heat and/or sulfide as settling cues. Life span unknown.

CONSERVATION STATUS
Not listed by the IUCN. Relatively inaccessible; observed solely from deep-diving submersibles.

SIGNIFICANCE TO HUMANS
None known. ◆

Northern quahog
Mercenaria mercenaria

ORDER
Veneroida

FAMILY
Veneridae

TAXONOMY
Mercenaria mercenaria (Linnaeus, 1758), Pennsylvania. One recognized brown-zigzagged color variety, *M. m.* var *notata* (Say, 1822), coast of the United States.

OTHER COMMON NAMES
English: Cherrystone clam, chowder clam, hard clam, hardshell clam, littleneck clam, quahog; French: Palourde américaine, praire, quahaug; German: Ostamerikanische Venusmuschel; Spanish: Almeja.

PHYSICAL CHARACTERISTICS
Shell is ovate-trigonal in shape, with the umbo curved toward the anterior end; moderately inflated, smooth with concentric growth lines, not gaping. Color is white or gray (with brown zigzag markings in form *notata*). Grows as long as 6 in (15 cm). Interior is porcelain white with a purple stain at posterior end; moderately deep pallial sinus (to contain retracted siphons). Soft body with muscular foot; mantle is fused to form siphons.

DISTRIBUTION
Eastern North America from Canada to Florida and the Gulf of Mexico. Introduced to California, England, and France. Raised in aquaculture in France and the southeastern United States.

HABITAT
Infaunal, buried in soft shelly substrates containing shells; less abundant in sand and mud. Found in depths from the intertidal zone to 50 ft (15 m). Tolerant of estuarine conditions.

BEHAVIOR
Generally sedentary.

FEEDING ECOLOGY AND DIET
Filter feeder.

REPRODUCTIVE BIOLOGY
Protandric hermaphrodite; broadcast spawner, with planktotrophic larvae settling after two weeks. Reaches reproductive maturity at one year, with maximum life span perhaps 40 years.

CONSERVATION STATUS
Not listed by the IUCN. Collection for human consumption regulated by fisheries and public health agencies in most areas.

SIGNIFICANCE TO HUMANS
Human food source, both through wild harvest and aquaculture. Commercial categories (in order of increasing size) include seed clams, beans, buttons, littlenecks, topnecks, cherrystones, and chowders. Esteemed as food item, reflected in its status as the official state shell of Rhode Island. American aquaculture groups stress the brown-marked *notata* form, which bears the visible "brand" of a farm-raised product. Drilled shell pieces were historically produced as wampum beads by Native Americans in the eastern United States. Wampum beads were used within tribes for gifts, exchange, and ornaments but not as money; wampum served as currency, however, for European settlers in the original thirteen colonies as late as 1701. *Mercenaria mercenaria* is presently being investigated for the anticancer activity of its digestive gland. Also used in college-level zoology courses to study invertebrate anatomy. ◆

Resources

Books

Gosling, Elizabeth, ed. *The Mussel* Mytilus: *Ecology, Physiology, Genetics and Culture.* Amsterdam: Elsevier Science Publishers, 1992.

Harper, E. M., J. D. Taylor, and J. A. Crame, eds. *The Evolutionary Biology of the Bivalvia.* London: The Geological Society, 2000.

Kennedy, Victor S., Roger I. E. Newell, and Albert F. Ebel, eds. *The Eastern Oyster:* Crassostrea virginica. College Park: Maryland Sea Grant College, 1996.

Landman, Neil H., Paula M. Mikkelsen, Rüdiger Bieler, and Bennett Bronson. *Pearls: A Natural History.* New York: Harry N. Abrams, 2001.

Morton, Brian. "The Evolutionary History of the Bivalvia." In *Origin and Evolutionary Radiation of the Mollusca,* edited by John D. Taylor. Oxford, U.K.: Oxford University Press, 1996.

Morton, Brian, Robert S. Prezant, and Barry Wilson. "Class Bivalvia." In *Mollusca: The Southern Synthesis, Fauna of Australia,* Vol. 5, Part A. Melbourne, Australia: CSIRO Publishing, 1998.

Nalepa, Thomas F., and Donald W. Schloesser, eds. *Zebra Mussels: Biology, Impacts, and Control.* Boca Raton, FL: Lewis Publishers, 1993.

Turner, Ruth D. *A Survey and Illustrated Catalogue of the Teredinidae.* Cambridge, MA: Museum of Comparative Zoology, Harvard University, 1966.

Periodicals

Adamkewicz, S. Laura, and Miroslav G. Harasewych. "Systematics and Biogeography of the Genus *Donax* (Bivalvia: Donacidae) in Eastern North America." *American Malacological Bulletin* 13, no. 1–2 (1996): 97–103.

Boss, Kenneth J., and Ruth D. Turner. "The Giant White Clam from the Galapagos Rift, *Calyptogena magnifica* species novum." *Malacologia* 20, no. 1 (1980): 161–194.

Carlos, A. A., B. K. Baillie, and T. Maruyama. "Diversity of Dinoflagellate Symbionts (Zooxanthellae) in a Host Individual." *Marine Ecology Progress Series* 195 (2000): 93–100.

Carlton, James T. "Introduced Marine and Estuarine Mollusks of North America: An End-of-the-20th-Century Perspective." *Journal of Shellfish Research* 11, no. 2 (1992): 489–505.

Giacobbe, Salvatore. "Epibiontic Mollusc Communities on *Pinna nobilis* L. (Bivalvia, Mollusca)." *Journal of Natural History* 36 (2002): 1385–1396.

Giribet, Gonzalo, and Ward Wheeler. "On Bivalve Phylogeny: A High-Level Analysis of the Bivalvia (Mollusca) Based on Combined Morphology and DNA Sequence Data." *Invertebrate Biology* 121, no. 4 (2002): 271–324.

Johnson, Claudia C. "The Rise and Fall of Rudist Reefs." *American Scientist* 90, no. 2 (2002): 148–153.

Mikkelsen, Paula M., and Rüdiger Bieler. "Biology and Comparative Anatomy of *Divariscintilla yoyo* and *D. troglodytes,* Two New Species of Galeommatidae (Bivalvia) from Stomatopod Burrows in Eastern Florida." *Malacologia* 31, no. 1 (1989): 175–195.

Morton, Brian. "Biology and Functional Morphology of the Watering Pot Shell *Brechites vaginiferus* (Bivalvia: Anomalodesmata: Clavagelloidea)." *Journal of Zoology (London)* 257 (2002): 545–562.

Rosewater, Joseph. "The Family Tridacnidae in the Indo-Pacific." *Indo-Pacific Mollusca* 1, no. 6 (1965): 347–396.

Smith, Brian J. "Revision of the Recent Species of the Family Clavagellidae (Mollusca, Bivalvia)." *Journal of the Malacological Society of Australia* 3, no. 3–4 (1976): 187–209.

Wilson, James G. "Population Dynamics and Energy Budget for a Population of *Donax variabilis* (Say) on an Exposed South Carolina Beach." *Journal of Experimental Marine Biology and Ecology* 239, no. 1 (1999): 61–83.

Organizations

American Malacological Society. Web site: <<http://erato.acnatsci.org/ams/>

Conchologists of America. Web site: <<http://coa.acnatsci.org/conchnet/>

Freshwater Mollusk Conservation Society. Web site: <<http://ellipse.inhs.uiuc.edu/FMCS/index.html>

Other

"NS&T Mussel Watch Project." National Status and Trends Program, National Ocean Service, National Oceanographic and Atmospheric Administration. [11 Aug. 2003]. <http://ccma.nos.noaa.gov/NSandT/NS%26TMW.html>.

"Zebra Mussel Information." United States Geological Survey (USGS). 9 July 2003 [11 Aug. 2003]. <http://nas.er.usgs.gov/zebra.mussel/>.

Paula M. Mikkelsen, PhD

Scaphopoda
(Tusk shells)

Phylum Mollusca

Class Scaphopoda

Number of families 14

Thumbnail description
Single-shelled tusk-shaped mollusks, ranging from about an eighth of an inch (a few mm) to 7.8 in (20 cm) in length. Burrow into exclusively marine sediments with a muscular foot; feed on microorganisms with thin tentacular captacula (threadlike cilia-bearing organs).

Photo: Three species of scaphopod mollusks. The dark colored specimen is *Dentalium priseum,* found in Carboniferous strata of Scotland. At center are specimens of *D. sexangulare,* from Miocene rocks in Tuscany, Italy. The tusklike shells, right, are of *D. striatum* from Eocene rocks in Hampshire, United Kingdom. (Photo by Biophoto Associates/ Photo Researchers, Inc. Reproduced by permission.)

Evolution and systematics

Scaphopoda is the most recent class of mollusks to appear in the fossil record, dating from at least the Mississippian Carboniferous period about 360 million years ago (mya). Several earlier fossils, however, have been assigned to the Scaphopoda, but not without some debate as to their relationship to extant scaphopods. These earlier specimens date from the Devonian into the Ordovician period (about 450 mya). The earliest putative representatives cited are *Plagioglypta iowensis* and *Rhytiodentalium kentuckyensis*. The most recent classification system places scaphopods into two orders, Dentaliida and Gadilida. Within these orders, there are 14 families and 60 genera, 46 of which are extant.

There is little agreement on the relationship of scaphopods to other mollusks. While scaphopods certainly belong among the uni- or bivalved mollusks in the subphylum Conchifera, studies in the past several years have identified many different taxa as the closest scaphopod relative, including cephalopods, bivalves, gastropods, and the extinct class of bivalved mollusks known as Rostroconchia. Historically, this unclear picture has likely been the result of difficulties in comparing the morphology (form or structure) of scaphopods to that of other mollusks. Recent molecular studies, however, have brought new evidence to bear on the problem. Expression of the *engrailed* gene in the early development of the scaphopod shell indicates an affinity with univalved cephalopods or gastropods rather than with bivalves. Moreover, 18S RNA sequence data indicate that scaphopods are most closely related to cephalopods among living mollusks.

Physical characteristics

Scaphopods get their name from the keel-shaped curvature of their shell. The shell is wider at the anterior end in most species, which explains the group's common English name of tusk or tooth shell. In some scaphopods, the widest part of the shell is about one-third of the way back from the anterior opening. The shells of most scaphopods, which range from about an eighth of an inch to upwards of 7.8 in (20 cm) in length, are smooth or have faint circumferential growth lines. Some species have longitudinal or annular (ringlike) ribbing. Generally, the shell is white, but a few scaphopod species have strongly colored shells that are green or banded in red and yellow. In species with thin translucent shells, the pink or yellow color of the gonads (eggs) may show through.

A muscular foot and numerous thin feeding tentacles, or captacula, extend from the front opening of a typical scaphopod. The narrower posterior end of the animal is also open, providing access to the mantle cavity for respiratory currents and the outflow of gametes and waste products. A small siphon-like extension of the mantle edge extends to and sometimes slightly beyond the posterior opening. The mantle cavity extends the length of the shell, partially encircling the body but otherwise quite narrow in its dorsoventral dimension. The restricted space has led to the loss of gills and the osphradium (chemosensory organ), which are structures found in the mantle cavity of most other mollusks. In turn, the loss of gills in scaphopods is reflected in the absence of auricles in the heart; in addition, the ventricle is much reduced.

Distribution

Scaphopods are found worldwide throughout the marine environment but not in brackish or estuarine systems. In terms of bathymetric distribution, they have been found from the intertidal zone in Australia to depths of about 23,000 ft (7,000 m) in the Sunda Trench, and are generally considered to be relatively well represented in deep-sea benthic (seabed) communities. Latitudinal distribution is similarly broad; scaphopods have been discovered well within both the Arctic and Antarctic Circles, and are found throughout temperate and tropical waters. While researchers have just begun to study the global distributional patterns of scaphopods, they know that the diversity of these animals in the Indo-Pacific region is relatively high, mirroring the distribution of diversity of many other marine taxa.

Habitat

Scaphopods are found exclusively in soft silty marine sediments, into which they burrow in search of food. Some species burrow only a short distance into the mud, leaving their narrow apical end protruding from the sediment surface; this position has become a classic illustration in many treatments of the group. Many species, such as *Gadila aberrans*, however, burrow to depths several times their body length.

While most species favor foraminiferans as food, and thereby have a preference for habitats with high densities of these shelled protists, some species successfully inhabit other areas by broadening the variety of microorganisms they eat. While scaphopods are considered relatively uncommon members of the benthic communities in which they live, in some areas they can reach relatively high densities. For example, *Rhabdus rectius* reaches almost 60 individuals per square yard (0.76 square meter) around Sandford Island off the western coast of Vancouver Island, British Columbia.

Behavior

While all scaphopods burrow into soft sediments, some species may not dig deeply enough to completely cover the apical shell, which protrudes slightly from the seafloor surface. This behavior occurs naturally in at least some of the larger dentaliids, although it is also an artifact of laboratory observation that led to textbook overgeneralizations about the group over the years. In contrast, some gadilids have been reported to burrow quickly and go much deeper into the seabed, as far as 16 in (40 cm) at rates of 0.40 in/sec (1 cm/sec).

Scaphopods test the respiratory currents entering the mantle cavity with sensory receptors located around the mantle edge. If the current is fouled, the animal usually makes an escape response of deeper burrowing or withdrawal of the foot into the shell.

Feeding ecology and diet

Scaphopods feed on microorganisms, their diet dominated to a greater or lesser extent by foraminiferans. Some species, however, also feed on mite and other eggs; kinorhynchs (microscopic marine invertebrates); ostracods (tiny crustaceans); and bivalves. As such, they can be considered microcarnivores or microomnivores. They select their prey by using thin, ciliated muscular tentacles called captacula, which have an adhesive gland complex in their tip that can transport food items to the mouth inside the shell. Depending on the species, the tentacles either extend out through the sediment or search the wall of a feeding cavity created by the scaphopod's foot. Foraminiferans have a shell or test, and scaphopods deal with these protective structures by crushing food items with their relatively large radula, a ribbon of teeth found in most mollusks. The scaphopod radula is short, stout, and heavily mineralized with iron. Food passes into the stomach and through a U-shaped gut; feces are released into the mantle cavity and discharged through the posterior aperture.

Scaphopods are preyed upon by all marine organisms that plow the sea floor mud for food. Their predators include such fishes as the ratfish. Naticid snails also feed on scaphopods, leaving behind the beveled bore-hole as the characteristic evidence of their predation.

Reproductive biology

Hermaphroditic scaphopods have been reported but are rare. Most members of the group release their eggs or sperm into the water column, either through the anterior or posterior aperture via the right kidney. Fertilization therefore occurs externally in the water; there is no courtship behavior or parental care of young. Few studies have been published on the reproductive ecology of scaphopods, although these do indicate some seasonality in gamete production. The eggs of scaphopods, particularly those of *Antalis entalis* and related species, have been used for over a century as models for the experimental study of early development of embryos.

Scaphopods have a free-swimming larval stage that develops in the water. The larvae have a general resemblance to the trochophore and veliger larvae of other mollusks. The orientation of the larval body eventually changes into the typical scaphopod body shape. It is noteworthy that the first shell secreted by the scaphopod, the larval *praetubulus*, is soon shed as growth continues at the larger anterior end of the animal—a pattern of shell secretion and dissolution that continues throughout the scaphopod's life.

Conservation status

Little is known about the conservation status of scaphopods. With presumably wide-ranging but ill-defined distributions of individual species, there are few instances where population estimates can be made, and where monitoring of those populations can have significant consequences for conservation measures. As members of shallow water marine ecosystems, scaphopods are usually only a small component of local diversity, and so are not the focus of key-species studies. No species is listed by the IUCN.

Significance to humans

Scaphopods were important to Native Americans of the Pacific Northwest. Their use as a form of currency extended well into central Canada and south to California, despite only a few source sites on the west coast of Vancouver Island, where collection was difficult. Accounts of early European travelers indicate that the value of blankets could be measured in scapho-pod shells. Naturally the shells were worn as displays of wealth, as can be seen in photographs by Edward Curtis as well as in the displays at such museums as the Museum of Anthropology at the University of British Columbia in Vancouver, the Royal British Columbia Museum in Victoria, and the Field Museum in Chicago. As of 2003, scaphopod shells are frequently sold in shell shops and often made into jewelry.

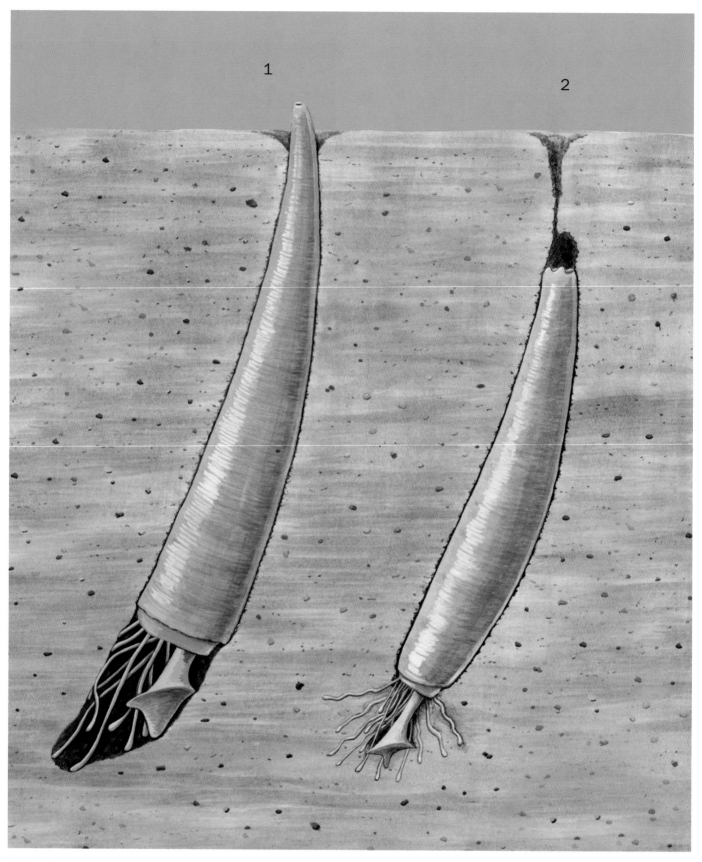

1. Tusk shell (*Antalis entalis*); 2. Aberrant tooth shell (*Gadila aberrans*). (Illustration by Dan Erickson)

Species accounts

Tusk shell
Antalis entalis

ORDER
Dentaliida

FAMILY
Dentaliidae

TAXONOMY
Dentalium entale Linnaeus, 1758, European, Indian Oceans.

OTHER COMMON NAMES
English: Entale tusk.

PHYSICAL CHARACTERISTICS
Smooth white shell, sometimes with faint longitudinal ribs toward the apex. Can reach an inch or more (several centimeters) in length, with the maximum diameter at the anterior opening. The posterior opening often has a shallow notch on its concave side.

DISTRIBUTION
Northeastern Atlantic from Ireland to the Canary Islands; one subspecies, *A. entalis stimpsoni*, is found in the northwestern Atlantic from Nova Scotia to Cape Cod.

HABITAT
Soft sediments at depths of several feet to about 6,600 ft (2,000 m).

BEHAVIOR
Often depicted with posterior opening protruding from the sediment surface.

FEEDING ECOLOGY AND DIET
Microomnivorous but prefers foraminiferans.

REPRODUCTIVE BIOLOGY
Sexes are separate; eggs and sperm are released into the water column for fertilization and development. Used as a model organism for studies of experimental embryology.

CONSERVATION STATUS
Not threatened.

SIGNIFICANCE TO HUMANS
None known. ◆

Aberrant tooth shell
Gadila aberrans

ORDER
Gadilida

FAMILY
Gadilidae

TAXONOMY
Cadulus aberrans Whiteaves, 1887, Quatsino Sound, Vancouver Island, British Columbia, Canada.

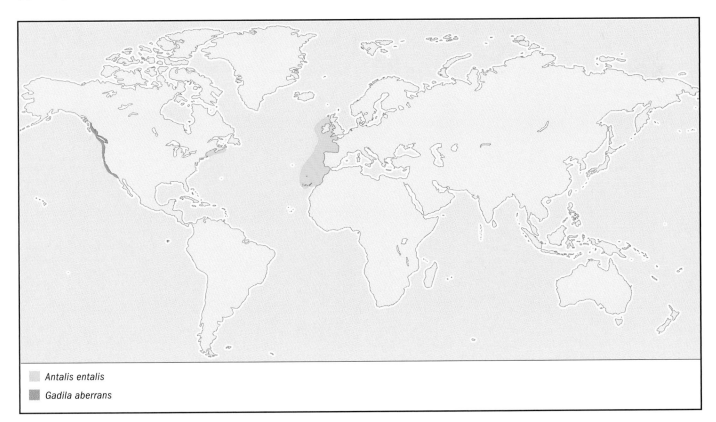

Antalis entalis
Gadila aberrans

OTHER COMMON NAMES
None known.

PHYSICAL CHARACTERISTICS
Smooth translucent shell, white or cream in color. Grows as long as 0.8 in (2 cm). Maximum diameter near anterior opening, which is slightly constricted; posterior opening often has four notches.

DISTRIBUTION
Northeastern Pacific Ocean from British Columbia to southern California.

HABITAT
Soft sediments at depths of 10–165 ft (3–50 m).

BEHAVIOR
Usually burrows several body lengths beneath the sediment surface.

FEEDING ECOLOGY AND DIET
Microcarnivorous, but prefers foraminiferans.

REPRODUCTIVE BIOLOGY
Sexes are separate; eggs and sperm are spawned into the water column where fertilization and development occur.

CONSERVATION STATUS
Not threatened.

SIGNIFICANCE TO HUMANS
None known. ◆

Resources

Books

Shimek, R. L., and G. Steiner. "Scaphopoda." In *Microscopic Anatomy of Invertebrates.* Volume 6B, *Mollusca II*, edited by F.W. Harrison and A. J. Kohn. New York: Wiley-Liss, 1997.

Steiner, G., and P. D. Reynolds. "Molecular Systematics of the Scaphopoda." In *Molecular Systematics and Phylogeography of Mollusks*, edited by C. Lydeard and D. R. Lindberg. Washington DC: Smithsonian Institution Press, in press.

Periodicals

Lacaze-Duthiers, H. "Histoire de l'organisation et du développement du Dentale." *Annales des Sciences Naturelles* Quatrième Serie, Paris (1856-57): Tome 6, 225–281, pls. 8–10; 319–385, pls. 11–13; Tome 7, 5–51, pls. 2 4; 171–255, pls. 5–9.; Tome 8, 18–44.

Reynolds, P. D. "The Scaphopoda." *Advances in Marine Biology* 42 (2002): 137–236.

Steiner, G. and A. R. Kabat. "Catalogue of Supraspecific Taxa of Scaphopoda (Mollusca)." *Zoosystema* 23 (2001): 433–460.

Patrick D. Reynolds, PhD

Cephalopoda

(Nautilids, octopods, cuttlefishes, squids, and relatives)

Phylum Mollusca

Class Cephalopoda

Number of families About 45

Thumbnail description
Mollusks bearing a radula (toothed tongue) and well-developed heads; mouth characterized by a dorsoventral pair of horny jaws known as beaks and encircled by the bases of 8–ca. 60 grasping appendages; single pair of lateral, image-forming eyes; well-developed brain and peripheral nervous system

Photo: A greater blue ringed octopus (*Hapalochlaena lunulata*) takes a crab as it emerges from its hole. (Photo ©Tony Wu/www.silent-symphony.com. Reproduced by permission.)

Evolution and systematics

Cephalopod shells are very well represented in the fossil record as far back as the Upper Cambrian period, about 505 million years ago. Soft-tissue fossils, however, are extremely rare. Two very different subclasses of cephalopods live in modern seas: (1) the nautilids (subclass Nautiloidea), represented by two genera and about six species that have two pairs of gills and external shells into which they can withdraw; and (2) the neocoleoids (a division of subclass Coleoidea), sometimes called dibranchiates because they have a single pair of gills, comprising all other living species of cephalopods (fewer than a thousand). The neocoleoids include the familiar squids, cuttlefishes, and octopods.

There are about 44 families of extant neocoleoids, plus one family of nautilids. Only eight families have more than 20 species; many families consist of only one species or one genus. Although the familial relationships of living cephalopods are fairly stable, researchers have not completely resolved many relationships at higher and lower levels of classification (orders and genera). Eight distinctive groups of neocoleoids can be defined, however: Incirrata (common octopods); Cirrata (finned octopods); Vampyromorpha (the vampire squid, one species); Sepiida (cuttlefishes), Spirulida (the ram's horn squid, one species); Sepiolida (bobtail and bottle squids); Myopsida

(inshore squids), and Oegopsida (oceanic squids). There is no consensus regarding the taxonomic level of these groups, and some families cannot be assigned with confidence to any of them. The first three groups listed comprise the Octopodiformes; the remaining five are grouped together as the Decapodiformes, commonly called decapods.

Physical characteristics

With regard to general structure, cephalopods have three easily distinguished regions. From front to rear, these regions are: (1) the brachial crown (arms and tentacles) surrounding the mouth; (2) the head, with prominent lateral eyes; and (3) the mantle, which may have a pair of fins on the sides. This overall three-part structure is less distinct in nautilids but is still recognizable. Nautilids have approximately 60 arms (sometimes called tentacles or cirri) arranged in two rings around the mouth. Neocoleoids have a single ring of either eight or ten appendages surrounding the mouth. On those that have ten, two appendages are modified into either ventrolateral tentacles (decapods) or dorsolateral velar filaments (vampires). Thus, all living cephalopods other than the nautilids have eight arms, and some have either two additional tentacles or two filaments. Although some nonspecialists refer to all cephalopod appendages as tentacles, one should

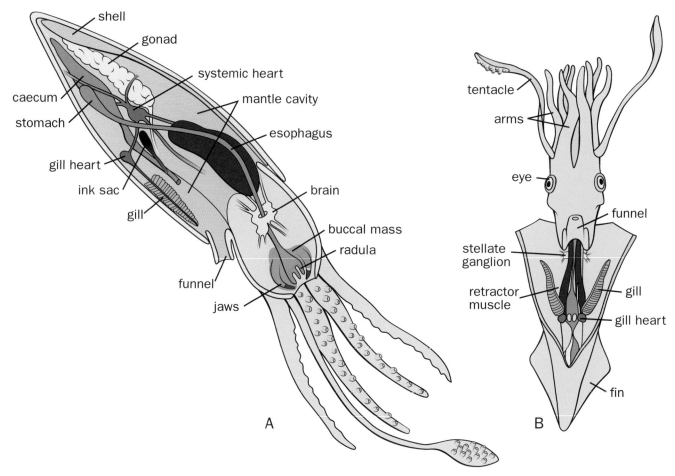

Cephalopod anatomy. A. Lateral view; B. Ventral view. (Illustration by Patricia Ferrer)

avoid this usage because it confuses the differentiation of the specialized appendages in decapods and vampire squids.

Cephalopods range in size from the giant squids, *Architeuthis* spp., which are commonly longer than 6.56 ft (2 m) in mantle length (ML) and reportedly reach 16.4 ft (5 m) in ML, 59 ft (18 m) in total length, and 661.3 pounds (300 kg) in weight, to tiny species like *Idiosepius* spp., decapods which mature at an ML of about 0.23–0.31 in (6–8 mm); or the sexually dimorphic pelagic octopods, *Argonauta* spp., in which mature males are only 0.39 in (1 cm) in ML. Several squid species and at least two species of octopods reach sizes larger than an adult human. They include the true giant squid as well as the muscular (and dangerous) ommastrephids, the weakly muscled cranchiids and chiroteuthids, and the giant Pacific octopus or "devilfish."

Many species of squids, cuttlefishes, and octopods can radically change their appearance within a fraction of a second. This ability to transform themselves so rapidly is the reason that cephalopods have been called masters of disguise. These remarkable transformations result from interactions among brown, red and yellow pigment-filled chromatophore (color-producing) organs under the control of the animal's nervous system; reflective iridophores (cells that produce a silvery or iridescent pigment); fixed white leucophores (pigment cells containing guanine); fixed tubercles (small nodules); and erectile muscular flaps and papillae in the animal's skin. Furthermore, some cephalopods have light-producing organs called photophores. Cephalopod photophores come in two fundamentally different types. In the first type, known as intrinsic photophores, the light is produced biochemically by the squid or octopod. The second type, called bacterial photophores, makes use of symbiotic photogenic bacteria that grow in special chambers associated with the ink sac in the host. The light produced by both types of photophores is usually blue-green in color; the color can be altered, however, by structures associated with the photophore.

Other features include an ink gland and ink sac associated with the intestine that is characteristic of neocoleoids. Complex, highly developed nervous and sensory systems are also typical of living cephalopods, although less so for the nautilids than for the others. Especially noteworthy are the image-forming eyes, with lenses in the neocoleoids; and the complex brain developed from the nerve ring surrounding the esophagus. Other noteworthy peculiarities, including a muscular hydrostatic skeleton and a pattern of muscle fiber alignment known as oblique striation, also are characteristics of cephalopods.

Distribution

All cephalopods are limited to marine environments. Only a few species can tolerate the low levels of salt in the waters of estuaries and fjords, the minimum being 17.5 PSU (practical salinity units), or about half the saltiness of full-strength seawater. No cephalopods can live in freshwater. Some octopods are known to crawl from one tidal pool to another but cannot remain out of the water for long.

Habitat

All cephalopods are mobile throughout their lives, except for the egg stages that in many species are attached to various substrates. As far as we know, all cephalopods are either carnivorous (the neocoleoids) or scavengers (the nautilids) throughout their life cycles after hatching. Some researchers have proposed that some cephalopod paralarvae feed on phytoplankton, but the evidence in support of this hypothesis currently is not very strong.

The life cycles of most neocoleoid cephalopods are very different from that of *Nautilus*. Whereas the latter is long-lived—it may live for 20 years or more—the lifestyles of other cephalopods have been characterized as "live fast and die young." Their life spans seem to range from a few months for small species to a few years for larger species. Many of the generalizations that have been made for neocoleoid cephalopods, however, have been based on observations of a few coastal species whose habitats are convenient for research.

Some cephalopod species, especially squids, can be very abundant; they are occasionally among the dominant organisms in their ecosystems. Because they are important food for larger animals, in addition to being voracious predators, such species are key members of some marine food webs. There are, however, significant gaps in current information about the life history and ecology of these organisms. Furthermore, many generalizations about cephalopods based on one or a few species are turning out to be either questionable or wrong. For example, because the common European octopus, *Octopus vulgaris*, and the California market squid, *Loligo opalescens*, spawn once and then die, their reproductive cycle has been widely regarded as the general pattern for cephalopods. Researchers are accumulating evidence, however, that many species of squids and octopods spawn many times and continue to live and feed after laying their eggs. Squids have been thought not to care for their eggs because the coastal species that have been most closely observed, as well as a few oceanic species, seem to release their egg masses and then depart either by moving away or by dying. Recently, though, some gonatid squids have been found to carry their egg masses around after releasing them. We know so little about most cephalopod species, especially oceanic and deep-sea species, that many such surprises likely await discovery. In short, many generalizations about the group may need to be modified.

Cephalopods live in a range of water depths from intertidal levels to over 16,400 ft (5,000 m). Different marine ecosystems have very different cephalopod faunas. For example, no cuttlefishes are found in American waters. Many octopod species have recently been described from Antarctic Ocean waters, whereas the Arctic appears to have few octopod species. The deep sea is home to both primitive vampire squids and morphologically similar finned octopods as well as some strange oegopsid squids. Oegopsid squids tend to be dominant in epipelagic and mesopelagic oceanic waters (the upper layer of water that admits enough light for photosynthesis to occur, and the "twilight zone" just below it); whereas myopsid squids share the continental shelves with incirrate octopods and cuttlefishes. Some incirrate octopods are found in deep-sea habitats on the ocean bottom whereas others are entirely pelagic. Because cephalopods are so widespread throughout marine habitats, their patterns of habitat utilization also vary widely.

Behavior

Cephalopods are renowned for their large brains, well-developed eyes, and complex behavior. Their social organization varies from the solitary life of octopods through small schools of cuttlefish to very large shoals of oceanic oegopsids. Some species of cuttlefishes and squids are known to form seasonal aggregations or gatherings for mating and spawning.

The sexes are separate in cephalopods. Courtship patterns vary from simple male grasping of the female followed by the implanting of spermatophores to complex courtship rituals in which both individuals display elaborate forms of touching prior to mating. Several species produce small "sneaker males" that resemble females; these sneaker males have been shown to be an important alternative to large, dominant mate-guarding males. The mating patterns of most oceanic cephalopods are largely unknown.

In addition to social organization and mating behavior, the use of visual communication among cephalopods has been a subject of debate. Whereas most observers who have watched cephalopods will agree that they perform complex visual signaling, these researchers debate whether such signaling can be considered a language. Cephalopod visual signaling involves a variety of behaviors, some of which are regularly directed toward conspecifics (e.g., mates or potential rivals), and others toward members of other species (e.g., prey or potential predators). The use of ink or other chemicals as an alarm signal to warn conspecifics has been demonstrated but not thoroughly investigated.

As with so many subjects about cephalopods, display patterns have been studied in detail for only a few easily accessible species. In general, these patterns include both acute patterns, lasting for a few seconds to a few minutes; and chronic patterns, lasting several minutes to several hours. The display patterns comprise chromatic (both dark and light), textural, postural, and movement changes. Commonly recorded displays include crypsis (hiding), in addition to what Hanlon and Messenger (1996) refer to as deimatic (threatening, startling, frightening, or bluffing) behaviors and protean (unpredictable or erratic) escape maneuvers.

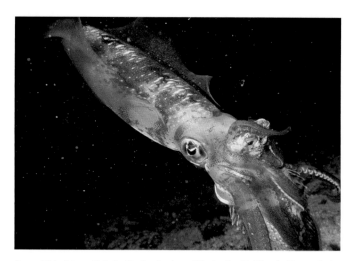

A squid holds a fish in its tentacles. (Photo by B. Wood. Bruce Coleman, Inc. Reproduced by permission.)

Although some nearshore benthic octopods are known to occupy and defend territories for limited time periods, cephalopods generally are not territorial.

The primary variable in the activity patterns of cephalopods is the 24-hour diel cycle. Some neritic (coastal zone) species forage primarily by daylight, but most are nocturnal. Some species are crepuscular, which means that they are most active at dawn and dusk. Diel activities of oceanic species primarily involve vertical migration. The activity patterns of most deep-sea benthic species are unknown.

Many species of oceanic cephalopods undergo diel vertical migrations, wherein they live at depths of about 1,310–3,280 ft (400–1,000 m) during the day and then ascend into the uppermost 656 ft (200 m) of water during the night. Shifts in distribution over the course of a cephalopod's life history are common. Some oegopsid squids are believed to undertake life-cycle migrations over large geographic distances covering hundreds of miles.

Feeding ecology and diet

Neocoleoid cephalopods are active predators that feed upon shrimps, crabs, fishes, other cephalopods, planktonic crustaceans, and, in the case of octopods, on other mollusks. Nautilids are scavengers.

Foraging behavior varies widely among species in various taxonomic groups. Some examples include rapid raptorial attacks (oegopsid and myopsid squids); raptorial attacks after stalking (cuttlefishes); drifting with dangling tentacles (some oegopsid squids); ambush from a hiding place (some bobtail squids, cuttlefishes, and benthic incirrate octopods); and perhaps even the use of mucus to entangle small prey (a cirrate octopod). Special foraging adaptations also vary widely among cephalopod species.

Cephalopods are major food items in the diets of many marine vertebrates, including toothed whales; seals; pelagic birds (penguins, petrels, albatrosses, etc.); and both benthic and pelagic fishes (e.g., sea basses, lancetfishes, tunas, billfishes).

Reproductive biology

The sexes are separate in all cephalopods. The unpaired gonads, either ovary or testis, are always located in the posterior region of the mantle cavity. Females may have one or two oviducts depending on their major taxonomic group. Each oviduct has an oviducal gland that secretes a primary coating around the egg. Many female cephalopods have an additional nidimental gland that adds extra layers of protective coating to the egg. This gland is found beyond the end of the oviduct.

In males, the sperm passes through the spermatophoric gland complex, which packages the sperm into a packet known as a spermatophore. The hectocotylus is a modified arm on the males of many types of cephalopods that is used to transfer spermatophores to the female. The modifications of the arm take many forms, some quite bizarre. The males of some species also exhibit modifications of other arms in addition to the hectocotylus. Females of some species also develop such modified structures as arm-tip photophores when mature. Other forms of sexual dimorphism vary among species within families. These forms include such characters as greatly elongated arm pairs or tails; enlarged suckers; and special photophores. Although researchers have speculated about the functions of these features, little is known about them as of 2003.

Spermatophores are implanted in specific locations on the females by the male hectocotylus in species possessing one, or by an elongated penis (the terminal organ of the sperm duct) in nonhectocotylized species. The time required for the process ranges from seconds in some squids to hours in some octopods. The spermatophores may be implanted in the female's oviducal gland; inside the mantle cavity; around the mantle opening on the neck; in a pocket under the eye; or around the mouth. The mode of reproduction and egg-laying for many cephalopods is unknown, especially oceanic and deep-sea species.

All cephalopod eggs have substantial amounts of yolk. Embryonic development is unlike that of any other mollusk. Nautilid eggs are as much as 1.14 in (2.9 cm) in length. Neocoleoid eggs vary in size from about 1.6 in (4.2 cm) in some *Graneledone* and *Megaledone* species (among the largest of any invertebrate) to 0.03 in (0.8 mm) long in *Argonauta*. The eggs have one or more layers of protective coatings and generally are laid as egg masses. Egg masses may be benthic (laid on the ocean bottom) or pelagic, varying among major taxonomic groups. The time span of embryonic development also varies widely from a few days to many months, depending on the species and temperature conditions. Hatching may occur synchronously from a single clutch or extend over a period of 2–3 weeks.

Except for nautilids, which reproduce repeatedly over a period of years, extant cephalopods apparently mate only once. Their reproductive period, however, may comprise the brief terminus of their life span, may extend for a considerable period, or may be intermediate between these extremes. Different species, even within a single family, are found at different points along this continuum but generally cluster at the two extremes. Similarly, the number of eggs produced by a female range from a few dozen to hundreds of thousands. Hatchlings from benthic eggs may either be benthic or temporarily planktonic, eventually settling back to the adult benthic habitat. Pelagic hatchlings are planktonic.

The development of cephalopod embryos is direct, which means that the embryos do not have true metamorphic stages. The hatchlings of species with large eggs look like miniatures of the adult, whereas hatchlings of species with small eggs undergo changes in body proportion during development. The young of some species differ conspicuously in the proportion of their body parts and development of specialized structures (e.g., photophores; modification of suckers into hooks) from the adults. Thus, researchers have coined the term "paralarva" for early stages of cephalopods that differ morphologically and ecologically from later stages.

Most incirrate octopods and, as has recently been discovered, some squids care for their egg masses until the eggs hatch. Pelagic species that exhibit such behavior swim while holding the egg mass in their arms. Benthic octopods use such structures as the interiors of mollusk shells, rock crevices, or discarded bottles as dens. They manipulate the eggs to keep them free of detritus and blow water across them, presumably to aerate them.

Reproduction, at least among coastal species, is seasonal, although there may be either one or two peak periods of seasonal reproduction. Because of their short life spans, neocoleoid cephalopods tend to exhibit strong seasonality in the occurrence of their different life-history stages.

Conservation status

Distribution and conservation status vary among cephalopod species. Some coastal species are quite common whereas others, especially among the oceanic fauna, are rarely encountered. Small-scale endemism (confinement to a specific locality) is uncommon, although endemism is established among cephalopods within ocean basins.

No cephalopods appear on the IUCN Red List. In addition, no cephalopods appear on any United States regional listings as of 2003, although there has been discussion of proposing nautilids for listing under CITES. Most large declines in cephalopod abundance are probably cyclic in nature, although overfishing has been implicated in the declines of commercially important species. The decline in abundance of nautilids probably results from human harvesting of them for both shells and meat.

The abundance of cephalopods varies (depending on group, habitat, and season) from isolated territorial individuals (primarily benthic octopods and sepioids) through small schools with a few dozen individuals to huge schools with millions of individuals.

Significance to humans

Cephalopods have figured in the art and literature of many human cultures. Stories of large or intelligent cephalopods are well known from many ocean-oriented cultures, including Japanese, Polynesian, and Western European. Octopods are depicted on classical Greco-Roman pottery, frescoes, and similar artifacts. Medieval Germanic legends about "Kraken" or sea monsters probably refer to giant squids. Even at the present time, large squids and octopods continue to be subjects of much public interest. For example, the giant squid was described in a special issue of *Time* magazine on ocean exploration as "the last bona fide sea monster." This interest has been sustained by a series of popular novels and movies.

Many species of squid, cuttlefish, and octopod are very good food when prepared properly. Many people around the world enjoy dishes made from certain species of cephalopods. Fishermen also often equate cephalopods with "good bait" because other marine species like to eat them as well. These culinary considerations, both human and otherwise, reflect the importance of cephalopods in commercial fisheries and marine food webs. Although worldwide fisheries for squids, cuttlefishes, and octopods are dwarfed in comparison with those for shrimps, bivalves, and many fishes, catching and marketing cephalopods can dominate such local economies as that of the Falkland Islands. The total cephalopod catch officially reported for the year 2000 was about 4.0 million tons (3.6 million metric tons); this figure represents about 4.2% of the world total marine catch for the same year. Some statistics indicate that as humans reduce the populations of competing and predatory fishes and mammals, regional populations of cephalopods are increasing.

Octopods occasionally are considered to be pests in trap fisheries for mollusks (e.g., whelk) and crustaceans (e.g., lobsters) because the octopods enter the traps and eat the catch.

The rediscovery of squid giant axons by J. Z. Young in the 1930s allowed experiments that demonstrated much of what is known about the basic functioning of nerves. The giant axons of squid nerves have been widely used as one of the primary bases of neurobiology. Cephalopods are also being used increasingly as models in other biomedical fields, including sensory biology, information processing, and biochemistry. They are even being used as a source of information regarding the detoxification of nerve gas. The results of an electronic search of scientific literature for the word "squid" may be dominated by biomedical studies based on inshore squid as a convenient and interesting model.

Cephalopod bites, especially by octopods, can be painful at the least. Some are poisonous because of the injection of salivary secretions, and may even be lethal on rare occasions. Poisonous bites from the small blue-ringed octopus, *Hapalochlaena* spp., which secretes a substance known as cephalotoxin, have resulted in documented human deaths. Furthermore, schools of the large Humboldt squid, *Dosidicus gigas*, are reported to attack scuba divers and fishermen who have fallen into the water.

Many marine animals of particular interest to people, like whales, specialize in eating cephalopods. Among the animals with large brains that people consider intelligent, the cephalopods are unique because they are not vertebrates. The cephalopod brains and their remarkably familiar eyes have developed from an early precursor completely unlike the forerunners of dogs, dolphins, parrots, lizards, and fishes. This evolutionary history makes cephalopods the closest thing to an alien intelligence that humans have ever encountered.

Because cephalopods are important to biomedical researchers and fisheries, cephalopod biology is comparatively well-known from a few inshore species that can be caught close to such major marine biological laboratories as those in the northeastern United States; Plymouth, England; Naples, Italy; and Hakodate, Japan.

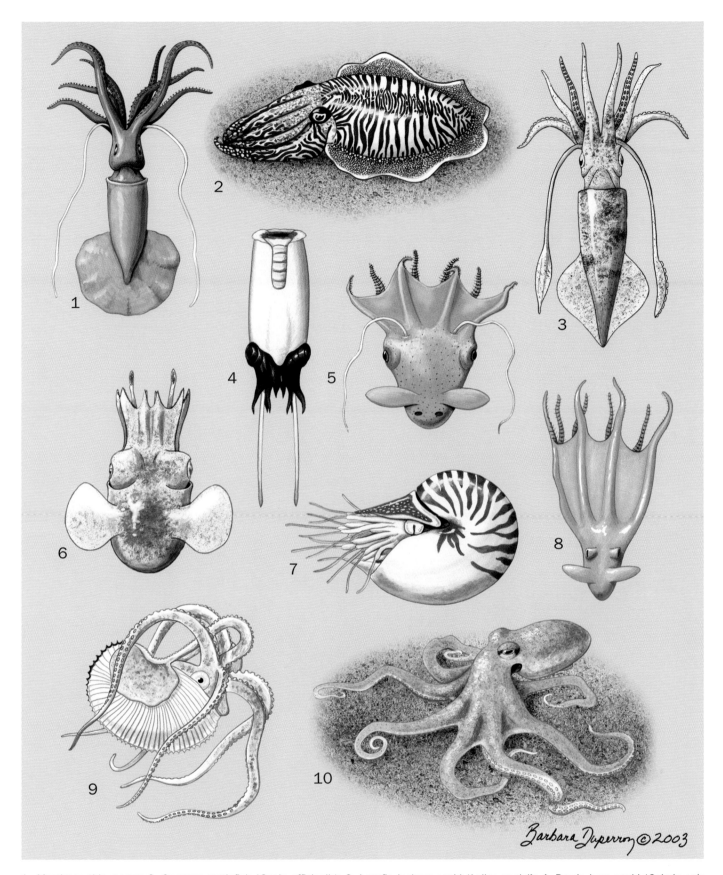

1. *Mastigoteuthis magna*; 2. Common cuttlefish (*Sepia officinalis*); 3. Longfin inshore squid (*Loligo pealeii*); 4. Ram's horn squid (*Spirula spirula*); 5. Vampire squid (*Vampyroteuthis infernalis*); 6. Butterfly bobtail squid (*Stoloteuthis leucoptera*); 7. Pearly nautilus (*Nautilus pompilius*); 8. *Stauroteuthis syrtensis*; 9. Greater argonaut (*Argonauta argo*); 10. Common octopus (*Octopus* cf. *vulgaris*). (Illustration by Barbara Duperron)

Species accounts

Longfin inshore squid
Loligo pealeii

ORDER
Myopsida

FAMILY
Loliginidae

TAXONOMY
Loligo pealeii LeSueur, 1821.

OTHER COMMON NAMES
English: Longfinned squid; French: Calmar.

PHYSICAL CHARACTERISTICS
A transparent skin (corneal membrane) covers eye lens. Funnel-locking apparatus is a simple, straight groove and ridge. Fins attach to lateral regions of mantle. Arms have suckers in two series. Tentacular club with suckers in four series. Hooks are never present. Buccal connectives attach to ventral margins of ventral arms. Seven buccal lappets possess small suckers. Mantle is long, moderately slender, cylindrical, with a bluntly pointed posterior end; the fins are rhomboid with nearly straight sides. The gladius is long, rather wide, and feather-shaped; the ratio of greatest width of the vane of the gladius to greatest width of the rhachis (the central spine of the gladius) is 2.7–3.7 in females, 2.4–2.9 in males. The edge of the vane is curved (sometimes straight in males), thin, and rarely ribbed. Eyes are not unusually large; the diameter of the externally visible eyeball is 8–18% ML, and the diameter of dissected lens is 2–6% ML. The left ventral arm of mature males is hectocotylized by modification of the outermost third to fourth of arm, but the modification does not extend to arm tip. Fewer than 12 of the suckers in dorsal row usually smaller than half the size of their counterparts in the ventral row; the bases or pedicels of some of the modified suckers are rounded and narrowly triangular.

DISTRIBUTION
Western Atlantic continental shelf and upper slope waters from Nova Scotia to Venezuela, including the Gulf of Mexico and the Caribbean Sea. Not found around islands except as rare strays at islands close to the continental shelf or slope.

HABITAT
Restricted in summer to surface and shallow water, but lives at a depth of 92–1,200 ft (28–366 m) in winter; peak concentrations occur at 328–633 ft (100–193 m); adults are found on the bottom during day but leave the bottom at night, dispersing into the water column. They may appear at the water surface in summer or in warm water.

BEHAVIOR
Longfin inshore squid have a summer, inshore-northerly spawning migration to shallow coastal and shelf waters, fol-

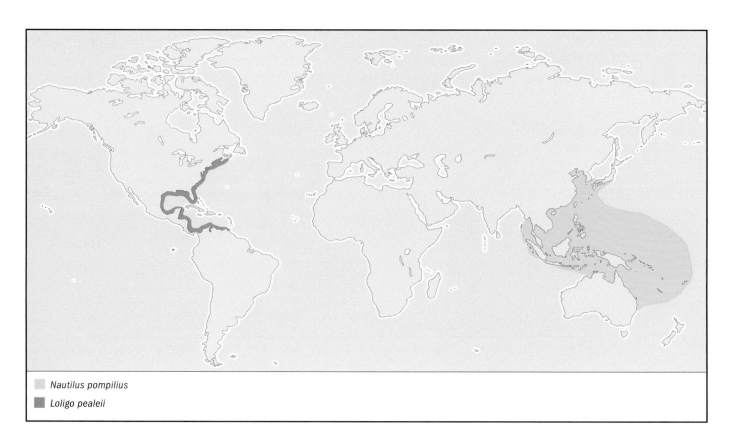

Nautilus pompilius
Loligo pealeii

lowed by an offshore-southerly retreat in fall and winter to the waters of the continental slope.

FEEDING ECOLOGY AND DIET
Food includes crustaceans, fishes, and squids.

REPRODUCTIVE BIOLOGY
Eggs are laid in gelatinous finger-like strands, many of which are attached together in large masses ("sea mops") to a solid substrate at depths from a few feet to 820 ft (250 m). The paralarvae are planktonic.

CONSERVATION STATUS
Not listed by the IUCN. Occasionally of concern because of potential overfishing.

SIGNIFICANCE TO HUMANS
A very important species for commercial fisheries. Has been used extensively in behavioral and neurological experiments. ◆

Pearly nautilus
Nautilus pompilius

ORDER
Nautilida

FAMILY
Nautilidae

TAXONOMY
Nautilus pompilius Linnaeus, 1758.

OTHER COMMON NAMES
English: Chambered nautilus, common nautilus, emperor nautilus, flame nautilus; French: Nautile chambré, nautile naturel; German: Perlboot, Schiffsboot; Italian: Nautilo.

PHYSICAL CHARACTERISTICS
External chambered shell, umbilicus filled in with a concretion; eyes form images via a pinhole opening; funnel formed by overlapping flaps; flame-striped color pattern extending across nearly entire shell.

DISTRIBUTION
Indo-West Pacific.

HABITAT
Continental shelf and slope from near the surface to a depth of about 2,460 ft (750 m). Associated with hard ocean bottoms, particularly coral reefs. Undergoes diel vertical migrations.

BEHAVIOR
Activity primarily nocturnal.

FEEDING ECOLOGY AND DIET
Feeds on sedentary benthic prey and carrion (dead or decaying animal flesh).

REPRODUCTIVE BIOLOGY
Reproduces repeatedly over a period of years; attaches large eggs with irregularly shaped coverings to hard substrates.

CONSERVATION STATUS
Not listed by the IUCN. Of concern because of pressure from locally heavy fishing.

SIGNIFICANCE TO HUMANS
Fished for meat and shells, which are valuable to collectors. ◆

Greater argonaut
Argonauta argo

ORDER
Octopoda

FAMILY
Argonautidae

TAXONOMY
Argonauta argo Linnaeus, 1758.

OTHER COMMON NAMES
English: Paper nautilus; French: Argonaute, argonaute papier; German: Papierboot; Spanish: Argonauta comùn.

PHYSICAL CHARACTERISTICS
Has a locking apparatus for the funnel and mantle that consists of a knob and pit. Lacks water pores. Mature females produce an external shell-like egg case. Females have a flag-like expansion of the web of the dorsal arms that contains shell-secreting glands. The male hectocotylus develops in a sac beneath the eye; lacks a lateral papillate fringe.

DISTRIBUTION
Open ocean in tropical and subtropical regions.

HABITAT
Surface waters; rarely encountered near shore.

BEHAVIOR
Has been found attached to jellyfish that it uses as a source of food and protection. Males have been reported living within salps (free-swimming tunicates).

FEEDING ECOLOGY AND DIET
Feeds on crustaceans and jellyfish.

REPRODUCTIVE BIOLOGY
The male argonaut is a dwarf, about 10% of the length of the female. The entire third right arm is hectocotylized and carried in a special sac. At mating, the hectocotylus, which carries one large spermatophore, breaks out of its sac and free from the male body. It then invades or is deposited in the female's mantle cavity, where it remains viable and active for some time. Females brood many thousands of very small eggs deposited in an external shell-like egg case.

CONSERVATION STATUS
Not listed by the IUCN.

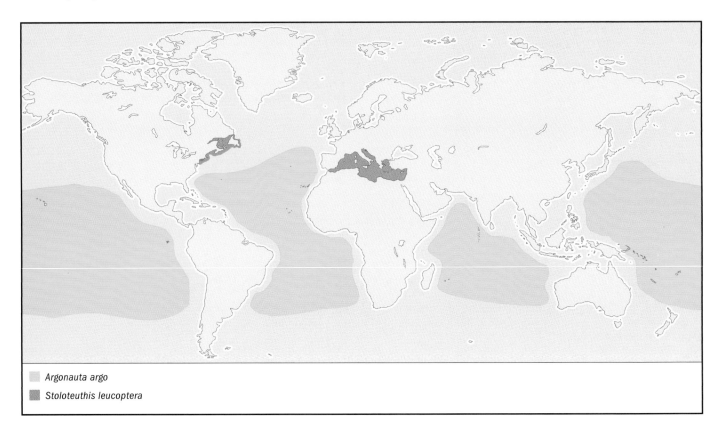

Argonauta argo

Stoloteuthis leucoptera

SIGNIFICANCE TO HUMANS
The egg cases are valuable to shell collectors. ◆

Common octopus
Octopus cf *vulgaris*

ORDER
Octopoda

FAMILY
Octopodidae

TAXONOMY
Octopus vulgaris Cuvier, 1797. Although *Octopus vulgaris* has been reported to be widely distributed around the world, researchers interested in cephalopod systematics have long known that these reports represent a species complex. Unfortunately, the relationships among the various populations within this complex have not yet been resolved.

OTHER COMMON NAMES
English: Devilfish, scuttle; French: Pieuvre, poulpe commun; German: Gewöhnlicher Krake, Oktopus, Seepolyp; Portuguese: Polvo; Spanish: Pulpo común, pulpo de roca.

PHYSICAL CHARACTERISTICS
Has two series of suckers. Structures (when present) on dorsal surfaces of mantle, head, arms not star-shaped cartilaginous tubercles. Funnel organ is W-shaped. Ink sac present. No ocelli between eye and bases of lateral arms. Has 9–11 gill lamellae on each outer demibranch. Mantle opening is wide. Dorsal arms are shortest; lateral arms longer than ventral arms; ventrolateral arms only slightly longer than dorsolateral arms.

Character states for distinguishing among species in this group await publication of redescriptions.

DISTRIBUTION
This species group is found worldwide in temperate and tropical waters; the true *O. vulgaris* is probably restricted to European and northern African waters.

HABITAT
A benthic species occurring from the coastline to the outer edge of the continental shelf in depths from 0–656 ft (0–200 m), where it is found in a variety of habitats, including rocks, coral reefs, and seagrass beds. The paralarvae are planktonic.

BEHAVIOR
The complex behavior of these octopods has been described in great detail. Although researchers disagree about the level of intelligence of these mollusks, or even how to measure it, they have referred to these animals as the closest thing to alien intelligence encountered by humans.

FEEDING ECOLOGY AND DIET
Feeds primarily on crustaceans (crabs and lobsters) and mollusks (clams and snails). Females may produce between 120,000 and 400,000 eggs little longer than 0.07 in (2 mm), which they deposit in strings in crevices or holes, usually in shallow waters. Spawning may take as long as a month. During the brooding period (25–65 days, depending on water temperature), the females almost stop feeding; many die after the hatching of the paralarvae.

REPRODUCTIVE BIOLOGY
Two peak periods of spawning, in the spring and autumn of each year.

CONSERVATION STATUS
Not listed by the IUCN. Occasionally of concern because of potential overfishing.

SIGNIFICANCE TO HUMANS
A very important species for commercial fisheries. Has also been used extensively in behavioral experiments. ◆

No common name
Stauroteuthis syrtensis

ORDER
Octopoda

FAMILY
Stauroteuthidae

TAXONOMY
Stauroteuthis syrtensis Verrill, 1879.

OTHER COMMON NAMES
None known.

PHYSICAL CHARACTERISTICS
Fins present. Cirri on arms. One series of suckers along the entire length of the arm. No filamentous appendages in pouches between the bases of the dorsal and dorsolateral arms. Light organ is present at base of each sucker. Contractile intermediate membrane or secondary web is present between each arm and the primary web. Shell is cartilaginous and has a simple U-shape. Gills are sepioid.

DISTRIBUTION
North Atlantic Ocean.

HABITAT
Found near bottom at depths of 1,640–13,125 ft (500–4,000 m) with maximum abundance between 4,920–8,200 ft (1,500–2,500 m).

BEHAVIOR
Has been observed drifting passively with the arms and web held in a bell-shaped posture. Swims by flapping its fins rather than jetting.

FEEDING ECOLOGY AND DIET
Feeds on small crustaceans that may be entangled in mucus produced by glands around the mouth.

REPRODUCTIVE BIOLOGY
Eggs of various sizes in ovary indicate slow and prolonged spawning rather than spawning of synchronous batches.

CONSERVATION STATUS
Not listed by the IUCN.

SIGNIFICANCE TO HUMANS
None known. ◆

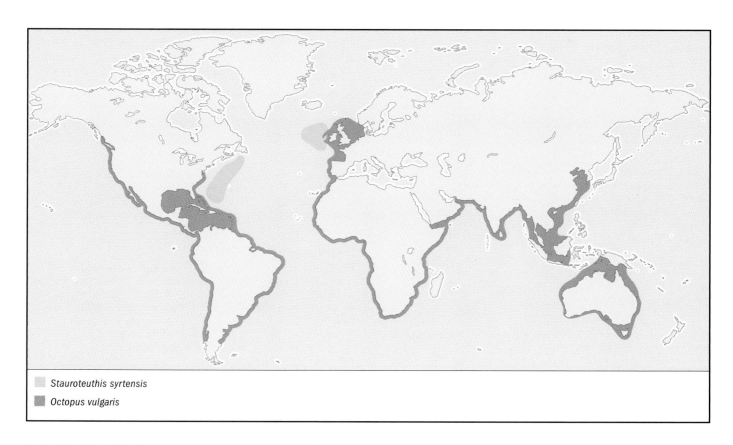

▢ *Stauroteuthis syrtensis*
▪ *Octopus vulgaris*

No common name
Mastigoteuthis magna

ORDER
Oegopsida

FAMILY
Mastigoteuthidae

TAXONOMY
Mastigoteuthis magna Joubin, 1913.

OTHER COMMON NAMES
None known.

PHYSICAL CHARACTERISTICS
Ventral arms are elongated. Tentacles are vermiform (worm-like); clubs not expanded or only slightly expanded; moderate in length to very elongate with minute suckers in many series. Funnel locking apparatus is oval with knobs affecting the shape. Fins are very large and positioned mostly behind the muscular part of the mantle. This cephalopod is weakly muscled and reddish in color; much of the red pigment is not located in chromatophore organs but is dispersed among other cells in the organism's skin.

DISTRIBUTION
Tropical and northern subtropical Atlantic; also reported from the Indian Ocean and Tasman Sea.

HABITAT
Bathypelagic and bathyal depths (between 600–13,000 ft or 200–4,000 m).

BEHAVIOR
Has been observed from submersibles drifting in a vertical posture just above the ocean floor. The organism's ventral arms were spread and held the tentacles, which dangled downward very close to the bottom.

FEEDING ECOLOGY AND DIET
Feeds on small planktonic crustaceans.

REPRODUCTIVE BIOLOGY
Not studied.

CONSERVATION STATUS
Not listed by the IUCN.

SIGNIFICANCE TO HUMANS
Important part of food webs on the continental slope that support commercial finfish fisheries. ◆

Common cuttlefish
Sepia officinalis

ORDER
Sepioidea

FAMILY
Sepiidae

TAXONOMY
Sepia officinalis Linnaeus, 1758.

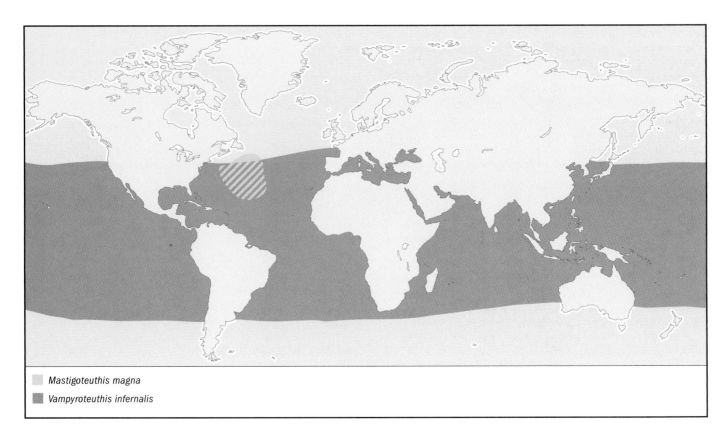

Mastigoteuthis magna
Vampyroteuthis infernalis

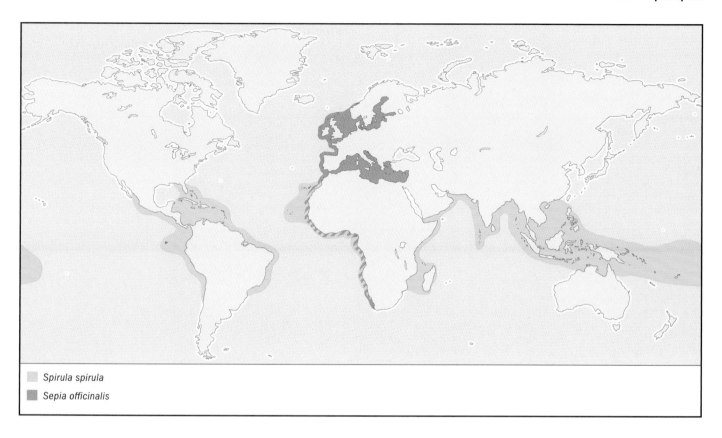

Spirula spirula

Sepia officinalis

OTHER COMMON NAMES
French: Casseron, seiche commune; German: Gemeine Sepie,
Gcmcinc Tintcnfisch; Italian: Scppia comunc; Spanish: Casta-
fuela, sepia común.

PHYSICAL CHARACTERISTICS
Internal shell is a cuttlebone. Cuttlebone length approximately
equals mantle length. Mantle is elongated and oval in shape.
Fins extend almost the full length of mantle. Tentacles can be
completely withdrawn into pockets. Tentacular club is short,
narrow, and crescent-shaped. Eye lens is covered by a cornea.

DISTRIBUTION
Eastern Atlantic: from the Baltic and North Seas to South
Africa; Mediterranean Sea.

HABITAT
A demersal (swimming near bottom) species occurring predominantly on sandy to muddy bottoms from the coastline to
about 656 ft (200 m) depth, but most abundant at less than 328
ft (100 m).

BEHAVIOR
As with *O. vulgaris*, the behavior of cuttlefish is very complex
and has been studied in great detail.

FEEDING ECOLOGY AND DIET
Feeds on small mollusks, crabs, shrimps, other cuttlefishes, and
juvenile demersal fishes. Cannibalism is common and has been
interpreted as a strategy to overcome temporary shortages of
adequately sized prey.

REPRODUCTIVE BIOLOGY
Males may carry as many as 1,400 spermatophores, and females
produce between 150–4,000 eggs, depending on their size. The
eggs measure from 0.31–0.39 in (8–10 mm) in diameter and

are attached in grapelike clusters to seaweeds, debris, shells and
other substrates. They hatch after 30–90 days depending on
temperature 70.7 59°F (21.5 15°C, respectively). Hatchlings
are large (0.27–0.31 in or 7–8 mm) dorsal mantle length and
look like miniature versions of adults.

CONSERVATION STATUS
Not listed by the IUCN. Occasionally of concern because of
potential overfishing.

SIGNIFICANCE TO HUMANS
A very important species for commercial fisheries. Has also
been used extensively in behavioral experiments. At one time
the ink produced by the cuttlefish was used as a dye. ◆

Butterfly bobtail squid
Stoloteuthis leucoptera

ORDER
Sepioidea

FAMILY
Sepiolidae

TAXONOMY
Stoloteuthis leucoptera Verrill, 1878.

OTHER COMMON NAMES
None known.

PHYSICAL CHARACTERISTICS
Mantle is short, and rounded toward the rear. Fins are broadly
separated toward the rear, with free anterior and posterior

lobes. The gladius is rudimentary. There is a median mantle septum with strong adductor muscles. The eye is covered by a cornea; there is a ventral eyelid present. The arms lack protective membranes. The dorsal mantle is fused to the head; the area of fusion is narrower than the width of the head. Posterior edges of fins do not extend beyond posterior mantle. Ventral shield of mantle does not extend beyond middle of eyes. Males have enlarged suckers on their dorsolateral arms.

DISTRIBUTION
North Atlantic off the eastern United States; Mediterranean Sea.

HABITAT
Continental shelf and slope at depths of 575–1,115 ft (175–340 m).

BEHAVIOR
Has been observed hovering in midwater far above the bottom, rapidly beating its fins.

FEEDING ECOLOGY AND DIET
Not studied; probably feeds on small planktonic crustaceans.

REPRODUCTIVE BIOLOGY
Nothing is known.

CONSERVATION STATUS
Not listed by the IUCN.

SIGNIFICANCE TO HUMANS
None known. ◆

Ram's horn squid
Spirula spirula

ORDER
Sepioidea

FAMILY
Spirulidae

TAXONOMY
Spirula spirula Linnaeus, 1758. Some experts assign this species to a separate order, Spirulida.

OTHER COMMON NAMES
French: Spirule; German: Posthorn.

PHYSICAL CHARACTERISTICS
Partially internal shell is curved ventrally in an open coil; each coil is round in cross section and has transverse septa (dividers) with a siphuncle. There are four series of suckers on the arms. Both ventral arms are hectocotylized in males. There are tentacular clubs with suckers in 16 series; not divided into manus and dactylus. Eyes lack a cornea. Fins are separate, terminal, and lie in a plane nearly transverse to body axis. Large photophore at rear end of body.

DISTRIBUTION
Mesopelagic waters of open tropical oceans.

HABITAT
Found at depths between 1,800–3,280 ft (550–1,000 m), mostly over the slopes of continents or islands where the ocean bottom lies between 3,280–6,560 ft (1,000–2,000 m).

BEHAVIOR
Captive *Spirula* hang head-downward in the water. This species is able to withdraw its head and arms completely within the mantle; the mantle opening can then be closed by folding over the large dorsal and ventrolateral extensions of the mantle margin. The photophore at the posterior end of the body is known to glow for hours at a time. When the animal is swimming slowly downward, head first, its terminal fins are pointed upward (i.e. posteriorly) and move with a rapid "waving or fluttering motion" that propels it downward (Bruun, 1943).

FEEDING ECOLOGY AND DIET
Nothing is known.

REPRODUCTIVE BIOLOGY
It has been suggested that *Spirula* lays its eggs on the ocean floor; the capture of very small animals in deep water supports this hypothesis.

CONSERVATION STATUS
Not listed by the IUCN.

SIGNIFICANCE TO HUMANS
The shells, which commonly wash ashore in some areas, are prized by collectors. ◆

Vampire squid
Vampyroteuthis infernalis

ORDER
Vampyromorpha

FAMILY
Vampyroteuthidae

TAXONOMY
Vampyroteuthis infernalis Chun, 1903.

OTHER COMMON NAMES
French: Vampire des enfers; German: Vampir-Tintenfisch.

PHYSICAL CHARACTERISTICS
Has retractile filaments extending from pockets between the dorsal and dorsolateral arms. Fins present. Large circular, lidded photophores present behind each adult fin ("fin-base" organs); numerous small photophores distributed over the lower surfaces of mantle, funnel, head and the aboral surface of the arms and web ("skin-nodule" organs). A gladius, or internal remnant of an original external shell, is present with a broad median field and conus (cup-shaped tip). Cirri are present over the entire length of the arm; suckers lack a cuticular lining and are present only on the outer half of arms.

DISTRIBUTION
Found throughout tropical and temperate oceans.

HABITAT
Meso to bathypelagic depths, generally 1,965–4,920 ft (600–1,500 m).

BEHAVIOR
Can swim surprisingly fast for a gelatinous animal. Arms are sometimes spread forward to form, along with the web, an umbrella-like or bell-shaped posture. Filaments appear to be tactile sense organs. Uses combinations of photophores for complex luminescent displays.

FEEDING ECOLOGY AND DIET
Poorly understood; diet includes gelatinous megaplankton.

REPRODUCTIVE BIOLOGY
No hectocotylus. Early development passes through stages with (1) a single pair of larval fins; (2) two pairs of fins, larval plus adult; and (3) a single pair of adult fins.

CONSERVATION STATUS
Not listed by the IUCN. Only a single extant species has been described in this order.

SIGNIFICANCE TO HUMANS
Frequently featured in natural-history television programs. ◆

Resources

Books

Abbott, J., R. Williamson, and L. Maddock, eds. *Cephalopod Neurobiology.* New York: Oxford University Press, 1995.

Boyle, P. R., ed. *Cephalopod Life Cycles.* Vol. I, *Species Accounts.* London: Academic Press, 1983.

———. *Cephalopod Life Cycles.* Vol. II, *Comparative Reviews.* London: Academic Press, 1987.

Budelman, B. U., R. Schipp, and S. von Boltezky. "Cephalopoda." In *Microscopic Anatomy of Invertebrates.* Vol. 6A, *Mollusca II*, edited by F. W. Harrison and A. J. Kohn. New York: Wiley-Liss, Inc., 1997.

Clarke, M. R., ed. *A Handbook for the Identification of Cephalopod Beaks.* Oxford, U.K.: The Clarendon Press, 1986.

Gilbert, D. L., W. J. Adelman, and J. M. Arnold, eds. *Squid as Experimental Animals.* New York: Plenum Press, 1990.

Hanlon, R. T., and J. B. Messenger. *Cephalopod Behaviour.* Cambridge, U.K.: Cambridge University Press, 1996.

Kinne, O., ed. *Diseases of Marine Animals*, Vol. III, *Cephalopoda through Urochordata.* Hamburg, Germany: Biologische Anstalt Helgoland, 1990.

Mangold, K. *Traité de Zoologie. Anatomie, Systématique, Biologie.* Tome 5, fascicule 4, *Céphalopodes.* Paris: Masson et Cie, 1989.

Nesis, K. N. *Cephalopods of the World: Squids, Cuttlefishes, Octopuses and Allies.* English translation. Neptune City, NJ: T. F. H. Publications, 1987.

Okutani, T. *Cuttlefish and Squids of the World in Color.* Tokyo: National Cooperative Association of Squid Processors, 1995.

Saunders, W. B., and N. H. Landman, eds. *Nautilus. The Biology and Paleobiology of a Living Fossil.* New York: Plenum Press, 1987.

Ward, P. D. *Natural History of Nautilus.* London: Allen and Unwin, 1987.

Periodicals

Clarke, M. R., ed. "The Role of Cephalopods in the World's Oceans." *Philosophical Transactions of the Royal Society, London* 351, no. 1343 (1996): 977–1112.

Roper, C. F. E., and M. Vecchione, eds. "The Gilbert L. Voss Memoral Issue: Systematics, Fisheries and Biology of Cephalopods." *Bulletin of Marine Science* 49, nos. 1–2 (1991): 1–670.

Sweeney, M. J., et al., eds. "'Larval' and Juvenile Cephalopods: A Manual for Their Identification." *Smithsonian Contributions to Zoology* 513 (1992): 1–282.

Voss, N. A., et al., eds. "Systematics and Biogeography of Cephalopods." *Smithsonian Contributions to Zoology* 586 (1998): 1–599.

Organizations

Cephalopod International Advisory Council (CIAC). Dr. Michael Vecchione, President. NOAA/NMFS National Systematics Laboratory, National Museum of Natural History, MRC-153, Washington, DC 20013-7012.

Other

"Cephalopoda Cuvier, 1797." Tree of Life Web Project. 1996 (21 July 2003). <http://tolweb.org/tree/eukaryotes/animals/mollusca/cephalopoda/cephalopoda.html>.

"Cephalopods at the National Museum of Natural History." Smithsonian National Museum of Natural History. 28 Jan. 2002 (21 July 2003). <http://www.nmnh.si.edu/cephs/>.

CephBase. 18 July 2002 (21 July 2003). <http://www.cephbase.utmb.edu/>.

Fossil Coleoidea Page. 1998 (21 July 2003). <http://userpage.fu-berlin.de/palaeont/fossilcoleoidea/welcome.html>.

Michael Vecchione, PhD

Phoronida
(Phoronids)

Phylum Phoronida
Number of families 1

Thumbnail description
Sedentary, infaunal, benthic suspension-feeders with a vermiform (worm-like) body that bears a lophophore and is enclosed in a slender tube in which the animal moves freely and is anchored by an ampulla.

Photo: A golden phoronid (*Phoronopsis californica*) in Madeira (São Pedro, southeast coast, about 100 ft [30 m] depth) showing the helicoidal shape of the lophophore. (Photo by Peter Wirtz. Reproduced by permission.)

Evolution and systematics

The phylum Phoronida is known to have existed since the Devonian, but there is a poor fossil record of burrows and borings attributed to phoronids. Many scientists now regard the Phoronida as a class within the phylum Lophophorata, along with the Brachiopoda and perhaps the Bryozoa. Phoronida consists of two genera, *Phoronis* and *Phoronopsis*, which are characterized by the presence of an epidermal collar fold at the base of the lophophore. The group takes its name from the genus name *Phoronis*, one of the numerous epithets of the Egyptian goddess Isis.

The phoronid larva, commonly called an actinotroch, retains a separate "generic" name, *Actinotrocha*, which authorities still consider different from the adult species.

There are 10 well-defined species of phoronid: *Phoronis ovalis* (creeping larva), *P. hippocrepia* (larva, *Actinotrocha vancouverensis*), *P. ijimai* (larva, *A. vancouverensis*), *P. australis* (larva, unknown), *P. muelleri* (larva, *A. branchiata*), *P. psammophila* (larva, *A. sabatieri*), *P. pallida* (larva, *A. pallida*), *Phoronopsis albomaculata* (larva, unknown), *Phoronopsis harmeri* (larva, *Actinotrocha harmeri*), and *Phoronopsis californica* (larva, unknown). There is no class or order designation in this phylum.

Physical characteristics

Body length varies from several millimeters to more than 18 in (45 cm) and corresponds to the extended size, which is generally approximately five times longer than the contracted size. Body diameter varies from 0.006 in (0.15 mm) to approximately 0.2 in (5 mm). Color is generally fleshy, sometimes brownish to black.

The other features are anatomical. Phoronids have at least two body regions in larval and adult forms, each containing its own coelom. A lophophore, defined as a tentacular extension of the mesosome (and of its coelomic cavity, the mesocoelom) embraces the mouth but not the anus. The main functions of the lophophore are feeding, respiration, and protection. The site and shape of the lophophore are proportional to body size, ranging from oval to horseshoe to helicoidal in relation to an increase in the number of tentacles. Phoronids have a U-shaped digestive tract. A nervous center is present between the mouth and anus, as is a ring nerve at the base of the lophophore, and the animal has one or two giant nerve fibers. Phoronids have one pair of nephridia and a closed circulatory system with red blood corpuscles.

Distribution

Phoronids are found in all oceans and seas except the Antarctic Ocean. They are present in the intertidal zone to approximately 1,310 ft (400 m). All species have a wide geographical range; most are cosmopolitan.

Habitat

Phoronids secrete characteristic rigid tubes consisting of layers of chitin to which adhere particles of sediment and debris from the animal's immediate environment. Phoronids occur singly, vertically embedded in soft sediment (sand, mud, or fine gravel) or form tangled masses of many individuals buried in, or encrusting, rocks and shells. *Phoronis australis* is associated with cerianthid anemones in the tube in which it buries. In some habitats phoronids are very abundant, reaching several tens of thousand of individuals per 11 ft^2 (1 m^2).

Predators of phoronids are not well known, but they include fishes, gastropods, and nematodes.

The larval actinotroch is a familiar component of plankton.

A phoronid worm (*Phoronopsis californica*). (Photo by Lawrence Naylor/ Photo Researchers, Inc. Reproduced by permission.)

Behavior

The reaction of a phoronid to a predator is very rapid retraction of the animal down into the tube out of harm's way. Should the lophophore be taken by a predator, a phoronid can regenerate the lost parts within two or three days. Thus phoronids may provide an important food source for grazers.

Feeding ecology and diet

Phoronids are suspension feeders, capturing algae, diatoms, flagellates, peridinians, small invertebrate larvae, and detritus from the water by means of the lophophore. Phoronids orient their lophophores into the prevailing water current, and when currents change direction, phoronids can rapidly reorient to maintain the food-catching surface of the lophophore into the water flow. The cilia on the tentacles create a feeding current, which transports the food particles along the frontal surface of the animal down to the mouth at the bottom of the lophophoral cavity.

Direct uptake of amino acids through the epidermis displays seasonal variation, the maximum occurring in summer.

Several types of parasite may be present in phoronids: progenetic trematode metacercariae and cysts with spores in the coelomic cavities; unidentified gregarines in the digestive tract; and an ancistrocomid ciliate parasite, *Heterocineta*, in the tentacles.

The life span of phoronids is approximately one year.

Reproductive biology

Phoronids generally breed between spring and autumn over a fairly long period. These organisms are hermaphroditic or dioecious. Gametes are released through the nephridia, which serve as gonoducts. Sperm is disseminated by means of spermatophores formed in two lophophoral organs. Fertilization is internal, and cross-fertilization takes place in hermaphroditic species. Egg cleavage is total, equal, and typically radial.

Three types of developmental patterns occur. Species with small eggs undergo complete planktonic development, and species with larger eggs brood either in nidamental glands within the lophophore concavity or within the parental tube until the first larval stages and then undergo planktotrophic development. Embryonic development leads to a characteristic free-swimming larva, called the *actinotroch*, that bears an anterior preoral lobe on the top of which is located the nervous ganglion, a tentacular ridge, a pair of protonephridia, and, in the posterior, a ciliated ring around the anus. The larvae undergo planktotrophic development and settle after approximately 20 days. Metamorphosis is "catastrophic," occurring in less than 30 minutes and leading to a slender young phoronid, which burrows into the substratum.

Phoronids also reproduce asexually, usually by transverse fission around the middle of the trunk.

Conservation status

No species are listed by the IUCN.

Significance to humans

None known.

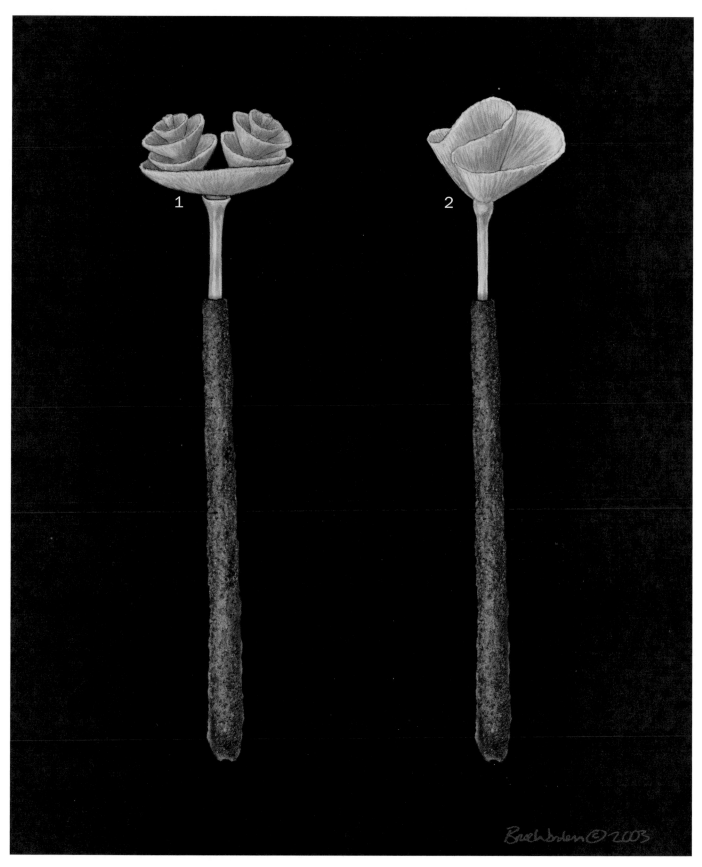

1. *Phoronopsis harmeri*; 2. *Phoronis ijimai*. (Illustration by Bruce Worden)

Species accounts

No common name
Phoronis ijimai

FAMILY
None

TAXONOMY
Phoronis ijimai Oka, 1897, Akkeshi Bay, Tokyo Bay, Japan.
Synonym: *Phoronis vancouverensis* Pixell, 1912, Vancouver
Island, Canada.

OTHER COMMON NAMES
None known.

PHYSICAL CHARACTERISTICS
Extended specimens of phoronids reach a length of 5 in (120
mm) and a diameter of 0.02–0.08 in (0.5–2 mm). The color in
life is fleshy to transparent. The lophophore sometimes is
transparent and has white pigment spots. The lophophore is
horseshoe shaped or spiral with approximately one coil and as
many as 230 tentacles 0.08–0.2 in (2–5 mm) long.

DISTRIBUTION
Eastern and western Pacific Ocean, northwestern Atlantic
Ocean. (Specific distribution map not available.)

HABITAT
Phoronids encrust on hard surfaces (rocks, wood) or live on
soft substrate in turf-like masses with many intertwined tubes
in quiet environments. Some burrow in hard substrate (calcareous rocks and algae, coral, mollusc shells) under strong water
motion. The range of this species extends from the intertidal
zone to a depth of approximately 33 ft (10 m).

BEHAVIOR
Reacts to predators with rapid retraction into the tube. Can
regenerate lost parts within two or three days. Thus may provide an important food source for grazers.

FEEDING ECOLOGY AND DIET
Suspension feeder, capturing algae, diatoms, flagellates, peridinians, small invertebrate larvae, and detritus from the water
by means of the lophophore. Orients the lophophore into the
prevailing water current.

REPRODUCTIVE BIOLOGY
Hermaphroditic. The embryos are brooded in two masses by
the nidamental glands of A type (on the floor of the
lophophoral concavity and on the inner surface of the internal
tentacles). The lophophoral organs, which produce the spermatophores, are small. Asexual reproduction is by transverse
fission. The larva is called *Actinotrocha vancouverensis* Zimmer,
1964. It is approximately 0.03 (0.8 mm) long when ready to
metamorphose.

CONSERVATION STATUS
Not listed by the IUCN.

SIGNIFICANCE TO HUMANS
None known. ◆

No common name
Phoronopsis harmeri

FAMILY
None

TAXONOMY
Phoronopsis harmeri Pixell, 1912, Vancouver Island, Canada.
Synonyms: *Phoronopsis striata* Hilton, 1930; *Phoronopsis viridis*
Hilton, 1930; *Phoronopsis malakhovi* Temereva, 2000.

OTHER COMMON NAMES
None known.

PHYSICAL CHARACTERISTICS
Collar fold below the lophophore (genus characteristic) is well
marked. Extended specimens reach a length of 9 in (220 mm)
and a diameter of 0.02–0.16 in (0.6–4 mm). The body color in
life is pink to greenish, the lophophore being transparent or
sometimes white pigmented. The lophophore is spiral with
1 2.5 coils on each side. This animal has as many as 400 tentacles 0.08–0.2 in (2–5 mm) long.

DISTRIBUTION
Pacific Ocean, North Atlantic Ocean, and western Mediterranean Sea. (Specific distribution map not available.)

HABITAT
The tubes are embedded vertically in soft sediment from sand
to muddy sand, sometimes with a coarse grain fraction. Depth
ranges from the intertidal zone to approximately 328 ft (100
m) with a common range of 0–66 ft (0–20 m). Density can
reach 28,000 individuals per 11 ft^2 (1 m^2).

BEHAVIOR
Reacts to predators with rapid retraction into the tube. Can regenerate lost parts within two or three days. Thus may provide
an important food source for grazers.

FEEDING ECOLOGY AND DIET
Suspension feeder, capturing algae, diatoms, flagellates, peridinians, small invertebrate larvae, and detritus from the water
by means of the lophophore. Orients the lophophore into the
prevailing water current.

REPRODUCTIVE BIOLOGY
Dioecious. Females shed the ova directly into sea water. Males
have large, membranous lophophoral organs. The larva is
Actinotrocha harmeri Zimmer, 1964. Approximately 0.04 in
(1 mm) long when ready to metamorphose. Asexual reproduction by transverse fission.

CONSERVATION STATUS
Not listed by the IUCN.

SIGNIFICANCE TO HUMANS
None known. ◆

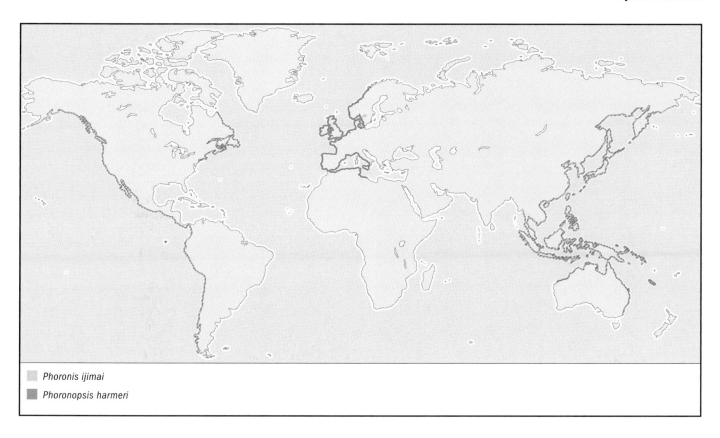

Phoronis ijimai

Phoronopsis harmeri

Resources

Periodicals

Cohen B. L. "Monophyly of Brachiopods and Phoronids: Reconciliation of Molecular Evidence with Linnaean Classification (the Subphylum Phoroniformea nov.)." *Proceedings of the Royal Society, London* Series B, 267 (2000): 225-31.

Emig C. C. "The Biology of Phoronida." *Advances in Marine Biology* 19 (1982): 1-89.

———. "Les Lophophorates Constituent: Ils un Embranchement?" *Bulletin de la Société zoologique de France* 122 (1997): 279-88.

Riisgård H. U. "Methods of Ciliary Filter Feeding in Adult *Phoronis muelleri* (Phylum Phoronida) and in Its Free-swimming Actinotroch Larva." *Marine Biology* 141 (2002): 75-87.

Other

Phoronid@2003. 22 July 2003 [29 July 2003]. <http://www.com.univ-mrs.fr/DIMAR/Phoro/>.

Christian C. Emig, Dr es-Sciences

Phylactolaemata

(Freshwater bryozoans)

Phylum Ectoprocta

Class Phylactolaemata

Number of families 5

Thumbnail description
Filter-feeding invertebrate animals composed of many identical zooids joined seamlessly as a colony, each zooid capable of independent feeding, digestion, and reproduction

Illustration: *Pectinatella magnifica.* (Illustration by Bruce Worden)

Evolution and systematics

Class Phylactolaemata includes five families and approximately 80 known species. Families include Cristatellidae, Fredericellidae, Lophopodidae, Pectinatellidae, and Plumatellidae.

Physical characteristics

Freshwater bryozoans appear in a variety of forms. The most conspicuous colonies are large, gelatinous masses, but other species are small, moss-like growths or patches of branching tubules. At the microscopic level the basic design is the same: every colony is composed of many identical zooids the anatomical features of which are devoted mostly to feeding and digestion. A horseshoe-shaped lophophore of ciliated tentacles (lophophore shape is circular in Fredericellidae) generates a current of water that draws suspended food particles toward the mouth. An elongated U-shaped gut mixes and digests the food then packs undigested particles into small pellets, which are expelled through the anus. Circular muscles within the colony wall maintain slight hydrostatic pressure, which enables the feeding apparatus to project into the water while long retractor muscles pull the lophophore unit back into the colony interior. Each zooid has a single nerve ganglion between mouth and anus, and nerve tracts extend into the lophophore and to the gut. No nerves pass between zooids. A cord of tissue called a *funiculus* connects the tip of the gut to the inner colony wall and functions in sexual and asexual reproduction. Each zooid is capable of budding additional zooids and forming dormant statoblasts.

Distribution

Every continent except Antarctica. Common in lakes, ponds, rivers, and streams growing on a variety of solid, submerged surfaces. Colonies and dormant statoblasts are easily carried to new sites by waterfowl or transported on aquatic vegetation.

Habitat

All species can be found in shallow water, where they grow on the sides or undersides of submerged rocks, wood, vegetation, rubber tires, plastic trash, and other solid, stable materials. These animals avoid rotting, oily, or marly substrata, and generally do not occur on wave-tossed materials. Most species thrive at a pH greater than 6 and a temperature range of 41–86°F (5–30°C).

Behavior

Nothing is known.

Feeding ecology and diet

Phylactolaemate bryozoans appear to be indiscriminate in the materials they eat, rejecting only particles that are too bulky or active. Ingested materials are reported to include a variety of plankton and detritus. Retention time in the gut varies with the rate of ingestion, which depends on the concentration of suspended particles in the water. During active feeding many of the living organisms taken into the gut emerge from the anus still alive. Studies have shown ingested bacteria to be an important source of nutrition, and it is possible that many ingested particles are simply carriers of bacterial food.

Reproductive biology

The life cycle includes both sexual and asexual reproduction. During the brief season of sexual activity, sperm are formed on the funiculus and then detach to circulate freely within the colony. Grape-like clusters of eggs appear on the inner colony wall near the tip of certain zooids. Internal fertilization appears possible, but there is also evidence of outcrossing. The mechanism for sperm exchange between colonies is unknown. The fertilized egg develops within a special embryo sac, becoming a motile stage composed of one or two fully formed zooids surrounded by a ciliated mantle. Once released from the colony, this larva-like structure can swim actively for several hours before settling on a substrate. The mantle pulls back, and the new zooids begin feeding almost immediately.

In some species asexual reproduction may be as simple as fragments of the colony lodging in a new location. In some globular species the colonies may pinch off daughter colonies, which then glide slowly away. However, all species also engage in asexual production of dormant statoblasts capable of withstanding drought, freezing temperatures, and other unfavorable conditions. Many statoblasts contain air chambers that provide buoyancy, so on release from the colony, they can drift considerable distances or adhere to feathers of waterfowl. Another type of statoblast is attached directly to the substratum, where it remains long after the colony has disintegrated. The period of obligate dormancy can last several months, and some statoblasts remain viable for years. When favorable conditions resume, the statoblast germinates to produce a single zooid capable of forming a new colony.

Conservation status

Freshwater bryozoans in general are common and abundant, although certain species are considered rare, especially in tropical regions. No species are listed by the IUCN.

Significance to humans

In their natural habitat freshwater bryozoans contribute to nutrient cycling, and they are grazed upon by fish and certain invertebrates. However, they also are a nuisance when they grow inside pipelines and filters, blocking or seriously disrupting the flow of water in irrigation, wastewater, and cooling water systems. Any pipeline that carries untreated water from a lake or river is at risk of becoming fouled with bryozoan colonies.

1. *Pectinatella magnifica* individual and colony; 2. *Fredericella sultana* individual and colony; 3. *Plumatella fungosa* individual and colony. (Illustration by Bruce Worden)

Species accounts

No common name
Fredericella sultana

ORDER
Plumatellida

FAMILY
Fredericellidae

TAXONOMY
Fredericella sultana Blumenbach, 1779, canals in Göttingen, Germany.

OTHER COMMON NAMES
None known.

PHYSICAL CHARACTERISTICS
Colony composed of thin, brown, branching tubules, some attached to a submerged object and others extending freely into the water, easily mistaken for small plant roots. Lophophore has 18–25 tentacles arranged in a circle around the mouth. Statoblast resembles a tiny bean 0.016 in (0.4 mm) long.

DISTRIBUTION
Every continent except South America and Antarctica, common in Europe and Asia. Two slightly different species also occur in North and South America.

HABITAT
Clean water in a variety of habitats. Thrives in flowing water. Tolerates a wide range of temperature and pH. Capable of overwintering under ice.

BEHAVIOR
Nothing is known.

FEEDING ECOLOGY AND DIET
Feeds continuously, ingesting suspended particles from the water, presumably deriving most nutrition from bacteria and detritus.

REPRODUCTIVE BIOLOGY
The fertilized egg develops into a motile stage containing a single zooid. Seasonality of sexual reproduction has not been documented. Statoblasts are relatively sparse, occurring mostly in colony branches adjacent to substratum.

CONSERVATION STATUS
Not listed by the IUCN.

SIGNIFICANCE TO HUMANS
An important host for *Tetracapsuloides bryosalmonae*, the causative agent of proliferative kidney disease in salmonid fish, often resulting in significant economic loss to fish farmers. ◆

Pectinatella magnifica
Fredericella sultana

No common name
Pectinatella magnifica

ORDER
Plumatellida

FAMILY
Pectinatellidae

TAXONOMY
Pectinatella magnifica Leidy, 1851, stream near Philadelphia, Pennsylvania, United States.

OTHER COMMON NAMES
None known.

PHYSICAL CHARACTERISTICS
The young colony is flat and compact, eventually forming a small mound that enlarges to the size of a football. Zooids cover the entire outer surface, producing slimy mucus with a distinctive, sharp odor. The interior of the colony is a clear, gelatinous mass consisting mostly of water. Individual zooids have red pigment around the mouth and two conspicuous white spots on the lophophore. Only one type of statoblast is formed: buoyant, discoid, approximately 0.04 in (1 mm) in diameter with hooked spines radiating from the periphery.

DISTRIBUTION
United States and southern Canada. Scattered sites in Germany, Poland, Czechoslovakia, and France. In the 1960s invaded Japan and Korea, where the presence of colonies longer than 6 ft (2 m) has been documented.

HABITAT
Thrives in warm, productive habitats. Prefers quiet or slowly moving water, often found on submerged logs, tree stumps, and dock pilings.

BEHAVIOR
Nothing is known.

FEEDING ECOLOGY AND DIET
Feeds continuously, ingesting suspended particles from the water, presumably deriving most nutrition from bacteria and detritus.

REPRODUCTIVE BIOLOGY
Throughout most of the range, sexual activity occurs early in the summer, the motile stages released in June or early July. Buoyant statoblasts are formed and released in September and October.

CONSERVATION STATUS
Not listed by the IUCN.

SIGNIFICANCE TO HUMANS
Despite frequent assumptions to the contrary, *pectinatella* is a normal part of the freshwater animal community. Its presence does not indicate pollution or any other water quality problems. However, at the end of the season, large colonies sometimes detach from their substrate and drift toward shore, startling fishermen and clogging water intake pipes. ◆

No common name
Plumatella fungosa

ORDER
Plumatellida

FAMILY
Plumatellidae

TAXONOMY
Plumatella fungosa Pallas, 1768, Stariza River, Russia.

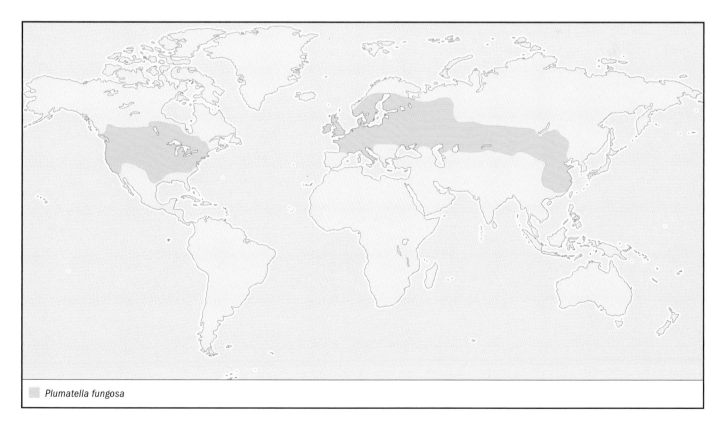

Plumatella fungosa

OTHER COMMON NAMES
None known.

PHYSICAL CHARACTERISTICS
Young colony composed of branched tubules spreading across submerged substratum. Zooids initially are oriented along the substratum but become erect when crowded. Older colonies reach a thickness up to 2 in (50 mm). As in all plumatellids, two types of statoblasts are produced: Floatoblasts have a ring of air-filled chambers for buoyancy and are released freely from the living colony. Sesoblasts lack air chambers and are cemented directly to the substratum.

DISTRIBUTION
North America, Europe, and northern Asia.

HABITAT
Still or gently flowing water. Thrives in eutrophic conditions and tolerates organic pollution better than other bryozoan species.

BEHAVIOR
Nothing is known.

FEEDING ECOLOGY AND DIET
Feeds continuously, ingesting suspended particles from the water, presumably deriving most nutrition from bacteria and detritus.

REPRODUCTIVE BIOLOGY
Sexual activity begins early in the season. Motile stages are released in late June and early July. Statoblasts are produced throughout the season in great abundance.

CONSERVATION STATUS
Not listed by the IUCN.

SIGNIFICANCE TO HUMANS
Colonies can clog pipes and filters in irrigation and water treatment systems. ◆

Resources

Books

Lacourt, Adrianus W. *A Monograph of the Freshwater Bryozoa: Phylactolaemata.* Leiden, The Netherlands: Zoologische Verhandelingen, 1968.

Ryland, John S. *Bryozoans.* London: Hutchinson University Library, 1970.

Smith, Douglas G. *Pennak's Freshwater Invertebrates of the United States: Porifera to Crustacea.* New York: Wiley and Sons, 2001.

Wood, Timothy S. "Bryozoans." In *Ecology and Classification of North American Freshwater Bryozoans.* 2nd edition, edited by John Thorp and Alan Covich. San Diego, CA: Academic Press, 2001.

———. *Ectoproct Bryozoans of Ohio.* Columbus: Ohio Biological Survey, 1989.

Wood, Timothy S., and Beth Okamura. *A New Key to the Freshwater Bryozoans of Britain, Ireland, and Continental Europe.* Ambleside, Cumbria, U.K.: Freshwater Biological Association, 2004.

Organizations

International Bryozoology Association. Web site: <http://petralia.civgeo.rmit.edu.au/bryozoa/iba.html>

Timothy S. Wood, PhD

Stenolaemata
(*Marine bryozoans*)

Class Stenolaemata

Number of families Approximately 25

Thumbnail description
Colonial marine animals, superficially plantlike in appearance, made up of many units that feed or perform other vital functions, and having rigid, calcified supportive skeletons.

Photo: *Crisulipora occidentalis*. (Photo by E. R. Degginger/Photo Researchers, Inc. Reproduced by permission.)

Evolution and systematics

Phylum Bryozoa, also called phylum Ectoprocta, sorts itself into three classes: Stenolaemata, Gymnolaemata, and Phylactolaemata. Stenolaemata are exclusively marine bryozoans, while Gymnolaemate bryozoans are mostly marine, plus a few freshwater types, and Phylactolaemates bryozoans are entirely reshwater.

The fossil record of class Stenolaemata extends back as far as 400 million years, into the early Ordovician Era. The Stenolaemata have passed their peak of diversification and are now in decline. Stenolaemata comprises one extant order, Cyclostomata (also called Tubuliporida), which first appeared in the Ordovician and scraped by with small colonies and little diversification through the Paleozoic, then exploded into extraordinary diversity during the Cretaceous. Its diversity declined toward the end of the Cretaceous, the number of genera was whittled down from about 175 to about 50, and the order has maintained that level of diversity into the present. There are about 500 extant species.

Physical characteristics

Stenolaemata, like bryozoans in general, are colonial animals, with many species superficially resembling seaweeds or corals. A common sort of stenolaemate colony consists of a branching array of hollow, calcified tubes, which serves as a skeleton, the tubes studded along the lengths of the tubes with individual, functional units called "zooids." Stenolaemate colonies can take a variety of forms, including crustlike, rosette, cup-shaped, branching, lumpy, and leaf-shaped. Colonies may bear calcified structures arising from the support skeleton, called "maculae," that function in colony hygiene, directing "used" water, or water already sifted for food particles and carrying wastes, away from the colony for hygiene purposes.

A single colony may be less than a millimeter high, or grow as high as a few feet (1 m) or more. The individual zooids are microscopic or nearly so. Microscopes are standard and necessary tools for studying and identifying bryozoans.

Stenolaemata zooids are characteristically elongated and cylindrical. Each zooid sits in a chamber called a "zoecium," encased in a calcified cystid for protection. Zooids within a single colony may differ considerably in form and function (polymorphism). Most numerous are the feeding zooids, or autozooids, each of which has a coelom, or inner body cavity, formed from the mesoderm, as well as a digestive tract, nervous system, and muscles. Although an autozooid feeds independently, it connects and communicates with the entire colony. Non-feeding zooids, or heterozooids, in Stenolaemata include gonozooids (reproductive) and kenozooids, which lack an internal organ system and serve as extra support for the colonial skeleton.

To feed, an autozooid extends an organ called a lophophore, consisting of a tentacle sheath and a ball of rolled-up, hollow tentacles. The tentacles open up into a graceful, flowerlike, bell-shaped structure. Cilia studding the tentacles draw currents of water into the mouth, to the center of the lophophore, and into the digestive tract and gut, with a caecum that grinds up food particles with peristaltic motions. The tract communicates directly, via pores, with another tract running the length of the supporting skeletal tube, allowing ingested food to be shared throughout the colony. A zooid passes its wastes through an anus, on the tentacle sheath below the tentacle ring.

Distribution

Stenolaemate bryozoans, like marine bryozoans in general, are found from the equator well into the arctic and Antarctic

Ocean regions, but their greatest diversity occurs in the cool waters of the temperate latitudes and the cold waters of the near-polar regions.

Habitat

Stenolaemate bryozoans live from shallow, near-shore waters to abyssal depths, but prefer shallow waters. A colony, depending on species, may attach or root itself in its formative stages to nearly any sort of substrate, including rocks, seashells, seaweed, and wooden piers.

Behavior

Stenolaemate bryozoans live passive, stationary lifestyles, similar to corals.

Feeding ecology and diet

Stenolaemata feed themselves via their zooids, which draw water, by the action of their ciliated lophopores, through their bodies, snagging microscopic algae, animals, and organic detritus. A tiny stomach churns up the ingested food with peristaltic contractions, digests it, then absorbs some and passes some on to the rest of the colony.

Reproductive biology

Stenolaemata, like bryozoans in general, reproduce both sexually and asexually. Sexually, most stenolaemata are hermaphroditic. Different species may be simultaneous hermaphrodites, producing sperm and eggs at the same time, or protandric hermaphrodites, producing eggs and sperm at different times. Ovaries and testes form from modified autozooids. Most stenolaemata species temporarily keep their fertilized eggs within brood chambers modified from the coela of feeding zooids. Free-swimming sperm are snatched from open water by the autozooids and passed on to the eggs for fertilization.

The fertilized eggs undergo cleavage and become free-swimming larvae that force their way out of the brood chamber and become independent, free-swimming organisms. In a few days, they settle down on a growth-friendly substrate, each larva changing into a rootlike form, the ancestrula, which commences to grow asexually into an entire colony, building the skeleton and budding the zooids; the colony continues to grow in this fashion. If a piece of a colony is broken off from the rest, it can settle and build itself into a new, complete colony.

Conservation status

No species are listed by the IUCN.

Significance to humans

Stenolaemata are of no special significance to humans. In time, they may, since research is being done on bryozoans to find possible biochemicals within them that may have medical benefits.

1. *Idmidronea atlantica*; 2. Joint-tubed bryozoan (*Crisia eburna*); 3. *Disporella hispida*. (Illustration by Dan Erickson)

Species accounts

Joint-tubed bryozoan

Crisia eburna

ORDER
Cyclostomata

FAMILY
Crisiidae

TAXONOMY
Crisia eburna Linnaeus, 1758.

OTHER COMMON NAMES
None known.

PHYSICAL CHARACTERISTICS
A colony takes the form of an upright, white miniature shrub or tuft, with many calcified branches. Colonies grow from 0.31–0.98 in (8–25 mm) high. Zooecia are arranged in two alternate rows on each branch.

DISTRIBUTION
Temperate to subtropical marine. Common in Woods Hole in Massachusetts, San Francisco Bay, and offshore Florida, United States.

HABITAT
Marine, living offshore and in brackish estuaries.

BEHAVIOR
Like most bryozoa, it is a stationary, colonial organism; colony feeds by means of autozooids.

FEEDING ECOLOGY AND DIET
Feeds on algae and other marine microorganisms by means of autozooids.

REPRODUCTIVE BIOLOGY
Observations lean toward it being hermaphroditic and protandrous. Breeding in San Francisco Bay runs from late February–May. Mature fertilized eggs are found in partially formed zooid buds that metamorphose into gonozoids, or brooding zooids, each nurturing a single embryo. Gonozooid is shaped like a vase, with a fat mid-region and a short neck, and is attached to the main colonial skeleton by a narrow stalk. The neck provides an opening for the escape of free-swimming larvae.

The eggs are fertilized internally, but the details of maturation and fertilization are not known. Fertilization is internal. The egg undergoes cleavage and develops into a 200-cell sphere, and then it goes on to undertake a process found only in stenolaemate bryozoans, producing buds that become independent, secondary embryos. These develop into free-swimming larvae and leave the gonozooid through its open neck. These settle, become ancestrules, and produce new colonies.

CONSERVATION STATUS
Not listed by the IUCN.

SIGNIFICANCE TO HUMANS
None known. ◆

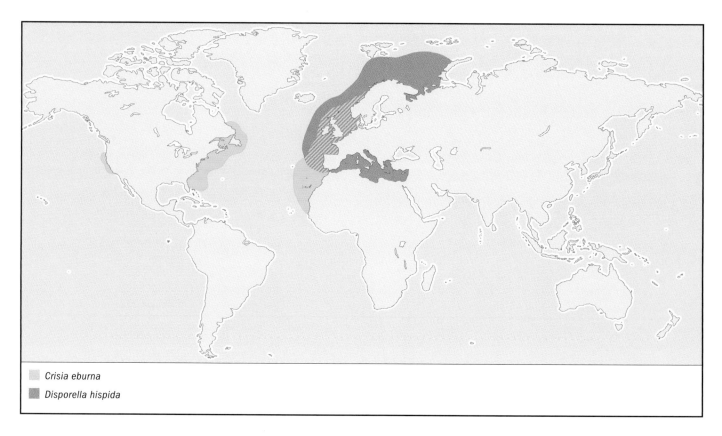

Crisia eburna
Disporella hispida

No common name
Disporella hispida

ORDER
Cyclostomata

FAMILY
Lichenoporidae

TAXONOMY
Disporella hispida Fleming, 1828.

OTHER COMMON NAMES
None known.

PHYSICAL CHARACTERISTICS
Known for wide variation in form among its colonies, which may be round or oval, flat, concave or cup-shaped, convex or mound-shaped; or colonies may form as sheets with several centers of growth. Colonies often form daughter colonies at their peripheries, while separate colonies may grow into and merge with one another. The calcified skeletons have complex forms. Autozooids in the colonies are arranged in rows radiating from the circular center.

DISTRIBUTION
Throughout the North Atlantic from Barents Sea to the eastern Mediterranean on the European side, and from Arctic Canada to Florida on the North American side.

HABITAT
Lives in offshore marine waters, to the edge of the continental shelf, at home on a wide array of substrates, including algae, rocks, shells, crustacean exoskeletons, and even other bryozoans.

BEHAVIOR
Stationary and passive, feeding on marine microorganisms.

FEEDING ECOLOGY AND DIET
Feeds on marine microorganisms by means of its autozooids.

REPRODUCTIVE BIOLOGY
Follows the general pattern for stenolaemate bryozoa.

CONSERVATION STATUS
Not listed by the IUCN.

SIGNIFICANCE TO HUMANS
None known. ◆

No common name
Idmidronea atlantica

ORDER
Cyclostomata

FAMILY
Tubuliporidae

TAXONOMY
Idmidronea atlantica Johnston, 1847.

OTHER COMMON NAMES
None known.

PHYSICAL CHARACTERISTICS
Colonies are erect with dichotomous, symmetrical branching, forming fanlike structures, but shapes and structures of

Idmidronea atlantica

individual colonies vary according to the immediate environment, also showing palmlike and elongated, irregularly branched forms. Zooids are placed alternately along axes of skeletal tubes.

DISTRIBUTION
Found from northern European Atlantic Ocean south to and throughout the Mediterranean; also in the Gulf of Mexico. Its full range is still unknown.

HABITAT
Marine species, living offshore on rocky and broken-shell gravel substrates.

BEHAVIOR
Sedentary, feeding on marine microorganisms by means of autozooids.

FEEDING ECOLOGY AND DIET
Feeds on marine microorganisms by means of autozooids.

REPRODUCTIVE BIOLOGY
Follows the general stenlaemate pattern, as exemplified in *Crisia eburna*.

CONSERVATION STATUS
Not listed by the IUCN.

SIGNIFICANCE TO HUMANS
None known. ◆

Resources

Books

Boardman, Richard S. *Reflections on the Morphology, Anatomy, Evolution, and Classification of the Class Stenolaemata (Bryozoa).* Smithsonian Contributions to Palaeobiology 86. Washington, DC: Smithsonian Institution Press, 1998.

Hayward, P. J., and J. S. Ryland. *Cyclostome Bryozoans: Keys and Notes for the Identification of the Species.* Leiden, The Netherlands: The Linnean Society of London, 1985.

McKinney, F. K, and J. B. C. Jackson. *Bryozoan Evolution.* Boston: Unwin Hyman, 1989.

Woollacott, R. M., and R. L. Zimmer, eds. *Bryozoans.* New York: Academic Press, Inc., 1977.

Periodicals

Faurie, A. S., E. R. Dempster, and M. R. Perrin. "Physiology and Ecology of Marine Bryozoans." *Advances in Marine Biology* no. 14 (1976): 285–443.

McKinney, F. K. "Feeding and Associated Colonial Morphology in Marine Bryozoans." *Reviews in Aquatic Sciences* no. 2 (1989): 255–280.

Ostrovsky, A. N. "Rejuvenation in Colonies of Some Antarctic Tubuliporids (Bryozoa: Stenolaemata)." *Ophelia* 46, no. 3 (1997): 175–185.

Organizations

International Bryozoology Association. E-mail: msj@nhm .ac.uk Web site: <http://www.nhm.ac.uk/hosted_sites/ iba/assoc.html>

Other

"Bryozoa Home Page." July 3, 2003 [July 30, 2003]. <http:// www.civgeo.rmit.edu.au/bryozoa/default.html>.

"Bryozoa Information and Links." August 6, 2001 [July 30, 2003]. <http://oliver.geology.adelaide.edu.au/grad/ rschmidt/rschmidt/rolfbryo.html>.

Kevin F. Fitzgerald, BS

Gymnolaemata
(Marine bryozoans)

Phylum Ectoprocta

Class Gymnolaemata

Number of families About 130

Thumbnail description
Colonial bryozoans, usually polymorphic zooids bearing a lophophore that protrudes through the action of muscles pulling on a frontal wall of the body

Photo: Lace corals or moss animals such as *Idodictyum* sp. belong to their own phylum of lophophorate animals (use a lophophore for feeding) with individuals living in chambers. Each animal extends out of its box to filter feed and is extremely common in all marine habitats; shown here at Heron Island, southern Great Barrier Reef, Australia. (Photo by A. Flowers and L. Newman. Reproduced by permission.)

Evolution and systematics

Uncalcified gymnolaemates are known as fossils from the late Ordovician on, almost exclusively as distinctive borings in carbonate substrates such as shells. Nonboring, noncalcified gymnolaemate bryozoans are extremely rare as fossils and are known from the Jurassic and Cretaceous only. Calcareous gymnolaemates did not appear in the oceans until the Cretaceous, during which they diversified rapidly. There were very few species of gymnolaemates in the early Cretaceous, but by the end of the period, there were more than 100 genera. Gymnolaemates continued to diversify in the Cenozoic era. There are two gymnolaemate orders—Ctenostomata and Cheilostomatida—and more than 1,000 genera. The Ctenostomata are stoloniferous or compact colonies in which the uncalcified exoskeleton is membranous, chitinous, or gelatinous and the usually terminal orifices lack an operculum. Order Cheilostomatida contains colonies composed of boxlike zooids that are adjacent but have separate calcareous walls. The orifice of cheilostomes is covered with an operculum.

Physical characteristics

Gymnolaemates are morphologically varied. The simplest have no skeletons, no polymorphs, and no change in zooidal form during colony development. The most complex gymnolaemates have elaborately calcified skeletons and at least two kinds of polymorphic zooids and display considerable variation in zooidal form during colony development (astogeny). Zooids are cylindrical or flattened. The zoecium, which covers the zooid, consists of an organic cuticle composed of protein and chitin or of cuticle overlying calcium carbonate. In many species, the zoecium is heavy and rigid. Some impregnation of the chitinous layer with calcium carbonate may be present, even when a calcareous layer is absent. An orifice enables the lophophore (food-catching organ) to protrude. The lophophore is circular and consists of a simple ridge bearing eight to 30 or more tentacles. Just within the orifice is a chamber, the atrium. An epistome and intrinsic musculature in the body wall are lacking. Protrusion of the lophophore depends on body wall deformation. Within the body wall is a large coelom surrounding the U-shaped digestive tract. The mouth at the center of the lophophore opens into the digestive tract. The anus opens through the dorsal side of the tentacular sheath outside the lophophore, hence the name Ectoprocta ("outside anus"). Interzooidal communication occurs through a funicular network of tissue-plugged pores in vertical walls.

Gymnolaemates are colonial, sessile animals. Individuals composing the colonies usually are approximately 0.02 in (0.5 mm) long.

Distribution

All seas; attached to rocks, pilings, shells, algae, and other animals in coastal waters. Some species have been found in depths as great as 26,900 ft (8,200 m).

Habitat

Any type of hard surfaces (rock, shells, coral, and wood) and Gymnolaemates are capable of using extremely restricted spaces.

Behavior

Nothing is known.

Lace coral (*Distichopora violacea*) seen off the coast of Fiji. (Photo by A. Flowers & L. Newman. Reproduced by permission.)

Feeding ecology and diet

Gymnolaemates are suspension feeders, although they sometimes use supplemental methods.

Reproductive biology

Asexual reproduction is responsible for colony growth and regeneration of zooids. Each colony begins from a single, sexually produced larva that settles and metamorphoses into a founding zooid, called the *ancestrula*. The ancestrula undergoes budding to produce a group of daughter zooids, which themselves subsequently form more buds, as do succeeding generations. Budding involves only elements of the body wall. The developing bud originates from the parent zooid. The bud initially includes only components of the body wall, or cystid, and an internal coelomic compartment. A new polypide is then generated from the living tissues of the bud, that is, the epidermis and the peritoneum. The epidermis and peritoneum invaginate, the former producing the lophophore and the gut. The peritoneum produces all of the new coelomic linings and the funiculus. Most gymnolaemates are hermaphroditic. Testes and ovaries develop either within the same zooid (zooidal hermaphroditism) or in different zooids within the same colony (zooidal gonochorism). In species that exhibit zooidal gonochorism, the colonies may be protandrous, protogynous, or simultaneous hermaphrodites, and the male and female zooids usually exhibit sexual dimorphism. Sexual polymorphism in hermaphroditic zooids

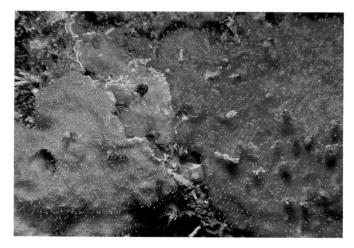

A variety of lace corals with their zooids extended, growing under a ledge near Heron Island, Great Barrier Reef. (Photo by A. Flowers & L. Newman. Reproduced by permission.)

usually is a seasonal specialization involving the polypide. In gonochoric zooids, sexual polymorphism is permanent and involves specialization of the cystid, zoecium, and sometimes the polypide.

Gymnolaemates have a variety of brooding methods, usually involving formation of an external brooding area called an *ovicell*, or *ooecium*.

Conservation status

No species are listed by the IUCN.

Significance to humans

Members of this class are used for medicinal and research purposes.

1. *Bugula turbinata*; 2. Sea mat (*Electra pilosa*). (Illustration by Bruce Worden)

Species accounts

No common name
Bugula turbinata

ORDER
Cheliostomata

FAMILY
Bugulidae

TAXONOMY
Bugula turbinata Alder, 1857

OTHER COMMON NAMES
None known.

PHYSICAL CHARACTERISTICS
Bugula turbinata forms an erect, orange to brown, tufted colony approximately 1.25–2.5 in (3–6 cm) high. The branches are arranged spirally around the main axis. There are two proximal rows of zooids and three or four distal rows. Individual zooids are rectangular, narrowing slightly at the proximal end and bearing a single short spine at each corner of the distal end. The front of the zooid is almost entirely membranous. The polypide bears 13 tentacles. Short, plump avicularia originate just below the spines. These structures resemble birds' heads with rectangularly hooked beaks. Inner avicularia are smaller than marginal avicularia. Brood chambers (ooecia) are globular and conspicuous. Colonies are attached to the substratum by extensions of the basal zooids (rhizoids).

DISTRIBUTION
Britain to Mediterranean Sea (eastern Atlantic).

HABITAT
Walls of gullies. Under boulders on lower shore and on bedrock, boulders, stones, and shells in the shallow subtidal zone.

BEHAVIOR
Nothing is known.

FEEDING ECOLOGY AND DIET
Phytoplankton, macroalgal spores, detritus, and bacteria.

REPRODUCTIVE BIOLOGY
Reproduction is by protogynous hermaphroditism. The developmental mechanism is lecithotrophic viviparity (no care). *Bugula* species are placental ovicell brooders, producing small embryos that are brooded in conspicuous hyperstomial ovicells. The size of embryos increases considerably during development owing to nutrition derived from the inside of the ovicell, which functions as a placenta.

CONSERVATION STATUS
Not listed by the IUCN.

SIGNIFICANCE TO HUMANS
Medicinal and research use. ◆

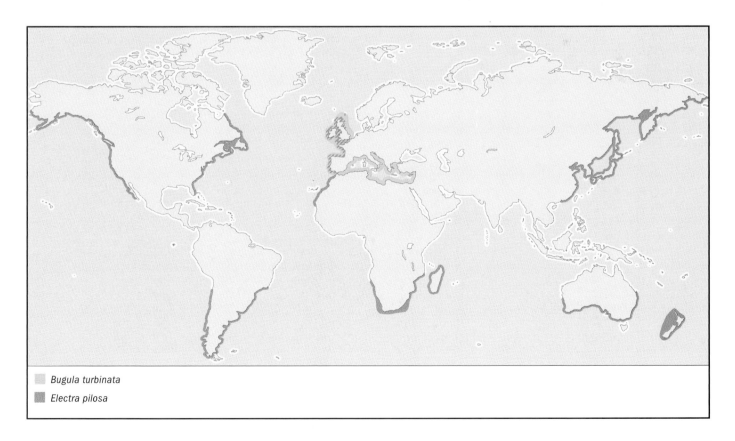

Bugula turbinata
Electra pilosa

Sea mat
Electra pilosa

ORDER
Cheliostomata

FAMILY
Electridae

TAXONOMY
Electra pilosa Linnaeus, 1767

OTHER COMMON NAMES
None known.

PHYSICAL CHARACTERISTICS
A calcareous encrusting species, sea mats form star-shaped or broad sheet colonies on the fronds of large algae, small irregular patches on stones and shells, narrow tufts (independent of the substratum), or cylindrical encrustations around the fronds of small red algae. The zooids are ovate to oblong. Approximately one half of the front of the zooid is calcified but translucent and perforated by large pores that leave a distal oval, membranous, frontal area surrounded by four to 12 (often nine) spines. Spines vary in length, but the median, proximal spine is always present and usually larger than the others. In some cases the proximal spine becomes well developed and longer than the zooid, giving the colony a hairy appearance.

DISTRIBUTION
All temperate seas; cosmopolitan.

HABITAT
A variety of substrata in marine habitats from low water into the shallow sublittoral zone (living at maximum water depths of approximately 164 ft [50 m]). Particularly common on macroalgae, such as *Fucus serratus* and laminaran kelp. Sea mats also encrust shellfish such as mussels and are a common fouling organism.

BEHAVIOR
Nothing is known.

FEEDING ECOLOGY AND DIET
Phytoplankton, algal spores.

REPRODUCTIVE BIOLOGY
Reproduction is by budding and hermaphroditism. The developmental mechanism is planktotrophic. The maternal lophophore may actively collect sperm. The ovaries produce up to 31 oocytes, which are released into the coelomic cavity. Fertilization is internal. Eggs come into contact with sperm (either as aggregates or singly) in the coelomic cavity, and fertilization occurs at or near ovulation. Embryos are shed into the water and develop into planktonic cyphonautes larvae.

CONSERVATION STATUS
Not listed by the IUCN.

SIGNIFICANCE TO HUMANS
Nothing is known. ◆

Resources

Books

Brusca, Richard C., and Gary J. Brusca. *Invertebrates.* Sunderland, MA: Sinauer Associates, 1990.

McKinney, F. K., and J. B. C. Jackson. *Bryozoan Evolution.* Chicago: University of Chicago Press, 1989.

Reed, Christopher G. "Bryozoa." In *Reproduction of Marine Invertebrates.* Vol. 6, edited by A. C. Giese, J. S. Pearse, and V. B. Pearse. Pacific Grove, CA: Boxwood Press, 1991.

Ryland, J. S. *Bryozoans.* London: Hutchinson University, 1970.

Ryland, J. S., and P. J. Hayward. "British Anascan Bryozoans." In *Synopses of British Fauna.* London: Academic Press, 1977.

Periodicals

Bayer, M. M., and C. D. Todd. "Evidence for Zooid Senescence in the Marine Bryozoan *Electra pilosa.*" *Invertebrate Biology* 116 (1997): 331-340.

Hermansen, P., P. S. Larsen, and H. U. Riisgard. "Colony Growth Rate of Encrusting Marine Bryozoans (*Electra pilosa* and *Celleporella hyalina*)." *Journal of Experimental Marine Biology and Ecology* 263 (2001): 1-23.

Other

Bryozoan Orders. RMIT University, Civil and Geological Engineering. 6 July 2000 [7 July 2003]. <http://www .civgeo.rmit.edu.au/bryozoa/orders.html>.

"Cheilostome Bryozoans Taxonomic List." *NMITA: Neogene Biota of Tropical America.* Department of Geoscience, University of Iowa. 5 Aug. 1998 [7 July 2003]. <http:// porites.geology.uiowa.edu/database/bryozoa/systemat/ bryotaxa.htm#cupula>.

Tyler, W. "A Sea Mat: *Electra pilosa.* " *MarLIN: The Marine Life Information Network for Britain and Ireland.* Marine Biological Association of the United Kingdom. 2003 [7 July 2003].<http://www.marlin.ac.uk/species/ Elepil.htm>.

Tyler-Waters, H. "An Erect Bryozoan: *Bugula turbinata.*" *MarLIN: The Marine Life Information Network for Britain and Ireland.* Marine Biological Association of the United Kingdom. 2002 [7 July 2003]. <http://www.marlin.ac.uk/ index2.htm?species/Bugtur.htm>.

Michela Borges, MSc

Inarticulata
(*Nonarticulate lampshells*)

Phylum Brachiopoda
Class Inarticulata
Number of families 3

Thumbnail description
Exclusively marine group of lophophorate animals that are suspension feeders attached at the base to the ocean bottom; they are called "inarticulated" because their shells lack articulation

Photo: A brachiopod lampshell of the genus *Lingula*, showing its stalk and shelled body. (Photo by Biophoto Associates/Photo Researchers, Inc. Reproduced by permission.)

Evolution and systematics

Nonarticulate lampshells, also known as inarticulated brachiopods, have been known since the Lower Cambrian period, about 550–600 million years ago. The traditional placement of these animals is in the class Inarticulata, but a more recent classification divides them into two subphyla: the Linguliformea and the Craniiformea. The living representatives of the Linguliformea are divided into two families: the Lingulidae with two genera, *Lingula* and *Glottidia*; and the Discinidae with four extant genera, *Discina*, *Discinisca*, *Discradisca*, and *Pelagodiscus*. The Craniiformea have one family, the Craniidae, with four genera: *Craniscus*, *Neoancistrocrania*, *Neocrania*, and *Valdiviathyris*. The inarticulated brachiopods are related to the Phoronida (horseshoe worms).

Physical characteristics

The shells of organisms in this group may grow as large as 2.75 in (7 cm) in the lingulides and 0.78 in (2 cm) in the discinids and craniids. These shells have a ventral and a dorsal valve; the muscles that close the shell include single or paired posterior adductor muscles as well as paired anterior adductor muscles. The oblique musculature, generally three or four pairs of muscles, controls the rotations and sliding movements of the valves. The lateral muscles of the trunk wall control the hydraulic mechanism that opens the shell. This mechanism involves changes in the pressure inside the organism's coelom, or body cavity. The lophophore or feeding organ is spirolophous (has elongated and coiled lateral lobes), except in *Pelagodiscus*, which has a schizolophous (lobes with relatively few filaments along its edges) lophophore. The lophophore has no supportive skeleton. The digestive tract ends in an anus. The ring of nerves around the animal's esophagus has either a single or paired dorsal and ventral ganglion (group or cluster of nerve cells).

Distribution

Lingulidae are exclusively infaunal (living in the sediment at the bottom of the ocean) in soft substrates from the intertidal zone to a depth of about 1,315 ft (400 m); they have a worldwide distribution within the 40° belt from temperate to equatorial areas. *Glottidia* occurs only in the Americas and *Lingula* in the waters of the other continents. The Discinidae are epifaunal (living on top of the bottom surface) on hard substrates and have a warm temperate to tropical distribution. They are found mainly on the continental shelf, except for the cosmopolitan deep-sea *Pelagodiscus*, which occurs from northern to southern high latitudes at depths between about 328 ft to about 18,044 ft (100 m to about 5500 m). The Craniidae are epifaunal on hard substrates and are widely distributed from northern to southern high latitudes, from shallow waters to a depth of about 7,546 ft (2,300 m) on the bathyal slope.

The life span of most animals in this group of brachiopods appears to be from 14 months to less than two years for *Glottidia*, to 6–10 years for *Lingula* and the discinids.

Habitat

The lingulides live in vertical burrows built within compact and stable sandy sediments under the influence of moderate water currents close to the bottom of the sea. The total length of the burrow is about ten times the length of the shell.

The distal bulb of the pedicle (footlike base of the organism) is firmly anchored into the substrate at the bottom of the burrow and is surrounded by a mass of sand and organic particles held together by sticky mucus.

The discinids live on hard substrates swept by currents under weak sedimentation. They occur singly or in clusters of many individuals attached to rocky surfaces or the shells of mollusks and brachiopods.

The craniids lack a pedicle; their ventral valves are cemented directly to hard substrates.

Behavior

Lingulides position their shells at the top of their burrows. The shell is held there by hydrostatic pressure exerted by the body cavity and the valves on the walls of the burrow; the pedicle plays no role in keeping the animal in this position. Lingulides cannot live in coarse or muddy substrata because the walls of the burrow do not support the shell in its normal filtering position. Rapid withdrawal into the burrow is an escape reflex or protective reaction that may be observed if the setae (bristles) on the front margin of the animal are touched or if there is a sudden change in light intensity.

The ventral valve of the discinids is always oriented towards the surface on which it lives. This positioning is related to the orientation of the larva when it settles on the substrate. Discinids are attached by a highly muscular pedicle to hard substrates.

The craniids are generally gregarious (living in groups), preferring to live on hard flat surfaces. They are cemented directly to the substrate by their ventral valves.

Feeding ecology and diet

Brachiopods are suspension feeders. The lophophore, which is suspended freely in the mantle cavity, functions as a feeding organ in capturing suspended particles. It also maintains the supply of oxygen to and removal of waste products from the mantle cavity by internal currents produced by the beating of its cilia.

Food items include plankton (mainly phytoplankton as diatoms); superficial meiobenthos (microscopic metazoans); colloidal (suspended) organic material; and dissolved organic material. This group of brachiopods also derives nourishment from direct absorption of dissolved nutrients.

Reproductive biology

The onset of breeding and the length of the spawning season depend primarily on water temperature, together with latitudinal and seasonal effects. In temperate waters, the breeding period takes place in midsummer and lasts about a month and a half; in the tropics, however, breeding may take place throughout the year.

Inarticulated brachiopods are dioecious. The sex ratio, at least in lingulides, is one to one. There is a single continuous gonad mass on each side of the visceral cavity in lingulides. Discinids have two gonads in the posterior part of the visceral cavity; males sometimes possess an additional single U-shaped testis. In the craniids, the gonads are in six separate groups, two in the visceral cavity and two in each lobe of the mantle. Fertilization is external in the lingulides, with synchronization of spawning in the two sexes. Ova and sperm are discharged through the metanephridia (primitive excretory organs) acting as gonoducts into the current of water that leaves through the lophophore. The occurence of fertilization is unknown in the discinids and craniids.

The larvae in lingulides and discinids have two parts: the lobe at the upper end or apex, which will form the future body; and the mantle lobe, which will develop into the future mantle lobes. The pedicle arises from the ventral mantle lobe during the larva's long planktotrophic stage. The larva grows in size and complexity as it feeds on phytoplankton. The duration of this stage ranges from 3–6 weeks. The lecithotrophic larva of *Neocrania* has a short larval stage of about 4–6 days after fertilization.

There is no true metamorphosis in the inarticulated brachiopods. The larval organs either atrophy (wither away) or are detached after the animal settles on the substrate.

Conservation status

No species are listed by the IUCN.

Significance to humans

Lingula species are eaten in the eastern Pacific islands, from Japan to New Caledonia. Specimens can be purchased at local markets.

1. *Glottidia pyramidata*; 2. *Neocrania anomala*. (Illustration by Bruce Worden)

Species accounts

No common name
Neocrania anomala

ORDER
Craniida

FAMILY
Craniidae

TAXONOMY
Neocrania anomala (Müller), 1776, North Atlantic Ocean.

OTHER COMMON NAMES
None known.

PHYSICAL CHARACTERISTICS
The shell is calcareous, punctuated, and rarely exceeds 0.78 in (20 mm) in width. The ventral valve is cemented directly to a hard substrate, the dorsal valve is smooth and conical. There is no pedicle. The lophophore is spirolophous with the spires directed upward; the main mantle canals (*vascula lateralia*) branch and radiate outward.

DISTRIBUTION
Subtidal to bathyal zone in the eastern North Atlantic Ocean and in the Mediterranean Sea. Known since the Pliocene period (1.8 million years ago).

HABITAT
This species lives on boulders and various hard substrates, cemented directly to them by its ventral valve. It is sometimes found in underwater caves and other cryptic (hidden) habitats. Its depth ranges from about 9.8 ft (3 m) to 4,920 ft (1,500 m).

BEHAVIOR
The ventral valve always faces toward the substrate and is directly cemented to it. *Neocrania* is generally gregarious and prefers rather flat hard surfaces.

FEEDING ECOLOGY AND DIET
Neocrania species are suspension feeders. The spirolophous lophophore collects food items—primarily phytoplankton, diatoms, superficial meiobenthos, suspended organic material, and dissolved organic material.

REPRODUCTIVE BIOLOGY
The eggs are freely spawned; they undergo radial cleavage and embolic gastrulation. The body cavity develops from a modification of the embryonic gut. The larvae are lecithotrophic (live on stored nutrients) and can settle immediately after complete embryogenesis in about a week.

CONSERVATION STATUS
Not listed by IUCN.

SIGNIFICANCE TO HUMANS
None known. ◆

☐ *Glottidia pyramidata*
■ *Neocrania anomala*

No common name
Glottidia pyramidata

ORDER
Lingulida

FAMILY
Lingulidae

TAXONOMY
Glottidia pyramidata (Stimpson, 1860), Beaufort, South Carolina, United States.

OTHER COMMON NAMES
None known.

PHYSICAL CHARACTERISTICS
The phosphatic shell is roughly circular, oblong, or oval in shape; it may grow as large as 1.2 in (3 cm). The single septum (dividing wall) in the dorsal valve reaches about 25–30% of the valve length and the two divergent septa in the ventral valve about 30–38%. The mantle papillae that are characteristic of this genus occur along the main mantle canals.

DISTRIBUTION
Littoral zone from the French West Indies to the eastern coast of Virginia.

HABITAT
Glottidia lives vertically in fine sand, sometimes covered by seagrass, in a burrow in which it is anchored by its pedicle. It is found in shallow water from the intertidal zone to about 246 ft (75 m).

BEHAVIOR
The animal lives in the upper part of its burrow. Only the three aligned pseudosiphons formed by the setae of the anterior mantle edges are visible on the surface of the sediment: the two lateral openings take in water and the central opening discharges it. These setae are sensitive to touch; they trigger a protective closure of the shell accompanied by contractions of the pedicle, which draws the shell downwards into the burrow.

FEEDING ECOLOGY AND DIET
Food particles enter through the two intake currents and are filtered by the lophophore. These food particles consist of plant and animal matter, phytoplankton, and gastropod larvae. Waste products are discharged into the outgoing current of water.

REPRODUCTIVE BIOLOGY
The reproductive period extends over 7–9 months. *Glottidia* are dioecious, fertilization is external and eggs undergo equal and radial cleavage. Hatching occurs when the embryo has developed the first tentacles. The larva is planktotrophic for about 20 days, after which it settles to the bottom for a sedentary infaunal mode of life.

CONSERVATION STATUS
Not listed by IUCN.

SIGNIFICANCE TO HUMANS
None known. ◆

Resources

Books

Emig, C. C. "Biogeography of the Inarticulated Brachiopods." In *Treatise on Invertebrate Paleontology*, Part H revised. *Brachiopoda*, vol. 1, edited by R. L. Kaesler. Boulder, CO, and Lawrence, KS: Geological Society of America and University of Kansas, 1997.

———. "Ecology of the Inarticulated Brachiopods." In *Treatise on Invertebrate Paleontology*, *Part H revised, Brachiopoda*, vol. 1, edited by R. L. Kaesler. Boulder, CO, and Lawrence, KS: Geological Society of America and University of Kansas, 1997.

Williams, M. A. James, C. C. Emig, S. Mackay, and M. C. Rhodes. "Anatomy." In *Treatise on Invertebrate Paleontology, Part H revised, Brachiopoda*, vol. 1, edited by R. L. Kaesler. Boulder, CO, and Lawrence, KS: Geological Society of America and University of Kansas, 1997.

Periodicals

Blochmann, F. "Untersuchungen über den Bau der Brachiopoden." Die Anatomie von *Crania anomala* O. F. M." *Bibliographia Zoologica, Jena* 1 (1892): 1–65, pl. 1–7.

Chuang, S. H. "Observations on the Ciliary Feeding Mechanism of the Brachiopod *Crania anomala*." *Journal of Zoology, London* 173 (1974): 441–449.

Cohen, B. L., A. B. Gawthrop, and T. Cavalier-Smith. "Molecular Phylogeny of Brachiopods and Phoronids Based on Nuclear-Encoded Small Subunit Ribosomal RNA Gene Sequences." *Philosophical Transactions of the Royal Society B* 353 (1998): 2039–2061.

Emig, C. C. "Taxonomie du genre *Glottidia* (Brachiopodes Inarticulés)." *Bulletin du Muséum National d'Histoire naturelle de Paris* 4, no. 5, sect. 4 (1983): 469–489.

———. "Taxonomie du genre *Lingula* (Brachiopodes, Inarticulés)." *Bulletin du Muséum National d'Histoire naturelle de Paris* 4, no. 4, sect. A (1982): 337–367.

———. "Terrier et position des Lingules (Brachiopodes, Inarticulés)." *Bulletin de la Société zoologique de France* 107 (1982): 185–194.

Freeman, G. "Regional Specification During Embryogenesis in the Craniiform Brachiopod *Crania anomala*." *Developmental Biology* 227 (2000): 219–238.

———. "Regional Specification During Embryogenesis in the Inarticulate Brachiopod *Glottidia*." *Developmental Biology* 172 (1999): 15–36.

Lee, D. E., and C. H. C. Brunton. "*Neocrania* n. gen., and a Revision of Cretaceous-Recent Brachiopod Genera in the Family Craniidae." *Bulletin of the British Museum of Natural History (Geology)* 40 (1986): 141–160.

Paine, R. T. "Ecology of the Brachiopod *Glottidia pyramidata*." *Ecological Monographs* 33 (1963): 187–213.

Other

"Brachiopoda and Brachiopodologists." *BrachNet*. 22 July 2003 (23 July 2003). <http://paleopolis.rediris.es/BrachNet/>.

Christian C. Emig, Dr es-Sciences

Articulata

(*Articulate lampshells*)

Phylum Brachiopoda

Class Articulata

Number of families 20

Thumbnail description
Brachiopods that live within a rounded, hinged, and mostly calcareous shell composed of two bilaterally symmetrical but dissimilar valves, and that generally attach themselves to hard substrates with a pedicle (foot-like structure) supported by connective tissue

Photo: Lampshell larvae seen near Heron Island in the Great Barrier Reef. (Photo by A. Flowers & L. Newman. Reproduced by permission.)

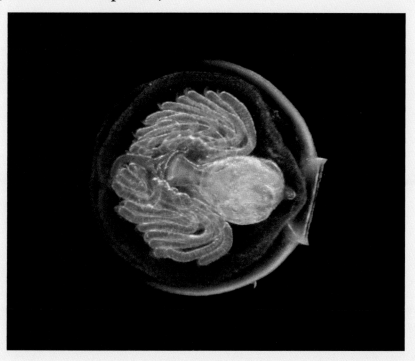

Evolution and systematics

Lampshells were a dominant life form during the Cambrian period (570–500 million years ago [mya]). They flourished during the Paleozoic (570–240 mya) and Mesozoic eras (240–65 mya). The brachiopods reached the zenith of their evolutionary diversification during the Ordovician period (500–435 mya). At that time, the articulate lampshells appeared and underwent their most important phase of evolution. Since the close of the Paleozoic era, however, when their populations were decimated by mass extinctions from unknown causes, their numbers have been steadily decreasing.

Fossil remains have shown that articulates were once widely distributed and abundant throughout the marine world. Although over 30,000 species are known from fossil records between the Cambrian period and the present day, only 250 to 325 species are thought to be extant. The majority of present-day brachiopod species are assigned to the class Articulata rather than the class Inarticulata. The two classes are distinguished primarily by the way in which the two valves are attached along the rear line of the organism, and their method of contact. As of 2003, there are three living orders of articulates and four extinct ones.

Physical characteristics

Most articulate lampshells are less than 1 in (25 mm) across; in some instances, however, they can grow as much as 2 in (50 mm) in width. They contain two very conspicuous hinged valves (or shells) that are composed of scleroproteins and calcite (a form of calcium carbonate). The shells are segmented in a radial pattern; for the most part, they are usually brown in color, or sometimes greenish when surrounded by algae. Only a few lampshells are naturally white. The upper (dorsal) valve is smaller than the lower (ventral) valve. The valves are connected by a joint that can articulate or pivot when teeth located on the ventral valve are inserted into sockets on the dorsal valve along a hinge line. In this manner, the shell can close in front of the hinge and open from behind the hinge, thus locking itself at the rear. The hinge restricts the organism's movements to simple opening and closing. Only a slight anterior gap of about 10° is apparent when the valves are locked in place. This gap at the apex of the two shells resembles the shape of the spout of ancient Roman and Greek oil lamps—whence the English name "lampshell."

The space inside the shell is divided approximately into two halves, the mantle cavity and the body or coelomic cavity. The body of an articulate lies between the shell valves, sheathed in mantle tissue that secretes the shell valves. The coelom lies between the body wall and the gut, and surrounds the lophophore and its tentacles. The lophophore is a crown-shaped structure of hollow tentacles with a pair of arms or brachia (from which the name *brachiopod* is derived), one brachium on each side of the mouth. The lateral arms of the lophophore are partially supported by a cartilaginous axis on the dorsal valve and a fluid-filled canal within each brachium. The arms climb in a spiral pattern into each half of the mantle cavity.

Northern lampshells (*Terebratulina septentrionalis*) seen in the Gulf of Maine. (Photo by Andrew J. Martinez/Photo Researchers, Inc. Reproduced by permission.)

Articulate lampshells have three sets of muscles that control the movements of their valves. The largest are the adductor muscles, used to close the shell. These muscles usually have two, occasionally four, closely spaced foundations on the ventral valve and four distinctly separated foundations on the dorsal valve. The posterior adductors close the shell rapidly, while the anterior adductors hold the shell tightly closed over longer periods of time. The diductor muscles are used to open the shell for feeding; they have two large foundations on the ventral valve and an attachment area on the rear tip of the dorsal valve. The adjustor muscles, positioned for moving the shell relative to the pedicle, have two attachments near the posterior end of each valve. The pedicle itself is a cylindrical stalk of horny material used to anchor the organism on such substrates as pebbles, stone particles, or coral. The pedicle in articulates ranges from absent in a few species to short and nonmuscular to very long, flexible, muscular, and branched. The pedicle does not expand where it leaves the shell but emerges through a slit or notch on the ventral valve. The pedicle lacks internal muscles and a coelomic lumen; is supported by connective tissue; and is composed of short papillae on its distal end that may branch, wrap around a substrate, or penetrate a soft substrate with acid secretions.

The short U-shaped intestine of articulates begins at the lower end of the stomach and ends in a blind pouch with no anus. The excretory system consists of segmental organs known as metanephridia, which are short tubes ending in large nephrostomes. Articulate brachiopods have a simple circulatory system containing colorless blood with only a few cells. The system includes a dorsal vessel that functions as a primitive pumping heart. The simple nervous system consists of a small dorsal ganglion, or group of nerve cells, behind the mouth and a large ventral ganglion in front of it. A nerve ring around the esophagus connects the various parts of the nervous system, including a nerve running into each tentacle. Articulates do not possess special sense organs, but the mantle margin is most likely an important area of sensory reception.

The setae (bristles) on the mantle are thought to transmit stimuli to receptors in the mantle epidermis.

Distribution

Articulate lampshells are scattered around the world in marine environments, mostly in colder waters. Where they are found, they usually occur in dense numbers.

Habitat

Lampshells are found at depths ranging from the intertidal zone (between high and low tide levels in coastal areas) to the deep sea as far as 17,410 ft (5,300 m). Most articulate lampshells live in moderately deep water as low as 1,500 ft (450 m) in polar waters and cryptic (hidden) environments. They are classified as epifaunal; that is, they live on the sea floor or attached to other animals or objects underwater. Most species live on rocks or other solid substrates, but some dig vertical burrows in sandy or muddy bottoms.

Bchavior

Articulate lampshells are sedentary or sessile in their lifestyle, attaching themselves to a substrate by the use of a cord-like pedicle on their lower valves. The lower valve is usually positioned on top. Lampshells with pedicles tend to orient themselves with regard to the water so that the front-to-back axis of the shell is at right angles to the current, with the commissural plane parallel to the current. The commissure is the line of junction between the edges of two valves. The few species of articulates that lack pedicles attach themselves to the substrate by the lower valve.

Articulate lampshells protect themselves by closing their shells. When disturbed, they will also contract their pedicles and pull themselves downward toward the substrate. They are capable of only limited lateral motion. Most of their movement involves opening the shell for feeding and closing it for protection.

Feeding ecology and diet

Articulates are generally suspension feeders, taking in minute particles of nourishment from the water. The lophophore is primarily a feeding and respiratory organism. To feed, the lampshell valves open a small gap in the front to allow water to flow over the lophophore. The hollow tentacles, with cilia on the outside, can reach to the front of the valves, while being held near the upper valve. Suspended microorganisms and food particles are trapped in mucus-covered loops and swept by the beating cilia with the intake current down to the mouth through a brachial groove. The mouth leads into an esophagus and stomach, which digests the food. Coelomocyte cells collect nitrogenous material throughout the body and release these wastes into the nephridia, or simple kidneys. The two nephridia have a thin duct leading into the mantle cavity from the coelom so that the wastes can be passed into special outflow currents that carry them away through the animal's mouth. The continual flow of water over the lophophoral arms allows for respiratory exchange of oxygen and carbon dioxide.

Reproductive biology

Most articulates have separate sexes. In addition, most species have no special breeding season, although some breed only at certain times of the year. There are usually four gonads, which are masses of developing gametes located beneath the mesothelium of the coelom. The ova or sperm are found in patches on the coelomic epithelium. They pass into the coelom and shell cavity through tubes in the segmental organs. In most species, large numbers of gametes are produced and discharged into the water, where fertilization takes place; in a few species, however, embryonic development takes place within the parent's shell, sometimes in a brood pouch.

Free-swimming larvae develop from the fertilized ova, propelling themselves through the water by rows of cilia on their three-lobed bodies. After a few days they sink to the bottom and undergo metamorphosis. At this point, the larvae have three developing regions: body, mantle, and pedicle lobes. No shell appears until the post-larval stage. The posterior lobe becomes elongated into a stalk, the middle lobe secretes a shell, and the remaining lobe forms the lophophore.

Conservation status

Articulate lampshells have diminished dramatically from the height of their diversity over 450 million years ago. Extant species are not, however, considered threatened. No species are listed by the IUCN.

Significance to humans

Articulate brachiopods have little economic significance but are important in understanding the fossil record and the evolutionary history of invertebrates.

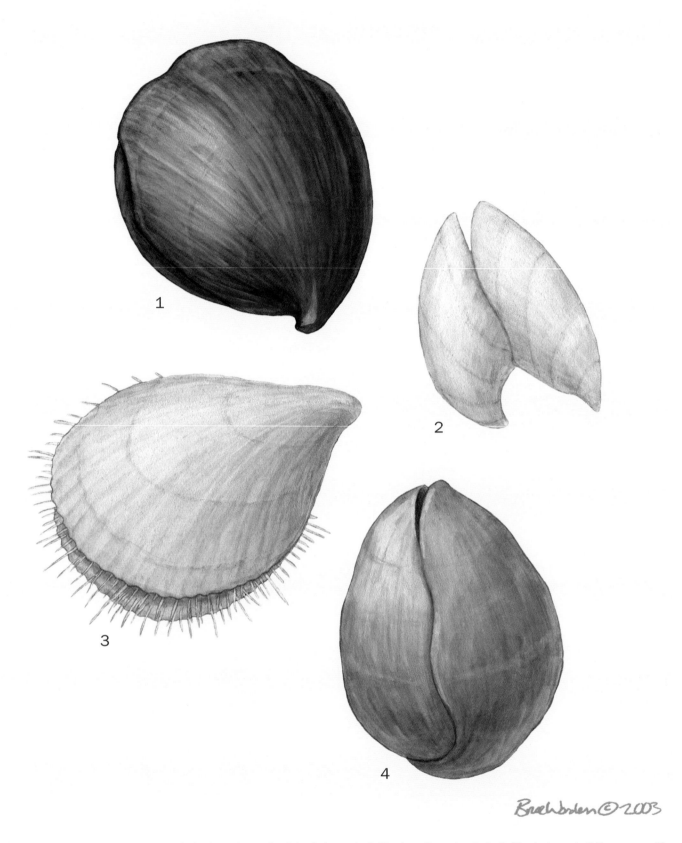

1. Black lampshell (*Hemithyris psittacea*); 2. *Argyrotheca cistellula*; 3. Lampshell (*Terebratulina retusa*); 4. California lampshell (*Laqueus californianus*). (Illustration by Bruce Worden)

Species accounts

Black lampshell
Hemithyris psittacea

ORDER
Rhynchonellida

FAMILY
Hemithyrididae

TAXONOMY
Hemithyris psittacea Gmelin, 1791.

OTHER COMMON NAMES
None known.

PHYSICAL CHARACTERISTICS
Black lampshells possess a shell that is small to medium in size, with a maximum adult width of 1.38 in (35 mm). The shell is moderately thick with a highly convex shape and a brownish-purple color. The ventral valve narrows to a long and curved beak. There are two spirolophous lophophore spirals, which are dorsally directed and partially supported by crura (hard processes extending forward from the socket region of the dorsal valve) and two pairs of metanephridia. The dorsal floor contains a low median ridge. The slender curved radulifer crura are attached to small outer hinge plates. The pedicle is somewhat short, almost as broad as the foramen, and not used for locomotion. The intestine is curved distally and is enlarged at its end. The adductor muscles possess two attachments on the ventral valve. The mantle canals contain two primary trunks in each mantle lobe. There are no spicules in the soft parts of this lampshell.

DISTRIBUTION
Circumpolar, primarily in the Pacific Ocean and Atlantic Ocean.

HABITAT
Black lampshells inhabit areas with tropical to cold surface waters. They are found at depths ranging from intertidal zones at about 33 ft (10 m) to deeper waters at about 4,265 ft (1,300 m).

BEHAVIOR
Sessile lifestyle; generally attached to substrates by a pedicle.

FEEDING ECOLOGY AND DIET
Feed on dissolved organic matter.

REPRODUCTIVE BIOLOGY
The posterior portion of each mantle lobe contains a specialized pair of pillared areas for the gonads.

CONSERVATION STATUS
Not listed by the IUCN.

SIGNIFICANCE TO HUMANS
None known. ◆

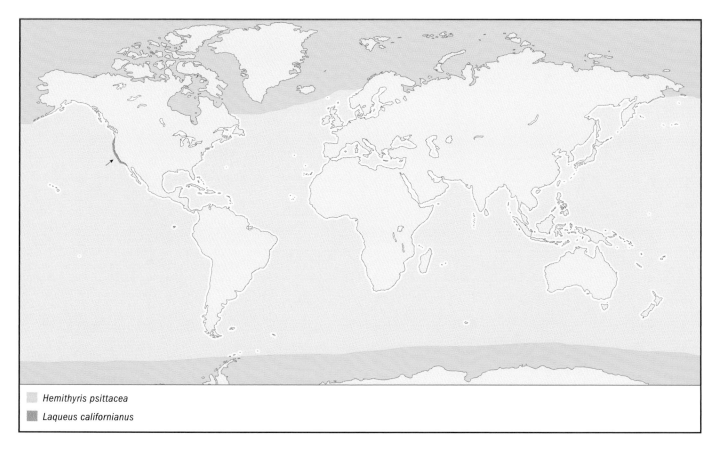

☐ *Hemithyris psittacea*
■ *Laqueus californianus*

No common name
Argyrotheca cistellula

ORDER
Terebratulida

FAMILY
Argyrotheca

TAXONOMY
Argyrotheca cistellula Searles-Wood, 1841.

OTHER COMMON NAMES
None known.

PHYSICAL CHARACTERISTICS
Argyrotheca cistellula has a shell width of about 0.59 in (15 mm). The shell contains a yellowish brown, four-cornered ventral valve and a five-cornered dorsal valve. The lophophore loops are schizolophous, which means that they consist of lobed discs with only a small number of marginal filaments. The peduncle is very small. This species generally lacks setae.

DISTRIBUTION
Argyrotheca cistellula is found off the coast of Europe and is common in the Mediterranean Sea.

HABITAT
This species is believed to occur over a wide range of depths, but is most frequently found between 7–200 ft (2–60 m).

BEHAVIOR
Sessile lifestyle; generally attached to substrates by the pedicle.

FEEDING ECOLOGY AND DIET
Feeds on dissolved organic matter.

REPRODUCTIVE BIOLOGY
Little is known about the reproductive cycle of *A. cistellula*. It is thought to breed continuously, on the basis of the maturing eggs and larvae that are found in its brood pouches throughout the year. Each embryo goes through a gastrula stage as well as two-lobed and three-lobed stages. Before leaving the brood

pouch, the larvae have an apical lobe with a ring of long cilia, a mantle lobe with a mid-ventral strip of cilia, and a pedicle lobe that lacks cilia. The larvae lack setae. Unlike most lampshells, *A. cistellula* is hermaphroditic, which means that each individual possesses both male and female gonads.

CONSERVATION STATUS
Not listed by IUCN.

SIGNIFICANCE TO HUMANS
None known. ◆

Lampshell
Terebratulina retusa

ORDER
Terebratulida

FAMILY
Cancellothyrididae

TAXONOMY
Terebratulina retusa Linnaeus, 1758.

OTHER COMMON NAMES
None known.

PHYSICAL CHARACTERISTICS
Adults of this species are small to moderately large brachiopods, with a maximum shell width of 1.18 in (30 mm). The shell is roughly oval in shape, and the anterior commissure is slightly grooved. Externally, the shells have numerous and somewhat coarse radial ribs. These ribs become knobby, forming smooth, rounded tubercles (nodules) at the rear and side margins of both valves, although they are most developed on the sides of the ventral beak or umbo. The beak represents the initial point of growth of a valve. A wide number of concentric growth lines are present. Shell color ranges from white to yellowish gray. The small deltidial plates are separated. The species is endopunctate, which means that there are tiny canals called endopunctae extending from the inner surface of the valve almost to the outside. In *T. retusa*, the endopunctae are usually arranged radially (corresponding to the external ribs) on the inside of the valves. Spicules, or hard calcareous plates, are found in the body tissue near the lophophore as well as in the tissue covering the mantle canals. The spicules help to support the soft body tissues. The tip of the pedicle divides into short rootlets that are able to penetrate the substrate.

DISTRIBUTION
Terebratulina retusa is found in the eastern part of the Atlantic Ocean around the British Isles, excluding the eastern coast of England and the southern Irish Sea. The species also occurs in the North Atlantic from Scandinavia to the eastern coast of Greenland, and as far south as the Mediterranean Sea.

HABITAT
Terebratulina retusa is found in tropical to cold surface waters, at depths ranging from the intertidal zone (about 50 ft [15 m]) to the deep sea (about 4,850 ft [1,480 m]).

BEHAVIOR
This species is generally sedentary or sessile in its lifestyle, generally attaching itself to substrates by the use of the pedicle.

FEEDING ECOLOGY AND DIET
They primarily eat dissolved organic matter.

◼ *Terebratulina retusa*
◼ *Argyrotheca cistellula*

REPRODUCTIVE BIOLOGY
The shell of *T. retusa* often acquires a brighter color during the breeding season, as the yellow or orange of the ripe gonads in females or the cream color of the male gonads often show through the external surface. Female gonads are almost always orange, red, or brown in color, however, even when the shell has no coloration.

CONSERVATION STATUS
Not listed by IUCN.

SIGNIFICANCE TO HUMANS
None known. ◆

California lampshell
Laqueus californianus

ORDER
Terebratulida

FAMILY
Laqueidae

TAXONOMY
Laqueus californianus Koch, 1848.

OTHER COMMON NAMES
English: Smooth lampshell.

PHYSICAL CHARACTERISTICS
The shell of California lampshells is medium to moderately large in size, with an adult dimension of 0.79–1.97 in (20–50 mm). The tan-colored shell is biconvex and may be smooth or ribbed. The pedicle is short to moderately long. The loop that supports the lophophore passes through one or more bilacunar phases. The lophophore is plectolophous, which means that it has longer lateral lobes and a coiled median lobe. Spicules occasionally occur scattered over the mantle canals. Dental plates are present. The ventral median septum (partition) is weak. Hinge plates are variably developed.

DISTRIBUTION
California lampshells are found near the break between the continental shelf and slope along the western coast of the United States.

HABITAT
California lampshells live in the continental shelf zone at depths of 325–650 ft (100–200 m). They are also found in tropical to cold surface waters and in deeper waters.

FEEDING ECOLOGY AND DIET
California lampshells are believed to eat dissolved silt as well as organic matter, diatoms, dinoflagellates, and phytoplankton, but may not be limited to these items.

BEHAVIOR
California lampshells are generally sedentary or sessile in their lifestyle, attaching themselves to substrates by their pedicles. They are usually found with the ventral shell uppermost and the valves slightly apart.

REPRODUCTIVE BIOLOGY
The eggs of California lampshells are 0.00512–0.00551 in (130–140 μm) in diameter, and the sperm are unmodified. At 50°F (10°C) an embryo develops within 72 hours, and gastrulation occurs within 24–38 hours. A three-lobed larva with an attachment disk develops in about seven days. The larvae, however, die within one day at 77°F (25°C). At 68°F (20°C), development is normal but may result in abnormal settlement of larvae. At 59°F (15°C), 50°F (10°C), and 41°F (5°C), most larvae achieve competence in five, seven, and nine days, respectively. Settlement and metamorphosis can occur within one day after the larvae make contact with the substrate.

CONSERVATION STATUS
Not listed by IUCN.

SIGNIFICANCE TO HUMANS
None known. ◆

Resources

Books

Audubon Society Encyclopedia of Animal Life. New York: Clarkson N. Potter, 1982.

Banister, Keith, and Andrew Campbell. *Encyclopedia of Aquatic Life.* New York: Facts on File, 1985.

Barnes, R. S. K., et al. *The Invertebrates: A Synthesis.* Malden, MA: Blackwell Science, 2001.

Brunton, C. Howard, L. Robin M. Cocks, and Sarah L. Long, eds. *Brachiopods Past and Present.* London and New York: Taylor and Francis, 2001.

Buchsbaum, Ralph Morris. *Animals Without Backbones: An Introduction to the Invertebrates.* 2nd ed. Chicago: University of Chicago Press, 1976.

Harrison, Frederick W., and Robert M. Woollacott, eds. *Microscopic Anatomy of Invertebrates.* Vol. 13. New York: Wiley-Liss, 1997.

New Larousse Encyclopedia of Animal Life. New York: Bonanza Books, 1981.

Nybakken, James W. *Diversity of the Invertebrates: A Laboratory Manual.* Pacific Coast Version. Dubuque, IA: Wm. C. Brown Publishers, 1996.

Ruppert, Edward E., and Robert D. Barnes. *Invertebrate Zoology.* 6th ed. Fort Worth, TX: Saunders College Publishing, 1994.

Stachowitsch, Michael. *The Invertebrates: An Illustrated Glossary.* New York: Wiley-Liss, 1992.

William Arthur Atkins

For further reading

Audubon Society Encyclopedia of Animal Life. New York: Clarkson N. Potter, 1982.

Abbott, J., R. Williamson, and L. Maddock, eds. *Cephalopod Neurobiology.* New York: Oxford University Press, 1995.

Abbott, R. Tucker. *A Compendium of Landshells.* Melbourne, FL: American Malacologists, Inc., 1989.

Abramson, C. I. *Invertebrate Learning: A Laboratory Manual and Source Book.* Washington, DC: American Psychological Association, 1990.

———. *A Primer of Invertebrate Learning: The Behavioral Perspective.* Washington, DC: American Psychological Association, 1994.

Abramson, C. I., and I. S. Aqunio. *A Scanning Electron Microscopy Atlas of the Africanized Honey Bee* (Apis mellifera): *Photographs for the General Public.* Campina Grande, Brazil: Arte Express, 2002.

Abramson, C. I., Z. P. Shuranova, and Y. M. Burmistrov, eds. *Contributions to Invertebrate Behavior.* (In Russian.) Westport, CT: Praeger, 1996.

Ahmadjian, Vernon, and Surindar Paracer. *Symbiosis: An Introduction to Biological Associations.* Hanover, NH, and London: University Press of New England, 1986.

Alikhan, M. A., ed. *Crustacean Issues 9: Terrestrial Isopod Biology.* Rotterdam, The Netherlands: A. A. Balkema, 1995.

Allan, J. D. *Stream Ecology: Structure and Function of Running Waters.* New York: Chapman and Hall, 1995.

Arthur, Wallace. *The Origin of Animal Body Plans: A Study in Evolutionary Developmental Biology.* Cambridge, U.K.: Cambridge University Press, 1997.

Bacescu, M. *Cumacea I: Crustaceorum Catalogus.* Part 7. The Hague, The Netherlands: SPB Academic Publishing, 1988.

———. *Cumacea II: Crustaceorum Catalogus.* Part 8. The Hague, The Netherlands: SPB Academic Publishing, 1992.

Baer, Jean G. *Animal Parasites.* London: World University Library, 1971.

Banister, Keith, and Andrew Campbell. *Encyclopedia of Aquatic Life.* New York: Facts on File, 1985.

Barker, G. M., ed. *The Biology of Terrestrial Molluscs.* Oxford, U.K.: CABI Publishing, 2001.

———. *Molluscs as Crop Pests.* Oxford, U.K.: CABI Publishing, 2002.

Barnes, R. S. K., et al. *The Invertebrates: A Synthesis.* Malden, MA: Blackwell Science, 2001.

Barnes, Robert D. *Invertebrate Zoology,* 4th ed. Philadelphia: Saunders College Publishing, 1980.

Bauer, Raymond T., and Joel W. Martin, eds. *Crustacean Sexual Biology.* New York: Columbia University Press, 1991.

Beacham, Walton, Frank V. Castronova, and Suzanne Sessine, eds. *Beacham's Guide to the Endangered Species of North America.* Detroit, MI: Gale Group, 2001.

Beesley, Pamela L., Graham J. B. Ross, and Alice Wells. *Mollusca—The Southern Synthesis, Part A.* Canberra, Australia: Australian Biological Resources Study, 1998.

Benson, R. H., et al. *Treatise on Invertebrate Paleontology, Part Q, Arthropoda 3.* Lawrence, KS: Geological Society of America and University of Kansas Press, 1961.

Bertolani, Roberto. *Tardigradi. Guide per il Riconnoscimento delle Specie Animali delle Acque Interne Italiane.* Verona, Italy: Consiglio Nazionale Delle Ricerche, 1982.

Bliss, Dorothy E. *Biology of the Crustacea.* New York: Academic Press, 1982–1985.

———. *Shrimps, Lobsters and Crabs.* New York: Columbia University Press, 1989.

Boardman, Richard S. *Reflections on the Morphology, Anatomy, Evolution, and Classification of the Class Stenolaemata (Bryozoa).* Smithsonian Contributions to Palaeobiology 86. Washington, DC: Smithsonian Institution Press, 1998.

Bosik, J. J. *Common Names of Insects and Related Organisms.* Lanham, MD: Entomological Society of America, 1997.

Boyle, P. R., ed. *Cephalopod Life Cycles,* Vol. I, *Species Accounts.* London: Academic Press, 1983.

———. *Cephalopod Life Cycles,* Vol. II, *Comparative Reviews.* London: Academic Press, 1987.

Brunton, C. Howard, L. Robin M. Cocks, and Sarah L. Long, eds. *Brachiopods Past and Present.* London and New York: Taylor and Francis, 2001.

Brusca, R. C., and G. J. Brusca. *Invertebrates.* New York: Sinauer Associates Inc., 1990, 2nd ed., 2003.

Buchsbaum, Ralph Morris. *Animals Without Backbones: An Introduction to the Invertebrates,* 2nd ed. Chicago: University of Chicago Press, 1976.

Burggren, Warren W., and Brian R. McMahon. *Biology of the Land Crabs.* Cambridge, U.K.: Cambridge University Press, 1988.

Child, C. A. *Marine Fauna of New Zealand: Pycnogonida (Sea Spiders).* Wellington, New Zealand: National Institute of Water and Atmospheric Research, 1998.

Clarke, M. R., ed. *A Handbook for the Identification of Cephalopod Beaks.* Oxford, U.K.: The Clarendon Press, 1986.

Clarkson, Euan N. K. *Invertebrate Palaeontology and Evolution,* 4th ed. Oxford, U.K.: Blackwell Science, Ltd., 1999.

Cloudsley-Thompson, J. L. *Spiders, Scorpions, Centipedes and Mites.* Oxford, U.K.: Pergamon Press, 1968.

Conn, David Bruce. *Atlas of Invertebrate Reproduction and Development,* 2nd ed. New York: Wiley-Liss, 2000.

Conn, David Bruce, Richard A. Lutz, Ya-Ping Hu, and Victor S. Kennedy. *Guide to the Identification of Larval and Postlarval Stages of Zebra Mussels,* Dreissena *spp. and the Dark False Mussel,* Mytilopsis leucophaeata. Stony Brook, NY: New York Sea Grant Institute, 1993.

Crane, Jocelyn. *Fiddler Crabs of the World.* Princeton, NJ: Princeton University Press, 1975.

Cutler, Edward, B. *The Sipuncula. Their Systematics, Biology, and Evolution.* Ithaca, NY: Cornell University Press, 1994.

Desbruyères, D., and M. Segonzac, eds. *Handbook of Deep-Sea Hydrothermal Vent Fauna.* Brest, France: IFREMER, 1997.

Douglas, Angela E. *Symbiotic Interactions.* Oxford, U.K.: Oxford University Press, 1994.

Dumont, H. J., and S. V. Negrea. *Introduction to the Class Branchiopoda.* Leiden, The Netherlands: Backhuys Publishers, 2002.

Emberlin, J. C. *Introduction to Ecology.* Plymouth, MA: Macdonald and Evans Handbooks, 1983.

Ensminger, Peter J. *Life Under the Sun.* London and New York: Academic Press, 1997.

Erikson, C., and D. Belk. *Fairy Shrimp of California's Puddles, Pools and Playas.* Eureka, CA: Mad River Press, 1999.

Factor, Jan Robert. *Biology of the Lobster.* New York: Academic Press, 1995.

Felsenstein, Joseph. *Inferring Phylogenies.* Sunderland, MA: Sinauer Associates, 2003.

Foelix, Rainer F. *Biology of Spiders.* Cambridge: Harvard University Press, 1982.

Fretter, Vera, and A. Graham. *A Functional Anatomy of Invertebrates.* London: Academic Press, 1976.

Fretter, Vera and J. Peake, eds. *Pulmonates: Functional Anatomy and Physiology,* vol. 1. London: Academic Press. 1975.

———. *Pulmonates: Economic Malacology,* vol. 2B. London: Academic Press, 1979.

Fusetani, Nobuhiro, ed. *Drugs from the Sea.* Basel, Switzerland: Karger, 2000.

Futuyama, Douglas J. *Evolutionary Biology,* 2nd ed. Sunderland, MA: Sinauer Associates, 1998.

Geise, Arthur C., and John S. Pearse. *Reproduction of Marine Invertebrates.* Vol. V, *Molluscs: Pelecypods and Lesser Classes.* London: Academic Press, 1975.

Gilbert, D. L., W. J. Adelman, and J. M. Arnold, eds. *Squid as Experimental Animals.* New York: Plenum Press, 1990.

Gilbert, Scott F., and Anne M. Raunio, eds. *Embryology: Constructing the Organism.* Sunderland, MA: Sinauer Associates, 1997.

Giller, P. S., and B. Malmqvist. *The Biology of Streams and Rivers.* Oxford, U.K.: Oxford University Press, 1998.

Gosling, Elizabeth, ed. *The Mussel* Mytilus: *Ecology, Physiology, Genetics and Culture.* Amsterdam: Elsevier Science Publishers, 1992.

Grassé, P. P. *Traitè de Zoologie,* Vol. 5. Paris: Masson et Cie., 1959.

Green, N. P. O., G. W. Stout, D. J. Taylor, and R. Soper. *Biological Science, Organisms, Energy and Environment.* Cambridge, U.K.: Cambridge University Press, 1984.

Greven, Hartmut. *Die Bärtierschen. Die NeueBrehm-Bucherei,* Volume 537. Wittenberg, Germany: A. Ziemsen Verlag, 1980.

Hanlon, R. T., and J. B. Messenger. *Cephalopod Behaviour.* Cambridge, U.K.: Cambridge University Press, 1996.

Harding, M. A., and J. P. Moore. *The Fauna of British India: Hirudinea.* London: Taylor and Francis, 1927.

Harper, E. M., J. D. Taylor, and J. A. Crame, eds. *The Evolutionary Biology of the Bivalvia.* London: The Geological Society, 2000.

Harrison, Frederick W., and Robert M. Woollacott, eds. *Microscopic Anatomy of Invertebrates,* vol. 13. New York: Wiley-Liss, 1997.

Hartmann, G., and M.-C. Guillaume. *Classe des Ostracodes. Traité de Zoologie. VII. Crustaces,* Fasc. 2. Paris: Masson, 1996.

Hayward, P. J., and J. S. Ryland. *Cyclostome Bryozoans: Keys and Notes for the Identification of the Species.* Leiden, The Netherlands: The Linnean Society of London, 1985.

————. *Handbook of the Marine Fauna of North-West Europe.* Oxford, U.K.: Oxford University Press, 1995.

Hopkin, S. P., and H. J. Read. *The Biology of Millipedes.* Oxford, U.K.: Oxford University Press, 1992.

Huys, R., and G. A. Boxshall. *Copepod Evolution.* London: The Ray Society, 1991.

Ivanov, A. V. *Pogonophora.* London: Academic Press, 1963.

Kabata, Z. *Parasitic Copepoda of British Fishes.* London: The Ray Society, 1979.

Kaestner, A. *Invertebrate Zoology.* Vol. 3, *Crustacea.* New York: Wiley-Interscience, 1970.

Keegan, H. L., S. Tashioka, and H. Suzuki. *Bloodsucking Asian Leeches of Families Hirudinidae and Haemadipsidae.* Tokyo: United States Army Medical Command, Japan, 1968.

Kennedy, Victor S., Roger I. E. Newell, and Albert F. Ebel, eds. *The Eastern Oyster:* Crassostrea virginica. College Park, MD: Maryland Sea Grant College, 1996.

Kinchin, Ian M. *The Biology of Tardigrades.* London: Portland, 1994.

Kinne, O., ed. *Diseases of Marine Animals,* Vol. III, *Cephalopoda through Urochordata.* Hamburg, Germany: Biologische Anstalt Helgoland, 1990.

Lacourt, Adrianus W. *A Monograph of the Freshwater Bryozoa: Phylactolaemata.* Leiden, The Netherlands: Zoologische Verhandelingen, 1968.

Lalli, Carol M., and Ronald W. Gilmore. *Pelagic Snails: The Biology of Holoplanktonic Gastropod Mollusks.* Stanford, CA: Stanford University Press, 1989.

Lalli, Carol M., and T. R. Parsons. *Biological Oceanography, An Introduction.* Oxford, U.K.: Elsevier Science, 1994.

Landman, Neil H., Paula M. Mikkelsen, Rüdiger Bieler, and Bennett Bronson. *Pearls: A Natural History.* New York: Harry N. Abrams, 2001.

Lerman, M. *Marine Biology: Environment, Diversity, and Ecology.* Redwood City, CA: Benjamin Cummings Publishing Company, 1986.

Levin, Simon Asher, ed. *Encyclopedia of Biodiversity.* San Diego, CA: Academic Press, 2001.

Lewis, J. G. E. *The Biology of Centipedes.* Cambridge, U.K.: Cambridge University Press, 1981.

Lutz, P. E. *Invertebrate Zoology.* Menlo Park, CA: Benjamin/Cummings Publishing Company, Inc., 1986.

McDaniel, Burruss. *How to Know the Mites and Ticks.* Dubuque, IA: William C. Brown Company Publishers, 1979.

McKinney, F. K., and J. B. C. Jackson. *Bryozoan Evolution.* Chicago: University of Chicago Press, 1989.

McLaughlin, Patsy A. *Comparative Morphology of Recent Crustacea.* San Francisco: W. H. Freeman and Company, 1980.

McLusky, D. S. *The Estuarine Ecosystem.* New York: Halsted Press, 1981.

Mangold, K. *Traité de Zoologie. Anatomie, Systématique, Biologie.* Tome 5, fascicule 4, *Céphalopodes.* Paris: Masson, 1989.

Mann, K. H. *Leeches (Hirudinea): Their Structure, Physiology, Ecology and Embryology.* London: Pergamon Press, 1962.

————. *Ecology of Coastal Waters with Implications for Management.* Malden, MA: Blackwell Science, 2000.

Mauchline, J. *The Biology of Mysids and Euphausiids,* Part I, *The Biology of Mysids.* London: Academic Press, 1980.

————. *The Biology of Calanoid Copepods. Advances in Marine Biology.* New York, London: Academic Press, 1998.

Mead, Albert R. *The Giant African Land Snail: A Problem in Economic Malacology.* Chicago: University of Chicago Press, 1961.

Mehlhorn, H. *Parasitology in Focus.* Berlin: Springer Verlag, 1988.

Milne, Lorus, and Susan Rayfield. *The Audubon Society Field Guide to North American Insects and Spiders.* New York: Knopf, 1980.

Minelli, A., ed. *Proceedings of the Seventh International Congress of Myriapodology.* Leiden, The Netherlands: E. J. Brill, 1990.

Morgan, M. D. *Ecology of Mysidacea.* Vol. 10, *Developments in Hydrobiology.* The Hague, The Netherlands: W. Junk Publishers, 1982.

Morris, R. H., D. P. Abbott, and E. C. Haderlie. *Intertidal Invertebrates of California.* Stanford, CA: Stanford University Press, 1980.

Morton, Bryan. *Partnerships in the Sea: Hong Kong's Marine Symbioses.* Leiden, The Netherlands: E. J. Brill, 1989.

Müller, H.-G. *World Catalogue and Bibliography of the Recent Mysidacea.* Wetzlar, Germany: Wissenschaftler Verlag, Laboratory for Tropical Ecosystems, Research and Information Service, 1993.

Nalepa, Thomas F., and Donald W. Schloesser, eds. *Zebra Mussels: Biology, Impacts, and Control.* Boca Raton, FL: Lewis Publishers, 1993.

Nesis, K. N. *Cephalopods of the World: Squids, Cuttlefishes, Octopuses and Allies.* English translation, Neptune City, NJ: T. F. H. Publications, 1987.

Nesler, T. P., and E. P. Bergerson, eds. *Mysids in Fisheries: Hard Lessons from Headlong Introductions.* Bethesda, MD: American Fisheries Society, 1991.

Nicol, Stephen. *Krill Fisheries of the World,* Technical Paper 167. Rome: Food and Agriculture Organization of the United Nations, 1997.

Nielsen, Claus. *Animal Evolution, Interrelationships of the Living Phyla.* Oxford, U.K.: Oxford University Press, 1995.

Noble, Elmer R., and Glenn A. Noble. *Parasitology: The Biology of Animal Parasites.* Philadelphia: Lea and Febiger, 1982.

Nouvel, H., J.-P. Casanova, and J. P. Lagardère. *Ordre des Mysidacés. Mémoires de l'institut océanographique.* Monaco: [n.p.], 1999.

Nybakken, James W. *Diversity of the Invertebrates: A Laboratory Manual,* Pacific Coast Version. Dubuque, IA: Wm. C. Brown Publishers, 1996.

Okutani, T. *Cuttlefish and Squids of the World in Color.* Tokyo: National Cooperative Association of Squid Processors, 1995.

Parker, Sybil P., ed. *Synopsis and Classification of Living Organisms.* New York: McGraw-Hill Book Company, 1982.

Parks, P. *The World You Never See: Underwater Life.* Skokie, IL: Rand McNally, 1976.

Pennak, R. W. *Fresh-Water Invertebrates of the United States,* 3rd ed. New York: John Wiley and Sons, 1989.

Persoone, G., et al. *The Brine Shrimp* Artemia, 3 vols. Wetteren, Belgium: Universal Press, 1980.

Polis, Gary A., ed. *The Biology of Scorpions.* Stanford, CA: Stanford University Press, 1990.

Preston-Mafham, R., and K. Preston-Mafham. *The Encyclopedia of Land Invertebrate Behavior.* Cambridge, MA: The MIT Press, 1993.

Price, P. W. *Insect Ecology.* New York: John Wiley and Sons, 1997.

Purchon, R. D. *The Biology of Mollusca.* Oxford, U.K.: Pergamon Press, Ltd., 1968.

Reaka, M. L., ed. *The Ecology of Coral Reefs.* Rockville, MD: NOAA Undersea Research Program, 1985.

Robinson, M. A., and J. F. Wiggins. *Animal Types 1, Invertebrates.* London: Stanley Thornes, 1988.

Rouse, Greg, and Fredrik W. Pleijel. *Polychaetes.* New York: Oxford University Press Inc., 2001.

Ruppert, Edward E., and Robert D. Barnes. *Invertebrate Zoology,* 6th ed. Fort Worth, TX: Saunders College Publishing, 1994.

Ryland, John S. *Bryozoans.* London: Hutchinson University Library, 1970.

Sars, G. O. *Cumacea: An Account of the Crustacea of Norway.* Christiania, Norway: Cammermeyers Forlag, 1899–1900.

Saunders, W. B., and N. H. Landman, eds. *Nautilus. The Biology and Paleobiology of a Living Fossil.* New York: Plenum Press, 1987.

Savory, Theodore. *Arachnida,* 2nd ed. London, U.K., and New York: Academic Press, 1977.

Sawyer, R. T. *Leech Biology and Behaviour.* Oxford, U.K.: The Clarendon Press, 1986.

Schmitt, W. L. *Crustaceans.* Ann Arbor: University of Michigan Press, 1965.

Schram, Frederick R. *Crustacea.* Oxford, U.K.: Oxford University Press, 1986.

Smith, Douglas G. *Pennak's Freshwater Invertebrates of the United States: Porifera to Crustacea.* New York: Wiley, 2001.

Smith, R. I., and J. T. Carlton. *Light's Manual: Intertidal Invertebrates of the Central California Coast.* Berkeley: University of California Press, 1976.

Solem, Alan. *The Shell Makers.* New York: John Wiley and Sons, 1974.

South, A. *Terrestrial Slugs: Biology, Ecology, and Control.* London: Chapman and Hall, 1992.

Stachowitsch, Michael. *The Invertebrates: An Illustrated Glossary.* New York: Wiley-Liss, 1992.

Stephen, A. C., and S. J. Edmonds. *The Phyla Sipuncula and Echiura.* London: Trustees of the British Museum (Natural History), 1972.

Strathmann, Megumi F., ed. *Reproduction and Development of Marine Invertebrates of the Northern Pacific Coast.* Seattle: University of Washington Press, 1987.

Tanacredi, John T., ed. *Limulus in the Limelight: A Species 350 Million Years in the Making and in Peril?* New York: Kluwer Academic/Plenum, 2001.

Thorp, J. H., and A. P. Covich, eds. *Ecology and Classification of North American Freshwater Invertebrates,* 2nd ed. New York: Academic Press, 2001.

Tudge, Colin. *The Variety of Life: A Survey and a Celebration of All the Creatures That Have Ever Lived.* Oxford, U.K., and New York: Oxford University Press, 2000.

Turner, Ruth D. *A Survey and Illustrated Catalogue of the Teredinidae.* Cambridge, MA: Museum of Comparative Zoology, Harvard University, 1966.

Van Dover, C. L. *The Ecology of Deep-Sea Hydrothermal Vents.* Princeton, NJ: Princeton University Press, 2000.

Vaught, Kay Cunningham. *A Classification of the Living Mollusca.* Melbourne, FL: American Malacologists Inc., 1989.

Warburg, M. R. *Evolutionary Biology of Land Isopods.* Berlin: Springer-Verlag, 1993.

Ward, P. D. *Natural History of Nautilus.* London: Allen and Unwin, 1987.

Warner, G. F. *The Biology of Crabs.* New York: Van Nostrand Reinhold Company, 1977.

Weigle, Marta. *Spiders and Spinsters: Women and Mythology.* Albuquerque, NM: University of New Mexico Press, 2001.

Williams, Austin B. *Shrimps, Lobsters, and Crabs of the Atlantic Coast of the Eastern United States, Maine to Florida.* Washington, DC: Smithsonian Institution Press, 1984.

Wilson, W. H., Stephen A. Sticker, and George L. Shinn, eds. *Reproduction and Development of Marine Invertebrates.* Baltimore: Johns Hopkins University Press, 1994.

Wood, Timothy S. *Ectoproct Bryozoans of Ohio.* Columbus: Ohio Biological Survey, 1989.

Wood, Timothy S., and Beth Okamura. *A New Key to the Freshwater Bryozoans of Britain, Ireland, and Continental Europe.* Ambleside, Cumbria, U.K.: Freshwater Biological Association, 2004.

Woollacott, R. M., and R. L. Zimmer, eds. *Bryozoans.* New York: Academic Press, Inc., 1977.

Woolley, Tyler A. *Acarology: Mites and Human Welfare.* New York: John Wiley and Sons, 1988.

Yamaji, I. *Illustrations of the Marine Plankton of Japan.* Osaka, Japan: Hoikusha Publishing Co., 1976.

Young, Craig M., M. A. Sewell, and Mary E. Rice, eds. *Atlas of Marine Invertebrate Larvae.* San Diego, CA: Academic Press, 2002.

FOR FURTHER READING

Organizations

Agricultural Research Service
<http://www.ars.usda.gov/is/AR/archive/may01/worms0501.htm>

American Society of Parasitologists
<http://asp.unl.edu>

American Zoo and Aquarium Association
8403 Colesville Road, Suite 710
Silver Spring, MD 20910 USA
<http://www.aza.org>

Australian Regional Association of Zoological Parks and Aquaria
PO Box 20
Mosman, NSW 2088
Australia
Phone: 61 (2) 9978-4797
Fax: 61 (2) 9978-4761
<http://www.arazpa.org>

British and Irish Graptolite Group
c/o Dr. A. W. A. Rushton
The Natural History Museum, Cromwell Rd.
London SW7 5BD United Kingdom
<http://www.graptolites.co.uk/>

British Myriapod and Isopod Group
E-Mail: steve.gregory@northmoortrust.co.uk
<http://www.salticus.demon.co.uk/bmig/html/info.htm>

Cephalopod International Advisory Council (CIAC)
Dr. Michael Vecchione, President
NOAA/NMFS National Systematics Laboratory
National Museum of Natural History, MR C-153
Washington, DC 20013-7012

European Association of Zoos and Aquaria
PO Box 20164
1000 HD Amsterdam
The Netherlands
<http://www.eaza.net>

The Graptolite Working Group of the International Palaeontological Association
c/o Dr. Charles E. Mitchell, Department of Geology
State University of New York at Buffalo
Buffalo, NY 14260-3050
E-mail: cem@acsu.buffalo.edu
<http://www.geology.buffalo.edu/gwg/index.htm>

Helminthological Society of Washington
Allen Richards, Ricksettsial Disease Department
Naval Medical Research Center
503 Robert Grant Ave.
Silver Spring, MD 20910-7500 USA

Inland Water Crustacean Specialist Group
Denton Belk, 840 E. Mulberry Ave.
San Antonio, TX 78212-3194 USA
Phone: (210) 732-8809
Fax: (210) 732-3943
<http://www.iucn.org/themes/ssc/pubs/sgnewsl.htm>

International Bryozoology Association
<http://petralia.civgeo.rmit.edu.au/bryozoa/iba.html>

International Isopod Research Group
<http://www.uni-kiel.de/zoologie/institut/limnologie/IIRG.htm>

International Society of Endocytobiology
<http://www.endocytobiology.org/>

International Symbiosis Society
<http://www.ma.psu.edu/lkh1/iss/>

Monterey Bay Aquarium Research Institute
7700 Sandholdt Road
Moss Landing, CA 95039 USA
Phone: (831) 775-1700
Fax: (831) 775-1620
E-mail: mage@mbari.org
<http://www.mbari.org>

Species Survival Commission, IUCN—The World Conservation Union
Rue Mauverney 28
Gland CH-1196 Switzerland
Phone: +41-22-999-0152
Fax: +41-22-999-0015
E-mail: ssc@iucn.org
<http://www.iucn.org>

Xiamen Rare Marine Creatures Conservation Areas
<http://ois.xmu.edu.cn/cbcm/english/resource/save/save04.htm>

Contributors to the first edition

The following individuals contributed chapters to the original edition of Grzimek's Animal Life Encyclopedia, *which was edited by Dr. Bernhard Grzimek, Professor, Justus Liebig University of Giessen, Germany; Director, Frankfurt Zoological Garden, Germany; and Trustee, Tanzanian National Parks, Tanzania.*

Dr. Fritz Dieterlen
Zoological Research Institute, A.
Koenig Museum
Bonn, Germany

Dr. Rolf Dircksen
Professor, Pedagogical Institute
Bielefeld, Germany

Josef Donner
Instructor of Biology
Katzelsdorf, Austria

Dr. Jean Dorst
Professor, National Museum of
Natural History
Paris, France

Dr. Gerti DÜcker
Professor and Chief Curator,
Zoological Institute, University of
Münster
Münster, Germany

Dr. Michael Dzwillo
Zoological Institute and Museum,
University of Hamburg
Hamburg, Germany

Dr. Irenäus Eibl-Eibesfeldt
Professor and Director, Institute of
Human Ethology, Max Planck
Institute for Behavioral Physiology
Percha/Starnberg, Germany

Dr. Martin Eisentraut
Professor and Director, Zoological
Research Institute and A. Koenig
Museum
Bonn, Germany

Dr. Eberhard Ernst
Swiss Tropical Institute
Basel, Switzerland

R. D. Etchecopar
Director, National Museum of
Natural History
Paris, France

Dr. R. A. Falla
Director, Dominion Museum
Wellington, New Zealand

Dr. Hubert Fechter
Curator, Lower Animals, Zoological
Collection of the State of Bavaria
Munich, Germany

Dr. Walter Fiedler
Docent, University of Vienna, and
Director, Schönbrunn Zoo
Vienna, Austria

Wolfgang Fischer
Inspector of Animals, Animal Park
Berlin, Germany

Dr. C. A. Fleming
Geological Survey Department of
Scientific and Industrial Research
Lower Hutt, New Zealand

Dr. Hans Frädrich
Zoological Garden
Berlin, Germany

Dr. Hans-Albrecht Freye
Professor and Director, Biological
Institute of the Medical School
Halle a.d.S., Germany

Günther E. Freytag
Former Director, Reptile and
Amphibian Collection, Museum of
Cultural History in Magdeburg
Berlin, Germany

Dr. Herbert Friedmann
Director, Los Angeles County
Museum of Natural History
Los Angeles, California, U.S.A.

Dr. H. Friedrich
Professor, Overseas Museum
Bremen, Germany

Dr. Jan Frijlink
Zoological Laboratory, University of
Amsterdam
Amsterdam, The Netherlands

Dr. H.C. Karl Von Frisch
Professor Emeritus and former
Director, Zoological Institute,
University of Munich
Munich, Germany

Dr. H. J. Frith
C.S.I.R.O. Research Institute
Canberra, Australia

Dr. Ion E. Fuhn
Academy of the Roumanian Socialist
Republic, Trajan Savulescu Institute of
Biology
Bucharest, Romania

Dr. Carl Gans
Professor, Department of Biology,
State University of New York at
Buffalo
Buffalo, New York, U.S.A.

Dr. Rudolf Geigy
Professor and Director, Swiss Tropical
Institute
Basel, Switzerland

Dr. Jacques Gery
St. Genies, France

Dr. Wolfgang Gewalt
Director, Animal Park
Duisburg, Germany

Dr. H.C. Viktor Goerttler
Professor Emeritus, University of Jena
Jena, Germany

Dr. Friedrich Goethe
Director, Institute of Ornithology,
Heligoland Ornithological Station
Wilhelmshaven, Germany

Dr. Ulrich F. Gruber
Herpetological Section, Zoological
Research Institute and A. Koenig
Museum
Bonn, Germany

Dr. H. R. Haefelfinger
Museum of Natural History
Basel, Switzerland

Dr. Theodor Haltenorth
Director, Mammalology, Zoological
Collection of the State of Bavaria
Munich, Germany

Barbara Harrisson
Sarawak Museum, Kuching, Borneo
Ithaca, New York, U.S.A.

Dr. Francois Haverschmidt
President, High Court (retired)
Paramaribo, Surinam

Dr. Heinz Heck
Director, Catskill Game Farm
Catskill, New York, U.S.A.

Dr. Lutz Heck
Professor (retired), and Director,
Zoological Garden, Berlin
Wiesbaden, Germany

Dr. H.C. Heini Hediger
Director, Zoological Garder
Zurich, Switzerland

Dr. Dietrich Heinemann
Director, Zoological Garden, Münster
Dörnigheim, Germany

Dr. Helmut Hemmer
Institute for Physiological Zoology,
University of Mainz
Mainz, Germany

Dr. W. G. Heptner
Professor, Zoological Museum,
University of Moscow
Moscow, Russia

Dr. Konrad Herter
Professor Emeritus and Director
(retired), Zoological Institute, Free
University of Berlin
Berlin, Germany

Dr. Hans Rudolf Heusser
Zoological Museum, University of
Zurich
Zurich, Switzerland

Dr. Emil Otto Höhn
Associate Professor of Physiology,
University of Alberta
Edmonton, Canada

Dr. W. Hohorst
Professor and Director, Parasitological
Institute, Farbwerke Hoechst A.G.
Frankfurt-Höchst, Germany

Dr. Folkhart Hückinghaus
Director, Senckenbergische Anatomy,
University of Frankfurt a.M.
Frankfurt a.M., Germany

Francois Hüe
National Museum of Natural History
Paris, France

Dr. K. Immelmann
Professor, Zoological Institute,
Technical University of Braunschweig
Braunschweig, Germany

Dr. Junichiro Itani
Kyoto University
Kyoto, Japan

Dr. Richard F. Johnston
Professor of Zoology, University of
Kansas
Lawrence, Kansas, U.S.A.

Otto Jost
Oberstudienrat, Freiherr-vom-Stein
Gymnasium
Fulda, Germany

Dr. Paul Kähsbauer
Curator, Fishes, Museum of Natural
History
Vienna, Austria

Dr. Ludwig Karbe
Zoological State Institute and
Museum
Hamburg, Germany

Dr. N. N. Kartaschew
Docent, Department of Biology,
Lomonossow State University
Moscow, Russia

Dr. Werner Kästle
Oberstudienrat, Gisela Gymnasium
Munich, Germany

Dr. Reinhard Kaufmann
Field Station of the Tropical Institute,
Justus Liebig University, Giessen,
Germany
Santa Marta, Colombia

Dr. Masao Kawai
Primate Research Institute, Kyoto
University
Kyoto, Japan

Dr. Ernst F. Kilian
Professor, Giessen University and
Catedratico Universidad Austral,
Valdivia-Chile
Giessen, Germany

Dr. Ragnar Kinzelbach
Institute for General Zoology,
University of Mainz
Mainz, Germany

Dr. Heinrich Kirchner
Landwirtschaftsrat (retired)
Bad Oldesloe, Germany

Dr. Rosl Kirchshofer
Zoological Garden, University of
Frankfort a.M.
Frankfurt a.M., Germany

Dr. Wolfgang Klausewitz
Curator, Senckenberg Nature
Museum and Research Institute
Frankfurt a.M., Germany

Dr. Konrad Klemmer
Curator, Senckenberg Nature
Museum and Research Institute
Frankfurt a.M., Germany

Dr. Erich Klinghammer
Laboratory of Ethology, Purdue
University
Lafayette, Indiana, U.S.A.

Dr. Heinz-Georg Klös
Professor and Director, Zoological
Garden
Berlin, Germany

Ursula Klös
Zoological Garden
Berlin, Germany

Dr. Otto Koehler
Professor Emeritus, Zoological
Institute, University of Freiburg
Freiburg i. BR., Germany

Dr. Kurt Kolar
Institute of Ethology, Austrian
Academy of Sciences
Vienna, Austria

Dr. Claus König
State Ornithological Station of Baden-
Württemberg
Ludwigsburg, Germany

Dr. Adriaan Kortlandt
Zoological Laboratory, University of
Amsterdam
Amsterdam, The Netherlands

Dr. Helmut Kraft
Professor and Scientific Councillor,
Medical Animal Clinic, University of
Munich
Munich, Germany

Dr. Helmut Kramer
Zoological Research Institute and A.
Koenig Museum
Bonn, Germany

Dr. Franz Krapp
Zoological Institute, University of
Freiburg
Freiburg, Switzerland

Dr. Otto Kraus
Professor, University of Hamburg,
and Director, Zoological Institute and
Museum
Hamburg, Germany

Dr. Hans Krieg
Professor and First Director (retired),
Scientific Collections of the State of
Bavaria
Munich, Germany

Dr. Heinrich Kühl
Federal Research Institute for
Fisheries, Cuxhaven Laboratory
Cuxhaven, Germany

Dr. Oskar Kuhn
Professor, formerly University
Halle/Saale
Munich, Germany

Dr. Hans Kumerloeve
First Director (retired), State
Scientific Museum, Vienna
Munich, Germany

Dr. Nagamichi Kuroda
Yamashina Ornithological Institute,
Shibuya-Ku
Tokyo, Japan

Dr. Fred Kurt
Zoological Museum of Zurich
University, Smithsonian Elephant
Survey
Colombo, Ceylon

Dr. Werner Ladiges
Professor and Chief Curator,
Zoological Institute and Museum,
University of Hamburg
Hamburg, Germany

Leslie Laidlaw
Department of Animal Sciences,
Purdue University
Lafayette, Indiana, U.S.A.

Dr. Ernst M. Lang
Director, Zoological Garden
Basel, Switzerland

Dr. Alfredo Langguth
Department of Zoology, Faculty of
Humanities and Sciences, University
of the Republic
Montevideo, Uruguay

Leo Lehtonen
Science Writer
Helsinki, Finland

Bernd Leisler
Second Zoological Institute
University of Vienna
Vienna, Austria

Dr. Kurt Lillelund
Professor and Director, Institute for
Hydrobiology and Fishery Sciences,
University of Hamburg
Hamburg, Germany

R. Liversidge
Alexander MacGregor Memorial
Museum
Kimberley, South Africa

Dr. Konrad Lorenz
Professor and Director, Max Planck
Institute for Behavioral Physiology
Seewiesen/Obb., Germany

Dr. Martin Lühmann
Federal Research Institute for the
Breeding of Small Animals
Celle, Germany

Dr. Johannes Lüttschwager
Oberstudienrat (retired)
Heidelberg, Germany

Dr. Wolfgang Makatsch
Bautzen, Germany

Dr. Hubert Markl
Professor and Director, Zoological
Institute, Technical University of
Darmstadt
Darmstadt, Germany

Basil J. Marlow, B.SC. (Hons)
Curator, Australian Museum
Sydney, Australia

Dr. Theodor Mebs
Instructor of Biology
Weissenhaus/Ostsee, Germany

Dr. Gerlof Fokko Mees
Curator of Birds, Rijks Museum of
Natural History
Leiden, The Netherlands

Hermann Meinken
Director, Fish Identification Institute,
V.D.A.
Bremen, Germany

Dr. Wilhelm Meise
Chief Curator, Zoological Institute
and Museum, University of Hamburg
Hamburg, Germany

Dr. Joachim Messtorff
Field Station of the Federal Fisheries
Research Institute
Bremerhaven, Germany

Dr. Marian Mlynarski
Professor, Polish Academy of
Sciences, Institute for Systematic and
Experimental Zoology
Cracow, Poland

Dr. Walburga Moeller
Nature Museum
Hamburg, Germany

Dr. H. C. Erna Mohr
Curator (retired), Zoological State
Institute and Museum
Hamburg, Germany

Dr. Karl-Heinz Moll
Waren/Müritz, Germany

Dr. Detlev Müller-Using
Professor, Institute for Game
Management, University of Göttingen
Hannoversch-Münden, Germany

Werner Münster
Instructor of Biology
Ebersbach, Germany

Dr. Joachim Münzing
Altona Museum
Hamburg, Germany

Dr. Wilbert Neugebauer
Wilhelma Zoo
Stuttgart-Bad Cannstatt, Germany

Dr. Ian Newton
Senior Scientific Officer, The Nature
Conservancy
Edinburgh, Scotland

Dr. Jürgen Nicolai
Max Planck Institute for Behavioral
Physiology
Seewiesen/Obb., Germany

Dr. Günther Niethammer
Professor, Zoological Research
Institute and A. Koenig Museum
Bonn, Germany

Dr. Bernhard Nievergelt
Zoological Museum, University of
Zurich
Zurich, Switzerland

Dr. C. C. Olrog
Institut Miguel Lillo San Miguel de
Tucuman
Tucuman, Argentina

Alwin Pedersen
Mammal Research and aRctic
Explorer
Holte, Denmark

Dr. Dieter Stefan Peters
Nature Museum and Senckenberg
Research Institute
Frankfurt a.M., Germany

Dr. Nicolaus Peters
Scientific Councillor and Docent,
Institute of Hydrobiology and
Fisheries, University of Hamburg
Hamburg, Germany

Dr. Hans-Günter Petzold
Assistant Director, Zoological Garden
Berlin, Germany

Dr. Rudolf Piechocki
Docent, Zoological Institute,
University of Halle
Halle a.d.S., Germany

Dr. Ivo Poglayen-Neuwall
Director, Zoological Garden
Louisville, Kentucky, U.S.A.

Dr. Egon Popp
Zoological Collection of the State of
Bavaria
Munich, Germany

Dr. H.C. Adolf Portmann
Professor Emeritus, Zoological
Institute, University of Basel
Basel, Switzerland

Hans Psenner
Professor and Director, Alpine Zoo
Innsbruck, Austria

Dr. Heinz-Siburd Raethel
Oberveterinärrat
Berlin, Germany

Dr. Urs H. Rahm
Professor, Museum of Natural History
Basel, Switzerland

Dr. Werner Rathmayer
Biology Institute, University of
Konstanz
Konstanz, Germany

Walter Reinhard
Biologist
Baden-Baden, Germany

Dr. H. H. Reinsch
Federal Fisheries Research Institute
Bremerhaven, Germany

Dr. Bernhard Rensch
Professor Emeritus, Zoological
Institute, University of Münster
Münster, Germany

Dr. Vernon Reynolds
Docent, Department of Sociology,
University of Bristol
Bristol, England

Dr. Rupert Riedl
Professor, Department of Zoology,
University of North Carolina
Chapel Hill, North Carolina, U.S.A.

Dr. Peter Rietschel
Professor (retired), Zoological
Institute, University of Frankfurt a.M.
Frankfurt a.M., Germany

Dr. Siegfried Rietschel
Docent, University of Frankfurt;
Curator, Nature Museum and
Research Institute Senckenberg
Frankfurt a.M., Germany

Herbert Ringleben
Institute of Ornithology, Heligoland
Ornithological Station
Wilhelmshaven, Germany

Dr. K. Rohde
Institute for General Zoology, Ruhr
University
Bochum, Germany

Dr. Peter Röben
Academic Councillor, Zoological
Institute, Heidelberg University
Heidelberg, Germany

Dr. Anton E. M. De Roo
Royal Museum of Central Africa
Tervuren, South Africa

Dr. Hubert Saint Girons
Research Director, Center for
National Scientific Research
Brunoy (Essonne), France

Dr. Luitfried Von Salvini-Plawen
First Zoological Institute, University
of Vienna
Vienna, Austria

Dr. Kurt Sanft
Oberstudienrat, Diesterweg-
Gymnasium
Berlin, Germany

Dr. E. G. Franz Sauer
Professor, Zoological Research
Institute and A. Koenig Museum,
University of Bonn
Bonn, Germany

Dr. Eleonore M. Sauer
Zoological Research Institute and A.
Koenig Museum, University of Bonn
Bonn, Germany

Dr. Ernst Schäfer
Curator, State Museum of Lower
Saxony
Hannover, Germany

Dr. Friedrich Schaller
Professor and Chairman, First
Zoological Institute, University of
Vienna
Vienna, Austria

Dr. George B. Schaller
Serengeti Research Institute, Michael
Grzimek Laboratory
Seronera, Tanzania

Dr. Georg Scheer
Chief Curator and Director,
Zoological Institute, State Museum of
Hesse
Darmstadt, Germany

Dr. Christoph Scherpner
Zoological Garden
Frankfurt a.M., Germany

Dr. Herbert Schifter
Bird Collection, Museum of Natural
History
Vienna, Austria

Dr. Marco Schnitter
Zoological Museum, Zurich University
Zurich, Switzerland

Dr. Kurt Schubert
Federal Fisheries Research Institute
Hamburg, Germany

Eugen Schuhmacher
Director, Animals Films, I.U.C.N.
Munich, Germany

Dr. Thomas Schultze-Westrum
Zoological Institute, University of
Munich
Munich, Germany

Dr. Ernst Schüt
Professor and Director (retired), State
Museum of Natural History
Stuttgart, Germany

Dr. Lester L. Short , Jr.
Associate Curator, American Museum
of Natural History
New York, New York, U.S.A.

Dr. Helmut Sick
National Museum
Rio de Janeiro, Brazil

Dr. Alexander F. Skutch
Professor of Ornithology, University
of Costa Rica
San Isidro del General, Costa Rica

Dr. Everhard J. Slijper
Professor, Zoological Laboratory,
University of Amsterdam
Amsterdam, The Netherlands

Bertram E. Smythies
Curator (retired), Division of Forestry
Management, Sarawak-Malaysia
Estepona, Spain

Dr. Kenneth E. Stager
Chief Curator, Los Angeles County
Museum of Natural History
Los Angeles, California, U.S.A.

Dr. H. C. Georg H. W. Stein
Professor, Curator of Mammals,
Institute of Zoology and Zoological
Museum, Humboldt University
Berlin, Germany

Dr. Joachim Steinbacher
Curator, Nature Museum and
Senckenberg Research Institute
Frankfurt a.M., Germany

Dr. Bernard Stonehouse
Canterbury University
Christchurch, New Zealand

Dr. Richard Zur Strassen
Curator, Nature Museum and
Senckenberg Research Institute
Frandfurt a.M., Germany

Dr. Adelheid Studer-Thiersch
Zoological Garden
Basel, Switzerland

Dr. Ernst Sutter
Museum of Natural History
Basel, Switzerland

Dr. Fritz Terofal
Director, Fish Collection, Zoological
Collection of the State of Bavaria
Munich, Germany

Dr. G. F. Van Tets
Wildlife Research
Canberra, Australia

Ellen Thaler-Kottek
Institute of Zoology, University of
Innsbruck
Innsbruck, Austria

Dr. Erich Thenius
Professor and Director, Institute of
Paleontolgy, University of Vienna
Vienna, Austria

Dr. Niko Tinbergen
Professor of Animal Behavior,
Department of Zoology, Oxford
University
Oxford, England

Alexander Tsurikov
Lecturer, University of Munich
Munich, Germany

Dr. Wolfgang Villwock
Zoological Institute and Museum,
University of Hamburg
Hamburg, Germany

Zdenek Vogel
Director, Suchdol Herpetological
Station
Prague, Czechoslovakia

Dieter Vogt
Schorndorf, Germany

Dr. Jiri Volf
Zoological Garden
Prague, Czechoslovakia

Otto Wadewitz
Leipzig, Germany

Dr. Helmut O. Wagner
Director (retired), Overseas Museum,
Bremen
Mexico City, Mexico

Dr. Fritz Walther
Professor, Texas A & M University
College Station, Texas, U.S.A.

John Warham
Zoology Department, Canterbury
University
Christchurch, New Zealand

Dr. Sherwood L. Washburn
University of California at Berkeley
Berkeley, California, U.S.A.

Eberhard Wawra
First Zoological Institute, University
of Vienna
Vienna, Austria

Dr. Ingrid Weigel
Zoological Collection of the State of
Bavaria
Munich, Germany

Dr. B. Weischer
Institute of Nematode Research,
Federal Biological Institute
Münster/Westfalen, Germany

Herbert Wendt
Author, Natural History
Baden-Baden, Germany

Dr. Heinz Wermuth
Chief Curator, State Nature Museum,
Stuttgart
Ludwigsburg, Germany

Dr. Wolfgang Von Westernhagen
Preetz/Holstein, Germany

Dr. Alexander Wetmore
United States National Museum,
Smithsonian Institution
Washington, D.C., U.S.A.

Dr. Dietrich E. Wilcke
Röttgen, Germany

Dr. Helmut Wilkens
Professor and Director, Institute of
Anatomy, School of Veterinary
Medicine
Hannover, Germany

Dr. Michael L. Wolfe
Utah, U.S.A.

Hans Edmund Wolters
Zoological Research Institute and A.
Koenig Museum
Bonn, Germany

Dr. Arnfrid Wünschmann
Research Associate, Zoological Garden
Berlin, Germany

Dr. Walter Wüst
Instructor, Wilhelms Gymnasium
Munich, Germany

Dr. Heinz Wundt
Zoological Collection of the State of
Bavaria
Munich, Germany

Dr. Claus-Dieter Zander
Zoological Institute and Museum,
University of Hamburg
Hamburg, Germany

Dr. Fritz Zumpt
Director, Entomology and
Parasitology, South African Institute
for Medical Research
Johannesburg, South Africa

Dr. Richard L. Zusi
Curator of Birds, United States
National Museum, Smithsonian
Institution
Washington, D.C., U.S.A.

Glossary

4d cell—Mesentoblast; a blastomere cell that results from zygotes that have spiral cleavage divisions, and contains an unidentified cytoplasmic factor that causes the cell and its progeny to form mesoderm.

Abdomen—The posterior of the main body divisions.

Abyssal—Of, or relating to the deepest regions of the ocean.

Acanthor—First larval stage of acanthocephalans.

Aciculum—Small needlelike structure resembling a rod that supports the divisions of the parapodium.

Acoelomate—An organism, particularly an invertebrate, lacking a coelom that is characterized by bilateral symmetry.

Actinotrocha—Tentacle-like ciliated larva of phoronids.

Aestivation—A period of dormancy that is entered into when conditions are not favorable, particularly during very warm or very dry seasons.

Aflagellate—An organism that lacks a flagella.

Agamete—Nucleus within the plasmodium that divides mitotically and gives rise to a sexual adult.

Ametabolous—Development in which little or no external metamorphic changes are noticeable in the larval to adult transition.

Anal—Relating to or being close to the anus.

Anamorphic—Development in which only part of the adult segments are present in recently hatched young.

Ancestrula—Zooid that develops from an egg.

Anecic—An earthworm known for burying leaf litter in the soil and pulling it into underground burrows for consumption.

Antibiosis—A provocative association between organisms that is detrimental, inhibitive, and preventative to one or more of them but produces a metabolic product in another.

Aphotic zone—Region of the ocean where no sun light reaches and exists in complete darkness.

Apical field—An area inside the circumapical band of rotifers that is devoid of cilia.

Arboreal—An organism that lives in, on, or among trees.

Ascidiologists—Scientists who study the Ascidiacea.

Auricularia—Primary larval stage in holothuroid development.

Benthic—An organism that lives on the bottom of the ocean floor.

Bipinnaria—Free-swimming larval stage of asteroids.

Biramous—Having two branches, such as the two appendages in crustaceans.

Bivoltine—The production of two broods or generations in a season or year.

Blastomere—Zygote cleavage divisions that result in a cell.

Blastopore—The first opening of the early digestive tract.

Blastula—Spere of blastomeres.

Brachiolaria—Second stage of asteroid larva.

Brood—When the care of eggs takes place outside or inside of the mother's body for at least the early part of development.

Buccal cavity—A cavity that is present within the mouth.

Bud—The development of new progeny cells or new outgrowth.

Caudal—Referring or pertaining to the posterior end of the body.

Cephalic—Referring or pertaining to the anterior end of the body.

Cephalothorax—The body region that consists of the head and thoracic segments.

Chelicera—Pair of appendages present in the anterior body of arachnids.

Chorion—The shell or covering of an egg.

Cilia—Outgrowth present on the cell surface that is short and produces a lashing movement capable of creating locomotion.

Cloaca—Chamber into which the intestinal and urogenital tracts discharge.

Coelom—The epithelium-lined space between the body wall and the digestive tract.

Colony—Body composed of zooids that share resources.

Commensalism—Symbiotic relationship between two or more species in which no group is injured, and at least one group benefits.

Commercial fishery—The industry of catching a certain species for sale.

Communal—Cooperation between females of one species in production and building, but not in caring for the brood.

Conspecific—Belonging to the same species.

Coracidium—Ciliated free-swimming stage of cestode.

Cosmopolitan—Occurring throughout most of the world.

Cuticle—The noncellular outer layers of the body.

Cydippid—Free-swimming ctenophore larva.

Cyphonarutes—Planktonic larva of some nonbrooding gymnolaemate bryozoans.

Definitive host—See Primary host.

Demersal—Aquatic animals that live near, are deposited on, or sink to the bottom of the sea.

Dentate—Having teeth, or structures that function, or as derived from, as teeth.

Denticles—Teeth, or structures that function as teeth.

Deposit feeders—Animals that feed upon matter that has settled on the substrate.

Detritus—Fragments of plant, animal, or waste remnants.

Deuterostome—Division of the animal kingdom that includes animals that are bilaterally symmetrical, have indeterminate cleavage and a mouth that does not arise from the blastopore.

Diapause—A period of time in which development is suspended or arrested and the body is dormant.

Dioecious—Organisms that have male reproductive organs in one individual and female in another.

Doliolaria—Barrel-shaped larval stage.

Ecdysis—Molting or shedding of the exoskeleton.

Ectoparasite—A parasite that lives on the outside of a host.

Endemic—Belonging to or from a particular geographical region.

Endocuticle—The innermost layer of the cuticle.

Endogeic—An earthworm that primarily feeds on soil and plant roots.

Endoparasite—A parasite that lives inside the body of its host.

Endosymbiont—Symbiotic relationship in which a symbiont dwells within the body of its symbiotic partner.

Enterocoely—Development of the coelom (body cavity) from the embryonic gut (archenteron) occurring in dueterostomes.

Epicuticle—The surface layers of the cuticle.

Epigeic—An earthworm that lives primarily in leaf litter above soil and feeds on surrounding plant debris.

Epiphragm—Temporary mucus door over the aperture (opening) that hardens to seal the snail inside.

Epizoic—An animal or plant that lives on another animal or plant.

Estuary—A semi-enclosed body of water that is diluted by freshwater input and has an open connection to the sea. Typically, there is a mixing of sea and fresh water, and the influx of nutrients from both sources results in high productivity.

Eurybathic—An animal that occurs in a wide range of depths.

Euryhaline—An animal that occurs in a wide variety of salinities.

Eurythermic—An animal that occurs in a wide range of temperatures.

Eversible—Capable of being turned inside out.

Exocuticle—Hard and darkened layer of the cuticle lying between the endocuticle and epicuticle.

Exoskeleton—The external plates of the body wall.

Fibrillae—Small filaments, hairs, or fibers.

Fishery—The industry of catching fish, crustaceans, mollusks or other aquatic animals for commercial, recreational, subsistence or aesthetic purposes.

Furca—An appendage that is forked.

Fusiform—Having a shape that tapers toward each end.

Ganglion—A nerve tissue mass containing cell bodies of neurons external to the brain or spinal cord.

Girdle—Outer mantle of the polyplacophoran that is thick and stiff, extending out from the shell plate.

Glycocalyx—Protein and carbohydrate surface coat in cells.

Gonochoric—An animal with separate sexes.

Gonopore—Reproductive aperture or pore present in the genital area.

Gynandromorph—An individual that exhibits both male and female characteristics.

Hematophagous—A group that feeds or subsides on blood.

Hemitransparent—Half or partially transparent.

Hermaphrodite—An organism that has both male and female sexual organs.

Heterothermic springs—Springs that may freeze in the winter.

Higgins larva—Loriciferan larval stage.

Holoplankton—An animal that lives in plankton all of its life.

Homothermic springs—Those with a constant temperature throughout the year.

Host—The organism in or on which a parasite lives.

Hyaline—Transparent, clear, and colorless.

Hydromedusa—Medusa of the hydrozoans.

Hyperparasite—A parasitic organism whose host is another parasite.

Infauna—An animal that lives among sediment.

Inquiline—Animal that lives in the nests or abode of another species.

Integument—A layer of skin, membrane, or cuticle that envelops an organism or one of its parts.

Intermediate host—Host for the larval stage of a parasitic organism.

Intromittent—Used in copulation; often used to describe the external reproductive organs of males.

Kinesis—A movement that lacks directional orientation and depends upon the intensity of stimulation.

Lamina—Thin, parallel plates of soft vascular sensitive tissue.

Larva—An immature development stage.

Larviparous—Eggs brooded within the female that are later released as larvae.

Lecithotrophic—Larvae that do not feed, but rather derive nutrition from yolk.

Lorica—Specialized girdle-like structure made of a set of hardened parts that protect the body, named for the segmented corselet of armor worn by Roman soldiers.

Lumen—Cavity of a tubular organ.

Mandible—The jaw.

Manubria—Tube that bears the mouth and hangs down from the subumbrella or medusae.

Maxilla—One of two components of the mouth immediately behind the mandibles.

Medusae—Well-developed cnidarian that is gelatinous and free-swimming.

Meiosis—Cellular process that results in the number of chromosomes in gamete-producing cells (usually sex cells) being reduced to one half.

Mesoderm—Tissue derived from the three primary embryonic germ layers, and the source of many bodily tissues and structures.

Metachronous—using coordinated waves, as in bands of cilia beating metachronously.

Metamorphosis—A change in physical form or substance.

Miracidium—Free-swimming first larva of trematodes that is ciliated.

Mitosis—A process that takes place in the nucleus of a dividing cell that results in the formation of two new nuclei having the same number of chromosomes as the parent nucleus.

Moult—The shedding of the exoskeleton.

Mutualism—Symbiotic relationship in which both members of the relationship benefit.

Myoepithelial—Cells of the epithelium.

Nauplius larva—Name given to crustacean larvae.

Nematocyst—Stingers or stinging cnida of cnidarians.

Neritic—An organism that inhabits the region of shallow water adjoining the seacoast.

Nocturnal—An organism that is active mostly at night.

Obligate ectoparasites—External parasites that cannot complete their cycle when removed from their host.

Oocyte—The egg before it has reached maturation.

Ootheca—The cover or case that surrounds a mass of eggs.

Oral lamella—Oral membrane or layer.

Ovigerous—A female that carries developing eggs until they hatch.

Oviparous—An organism that lays eggs.

Ovipositor—The apparatus through which the female lays eggs.

Ovoviviparous—An organism that produces young that hatch out of their egg while still within their mother.

Parapodium—Appendage present on annelids that resembles a paddle.

Parasite—An organism that lives in or on the body of another living organism, feeding off of its host.

Parenchymula—Larval sponge.

Parthenogenetic—Development of an egg without fertilization.

Pelagic—Organisms that live in the open sea, above the ocean floor.

Pelagosphera—Second planktotrophic larva of sipunculans.

Petancula—Stage of metamorphosis for holothuroids.

Phoresy—Nonparasitic relationship between two organisms in which one uses the other as a means of transportation.

Photokinesis—Activity induced by the presence of light.

Photophore—Cell or group of cells that produce light.

Phytophagous—An organism that solely feeds upon plants.

Pilidium—Free-swimming, planktotrophic larva of heteronemerteans.

Pinnules—Small branches.

Planktotrophic larvae—Larvae that feed during their planktonic phase.

Planulae—Larval cnidarians.

Pleonite—Also known as abdominal somite. The single division of a body after the thorax.

Pleopod—An appendage originating from the abdomen.

Plerocercoid—Last larval lifestage of tapeworms.

Polyembryony—The production of several embryos from a single egg.

Polyp—Cnidarian form that is sessile.

Polyphagous—An organism that consumes a variety of foods.

Positively phototactic—Movement toward light.

Predaceous—An organism that preys on other organisms.

Predator—An animal that attacks and feeds on other animals.

Primary host—An organism that acts as the host for an adult stage of a parasite. Also called a definitive host.

Protandric hermaphrodites—Animals that hatch as males and later develop into females.

Protonephridia—Ciliated excretory tube that is specialized for filtration.

Protonymph—The second instar of a mite.

Protostome—Bilateral metazoans characterized by determinate and spiral cleavage, the formation of a mouth and anus directly from the blastopore, and the formation of the coelom by the embryonic mesoderm having split.

Pseudovipositor—Terminal abdominal segment of females from which eggs are layed.

Radial symmetry—The exact arrangement of parts or organs around a central axis.

Ramate—An animal or organism with branches.

Raptorial—An organism that has specially adapted the ability to seize and grasp prey.

Rhagon—Stage of development in demosponge larva.

Rostrum— The beak, snout, spine, proboscis, or anterior median prolongation of the carapace or head of an organism.

Saprophytic—An organism that lives on dead or decaying organic matter.

Scalids—Sets of complex spines that allow the organism to move, capture food, or sense changes in its environment.

Scyphistoma—Scyphozoan polyp.

Sclerites—Thick layer of the exoskeleton.

Segment—A rings or subdivisions of the body.

Sensu stricto—In the "strict sense."

Seta—A bristle.

Somites—The similar or identical segments that divide an animal (especially invertebrates) longitudinally.

Spermatophore—Packet of sperm that is usually transferred from one individual to another during mating.

Spiral cleavage—Cleavage pattern in which spindles or places are oblique to the axis of the egg.

Spiralians—Animal groups that show spiral cleavage patterns.

Spirocyst—Adhesive threads present on Cnidarians that capture prey and attach to immobile objects.

Stock—A biologically distinct and interbreeding population within a species of aquatic animals.

Stoloniferous—An organism that bears or develops a branch from its base to produce new plants from buds, or an extension of the body wall that develops buds giving rise to new zooids.

Strobilation—Asexual reproduction by division into body segments.

Subsistence fishery—A fishery in which the harvested resource is used directly by the fisher.

Symbiont—An organism living in a symbiotic relationship with another organism.

Symbiosis—An intimate association, union, or living arragement between two dissimilar organisms in which at least one of the organisms is dependent upon the other.

Synanthropic—Associated with human habitation.

Syncytial—Multinucleate mass of cytoplasm resulting from the fusion of cells.

Taxis—Reflex movement by an organism in relation to a source of stimulation.

Tegument—Outer, nonciliated layer of the body wall of platyhelminth parasites.

Test—Shell-like encasement or skeleton.

Triploblastic—Embryos with three germ layers.

Trochophore—Larva that has a girdle ring of cilia.

Troglophilous—An organism that lives in caves.

Unci—Hooked anatomical structure.

Uncinus—Miniature hooked anatomical structure.

Uniramous—Having one branch, such as only one appendage in crustaceans. Typically results from loss of the appendage.

Univoltine—A group that produces only one generation per year.

Velum—Shelf present under the umbrella of most hydromedusae, or a ciliated growth with which larva swim.

Vermiform larva—A legless, worm-like larva without a well-developed head.

Vibrissae—A pair of large bristles that is present just above the mouth in some organisms.

Vitellarium—Part of the ovary that produces yolk-filled nurse cells.

Viviparous—An organism that produces live young.

Zoea—Second to last larval stage of many crustaceans.

Zooid—Individual invertebrate that reproduces nonsexually by budding or splitting, especially one that lives in a colony in which each member is joined to others by living material, for example, a coral.

Zooplankton—Free-swimming, microscopic planktonic animals present in lakes and oceans.

Protostomes order list

Annelida [Phylum]
  Polychaeta [Class]
    Amphinomida [Order]
    Capitellida
    Chaetopterida
    Cirratulida
    Cossurida
    Ctenodrilida
    Dinophilida
    Eunicida
    Flabelligerida
    Magelonida
    Nerillida
    Opheliida
    Orbiniida
    Oweniida
    Phyllodocida
    Poeobiida
    Polygordiida
    Protodrilida
    Psammodrilida
    Sabellariida
    Sabellida
    Spintherida
    Spionida
    Sternaspida
    Terebellida

  Myzostomida [Class]
    Pharyngidea [Order]
    Proboscidea

  Oligochaeta [Class]
    Haplotaxida [Order]
    Lumbriculida
    Moniligastrida
    Opisthopora

  Hirudinea [Class]
    Arhynchobdellae [Order]
    Rhynchobdellida

  Pogonophora [Class]
    Athecanephria [Order]
    Thecanephria

Vestimentifera [Phylum]
  Basibranchia [Order]
  Riftiidae

Sipuncula [Phylum]
  Aspidosiphoniformes [Order]
  Golfingiaformes
  Phascolosomatiformes
  Sipunculiformes

Echiura [Phylum]
  Bonellioinea [Order]
  Echiuroinea
  Heteromyota
  Xenopneusta

Onychophora [Phylum]
  No order designations

Tardigrada [Phylum]
  Arthrotardigrada [Order]
  Echiniscoidea
  Parachela
  Apochela
  Mesotardigrada

Arthropoda [Phylum]
  Crustacea [Subphylum]
    Remipedia [Class]
      Nectiopoda [Order]

    Cephalocarida [Class]
      Brachypoda [Order]

    Branchiopoda [Class]
      Anostraca [Order]
      Notostraca
      Conchostraca
      Cladocera

    Malacostraca [Class]
      Phyllocarida [Subclass]
        Leptostraca [Order]

      Eumalacostraca [Subclass]
        Stomatopoda [Order]
        Bathynellacea

        Anaspidacea
        Euphausiacea
        Amphionidacea
        Decapoda
        Mysida
        Lophograstrida
        Cumacea
        Tanaidacea
        Mictacea
        Spelaeogriphacea
        Thermosbaenacea
        Isopoda
        Amphipoda

    Maxillopoda [Class]
      Thecostraca [Subclass]
        Acrothoracica [Order]
        Ascothoracica
        Rhizocephala
        Thoracica

      Tantulocarida [Subclass]
        No order designations

      Branchiura [Subclass]
        Arguloida [Order]

      Mystacocarida [Subclass]
        No order designations

      Copepoda [Subclass]
        Calanoida [Order]
        Cyclopoida
        Gelyelloida
        Harpacticoida
        Misophrioida
        Monstrilloida
        Mormonilloida
        Platycopioida
        Siphonostomatoida

      Ostracoda [Subclass]
        Myodocopida [Order]
        Palaeocopida
        Podocopida

Pentastomida [Class]
Cephalobaenida [Order]
Porocephalida

Cheliceriformes [Subphylum]
Pycnogonida [Class]
Pantopoda [Order]

Chelicerata [Class]
Merostomata [Subclass]
Xiphosura [Order]

Arachnida [Subclass]
Acari [Order]
Amblypygi
Araneae
Opiliones
Palpigradi
Pseudoscorpiones
Ricinulei
Schizomida
Scorpionida
Solpugida
Uropygi

Uniramia [Subphylum]
Myriapoda [Class]
Chilopoda [Subclass]
Craterostigmomophora [Order]
Geophilomorpha
Lithobiomorpha
Scolopendromorpha
Scutigeromorpha

Diplopoda [Subclass]
Callipodida [Order]
Chordeumatida
Glomerida
Glomeridesmida
Julida
Platydesmida
Polydesmida
Polyxenida
Polyzoniida
Siphoniulida
Siphonophorida
Sphaerotheriida
Spirobolida

Spirostreptida
Stemmiulida

Symphyla [Subclass]
Symphyla [Order]

Pauropoda [Subclass]
Hexamerocerata [Order]
Tetramerocerata

Mollusca [Phylum]
Aplacophora [Class]
Cavibelonia [Order]
Neomeniomorpha
Pholidoskepia

Monoplacophora [Class]
Tryblidioidea [Order]

Polyplacophora [Class]
Acanthochitonida [Order]
Ischnochitonida
Lepidopleurida

Gastropoda [Class]
Opisthobranchia [Subclass]
Acochlidioidea [Order]
Anaspidea
Cephalaspidea
Gymnosomata
Notaspidea
Nudibranchia
Runcinoidea
Sacoglossa
Thecosomata

Pulmonata [Subclass]
Actophila [Order]
Basommatophora
Stylommatophora
Systellommatophora
Patellogastropoda
Vetigastropoda [Superorder]
Haliotoidea
Lepetodriloidea
Vetigastropoda
Cocculiniformia
Neritopsina
Caenogastropoda

Bivalvia [Class]
Arcoida [Order]
Hippuritoida
Limoida
Myoida
Mytiloida
Nuculoida
Ostreoida
Pholadomyoida
Pterioida
Solemyoida
Trigonioida
Unionoida
Veneroida

Scaphopoda [Class]
Dentaliida [Order]
Gadilida

Cephalopoda [Class]
Nautilida [Order]
Octopoda
Sepioidea
Spirulida
Teuthoidea
Vampyromorphida

Phoronida [Phylum]
No order designations

Ectoprocta [Phylum]
Phylactolaemata [Class]
No order designations

Stenolaemata [Class]
Cyclostomata [Order]

Gymnolaemata [Class]
Cheilostomatida [Order]
Ctenostomata

Brachiopoda [Phylum]
Inarticulata [Class]
Acrotretida [Order]
Lingulida

Articulata [Class]
Rhynchonellida [Order]
Terebratulida
Thecideidina

A brief geologic history of animal life

A note about geologic time scales: A cursory look will reveal that the timing of various geological periods differs among textbooks. Is one right and the others wrong? Not necessarily. Scientists use different methods to estimate geological time—methods with a precision sometimes measured in tens of millions of years. There is, however, a general agreement on the magnitude and relative timing associated with modern time scales. The closer in geological time one comes to the present, the more accurate science can be—and sometimes the more disagreement there seems to be. The following account was compiled using the more widely accepted boundaries from a diverse selection of reputable scientific resources.

Geologic time scale

Era	Period	Epoch	Dates	Life forms
Proterozoic			2,500-544 mya*	First single-celled organisms, simple plants, and invertebrates (such as algae, amoebas, and jellyfish)
Paleozoic	Cambrian		544-490 mya	First crustaceans, mollusks, sponges, nautiloids, and annelids (worms)
	Ordovician		490-438 mya	Trilobites dominant. Also first fungi, jawless vertebrates, starfishes, sea scorpions, and urchins
	Silurian		438-408 mya	First terrestrial plants, sharks, and bony fishes
	Devonian		408-360 mya	First insects, arachnids (scorpions), and tetrapods
	Carboniferous	Mississippian	360-325 mya	Amphibians abundant. Also first spiders, land snails
		Pennsylvanian	325-286 mya	First reptiles and synapsids
	Permian		286-248 mya	Reptiles abundant. Extinction of trilobytes. Most modern insect orders
Mesozoic	Triassic		248-205 mya	Diversification of reptiles: turtles, crocodiles, therapsids (mammal-like reptiles), first dinosaurs, first flies
	Jurassic		205-145 mya	Insects abundant, dinosaurs dominant in later stage. First mammals, lizards, frogs, and birds
	Cretaceous		145-65 mya	First snakes and modern fish. Extinction of dinosaurs and ammonites, rise and fall of toothed birds
Cenozoic	Tertiary	Paleocene	65-55.5 mya	Diversification of mammals
		Eocene	55.5-33.7 mya	First horses, whales, monkeys, and leafminer insects
		Oligocene	33.7-23.8 mya	Diversification of birds. First anthropoids (higher primates)
		Miocene	23.8-5.6 mya	First hominids
		Pliocene	5.6-1.8 mya	First australopithecines
	Quaternary	Pleistocene	1.8 mya-8,000 ya	Mammoths, mastodons, and Neanderthals
		Holocene	8,000 ya-present	First modern humans

*Millions of years ago (mya)

Index

Bold page numbers indicate the primary discussion of a topic; page numbers in italics indicate illustrations.

A

Abalone, 2:431, *2:432*, *2:433*
Aberrant tooth shells, 2:470, *2:472*, *2:473–474*
Abiotic environments, 2:25–26
  See also Habitats
Acanthocephala, 2:33
Acanthodrilidae, 2:65
Acari, 2:333
Acartia clausi, *2:303*, 2:304
Achatina spp., 2:414
Achatinella mustelina. *See* Agate snails
Achatinidae, 2:411
Acmaeiadae, 2:423
Acmaeina, 2:423
Acoels, 2:12, 2:13, 2:35
Acorn barnacles. *See* Rock barnacles
Acorn shells. *See* Rock barnacles
Acrothoracica, 2:273, 2:274
Actinotrocha spp. *See* Actinotrochs
Actinotrocha branchiata, 2:491
Actinotrocha harmeri, 2:491, 2:494
Actinotrocha pallida, 2:491
Actinotrocha sabatieri, 2:491
Actinotrocha vancouverensis, 2:491, 2:494
Actinotrochs, 2:491, *2:492*
Active mimicry, 2:38
  See also Behavior
Actophila, 2:412, 2:413, 2:414
Acyphoderes sexualis, 2:38
Addisonia excentrica, 2:435
Aega spp., 2:252
Aegism, 2:32
  See also Behavior
Aeolidiella sanguinea, 2:406, *2:408*
Aequipecten opercularis. *See* Queen scallops
Aesop's Fables, 2:42
Aethomerus spp. *See* Longhorn beetles
Afrauropodidae, 2:375
African land snails, giant. *See* Giant African land snails
African worms, 2:68, *2:69*, *2:70*
Agastoschizomus lucifer, *2:338*, *2:340*, 2:348–349
Agate snails, 2:416, *2:418*
Agema spp., 2:317
Ailoscolecidae, 2:65
Alaska king crabs. *See* Red king crabs
Alaysia spp., 2:91
Alcyonidium spp., 2:25
Allopauropus carolinensis, *2:375*, *2:377*
Allopauropus gracilis, *2:376*
Alluroididae, 2:65
Almidae, 2:65

Alofia spp., 2:317, 2:320
Alphaeidae, 2:200
Amathimysis trigibba, *2:218*, *2:220*
Amblypygi, 2:333
Ambulacralia, 2:4
American horseshoe crabs, *2:327*, *2:328*, *2:330*, *2:331–332*
American king crabs. *See* Red king crabs
American leeches. *See* North American medicinal leeches
American oysters. *See* Eastern American oysters
American tadpole shrimps. *See* Longtail tadpole shrimps
Americobdellidae, 2:76
Ammonites, *2:10*
Amphibolidae, 2:413
Amphibolus nebulosus, 2:117
Amphionides spp. *See Amphionides reynaudii*
Amphionides reynaudii, 2:195
Amphionididae. *See Amphionides reynaudii*
Amphionids, **2:195–196**
Amphipods, **2:261–272**, *2:263*, *2:264*
  behavior, 2:261
  conservation status, 2:262
  distribution, 2:261
  evolution, 2:261
  feeding ecology, 2:29, 2:261
  habitats, 2:261
  humans and, 2:262
  physical characteristics, 2:261
  reproduction, 2:261–262
  species of, 2:265–271
  taxonomy, 2:261
Amphiprion spp. *See* Anemonefishes
Ampullaria canaliculata. *See* Apple snails
Ampullarioidea, 2:445, 2:446
Amynthas corticis, *2:69*, 2:71
Anaea spp., 2:38
Anaspidaceans, **2:181–183**
Anaspides tasmaniae, *2:181*, *2:183*
Anaspididae, 2:181
Anaspidinea, 2:181
Ancistrocerus inflictus. *See* Katydids
Anecic earthworms, 2:66
Anemonefishes, 2:34, *2:34*
Anemones, *2:34*, 2:36
Anilocra laticaudata, 2:252
Animalia, 2:8
  See also Taxonomy; specific species
Annelida, 2:3, 2:25, 2:45, 2:59
  evolution, 2:14
  reproduction, 2:16, 2:17, 2:21, 2:22–23
  taxonomy, 2:35
  See also specific annelids

Anomalodesmata, 2:451
Anomura, 2:197, 2:198, 2:200
Anoplodactylus evansi, *2:323*, *2:324–325*
Anostraca. *See* Fairy shrimps
Anostracina, 2:135
Antalis entalis. *See* Tusk shells
Antarctic krill, *2:185*, *2:188*, 2:189–191, *1:190*
Anthozoa, 2:13
Anthracocaridomorpha, 2:235
Anticoagulants, 2:76
Antillesoma antillarum, *2:99*, *2:100*
Antrobathynella stammeri, *2:179*
Ants, 2:23, 2:35, 2:36, 2:37
Anus, 2:3, 2:21
  See also Physical characteristics
Aphids, 2:25, 2:37
Aphroditidae, 2:45
Apicomplexa, 2:11
Apis spp. *See* Bees
Aplacophorans, 2:14, **2:379–385**, *2:381*, *2:382*
Apochela, 2:115
Apomorphies, 2:13
Aporrectodea caliginosa. *See* Common field worms
Apple snails, *2:448*, *2:449*
Apseudes intermedius, *2:237*, *2:238*
Apseudes spectabilis, 2:236
Apseudomorpha, 2:235
Aquaculture, 2:41
  See also Humans
Aquatic leaf leeches, 2:75
Arachnida, 2:25, *2:337*, *2:338*
  behavior, 2:37, 2:335
  conservation status, 2:336
  distribution, 2:335
  evolution, 2:333
  feeding ecology, 2:335
  habitats, 2:335
  humans and, 2:336
  physical characteristics, 2:333–*334*
  reproduction, 2:23, 2:335
  species of, 2:339–352
  taxonomy, 2:333
  See also specific arachnids
Araneae, 2:333
Archaebacteria, 2:8
Archaeophiala spp., 2:387
Archea. *See* Archaebacteria
Archenteron, 2:4, 2:21
Archispirostreptus syriacus, 2:364
Architaenioglossa, 2:445, 2:446
Architeuthis spp. *See* Giant squids
Arcovestia spp., 2:91
Arctiid moths, 2:37

D

INDEX

INDEX

INDEX

INDEX